Lecture Notes in Computer Science

Founding Editors

Gerhard Goos
Juris Hartmanis

Editorial Board Members

Elisa Bertino, *Purdue University, West Lafayette, IN, USA*
Wen Gao, *Peking University, Beijing, China*
Bernhard Steffen , *TU Dortmund University, Dortmund, Germany*
Moti Yung , *Columbia University, New York, NY, USA*

The series Lecture Notes in Computer Science (LNCS), including its subseries Lecture Notes in Artificial Intelligence (LNAI) and Lecture Notes in Bioinformatics (LNBI), has established itself as a medium for the publication of new developments in computer science and information technology research, teaching, and education.

LNCS enjoys close cooperation with the computer science R & D community, the series counts many renowned academics among its volume editors and paper authors, and collaborates with prestigious societies. Its mission is to serve this international community by providing an invaluable service, mainly focused on the publication of conference and workshop proceedings and postproceedings. LNCS commenced publication in 1973.

Aleš Leonardis · Elisa Ricci · Stefan Roth ·
Olga Russakovsky · Torsten Sattler · Gül Varol
Editors

Computer Vision – ECCV 2024

18th European Conference
Milan, Italy, September 29–October 4, 2024
Proceedings, Part IV

Springer

Editors
Aleš Leonardis
University of Birmingham
Birmingham, UK

Stefan Roth
Technical University of Darmstadt
Darmstadt, Germany

Torsten Sattler
Czech Technical University in Prague
Prague, Czech Republic

Elisa Ricci
University of Trento
Trento, Italy

Olga Russakovsky
Princeton University
Princeton, NJ, USA

Gül Varol
École des Ponts ParisTech
Marne-la-Vallée, France

ISSN 0302-9743　　　　　　　　ISSN 1611-3349　(electronic)
Lecture Notes in Computer Science
ISBN 978-3-031-73234-8　　　　ISBN 978-3-031-73235-5　(eBook)
https://doi.org/10.1007/978-3-031-73235-5

© The Editor(s) (if applicable) and The Author(s), under exclusive license
to Springer Nature Switzerland AG 2025

This work is subject to copyright. All rights are solely and exclusively licensed by the Publisher, whether the whole or part of the material is concerned, specifically the rights of translation, reprinting, reuse of illustrations, recitation, broadcasting, reproduction on microfilms or in any other physical way, and transmission or information storage and retrieval, electronic adaptation, computer software, or by similar or dissimilar methodology now known or hereafter developed.
The use of general descriptive names, registered names, trademarks, service marks, etc. in this publication does not imply, even in the absence of a specific statement, that such names are exempt from the relevant protective laws and regulations and therefore free for general use.
The publisher, the authors and the editors are safe to assume that the advice and information in this book are believed to be true and accurate at the date of publication. Neither the publisher nor the authors or the editors give a warranty, expressed or implied, with respect to the material contained herein or for any errors or omissions that may have been made. The publisher remains neutral with regard to jurisdictional claims in published maps and institutional affiliations.

This Springer imprint is published by the registered company Springer Nature Switzerland AG
The registered company address is: Gewerbestrasse 11, 6330 Cham, Switzerland

If disposing of this product, please recycle the paper.

Foreword

Welcome to the proceedings of the European Conference on Computer Vision (ECCV) 2024, the 18th edition of the conference, which has been held biennially since its founding in France in 1990. We are holding the event in Italy for the third time, in a convention centre where we can accommodate nearly 7000 in-person attendees.

The journey to ECCV 2024 started at CVPR 2019, where the general chairs met at a sketchy bar in Long Beach to discuss ideas on how to organize an amazing future ECCV. After that, Covid came and changed conferences probably forever, but we maintained our excitement to organize an awesome in-person conference in 2024, particularly having seen the wonderful event that was ECCV 2022 in Tel Aviv. And here we are, 5 years later, delighted that we finally get to welcome all of you to Milan!

First and foremost, ECCV is about the exchange of scientific ideas, and hence the most important organizational task is the selection of the programme. To our Program Chairs, Aleš Leonardis, Elisa Ricci, Gül Varol, Olga Russakovsky, Stefan Roth, and Torsten Sattler, our entire community owe a huge debt of gratitude. As you will read in their Preface to the proceedings, they dealt with over 8,500 submissions, and their diligence, thoroughness, and sheer effort has been inspirational. From the early stages of designing the call for papers, through recruiting and selecting area chairs and reviewers, to ensuring every paper has a fair and thorough assessment, their professionalism, commitment, and attention to detail has been exemplary. From all the choices that we made to organize ECCV, choosing this amazing team of Program Chairs has been, without a doubt, our very best decision.

The Program Chairs were advised and supported by an Ethics Review Committee, Chloé Bakalar, Kate Saenko, Remi Denton, and Yisong Yue, who provided invaluable input on papers where reviewers or Area Chairs had raised ethical concerns.

The Publication Chairs, Jovita Lukasik, Michael Möller, François Brémond, and Mahmoud Ali, did great work in assembling the camera-ready papers into these Springer volumes, dealing with numerous issues that arose promptly and professionally.

With the ever-expanding importance of workshops, tutorials, and demos at our major conferences comes a corresponding increase in the challenge of organizing over 70 workshops and 9 tutorials attached to the main conference. The Workshop and Tutorial Chairs, Alessio Del Bue, Jordi Pont-Tuset, Cristian Canton, and Tatiana Tommasi, and Demo Chairs, Hyung Chang and Marco Cristani, worked tirelessly to coordinate the very varied demands and structures of these different events, yielding an extremely rich auxiliary program which will enhance the conference experience for everyone who attends.

The Industry Chairs, Shaogang Gong and Cees Snoek, working with Limor Urfaly and Lior Gelfand, managed to attract numerous companies and organizations from all around the world to the industrial Expo, making ECCV 2024 an international event that not only showcases excellent foundational research, but also presents how such research can be transformed in real products. It is notable that together with the big companies, several start-ups chose ECCV to present their business ideas and activities.

As usual in the recent main conference, we also organized Speed Mentoring and Doctoral Consortium events. While the latter is an institutional occasion allowing senior PhDs to show their work, the former is especially important to answer to the many doubts and uncertainties young scholars have while undertaking a career in Computer Vision and Artificial Intelligence. We warmly thank our Social Activities Chairs, Raffaella Lanzarotti, Simone Bianco and Giovanni Farinella, and the Doctoral Consortium Chairs, Cigdem Beyan and Or Litany, for their dedication to organize such events, so important for our young colleagues, as well as all the mentors who were available to share their experience.

As the conference becomes larger and larger, the overall budget increases to a level where uncertainties in estimates of attendance can result in significant losses, perhaps enough to significantly damage the prospects of running the conference in the future. This risk imposed the need to require authors to commit to full conference registrations, with the undesirable consequent possibility to cause hardship to authors, particularly student authors, from organizations where funding attendance is difficult. To mitigate this effect, we allocated travel grants to students who might otherwise have had difficulty in attending. Our Diversity Chairs, David Fouhey and Rita Cucchiara, put tremendous effort into a wonderfully careful and thoughtful programme, balancing diversity in all its forms with a changing budget landscape, allowing us to support the participation of many people who might not otherwise have been able to attend.

This year's conference also marks the move to a unified, multi-year conference website, based on that used for other leading computer vision and AI conferences. This was an effort that we decided to take on to make the ECCV website more familiar to the community and support the organization of the future ECCV editions. Our thanks go to Lee Campbell of Eventhosts who managed the entire design and development of the new website and provided responsive support and updates throughout the conference organization timeline.

Our Finance Chairs, Gérard Medioni and Nicole Finn, provided invaluable advice and assistance on all aspects of the financial planning of the conference, in conjunction with the professional conference organizers AIM Group, where Lavinia Ricci led a fantastic team that managed thousands of details from space planning to food to registrations. In particular, Elder Bromley and Mara Carletti were a tremendous support to the entire conference organization. The conference centre, Mico DMC, also provided an excellent service, showing considerable flexibility in adapting the venue and processes to the particular demands of an academic-focused conference.

Finally, we cannot forget to thank our Social Media Chair, Kosta Derpanis, supported by Abby Stylianou and Jia-Bin Huang, who, with their presence on social media, spread ECCV news and responded to small and big questions in a timely manner, helping the community to promptly receive information and fix their problems.

October 2024
Andrew Fitzgibbon
Laura Leal-Taixé
Vittorio Murino

Preface

ECCV 2024 saw a remarkable 48% increase in submissions, with 8585 valid papers submitted, compared to 5804 in ECCV 2022, 5150 in 2020, and 2439 in 2018. Of these, 2387 papers (27.9%) were accepted for publication, with 200 (2.3% overall) selected for oral presentations.

A total of 173 complete submissions were desk-rejected for various reasons. Common issues included revealing author identities in the paper or supplementary material (for example: by including the author names, referring to specific grants in the acknowledgments, including links to GitHub accounts with visible author information, etc.), exceeding the page limit or otherwise significantly altering the submission template, or posting an arXiv preprint explicitly mentioning the work as an ECCV 2024 submission. The desk-rejected submissions also include 14 papers identified as dual submissions to other concurrent conferences, such as ICML, NeurIPS, and MICCAI. Seven papers were rejected among the accepted papers due to plagiarism issues, after the Program Chairs conducted a thorough plagiarism check with the iThenticate software followed by manual inspection.

The double-blind review process was managed using the CMT system, ensuring anonymity between authors and Area Chairs/reviewers. Each paper received at least three reviews, amounting to over 26,000 reviews in total. To maintain the quality and the fairness of the reviews, we recruited 469 Area Chairs (ACs) and more than 8200 reviewers. Of the recruited reviewers, 7293 ultimately participated.

Area Chairs were chosen for their technical expertise and reputation; many of them had served in similar roles at other top conferences. We also recruited additional Area Chairs through an open call for self-nominations. Among the selected ACs, around 18% were women. Geographical distribution was taken into account during the selection process, with 31% of ACs working in Europe, 38% in the Americas, 27% in Asia, and 4% in the rest of the world. Despite our best efforts to consider diversity factors in the selection process, the gender diversity and geographic diversity of ACs is still below what we were hoping for.

To ensure a smooth decision-making process, we grouped the ACs into triplets. Upon accepting their role, ACs were requested to enter into the CMT system domain conflicts and subject areas, as well as information about their DBLP, TPMS (Toronto Paper Matching System), and OpenReview profiles. These were used to determine whether different subject areas were sufficiently covered, to detect conflicts, and to automatically form AC triplets. An AC triplet was a group of three ACs who worked together throughout the process, which helped to facilitate discussion, calibrate the decision-making process, and provide support for first-time ACs. AC triplets were manually refined by Program Chairs to ensure no triplet contained more than two major time zones to enable easier coordination for virtual AC meetings. Paper assignments were made such that no AC within the triplet was conflicted with any of the papers. Each AC was responsible for overseeing 18 papers on average.

In each triplet, we identified a Lead AC who was an experienced member of the computer vision community willing to take on additional administrative duties. The primary responsibilities of Lead ACs included coordinating the AC triplet to ensure deadlines were met and offering guidance to less experienced ACs. They also served as the main contact person with the Program Chairs and helped in handling papers if an original AC was unable to complete their tasks.

Reviewers were invited from reviewer pools of previous conferences and also selected from the pool of authors. We asked experienced ACs to recommend more potential reviewers. We further recruited reviewers through self-nominations, where researchers volunteered to review for ECCV 2024 via an open call. The self-nominations were then filtered, selecting those reviewers with at least two papers published in related conferences such as ECCV, ICCV, CVPR, ICLR, NeurIPS, etc.

Similarly to ACs, reviewers were also asked to update their CMT, Toronto Paper Matching System (TPMS), and OpenReview profiles. Reviewers had the option to classify themselves as either senior researchers or junior researchers and students, with the number of assigned papers varying accordingly (a quota of 6 for senior researchers with 3 or more times reviewing experience, and a quota of 4 otherwise). On average, each reviewer was assigned about 4 papers.

Conflicts of interest among authors, ACs, and reviewers were managed automatically via conflict domains and co-authorship information from DBLP. Paper assignments to ACs and reviewers were determined using an algorithm that combined subject-area affinity scores from CMT, TPMS, and OpenReview, along with additional criteria such as ensuring that every paper had at least one non-student reviewer. For AC assignments, we additionally incorporated a newly built affinity score based on embedding similarity with the papers on the Google Scholar profile of the ACs. Once the assignments were made, ACs and reviewers were asked to promptly report any undetected conflicts of interest or other concerns that would preclude them from handling the assigned papers to ensure immediate reassignment, if necessary.

Given the widespread use of Large Language Models (LLMs), ECCV 2024, like previous conferences, implemented a special policy regulating their use. Authors were permitted to use any tools, including LLMs, in preparing their papers, but they remained fully responsible for any misrepresentation, factual inaccuracies, or plagiarism. Reviewers were also allowed to use LLMs to refine the wording of their reviews; however, they were held accountable for the accuracy of their content and were strictly prohibited from inputting submissions into an LLM. One challenge we ran into was that tools that detect LLM-generated content were not able to distinguish between text polished vs. text written from scratch by an LLM. Thus, it became quite difficult to enforce the policy in practice.

Overall, the challenges encountered during the review process were consistent with those faced at previous computer vision conferences. For instance, a small number of reviewers ultimately did not submit their reviews or provided brief, uninformative reviews, necessitating last-minute reassignment to emergency reviewers. As the community expands and the number of submissions rises, securing enough qualified reviewers and performing a quality check of the received reviews becomes increasingly difficult.

The review process included a rebuttal phase where authors were given a week to address any concerns raised by the reviewers. Each response was limited to a single PDF page using a predefined template. Rebuttals with formatting or anonymity violations were removed by the Program Chairs. ACs then facilitated discussions with reviewers to evaluate the merits of each submission. While the goal was to reach a consensus, the final decision was left to the ACs in the triplet to ensure fairness. As is common, the ACs within each triplet were tasked with organizing meetings to discuss papers, especially borderline cases. The entire process was conducted online, with no in-person meetings. For each paper, a primary and a secondary AC were assigned from the AC triplet. The primary AC was responsible for entering the decisions and meta-reviews into the CMT system, which were then reviewed and approved by the secondary ACs. Besides making accept/reject decisions, the ACs were also required to provide recommendations regarding oral/poster presentations and award nominations.

Reviewers and ACs were asked to flag papers with potential ethical concerns. An ethics review committee, consisting of four leading researchers from both industry and academia, was appointed. The role of the committee was to evaluate papers escalated by the AC for further investigation. In total, 31 papers were flagged to the committee. The committee requested that the authors of 15 papers address specific concerns in their camera-ready submissions. The authors all complied with the requests.

Following the recent IEEE decision to ban the use of the photograph of Lena Forsén, we automatically identified all submissions that included this image. Although the ban did not officially apply to ECCV, we nevertheless shared the context with the authors of the affected papers and suggested that they remove the photograph in the camera-ready submission. All authors complied.

To help the ACs to track the overall progress of the reviewing and decision-making processes, we created AC triplet activity dashboards for each AC triplet (via Google sheets), which were updated on a regular basis. During the review process, the activity dashboard enabled ACs to track the progress of incoming reviews for each paper and promptly identify papers that necessitated attention, e.g., papers potentially requiring the invitation of emergency reviewers, or papers with suspiciously short reviews. During the decision-making process, the dashboard was also used by the Program Chairs to identify papers with incomplete review information, e.g., for which reviewers did not enter their final recommendation, the AC did not enter their final decision or meta-review, the secondary AC did not confirm the recommendation, etc.

At the end of the decision-making progress, the Program Chairs carefully examined the very small number of cases where ACs overruled a unanimous reviewer recommendation for acceptance or rejection. In such instances, it was crucial for the primary and secondary ACs to explain in detail the motivation behind the decision. To ensure a fair evaluation, an additional Area Chair, not originally part of the decision-making triplet, was appointed to provide an independent opinion and advise the Program Chairs in determining whether to uphold or reverse the overturn.

Inevitably, after decisions were released, some authors were dissatisfied. The authors were given the possibility to appeal the final decision in case of procedural issues encountered during the review process by filling out a form to directly contact the Program Chairs. Valid reasons for appealing a decision included policy errors (e.g., reviewers or ACs enforcing a non-existent policy), clerical errors (e.g., the meta-review clearly indicated an intention to accept a paper, but it was mistakenly rejected), and significant misunderstandings by the reviewers or ACs. In total, we received 59 appeals and upheld 6 of them. One was due to a clear policy error by an AC, and the paper decision changed from reject to accept following a successful appeal. The other 5 papers were already conditionally accepted – but the AC/reviewers wrongly flagged them as papers for which the associated dataset was required to be released prior to final acceptance; this condition was removed following the appeal.

After the camera-ready deadline, additional checks were conducted, particularly for papers contributing a new dataset and for those flagged for potential ethical concerns. Specifically, during the submission phase, authors were required to specify if the primary contribution of their papers was the release of a new dataset, and for these papers, a valid dataset link had to be provided by the camera-ready deadline. A total of 435 submissions included a dataset as part of their contribution, and all complied with the requirement to release the dataset by the camera-ready deadline.

We sincerely thank all the ACs and reviewers for their invaluable contributions to the review process. Their dedication and expertise were crucial in maintaining the high standards of the conference. We would also like to thank our Technical Program Chair, Sascha Hornauer, who did tremendous work behind the scenes, and to the entire CMT team for their prompt support with any technical issues.

October 2024

Aleš Leonardis
Elisa Ricci
Stefan Roth
Olga Russakovsky
Torsten Sattler
Gül Varol

Organization

General Chairs

Andrew Fitzgibbon Graphcore, UK
Laura Leal-Taixé NVIDIA, Italy
Vittorio Murino University of Verona & University of Genoa, Italy

Program Chairs

Aleš Leonardis University of Birmingham, UK
Elisa Ricci University of Trento, Italy
Stefan Roth Technical University of Darmstadt, Germany
Olga Russakovsky Princeton University, USA
Torsten Sattler Czech Technical University in Prague, Czech Republic
Gül Varol Ecole des Ponts Paris Tech, France

Technical Program Chair

Sascha Hornauer Mines Paris – PSL, France

Industrial Liaison Chairs

Cees Snoek University of Amsterdam, Netherlands
Shaogang Gong Queen Mary University of London, UK

Publication Chairs

Francois Bremond Inria, France
Mahmoud Ali Inria, France
Jovita Lukasik University of Siegen, Germany
Michael Moeller University of Siegen, Germany

Poster Chairs

Aljoša Ošep   Carnegie Mellon University, USA
Zuzana Kukelova   Czech Technical University in Prague, Czech Republic

Diversity Chairs

David Fouhey   New York University, USA
Rita Cucchiara   Università di Modena e Reggio Emilia, Italy

Local Chairs

Raffaella Lanzarotti   Università degli Studi di Milano, Italy
Simone Bianco   University of Milano-Bicocca, Italy

Conference Ombuds

Georgia Gkioxari   California Institute of Technology, USA
Greg Mori   Borealis AI & Simon Fraser University, Canada

Ethics Review Committee

Chloé Bakalar   Meta, USA
Kate Saenko   Boston University, USA
Remi Denton   Google Research, USA
Yisong Yue   Caltech, Asari AI & Latitude AI, USA

Workshop and Tutorial Chairs

Alessio Del Bue   Istituto Italiano di Tecnologia, Italy
Cristian Canton   Meta AI, USA
Jordi Pont-Tuset   Google DeepMind, Switzerland
Tatiana Tommasi   Politecnico di Torino, Italy

Finance Chairs

Gerard Medioni   Amazon, USA
Nicole Finn   c to c events, USA

Demo Chairs

Hyung Chang   University of Birmingham, UK
Marco Cristani   University of Verona, Italy

Publicity and Social Media Chairs

Konstantinos Derpanis   York University & Samsung AI Centre Toronto, Canada
Jia-Bin Huang   University of Maryland College Park, USA
Abby Stylianou   Saint Louis University, USA

Social Activities Chairs

Giovanni Maria Farinella   University of Catania, Italy
Raffaella Lanzarotti   Università degli Studi di Milano, Italy
Simone Bianco   University of Milano-Bicocca, Italy

Doctoral Consortium Chairs

Cigdem Beyan   University of Verona, Italy
Or Litany   NVIDIA & Technion, USA

Web Developer

Lee Campbell   Eventhosts, USA

Area Chairs

Ehsan Adeli	Stanford University, USA
Aishwarya Agrawal	University of Montreal & Mila & DeepMind, Canada
Naveed Akhtar	University of Western Australia, Australia
Yagiz Aksoy	Simon Fraser University, Canada
Karteek Alahari	Inria, France
Jose Alvarez	NVIDIA, USA
Djamila Aouada	Interdisciplinary Centre for Security, Reliability and Trust, University of Luxembourg, Luxembourg
Andre Araujo	Google, Brazil
Pablo Arbelaez	Universidad de los Andes, Colombia
Iro Armeni	Stanford University, USA
Yuki Asano	University of Amsterdam, Netherlands
Karl Åström	Lund University, Sweden
Mathieu Aubry	Ecole des Ponts ParisTech, France
Shai Bagon	Weizmann Institute of Science, Israel
Song Bai	ByteDance, Singapore
Xiang Bai	Huazhong University of Science and Technology, China
Yang Bai	Tencent, China
Lamberto Ballan	University of Padova, Italy
Linchao Bao	University of Birmingham, UK
Ronen Basri	Weizmann Institute of Science, Israel
Sara Beery	Massachusetts Institute of Technology, USA
Vasileios Belagiannis	University of Erlangen–Nuremberg, Germany
Serge Belongie	University of Copenhagen, Denmark
Sagie Benaim	Hebrew University of Jerusalem, Israel
Rodrigo Benenson	Google, Switzerland
Cigdem Beyan	University of Bergamo, Italy
Bharat Bhatnagar	Meta Reality Labs Research, Switzerland
Tolga Birdal	Imperial College London, UK
Matthew Blaschko	KU Leuven, Belgium
Federica Bogo	Meta Reality Labs Research, Switzerland
Timo Bolkart	Google, Switzerland
Katherine Bouman	California Institute of Technology, USA
Lubomir Bourdev	Primepoint, USA
Edmond Boyer	Inria, France
Yuri Boykov	University of Waterloo, Canada
Eric Brachmann	Niantic, Germany

Wieland Brendel	University of Tübingen, Germany
Gabriel Brostow	University College London, UK
Michael Brown	York University, Canada
Andrés Bruhn	University of Stuttgart, Germany
Andrei Bursuc	valeo.ai, France
Benjamin Busam	Technical University of Munich, Germany
Holger Caesar	TU Delft, Netherlands
Jianfei Cai	Monash University, Australia
Simone Calderara	University of Modena and Reggio Emilia, Italy
Octavia Camps	Northeastern University, Boston, USA
Joao Carreira	DeepMind, UK
Silvia Cascianelli	University of Modena and Reggio Emilia, Italy
Ayan Chakrabarti	Google Research, USA
Tat-Jen Cham	Nanyang Technological University, Singapore
Antoni Chan	City University of Hong Kong, China
Manmohan Chandraker	University of California, San Diego, USA
Xiaojun Chang	University of Technology Sydney, Australia
Devendra Singh Chaplot	Mistral AI, USA
Liang-Chieh Chen	TikTok, USA
Long Chen	Hong Kong University of Science and Technology, China
Mei Chen	Microsoft, USA
Shizhe Chen	Inria, France
Shuo Chen	RIKEN, Japan
Xilin Chen	Institute of Computing Technology, Chinese Academy of Sciences, China
Anoop Cherian	Mitsubishi Electric Research Labs, USA
Tat-Jun Chin	University of Adelaide, Australia
Minsu Cho	Pohang University of Science and Technology, Korea
Hisham Cholakkal	Mohamed bin Zayed University of Artificial Intelligence, United Arab Emirates
Yung-Yu Chuang	National Taiwan University, Taiwan
Ondrej Chum	Czech Technical University in Prague, Czech Republic
Marcella Cornia	University of Modena and Reggio Emilia, Italy
Marco Cristani	University of Verona, Italy
Cristian Canton	Meta AI, USA
Zhaopeng Cui	Zhejiang University, China
Angela Dai	Technical University of Munich, Germany
Jifeng Dai	Tsinghua University, China
Yuchao Dai	Northwestern Polytechnical University, China

Dima Damen	University of Bristol, UK
Kostas Daniilidis	University of Pennsylvania, USA
Antitza Dantcheva	Inria, France
Raoul de Charette	Inria, France
Shalini De Mello	NVIDIA Research, USA
Tali Dekel	Weizmann Institute of Science, Israel
Ilke Demir	Intel Corporation, USA
Joachim Denzler	Friedrich Schiller University Jena, Germany
Konstantinos Derpanis	York University, Canada
Jose Dolz	École de technologie supérieure Montreal, Canada
Jiangxin Dong	Nanjing University of Science and Technology, China
Hazel Doughty	Leiden University, Netherlands
Iddo Drori	Boston University and Columbia University, USA
Enrique Dunn	Stevens Institute of Technology, USA
Thibaut Durand	Borealis AI, Canada
Sayna Ebrahimi	Google, USA
Mohamed Elhoseiny	King Abdullah University of Science and Technology, Saudi Arabia
Bin Fan	University of Science and Technology Beijing, China
Yi Fang	New York University, USA
Giovanni Maria Farinella	University of Catania, Italy
Ryan Farrell	Brigham Young University, USA
Alireza Fathi	Google, USA
Christoph Feichtenhofer	Meta & FAIR, USA
Rogerio Feris	MIT-IBM Watson AI Lab & IBM Research, USA
Basura Fernando	Agency for Science, Technology and Research (A*STAR), Singapore
David Fouhey	New York University, USA
Katerina Fragkiadaki	Carnegie Mellon University, USA
Friedrich Fraundorfer	Graz University of Technology, Austria
Oren Freifeld	Ben-Gurion University, Israel
Huan Fu	University of Sydney, China
Yasutaka Furukawa	Simon Fraser University, Canada
Andrea Fusiello	University of Udine, Italy
Fabio Galasso	Sapienza University, Italy
Jürgen Gall	University of Bonn, Germany
Orazio Gallo	NVIDIA Research, USA
Chuang Gan	MIT-IBM Watson AI Lab, USA
Lianli Gao	University of Electronic Science and Technology of China, China

Shenghua Gao	University of Hong Kong, China
Sourav Garg	University of Adelaide, Australia
Efstratios Gavves	University of Amsterdam, Netherlands
Peter Gehler	Zalando, Germany
Theo Gevers	University of Amsterdam, Netherlands
Golnaz Ghiasi	Google DeepMind, USA
Rohit Girdhar	Carnegie Mellon University, USA
Xavier Giro-i-Nieto	Universitat Politecnica de Catalunya, Spain
Ross Girshick	Allen Institute for Artificial Intelligence, USA
Zan Gojcic	NVIDIA, Switzerland
Vladislav Golyanik	Max Planck Institute for Informatics, Germany
Stephen Gould	Australian National University, Australia
Liangyan Gui	University of Illinois Urbana-Champaign, USA
Fatma Guney	Koc University, Turkey
Yulan Guo	Sun Yat-sen University, China
Yunhui Guo	University of Texas at Dallas, USA
Mohit Gupta	University of Wisconsin-Madison, USA
Minh Ha Quang	RIKEN Center for Advanced Intelligence Project, Japan
Bumsub Ham	Yonsei University, Korea
Bohyung Han	Seoul National University, Korea
Kai Han	University of Hong Kong, China
Xiaoguang Han	Shenzhen Research Institute of Big Data & Chinese University of Hong Kong (Shenzhen), China
Tatsuya Harada	University of Tokyo & RIKEN, Japan
Tal Hassner	Weir AI, USA
Xuming He	ShanghaiTech University, China
Felix Heide	Princeton & Algolux, USA
Anthony Hoogs	Kitware, USA
Timothy Hospedales	Edinburgh University, UK
Mahdi Hosseini	Concordia University, Canada
Peng Hu	College of Computer Science, Sichuan University, China
Gang Hua	Wormpex AI Research, USA
Jia-Bin Huang	University of Maryland College Park, USA
Qixing Huang	University of Texas at Austin, USA
Sharon Xiaolei Huang	Pennsylvania State University, USA
Junhwa Hur	Google, USA
Nazli Ikizler-Cinbis	Hacettepe University, Turkey
Eddy Ilg	Saarland University, Germany
Ahmet Iscen	Google, France

Nathan Jacobs	Washington University in St. Louis, USA
Varun Jampani	Stability AI, USA
C. V. Jawahar	International Institute of Information Technology - Hyderabad, India
Laszlo Jeni	Carnegie Mellon University, USA
Jiaya Jia	Chinese University of Hong Kong, China
Qin Jin	Renmin University of China, China
Jungseock Joo	NVIDIA & University of California, Los Angeles, USA
Fredrik Kahl	Chalmers, Sweden
Yannis Kalantidis	NAVER LABS Europe, France
Vicky Kalogeiton	Ecole Polytechnique, IP Paris, France
Evangelos Kalogerakis	University of Massachusetts Amherst, USA
Hiroshi Kawasaki	Kyushu University, Japan
Margret Keuper	University of Mannheim & Max Planck Institute for Informatics, Germany
Salman Khan	Mohamed bin Zayed University of Artificial Intelligence, United Arab Emirates
Anna Khoreva	Bosch Center for Artificial Intelligence, Germany
Gunhee Kim	Seoul National University, Korea
Junmo Kim	Korea Advanced Institute of Science and Technology, Korea
Min H. Kim	Korea Advanced Institute of Science and Technology, Korea
Seon Joo Kim	Yonsei University, Korea
Tae-Kyun Kim	Korea Advanced Institute of Science and Technology & Imperial College London, Korea
Benjamin Kimia	Brown University, USA
Hedvig Kjellström	KTH Royal Institute of Technology, Sweden
Laurent Kneip	ShanghaiTech University, China
A. Sophia Koepke	University of Tübingen, Germany
Iasonas Kokkinos	University College London, UK
Wai-Kin Adams Kong	Nanyang Technological University, Singapore
Piotr Koniusz	Data61/CSIRO & Australian National University, Australia
Jana Kosecka	George Mason University, USA
Adriana Kovashka	University of Pittsburgh, USA
Hilde Kuehne	University of Bonn, Germany
Zuzana Kukelova	Czech Technical University in Prague, Czech Republic
Ajay Kumar	Hong Kong Polytechnic University, China
Kiriakos Kutulakos	University of Toronto, Canada

Suha Kwak	Pohang University of Science and Technology, Korea
Zorah Laehner	University of Bonn, Germany
Shang-Hong Lai	National Tsing Hua University, Taiwan
Iro Laina	University of Oxford, UK
Jean-Francois Lalonde	Université Laval, Canada
Loic Landrieu	École des Ponts ParisTech, France
Diane Larlus	Naver Labs Europe, France
Viktor Larsson	Lund University, Sweden
Stéphane Lathuilière	Telecom-Paris, France
Gim Hee Lee	National University of Singapore, Singapore
Yong Jae Lee	University of Wisconsin-Madison, USA
Bastian Leibe	RWTH Aachen University, Germany
Ales Leonardis	University of Birmingham, UK
Vincent Lepetit	Ecole des Ponts ParisTech, France
Fuxin Li	Oregon State University, USA
Hongdong Li	Australian National University, Australia
Xi Li	Zhejiang University, China
Yin Li	University of Wisconsin-Madison, USA
Yu Li	International Digital Economy Academy, Singapore
Xiaodan Liang	Sun Yat-sen University, China
Shengcai Liao	Inception Institute of Artificial Intelligence, United Arab Emirates
Jongwoo Lim	Seoul National University, Korea
Ser-Nam Lim	University of Central Florida, USA
Stephen Lin	Microsoft Research, China
Tsung-Yi Lin	NVIDIA Research, USA
Yen-Yu Lin	National Yang Ming Chiao Tung University, Taiwan
Yutian Lin	Wuhan University, China
Zhe Lin	Adobe Research, USA
Haibin Ling	Stony Brook University, USA
Or Litany	NVIDIA, USA
Jiaying Liu	Peking University, China
Jun Liu	Lancaster University, UK
Miaomiao Liu	Australian National University, Australia
Si Liu	Beihang University, China
Sifei Liu	NVIDIA, USA
Wei Liu	Tencent, China
Yanxi Liu	Pennsylvania State University, USA
Yaoyao Liu	Johns Hopkins University, USA

Yebin Liu	Tsinghua University, China
Zicheng Liu	Microsoft, USA
Ziwei Liu	Nanyang Technological University, Singapore
Ismini Lourentzou	University of Illinois, Urbana-Champaign, USA
Chen Change Loy	Nanyang Technological University, Singapore
Feng Lu	Beihang University, China
Le Lu	Alibaba Group, USA
Oisin Mac Aodha	University of Edinburgh, UK
Luca Magri	Politecnico di Milano, Italy
Michael Maire	University of Chicago, USA
Subhransu Maji	University of Massachusetts, Amherst, USA
Atsuto Maki	KTH Royal Institute of Technology, Sweden
Yasushi Makihara	Osaka University, Japan
Massimiliano Mancini	University of Trento, Italy
Kevis-Kokitsi Maninis	Google Research, Switzerland
Kenneth Marino	Google DeepMind, USA
Renaud Marlet	Ecole des Ponts ParisTech & Valeo, France
Niki Martinel	University of Udine, Italy
Jiri Matas	Czech Technical University in Pragu, Czech Republic
Yasuyuki Matsushita	Osaka University, Japan
Simone Melzi	University of Milano-Bicocca, Italy
Thomas Mensink	Google Research, Netherlands
Michele Merler	IBM Research, USA
Pascal Mettes	University of Amsterdam, Netherlands
Ajmal Mian	University of Western Australia, Australia
Krystian Mikolajczyk	Imperial College London, UK
Ishan Misra	Facebook AI Research, USA
Niloy Mitra	University College London, UK
Anurag Mittal	Indian Institute of Technology Madras, India
Davide Modolo	Amazon, USA
Davide Moltisanti	University of Bath, UK
Philippos Mordohai	Stevens Institute of Technology, USA
Francesc Moreno	Institut de Robotica i Informatica Industrial, Spain
Pietro Morerio	Istituto Italiano di Tecnologia, Italy
Greg Mori	Simon Fraser University & Borealis AI, Canada
Roozbeh Mottaghi	FAIR @ Meta, USA
Yadong Mu	Peking University, China
Vineeth N. Balasubramanian	Indian Institute of Technology, India
Hajime Nagahara	Osaka University, Japan
Vinay Namboodiri	University of Bath, UK
Liangliang Nan	Delft University of Technology, Netherlands

Michele Nappi	University of Salerno, Italy
Srinivasa Narasimhan	Carnegie Mellon University, USA
P. J. Narayanan	International Institute of Information Technology, Hyderabad, India
Nassir Navab	Technical University of Munich, Germany
Ram Nevatia	University of Southern California, USA
Natalia Neverova	Facebook AI Research, France
Juan Carlos Niebles	Salesforce & Stanford University, USA
Michael Niemeyer	Google, Germany
Simon Niklaus	Adobe Research, USA
Ko Nishino	Kyoto University, Japan
Nicoletta Noceti	MaLGa-DIBRIS & Università degli Studi di Genova, Italy
Matthew O'Toole	Carnegie Mellon University, USA
Jean-Marc Odobez	Idiap Research Institute, École polytechnique fédérale de Lausanne, Switzerland
Francesca Odone	Università di Genova, Italy
Takayuki Okatani	Tohoku University & RIKEN Center for Advanced Intelligence Project, Japan
Carl Olsson	Lund University, Sweden
Vicente Ordonez	Rice University, USA
Aljosa Osep	Technical University of Munich, Germany
Martin R. Oswald	University of Amsterdam, Netherlands
Andrew Owens	University of Michigan, USA
Tomas Pajdla	Czech Technical University in Prague, Czechia
Manohar Paluri	Meta, USA
Jinshan Pan	Nanjing University of Science and Technology, China
In Kyu Park	Inha University, Korea
Jaesik Park	Seoul National University, Korea
Despoina Paschalidou	Stanford, USA
Georgios Pavlakos	University of Texas at Austin, USA
Vladimir Pavlovic	Rutgers University, USA
Marcello Pelillo	University of Venice, Italy
David Picard	École des Ponts ParisTech, France
Hamed Pirsiavash	University of California, Davis, USA
Bryan Plummer	Boston University, USA
Thomas Pock	Graz University of Technology, Austria
Fabio Poiesi	Fondazione Bruno Kessler, Italy
Marc Pollefeys	ETH Zurich & Microsoft, Switzerland
Jean Ponce	Ecole normale supérieure - PSL, France
Gerard Pons-Moll	University of Tübingen, Germany

Jordi Pont-Tuset   Google, Switzerland
Fatih Porikli   Qualcomm R&D, USA
Brian Price   Adobe, USA
Guo-Jun Qi   Westlake University, USA
Long Quan   Hong Kong University of Science and
                                 Technology, China
Venkatesh Babu Radhakrishnan   Indian Institute of Science, India
Hossein Rahmani   Lancaster University, UK
Deva Ramanan   Carnegie Mellon University, USA
Vignesh Ramanathan   Facebook, USA
Vikram V. Ramaswamy   Princeton University, USA
Nalini Ratha   University at Buffalo, State University of New
                                 York, USA
Yogesh Rawat   University of Central Florida, USA
Hamid Rezatofighi   Monash University, Australia
Helge Rhodin   University of British Columbia, Canada
Stephan Richter   Apple, Germany
Anna Rohrbach   Technical University of Darmstadt, Germany
Marcus Rohrbach   Technical University of Darmstadt, Germany
Gemma Roig   Goethe University Frankfurt, Germany
Negar Rostamzadeh   Google, Canada
Paolo Rota   University of Trento, Italy
Stefan Roth   Technical University of Darmstadt, Germany
Carsten Rother   University of Heidelberg, Germany
Subhankar Roy   University of Trento, Italy
Michael Rubinstein   Google, USA
Christian Rupprecht   University of Oxford, UK
Olga Russakovsky   Princeton University, USA
Bryan Russell   Adobe Research, USA
Michael Ryoo   Stony Brook University & Google, USA
Reza Sabzevari   Delft University of Technology, Netherlands
Mathieu Salzmann   École polytechnique fédérale de Lausanne,
                                 Switzerland
Dimitris Samaras   Stony Brook University, USA
Aswin Sankaranarayanan   Carnegie Mellon University, USA
Sudeep Sarkar   University of South Florida, USA
Yoichi Sato   University of Tokyo, Japan
Manolis Savva   Simon Fraser University, Canada
Simone Schaub-Meyer   Technical University of Darmstadt, Germany
Bernt Schiele   Max Planck Institute for Informatics, Germany
Cordelia Schmid   Inria & Google, France
Nicu Sebe   University of Trento, Italy

Laura Sevilla-Lara	University of Edinburgh, UK
Fahad Shahbaz Khan	Mohamed bin Zayed University of Artificial Intelligence, United Arab Emirates
Qi Shan	Apple, USA
Shiguang Shan	Institute of Computing Technology, Chinese Academy of Sciences, China
Viktoriia Sharmanska	University of Sussex & Imperial College London, UK
Eli Shechtman	Adobe Research, USA
Evan Shelhamer	DeepMind, UK
Lu Sheng	Beihang University, China
Boxin Shi	Peking University, China
Jianbo Shi	University of Pennsylvania, USA
Mike Zheng Shou	National University of Singapore, Singapore
Abhinav Shrivastava	University of Maryland, USA
Aliaksandr Siarohin	Snap, USA
Kaleem Siddiqi	McGill University, Canada
Leonid Sigal	University of British Columbia, Canada
Oriane Siméoni	valeo.ai, France
Krishna Kumar Singh	Adobe Research, USA
Richa Singh	Indian Institute of Technology Jodhpur, India
Yale Song	Facebook AI Research, USA
Concetto Spampinato	University of Catania, Italy
Srinath Sridhar	Brown University, USA
Bjorn Stenger	Rakuten, Japan
Abby Stylianou	Saint Louis University, USA
Akihiro Sugimoto	National Institute of Informatics, Japan
Chen Sun	Brown University, USA
Deqing Sun	Google, USA
Jian Sun	Xi'an Jiaotong University, China
Min Sun	National Tsing Hua University, Taiwan
Qianru Sun	Singapore Management University, Singapore
Yu-Wing Tai	Dartmouth College, China
Ayellet Tal	Technion, Israel
Ping Tan	Hong Kong University of Science and Technology, China
Robby Tan	National University of Singapore, Singapore
Chi-Keung Tang	Hong Kong University of Science and Technology, China
Makarand Tapaswi	International Institute of Information Technology, Hyderabad & Wadhwani AI, India
Justus Thies	Technical University of Darmstadt, Germany

Radu Timofte	University of Wurzburg & ETH Zurich, Germany
Giorgos Tolias	Czech Technical University in Prague, Czechia
Federico Tombari	Google & Technical University of Munich, Switzerland
Tatiana Tommasi	Politecnico di Torino, Italy
James Tompkin	Brown University, USA
Matthew Trager	AWS AI Labs, USA
Anh Tran	VinAI, Vietnam
Du Tran	Google, USA
Shubham Tulsiani	Carnegie Mellon University, USA
Sergey Tulyakov	Snap, USA
Dimitrios Tzionas	University of Amsterdam, Netherlands
Maria Vakalopoulou	CentraleSupelec, France
Joost van de Weijer	Computer Vision Center, Spain
Laurens van der Maaten	Meta, USA
Jan van Gemert	Delft University of Technology, Netherlands
Sebastiano Vascon	Ca' Foscari University of Venice & European Centre for Living Technology, Italy
Nuno Vasconcelos	University of California, San Diego, USA
Mayank Vatsa	Indian Institute of Technology Jodhpur, India
Javier Vazquez-Corral	Autonomous University of Barcelona, Spain
Andrea Vedaldi	Oxford University, UK
Olga Veksler	University of Waterloo, Canada
Luisa Verdoliva	University Federico II of Naples, Italy
Rene Vidal	University of Pennsylvania, USA
Bo Wang	CtrsVision, USA
Heng Wang	TikTok, USA
Jingdong Wang	Baidu, China
Jingya Wang	ShanghaiTech University, China
Ruiping Wang	Institute of Computing Technology, Chinese Academy of Sciences, China
Shenlong Wang	University of Illinois, Urbana-Champaign, USA
Ting-Chun Wang	NVIDIA, USA
Wenguan Wang	Zhejiang University, China
Xiaolong Wang	University of California, San Diego, USA
Xiaoqian Wang	Purdue University, USA
Xiaoyu Wang	Hong Kong University of Science and Technology, USA
Xin Wang	Microsoft Research, USA
Xinchao Wang	National University of Singapore, Singapore
Yang Wang	Concordia University, Canada
Yiming Wang	Fondazione Bruno Kessler, Italy

Yu-Chiang Frank Wang	National Taiwan University, Taiwan
Yu-Xiong Wang	University of Illinois, Urbana-Champaign, USA
Zhangyang Wang	University of Texas at Austin, USA
Olivia Wiles	DeepMind, UK
Christian Wolf	Naver Labs Europe, France
Michael Wray	University of Bristol, UK
Jiajun Wu	Stanford University, USA
Jianxin Wu	Nanjing University, China
Shangzhe Wu	Stanford University, USA
Ying Wu	Northwestern University, USA
Yu Wu	Wuhan University, China
Jianwen Xie	Baidu Research, USA
Saining Xie	New York University, USA
Weidi Xie	Shanghai Jiao Tong University, China
Dan Xu	Hong Kong University of Science and Technology, China
Huijuan Xu	Pennsylvania State University, USA
Xiangyu Xu	Xi'an Jiaotong University, China
Yan Yan	Illinois Institute of Technology, USA
Jiaolong Yang	Microsoft Research, China
Kaiyu Yang	California Institute of Technology, USA
Linjie Yang	ByteDance AI Lab, USA
Ming-Hsuan Yang	University of California, Merced, USA
Yanchao Yang	University of Hong Kong, China
Yi Yang	Zhejiang University, China
Angela Yao	National University of Singapore, Singapore
Mang Ye	Wuhan University, China
Kwang Moo Yi	University of British Columbia, Canada
Xi Yin	Facebook, USA
Chang D. Yoo	Korea Advanced Institute of Science and Technology, Korea
Shaodi You	University of Amsterdam, Netherlands
Jingyi Yu	Shanghai Tech University, China
Ning Yu	Netflix Eyeline Studios, USA
Qi Yu	Rochester Institute of Technology, USA
Stella Yu	University of Michigan, USA
Yizhou Yu	University of Hong Kong, China
Junsong Yuan	University at Buffalo, State University of New York, USA
Lu Yuan	Microsoft, USA
Sangdoo Yun	NAVER AI Lab, Korea
Hongbin Zha	Peking University, China

xxvi   Organization

Jing Zhang   Australian National University, Australia
Lei Zhang   Hong Kong Polytechnic University, China
Ning Zhang   Facebook AI, USA
Richard Zhang   Adobe, USA
Tianzhu Zhang   University of Science and Technology of China, China

Hengshuang Zhao   University of Hong Kong, China
Qi Zhao   University of Minnesota, USA
Sicheng Zhao   Tsinghua University, China
Liang Zheng   Australian National University, Australia
Wei-Shi Zheng   Sun Yat-sen University, China
Zhun Zhong   University of Nottingham, UK
Bolei Zhou   University of California, Los Angeles, USA
Kaiyang Zhou   Hong Kong Baptist University, China
Luping Zhou   University of Sydney, Australia
S. Kevin Zhou   University of Science and Technology of China, China
Lei Zhu   Hong Kong University of Science and Technology (Guangzhou), China
Linchao Zhu   Zhejiang University, China
Andrew Zisserman   University of Oxford, UK
Daniel Zoran   DeepMind, UK
Wangmeng Zuo   Harbin Institute of Technology, China

Technical Program Committee

Sathyanarayanan N. Aakur
Andrea Abate
Wim Abbeloos
Amos L. Abbott
Rameen Abdal
Salma Abdel Magid
Rabab Abdelfattah
Ahmed Abdelkader
Eslam Mohamed Abdelrahman
Ahmed Abdelreheem
Akash Abdu Jyothi
Kfir Aberman
Shahira Abousamra
Yazan Abu Farha
Shady Abu-Hussein

Abulikemu Abuduweili
Hanno Ackermann
Chandranath Adak
Aikaterini Adam
Lukáš Adam
Kamil Adamczewski
Donald Adjeroh
Mohammed Adnan
Mahmoud Afifi
Triantafyllos Afouras
Akshay Agarwal
Madhav Agarwal
Nakul Agarwal
Sharat Agarwal
Shivang Agarwal
Vatsal Agarwal

Abhinav Agarwalla
Shivam Aggarwal
Sara Aghajanzadeh
Shashank Agnihotri
Harsh Agrawal
Antonio Agudo
Shai Aharon
Saba Ahmadi
Waqar Ahmed
Byeongjoo Ahn
Chanho Ahn
Jaewoo Ahn
Namhyuk Ahn
Seoyoung Ahn
Nilesh A. Ahuja
Yihao Ai

Abhishek Aich
Emanuele Aiello
Noam Aigerman
Kiyoharu Aizawa
Kenan Emir Ak
Reza Akbarian Bafghi
Sai Aparna Aketi
Derya Akkaynak
Ibrahim Alabdulmohsin
Yuval Alaluf
Md Ferdous Alam
Stephan Alaniz
Badour A. Sh AlBahar
Paul Albert
Isabela Albuquerque
Emanuel Aldea
Luca Alessandrini
Emma Alexander
Konstantinos P.
    Alexandridis
Stamatis Alexandropoulos
Saghir Alfasly
Mohsen Ali
Waqar Ali
Daniel Aliaga
Sadegh Aliakbarian
Garvita Allabadi
Thiemo Alldieck
Antonio Alliegro
Jurandy Almeida
Gal Alon
Hadi Alzayer
Ibtihel Amara
Giuseppe Amato
Sameer Ambekar
Niki Amini-Naieni
Shir Amir
Roberto Amoroso
Dongsheng An
Jie An
Junyi An
Liang An
Xiang An
Yuexuan An
Zhaochong An

Zhulin An
Saket Anand
Pavan Kumar Anasosalu
    Vasu
Codruta O. Ancuti
Cosmin Ancuti
Fernanda Andaló
Connor Anderson
Juan Andrade-Cetto
Bruno Andreis
Alexander Andreopoulos
Bjoern Andres
Shivangi Aneja
Dragomir Anguelov
Damith Kawshan
    Anhettigama
Aditya Annavajjala
Michel Antunes
Tejas Anvekar
Saeed Anwar
Evlampios Apostolidis
Srikar Appalaraju
Relja Arandjelović
Nikita Araslanov
Alexandre Araujo
Helder Araujo
Eric Arazo
Dawit Mureja Argaw
Anurag Arnab
Federica Arrigoni
Alexey Artemov
Aditya Arun
Nader Asadi
Kumar Ashutosh
Vishal Asnani
Mahmoud Assran
Amir Atapour-Abarghouei
Nikos Athanasiou
Ali Athar
ShahRukh Athar
Rowel O. Atienza
Benjamin Attal
Matan Atzmon
Nicolas Audebert
Romaric Audigier

Tristan
    Aumentado-Armstrong
Yannis Avrithis
Muhammad Awais
Md Rabiul Awal
Görkay Aydemir
Mehmet Aygün
Melika Ayoughi
Ali Ayub
Kumar Ayush
Mehdi Azabou
Basim Azam
Mohammad Farid
    Azampour
Hossein Azizpour
Vivek B. S.
Yunhao Ba
Mohammadreza Babaee
Deepika Bablani
Sudarshan Babu
Roman Bachmann
Inhwan Bae
Jeonghun Baek
Kyungjune Baek
Seungryul Baek
Piyush Nitin Bagad
Mohammadhossein Bahari
Gaetan Bahl
Sherwin Bahmani
Bing Bai
Haoyue Bai
Jiawang Bai
Jinbin Bai
Lei Bai
Qingyan Bai
Shutao Bai
Xuyang Bai
Yalong Bai
Yuanchao Bai
Yue Bai
Yunpeng Bai
Ziyi Bai
Daniele Baieri
Sungyong Baik
Ramon Baldrich

Coloma Ballester
Irene Ballester
Sonat Baltaci
Biplab Banerjee
Monami Banerjee
Sandipan Banerjee
Snehasis Banerjee
Soumya Banerjee
Sreya Banerjee
Jihwan Bang
Ankan Bansal
Nitin Bansal
Siddhant Bansal
Chong Bao
Hangbo Bao
Runxue Bao
Wentao Bao
Yiwei Bao
Zhipeng Bao
Zhongyun Bao
Amir Bar
Omer Bar Tal
Manel Baradad Jurjo
Lorenzo Baraldi
Lorenzo Baraldi
Tal Barami
Danny Barash
Daniel Barath
Adrian Barbu
Nick Barnes
Alina J. Barnett
German Barquero
Joao P. Barreto
Jonathan T. Barron
Rodrigo C. Barros
Luca Barsellotti
Myles Bartlett
Kristijan Bartol
Hritam Basak
Dina Bashkirova
Chaim Baskin
Lennart C. Bastian
Abhipsa Basu
Soumen Basu
Peyman Bateni

Anil Batra
David Bau
Zuria Bauer
Eduard Gabriel Bazavan
Federico Becattini
Nathan A. Beck
Jan Bednarik
Peter A. Beerel
Ardhendu Behera
Harkirat Singh Behl
Jens Behley
William Beksi
Soufiane Belharbi
Giovanni Bellitto
Ismail Ben Ayed
Abdessamad Ben Hamza
Yizhak Ben-Shabat
Guy Ben-Yosef
Robert Benavente
Assia Benbihi
Yasser Benigmim
Urs Bergmann
Amit H Bermano
Jesus Bermudez-Cameo
Florian Bernard
Stefano Berretti
Gabriele Berton
Petra Bevandić
Matthew Beveridge
Lalit Bhagat
Yash Bhalgat
Homanga Bharadhwaj
Skanda Bharadwaj
Ambika Saklani Bhardwaj
Goutam Bhat
Gaurav Bhatt
Neel P. Bhatt
Deblina Bhattacharjee
Anish Bhattacharya
Rajarshi Bhattacharya
Uttaran Bhattacharya
Apratim Bhattacharyya
Anand Bhattad
Binod Bhattarai
Snehal Bhayani

Brojeshwar Bhowmick
Aritra Bhowmik
Neelanjan Bhowmik
Amran Bhuiyan
Ankan Kumar Bhunia
Ayan Kumar Bhunia
Jing Bi
Qi Bi
Xiuli Bi
Michael Bi Mi
Hao Bian
Jia-Wang Bian
Shaojun Bian
Simone Bianco
Pia Bideau
Xiaoyu Bie
Roberto Bigazzi
Mario Bijelic
Bahri Batuhan Bilecen
Guillaume-Alexandre
 Bilodeau
Alexander Binder
Massimo Bini
Niccolò Bisagno
Carmen Bisogni
Anthony R. Bisulco
Sandika Biswas
Sanket Biswas
Ksenia Bittner
Kevin Blackburn-Matzen
Nathaniel Blanchard
Hunter Blanton
Emile Blettery
Hermann Blum
Deyu Bo
Janusz Bobulski
Marius Bock
Lea Bogensperger
Simion-Vlad Bogolin
Moritz Böhle
Piotr Bojanowski
Alexey Bokhovkin
Georg Bökman
Ladislau Boloni
Daniel Bolya

Lorenzo Bonicelli
Gianpaolo Bontempo
Vivek Boominathan
Adrian Bors
Damian Borth
Matteo Bortolon
Davide Boscaini
Laurie Nicholas Bose
Shirsha Bose
Mark Boss
Andrea Bottino
Larbi Boubchir
Alexandre Boulch
Terrance E. Boult
Amine Bourki
Walid Bousselham
Fadi Boutros
Nicolas C. Boutry
Thierry Bouwmans
Richard S. Bowen
Kevin W. Bowyer
Ivaylo Boyadzhiev
Aidan Boyd
Joseph Boyd
Aljaz Bozic
Behzad Bozorgtabar
Manuel Brack
Samarth Brahmbhatt
Nikolas Brasch
Cameron E. Braunstein
Toby P. Breckon
Mathieu Bredif
Romain Brégier
Carlo Bretti
Joel R. Brogan
Clifford Broni-Bediako
Andrew Brown
Ellis L. Brown
Thomas Brox
Qingwen Bu
Tong Bu
Shyamal Buch
Alvaro Budria
Ignas Budvytis
José M. Buenaposada

Kyle R. Buettner
Marcel C. Bühler
Nhat-Tan Bui
Adrian Bulat
Valay M. Bundele
Nathaniel J. Burgdorfer
Ryan D. Burgert
Evgeny Burnaev
Pietro Buzzega
Marco Buzzelli
Kwon Byung-Ki
Fabian Caba
Felipe Cadar
Martin Cadik
Bowen Cai
Changjiang Cai
Guanyu Cai
Haoming Cai
Hongrui Cai
Jiarui Cai
Jinyu Cai
Kaixin Cai
Lile Cai
Minjie Cai
Mu Cai
Pingping Cai
Rizhao Cai
Ruisi Cai
Ruojin Cai
Shen Cai
Shengqu Cai
Weidong Cai
Weiwei Cai
Wenjia Cai
Wenxiao Cai
Yancheng Cai
Yuanhao Cai
Yujun Cai
Yuxuan Cai
Zhaowei Cai
Zhipeng Cai
Zhixi Cai
Zhongang Cai
Zikui Cai
Salvatore Calcagno

Roberto Caldelli
Akin Caliskan
Lilian Calvet
Necati Cihan Camgoz
Tommaso Campari
Dylan Campbell
Guglielmo Camporese
Yigit Baran Can
Shaun Canavan
Gemma Canet Tarrés
Marco Cannici
Ang Cao
Anh-Quan Cao
Bing Cao
Bingyi Cao
Chenjie Cao
Dongliang Cao
Hu Cao
Jiahang Cao
Jiale Cao
Jie Cao
Jiezhang Cao
Jinkun Cao
Meng Cao
Mingdeng Cao
Peng Cao
Pu Cao
Qinglong Cao
Qingxing Cao
Shengcao Cao
Song Cao
Weiwei Cao
Xiangyong Cao
Xiao Cao
Xu Cao
Xu Cao
Yan-Pei Cao
Yang Cao
Yichao Cao
Ying Cao
Yuan Cao
Yuhang Cao
Yukang Cao
Yulong Cao
Yun-Hao Cao

Yunhao Cao
Yutong Cao
Zhangjie Cao
Zhiguo Cao
Zhiwen Cao
Ziang Cao
Giacomo Capitani
Luigi Capogrosso
Andrea Caraffa
Jaime S. Cardoso
Chris Careaga
Zachariah Carmichael
Giuseppe Cartella
Vincent Cartillier
Paola Cascante-Bonilla
Lucia Cascone
Pol Caselles Rico
Angela Castillo
Lluis Castrejon
Modesto
   Castrillón-Santana
Francisco M. Castro
Pedro Castro
Jacopo Cavazza
Oya Celiktutan
Luigi Celona
Jiazhong Cen
Fernando J. Cendra
Fabio Cermelli
Duygu Ceylan
Eunju Cha
Junbum Cha
Rohan Chabra
Rohan Chacko
Julia Chae
Nicolas Chahine
Junyi Chai
Lucy Chai
Tianrui Chai
Wenhao Chai
Zenghao Chai
Anish Chakrabarty
Goirik Chakrabarty
Anirban Chakraborty
Deep Chakraborty
Souradeep Chakraborty
Punarjay Chakravarty
Jacob Chalk
Loick Chambon
Chee Seng Chan
David Chan
Dorian Y. Chan
Kin Sun Chan
Matthew Chan
Adrien Chan-Hon-Tong
Sampath Chanda
Shivam Chandhok
Prashanth Chandran
Kenneth Chaney
Chirui Chang
Di Chang
Dongliang Chang
Hong Chang
Hyung Jin Chang
Ju Yong Chang
Matthew Chang
Ming-Ching Chang
MingFang Chang
Nadine Chang
Qi Chang
Seunggyu Chang
Wei-Yi Chang
Xiaojun Chang
Yakun Chang
Yi Chang
Zheng Chang
Sumohana S.
   Channappayya
Hanqing Chao
Pradyumna Chari
Korawat Charoenpitaks
Dibyadip Chatterjee
Moitreya Chatterjee
Pratik Chaudhari
Muawiz Chaudhary
Ritwick Chaudhry
Ruchika Chavhan
Sofian Chaybouti
Zhengping Che
Gullal Singh Cheema
Kunal Chelani
Rama Chellappa
Allison Y. Chen
Anpei Chen
Aozhu Chen
Bin Chen
Bin Chen
Bin Chen
Bohong Chen
Bor-Chun Chen
Bowei Chen
Brian Chen
Chang Wen Chen
Changan Chen
Changhao Chen
Changhuai Chen
Chaofan Chen
Chaofan Chen
Chaofeng Chen
Chen Chen
Cheng Chen
Chengkuan Chen
Chieh-Yun Chen
Chong Chen
Chun-Fu Richard Chen
Congliang Chen
Cuiqun Chen
Da Chen
Daoyuan Chen
Defang Chen
Dian Chen
Dian Chen
Ding-Jie Chen
Dingfan Chen
Dong Chen
Dongdong Chen
Fangyi Chen
Guang-Yong Chen
Guangyao Chen
Guangyi Chen
Guanying Chen
Guikun Chen
Guo Chen
Haibo Chen
Haiwei Chen

Hansheng Chen
Hanting Chen
Hanzhi Chen
Hao Chen
Haodong Chen
Haokun Chen
Haoran Chen
Haoxin Chen
Haoyu Chen
Haoyu Chen
Hong-You Chen
Honghua Chen
Honglin Chen
Hongming Chen
Huanran Chen
Hui Chen
Hwann-Tzong Chen
Jiacheng Chen
Jiajing Chen
Jianqiu Chen
Jiaqi Chen
Jiaxin Chen
Jiayi Chen
Jie Chen
Jie Chen
Jie Chen
Jieneng Chen
Jierun Chen
Jinghui Chen
Jingjing Chen
Jintai Chen
Jinyu Chen
Joya Chen
Jun Chen
Jun Chen
Jun Chen
Jun-Cheng Chen
Jun-Kun Chen
Junjie Chen
Kai Chen
Kai Chen
Kaifeng Chen
Kefan Chen
Kejiang Chen
Kuan-Wen Chen

Kunyao Chen
Li Chen
Liang Chen
Lianggangxu Chen
Liangyu Chen
Ling-Hao Chen
Linghao Chen
Liyan Chen
Min Chen
Min-Hung Chen
Minghao Chen
Minghui Chen
Nenglun Chen
Peihao Chen
Pengguang Chen
Ping Chen
Qi Chen
Qi Chen
Qi Chen
Qiang Chen
Qimin Chen
Qingcai Chen
Rongyu Chen
Rui Chen
Runjian Chen
Shang-Fu Chen
Shen Chen
Sherry X. Chen
Shi Chen
Shiming Chen
Shiyan Chen
Shizhe Chen
Shoufa Chen
Shu-Yu Chen
Shuai Chen
Si Chen
Siheng Chen
Simin Chen
Sixiang Chen
Sizhe Chen
Songcan Chen
Tao Chen
Tianshui Chen
Tianyi Chen
Tsai-Shien Chen

Wei Chen
Weidong Chen
Weihua Chen
Weijie Chen
Weikai Chen
Wuyang Chen
Xi Chen
Xi Chen
Xiang Chen
Xiangyu Chen
Xianyu Chen
Xiao Chen
Xiao Chen
Xiaohan Chen
Xiaokang Chen
Xiaotong Chen
Xin Chen
Xin Chen
Xin Chen
Xinghao Chen
Xingyu Chen
Xingyu Chen
Xinlei Chen
Xiuwei Chen
Xuanbai Chen
Xuanhong Chen
Xuesong Chen
Xuweiyi Chen
Xuxi Chen
Yanbei Chen
Yang Chen
Yaofo Chen
Yen-Chun Chen
Yi-Ting Chen
Yi-Wen Chen
Yilun Chen
Yinbo Chen
Yingcong Chen
Yingwen Chen
Yiting Chen
Yixin Chen
Yixin Chen
Yu Chen
Yuanhong Chen
Yuanqi Chen

Yuedong Chen	Ziyang Chen	Seunggeun Chi
Yueting Chen	An-Chieh Cheng	Zhixiang Chi
Yuhao Chen	Bowen Cheng	Naoya Chiba
Yuhao Chen	De Cheng	Julian Chibane
Yujin Chen	Erkang Cheng	Aditya Chinchure
Yukang Chen	Feng Cheng	Eleni Chiou
Yun-Chun Chen	Guangliang Cheng	Mia Chiquier
Yunjin Chen	Hao Cheng	Chiranjeev Chiranjeev
Yunlu Chen	Hao Cheng	Kashyap Chitta
Yuntao Chen	Harry Cheng	Hsu-kuang Chiu
Yushuo Chen	Ho Kei Cheng	Wei-Chen Chiu
Yuxiao Chen	Jingchun Cheng	Chee Kheng Chng
Zehui Chen	Jun Cheng	Shin-Fang Chng
Zerui Chen	Ka Leong Cheng	Donghyeon Cho
Zewen Chen	Kelvin B. Cheng	Gyusang Cho
Zeyuan Chen	Kevin Ho Man Cheng	Hanbyel Cho
Zeyuan Chen	Lechao Cheng	Hoonhee Cho
Zhang Chen	Shuo Cheng	Jae Won Cho
Zhao-Min Chen	Silin Cheng	Jang Hyun Cho
Zhaoliang Chen	Ta-Ying Cheng	Jungchan Cho
Zhaoxi Chen	Tianhang Cheng	Junhyeong Cho
Zhaoyu Chen	Weihao Cheng	Nam Ik Cho
Zhaozheng Chen	Wencan Cheng	Seokju Cho
Zhaozheng Chen	Wensheng Cheng	Seungju Cho
Zhe Chen	Xi Cheng	Suhwan Cho
Zhen Chen	Xize Cheng	Sunghyun Cho
Zhen Chen	Xuxin Cheng	Sungjun Cho
Zheng Chen	Yen-Chi Cheng	Sungmin Cho
Zhenghao Chen	Yi Cheng	Wonwoong Cho
Zhengyu Chen	Yihua Cheng	Nathaniel E. Chodosh
Zhenyu Chen	Yu Cheng	Jaesung Choe
Zhi Chen	Zesen Cheng	Junsuk Choe
Zhibo Chen	Zezhou Cheng	Changwoon Choi
Zhimin Chen	Zhanzhan Cheng	Chiho Choi
Zhineng Chen	Zhen Cheng	Ching Lam Choi
Zhiwei Chen	Zhi-Qi Cheng	Dongyoung Choi
Zhixiang Chen	Ziang Cheng	Dooseop Choi
Zhiyang Chen	Ziheng Cheng	Hongsuk Choi
Zhiyuan Chen	Soon Yau Cheong	Hyesong Choi
Zhuangzhuang Chen	Anoop Cherian	Jaewoong Choi
Zhuo Chen	Daniil Cherniavskii	Janghoon Choi
Zhuoxiao Chen	Ajad Chhatkuli	Jin Young Choi
Zihan Chen	Cheng Chi	Jinwoo Choi
Zijiao Chen	Haoang Chi	Jinyoung Choi
Ziwen Chen	Hyung-gun Chi	Jonghyun Choi

Jongwon Choi
Jooyoung Choi
Kanghyun Choi
Myungsub Choi
Seokeon Choi
Wonhyeok Choi
Bhavin Choksi
Jaegul Choo
Baptiste Chopin
Ayush Chopra
Gene Chou
Shih-Han Chou
Divya Choudhary
Siddharth Choudhary
Sunav Choudhary
Rohan Choudhury
Vasileios Choutas
Ka-Ho Chow
Pinaki Nath Chowdhury
Sanjoy Chowdhury
Sammy Christen
Fu-Jen Chu
Qi Chu
Ruihang Chu
Sanghyeok Chu
Wei-Ta Chu
Wen-Hsuan Chu
Wen-Hsuan Chu
Wei Qin Chuah
Andrei Chubarau
Ilya Chugunov
Sanghyuk Chun
Se Young Chun
Hyungjin Chung
Inseop Chung
Jaeyoung Chung
Jihoon Chung
Jiwan Chung
Joon Son Chung
Thomas A. Ciarfuglia
Umur A. Ciftci
Antonio E. Cinà
Ramazan Gokberk Cinbis
Anthony Cioppa
Javier Civera

Christopher A. Clark
James J. Clark
Ronald Clark
Brian S. Clipp
Mark Coates
Federico Cocchi
Felipe Codevilla
Cecília Coelho
Danny Cohen-Or
Forrester Cole
Julien Colin
John Collomosse
Marc Comino-Trinidad
Nicola Conci
Marcos V. Conde
Alexandru Paul
   Condurache
Runmin Cong
Tianshuo Cong
Wenyan Cong
Yuren Cong
Alessandro Conti
Andrea Conti
Enric Corona
John R. Corring
Jaime Corsetti
Jason J. Corso
Theo W. Costain
Guillaume Couairon
Robin Courant
Davide Cozzolino
Donato Crisostomi
Francesco Croce
Florinel Alin Croitoru
Ioana Croitoru
James L. Crowley
Steve Cruz
Gabriela Csurka
Alfredo Cuesta Infante
Aiyu Cui
Hainan Cui
Jiali Cui
Jiequan Cui
Justin Cui
Qiongjie Cui

Ruikai Cui
Yawen Cui
Ying Cui
Yuning Cui
Zhen Cui
Zhenyu Cui
Ziteng Cui
Benjamin J. Culpepper
Xiaodong Cun
Federico Cunico
Claudio Cusano
Sebastian Cygert
Guido Maria D'Amely di
   Melendugno
Moreno D'Incà
Feipeng Da
Mosam Dabhi
Rishabh Dabral
Konstantinos M. Dafnis
Matthew L. Daggitt
Anders B. Dahl
Bo Dai
Bo Dai
Haixing Dai
Mengyu Dai
Peng Dai
Peng Dai
Pingyang Dai
Qi Dai
Qiyu Dai
Wenrui Dai
Wenxun Dai
Zuozhuo Dai
Nicola Dall'Asen
Naser Damer
Bo Dang
Meghal Dani
Duolikun Danier
Donald G. Dansereau
Quan Dao
Son Duy Dao
Trung Tuan Dao
Mohamed Daoudi
Ahmad Darkhalil
Rangel Daroya

Abhijit Das
Abir Das
Anurag Das
Ayan Das
Nilotpal Das
Partha Das
Soumi Das
Srijan Das
Sukhendu Das
Swagatam Das
Ananyananda Dasari
Avijit Dasgupta
Gourav Datta
Achal Dave
Ishan Rajendrakumar Dave
Andrew Davison
Aram Davtyan
Axel Davy
Youssef Dawoud
Laura Daza
Francesca De Benetti
Daan de Geus
Alfredo S. De Goyeneche
Riccardo de Lutio
Maria De Marsico
Christophe De
   Vleeschouwer
Tonmoay Deb
Sayan Deb Sarkar
Dale Decatur
Thanos Delatolas
Fabien Delattre
Mauricio Delbracio
Ginger D. Delmas
Andong Deng
Bailin Deng
Bin Deng
Boyang Deng
Chaorui Deng
Cheng Deng
Congyue Deng
Jiacheng Deng
Jiajun Deng
Jiankang Deng
Jingjing Deng
Jinhong Deng
Kangle Deng
Kenan Deng
Lei Deng
Liang-Jian Deng
Shijian Deng
Weijian Deng
Xiaoming Deng
Xueqing Deng
Ye Deng
Yongjian Deng
Youming Deng
Yu Deng
Zhijie Deng
Zhongying Deng
Matteo Denitto
Mohammad Mahdi
   Derakhshani
Alakh Desai
Kevin Desai
Nishq P. Desai
Sai Vikas Desai
Jean-Emmanuel Deschaud
Frédéric Devernay
Rahul Dey
Arturo Deza
Helisa Dhamo
Amaya Dharmasiri
Ankit Dhiman
Vikas Dhiman
Yan Di
Zonglin Di
Ousmane A. Dia
Renwei Dian
Haiwen Diao
Xiaolei Diao
Gonçalo José Dias Pais
Abdallah Dib
Juan Carlos Dibene
   Simental
Sebastian Dille
Christian Diller
Mariella Dimiccoli
Anastasios Dimou
Or Dinari
Caiwen Ding
Changxing Ding
Choubo Ding
Fangqiang Ding
Guiguang Ding
Guodong Ding
Henghui Ding
Hui Ding
Jian Ding
Jian-Jiun Ding
Jianhao Ding
Li Ding
Liang Ding
Lihe Ding
Ming Ding
Runyu Ding
Shouhong Ding
Shuangrui Ding
Shuxiao Ding
Tianjiao Ding
Tianyu Ding
Wenhao Ding
Xiaohan Ding
Xinpeng Ding
Yaqing Ding
Yong Ding
Yuhang Ding
Yuqi Ding
Zheng Ding
Zhengming Ding
Zhipeng Ding
Zhuangzhuang Ding
Zihan Ding
Zihan Ding
Ziluo Ding
Anh-Dung Dinh
Markos Diomataris
Christos Diou
Alish Dipani
Alara Dirik
Ajay Divakaran
Abdelaziz Djelouah
Nemanja Djuric
Thanh-Toan Do
Tien Do

Anh-Dzung Doan
Bao Gia Doan
Khoa D. Doan
Carl Doersch
Andreea Dogaru
Nehal Doiphode
Csaba Domokos
Bowen Dong
Haoye Dong
Jiahua Dong
Jianfeng Dong
Jinpeng Dong
Junhao Dong
Junting Dong
Li Dong
Minjing Dong
Nanqing Dong
Peijie Dong
Qiaole Dong
Qihua Dong
Qiulei Dong
Runpei Dong
Shichao Dong
Shuting Dong
Sixun Dong
Siyan Dong
Tian Dong
Wei Dong
Wei Dong
Weisheng Dong
Xiaoyi Dong
Xiaoyu Dong
Xingping Dong
Yinpeng Dong
Zhenxing Dong
Jonathan C. Donnelly
Naren Doraiswamy
Gianfranco Doretto
Michael Dorkenwald
Muskan Dosi
Huanzhang Dou
Shuguang Dou
Yiming Dou
Zhiyang Frank Dou
Zi-Yi Dou
Arthur Douillard
Michail C. Doukas
Matthijs Douze
Amil Dravid
Vincent Drouard
Dawei Du
Jia-Run Du
Jinhao Du
Ruoyi Du
Weiyu Du
Xiaodan Du
Ye Du
Yingjun Du
Yong Du
Yuanqi Du
Yuntao Du
Zexing Du
Zhuoran Du
Radhika Dua
Haodong Duan
Haoran Duan
Huiyu Duan
Jiafei Duan
Jiali Duan
Peiqi Duan
Ye Duan
Yuchen Duan
Yueqi Duan
Zhihao Duan
Amanda Cardoso Duarte
Shiv Ram Dubey
Anastasia Dubrovina
Daniel Duckworth
Timothy Duff
Nicolas Dufour
Rahul Duggal
Shivam Duggal
Bardienus P. Duisterhof
Felix Dülmer
Matteo Dunnhofer
Chi Nhan Duong
Elona Dupont
Nikita Durasov
Zoran Duric
Mihai Dusmanu
Titir Dutta
Ujjal Kr Dutta
Isht Dwivedi
Sai Kumar Dwivedi
Sebastian Dziadzio
Thomas Eboli
Ramin Ebrahim Nakhli
Benjamin Eckart
Takeharu Eda
Johan Edstedt
Alexei A. Efros
Nikos Efthymiadis
Bernhard Egger
Max Ehrlich
Iván Eichhardt
Farshad Einabadi
Martin Eisemann
Peter Eisert
Mohamed El Banani
Mostafa El-Khamy
Alaa El-Nouby
Randa I. M. Elanwar
James Elder
Abdelrahman Eldesokey
Ismail Elezi
Ehsan Elhamifar
Moshe Eliasof
Sara Elkerdawy
Qing En
Ian Endres
Andreas Engelhardt
Francis Engelmann
Martin Engilberge
Kenji Enomoto
Chanho Eom
Guy Erez
Linus Ericsson
Maria C. Escobar
Marcos Escudero-Viñolo
Sedigheh Eslami
Valter Estevam
Ali Etemad
Martin Nicolas Everaert
Ivan Evtimov
Ralph Ewerth

Fevziye Irem Eyiokur
Cristobal Eyzaguirre
Gabriele Facciolo
Mohammad Fahes
Masud ANI Fahim
Fabrizio Falchi
Alex Falcon
Aoxiang Fan
Baojie Fan
Bin Fan
Chao Fan
David Fan
Haoqiang Fan
Hehe Fan
Heng Fan
Hongyi Fan
Huijie Fan
Junsong Fan
Lei Fan
Lei Fan
Lifeng Fan
Lue Fan
Mingyuan Fan
Qi Fan
Rui R. Fan
Xiran Fan
Yifei Fan
Yue Fan
Zejia Fan
Zhaoxin Fan
Zhenfeng Fan
Zhentao Fan
Zhipeng Fan
Zhiwen Fan
Zicong Fan
Sean Fanello
Chaowei Fang
Chuan Fang
Gongfan Fang
Guian Fang
Han Fang
Jiansheng Fang
Jiemin Fang
Jin Fang
Jun Fang
Kun Fang
Pengfei Fang
Rui Fang
Sheng Fang
Xiang Fang
Xiaolin Fang
Xiuwen Fang
Zhaoyuan Fang
Zhiyuan Fang
Eros Fanì
Matteo Farina
Farzan Farnia
Aiman Farooq
Azade Farshad
Mohsen Fayyaz
Arash Fayyazi
Ben Fei
Jingjing Fei
Xiaohan Fei
Zhengcong Fei
Michael Felsberg
Berthy T. Feng
Bin Feng
Brandon Y. Feng
Chao Feng
Chen Feng
Chengjian Feng
Chun-Mei Feng
Fan Feng
Guorui Feng
Hao Feng
Jianjiang Feng
Lan Feng
Lei Feng
Litong Feng
Mingtao Feng
Qianyu Feng
Ruicheng Feng
Ruili Feng
Ruoyu Feng
Ryan T. Feng
Shiwei Feng
Songhe Feng
Tuo Feng
Wei Feng
Weitao Feng
Weixi Feng
Xuelu Feng
Yanglin Feng
Yao Feng
Yue Feng
Zhanxiang Feng
Zhenhua Feng
Zunlei Feng
Clara Fernandez
Victoria Fernandez Abrevaya
Miguel-Ángel Fernández-Torres
Sira Ferradans
Claudio Ferrari
Ethan Fetaya
Mustansar Fiaz
Guénolé Fiche
Panagiotis P. Filntisis
Marco Fiorucci
Kai Fischer
Tobias Fischer
Tom Fischer
Volker Fischer
Robert B. Fisher
Alessandro Flaborea
Corneliu O. Florea
Georgios Floros
Alejandro Fontan
Tomaso Fontanini
Lin Geng Foo
Wolfgang Förstner
Niki M. Foteinopoulou
Simone Foti
Simone Foti
Victor Fragoso
Gianni Franchi
Jean-Sebastien Franco
Emanuele Frascaroli
Moti Freiman
Rafail Fridman
Sara Fridovich-Keil
Felix Friedrich
Stanislav Frolov

Andre Fu
Bin Fu
Changcheng Fu
Changqing Fu
Jianlong Fu
Jie Fu
Jingjing Fu
Keren Fu
Lan Fu
Lele Fu
Minghao Fu
Qichen Fu
Rao Fu
Taimeng Fu
Tianwen Fu
Yanwei Fu
Ying Fu
Yonggan Fu
Yun Fu
Yuqian Fu
Zehua Fu
Zhenqi Fu
Zipeng Fu
Wolfgang Fuhl
Yasuhisa Fujii
Yuki Fujimura
Katsuki Fujisawa
Kent Fujiwara
Takuya Funatomi
Christopher Funk
Antonino Furnari
Ryo Furukawa
Ryosuke Furuta
David Futschik
Akshay Gadi Patil
Siddhartha Gairola
Rinon Gal
Adrian Galdran
Silvio Galesso
Meirav Galun
Rofida Gamal
Weijie Gan
Yiming Gan
Yulu Gan
Vineet Gandhi
Kanchana Vaishnavi Gandikota
Aditya Ganeshan
Aditya Ganeshan
Swetava Ganguli
Sujoy Ganguly
Hanan Gani
Alireza Ganjdanesh
Harald Ganster
Roy Ganz
Angela F. Gao
Bin-Bin Gao
Changxin Gao
Chen Gao
Chongyang Gao
Daiheng Gao
Daoyi Gao
Difei Gao
Hang Gao
Hongchang Gao
Huiyu Gao
Junyu Gao
Junyu Gao
Kaifeng Gao
Katelyn Gao
Kuofeng Gao
Lin Gao
Ling Gao
Maolin Gao
Quankai Gao
Ruopeng Gao
Shanghua Gao
Shangqi Gao
Shangqian Gao
Shaobing Gao
Shenyuan Gao
Xiang Gao
Xiangjun Gao
Yang Gao
Yipeng Gao
Yuan Gao
Yue Gao
Yunhe Gao
Zhanning Gao
Zhi Gao
Zhihan Gao
Zhitong Gao
Zhongpai Gao
Ziteng Gao
Nicola Garau
Elena Garces
Noa Garcia
Nuno Cruz Garcia
Ricardo Garcia Pinel
Guillermo Garcia-Hernando
Saurabh Garg
Mathieu Garon
Risheek Garrepalli
Pablo Garrido
Quentin Garrido
Stefano Gasperini
Vincent Gaudilliere
Chandan Gautam
Shivam Gautam
Paul Gavrikov
Jonathan Z. Gazak
Chongjian Ge
Liming Ge
Runzhou Ge
Shiming Ge
Songwei Ge
Weifeng Ge
Wenhang Ge
Yanhao Ge
Yixiao Ge
Yunhao Ge
Yuying Ge
Zhiqi Ge
Zongyuan Ge
James Gee
Chen Geng
Daniel Geng
Xue Geng
Zhengyang Geng
Zichen Geng
Kyle Genova
Georgios Georgakis
Mariana-Iuliana Georgescu

Yiangos Georgiou
Markos Georgopoulos
Stamatios Georgoulis
Michal Geyer
Mina Ghadimi Atigh
Abhijay Ghildyal
Reza Ghoddoosian
Mohsen Gholami
Peyman Gholami
Enjie Ghorbel
Soumya Suvra Ghosal
Anindita Ghosh
Anurag Ghosh
Arnab Ghosh
Arthita Ghosh
Aurobrata Ghosh
Shreya Ghosh
Soham Ghosh
Sreyan Ghosh
Andrea Giachetti
Paris Giampouras
Alexander Gielisse
Andrew Gilbert
Guy Gilboa
Nate Gillman
Jhony H. Giraldo
Roger Girgis
Sharath Girish
Mario Valerio Giuffrida
Francesco Giuliari
Nikolaos Gkanatsios
Ioannis Gkioulekas
Alexi Gladstone
Derek Gloudemans
Aurele T. Gnanha
Hyojun Go
Akos Godo
Danilo D. Goede
Arushi Goel
Kratarth Goel
Rahul Goel
Shubham Goel
Vidit Goel
Tejas Gokhale
Vignesh Gokul

Aditya Golatkar
Micah Goldblum
S. Alireza Golestaneh
Gabriele Goletto
Lluis Gomez
Guillermo Gomez-Trenado
Alex Gomez-Villa
Nuno Gonçalves
Biao Gong
Boqing Gong
Chen Gong
Chengyue Gong
Dayoung Gong
Dong Gong
Jingyu Gong
Kaixiong Gong
Kehong Gong
Liyu Gong
Mingming Gong
Ran Gong
Rui Gong
Tao Gong
Xiaojin Gong
Xuan Gong
Xun Gong
Yifan Gong
Yu Gong
Yuanhao Gong
Yuning Gong
Yunye Gong
Cristina I. González
Adam Goodge
Nithin Gopalakrishnan Nair
Cameron Gordon
Dipam Goswami
Gaurav Goswami
Hanno Gottschalk
Chenhui Gou
Jianping Gou
Shreyank N. Gowda
Mohit Goyal
Julia Grabinski
Helmut Grabner
Patrick L. Grady

Alexandros Graikos
Eric Granger
Douglas R. Gray
Michael Alan Greenspan
Connor Greenwell
David Griffiths
Artur Grigorev
Alexey A. Gritsenko
Ivan Grubišić
Geonmo Gu
Jianyang Gu
Jiaqi Gu
Jiayuan Gu
Jinwei Gu
Peiyan Gu
Qiao Gu
Tianpei Gu
Xianfan Gu
Xiang Gu
Xiangming Gu
Xiao Gu
Xiuye Gu
Yanan Gu
Yiwen Gu
Yuchao Gu
Yuming Gu
Yun Gu
Zeqi Gu
Zhangxuan Gu
Banglei Guan
Junfeng Guan
Qingji Guan
Shanyan Guan
Tianrui Guan
Tongkun Guan
Ziqiao Guan
Ziyu Guan
Denis A. Gudovskiy
Antoine Guédon
Paul Guerrero
Ricardo Guerrero
Liangke Gui
Sadaf Gulshad
Manuel Günther
Chen Guo

Chenqi Guo
Chuan Guo
Guodong Guo
Haiyun Guo
Heng Guo
Hengtao Guo
Hongji Guo
Jia Guo
Jianwei Guo
Jianyuan Guo
Jiayi Guo
Jie Guo
Jie Guo
Jingcai Guo
Jingyuan Guo
Jinyang Guo
Lanqing Guo
Meng-Hao Guo
Mengxi China Guo
Minghao Guo
Pengfei Guo
Pengsheng Guo
Pinxue Guo
Qi Guo
Qin Guo
Qing Guo
Qingpei Guo
Saidi Guo
Shi Guo
Shuxuan Guo
Song Guo
Taian Guo
Tiantong Guo
Tianyu Guo
Wen Guo
Wenzhong Guo
Xianda Guo
Xiaobao Guo
Xiefan Guo
Yangyang Guo
Yanhui Guo
Yao Guo
Yichen Guo
Yong Guo
Yuan-Chen Guo

Yuliang Guo
Yuxiang Guo
Yuyu Guo
Zixian Guo
Ziyu Guo
Zujin Guo
Aarush Gupta
Akash Gupta
Akshita Gupta
Ankush Gupta
Anshul Gupta
Anubhav Gupta
Anup Kumar Gupta
Honey Gupta
Kunal Gupta
pravir singh gupta
Rohit Gupta
Saumya Gupta
Corina Gurau
Yeti Z. Gurbuz
Saket Gurukar
Siddharth Gururani
Vladimir Guzov
Matthew A. Gwilliam
Hyunho Ha
Ryo Hachiuma
Isma Hadji
Armin Hadzic
Daniel Haehn
Nicolai Haeni
Ronny Haensch
Meera Hahn
Oliver Hahn
Nguyen Hai
Yuval Haitman
Levente Hajder
Moayed Haji-Ali
Sina Hajimiri
Alexandros Haliassos
Peter M. Hall
Bumsub Ham
Ryuhei Hamaguchi
Abdullah J. Hamdi
Max Hamilton
Lars Hammarstrand

Hasan Abed Al Kader Hammoud
Beining Han
Boran Han
Dongyoon Han
Feng Han
Guangxing Han
Guangzeng Han
Hu Han
Jiaming Han
Jin Han
Jungong Han
Junlin Han
Kai Han
Kang Han
Kun Han
Ligong Han
Longfei Han
Mei Han
Mingfei Han
Muzhi Han
Sungwon Han
Tengda Han
Tengda Han
Wencheng Han
Xinran Han
Xintong Han
Yizeng Han
Yufei Han
Zhenjun Han
Zhi Han
Zhongyi Han
Zongbo Han
Zongyan Han
Ankur Handa
Tiankai Hang
Yucheng Hang
Asif Hanif
Param Hanji
Joëlle Hanna
Niklas Hanselmann
Nicklas A. Hansen
Fusheng Hao
Shaozhe Hao
Shengyu Hao

Shijie Hao
Tianxiang Hao
Yanbin Hao
Yu Hao
Zekun Hao
Takayuki Hara
Mehrtash Harandi
Sanjay Haresh
Haripriya Harikumar
Adam Harley
David M. Hart
Md Yousuf Harun
Md Rakibul Hasan
Connor Hashemi
Atsushi Hashimoto
Khurram Azeem Hashmi
Nozomi Hata
Ali Hatamizadeh
Ryuichiro Hataya
Timm Haucke
Joakim Bruslund Haurum
Stephen Hausler
Mohammad Havaei
Hideaki Hayashi
Zeeshan Hayder
Chengan He
Conghui He
Dailan He
Fan He
Fazhi He
Gaoqi He
Haoyu He
Hongliang He
Jiangpeng He
Jianzhong He
Jiawei He
Ju He
Jun-Yan He
Junfeng He
Lihuo He
Liqiang He
Nanjun He
Ruian He
Runze He
Ruozhen He

Shengfeng He
Shuting He
Songtao He
Tao He
Tianyu He
Tong He
Wei He
Xiang He
Xiangteng He
Xiangyu He
Xiaoxiao He
Xin He
Xingyi He
Xingzhe He
Xinlei He
Xinwei He
Yang He
Yang He
Yannan He
Yeting He
Yinan He
Ying He
Yisheng He
Yuhang He
Yuhang He
Zecheng He
Zewei He
Zexin He
Zhenyu He
Ziwen He
Ramya S. Hebbalaguppe
Peter Hedman
Deepti B. Hegde
Sindhu B. Hegde
Matthias Hein
Mattias Paul Heinrich
Aral Hekimoglu
Philipp Henzler
Byeongho Heo
Jae-Pil Heo
Miran Heo
Stephane Herbin
Alexander Hermans
Pedro Hermosilla Casajus
Jefferson E. Hernandez

Monica Hernandez
Charles Herrmann
Roei Herzig
Fabian Herzog
Georg Hess
Mauricio Hess-Flores
Robin Hesse
Chamin P. Hewa
 Koneputugodage
Masatoshi Hidaka
Richard E. L. Higgins
Mitchell K. Hill
Carlos Hinojosa
Tobias Hinz
Yusuke Hirota
Elad Hirsch
Shinsaku Hiura
Chih-Hui Ho
Man M. Ho
Tuan N. A. Hoang
Jennifer Hobbs
Jiun Tian Hoe
David T. Hoffmann
Derek Hoiem
Yannick Hold-Geoffroy
Lukas Höllein
Gregory I. Holste
Hiroto Honda
Cheeun Hong
Cheng-Yao Hong
Danfeng Hong
Deokki Hong
Fa-Ting Hong
Fangzhou Hong
Feng Hong
Guan Zhe Hong
Je Hyeong Hong
Ji Woo Hong
Jie Hong
Joanna Hong
Lanqing Hong
Lingyi Hong
Sangwoo Hong
Sungeun Hong
Sunghwan Hong

Susung Hong
Weixiang Hong
Xiaopeng Hong
Yan Hong
Yi Hong
Yuchen Hong
Julia Hornauer
Maxwell C. Horton
Eliahu Horwitz
Mir Rayat Imtiaz Hossain
Tonmoy Hossain
Mehdi Hosseinzadeh
Kazuhiro Hotta
Andrew Z. Hou
Fei Hou
Hao Hou
Ji Hou
Jian Hou
Jinghua Hou
Qibin Hou
Qiqi Hou
Ruibing Hou
Saihui Hou
Tingbo Hou
Xinhai Hou
Yuenan Hou
Yunzhong Hou
Zhi Hou
Lukas Hoyer
Petr Hruby
Jun-Wei Hsieh
Chih-Fan Hsu
Gee-Sern Hsu
Hung-Min Hsu
Yen-Chi Hsu
Anthony Hu
Benran Hu
Bo Hu
Dapeng Hu
Dongting Hu
Guosheng Hu
Haigen Hu
Hanjiang Hu
Hanzhe Hu
Hengtong Hu
Hezhen Hu
Jian Hu
Jian-Fang Hu
Jie Hu
Juan Hu
Junjie Hu
Kai Hu
Lianyu Hu
Lily Hu
MengShun Hu
Minghui Hu
Mu Hu
Panwen Hu
Panwen Hu
Ruizhen Hu
Shengshan Hu
Shiyu Hu
Shizhe Hu
Shu Hu
Tao Hu
Tao Hu
Vincent Tao Hu
Wenbo Hu
Xiaodan Hu
Xiaolin Hu
Xiaoling Hu
Xiaotao Hu
Xiaowei Hu
Xinting Hu
Xinting Hu
Xixi Hu
Xuefeng Hu
Yang Hu
Yaosi Hu
Yinlin Hu
Yueyu Hu
Zeyu Hu
Zhanhao Hu
zhengyu hu
Zhenzhen Hu
Zhiming Hu
Zhiming Hu
Zhongyun Hu
Zijian Hu
Andong Hua
Hang Hua
Miao Hua
Wei Hua
Mengdi Huai
Binbin Huang
Bo Huang
Buzhen Huang
Chao Huang
Chao-Tsung Huang
Chaoqin Huang
Ching-Chun Huang
Cong Huang
Danqing Huang
Di Huang
Feihu Huang
Haibin Huang
Haiyang Huang
Hao Huang
Hsin-Ping Huang
Huaibo Huang
Ian Y. Huang
Jiabo Huang
Jiahui Huang
Jiancheng Huang
Jiangyong Huang
Jiaxing Huang
Jin Huang
Jinfa Huang
Jing Huang
Jonathan Huang
Junkai Huang
Junwen Huang
Kejie Huang
Kuan-Chih Huang
Lei Huang
Libo Huang
Lin Huang
Linjiang Huang
Linyan Huang
linzhi Huang
Lun Huang
Luojie Huang
Mingzhen Huang
Qidong Huang
Qiusheng Huang

Runhui Huang
Ruqi Huang
Shaofei Huang
Sheng Huang
Sheng-Yu Huang
Shiyuan Huang
Shuaiyi Huang
Shuangping Huang
Siteng Huang
Siyu Huang
Siyuan Huang
Tao Huang
Thomas E. Huang
Tianyu Huang
Wenjian Huang
Wenke Huang
Xiaohu Huang
Xiaohua Huang
Xiaoke Huang
Xiaoshui Huang
Xiaoyang Huang
Xijie Huang
Xinyu Huang
Xuhua Huang
Yan Huang
Yaping Huang
Ye Huang
Yi Huang
Yi Huang
Yi Huang
Yi-Hua Huang
Yifei Huang
Yihao Huang
Ying Huang
Yizhan Huang
You Huang
Yue Huang
Yufei Huang
Yuge Huang
Yukun Huang
Yuyao Huang
Zaiyu Huang
Zehao Huang
Zeyi Huang
Zhe Huang

Zhewei Huang
Zhida Huang
Zhiqi Huang
Zhiyu Huang
Zhongzhan Huang
Ziling Huang
Zilong Huang
Ziqi Huang
Zixuan Huang
Ziyuan Huang
Jaesung Huh
Ka-Hei Hui
Le Hui
Tianrui Hui
Xiaofei Hui
Ahmed Imtiaz Humayun
Thomas Hummel
Jing Huo
Shuwei Huo
Yuankai Huo
Julio Hurtado
Rukhshanda Hussain
Mohamed Hussein
Noureldien Hussein
Chuong Minh Huynh
Tran Ngoc Huynh
Muhammad Huzaifa
Inwoo Hwang
Jaehui Hwang
Jenq-Neng Hwang
Sehyun Hwang
Seunghyun Hwang
Sukjun Hwang
Sunhee Hwang
Wonjun Hwang
Jeongseok Hyun
Sangeek Hyun
Ekaterina Iakovleva
Tomoki Ichikawa
A. S. M. Iftekhar
Masaaki Iiyama
Satoshi Ikehata
Sunghoon Im
Woobin Im
Nevrez Imamoglu

Abdullah Al Zubaer Imran
Tooba Imtiaz
Nakamasa Inoue
Naoto Inoue
Eldar Insafutdinov
Catalin Ionescu
Radu Tudor Ionescu
Nasar Iqbal
Umar Iqbal
Go Irie
Muhammad Zubair Irshad
Francesco Isgrò
Yasunori Ishii
Berivan Isik
Syed M. S. Islam
Mariko Isogawa
Vamsi Krishna K. Ithapu
Koichi Ito
Leonardo Iurada
Shun Iwase
Sergio Izquierdo
Sarah Jabbour
David Jacobs
David E. Jacobs
Yasamin Jafarian
Azin Jahedi
Achin Jain
Ayush Jain
Jitesh Jain
Kanishk Jain
Yash Jain
Nikita Jaipuria
Tomas Jakab
Muhammad Abdullah Jamal
Hadi Jamali-Rad
Stuart James
Nataraj Jammalamadaka
Donggon Jang
Sujin Jang
Young Kyun Jang
Youngjoon Jang
Steeven Janny
Paul Janson
Maximilian Jaritz

Ronnachai Jaroensri
Guillaume Jaume
Sajid Javed
Saqib Javed
Bhavin Jawade
Hirunima Jayasekara
Senthilnath Jayavelu
Guillaume Jeanneret
Pranav Jeevan
Rohit Jena
Tomas Jenicek
Simon Jenni
Hae-Gon Jeon
Jeimin Jeon
Seogkyu Jeon
Boseung Jeong
Jongheon Jeong
Yonghyun Jeong
Yoonwoo Jeong
Koteswar Rao Jerripothula
Artur Jesslen
Nikolay Jetchev
Ankit Jha
Sumit K. Jha
I-Hong Jhuo
Deyi Ji
Ge-Peng Ji
Hui Ji
Jianmin Ji
Jiayi Ji
Jingwei Ji
Naye Ji
Pengliang Ji
Shengpeng Ji
Wei Ji
Xiang Ji
Xiaopeng Ji
Xiaozhong Ji
Xinya Ji
Yu Ji
Yuanfeng Ji
Zhanghexuan Ji
Baoxiong Jia
Ding Jia
Jinrang Jia

Menglin Jia
Shan Jia
Shuai Jia
Tong Jia
Wenqi Jia
Xiaojun Jia
Xiaosong Jia
Xu Jia
Yiren Jian
Bo Jiang
Borui Jiang
Chen Jiang
Guangyuan Jiang
Hai Jiang
Haiyang Jiang
Haiyong Jiang
Hanwen Jiang
Hanxiao Jiang
Hao Jiang
Haobo Jiang
Huaizu Jiang
Huajie Jiang
Jianmin Jiang
Jianwen Jiang
Jiaxi Jiang
Jin Jiang
Jing Jiang
Kaixun Jiang
Kui Jiang
Li Jiang
Liming Jiang
Meirui Jiang
Ming Jiang
Ming Jiang
Peng Jiang
Peng-Tao Jiang
Runqing Jiang
Siyang Jiang
Tianjian Jiang
Tingting Jiang
Weisen Jiang
Wen Jiang
Xingyu Jiang
Xueying Jiang
Xuhao Jiang

Yanru Jiang
Yi Jiang
Yifan Jiang
Yingying Jiang
Yue Jiang
Yuming Jiang
Zeren Jiang
Zheheng Jiang
Zhenyu Jiang
Zhongyu Jiang
Zi-Hang Jiang
Zixuan Jiang
Ziyu Jiang
Zutao Jiang
Zutao Jiang
Jianbin Jiao
Jianbo Jiao
Licheng Jiao
Ruochen Jiao
Shuming Jiao
Wenpei Jiao
Zequn Jie
Dongkwon Jin
Gaojie Jin
Haian Jin
Haibo Jin
Kyong Hwan Jin
Lei Jin
Lianwen Jin
Linyi Jin
Peng Jin
Peng Jin
Sheng Jin
SouYoung Jin
Weiyang Jin
Xiao Jin
Xiaojie Jin
Xin Jin
Xin Jin
Yeying Jin
Yi Jin
Ying Jin
Zhenchao Jin
Chenchen Jing
Junpeng Jing

Mengmeng Jing
Taotao Jing
Jeonghee Jo
Yeonsik Jo
Younghyun Jo
Cameron A. Johnson
Faith M. Johnson
Maxwell Jones
Michael J. Jones
R. Kenny Jones
Ameya Joshi
Shantanu H. Joshi
Brendan Jou
Chen Ju
Wei Ju
Xuan Ju
Yan Ju
Felix Juefei-Xu
Florian Jug
Yohan Jun
Kim Jun-Seong
Masum Shah Junayed
Chanyong Jung
Claudio R. Jung
Hoin Jung
Hyunyoung Jung
Sangwon Jung
Steffen Jung
Sacha Jungerman
Hari Chandana K.
Prajwal K. R.
Berna Kabadayi
Anis Kacem
Anil Kag
Kumara Kahatapitiya
Bernhard Kainz
Ivana Kajic
Ioannis Kakogeorgiou
Niveditha Kalavakonda
Mahdi M. Kalayeh
Ajinkya Kale
Anmol Kalia
Sinan Kalkan
Jayateja Kalla
Tarun Kalluri

Uday Kamal
Sandesh Kamath
Chandra Kambhamettu
Meina Kan
Sai Srinivas Kancheti
Takuhiro Kaneko
Cuicui Kang
Dahyun Kang
Guoliang Kang
Hyolim Kang
Jaeyeon Kang
Juwon Kang
Li-Wei Kang
Mingon Kang
MinGuk Kang
Minjun Kang
Weitai Kang
Zhao Kang
Yash Mukund Kant
Yueying Kao
Saarthak Kapse
Aupendu Kar
Oğuzhan Fatih Kar
Ozgur Kara
Tamás Karácsony
Srikrishna Karanam
Neerav Karani
Mert Asim Karaoglu
Laurynas Karazija
Navid Kardan
Amirhossein Kardoost
Nour Karessli
Michelle Karg
Mohammad Reza Karimi
  Dastjerdi
Animesh Karnewar
Arjun M. Karpur
Shyamgopal Karthik
Korrawe Karunratanakul
Tejaswi Kasarla
Satyananda Kashyap
Yoni Kasten
Marc A. Kastner
Hirokatsu Kataoka
Isinsu Katircioglu

Kai Katsumata
Ilya Kaufman
Manuel Kaufmann
Chaitanya Kaul
Prannay Kaul
Prakhar Kaushik
Isaak Kavasidis
Ryo Kawahara
Yuki Kawana
Justin Kay
Evangelos Kazakos
Jingcheng Ke
Lei Ke
Tsung-Wei Ke
Wei Ke
Zhanghan Ke
Nikhil V. Keetha
Thomas Kehrenberg
Marilyn Keller
Rohit Keshari
Janis Keuper
Daniel Keysers
Hrant Khachatrian
Seyran Khademi
Wesley A. Khademi
Taras Khakhulin
Samir Khaki
Hasam Khalid
Umar Khalid
Amr Khalifa
Asif Hussain Khan
Faizan Farooq Khan
Muhammad Haris Khan
Naimul Khan
Qadeer Khan
Zeeshan Khan
Bishesh Khanal
Pulkit Khandelwal
Ishan Khatri
Muhammad Uzair Khattak
Vahid Reza Khazaie
Vaishnavi M. Khindkar
Rawal Khirodkar
Pirazh Khorramshahi
Sahil S. Khose

Kourosh Khoshelham
Sena Kiciroglu
Benjamin Kiefer
Kotaro Kikuchi
Mert Kilickaya
Benjamin Killeen
Beomyoung Kim
Boah Kim
Bumsoo Kim
Byeonghwi Kim
Byoungjip Kim
Changhoon Kim
Changick Kim
Chanho Kim
Chanyoung Kim
Dahun Kim
Dahyun Kim
Diana S. Kim
Dong-Jin Kim
Donggun Kim
Donghyun Kim
Dongkeun Kim
Dongwan Kim
Dongyoung Kim
Eun-Sol Kim
Eunji Kim
Euyoung Kim
Geeho Kim
Giseop Kim
Guisik Kim
Gwanghyun Kim
Hakyeong Kim
Hanjae Kim
Hanjung Kim
Heewon Kim
Howon Kim
Hyeongwoo Kim
Hyeongwoo Kim
Hyo Jin Kim
Hyung-Il Kim
Hyunwoo J. Kim
Insoo Kim
Jae Myung Kim
Jeongsol Kim
Jihwan Kim

Jinkyu Kim
Jinwoo Kim
Jongyoo Kim
Joonsoo Kim
Jung Uk Kim
Junho Kim
Junho Kim
Junsik Kim
Kangyeol Kim
Kunhee Kim
Kwang In Kim
Manjin Kim
Minchul Kim
Minji Kim
Minjung Kim
Namil Kim
Namyup Kim
Nayeong Kim
Sanghyun Kim
Seong Tae Kim
Seungbae Kim
Seungryong Kim
Seungwook Kim
Soo Ye Kim
Soohwan Kim
Sungnyun Kim
Sungyeon Kim
Sunnie S. Y. Kim
Tae Hyun Kim
Tae Hyung Kim
Taehoon Kim
Taehun Kim
Taehwan Kim
Taekyung Kim
Taeoh Kim
Taewoo Kim
Won Hwa Kim
Wonjae Kim
Woo Jae Kim
Young Min Kim
YoungBin Kim
Youngeun Kim
Youngseok Kim
Youngwook Kim
Akisato Kimura

Andreas Kirsch
Nikita Kister
Furkan Osman Kınlı
Marcus Klasson
Florian Kleber
Tzofi M. Klinghoffer
Jan P. Klopp
Florian Kluger
Hannah Kniesel
David M. Knigge
Byungsoo Ko
Dohwan Ko
Jongwoo Ko
Takumi Kobayashi
Muhammed Kocabas
Yeong Jun Koh
Kathlén Kohn
Subhadeep Koley
Nick Kolkin
Soheil Kolouri
Jacek Komorowski
Deying Kong
Fanjie Kong
Hanyang Kong
Hui Kong
Kyeongbo Kong
Lecheng Kong
Lingdong Kong
Linghe Kong
Lingshun Kong
Naejin Kong
Quan Kong
Shu Kong
Xianghao Kong
Xiangtao Kong
Xiangwei Kong
Xiaoyu Kong
Xin Kong
Youyong Kong
Yu Kong
Zhenglun Kong
Aishik Konwer
Nicholas C. Konz
Gwanhyeong Koo
Juil Koo

Julian F. P. Kooij
George Kopanas
Sanjeev J. Koppal
Bruno Korbar
Giorgos Kordopatis-Zilos
Dimitri Korsch
Adam Kortylewski
Divya Kothandaraman
Suraj Kothawade
Iuliia Kotseruba
Sasikanth Kotti
Alankar Kotwal
Shashank Kotyan
Alexandros Kouris
Petros Koutras
Rama Kovvuri
Dilip Krishnan
Praveen Krishnan
Ranganath Krishnan
Rohan M. Krishnan
Georg Krispel
Alexander Krull
Tianshu Kuai
Haowei Kuang
Zhengfei Kuang
Andrey Kuehlkamp
David Kügler
Arjan Kuijper
Anna Kukleva
Jonas Kulhanek
Peter Kulits
Akshay R. Kulkarni
Ashutosh C. Kulkarni
Kuldeep Kulkarni
Nilesh Kulkarni
Abhinav Kumar
Abhishek Kumar
Akash Kumar
Avinash Kumar
B. V. K. Vijaya Kumar
Chandan Kumar
Pulkit Kumar
Ratnesh Kumar
Sateesh Kumar
Satish Kumar

Suryansh Kumar
Yogesh Kumar
Nupur Kumari
Sudhakar Kumawat
Nilakshan Kunananthaseelan
Rohit Kundu
Souvik Kundu
Meng-Yu Jennifer Kuo
Weicheng Kuo
Shuhei Kurita
Yusuke Kurose
Takahiro Kushida
Uday Kusupati
Alina Kuznetsova
Jobin K. V.
Henry Kvinge
Ho Man Kwan
Hyeokjun Kweon
Donghyeon Kwon
Gihyun Kwon
Heeseung Kwon
Hyoukjun Kwon
Myung-Joon Kwon
Taein Kwon
YoungJoong Kwon
Cameron Kyle-Davidson
Christos Kyrkou
Jorma Laaksonen
Patrick Labatut
Yann Labbé
Manuel Ladron de Guevara
Florent Lafarge
Jean Lahoud
Bolin Lai
Farley Lai
Jian-Huang Lai
Shenqi Lai
Xin Lai
Yu-Kun Lai
Yung-Hsuan Lai
Zeqiang Lai
Zhengfeng Lai
Barath Lakshmanan
Rohit Lal

Rodney LaLonde
Hala Lamdouar
Meng Lan
Yushi Lan
Federico Landi
George V. Landon
Chunbo Lang
Jochen Lang
Nico Lang
Georg Langs
Raffaella Lanzarotti
Dong Lao
Yixing Lao
Yizhen Lao
Zakaria Laskar
Alexandros Lattas
Chun Pong Lau
Shlomi Laufer
Justin Lazarow
Svetlana Lazebnik
Duy Tho Le
Hieu Le
Hoang Le
Hoang Le
Thi-Thu-Huong Le
Trung Le
Trung-Nghia Le
Tung Thanh Le
Hoàng-Ân Lê
Herve Le Borgne
Guillaume Le Moing
Erik Learned-Miller
Tim Lebailly
Byeong-Uk Lee
Byung-Kwan Lee
Cheng-Han Lee
Chul Lee
Daeun Lee
Dogyoon Lee
Dong Hoon Lee
Eugene Eu Tzuan Lee
Eung-Joo Lee
Gyuseong Lee
Hsin-Ying Lee
Hwee Kuan Lee

Hyeongmin Lee
Hyungtae Lee
Jae Yong Lee
Jaeho Lee
Jaeseong Lee
Jaewon Lee
Jangho Lee
Jangwon Lee
Jihyun Lee
Jiyoung Lee
Jong-Seok Lee
Jongho Lee
Jongmin Lee
Joo-Ho Lee
Joon-Young Lee
Joonseok Lee
Jungbeom Lee
Jungho Lee
Jungwoo Lee
Junha Lee
Junhyun Lee
Junyong Lee
Kibok Lee
Kuan-Ying Lee
Kwonjoon Lee
Kwot Sin Lee
Kyungmin Lee
Kyungmoon Lee
Minhyeok Lee
Minsik Lee
Pilhyeon Lee
Saehyung Lee
Sangho Lee
Sanghyeok Lee
Sangmin Lee
Sehun Lee
Sehyung Lee
Seon-Ho Lee
Seong Hun Lee
Seongwon Lee
Seung Hyun Lee
Seung-Ik Lee
Seungho Lee
Seunghun Lee
Seungmin Lee

Seungyong Lee
Sohyun Lee
Suhyeon Lee
Sungho Lee
Sungmin Lee
Suyoung Lee
Taehyun Lee
Wooseok Lee
Yao-Chih Lee
Yi-Lun Lee
Yonghyeon Lee
Youngwan Lee
Leonidas Lefakis
Bowen Lei
Chenyang Lei
Chenyi Lei
Jiahui Lei
Na Lei
Qinqian Lei
Yinjie Lei
Thomas Leimkuehler
Abe Leite
Abdelhak Lemkhenter
Jiaxu Leng
Luziwei Leng
Zhiying Leng
Hendrik P. A. Lensch
Jan E. Lenssen
Ted Lentsch
Simon Lepage
Stefan Leutenegger
Filippo Leveni
Axel Levy
Ailin Li
Aixuan Li
Baiang Li
Baoxin Li
Bin Li
Bing Li
Bing Li
Bo Li
Bowen Li
Boying Li
Changlin Li
Changlin Li

Chao Li
Chenghong Li
Chenglin Li
Chenglong Li
Chengze Li
Chun-Guang Li
Daiqing Li
Dasong Li
Dian Li
Dong Li
Fangda Li
Feiran Li
Fenghai Li
Gen Li
Guanbin Li
Guangrui Li
Guihong Li
Guorong Li
Haifeng Li
Han Li
Hang Li
Hangyu Li
Hanhui Li
Hao Li
Hao Li
Haoang Li
Haoran Li
Haoxiang Li
Haoxin Li
He Li
Heng Li
Hengduo Li
Hongshan Li
Hongwei Bran Li
Hongxiang Li
Hongyang Li
Hongyu Li
Huafeng Li
Huan Li
Hui Li
Jiacheng Li
Jiahao Li
Jialu Li
Jiaman Li
Jiangmeng Li

Jiangtong Li
Jiangyuan Li
Jianing Li
Jianwei Li
Jianwu Li
Jiaqi Li
Jiaqi Li
Jiatong Li
Jiaxuan Li
Jiazhi Li
Jichang Li
Jie Li
Jin Li
Jinglun Li
Jingzhi Li
Jingzong Li
Jinlong Li
Jinlong Li
Jinpeng Li
Jinxing Li
Jun Li
Jun Li
Junbo Li
Juncheng Li
Junxuan Li
Junyi Li
Kai Li
Kaican Li
Kailin Li
Ke Li
Kehan Li
Keyu Li
Kun Li
Kunchang Li
Kunpeng Li
Lei Li
Lei Li
Li Li
Li Erran Li
Liang Li
Lin Li
Lincheng Li
Liulei Li
Liunian Harold Li
Lujun Li

Manyi Li
Maomao Li
Meng Li
Mengke Li
Mengtian Li
Mengtian Li
Ming Li
Ming Li
Minghan Li
Mingjie Li
Nannan Li
Nianyi Li
Peike Li
Peizhao Li
Peng Li
Pengpeng Li
Pengyu Li
Ping Li
Puhao Li
Qiang Li
Qing Li
Qingyong Li
Qiufu Li
Qizhang Li
Ren Li
Rong Li
Rongjie Li
Ru Li
Rui Li
Ruibo Li
Ruihui Li
Ruilong Li
Ruining Li
Ruixuan Li
Runze Li
Ruoteng Li
Shaohua Li
Shasha Li
Shigang Li
Shijie Li
Shikun Li
Shile Li
Shuai Li
Shuai Li
Shuang Li

Shuwei Li
Si Li
Siyao Li
Siyuan Li
Siyuan Li
Taihui Li
Tianye Li
Wanhua Li
Wanqing Li
Wei Li
Wei Li
Wei Li
Wei-Hong Li
Weihao Li
Weijia Li
Weiming Li
Wenbin Li
Wenbo Li
Wenhao Li
Wenjie Li
Wenshuo Li
Wentong Li
Wenxi Li
Xiang Li
Xiang Li
Xiang Li
Xiang Li
Xiang Li
Xiangtai Li
Xiangyang Li
Xianzhi Li
Xiao Li
Xiao Li
Xiaoguang Li
Xiaomeng Li
Xiaoming Li
Xiaoqi Li
Xiaoqiang Li
Xiaotian Li
Xiaoyu Li
Xin Li
Xin Li
Xin Li
Xinghui Li
Xingyi Li

Xingyu Li	Yuexiang Li	Ruyi Lian
Xinjie Li	Yuezun Li	Zhouhui Lian
Xinyu Li	Yuhang Li	Jin Lianbao
Xiu Li	Yuheng Li	Chao Liang
Xiujun Li	Yulin Li	Chia-Kai Liang
Xuan Li	Yumeng Li	Dingkang Liang
Xuanlin Li	Yunfan Li	Feng Liang
Xuelong Li	Yunheng Li	Gongbo Liang
Xuelu Li	Yunqiang Li	Hanxue Liang
Xueqian Li	Yunsheng Li	Hao Liang
Ya-Li Li	Yuyan Li	Hui Liang
Yanan Li	Yuyang Li	Jiadong Liang
Yang Li	Zejian Li	Jiajun Liang
Yang Li	Zekun Li	Jian Liang
Yangyan Li	Zekun Li	Jingyun Liang
Yanjing Li	Zhangheng Li	Jinxiu S. Liang
Yansheng Li	Zhangzikang Li	Junwei Liang
Yanwei Li	Zhaoshuo Li	Kaiqu Liang
Yanyu Li	Zhaowen Li	Ke Liang
Yaohui Li	Zhe Li	Kevin J. Liang
Yaowei Li	Zhe Li	Luming Liang
Yawei Li	Zhen Li	Mingfu Liang
Yi Li	Zhen Li	Pengpeng Liang
Yi Li	Zhen Li	Siyuan Liang
Yicong Li	Zheng Li	Xiaoxiao Liang
Yicong Li	Zhengqin Li	Xinran Liang
Yifei Li	Zhengyuan Li	Xiwen Liang
Yijin Li	Zhenyu Li	Yang Liang
Yijun Li	Zhichao Li	Yixun Liang
Yijun Li	Zhihao Li	Yongqing Liang
Yikang Li	Zhihao Li	Youwei Liang
Yimeng Li	Zhiheng Li	Yuanzhi Liang
Yiming Li	Zhiqi Li	Zhexin Liang
Yiming Li	Zhixuan Li	Zhihao Liang
Yingwei Li	Zhong Li	Zhixuan Liang
Yiting Li	Zhuoling Li	Kang Liao
Yixuan Li	Zhuowan Li	Liang Liao
Yize Li	Zhuowei Li	Minghui Liao
Yizhuo Li	Zhuoxiao Li	Ting-Hsuan Liao
Yong Li	Zihan Li	Wei Liao
Yong-Lu Li	Ziqiang Li	Xin Liao
Yongjie Li	Wen Qiao Li	Yinghong Liao
Yuanman Li	Dongze Lian	Yue Liao
Yuanming Li	Long Lian	Zhibin Liao
Yuelong Li	Qing Lian	Ziwei Liao

Benedetta Liberatori	Mingyuan Lin	Stefan P. Lionar
Daniel J. Lichy	Qiuxia Lin	Phillip Lippe
Maiko Lie	Shaohui Lin	Lahav O. Lipson
Qin Likun	Shih-Yao Lin	Joey Litalien
Isaak Lim	Sihao Lin	Ron Litman
Teck Yian Lim	Siyou Lin	Mattia Litrico
Bannapol Limanond	Tiancheng Lin	Dor Litvak
Baijiong Lin	Tsung-Yu Lin	Aishan Liu
Beibei Lin	Wanyu Lin	Ajian Liu
Cheng Lin	Wei Lin	Akide L. Y. Liu
Chenhao Lin	Wei Lin	Andrew Liu
Chia-Wen Lin	Wen-Yan Lin	Ao Liu
Chieh Hubert Lin	Wenbin Lin	Bang Liu
Chuang Lin	Xiangbo Lin	Benlin Liu
Chung-Ching Lin	Xianhui Lin	Bin Liu
Chunyu Lin	Xiaofan Lin	Bin Liu
Ci-Siang Lin	Xiaofeng Lin	Bing Liu
Di Lin	Xin Lin	Binghao Liu
Fanqing Lin	Xudong Lin	Bingyuan Liu
Feng Lin	Xue Lin	Bo Liu
Fudong Lin	Xuxin Lin	Bo Liu
Guangfeng Lin	Ya-Wei Eileen Lin	Bo Liu
Haotong Lin	Yan-Bo Lin	Boning Liu
Haozhe Lin	Yancong Lin	Bowen Liu
Hubert Lin	Yi Lin	Boxiao Liu
Hui Lin	Yijie Lin	Chang Liu
Jason Lin	Yiming Lin	Chang Liu
Jianxin Lin	Yiqi Lin	Chang Liu
Jiaqi Lin	Yiqun Lin	Chang Liu
Jiaying Lin	Yongliang Lin	Chao Liu
Jiehong Lin	Yu Lin	Chengxin Liu
Jierui Lin	Yuanze Lin	Chengxu Liu
Jintao Lin	Yuewei Lin	Chih-Ting Liu
Kai-En Lin	Zhi-Hao Lin	Chuanjian Liu
Ke Lin	Zhiqiu Lin	Chun-Hao Liu
Kevin Lin	ZhiWei Lin	Daizong Liu
Kevin Qinghong Lin	Zinan Lin	Decheng Liu
Kuan Heng Lin	Ziyi Lin	Di Liu
Kun-Yu Lin	David B. Lindell	Difan Liu
Kunyang Lin	Philipp Lindenberger	Dong Liu
Kwan-Yee Lin	Jingwang Ling	Dongnan Liu
Lijian Lin	Jun Ling	Fang Liu
Liqiang Lin	Yongguo Ling	Fang Liu
Liting Lin	Zhan Ling	Fangyi Liu
Luojun Lin	Alexander Liniger	Feng Liu

Fengbei Liu
Fenglin Liu
Fengqi Liu
Furui Liu
Fuxiao Liu
Haisong Liu
Han Liu
Hanwen Liu
Hanyuan Liu
Hao Liu
Haolin Liu
Haotian Liu
Haozhe Liu
Heshan Liu
Hong Liu
Hongbin Liu
Hongfu Liu
Hongyu Liu
Hsueh-Ti Derek Liu
Huidong Liu
Isabella Liu
Ji Liu
Jia-Wei Liu
Jiachen Liu
Jiaheng Liu
Jiahui Liu
Jiaming Liu
Jiancheng Liu
Jiang Liu
Jianmeng Liu
Jiashuo Liu
Jiawei Liu
Jiawei Liu
Jiayang Liu
Jiayi Liu
Jie Liu
Jie Liu
Jie Liu
Jihao Liu
Jing Liu
Jing Liu
Jing Liu
Jingyuan Liu
Jingyuan Liu
Jiuming Liu
Jiyuan Liu
Jun Liu
Kang-Jun Liu
Kangning Liu
Kenkun Liu
Kunhao Liu
Li Liu
Lijuan Liu
Lingbo Liu
Lingqiao Liu
Liu Liu
Liyang Liu
Meng Liu
Mengchen Liu
Miao Liu
Ming Liu
Minghao Liu
Minghua Liu
Mingxuan Liu
Mingyuan Liu
Nan Liu
Nian Liu
Ning Liu
Peidong Liu
Peirong Liu
Peiye Liu
Pengju Liu
Ping Liu
Qi Liu
Qiankun Liu
Qihao Liu
Qing Liu
Qingjie Liu
Richard Liu
Risheng Liu
Rui Liu
Ruicong Liu
Ruoshi Liu
Ruyu Liu
Shaohui Liu
Shaoteng Liu
Shaowei Liu
Sheng Liu
Shenglan Liu
Shikun Liu
Shilong Liu
Shuaicheng Liu
Shuaicheng Liu
Shuming Liu
Songhua Liu
Tao Liu
Tian Yu Liu
Tianci Liu
Tianshan Liu
Tongliang Liu
Tyng-Luh Liu
Wei Liu
Weifeng Liu
Weixiao Liu
Weiyu Liu
Wen Liu
Wenxi Liu
Wenyu Liu
Wenyu Liu
Wu Liu
Xian Liu
Xianglong Liu
Xianpeng Liu
Xiao Liu
Xiaohong Liu
Xiaoyu Liu
Xiaoyu Liu
Xihui Liu
Xin Liu
Xin Liu
Xinchen Liu
Xingtong Liu
Xingyu Liu
Xinhang Liu
Xinhui Liu
Xinpeng Liu
Xinwei Liu
Xinyu Liu
Xiulong Liu
Xiyao Liu
Xu Liu
Xubo Liu
Xudong Liu
Xueting Liu
Xueyi Liu

Yan Liu
Yanbin Liu
Yang Liu
Yang Liu
Yang Liu
Yang Liu
Yang Liu
Yanwei Liu
Yaojie Liu
Ye Liu
Yi Liu
Yihao Liu
Yingcheng Liu
Yingfei Liu
Yipeng Liu
Yipeng Liu
Yixin Liu
Yizhang Liu
Yong Liu
Yong Liu
Yonghuai Liu
Yongtuo Liu
Yu Liu
Yu-Lun Liu
Yu-Shen Liu
Yuan Liu
Yuang Liu
Yuanpei Liu
Yuanpeng Liu
Yuanwei Liu
Yuchen Liu
Yuchen Liu
Yuchi Liu
Yueh-Cheng Liu
Yufan Liu
Yuhao Liu
Yuliang Liu
Yun Liu
Yun Liu
Yun Liu
Yunfan Liu
Yunfei Liu
Yunze Liu
Yupei Liu
Yuqi Liu

Yuyang Liu
Yuyuan Liu
Zhaoqiang Liu
Zhe Liu
Zhe Liu
Zhen Liu
Zheng Liu
Zhenguang Liu
Zhi Liu
Zhihua Liu
Zhijian Liu
Zhili Liu
Zhuoran Liu
Ziquan Liu
Ziyi Liu
Zuxin Liu
Zuyan Liu
Josep Llados
Ling Lo
Shao-Yuan Lo
Liliana Lo Presti
Sylvain Lobry
Yaroslava Lochman
Fotios Logothetis
Suhas Lohit
Marios Loizou
Vishnu Suresh Lokhande
Cheng Long
Chengjiang Long
Fuchen Long
Guodong Long
Rujiao Long
Shangbang Long
Teng Long
Xiaoxiao Long
Zijun Long
Ivan Lopes
Vasco Lopes
Adrian Lopez-Rodriguez
Javier Lorenzo-Navarro
Yujing Lou
Brian C. Lovell
Weng Fei Low
Changsheng Lu
Chun-Shien Lu

Daohan Lu
Dongming Lu
Erika Lu
Fan Lu
Guangming Lu
Guo Lu
Hao Lu
Hao Lu
Hongtao Lu
Jiachen Lu
Jiaxin Lu
Jiwen Lu
Lewei Lu
Liying Lu
Quanfeng Lu
Shenyu Lu
Shun Lu
Tao Lu
Xiangyong Lu
Xiankai Lu
Xin Lu
Xuanchen Lu
Xuequan Lu
Yan Lu
Yang Lu
Yanye Lu
Yawen Lu
Yifan Lu
Yongchun Lu
Yongxi Lu
Yu Lu
Yu Lu
Yuzhe Lu
Zhichao Lu
Zhihe Lu
Zijia Lu
Tianyu Luan
Pauline Luc
Simon Lucey
Timo Lüddecke
Jonathon Luiten
Jovita Lukasik
Ao Luo
Cheng Luo
Chuanchen Luo

Donghao Luo
Fangzhou Luo
Gen Luo
Gongning Luo
Hongchen Luo
Jiahao Luo
Jiebo Luo
Jinqi Luo
Jinqi Luo
Jun Luo
Katie Z. Luo
Kunming Luo
Lei Luo
Mandi Luo
Mi Luo
Ruotian Luo
Sihui Luo
Tiange Luo
Wenhan Luo
Xiao Luo
Xiaotong Luo
Xiongbiao Luo
Xu Luo
Yadan Luo
Yawei Luo
Ye Luo
Yisi Luo
Yong Luo
You-Wei Luo
Yuanjing Luo
Zelun Luo
Zhengxiong Luo
Zhengyi Luo
Zhiming Luo
Zhipeng Luo
Zhongjin Luo
Zilin Luo
Ziyang Luo
Tung M. Luu
Diogo C. Luvizon
Jun Lv
Pei Lv
Yunqiu Lv
Zhaoyang Lv
Gengyu Lyu

Jiancheng Lyu
Jipeng Lyu
Junfeng Lyu
Mengyao Lyu
Mingzhi Lyu
Weimin Lyu
Xiaoyang Lyu
Xinyu Lyu
Yiwei Lyu
Youwei Lyu
Ailong Ma
Andy J. Ma
Benteng Ma
Bingpeng Ma
Chao Ma
Chuofan Ma
Cong Ma
Cuixia Ma
Fan Ma
Fangchang Ma
Fei Ma
Guozheng Ma
Haoyu Ma
Hengbo Ma
Huimin Ma
Jiahao Ma
Jianqi Ma
Jiawei Ma
Jiayi Ma
Kai Ma
Kede Ma
Lei Ma
Li Ma
Lin Ma
Liqian Ma
Lizhuang Ma
Mengmeng Ma
Ning Ma
Qianli Ma
Rui Ma
Shijie Ma
Shiqiang Ma
Shiqing Ma
Shuailei Ma
Sizhuo Ma

Tao Ma
Teli Ma
Wenxuan Ma
Wufei Ma
Xianzheng Ma
Xiaoxuan Ma
Xinyin Ma
Xinzhu Ma
Xu Ma
Yeyao Ma
Yifeng Ma
Yuexiao Ma
Yuexin Ma
Yunsheng Ma
Zhan Ma
Zhanyu Ma
Ziping Ma
Ziqiao Ma
Muhammad Maaz
Anish Madan
Neelu Madan
Spandan Madan
Sai Advaith Maddipatla
Rishi Madhok
Filippo Maggioli
Simone Magistri
Marcus Magnor
Sabarinath Mahadevan
Shweta Mahajan
Aniruddha Mahapatra
Sarthak Kumar Maharana
Behrooz Mahasseni
Upal Mahbub
Arif Mahmood
Kaleel Mahmood
Mohammed Mahmoud
Tanvir Mahmud
Jinjie Mai
Helena de Almeida Maia
Josef Maier
Shishira R. Maiya
Snehashis Majhi
Orchid Majumder
Sagnik Majumder
Ilya Makarov

Sina Malakouti
Hashmat Shadab Malik
Mateusz Malinowski
Utkarsh Mall
Srikanth Malla
Clement Mallet
Dimitrios Mallis
Abed Malti
Yunze Man
Oscar Mañas
Karttikeya Mangalam
Fabian Manhardt
Ioannis Maniadis Metaxas
Fahim Mannan
Rafal Mantiuk
Dongxing Mao
Jiageng Mao
Wei Mao
Weian Mao
Weixin Mao
Ye Mao
Yongsen Mao
Yunyao Mao
Yuxin Mao
Zhiyuan Mao
Emanuela Marasco
Matthew Marchellus
Alberto Marchisio
Diego Marcos
Alina E. Marcu
Riccardo Marin
Manuel J. Marín-Jiménez
Octave Mariotti
Dejan Markovic
Imad Eddine Marouf
Valerio Marsocci
Diego Martin Arroyo
Ricardo Martin-Brualla
Brais Martinez
Renato Martins
Damien Martins Gomes
Tetiana Martyniuk
Pierre Marza
David Masip
Carlo Masone

Timothée Masquelier
André G. Mateus
Minesh Mathew
Yusuke Matsui
Bruce A. Maxwell
Christoph Mayer
Prasanna Mayilvahanan
Amir Mazaheri
Amrita Mazumdar
Pratik Mazumder
Alessio Mazzucchelli
Amarachi B. Mbakwe
Scott McCloskey
Naga Venkata Kartheek
 Medathati
Henry Medeiros
Guofeng Mei
Haiyang Mei
Jie Mei
Jieru Mei
Kangfu Mei
Lingjie Mei
Xiaoguang Mei
Dennis Melamed
Luke Melas-Kyriazi
Iaroslav Melekhov
Yifang Men
Ricardo A. Mendoza-León
Depu Meng
Fanqing Meng
Jingke Meng
Lingchen Meng
Qier Meng
Qingjie Meng
Quan Meng
Yanda Meng
Zibo Meng
Otniel-Bogdan Mercea
Pablo Mesejo
Safa Messaoud
Nico Messikommer
Nando Metzger
Christopher Metzler
Vasileios Mezaris
Liang Mi

Zhenxing Mi
S. Mahdi H. Miangoleh
Bo Miao
Changtao Miao
Jiaxu Miao
Zichen Miao
Bjoern Michele
Christian Micheloni
Marko Mihajlovic
Zoltán Á. Milacski
Simone Milani
Leo Milecki
Roy Miles
Christen Millerdurai
Monica Millunzi
Chaerin Min
Cheol-Hui Min
Dongbo Min
Hyun-Seok Min
Jie Min
Juhong Min
Kyle Min
Yifei Min
Yuecong Min
Zhixiang Min
Matthias Minderer
Di Ming
Qi Ming
Xiang Ming
Riccardo Miotto
Aymen Mir
Pedro Miraldo
Parsa Mirdehghan
Seyed Ehsan Mirsadeghi
Muhammad Jehanzeb
 Mirza
Ashkan Mirzaei
Dmytro Mishkin
Anand Mishra
Ashish Mishra
Samarth Mishra
Shlok K. Mishra
Diganta Misra
Abhay Mittal
Gaurav Mittal

Surbhi Mittal
Trisha Mittal
Taiki Miyanishi
Daisuke Miyazaki
Hong Mo
Kaichun Mo
Sangwoo Mo
Sicheng Mo
Sicheng Mo
Zhipeng Mo
Michael Moeller
Peyman Moghadam
Hadi Mohaghegh Dolatabadi
Salman Mohamadi
Mirgahney H. Mohamed
Deen Dayal Mohan
Fnu Mohbat
Satyam Mohla
Tony C. W. Mok
Liliane Momeni
Pascal Monasse
Ajoy Mondal
Anindya Mondal
Mathew Monfort
Tom Monnier
Yusuke Monno
Eduardo F. Montesuma
Gyeongsik Moon
Taesup Moon
WonJun Moon
Dror Moran
Julie R. C. Mordacq
Deeptej S. More
Arthur Moreau
Davide Morelli
Luca Morelli
Pedro Morgado
Alexandre Morgand
Henrique Morimitsu
Matteo Moro
Lia Morra
Matteo Mosconi
Ali Mosleh

Sayed Mohammad Mostafavi Isfahani
Saman Motamed
Chong Mou
Dana Moukheiber
Pierre Moulon
Ramy A. Mounir
Théo Moutakanni
Fangzhou Mu
Jiteng Mu
Yao Mark Mu
Manasi Muglikar
Yasuhiro Mukaigawa
Amitangshu Mukherjee
Avideep Mukherjee
Prerana Mukherjee
Tanmoy Mukherjee
Anirban Mukhopadhyay
Soumik Mukhopadhyay
Yusuke Mukuta
Ravi Teja Mullapudi
Lea Müller
Norman Müller
Chaithanya Kumar Mummadi
Muhammad Akhtar Munir
Subrahmanyam Murala
Sanjeev Muralikrishnan
Ana C. Murillo
Nils Murrugarra-Llerena
Mohamed Adel Musallam
Damien Muselet
Josh David Myers-Dean
Byeonghu Na
Taeyoung Na
Muhammad Ferjad Naeem
Sauradip Nag
Pravin Nagar
Rajendra Nagar
Varun Nagaraja
Tushar Nagarajan
Seungjun Nah
Shu Nakamura
Gaku Nakano
Yuta Nakashima

Kiyohiro Nakayama
Mitsuru Nakazawa
Krishna Kanth Nakka
Yuesong Nan
Karthik Nandakumar
Paolo Napoletano
Syed S. Naqvi
Dinesh Reddy Narapureddy
Supreeth Narasimhaswamy
Kartik Narayan
Sriram Narayanan
Fabio Narducci
Erickson R. Nascimento
Muzammal Naseer
Kamal Nasrollahi
Lakshmanan Nataraj
Vishwesh Nath
Avisek Naug
Alexander Naumann
K. L. Navaneet
Pablo Navarrete Michelini
Shah Nawaz
Nazir Nayal
Niv Nayman
Amin Nejatbakhsh
Negar Nejatishahidin
Reyhaneh Neshatavar
Pedro C. Neto
Lukáš Neumann
Richard Newcombe
Alejandro Newell
Evonne Ng
Kam Woh Ng
Trung T. Ngo
Tuan Duc Ngo
Anh Nguyen
Anh Duy Nguyen
Cuong Cao Nguyen
Duc Anh Nguyen
Hoang Chuong Nguyen
Huy Hong Nguyen
Khai Nguyen
Khanh-Binh Nguyen

Khanh-Duy Nguyen
Khoi Nguyen
Khoi D. Nguyen
Kiet A. Nguyen
Ngoc Cuong Nguyen
Pha Nguyen
Phi Le Nguyen
Phong Ha Nguyen
Rang Nguyen
Tam V. Nguyen
Thao Nguyen
Thuan Hoang Nguyen
Toan Tien Nguyen
Trong-Tung Nguyen
Van Nguyen Nguyen
Van-Quang Nguyen
Thuong Nguyen Canh
Thien Trang Nguyen Vu
Haomiao Ni
Jiangqun Ni
Minheng Ni
Yao Ni
Zhen-Liang Ni
Zixuan Ni
Dong Nie
Hui Nie
Jiahao Nie
Lang Nie
Liqiang Nie
Qiang Nie
Ying Nie
Yinyu Nie
Yongwei Nie
Aditya Nigam
Kshitij N. Nikhal
Nick Nikzad
Jifeng Ning
Rui Ning
Xuefei Ning
Li Niu
Muyao Niu
Shuaicheng Niu
Wei Niu
Xuesong Niu
Yi Niu
Yulei Niu
Zhenxing Niu
Zhong-Han Niu
Shohei Nobuhara
Jongyoun Noh
Junhyug Noh
Nadhira Noor
Parsa Nooralinejad
Sotiris Nousias
Tiago Novello
Gal Novich
David Novotny
Slawomir Nowaczyk
Ewa M. Nowara
Evangelos Ntavelis
Valsamis Ntouskos
Leonardo Nunes
Oren Nuriel
Zhakshylyk Nurlanov
Simbarashe Nyatsanga
Lawrence O'Gorman
Anton Obukhov
Michael Oechsle
Ferda Ofli
Changjae Oh
Dongkeun Oh
Junghun Oh
Seoung Wug Oh
Youngtaek Oh
Hiroki Ohashi
Takehiko Ohkawa
Takeshi Oishi
Takahiro Okabe
Fumio Okura
Daniel Olmeda Reino
Suguru Onda
Trevine S. J. Oorloff
Michael Opitz
Roy Or-El
Jose Oramas
Jordi Orbay
Tribhuvanesh Orekondy
Evin Pınar Örnek
Alessandro Ortis
Magnus Oskarsson
Julian Ost
Daniil Ostashev
Mayu Otani
Naima Otberdout
Hatef Otroshi Shahreza
Yassine Ouali
Amine Ouasfi
Cheng Ouyang
Wanli Ouyang
Wenqi Ouyang
Xu Ouyang
Poojan B. Oza
Milind G. Padalkar
Johannes C. Paetzold
Gautam Pai
Anwesan Pal
Simone Palazzo
Avinash Paliwal
Cristina Palmero
Chengwei Pan
Fei Pan
Hao Pan
Jianhong Pan
Junting Pan
Liang Pan
Lili Pan
Linfei Pan
Liyuan Pan
Tai-Yu Pan
Xichen Pan
Xingjia Pan
Xinyu Pan
Yingwei Pan
Zhaoying Pan
Zhihong Pan
Zixuan Pan
Zizheng Pan
Rohit Pandey
Saurabh Pandey
Bo Pang
Guansong Pang
Lu Pang
Meng Pang
Tianyu Pang
Youwei Pang

Ziqi Pang
Omiros Pantazis
Juan J. Pantrigo
Hsing-Kuo Kenneth Pao
Marina Paolanti
Joao P. Papa
Samuele S. Papa
Dim P. Papadopoulos
Symeon Papadopoulos
George Papandreou
Toufiq Parag
Chethan Parameshwara
Foivos Paraperas
   Papantoniou
Shaifali Parashar
Alejandro Pardo
Jason R. Parham
Kranti K. Parida
Rishubh Parihar
Chunghyun Park
Daehee Park
Dongmin Park
Dongwon Park
Eunbyung Park
Eunhyeok Park
Eunil Park
Geon Yeong Park
Gyeong-Moon Park
Hyoungseob Park
Jae Sung Park
JaeYoo Park
Jin-Hwi Park
Jinhyung Park
Jinyoung Park
Jongwoo Park
JoonKyu Park
JungIn Park
Junheum Park
Kiru Park
Kwanyong Park
Seongsik Park
Seulki Park
Song Park
Sungho Park
Sungjune Park

Taesung Park
Yeachan Park
Gaurav Parmar
Paritosh Parmar
Maurizio Parton
Magdalini Paschali
Vito Paolo Pastore
Or Patashnik
Gaurav Patel
Maitreya Patel
Diego Patino
Suvam Patra
Viorica Patraucean
Badri Narayana Patro
Danda Pani Paudel
Angshuman Paul
Sneha Paul
Soumava Paul
Sudipta Paul
Sujoy Paul
Rémi Pautrat
Ioannis Pavlidis
Svetlana Pavlitska
Raju Pavuluri
Kim Steenstrup Pedersen
Marco Pedersoli
Adithya Pediredla
Pieter Peers
Jiju Peethambaran
Sen Pei
Wenjie Pei
Yuru Pei
Simone Alberto Peirone
Chantal Pellegrini
Latha Pemula
Abhirama Subramanyam
   V. B. Penamakuri
Adrian Penate-Sanchez
Baoyun Peng
Bo Peng
Can Peng
Cheng Peng
Chi-Han Peng
Chunlei Peng
Jie Peng

Jingliang Peng
Kebin Peng
Kunyu Peng
Liang Peng
Liangzu Peng
Pai Peng
Peixi Peng
Sida Peng
Songyou Peng
Wei Peng
Wen-Hsiao Peng
Xi Peng
Xiaojiang Peng
Yi-Xing Peng
Yuxin Peng
Zhiliang Peng
Ziqiao Peng
Matteo Pennisi
Or Perel
Gabriel Perez
Gustavo Perez
Juan C. Perez
Andres Felipe Perez
   Murcia
Eduardo Pérez-Pellitero
Neehar Peri
Skand Peri
Gabriel J. Perin
Federico Pernici
Chiara Pero
Elia Peruzzo
Marco Pesavento
Dmitry M. Petrov
Ilya A. Petrov
Mathis Petrovich
Vitali Petsiuk
Tomas Pevny
Shubham Milind Phal
Chau Pham
Hai X. Pham
Khoi Pham
Long Hoang Pham
Trong Thang Pham
Trung X. Pham
Tung Pham

Hoang Phan
Huy Phan
Minh Hieu Phan
Julien Philip
Stephen Phillips
Cheng Perng Phoo
Hao Phung
Shruti S. Phutke
Weiguo Pian
Yongri Piao
Luigi Piccinelli
A. J. Piergiovanni
Sara Pieri
Vipin Pillai
Wu Pingyu
Silvia L. Pintea
Francesco Pinto
Maura Pintor
Giovanni Pintore
Vittorio Pippi
Robinson Piramuthu
Fiora Pirri
Leonid Pishchulin
Francesca Pistilli
Francesco Pittaluga
Fabio Pizzati
Edward Pizzi
Benjamin Planche
Iuliia Pliushch
Chiara Plizzari
Ryan Po
GIovanni Poggi
Matteo Poggi
Kilian Pohl
Chandradeep Pokhariya
Ashwini Pokle
Matteo Polsinelli
Adrian Popescu
Teodora Popordanoska
Nikola Popović
Ronald Poppe
Samuele Poppi
Andrea Porfiri Dal Cin
Angelo Porrello

Pedro Porto Buarque de Gusmão
Rudra P. K. Poudel
Kossar Pourahmadi Meibodi
Hadi Pouransari
Ali Pourramezan Fard
Omid Poursaeed
Anish J. Prabhu
Mihir Prabhudesai
Aayush Prakash
Aditya Prakash
Shraman Pramanick
Mantini Pranav
B. H. Pawan Prasad
Meghshyam Prasad
Prateek Prasanna
Ekta Prashnani
Bardh Prenkaj
Derek S. Prijatelj
Véronique Prinet
Malte Prinzler
Victor Adrian Prisacariu
Federica Proietto Salanitri
Sergey Prokudin
Bill Psomas
Dongqi Pu
Mengyang Pu
Nan Pu
Shi Pu
Rita Pucci
Kuldeep Purohit
Senthil Purushwalkam
Waqas A. Qazi
Charles R. Qi
Chenyang Qi
Haozhi Qi
Jiaxin Qi
Lei Qi
Mengshi Qi
Peng Qi
Xianbiao Qi
Xiangyu Qi
Yuankai Qi
Zhangyang Qi

Guocheng Qian
Hangwei Qian
Jianing Qian
Qi Qian
Rui Qian
Shengsheng Qian
Shengyi Qian
Shenhan Qian
Wen Qian
Xuelin Qian
Yaguan Qian
Yijun Qian
Yiming Qian
Zhenxing Qian
Wenwen Qiang
Feng Qiao
Fengchun Qiao
Xiaotian Qiao
Yanyuan Qiao
Yi-Ling Qiao
Yu Qiao
Hangyu Qin
Haotong Qin
Jie Qin
Peiwu Qin
Siyang Qin
Wenda Qin
Xuebin Qin
Xugong Qin
Yang Qin
Yipeng Qin
Yongqiang Qin
Yuzhe Qin
Zequn Qin
Zeyu Qin
Zheng Qin
Zhenyue Qin
Ziheng Qin
Jiaxin Qing
Congpei Qiu
Haibo Qiu
Hang Qiu
Heqian Qiu
Jiayan Qiu
Jielin Qiu

Longtian Qiu
Mufan Qiu
Ri-Zhao Qiu
Weichao Qiu
Xuchong Qiu
Xuerui Qiu
Yuda Qiu
Yuheng Qiu
Zhongxi Qiu
Maan Qraitem
Chao Qu
Linhao Qu
Yanyun Qu
Kha Gia Quach
Ruijie Quan
Fabio Quattrini
Yvain Queau
Faisal Z. Qureshi
Rizwan Qureshi
Hamid R. Rabiee
Paolo Rabino
Ryan L. Rabinowitz
Petia Radeva
Bhaktipriya Radharapu
Krystian Radlak
Bodgan Raducanu
M. Usman Rafique
Francesco Ragusa
Sahar Rahimi Malakshan
Tanzila Rahman
Aashish Rai
Arushi Rai
Shyam Nandan Rai
Zobeir Raisi
Amit Raj
Kiran Raja
Sachin Raja
Deepu Rajan
Jathushan Rajasegaran
Gnana Praveen Rajasekhar
Ramanathan Rajendiran
Marie-Julie Rakotosaona
Gorthi Rama Krishna Sai Subrahmanyam
Sai Niranjan Ramachandran
Santhosh Kumar Ramakrishnan
Srikumar Ramalingam
Michaël Ramamonjisoa
Ravi Ramamoorthi
Shanmuganathan Raman
Mani Ramanagopal
Ashish Ramayee Asokan
Andrea Ramazzina
Jason Rambach
Sai Saketh Rambhatla
Sai Saketh Rambhatla
Clément Rambour
Francois Bernard Julien Rameau
Visvanathan Ramesh
Adín Ramírez Rivera
Haoxi Ran
Xuming Ran
Aakanksha Rana
Srinivas Rana
Kanchana N. Ranasinghe
Poorva G. Rane
Aneesh Rangnekar
Harsh Rangwani
Viresh Ranjan
Anyi Rao
Sukrut Rao
Yongming Rao
ZhiBo Rao
Carolina Raposo
Hanoona Abdul Rasheed
Amir Rasouli
Deevashwer Rathee
Christian Rathgeb
Avinash Ravichandran
Bharadwaj Ravichandran
Arijit Ray
Dripta S. Raychaudhuri
Sonia Raychaudhuri
Haziq Razali
Daniel Rebain
William T. Redman
Albert W. Reed
Aniket Rege
Christoph Reich
Christian Reimers
Simon Reiß
Konstantinos Rematas
Tal Remez
Davis Rempe
Bin Ren
Chao Ren
Chuan-Xian Ren
Dayong Ren
Dongwei Ren
Jiawei Ren
Jiaxiang Ren
Jing Ren
Mengwei Ren
Pengfei Ren
Pengzhen Ren
Qibing Ren
Shuhuai Ren
Sucheng Ren
Tianhe Ren
Weihong Ren
Wenqi Ren
Xuanchi Ren
Yanli Ren
Yihui Ren
Yixuan Ren
Yufan Ren
Zhenwen Ren
Zhihang Ren
Zhiyuan Ren
Zhongzheng Ren
Jose Restom
George Retsinas
Ambareesh Revanur
Ferdinand Rewicki
Manuel Rey Area
Md Alimoor Reza
Farnoush Rezaei Jafari
Hamed Rezazadegan Tavakoli
Rafael S. Rezende
Wonjong Rhee

Anthony D. Rhodes
Daniel Riccio
Alexander Richard
Christian Richardt
Luca Rigazio
Benjamin Risse
Dominik Rivoir
Luigi Riz
Mamshad Nayeem Rizve
Antonino M. Rizzo
Wes J. Robbins
Damien Robert
Jonathan Roberts
Joseph Robinson
Antonio Robles-Kelly
Mrigank Rochan
Chris Rockwell
Chris Rockwell
Ivan Rodin
Erik Rodner
Ranga Rodrigo
Andres C. Rodriguez
Cristian Rodriguez
Carlos Rodriguez-Pardo
Antonio J.
 Rodriguez-Sanchez
Barbara Roessle
Paul Roetzer
Alina Roitberg
Javier Romero
Meitar Ronen
Keran Rong
Xuejian Rong
Yu Rong
Marco Rosano
Bodo Rosenhahn
Gabriele Rosi
Candace Ross
Andreas Rössler
Giulio Rossolini
Mohammad Rostami
Edward Rosten
Daniel Roth
Karsten Roth
Mark S. Rothermel

Matthias Rottmann
Anastasios Roussos
Aniket Roy
Anirban Roy
Debaditya Roy
Shuvendu Roy
Sudipta Roy
Ahana Roy Choudhury
Amit Roy-Chowdhury
Aruni RoyChowdhury
Dávid Rozenberszki
Denys Rozumnyi
Lixiang Ru
Lingyan Ruan
Shulan Ruan
Viktor Rudnev
Daniel Rueckert
Nataniel Ruiz
Ewelina Rupnik
Evgenia Rusak
Chris Russell
Marc Rußwurm
Fiona Ryan
Dawid Damian Rymarczyk
DongHun Ryu
Sari Saba-Sadiya
Robert Sablatnig
Mohammad Sabokrou
Ragav Sachdeva
Ali Sadeghian
Arka Sadhu
Sadra Safadoust
Bardia Safaei
Ryusuke Sagawa
Avishkar Saha
Gobinda Saha
Oindrila Saha
Aditya Sahdev
Lakshmi Babu Saheer
Aadarsh Sahoo
Pritish Sahu
Aneeshan Sain
Nirat Saini
Saurabh Saini
Kuniaki Saito

Shunsuke Saito
Rahul Sajnani
Fumihiko Sakaue
Parikshit V. Sakurikar
Riccardo Salami
Soorena Salari
Mohammadreza Salehi
Leonard Salewski
Driton Salihu
Benjamin Salmon
Cristiano Saltori
Joel Saltz
Tim Salzmann
Sina Samangooei
Babak Samari
Nermin Samet
Fawaz Sammani
Leo Sampaio Ferraz
 Ribeiro
Shailaja Keyur Sampat
Alessio Sampieri
Jorge Sanchez
Pedro Sandoval-Segura
Nong Sang
Shengtian Sang
Patsorn Sangkloy
Depanshu Sani
Juan C. Sanmiguel
Hiroaki Santo
Joshua Santoso
Bikash Santra
Soubhik Sanyal
Hitesh Sapkota
Ayush Saraf
Nikolaos Sarafianos
István Sárándi
Kyle Sargent
Andranik Sargsyan
Josip Šarić
Mert Bulent Sariyildiz
Abhijit Sarkar
Anirban Sarkar
Chayan Sarkar
Michel Sarkis
Paul-Edouard Sarlin

Sara Sarto
Josua Sassen
Srikumar Sastry
Imari Sato
Takami Sato
Shin'ichi Satoh
Ravi Kumar Satzoda
Jack Saunders
Corentin Sautier
Mattia Savardi
Bogdan Savchynskyy
Mohamed Sayed
Marin Scalbert
Gianluca Scarpellini
Gerald Schaefer
Guilherme G. Schardong
David Schinagl
Phillip Schniter
Patrick Schramowski
Matthias Schubert
Peter Schüffler
Samuel Schulter
René Schuster
Klamer Schutte
Luca Scofano
Jesse Scott
Marcel Seelbach Benkner
Karthik Seemakurthy
Mattia Segù
Santi Seguí
Sinisa Segvic
Constantin Marc Seibold
Roman Seidel
Lorenzo Seidenari
Taiki Sekii
Yusuke Sekikawa
Matan Sela
Pratheba Selvaraju
Agniva Sengupta
Ahyun Seo
Jinhwan Seo
Junyoung Seo
Kwanggyoon Seo
Seonguk Seo
Seunghyeon Seo

Jinseok Seol
Hongje Seong
Ana F. Sequeira
Dario Serez
Dario Serez
David Serrano-Lozano
Pratinav Seth
Francesco Setti
Giorgos Sfikas
Mohammad Amin Shabani
Faisal Shafait
Anshul Shah
Chintan Shah
Jay Shah
Ketul Shah
Mubarak Shah
Viraj Shah
Mohamad Shahbazi
Muhammad Bilal B. Shaikh
Abdelrahman M. Shaker
Greg Shakhnarovich
Md Salman Shamil
Fahad Shamshad
Caifeng Shan
Dandan Shan
Hongming Shan
Xiaojun Shan
Chong Shang
Fanhua Shang
Jinghuan Shang
Lei Shang
Sifeng Shang
Wei Shang
Yuzhang Shang
Yuzhang Shang
Sukrit Shankar
Dian Shao
Mingwen Shao
Rui Shao
Ruizhi Shao
Shuai Shao
Shuwei Shao
Ron A. Shapira Weber
S. M. A. Sharif

Aashish Sharma
Avinash Sharma
Charu Sharma
Prafull Sharma
Prasen Kumar Sharma
Allam Shehata
Mark Sheinin
Sumit Shekhar
Oleksandr Shekhovtsov
Chuanfu Shen
Fei Shen
Fengyi Shen
Furao Shen
Hui-liang Shen
Jiajun Shen
Jianghao Shen
Jiangrong Shen
Jiayi Shen
Li Shen
Li-Yong Shen
Linlin Shen
Maying Shen
Qiu Shen
Qiuhong Shen
Shuai Shen
Shuhan Shen
Siqi Shen
Tianwei Shen
Tong Shen
Xiaolong Shen
Xiaoqian Shen
Yan Shen
Yanqing Shen
Yilin Shen
Ying Shen
Yiqing Shen
Yuan Shen
Yucong Shen
Yuhan Shen
Yunhang Shen
Zehong Shen
Zengming Shen
Zhijie Shen
Zhiqiang Shen
Hualian Sheng

Tao Sheng
Yichen Sheng
Zehua Sheng
Shivanand Venkanna
  Sheshappanavar
Ivaxi Sheth
Baoguang Shi
Botian Shi
Dachuan Shi
Daqian Shi
Haizhou Shi
Hengcan Shi
Jia Shi
Jing Shi
Jingang Shi
QingHongYa Shi
Ruoxi Shi
Tianyang Shi
Weishi Shi
Wu Shi
Wuxuan Shi
Xiaodan Shi
Xiaoshuang Shi
Xiaoyu Shi
Xingjian Shi
Xinyu Shi
Xuepeng Shi
Yichun Shi
Yujiao Shi
Zhenbo Shi
Zheng Shi
Zhensheng Shi
Zhenwei Shi
Zhihao Shi
Zifan Shi
Takashi Shibata
Meng-Li Shih
Yichang Shih
Dongseok Shim
Wataru Shimoda
Ilan Shimshoni
Changha Shin
Gyungin Shin
Hyungseob Shin
Inkyu Shin

Seungjoo Shin
Ukcheol Shin
Yooju Shin
Young Min Shin
Koichi Shinoda
Kaede Shiohara
Suprosanna Shit
Palaiahnakote
  Shivakumara
Sindi Shkodrani
Michal
  Shlapentokh-Rothman
Debaditya Shome
Hyounguk Shon
Sulabh Shrestha
Aman Shrivastava
Ayush Shrivastava
Gaurav Shrivastava
Aleksandar Shtedritski
Dong Wook Shu
Han Shu
Jun Shu
Xiangbo Shu
Xiujun Shu
Yang Shu
Bing Shuai
Hong-Han Shuai
Qing Shuai
Changjian Shui
Pushkar Shukla
Mustafa Shukor
Hubert P. H. Shum
Nina Shvetsova
Chenyang Si
Jianlou Si
Zilin Si
Mennatullah Siam
Sven Sickert
Désiré Sidibé
Ioannis Siglidis
Alberto Signoroni
Karan Sikka
Pedro Silva
Julio Silva-Rodríguez
Hyeonjun Sim

Jae-Young Sim
Chonghao Sima
Christian Simon
Martin Simon
Alessandro Simoni
Enis Simsar
Abhishek Singh
Apoorv Singh
Ashish Singh
Bharat Singh
Jasdeep Singh
Jaskirat Singh
Krishnakant Singh
Manish Kumar Singh
Mannat Singh
Nikhil Singh
Pravendra Singh
Rajat Vikram Singh
Simranjit Singh
Darshan Singh S.
Utkarsh Singhal
Dipika Singhania
Vasu Singla
Abhishek Kumar Sinha
Animesh Sinha
Sanjana Sinha
Saptarshi Sinha
Sudipta Sinha
Sophia A.
  Sirko-Galouchenko
Josef Sivic
Elena Sizikova
Geri Skenderi
Gregory Slabaugh
Habib Slim
Dmitriy Smirnov
James S. Smith
William Smith
Noah Snavely
Kihyuk Sohn
Bolivar E. Solarte
Mattia Soldan
Sobhan Soleymani
Samik Some
Nagabhushan Somraj

Jeany Son
Seung Woo Son
Byung Cheol Song
Chen Song
Guanglu Song
Jie Song
Jifei Song
Li Song
Liangchen Song
Lin Song
Luchuan Song
Mingli Song
Ran Song
Sibo Song
Sifan Song
Siyang Song
Weilian Song
Weinan Song
Wenfeng Song
Xiangchen Song
Xibin Song
Xinhang Song
Yafei Song
Yang Song
Yi-Yang Song
Yizhi Song
Yue Song
Zeen Song
Zhenbo Song
Zikai Song
Ekta Sood
Tomáš Souček
Rajiv Soundararajan
Albin Soutif-Cormerais
Jeremy Speth
Indro Spinelli
Jon Sporring
Manogna Sreenivas
Arvind Krishna Sridhar
Deepak Sridhar
Balaji Vasan Srinivasan
Pratul Srinivasan
Anuj Srivastava
Astitva Srivastava
Dhruv Srivastava

Koushik Srivatsan
Pierre-Luc St-Charles
Ioannis Stamos
Anastasis Stathopoulos
Colton Stearns
Jan Steinbrener
Jan-Martin O. Steitz
Sinisa Stekovic
Federico Stella
Michael Stengel
Alexandros Stergiou
Gleb Sterkin
Rainer Stiefelhagen
Noah Stier
Timo N. Stoffregen
Vladan Stojnić
Nick O. Stracke
Ombretta Strafforello
Julian Straub
Nicola Strisciuglio
Vitomir Struc
Yannick Strümpler
Joerg Stueckler
Chi Su
Hang Su
Hang Su
Kun Su
Rui Su
Shaolin Su
Sitong Su
Xingzhe Su
Xiu Su
Yao Su
Yiyang Su
Yongyi Su
Zhaoqi Su
Zhixun Su
Zhuo Su
Zhuo Su
Iago Suárez
Arulkumar Subramaniam
Sanjay Subramanian
A. Subramanyam
Swathikiran Sudhakaran
Yusuke Sugano

Masanori Suganuma
Yumin Suh
Mohammed Suhail
Xiuchao Sui
Yang Sui
Yao Sui
Heung-Il Suk
Pavel Suma
Baigui Sun
Baochen Sun
Bin Sun
Bo Sun
Changchang Sun
Che Sun
Cheng Sun
Chong Sun
Chunyi Sun
Gan Sun
Guofei Sun
Guoxing Sun
Haifeng Sun
Hanqing Sun
Haoliang Sun
He Sun
Heming Sun
Hongbin Sun
Huiming Sun
Jennifer J. Sun
Jian Sun
Jiande Sun
Jianhua Sun
Jiankai Sun
Jipeng Sun
Keqiang Sun
Lei Sun
Lichao Sun
Long Sun
Mingjie Sun
Peize Sun
Pengzhan Sun
Qiyue Sun
Shangquan Sun
Shanlin Sun
Shuyang Sun
Tao Sun

Tiancheng Sun
Wei Sun
Weiwei Sun
Weixuan Sun
Xianfang Sun
Xiaohang Sun
Xiaoshuai Sun
Xiaoxiao Sun
Ximeng Sun
Xuxiang Sun
Yanan Sun
Yasheng Sun
Yihong Sun
Ying Sun
Yixuan Sun
Yu Sun
Yuan Sun
Yuchong Sun
Zeren Sun
Zhanghao Sun
Zhaodong Sun
Zhaohui H. Sun
Zhicheng Sun
Zhicheng Sun
Haomiao Sun
Varun Sundar
Shobhita Sundaram
Minhyuk Sung
Kalyan Sunkavalli
Yucheng Suo
Indranil Sur
Saksham Suri
Naufal Suryanto
Vadim Sushko
David Suter
Roman Suvorov
Fnu Suya
Teppei Suzuki
Kunal Swami
Archana Swaminathan
Gurumurthy Swaminathan
Robin Swanson
Eran Swears
Alexander Swerdlow
Sirnam Swetha

Tabish A. Syed
Tanveer Syeda-Mahmood
Stanislaw K. Szymanowicz
Sethuraman T. V.
Calvin-Khang T. Ta
The-Anh Ta
Babak Taati
Samy Tafasca
Andrea Tagliasacchi
Haowei Tai
Yuan Tai
Francesco Taioli
Peng Taiying
Keita Takahashi
Naoya Takahashi
Jun Takamatsu
Nicolas Talabot
Hugues G. Talbot
Hossein Talebi
Davide Talon
Gary Tam
Toru Tamaki
Dipesh Tamboli
Andong Tan
Bin Tan
Cheng Tan
David Joseph New Tan
Fuwen Tan
Guang Tan
Jianchao Tan
Jing Tan
Jingru Tan
Lei Tan
Mingkui Tan
Mingxing Tan
Shuhan Tan
Shunquan Tan
Weimin Tan
Xin Tan
Zhentao Tan
Zhentao Tan
Masayuki Tanaka
Chen Tang
Chengzhou Tang
Chenwei Tang

Fan Tang
Feng Tang
Hao Tang
Haoran Tang
Jiajun Tang
Jiapeng Tang
Jiaxiang Tang
Jie Tang
Junshu Tang
Keke Tang
Luming Tang
Luyao Tang
Lv Tang
Ming Tang
Quan Tang
Shengji Tang
Sheyang Tang
Shitao Tang
Shixiang Tang
Tao Tang
Weixuan Tang
Xu Tang
Yang Tang
Yansong Tang
Yehui Tang
Yu-Ming Tang
Zheng Tang
Zhipeng Tang
Zitian Tang
Md Mehrab Tanjim
Julian Tanke
An Tao
Chaofan Tao
Chenxin Tao
Jiale Tao
Junli Tao
Keda Tao
Ming Tao
Ran Tao
Wenbing Tao
Xinhao Tao
Jean-Philippe G. Tarel
Laia Tarres
Laia Tarrés
Enzo Tartaglione

Keisuke Tateno
SaiKiran K. Tedla
Antonio Tejero-de-Pablos
Bugra Tekin
Purva Tendulkar
Minggui Teng
Ruwan Tennakoon
Andrew Beng Jin Teoh
Konstantinos Tertikas
Piotr Teterwak
Piotr Teterwak
Anh Thai
Kartik Thakral
Nupur Thakur
Sadbhawna Thakur
Balamurugan Thambiraja
Vikas Thamizharasan
Kevin Thandiackal
Sushil Thapa
Daksh Thapar
Jonas Theiner
Christian Theobalt
Spyridon Thermos
Fida Mohammad Thoker
Christopher L. Thomas
Diego Thomas
William Thong
Mamatha Thota
Mukund Varma Thottankara
Changyao Tian
Chunwei Tian
Jinyu Tian
Kai Tian
Lin Tian
Tai-Peng Tian
Xin Tian
Xinyu Tian
Yapeng Tian
Yu Tian
Yuan Tian
Yuesong Tian
Yunjie Tian
Yuxin Tian
Zhuotao Tian

Mert Tiftikci
Javier Tirado-Garín
Garvita Tiwari
Lokender Tiwari
Anastasia Tkach
Andrea Toaiari
Sinisa Todorovic
Pavel Tokmakov
Tri Ton
Adam Tonderski
Jinguang Tong
Peter Tong
Xin Tong
Zhan Tong
Francesco Tonini
Alessio Tonioni
Alessandro Torcinovich
Marwan Torki
Lorenzo Torresani
Fabio Tosi
Matteo Toso
Anh T. Tran
Hung Tran
Linh-Tam Tran
Minh-Triet Tran
Ngoc-Trung Tran
Phong Tran
Jonathan Tremblay
Alex Trevithick
Aditay Tripathi
Subarna Tripathi
Felix Tristram
Gabriele Trivigno
Emanuele Trucco
Prune Truong
Thanh-Dat Truong
Tomasz Trzcinski
Fu-Jen Tsai
Yu-Ju Tsai
Michael Tschannen
Tze Ho Elden Tse
Ethan Tseng
Yu-Chee Tseng
Shahar Tsiper
Hanzhang Tu

Rong-Cheng Tu
Yuanpeng Tu
Zhengzhong Tu
Zhigang Tu
Narek Tumanyan
Anil Osman Tur
Haithem Turki
Mehmet Ozgur Turkoglu
Daniyar Turmukhambetov
Victor G. Turrisi da Costa
Tinne Tuytelaars
Bartlomiej Twardowski
Radim Tylecek
Christos Tzelepis
Seiichi Uchida
Hideaki Uchiyama
Vishaal Udandarao
Mostofa Rafid Uddin
Kohei Uehara
Tatsumi Uezato
Nicolas Ugrinovic
Youngjung Uh
Norimichi Ukita
Amin Ullah
Markus Ulrich
Ardian Umam
Mesut Erhan Unal
Mathias Unberath
Devesh Upadhyay
Paul Upchurch
Shagun Uppal
Yoshitaka Ushiku
Anil Usumezbas
Yuzuko Utsumi
Roy Uziel
Anil Vadathya
Sharvaree Vadgama
Pratik Vaishnavi
Gregory Vaksman
Matias A. Valdenegro Toro
Lucas Valença
Eduardo Valle
Ernest Valveny
Laurens van der Maaten
Wouter Van Gansbeke

Nanne van Noord
Max W. F. van Spengler
Lorenzo Vaquero
Farshid Varno
Cristina Vasconcelos
Francisco Vasconcelos
Igor Vasiljevic
Florin-Alexandru
  Vasluianu
Subeesh Vasu
Arun Balajee Vasudevan
Vaibhav S. Vavilala
Kyle Vedder
Vijay Veerabadran
Ronny Xavier Velastegui
  Sandoval
Senem Velipasalar
Andreas Velten
Raviteja Vemulapalli
Deepika Vemuri
Edward Vendrow
Jonathan Ventura
Lucas Ventura
Jakob Verbeek
Dor Verbin
Eshan Verma
Manisha Verma
Monu Verma
Sahil Verma
Constantin Vertan
Eli Verwimp
Noranart Vesdapunt
Jordan J. Vice
Sara Vicente
Kavisha Vidanapathirana
Dat Viet Thanh Nguyen
Sudheendra
  Vijayanarasimhan
Sujal T. Vijayaraghavan
Deepak Vijaykeerthy
Elliot Vincent
Yael Vinker
Duc Minh Vo
Huy V. Vo
Khoa H. V. Vo

Romain Vo
Antonin Vobecky
Michele Volpi
Riccardo Volpi
Igor Vozniak
Nicholas Vretos
Vibashan V. S.
Ngoc-Son Vu
Tuan-Anh Vu
Khiem Vuong
Mårten Wadenbäck
Neal Wadhwa
Sophia J. Wagner
Muntasir Wahed
Nobuhiko Wakai
Devesh Walawalkar
Jacob Walker
Matthew Walmer
Matthew R. Walter
Bo Wan
Guancheng Wan
Jia Wan
Jin Wan
Jun Wan
Qiyang Wan
Renjie Wan
Wei Wan
Xingchen Wan
Yecong Wan
Zhexiong Wan
Ziyu Wan
Karan Wanchoo
Alex Jinpeng Wang
Angtian Wang
Baoyuan Wang
Benyou Wang
Biao Wang
Bin Wang
Bing Wang
Binghui Wang
Binglu Wang
Can Wang
Ce Wang
Changwei Wang
Chao Wang

Chaoyang Wang
Chen Wang
Chen Wang
Chen Wang
Chengrui Wang
Chien-Yao Wang
Chu Wang
Chuan Wang
Congli Wang
Dadong Wang
Di Wang
Dong Wang
Dong Wang
Dongdong Wang
Dongkai Wang
Dongqing Wang
Dongsheng Wang
X. Wang
Fan Wang
Fangfang Wang
Fangjinhua Wang
Fei Wang
Feng Wang
Feng Wang
Fu-Yun Wang
Gaoang Wang
Guangcong Wang
Guangming Wang
Guangrun Wang
Guangzhi Wang
Guanshuo Wang
Guo-Hua Wang
Guoqing Wang
Guoqing Wang
Haixin Wang
Haiyan Wang
Han Wang
Hanjing Wang
Hanyu Wang
Hao Wang
Hao Wang
Haobo Wang
Haochen Wang
Haochen Wang
Haohan Wang

Haoqi Wang
Haoran Wang
Haotao Wang
Haoxuan Wang
HaoYu Wang
Hengkang Wang
Hengli Wang
Hengyi Wang
Hesheng Wang
Hong Wang
Hongjun Wang
Hongxiao Wang
Hongyu Wang
Hongzhi Wang
Hua Wang
Huafeng Wang
Huan Wang
Huijie Wang
Huiyu Wang
Jiadong Wang
Jiahao Wang
Jiahao Wang
Jiahao Wang
Jiakai Wang
Jialiang Wang
Jiamian Wang
Jian Wang
Jiang Wang
Jiangliu Wang
Jianjia Wang
Jianyi Wang
Jianyuan Wang
Jiaqi Wang
Jiashun Wang
Jiayi Wang
Jiaze Wang
Jin Wang
Jinfeng Wang
Jingbo Wang
Jinghua Wang
Jingkang Wang
Jinglong Wang
Jinglu Wang
Jinpeng Wang
Jinqiao Wang

Jue Wang
Jun Wang
Jun Wang
Junjue Wang
Junke Wang
Junxiao Wang
Kai Wang
Kai Wang
Kai Wang
Kai Wang
Kaihong Wang
Kewei Wang
Keyan Wang
Kun Wang
Kup Wang
Lan Wang
Lanjun Wang
Le Wang
Lei Wang
Lei Wang
Lezi Wang
Liansheng Wang
Liao Wang
Lijuan Wang
Lijun Wang
Limin Wang
Lin Wang
Linwei Wang
Lishun Wang
Lixu Wang
Liyuan Wang
Lizhen Wang
Lizhi Wang
Longguang Wang
Luozhou Wang
Luting Wang
Mang Wang
Manning Wang
Mei Wang
Mengjiao Wang
Mengmeng Wang
Miaohui Wang
Min Wang
Naiyan Wang
Nannan Wang

Ning-Hsu Wang
Pei Wang
Peihao Wang
Peiqi Wang
Peng Wang
Pengfei Wang
Pengkun Wang
Pichao Wang
Pu Wang
Qi Wang
Qian Wang
Qiang Wang
Qiang Wang
Qiangchang Wang
Qianqian Wang
Qifei Wang
Qilong Wang
Qin Wang
Qing Wang
Qingzhong Wang
Qitong Wang
Qiufeng Wang
Ronggang Wang
Rui Wang
Rui Wang
Rui Wang
Ruibin Wang
Ruisheng Wang
Ruoyu Wang
Sai Wang
Sen Wang
Sen Wang
Shan Wang
Shaoru Wang
Sheng Wang
Sheng-Yu Wang
Shengze Wang
Shida Wang
Shijie Wang
Shipeng Wang
Shiping Wang
Shiyu Wang
Shizun Wang
Shuhui Wang
Shujun Wang

Shunli Wang
Shunxin Wang
Shuo Wang
Shuo Wang
Shuo Wang
Shuzhe Wang
Siqi Wang
Siwei Wang
Song Wang
Song Wang
Su Wang
Tan Wang
Tao Wang
Taoyue Wang
Teng Wang
Tengfei Wang
Tiancai Wang
Tianqi Wang
Tianyang Wang
Tianyu Wang
Tong Wang
Tsun-Hsuan Wang
Tuanfeng Wang
Tuanfeng Y. Wang
Wei Wang
Weihan Wang
Weikang Wang
Weimin Wang
Weiqiang Wang
Weixi Wang
Weiyao Wang
Weiyun Wang
Wen Wang
Wenbin Wang
Wenhao Wang
Wenjing Wang
Wenqian Wang
Wentao Wang
Wenxiao Wang
Wenxuan Wang
Wenzhe Wang
Xi Wang
Xi Wang
Xiang Wang
Xiao Wang

Xiao Wang
Xiao Wang
Xiaobing Wang
Xiaofeng Wang
Xiaohan Wang
Xiaosen Wang
Xiaosong Wang
Xiaoxing Wang
Xiaoyang Wang
Xijun Wang
Xijun Wang
Xinggang Wang
Xinghan Wang
Xinjiang Wang
Xinshao Wang
Xintong Wang
Xizi Wang
Xu Wang
Xuan Wang
Xuanhan Wang
Xue Wang
Xueping Wang
Xuyang Wang
Yali Wang
Yan Wang
Yan Wang
Yang Wang
Yangang Wang
Yangtao Wang
Yaohui Wang
Yaoming Wang
Yaxing Wang
Yaxiong Wang
Yi Wang
Yi Ru Wang
Yidong Wang
Yifan Wang
Yifeng Wang
Yifu Wang
Yikai Wang
Yilin Wang
Yilun Wang
Yin Wang
Yinggui Wang
Yingheng Wang

Yingqian Wang
Yipei Wang
Yiqun Wang
Yiran Wang
Yiwei Wang
Yixu Wang
Yizhi Wang
Yizhou Wang
Yizhou Wang
Yong Wang
Yu Wang
Yu-Shuen Wang
Yuan-Gen Wang
Yuchen Wang
Yude Wang
Yue Wang
Yuesong Wang
Yufei Wang
Yufu Wang
Yuguang Wang
Yuhan Wang
Yujia Wang
Yulin Wang
Yunke Wang
Yuting Wang
Yuxi Wang
YuXin Wang
Yuzheng Wang
Ze Wang
Zedong Wang
Zehan Wang
Zengmao Wang
Zeyu Wang
Zeyu Wang
Zhao Wang
Zhaokai Wang
Zhaowen Wang
Zhe Wang
Zhen Wang
Zhen Wang
Zhendong Wang
Zheng Wang
Zheng Wang
Zheng Wang
Zhengyi Wang

Zhennan Wang
Zhenting Wang
Zhenyi Wang
Zhenyu Wang
Zhenzhi Wang
Zhepeng Wang
Zhi Wang
Zhibo Wang
Zhihao Wang
Zhihui Wang
Zhijie Wang
Zhikang Wang
Zhixiang Wang
Zhiyong Wang
Zhongdao Wang
Zhonghao Wang
Zhouxia Wang
Zhu Wang
Zian Wang
Zifu Wang
Zihao Wang
Zijian Wang
Ziqiang Wang
Ziqin Wang
Ziqing Wang
Zirui Wang
Zirui Wang
Ziwei Wang
Ziyan Wang
Ziyang Wang
Ziyi Wang
Ziyun Wang
Frederik Warburg
Syed Talal Wasim
Daniel Watson
Jamie Watson
Ethan Weber
Silvan Weder
Jan Dirk Wegner
Chen Wei
Donglai Wei
Fangyin Wei
Fangyun Wei
Guoqiang Wei
Jia Wei

Jiacheng Wei
Kaixuan Wei
Kun Wei
Longhui Wei
Megan Wei
Mian Wei
Mingqiang Wei
Pengxu Wei
Ping Wei
Qiuhong Anna Wei
Shikui Wei
Tianyi Wei
Wei Wei
Wenqi Wei
Xian Wei
Xin Wei
Xing Wei
Xinyue Wei
Xiu-Shen Wei
Yi Wei
Yixuan Wei
Yunchao Wei
Yuxiang Wei
Yuxiang Wei
Zeming Wei
Zhipeng Wei
Zihao Wei
Zimian Wei
Jean-Baptiste Weibel
Luca Weihs
Martin Weinmann
Michael Weinmann
Bihan Wen
Bowen Wen
Chao Wen
Chenglu Wen
Chuan Wen
Congcong Wen
Jie Wen
Jing Wen
Qiang Wen
Rui Wen
Sijia Wen
Song Wen
Xiang Wen

Xin Wen
Yilin Wen
Youpeng Wen
Yuanbo Wen
Yuxin Wen
Chung-Yi Weng
Junwu Weng
Shuchen Weng
Wenming Weng
Yijia Weng
Zhenzhen Weng
Thomas Westfechtel
Christopher Johannes
 Wewer
Spencer Whitehead
Tobias Jan Wieczorek
Thaddäus Wiedemer
Julian Wiederer
Kevin Tirta Wijaya
Asiri Wijesinghe
Kimberly Wilber
Jeffrey R. Willette
Bryan M. Williams
Williem Williem
Christian Wilms
Benjamin Wilson
Richard Wilson
Felix Wimbauer
Vanessa Wirth
Scott Wisdom
Calden Wloka
Alex Wong
Chau-Wai Wong
Chi-Chong Wong
Ka Wai Wong
Kelvin Wong
Kok-Seng Wong
Kwan-Yee K. Wong
Yongkang Wong
Sangmin Woo
Simon S. Woo
Markus Worchel
Scott Workman
Marcel Worring
Safwan Wshah

Aming Wu
Bo Wu
Bojian Wu
Boxi Wu
Changguang Wu
Chaoyi Wu
Chen Henry Wu
Cheng-En Wu
Chenming Wu
Chenyan Wu
Chenyun Wu
Cho-Ying Wu
Chongruo Wu
Cong Wu
Dayan Wu
Di Wu
Dongming Wu
Fangzhao Wu
Fuxiang Wu
Gaojie Wu
Guanyao Wu
Guile Wu
Haiping Wu
Haiwei Wu
Haiyu Wu
Han Wu
Haoning Wu
Haoning Wu
Haotian Wu
Hefeng Wu
Huisi Wu
Jane Wu
Jay Zhangjie Wu
Jhih-Ciang Wu
Ji-Jia Wu
Jialian Wu
Jiaye Wu
Jimmy Wu
Jing Wu
Jing Wu
Jinjian Wu
Jiqing Wu
Jun Wu
Junfeng Wu
Junlin Wu
Junru Wu
Junyang Wu
Junyi Wu
Letian Wu
Lifang Wu
Lin Yuanbo Wu
Liwen Wu
Min Wu
Minye wu
Peng Wu
Penghao Wu
Qian Wu
Qiangqiang Wu
Qianyi Wu
Qingbo Wu
Rongliang Wu
Rui Wu
Rundi Wu
Shuang Wu
Shuzhe Wu
Tao Wu
Tao Wu
Tao Wu
Te-Lin Wu
Tianfu Wu
Tianhao Wu
Tianhao Wu
Ting-Wei Wu
Tong Wu
Tong Wu
Tsung-Han Wu
Tz-Ying Wu
Weibin Wu
Weijia Wu
Xian Wu
Xiao Wu
Xiaodong Wu
Xiaohe Wu
Xiaoqian Wu
Xiaoyang Wu
Xindi Wu
Xingjiao Wu
Xinxiao Wu
Xiuzhe Wu
Yang Wu
Yangzheng Wu
Yanze Wu
Yanzhao Wu
Yawen Wu
Yicheng Wu
Ying Nian Wu
Yingwen Wu
Yong Wu
Yuanwei Wu
Yue Wu
Yue Wu
Yuqun Wu
Yushu Wu
Yushuang Wu
Zhe Wu
Zheng Wu
Zhi-Fan Wu
Zhihao Wu
Zhijie Wu
Zhiliang Wu
Zhonghua Wu
Zijie Wu
Ziyi Wu
Zizhao Wu
Zongwei Wu
Zongyu Wu
Zongze Wu
Stefanie Wuhrer
Jamie M. Wynn
Monika Wysoczańska
Jianing Xi
Teng Xi
Bin Xia
Changqun Xia
Haifeng Xia
Jiaer Xia
Jiahao Xia
Kun Xia
Mingxuan Xia
Shihong Xia
Weihao Xia
Xiaobo Xia
Yan Xia
Ye Xia
Yifei Xia

Zhaoyang Xia
Zhihao Xia
Zhihua Xia
Zhuofan Xia
Zimin Xia
Chuhua Xian
Wenqi Xian
Yongqin Xian
Donglai Xiang
Jinhai Xiang
Liuyu Xiang
Tian-Zhu Xiang
Tiange Xiang
Wangmeng Xiang
Xiaoyu Xiang
Yuanbo Xiangli
Anqi Xiao
Aoran Xiao
Bei Xiao
Chunxia Xiao
Fanyi Xiao
Han Xiao
Jiancong Xiao
Jimin Xiao
Jing Xiao
Jing Xiao
Jun Xiao
Junbin Xiao
Junfei Xiao
Mingqing Xiao
Qingyang Xiao
Ruixuan Xiao
Taihong Xiao
Yang Xiao
Yanru Xiao
Yao Xiao
Yijun Xiao
Yuting Xiao
Zehao Xiao
Zeyu Xiao
Zihao Xiao
Binhui Xie
Chaohao Xie
Chi Xie
Christopher Xie

Chuanlong Xie
Fei Xie
Guo-Sen Xie
Haozhe Xie
Hongtao Xie
Jiahao Xie
Jiaxin Xie
Jin Xie
Jinheng Xie
Jiu-Cheng Xie
Jiyang Xie
Junyu Xie
Liuyue Xie
Ming-Kun Xie
Mingyang Xie
Qian Xie
Tingting Xie
Weicheng Xie
Xianghui Xie
Xiaohua Xie
Xudong Xie
Yichen Xie
Yiming Xie
You Xie
Yuan Xie
Yusheng Xie
Yutong Xie
Zeke Xie
Zhenda Xie
Zhenyu Xie
ZiYang Xie
Chaoyue Xing
Fuyong Xing
Jinbo Xing
Xiaoyan Xing
Xiaoying Xing
XiMing Xing
Xin Xing
Yazhou Xing
Yifan Xing
Yun Xing
Zhen Xing
Hongkai Xiong
Jingjing Xiong
Jinhui Xiong

Junwen Xiong
Peixi Xiong
Wei Xiong
Weihua Xiong
Yu Xiong
Yuanhao Xiong
Yuanjun Xiong
Yuwen Xiong
Zhexiao Xiong
Zhiwei Xiong
Yuliang Xiu
Alessio Xompero
An Xu
Angchi Xu
Baixin Xu
Bicheng Xu
Bo Xu
Chao Xu
Chenfeng Xu
Chenshu Xu
Chenxin Xu
Chi Xu
Dejia Xu
Dongli Xu
Feng Xu
Gangwei Xu
Haiming Xu
Haiyang Xu
Han Xu
Haofei Xu
Haohang Xu
Haoran Xu
Hongbin Xu
Hongmin Xu
Jiale Xu
Jianjin Xu
Jiaqi Xu
Jie Xu
Jilan Xu
Jinglin Xu
Jingyi Xu
Jun Xu
Kai Xu
Katherine Xu
Ke Xu

Kele Xu
Lan Xu
Lian Xu
Liang Xu
Linning Xu
Lumin Xu
Manjie Xu
Mengde Xu
Mengdi Xu
Mengmeng Frost Xu
Min Xu
Ming Xu
Mutian Xu
Peiran Xu
Peng Xu
Qi Xu
Qiang Xu
Qiangeng Xu
Qingshan Xu
Qingyang Xu
Qiuling Xu
Ran Xu
Renzhe Xu
Ruikang Xu
Runsen Xu
Runsheng Xu
Shichao Xu
Sirui Xu
Tongda Xu
Wanting Xu
Wei Xu
Weiwei Xu
Wenjia Xu
Wenju Xu
Wenqiang Xu
Xiang Xu
Xianghao Xu
Xiangyu Xu
Xiangyu Xu
Xiaogang Xu
Xiaohao Xu
Xin Xu
Xin Xu
Xin-Shun Xu
Xing Xu

Xinli Xu
Xinyu Xu
Xiuwei Xu
Xiyan Xu
Xudong Xu
Xuemiao Xu
Xun Xu
Yan Xu
Yan Xu
Yan Xu
Yangyang Xu
Yanwu Xu
Yating Xu
Yi Xu
Yi Xu
Yi Xu
Yihong Xu
YiKun Xu
Yinghao Xu
Yingyan Xu
Yinshuang Xu
Yiran Xu
Yixing Xu
Yongchao Xu
Yue Xu
Yufei Xu
Yunqiu Xu
Zexiang Xu
Zhan Xu
Zhe Xu
Zhengqin Xu
Zhenlin Xu
Zhiqiu Xu
Zhiyuan Xu
Zhongcong Xu
Zhuoer Xu
Zipeng Xu
Ziyue Xu
Zongyi Xu
Ziwei Xuan
Danna Xue
Fanglei Xue
Fei Xue
Feng Xue
Han Xue

Jianru Xue
Le Xue
Lixin Xue
Mingfu Xue
Nan Xue
Qinghan Xue
Shangjie Xue
Xiangyang Xue
Zihui Xue
Abhay Yadav
Amit Kumar Singh Yadav
Takuma Yagi
Tomas F Yago Vicente
I. Zeki Yalniz
Kota Yamaguchi
Shin'ya Yamaguchi
Burhaneddin Yaman
Toshihiko Yamasaki
Kohei Yamashita
Lee Juliette Yamin
Chaochao Yan
Hongyu Yan
Jiexi Yan
Kai Yan
Pei Yan
Qingan Yan
Qingsen Yan
Qingsong Yan
Rui Yan
Shaoqi Yan
Shi Yan
Siming Yan
Siming Yan
Siyuan Yan
Weilong Yan
Wending Yan
Xiangyi Yan
Xinchen Yan
Xingguang Yan
Xueting Yan
Yan Yan
Yichao Yan
Zhaoyi Yan
Zhiqiang Yan
Zhiyuan Yan

Zike Yan
Zizheng Yan
Keiji Yanai
Pinar Yanardag
Anqi Yang
Anqi Joyce Yang
Bangbang Yang
Baoyao Yang
Bin Yang
Binwei Yang
Bo Yang
Bo Yang
Boyu Yang
Changdi Yang
Chao Yang
Charig Yang
Cheng-Fu Yang
Cheng-Yen Yang
Chenhongyi Yang
Chuanguang Yang
De-Nian Yang
Dingcheng Yang
Dingkang Yang
Dong Yang
Erkun Yang
Fan Yang
Fan Yang
Fan Yang
Fan Yang
Fan Yang
Feng Yang
Fengting Yang
Fengxiang Yang
Fengyuan Yang
Fu-En Yang
Gang Yang
Gengshan Yang
Guandao Yang
Guanglei Yang
Haitao Yang
Hanqing Yang
Heran Yang
Honghui Yang
Huanrui Yang
Huiyuan Yang

Huizong Yang
Hunmin Yang
Jiange Yang
Jiaqi Yang
Jiawei Yang
Jiayu Yang
Jiazhi Yang
Jie Yang
Jie Yang
Jiewen Yang
Jihan Yang
Jing Yang
Jingkang Yang
Jinhui Yang
Jinlong Yang
Jinrong Yang
Jinyu Yang
Kaicheng Yang
Kailun Yang
Lan Yang
Le Yang
Lehan Yang
Lei Yang
Lei Yang
Lei Yang
Li Yang
Lihe Yang
Ling Yang
Lingxiao Yang
Linlin Yang
Lixin Yang
Longrong Yang
Lu Yang
Luwei Yang
Michael Ying Yang
Min Yang
Ming Yang
MingKun Yang
Mouxing Yang
Muli Yang
Peiyu Yang
Qi Yang
Qian Yang
Qiushi Yang
Ren Yang

Rui Yang
Ruihan Yang
Sejong Yang
Shan Yang
Shangrong Yang
Shiqi Yang
Shuai Yang
Shuai Yang
Shuang Yang
Shuo Yang
Shusheng Yang
Sibei Yang
Siwei Yang
Siyuan Yang
Siyuan Yang
Song Yang
Songlin Yang
Tianyu Yang
Tong Yang
Wankou Yang
Wenhan Yang
Wenhan Yang
Wenjie Yang
Wenqi Yang
William Yang
Xi Yang
Xi Yang
Xiangpeng Yang
Xiao Yang
Xiaofeng Yang
Xiaoshan Yang
Xin Jeremy Yang
Xingyi Yang
Xinlong Yang
Xitong Yang
Xiulong Yang
Xu Yang
Xuan Yang
Xue Yang
Xuelin Yang
Xun Yang
Yan Yang
Yan Yang
Yang Yang
Yaokun Yang

Yezhou Yang
Yiding Yang
Yijun Yang
Yijun Yang
Yin Yang
Yinfei Yang
Yixin Yang
Yongqi Yang
Yongqi Yang
Yue Yang
Yuewei Yang
Yuezhi Yang
Yujiu Yang
Yung-Hsu Yang
Yuwei Yang
Ze Yang
Ze Yang
Zetong Yang
Zhangsihao Yang
Zhaoyuan Yang
Zhen Yang
Zhenpei Yang
Zhibo Yang
Zhiwei Yang
Zhiwen Yang
Zhiyuan Yang
Zhuoqian Yang
Ziyan Yang
Ziyun Yang
Zongxin Yang
Zuhao Yang
Chengtang Yao
Cong Yao
Hantao Yao
Jiawen Yao
Lina Yao
Mingde Yao
Mingshuai Yao
Qingsong Yao
Shunyu Yao
Taiping Yao
Ting Yao
Xincheng Yao
Xinwei Yao
Xu Yao
Xufeng Yao
Yao Yao
Yazhou Yao
Yue Yao
Ziwei Yao
Sudhir Yarram
Rajeev Yasarla
Mohsen Yavartanoo
Botao Ye
Dengpan Ye
Fei Ye
Hanrong Ye
Jianglong Ye
Jiarong Ye
Jin Ye
Jingwen Ye
Jinwei Ye
Junjie Ye
Keren Ye
Maosheng Ye
Meng Ye
Meng Ye
Muchao Ye
Nanyang Ye
Peng Ye
Qi Ye
Qian Ye
Qinghao Ye
Qixiang Ye
Ruolin Ye
Shuquan Ye
Tian Ye
Vickie Ye
Wenqian Ye
Xinchen Ye
Yufei Ye
Moon Ye-Bin
Yousef Yeganeh
Chun-Hsiao Yeh
Raymond Yeh
Yu-Ying Yeh
Florence Yellin
Sriram Yenamandra
Tarun Yenamandra
Promod Yenigalla
Chandan Yeshwanth
Dong Yi
Hongwei Yi
Kai Yi
Ran Yi
Renjiao Yi
Xinyu Yi
Alper Yilmaz
Jonghwa Yim
Aoxiong Yin
Fei Yin
Fukun Yin
Jia-Li Yin
Ming Yin
Nan Yin
Ruihong Yin
Tianwei Yin
Wenzhe Yin
Xiaoqi Yin
Yingda Yin
Yu Yin
Yufeng Yin
Zhenfei Yin
Xianghua Ying
Xiaowen Ying
Naoto Yokoya
Chen YongCan
ByungIn Yoo
Innfarn Yoo
Jinsu Yoo
Sungjoo Yoo
Hee Suk Yoon
Jae Shin Yoon
Jihun Yoon
Sangwoong Yoon
Sejong Yoon
Sung Whan Yoon
Sung-Hoon Yoon
Sunjae Yoon
Youngho Yoon
Youngseok Yoon
Youngseok Yoon
Yuichi Yoshida
Ryota Yoshihashi
Yusuke Yoshiyasu

Chenyu You
Haoran You
Haoxuan You
Shan You
Yang You
Yingxuan You
Yurong You
Chan-Hyun Youn
Kim Youwang
Nikolaos-Antonios
    Ypsilantis
Baosheng Yu
Bei Yu
Bruce X. B. Yu
Chaohui Yu
Chunlin Yu
Cunjun Yu
Dahai Yu
En Yu
En Yu
Fenggen Yu
Gang Yu
Haibao Yu
Hanchao Yu
Hang Yu
Hao Yu
Hao Yu
Haojun Yu
Heng Yu
Hong-Xing Yu
Houjian Yu
Jianhui Yu
Jiashuo Yu
Jing Yu
Jiwen Yu
Jiyang Yu
Kaicheng Yu
Lei Yu
Lidong Yu
Lijun Yu
Mulin Yu
Peilin Yu
Qian Yu
Qihang Yu
Qing Yu

Rui Yu
Ruixuan Yu
Runpeng Yu
Shaozuo Yu
Shuzhi Yu
Sihyun Yu
Tan Yu
Tao Yu
Tianjiao Yu
Wei Yu
Weihao Yu
Wenwen Yu
Xi Yu
Xiaohan Yu
Xin Yu
Xin Yu
Xuehui Yu
Yingchen Yu
Yongsheng Yu
Yunlong Yu
Zehao Yu
Zhaofei Yu
Zhengdi Yu
Zhengdi Yu
Zhixuan Yu
Zhongzhi Yu
Zhuoran Yu
Zitong Yu
Chun Yuan
Chunfeng Yuan
Hangjie Yuan
Haobo Yuan
Jiakang Yuan
Jiangbo Yuan
Liangzhe Yuan
Maoxun Yuan
Shanxin Yuan
Shengming Yuan
Shuai Yuan
Shuaihang Yuan
Wentao Yuan
Xiaoding Yuan
Xiaoyun Yuan
Xin Yuan
Xin Yuan

Yixuan Yuan
Yu-Jie Yuan
Yuan Yuan
Yuan Yuan
Yuhui Yuan
Zheng Yuan
Zhuoning Yuan
Mehmet Kerim Yücel
Dongxu Yue
Haixiao Yue
Kaiyu Yue
Tao Yue
Xiangyu Yue
Zihao Yue
Zongsheng Yue
Heeseung Yun
Jooyeol Yun
Juseung Yun
Kimin Yun
Se-Young Yun
Sukwon Yun
Tian Yun
Raza Yunus
Ekim Yurtsever
Eloi Zablocki
Riccardo Zaccone
Martin Zach
Muhammad Zaigham
    Zaheer
Ilya Zakharkin
Egor Zakharov
Abhaysinh S. Zala
Pierluigi Zama Ramirez
Eduard Sebastian Zamfir
Amir Zamir
Luca Zancato
Yuan Zang
Yuhang Zang
Zelin Zang
Pietro Zanuttigh
Giacomo Zara
Samira Zare
Olga Zatsarynna
Denis Zavadski
Vitjan Zavrtanik

Jan Zdenek
Yanjie Ze
Bernhard Zeisl
John Zelek
Oliver Zendel
Ailing Zeng
Chong Zeng
Dan Zeng
Fangao Zeng
Haijin Zeng
Huimin Zeng
Jia Zeng
Jiabei Zeng
Kuo-Hao Zeng
Libing Zeng
Ling-An Zeng
Ming Zeng
Pengpeng Zeng
Runhao Zeng
Tieyong Zeng
Wei Zeng
Yan Zeng
Yanhong Zeng
Yawen Zeng
Yuyuan Zeng
Zilai Zeng
Ziyao Zeng
Kaiwen Zha
Ruyi Zha
Yaohua Zha
Bohan Zhai
Qiang Zhai
Runtian Zhai
Wei Zhai
YiKui Zhai
Yuanhao Zhai
Yunpeng Zhai
De-Chuan Zhan
Fangneng Zhan
Guanqi Zhan
Huangying Zhan
Huijing Zhan
Kun Zhan
Xueying Zhan
Aidong Zhang
Baochang Zhang
Baoheng Zhang
Baoming Zhang
Biao Zhang
Bingfeng Zhang
Binjie Zhang
Bo Zhang
Bo Zhang
Borui Zhang
Bowen Zhang
Can Zhang
Ce Zhang
Chang-Bin Zhang
Chao Zhang
Chao Zhang
Chen-Lin Zhang
Cheng Zhang
Cheng Zhang
Chenghao Zhang
Chenyangguang Zhang
Chi Zhang
Chongyang Zhang
Chris Zhang
Chuhan Zhang
Chunhui Zhang
Chuyu Zhang
Congyi Zhang
Daichi Zhang
Dan Zhang
Daoan Zhang
Daoqiang Zhang
David Junhao Zhang
Dexuan Zhang
Dingwen Zhang
Dingyuan Zhang
Dongsu Zhang
Fan Zhang
Fan Zhang
Fang-Lue Zhang
Feilong Zhang
Frederic Z. Zhang
Fuyang Zhang
Gang Zhang
Gengwei Zhang
Gengyu Zhang
Gengyuan Zhang
Gongjie Zhang
GuiXuan Zhang
Guofeng Zhang
Guozhen Zhang
Hang Zhang
Hang Zhang
Hanwang Zhang
Hao Zhang
Hao Zhang
Hao Zhang
Haokui Zhang
Haonan Zhang
Haotian Zhang
Hengrui Zhang
Hongguang Zhang
Hongrun Zhang
Hongyuan Zhang
Howard Zhang
Huaidong Zhang
Huaiwen Zhang
Hui Zhang
Hui Zhang
Jason Y. Zhang
Ji Zhang
Jiahui Zhang
Jiakai Zhang
Jiaming Zhang
Jian Zhang
Jianfu Zhang
Jiangning Zhang
Jianhua Zhang
Jianming Zhang
Jianpeng Zhang
Jianping Zhang
Jianrong Zhang
Jichao Zhang
Jie Zhang
Jie Zhang
Jie Zhang
Jimuyang Zhang
Jing Zhang
Jing Zhang
Jinghao Zhang
Jingyi Zhang

Jinlu Zhang
Jiqing Zhang
Jiyuan Zhang
Junbo Zhang
Junge Zhang
Junyi Zhang
Juyong Zhang
Kai Zhang
Kai Zhang
Kaidong Zhang
Kaihao Zhang
Kaipeng Zhang
Kaiyi Zhang
Ke Zhang
Ke Zhang
Kui Zhang
Le Zhang
Le Zhang
Lefei Zhang
Lei Zhang
Leo Yu Zhang
Li Zhang
Lianbo Zhang
Liang Zhang
Liangpei Zhang
Lin Zhang
Linfeng Zhang
Liqing Zhang
Lu Zhang
Malu Zhang
Manyuan Zhang
Mengmi Zhang
Mengqi Zhang
Mi Zhang
Min Zhang
Min-Ling Zhang
Mingda Zhang
Mingfang Zhang
Minghui Zhang
Mingjin Zhang
Mingyuan Zhang
Minjia Zhang
Ni Zhang
Pan Zhang
Peiyan Zhang

Pengze Zhang
Pingping Zhang
Qi Zhang
Qi Zhang
Qian Zhang
Qiang Zhang
Qijian Zhang
Qiming Zhang
Qing Zhang
Qing Zhang
Renrui Zhang
Rongyu Zhang
Ruida Zhang
Ruimao Zhang
Ruixin Zhang
Runze Zhang
Sanyi Zhang
Shan Zhang
Shanghang Zhang
Shaofeng Zhang
Sheng Zhang
Shengping Zhang
Shengyu Zhang
Shimian Zhang
Shiwei Zhang
Shizhou Zhang
Shu Zhang
Shuo Zhang
Siwei Zhang
Song-Hai Zhang
Tao Zhang
Tianyun Zhang
Ting Zhang
Tong Zhang
Weixia Zhang
Wendong Zhang
Wenlong Zhang
Wenqiang Zhang
Wentao Zhang
Wentian Zhang
Wenxiao Zhang
Wenxuan Zhang
Xi Zhang
Xiang Zhang
Xiang Zhang

Xianling Zhang
Xiao Zhang
Xiaohan Zhang
Xiaoming Zhang
Xiaoran Zhang
Xiaowei Zhang
Xiaoyun Zhang
Xikun Zhang
Xin Zhang
Xinfeng Zhang
Xingchen Zhang
Xingguang Zhang
Xingxuan Zhang
Xiong Zhang
Xiuming Zhang
Xu Zhang
Xuanyang Zhang
Xucong Zhang
Xuying Zhang
Yabin Zhang
Yabo Zhang
Yachao Zhang
Yahui Zhang
Yan Zhang
Yan Zhang
Yanan Zhang
Yang Zhang
Yanghao Zhang
Yawen Zhang
Yechao Zhang
Yi Zhang
Yi Zhang
Yi Zhang
Yi-Fan Zhang
Yifan Zhang
Yifei Zhang
Yifeng Zhang
Yihao Zhang
Yihua Zhang
Yimeng Zhang
Yiming Zhang
Yin Zhang
Yinan Zhang
Yinda Zhang
Ying Zhang

Yingliang Zhang
Yitian Zhang
Yixin Zhang
Yiyuan Zhang
Yongfei Zhang
Yonggang Zhang
Yonghua Zhang
Youjian Zhang
Youmin Zhang
Youshan Zhang
Yu Zhang
Yu Zhang
Yuan Zhang
Yuechen Zhang
Yuexi Zhang
Yufei Zhang
Yuhan Zhang
Yuhang Zhang
Yunchao Zhang
Yunhe Zhang
Yunhua Zhang
Yunpeng Zhang
Yunzhi Zhang
Yuxin Zhang
Yuyao Zhang
Zaixi Zhang
Zeliang Zhang
Zewei Zhang
Zeyu Zhang
Zhang Zhang
Zhao Zhang
Zhaoxiang Zhang
Zhen Zhang
Zheng Zhang
Zheng Zhang
Zhenyu Zhang
Zhenyu Zhang
Zheyuan Zhang
Zhicheng Zhang
Zhilu Zhang
Zhishuai Zhang
Zhitian Zhang
Zhiwei Zhang
Zhixing Zhang
Zhiyuan Zhang

Zhong Zhang
Zhongping Zhang
Zhongqun Zhang
Zicheng Zhang
Zicheng Zhang
Zihao Zhang
Ziming Zhang
Ziqi Zhang
Qilong Zhangli
Bin Zhao
Bingchen Zhao
Bingyin Zhao
Cairong Zhao
Can Zhao
Chen Zhao
Dong Zhao
Dongxu Zhao
Fang Zhao
Feng Zhao
Fuqiang Zhao
Gangming Zhao
Ganlong Zhao
Guiyu Zhao
Haimei Zhao
Hanbin Zhao
Handong Zhao
Jian Zhao
Jiaqi Zhao
Jie Zhao
Kai Zhao
Kaifeng Zhao
Lei Zhao
Liang Zhao
Lirui Zhao
Long Zhao
Luxi Zhao
Mingyang Zhao
Minyi Zhao
Na Zhao
Nanxuan Zhao
Pu Zhao
Qi Zhao
Qian Zhao
Qibin Zhao
Qingsong Zhao

Qingyu Zhao
Qinyu Zhao
Rongchang Zhao
Rui Zhao
Rui Zhao
Rui Zhao
Ruiqi Zhao
Shanshan Zhao
Shihao Zhao
Shiyu Zhao
Shizhen Zhao
Shuai Zhao
Siheng Zhao
Tianchen Zhao
TianHao Zhao
Tiesong Zhao
Wang Zhao
Wangbo Zhao
Weichao Zhao
Weiyue Zhao
Wenda Zhao
Wenliang Zhao
Xiangyun Zhao
Xiaoming Zhao
Xiaonan Zhao
Xiaoqi Zhao
Xin Zhao
Xin Zhao
Xingyu Zhao
Xu Zhao
Yajie Zhao
Yang Zhao
Yifan Zhao
Ying Zhao
Yiqun Zhao
Yiqun Zhao
Yizhou Zhao
Yizhou Zhao
Yucheng Zhao
Yue Zhao
Yunhan Zhao
Yuyang Zhao
Yuzhi Zhao
Zelin Zhao
Zengqun Zhao

Zhen Zhao	Siming Zheng	Dongzhan Zhou
Zhenghao Zhao	Tianhang Zheng	Fan Zhou
Zhengyu Zhao	Wenting Zheng	Hang Zhou
Zhou Zhao	Wenzhao Zheng	Hang Zhou
Zibo Zhao	Xiaozheng Zheng	Hao Zhou
Zimeng Zhao	Xiawu Zheng	Hao Zhou
Ziwei Zhao	Xu Zheng	Haoyi Zhou
Ziwei Zhao	Yajing Zheng	Hong-Yu Zhou
Zixiang Zhao	Yalin Zheng	Honglu Zhou
Jin Zhe	Yang Zheng	Huayi Zhou
Anling Zheng	Ye Zheng	Jiahuan Zhou
Ce Zheng	Yinglin Zheng	Jiaming Zhou
Chaoda Zheng	Yinqiang Zheng	Jian Zhou
Chengwei Zheng	Yu Zheng	Jianan Zhou
Chuanxia Zheng	Zangwei Zheng	Jiantao Zhou
Duo Zheng	Zehan Zheng	Jianxiong Zhou
Ervine Zheng	Zerong Zheng	JIngkai Zhou
Guangcong Zheng	Zhaoheng Zheng	Junsheng Zhou
Haitian Zheng	Zhedong Zheng	Kailai Zhou
Haiyong Zheng	Zhuo Zheng	Kaiyang Zhou
Hao Zheng	Zilong Zheng	Keyang Zhou
Haotian Zheng	Shuaifeng Zhi	Kun Zhou
Haoxin Zheng	Tiancheng Zhi	Lei Zhou
Huan Zheng	Bineng Zhong	Mingyang Zhou
Huan Zheng	Fangcheng Zhong	Mingyi Zhou
Jia Zheng	Fangwei Zhong	Mingyuan Zhou
Jian Zheng	Guoqiang Zhong	Mo Zhou
Jian-Qing Zheng	Nan Zhong	Pan Zhou
Jianqiao Zheng	Yaoyao Zhong	Peng Zhou
Jianwei Zheng	Yijie Zhong	Peng Zhou
Jin Zheng	Yiqi Zhong	Qianyi Zhou
Jingxiao Zheng	Yiran Zhong	Qianyu Zhou
Kaiwen Zheng	Yiwu Zhong	Qihua Zhou
Kecheng Zheng	Yunshan Zhong	Qin Zhou
Léon Zheng	Zichun Zhong	Qinqin Zhou
Meng Zheng	Ziming Zhong	Qunjie Zhou
Naishan Zheng	Brady Zhou	Sheng Zhou
Peng Zheng	Chong Zhou	Shenglong Zhou
Qi Zheng	Chu Zhou	Shijie Zhou
Qian Zheng	Chunluan Zhou	Shuchang Zhou
Rongkun Zheng	Da-Wei Zhou	Sihang Zhou
Shen Zheng	Dawei Zhou	Tao Zhou
Shuai Zheng	Dewei Zhou	Tianfei Zhou
Shuhong Zheng	Dingfu Zhou	Xiangdong Zhou
Shunyuan Zheng	Donghao Zhou	Xiaoqiang Zhou

Xin Zhou
Xingyi Zhou
Yan-Jie Zhou
Yang Zhou
Yang Zhou
Yanqi Zhou
Yi Zhou
Yi Zhou
Yichao Zhou
Yichen Zhou
Yin Zhou
Yiyi Zhou
Yu Zhou
Yucheng Zhou
Yufan Zhou
Yunsong Zhou
Yuqian Zhou
Yuxiao Zhou
Yuxuan Zhou
Zhenyu Zhou
Zijian Zhou
Zikun Zhou
Ziqi Zhou
Zixiang Zhou
Zongwei Zhou
Alex Z. Zhu
Benjin Zhu
Bin Zhu
Bin Zhu
Chenming Zhu
Chenyang Zhu
Deyao Zhu
Dongxiao Zhu
Fangrui Zhu
Fei Zhu
Feida Zhu
Fengqing Maggie Zhu
Guibo Zhu
Haidong Zhu
Hanwei Zhu
Hao Zhu
Hao Zhu
Heming Zhu
Jiachen Zhu
Jianke Zhu

Jiawen Zhu
Jiayin Zhu
Jinjing Zhu
Junyi Zhu
Kai Zhu
Ke Zhu
Lanyun Zhu
Lin Zhu
Linchao Zhu
Liyuan Zhu
Meilu Zhu
Muzhi Zhu
Qingtian Zhu
Ronghang Zhu
Rui Zhu
Rui Zhu
Rui-Jie Zhu
Ruizhao Zhu
Shengjie Zhu
Sijie Zhu
Siyu Zhu
Tyler Zhu
Wang Zhu
Weicheng Zhu
Wenwu Zhu
Xiangyu Zhu
Xiaofeng Zhu
Xiaoguang Zhu
Xiaosu Zhu
Xiaoyu Zhu
Xingkui Zhu
Xinxin Zhu
Xiyue Zhu
Yangguang Zhu
Yanjun Zhu
Yao Zhu
Ye Zhu
Feida Zhu
Fengqing Maggie Zhu
Guibo Zhu
Haidong Zhu
Hanwei Zhu
Hao Zhu
Hao Zhu
Heming Zhu

Jiachen Zhu
Jianke Zhu
Jiawen Zhu
Jiayin Zhu
Jinjing Zhu
Junyi Zhu
Kai Zhu
Ke Zhu
Lanyun Zhu
Lin Zhu
Linchao Zhu
Liyuan Zhu
Meilu Zhu
Muzhi Zhu
Qingtian Zhu
Ronghang Zhu
Rui Zhu
Rui Zhu
Rui-Jie Zhu
Ruizhao Zhu
Shengjie Zhu
Sijie Zhu
Siyu Zhu
Tyler Zhu
Wang Zhu
Weicheng Zhu
Wenwu Zhu
Xiangyu Zhu
Xiaofeng Zhu
Xiaoguang Zhu
Xiaosu Zhu
Xiaoyu Zhu
Xingkui zhu
Xinxin Zhu
Xiyue Zhu
Yangguang Zhu
Yanjun Zhu
Yao Zhu
Ye Zhu
Ye Zhu
Yichen Zhu
Yingying Zhu
Yousong Zhu
Yuansheng Zhu
Yurui Zhu

Zhen Zhu
Zhenwei Zhu
Zhenyao Zhu
Zhifan Zhu
Zhigang Zhu
Zhihong Zhu
Zihan Zhu
Zixin Zhu
Zunjie Zhu
Bingbing Zhuang
Jia-Xin Zhuang
Jiafan Zhuang
Peiye Zhuang
Wanyi Zhuang
Weiming Zhuang
Yihong Zhuang
Yixin Zhuang
Mingchen Zhuge
Tao Zhuo
Wei Zhuo

Yaoxin Zhuo
Bartosz Zieliński
Wojciech Zielonka
Filippo Ziliotto
Karel Zimmermann
Primo Zingaretti
Nikolaos Zioulis
Liu Ziyin
Mohammad Zohaib
Yongshuo Zong
Zhuofan Zong
Maria Zontak
Gaspard Zoss
Changqing Zou
Chuhang Zou
Danping Zou
Dongqing Zou
Haoming Zou
Longkun Zou
Shihao Zou

Xingxing Zou
Xueyan Zou
Yang Zou
Yuexian Zou
Yuli Zou
Yuliang Zou
Yunhao Zou
Zhiming Zou
Zihang Zou
Silvia Zuffi
Idil Esen Zulfikar
Maria A. Zuluaga
Ronglai Zuo
Xingxing Zuo
Xinxin Zuo
Yifan Zuo
Yiming Zuo
Reyer Zwiggelaar
Vlas Zyrianov

Contents – Part IV

LGM: Large Multi-view Gaussian Model for High-Resolution 3D Content Creation 1
   Jiaxiang Tang, Zhaoxi Chen, Xiaokang Chen, Tengfei Wang, Gang Zeng, and Ziwei Liu

Mahalanobis Distance-Based Multi-view Optimal Transport for Multi-view Crowd Localization 19
   Qi Zhang, Kaiyi Zhang, Antoni B. Chan, and Hui Huang

RAW-Adapter: Adapting Pre-trained Visual Model to Camera RAW Images ... 37
   Ziteng Cui and Tatsuya Harada

SLEDGE: Synthesizing Driving Environments with Generative Models and Rule-Based Traffic 57
   Kashyap Chitta, Daniel Dauner, and Andreas Geiger

AFreeCA: Annotation-Free Counting for All 75
   Adriano D'Alessandro, Ali Mahdavi-Amiri, and Ghassan Hamarneh

Adversarially Robust Distillation by Reducing the Student-Teacher Variance Gap 92
   Junhao Dong, Piotr Koniusz, Junxi Chen, and Yew-Soon Ong

LN3DIFF: Scalable Latent Neural Fields Diffusion for Speedy 3D Generation 112
   Yushi Lan, Fangzhou Hong, Shuai Yang, Shangchen Zhou, Xuyi Meng, Bo Dai, Xingang Pan, and Chen Change Loy

Hierarchical Temporal Context Learning for Camera-Based Semantic Scene Completion 131
   Bohan Li, Jiajun Deng, Wenyao Zhang, Zhujin Liang, Dalong Du, Xin Jin, and Wenjun Zeng

Equi-GSPR: Equivariant SE(3) Graph Network Model for Sparse Point Cloud Registration 149
   Xueyang Kang, Zhaoliang Luan, Kourosh Khoshelham, and Bing Wang

GTP-4o: Modality-Prompted Heterogeneous Graph Learning
for Omni-Modal Biomedical Representation 168
  *Chenxin Li, Xinyu Liu, Cheng Wang, Yifan Liu, Weihao Yu, Jing Shao,
  and Yixuan Yuan*

PromptCCD: Learning Gaussian Mixture Prompt Pool for Continual
Category Discovery 188
  Fernando Julio Cendra, Bingchen Zhao, and Kai Han

Sapiens: Foundation for Human Vision Models 206
  *Rawal Khirodkar, Timur Bagautdinov, Julieta Martinez, Su Zhaoen,
  Austin James, Peter Selednik, Stuart Anderson, and Shunsuke Saito*

Linearly Controllable GAN: Unsupervised Feature Categorization
and Decomposition for Image Generation and Manipulation 229
  Sehyung Lee, Mijung Kim, Yeongnam Chae, and Björn Stenger

Generating Human Interaction Motions in Scenes with Text Control 246
  *Hongwei Yi, Justus Thies, Michael J. Black, Xue Bin Peng,
  and Davis Rempe*

NOVUM: Neural Object Volumes for Robust Object Classification 264
  *Artur Jesslen, Guofeng Zhang, Angtian Wang, Wufei Ma, Alan Yuille,
  and Adam Kortylewski*

Align Before Collaborate: Mitigating Feature Misalignment for Robust
Multi-agent Perception 282
  *Kun Yang, Dingkang Yang, Ke Li, Dongling Xiao, Zedian Shao,
  Peng Sun, and Liang Song*

HIMO: A New Benchmark for Full-Body Human Interacting with Multiple
Objects 300
  *Xintao Lv, Liang Xu, Yichao Yan, Xin Jin, Congsheng Xu, Shuwen Wu,
  Yifan Liu, Lincheng Li, Mengxiao Bi, Wenjun Zeng, and Xiaokang Yang*

SAIR: Learning Semantic-Aware Implicit Representation 319
  Canyu Zhang, Xiaoguang Li, Qing Guo, and Song Wang

ColorMNet: A Memory-Based Deep Spatial-Temporal Feature
Propagation Network for Video Colorization 336
  Yixin Yang, Jiangxin Dong, Jinhui Tang, and Jinshan Pan

UNIC: Universal Classification Models via Multi-teacher Distillation 353
  *Mert Bülent Sarıyıldız, Philippe Weinzaepfel, Thomas Lucas,
  Diane Larlus, and Yannis Kalantidis*

Instance-Dependent Noisy-Label Learning with Graphical Model Based
Noise-Rate Estimation 372
  *Arpit Garg, Cuong Nguyen, Rafael Felix, Thanh-Toan Do,
  and Gustavo Carneiro*

Eliminating Warping Shakes for Unsupervised Online Video Stitching 390
  *Lang Nie, Chunyu Lin, Kang Liao, Yun Zhang, Shuaicheng Liu, Rui Ai,
  and Yao Zhao*

Vary: Scaling up the Vision Vocabulary for Large Vision-Language Model 408
  *Haoran Wei, Lingyu Kong, Jinyue Chen, Liang Zhao, Zheng Ge,
  Jinrong Yang, Jianjian Sun, Chunrui Han, and Xiangyu Zhang*

Merlin: Empowering Multimodal LLMs with Foresight Minds 425
  *En Yu, Liang Zhao, Yana Wei, Jinrong Yang, Dongming Wu,
  Lingyu Kong, Haoran Wei, Tiancai Wang, Zheng Ge, Xiangyu Zhang,
  and Wenbing Tao*

ViC-MAE: Self-supervised Representation Learning from Images
and Video with Contrastive Masked Autoencoders 444
  Jefferson Hernandez, Ruben Villegas, and Vicente Ordonez

E.T. the Exceptional Trajectories: Text-to-Camera-Trajectory Generation
with Character Awareness 464
  *Robin Courant, Nicolas Dufour, Xi Wang, Marc Christie,
  and Vicky Kalogeiton*

OphNet: A Large-Scale Video Benchmark for Ophthalmic Surgical
Workflow Understanding 481
  *Ming Hu, Peng Xia, Lin Wang, Siyuan Yan, Feilong Tang,
  Zhongxing Xu, Yimin Luo, Kaimin Song, Jurgen Leitner,
  Xuelian Cheng, Jun Cheng, Chi Liu, Kaijing Zhou, and Zongyuan Ge*

Author Index 501

LGM: Large Multi-view Gaussian Model for High-Resolution 3D Content Creation

Jiaxiang Tang[1]([✉]), Zhaoxi Chen[2], Xiaokang Chen[1], Tengfei Wang[3], Gang Zeng[1], and Ziwei Liu[2]

[1] National Key Lab of General AI, Peking University, Beijing, China
tjx@pku.edu.cn
[2] S-Lab, Nanyang Technological University, Singapore, Singapore
[3] Shanghai AI Lab, Shanghai, China

Abstract. 3D content creation has achieved significant progress in terms of both quality and speed. Although current feed-forward models can produce 3D objects in seconds, their resolution is constrained by the intensive computation required during training. In this paper, we introduce **Large Multi-View Gaussian Model (LGM)**, a novel framework designed to generate high-resolution 3D models from text prompts or single-view images. Our key insights are two-fold: **1) 3D Representation:** We propose multi-view Gaussian features as an efficient yet powerful representation, which can then be fused together for differentiable rendering. **2) 3D Backbone:** We present an asymmetric U-Net as a high-throughput backbone operating on multi-view images, which can be produced from text or single-view image input by leveraging multi-view diffusion models. Extensive experiments demonstrate the high fidelity and efficiency of our approach. Notably, we maintain the fast speed to generate 3D objects within 5 s while boosting the training resolution to 512, thereby achieving high-resolution 3D content generation. Our project page is available at https://me.kiui.moe/lgm/.

Keywords: 3D Generation · Gaussian Splatting · High Resolution

1 Introduction

Automatic 3D content creation has great potential in numerous fields such as digital games, virtual reality, and films. The fundamental techniques, like image-to-3D and text-to-3D, provide significant benefits by remarkably decreasing the requirement for manual labor among professional 3D artists, enabling those without expertise to participate in 3D asset creation.

J. Tang—Work done while visiting S-Lab, Nanyang Technological University.

Supplementary Information The online version contains supplementary material available at https://doi.org/10.1007/978-3-031-73235-5_1.

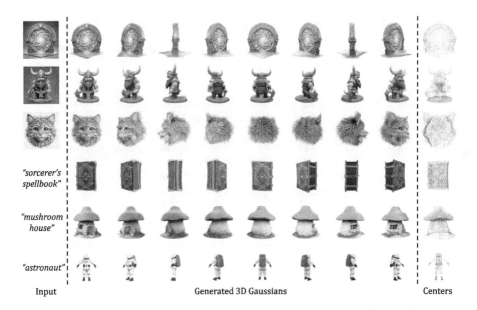

Fig. 1. Our method generates **high-resolution** 3D Gaussians in **5 s** from single-view images or texts.

Previous research on 3D generation has predominantly focused on score distillation sampling (SDS) [22,24,36,48] to lift 2D diffusion priors into 3D. These optimization-based methods can create highly detailed 3D objects from text or single-view image inputs, but they often face issues such as slow generation speed and limited diversity. Recent advancements have significantly decreased the time required to generate 3D objects using large reconstruction models from single-view or few-shot images [15,19,53,56,58]. These methods utilize transformers to directly regress triplane-based [2] neural radiance fields (NeRF) [32]. However, these methods cannot produce detailed textures and complicated geometry due to the low-resolution training. We argue that their bottlenecks are **1)** inefficient 3D representation, and **2)** heavily parameterized 3D backbone. For instance, given a fixed compute budget, the triplane representation of LRM [15] is limited to the resolution of 32, while the resolution of the rendered image is capped at 128 due to the online volume rendering. Despite this, these methods suffer from the computationally intensive transformer-based backbone, which also leads to a limited training resolution.

To address these challenges, we present a novel method to train a few-shot 3D reconstruction model without relying on triplane-based volume rendering or transformers [15]. Instead, our approach employs 3D Gaussian splatting [17] of which features are predicted by an asymmetric U-Net as a high-throughput backbone [41,47]. The motivation of this design is to achieve high-resolution 3D generation, which necessitates an expressive 3D representation and the ability to train at high resolutions. Gaussian splatting stands out for 1) the expressiveness of compactly representing a scene compared with a single triplane, and 2) rendering efficiency compared with heavy volume rendering, which facilitates high-resolution training. However, it requires a sufficient number of 3D Gaussians to accurately represent detailed 3D information. Inspired by splatter image [47], we found that U-Net is effective in generating a sufficient number of Gaussians from multiview pixels, which maintains the capacity for high-resolution training at the same time. Note that, compared to previous methods [15,63], our default model is capable of generating 3D models with up to 65,536 Gaussians and can be trained at a resolution of 512, while still maintaining the rapid generation speed of feed-forward regression models. As shown in Fig. 1, our model supports both image-to-3D and text-to-3D tasks, capable of producing high-resolution, richly detailed 3D Gaussians in approximately 5 s.

Our method adopts a multi-view reconstruction setting similar to Instant3D [19]. In this process, the image and camera embedding from each input view are transformed into a feature map, which can be decoded and fused as a set of Gaussians. Differentiable rendering is applied to render novel views from the fused 3D Gaussians, allowing end-to-end image-level supervision in high resolution. To enhance information sharing across all input views, attention blocks are integrated into the deeper layers of the U-Net. This enables us to train our network on multi-view image datasets [12] using only regressing objectives. During inference, our method leverages existing image or text to multi-view diffusion models [27,44,45,52] to produce multi-view images as inputs for our Gaussian fusion network. To overcome the domain gap between multi-view images rendered from actual 3D objects and synthesized using diffusion models, we further propose two proper data augmentations for robust training. Finally, considering the preference for polygonal meshes in downstream tasks, we design a general algorithm to convert generated 3D Gaussians to smooth and textured meshes.

In summary, our contributions are:

1. We propose a novel framework to generate high-resolution 3D Gaussians by fusing information from multi-view images, which can be generated from text prompts or single-view images.
2. We design an asymmetric U-Net based architecture for efficient end-to-end training with significantly higher resolution, investigate data augmentation techniques for robust training, and propose a general mesh extraction approach from 3D Gaussians.
3. Extensive experiments demonstrate the superior quality, resolution, and efficiency of our method in both text-to-3D and image-to-3D tasks.

2 Related Work

High-Resolution 3D Generation. Current approaches for generating high-fidelity 3D models mostly rely on SDS-based optimization techniques. It requires both an expressive 3D representation and high-resolution supervision to effectively distill detailed information from 2D diffusion models into 3D. Due to the significant memory consumption associated with high-resolution rendering of NeRF, Magic3D [22] first converts NeRF to DMTet [43] and subsequently trains a second stage for finer resolution refinement. The hybrid representation of DMTet geometry and hash grid [34] textures enables the capture of high-quality 3D information, which can be efficiently rendered using differentiable rasterization [18]. Fantasia3D [6] explores to directly train DMTet with disentangled geometry and appearance generation. Subsequent studies [8,20,21,37,48,50,55] also employ a similar mesh-based stage, enabling high-resolution supervision for enhanced detail. Another promising 3D representation is Gaussian splatting [17] for its expressiveness and efficient rendering capabilities. Nonetheless, achieving rich details with this method necessitates appropriate initialization and careful densification during optimization [10,60]. In contrast, our work investigates a feed-forward approach to directly generate a sufficient number of 3D Gaussians.

Efficient 3D Generation. In contrast to SDS-based optimization methods, feed-forward 3D native methods are able to generate 3D assets within seconds after training on large-scale 3D datasets [11,12]. Some works attempt to train text-conditioned diffusion models on 3D representations such as point clouds and volumes [1,5,9,16,26,33,35,54,59,62]. However, these methods either cannot generalize well to large datasets or only produce low-quality 3D assets with simple textures. Recently, LRM [15] first shows that a regression model can be trained to robustly predict NeRF from a single-view image in just 5 s, which can be further exported to meshes. Instant3D [19] trains a text to multi-view images diffusion model and a multi-view LRM to perform fast and diverse text-to-3D generation. The following works extend LRM to predict poses given multi-view images [53], combine with diffusion [58], and specialize on human data [56]. These feed-forward models can be trained with simple regression objectives and significantly accelerate the speed of 3D object generation. However, their triplane NeRF-based representation is restricted to a relatively low resolution and limits the final generation fidelity. Our model instead seeks to train a high-fidelity feed-forward model using Gaussian splatting and U-Net.

Gaussian Splatting for Generation. We specifically discuss recent methods in generation tasks using Gaussian splatting [4,7,23,39,57]. DreamGaussian [48] first combines 3D Gaussians with SDS-based optimization approaches to decrease generation time. GSGen [10] and GaussianDreamer [60] explore various densification and initialization strategies for text to 3D Gaussians generation. Despite the acceleration achieved, generating high-fidelity 3D Gaussians using these optimization-based methods still requires several minutes. Triplane-Gaussian [63] introduces Gaussian splatting into the framework of LRM. This method starts by predicting Gaussian centers as point clouds and then projects them onto a triplane for other features. Nonetheless, the number of Gaussians and the resolution of the triplane are still limited, affecting the quality of the generated Gaussians. Splatter image [47] proposes to predict 3D Gaussians as pixels on the output feature map using U-Net from single-view images. This approach mainly focuses on single-view or two-view scenarios, limiting its generalization to large-scale datasets. Similarly, PixelSplat [3] predicts Gaussian parameters for each pixel of two posed images from scene datasets. We design a 4-view reconstruction model combined with existing multi-view diffusion models for general text or image to high-fidelity 3D object generation.

3 Large Multi-view Gaussian Model

We first provide the background information on Gaussian splatting and multi-view diffusion models (Sect. 3.1). Then we introduce our high-resolution 3D content generation framework (Sect. 3.2), where the core part is an asymmetric U-Net backbone to predict and fuse 3D Gaussians from multi-view images (Sect. 3.3). We design careful data augmentation and training pipeline to enhance robustness and stability (Sect. 3.4). Finally, we describe an effective method for smooth textured mesh extraction from the generated 3D Gaussians (Sect. 3.5).

3.1 Preliminaries

Gaussian Splatting. As introduced in [17], Gaussian splatting employs a collection of 3D Gaussians to represent 3D data. Specifically, each Gaussian is defined by a center $\mathbf{x} \in \mathbb{R}^3$, a scaling factor $\mathbf{s} \in \mathbb{R}^3$, and a rotation quaternion $\mathbf{q} \in \mathbb{R}^4$. Additionally, an opacity value $\alpha \in \mathbb{R}$ and a color feature $\mathbf{c} \in \mathbb{R}^C$ are maintained for rendering, where spherical harmonics can be used to model view-dependent effects. These parameters can be collectively denoted by Θ, with $\Theta_i = \{\mathbf{x}_i, \mathbf{s}_i, \mathbf{q}_i, \alpha_i, \mathbf{c}_i\}$ representing the parameters for the i-th Gaussian. Rendering of the 3D Gaussians involves projecting them onto the image plane as 2D Gaussians and performing alpha composition for each pixel in front-to-back depth order, thereby determining the final color and alpha.

Multi-view Diffusion Models. Original 2D diffusion models [40,42] primarily focus on generating single-view images and do not support 3D viewpoint manipulation. Recently, several methods [20,27,44,45,52] propose to fine-tune multi-view diffusion models on 3D datasets to incorporate camera poses as an additional input. These approaches enable the creation of multi-view images of the same object, either from a text prompt or a single-view image. However, due to the absence of an actual 3D model, inconsistencies may still occur across the generated views.

3.2 Overall Framework

As illustrated in Fig. 2, we adopt a two-step 3D generation pipeline at inference. Firstly, we take advantage of off-the-shelf text or image to multi-view diffusion models to generate multi-view images. Specifically, we adopt MVDream [45] for text input and ImageDream [52] for image (and optionally text) input. Both models are designed to generate multi-view images at four orthogonal azimuths and a fixed elevation. In the second step, we use a U-Net based model to predict 3D Gaussians from these sparse view images. Specifically, our model is trained to

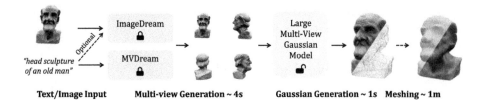

Fig. 2. Pipeline. Our model is trained to reconstruct 3D Gaussians from multi-view images, which can be synthesized by off-the-shelf models [45,52] at inference time from only text, or only image, or both input. Polygonal meshes can be extracted optionally.

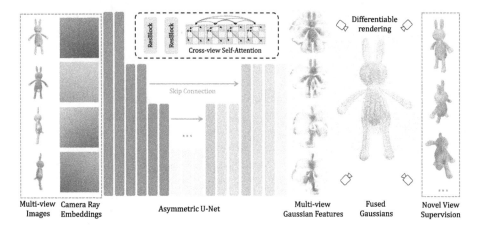

Fig. 3. Architecture of LGM. Our network adopts an asymmetric U-Net with cross-view self-attentions. We take four images with camera ray embeddings as the input, and output four feature maps of 14 channels (visualized using the RGB and opacity channels) which are interpreted as 3D Gaussians. The fused Gaussians are then rendered at novel views and supervised with ground truth images.

take four images with camera pose embeddings as input and predict four sets of Gaussians, which are fused to form the final 3D Gaussians. The generated Gaussians can be optionally converted to polygonal meshes using an extra conversion step, which is more suitable for downstream tasks.

3.3 Asymmetric U-Net for 3D Gaussians

At the core of our framework is an asymmetric U-Net to predict and fuse Gaussians from multi-view images. The network architecture is shown in Fig. 3. We take four images and corresponding camera poses as the input. Following previous works [58], we use the Plücker ray embedding to densely encode the camera poses. The RGB value and ray embedding are concatenated into a 9-channel feature map as the input to the first layer:

$$\mathbf{f}_i = \{\mathbf{c}_i, \mathbf{o}_i \times \mathbf{d}_i, \mathbf{d}_i\} \tag{1}$$

where $\mathbf{f}_i$ is the input feature for pixel i, $\mathbf{c}_i$ is the RGB value, $\mathbf{d}_i$ is the ray direction, and $\mathbf{o}_i$ is the ray origin.

The U-Net is built with residual layers [13] and self-attention layers [51] similar to previous works [14,31,47]. We only add self-attention at deeper layers where the feature map resolution is down-sampled to save memory. To propagate information across multiple views, we flatten the four image features and concatenate them before applying self-attention, similar to previous multi-view diffusion models [45,52].

Each pixel of the output feature map is treated as a 3D Gaussian inspired by splatter image [47]. Differently, our U-Net is designed to be asymmetric with a smaller output resolution compared to input, which allows us to use higher resolution input images and limit the number of output Gaussians. We discard the depth prediction required by explicit ray-wise camera projection in [47]. The output feature map contains 14 channels corresponding to the original attributes of each Gaussian Θ_i. To stabilize the training, we choose some different activation functions compared to the original Gaussian Splatting [17]. We clamp the predicted positions $\mathbf{x}_i$ into $[-1,1]^3$, and multiply the softplus-activated scales $\mathbf{s}_i$ with 0.1, such that the generated Gaussians at the beginning of training is close to the scene center. For each input view, the output feature map is transformed into a set of Gaussians. We simply concatenate these Gaussians from all four views as the final 3D Gaussians, which are used to render images at novel views for supervision.

3.4 Robust Training

Data Augmentation. We use multi-view images rendered from the Objaverse [12] dataset for training. However, at inference, we use synthesized multi-view images by diffusion models [45,52]. To mitigate the domain gap, we design two types of data augmentation for more robust training.

Grid Distortion. Synthesizing 3D consistent multi-view images using 2D diffusion models has been explored by many works [25,44,45,52]. However, since there is no underlying 3D representation, the generated multi-view images often suffer from subtle inconsistency across different views. We try to simulate such inconsistency using grid distortion. Except for the first input view, the other three input views are randomly distorted with a random grid during training. This makes the model more robust to inconsistent multi-view input images.

Orbital Camera Jitter. Another problem is that the synthesized multi-view images may not accurately follow the given camera poses. Following [15], we always normalized the camera poses at each training step such that the first view's camera pose is fixed. We therefore apply camera jitter to the last three input views during training.

Loss Function. To supervise the concatenated Gaussians, we use the differentiable renderer implementation from [17] to render them. At each training step,

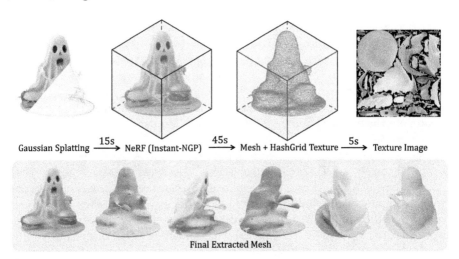

Fig. 4. Mesh Extraction Pipeline. We implement an efficient pipeline to convert the 3D Gaussians into smooth and textured meshes.

we render the RGB image and alpha image of eight views, including four input views and four novel views. Following [15], we apply mean square error loss and VGG-based LPIPS loss [61] to the RGB image:

$$\mathcal{L}_{\text{rgb}} = \mathcal{L}_{\text{MSE}}(I_{\text{rgb}}, I_{\text{rgb}}^{\text{GT}}) + \mathcal{L}_{\text{LPIPS}}(I_{\text{rgb}}, I_{\text{rgb}}^{\text{GT}}) \qquad (2)$$

We further apply mean square error loss on the alpha image for faster convergence of the shape:

$$\mathcal{L}_\alpha = \mathcal{L}_{\text{MSE}}(I_\alpha, I_\alpha^{\text{GT}}) \qquad (3)$$

3.5 Mesh Extraction

Since polygonal meshes are still the most widely used 3D representation in downstream tasks, we hope to further extract meshes from our generated Gaussians. Previous works [48] have tried to directly convert the opacity value of 3D Gaussians into an occupancy field for mesh extraction. However, we find this method dependent on aggressive densification during the optimization of 3D Gaussians to produce smooth occupancy field. On the contrary, the generated Gaussians in our method are usually sparse and cannot produce a suitable occupancy field, leading to an unsatisfactory surface with visible holes.

Instead, we propose a more general mesh extraction pipeline from 3D Gaussians as illustrated in Fig. 4. We first train an efficient NeRF [34] using the rendered images from 3D Gaussians on-the-fly, and then convert the NeRF to polygonal meshes [49]. Marching Cubes [28] is applied to extract a coarse mesh, which is then iteratively refined together with the appearance hash grid using differentiable rendering. Finally, we bake the appearance field onto the refined mesh to extract texture images. For more details, please refer to the supplementary materials and NeRF2Mesh [49]. With adequately optimized implementation, it takes only about 1 min to perform this Gaussians to NeRF to mesh conversion.

4 Experiments

4.1 Implementation Details

Datasets. We use a filtered subset of the Objaverse [12] dataset to train our model. Since there are many low-quality 3D models (*e.g.*, partial scans, missing textures) in the original Objaverse dataset, we filter the dataset by two empirical rules: (1) We examine the captions from Cap3D [30], and curate a list of words that usually appears in bad models (*e.g.*, 'resembling', 'debris', 'frame'), which is used to filter all models whose caption includes any of these words. (2) We discard models with mostly white color after rendering, which usually indicates missing texture. These lead to a final set of around 80K 3D objects. We render the RGBA image from 100 camera views at the resolution of 512×512 for training and validation.

Network Architecture. Our asymmetric U-Net model consists of 6 down blocks, 1 middle block, and 5 up blocks, with the input image at 256×256 and output Gaussian feature map at 128×128. We use 4 input views, so the number of output Gaussians is $128 \times 128 \times 4 = 65,536$. The feature channels for all blocks are $[64, 128, 256, 512, 1024, 1024]$, $[1024]$ and $[1024, 1024, 512, 256, 128]$ respectively. Each block contains a series of residual layers and an optional downsample or up-sample layer. For the last 3 of down blocks, the middle block, and the first 3 up blocks, we also insert cross-view self-attention layers after the residual layers. The final feature maps are processed by a 1×1 convolution layer to 14-channel pixel-wise Gaussian features. Following previous works [40, 47], we adopt `Silu` activation and group normalization for the U-Net.

Training. We train our model on 32 NVIDIA A100 (80G) GPUs for about 4 days. A batch size of 8 for each GPU is used, leading to an effective batch size of 256. For each batch, we randomly sample 8 camera views, with the first 4 views as the input, and all 8 views as the output for supervision. Similar to LRM [15], we transform the cameras of each batch such that the first input view is always the front view with an identity rotation matrix and fixed translation. The input images are assumed to have a white background. The output 3D Gaussians are rendered at 512×512 resolution for mean square error loss. We resize the images to 256×256 for LPIPS loss to save memory. The AdamW [29] optimizer is adopted with the learning rate of 4×10^{-4}, weight decay of 0.05, and betas of $(0.9, 0.95)$. The learning rate is cosine annealed to 0 during the training. We clip the gradient with a maximum norm of 1.0. The probability for grid distortion and camera jitter is set to 50%.

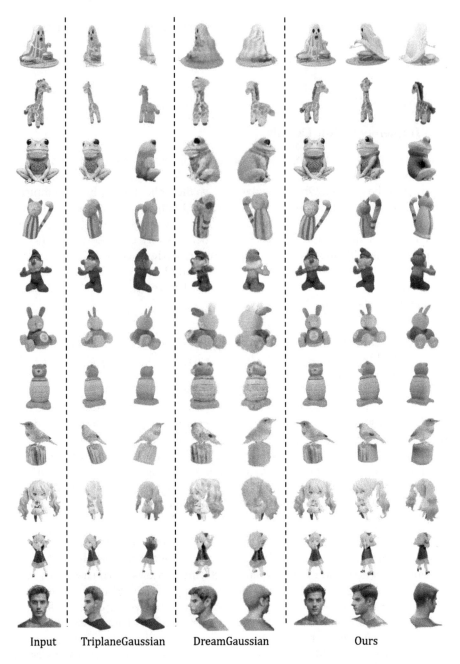

Fig. 5. Comparisons of generated 3D Gaussians for image-to-3D. Our method generates Gaussian splatting with better visual quality on various challenging images.

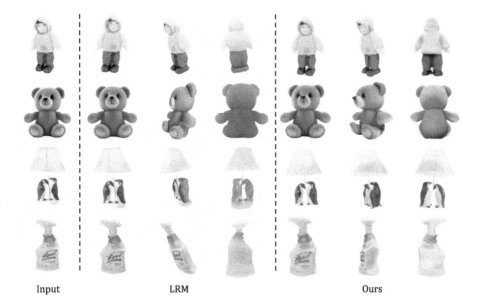

| Input | LRM | Ours |

Fig. 6. Comparisons with LRM for image-to-3D. We compare our method with available results from LRM [15].

Inference. Our whole pipeline, including two multi-view diffusion models, takes only about 10 GB of GPU memory for inference, which is friendly for deployment. For the multi-view diffusion models, we use a guidance scale of 5 for ImageDream [52] and 7.5 for MVDream [45] following the original paper. The number of diffusion steps is set to 30 using the DDIM [46] scheduler. The camera elevation is fixed to 0, and azimuths to $[0, 90, 180, 270]$ degree for the four generated views. For ImageDream [52], the text prompt is always left empty so the only input is a single-view image. Since the images generated by MVDream may contain various backgrounds, we apply background removal [38] and use white background.

4.2 Qualitative Comparisons

Image-to-3D. We first compare against recent methods [48,63] that are capable of generating 3D Gaussians. Figure 5 shows images rendered from the generated 3D Gaussians for comparison. The 3D Gaussians produced by our method have better visual quality and effectively preserve the content from the input view. Our high-resolution 3D Gaussians can be transformed into smooth textured meshes with minimal loss of quality in most cases. We also compare our results against LRM [15] using the available videos from their website in Fig. 6. Specifically, our multi-view setting successfully mitigates the issue of blurry back views and flat geometry, resulting in enhanced detail even in unseen views.

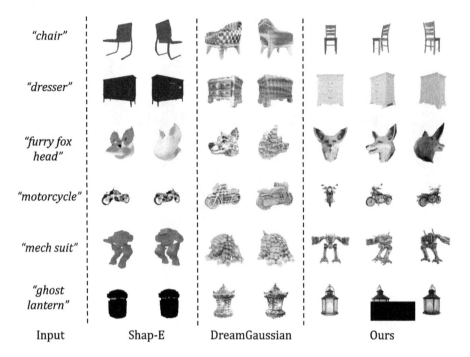

Fig. 7. Comparisons of generated 3D models for text-to-3D. Our method achieves better text alignment and visual quality.

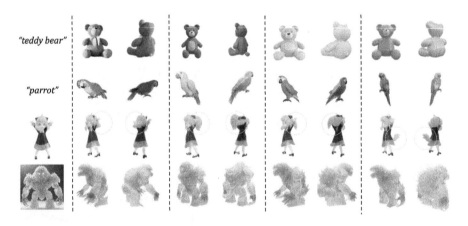

Fig. 8. Diversity of our 3D generation. We can generate diverse 3D models given an ambiguous text description or single-view image.

Text-to-3D. We then compare with recent methods [16,48] on text-to-3D tasks. We observe an enhanced quality and efficiency in our method, generating more realistic 3D objects, as illustrated in Fig. 7. Due to the multi-view diffusion models, our model is also free from multi-face problems.

Diversity. Notably, our pipeline exhibits high diversity in 3D generation, owing to the capability of multi-view diffusion model [45,52]. As shown in Fig. 8, with different random seeds, we can generate a variety of feasible objects from the same ambiguous text prompt or single-view image.

4.3 Quantitative Comparisons

We majorly conduct a user study to quantitatively evaluate our image-to-3D Gaussians generation performance. For a collection of 30 images, we render 360-degree rotating videos of the 3D Gaussians generated from DreamGaussain [48] (only the first stage), TriplaneGaussian [63], and ours. There are in total 90 videos for evaluation in our user study. Each volunteer is shown 30 samples from mixed random methods, and asked to rate in two aspects: image consistency and overall model quality. We collect results from 20 volunteers and get 600 valid scores in total. As shown in Table 1, our method is preferred as it aligns with the original image content and shows better overall quality.

Table 1. User Study on the quality of generated 3D Gaussians for image-to-3D tasks. The rating is of scale 1–5, the higher the better.

	Image Consistency ↑	Overall Quality ↑
DreamGaussian [48]	2.30	1.98
TriplaneGaussian [63]	3.02	2.67
LGM (Ours)	**4.18**	**3.95**

4.4 Ablation Study

Number of Views. We train an image-to-3D model with only one input views similar to splatter image [47], *i.e.*, without the multi-view generation step. The U-Net takes the single input view as input with self-attention, and outputs Gaussian features as in our multi-view model. To compensate the number of Gaussians, we predict two Gaussians for each pixel of the output feature maps, leading to $128 \times 128 \times 2 = 32,768$ Gaussians. As illustrated in the top-left part of Fig. 9, the single-view model can reconstruct faithful front-view, but fails to distinguish the back view and results in blurriness. This is as expected since the regressive U-Net is more suitable for reconstruction tasks, and it's hard to generalize to large datasets in our experiments.

Data Augmentation. We train a smaller model with or without applying data augmentation to validate its effectiveness. Although we observe a lower training loss for the model without data augmentation, the domain gap during inference leads to more floaters and worse geometry as shown in the bottom-left part of

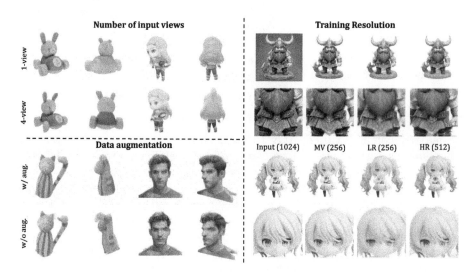

Fig. 9. Ablation Study. We carry out ablation study on designs of our method.

Fig. 9. The model with data augmentation strategy can better correct the 3D inconsistency and inaccurate camera poses in the generated multi-view images, leading to fewer floaters.

Training Resolution. Lastly, we train a model with a fewer number of Gaussians and smaller rendering resolution as in the right part of Fig. 9. We remove the last up block of the U-Net so the number of output Gaussians is $64 \times 64 \times 4 = 16,384$, and we render it at 256×256 for supervision. The model can still converge and successfully reconstruct 3D Gaussians, but the details are worse compared to the 256×256 input multi-view images. In contrast, our large resolution model at 512×512 can capture better details and generate Gaussians with higher resolution.

4.5 Limitations

Since our model is essentially a multi-view reconstruction model, the 3D generation quality highly depends on the quality of four input views. However, current multi-view diffusion models [45,52] are far from perfect: (1) There can be 3D inconsistency which misleads the reconstruction model to generate floaters. (2) The resolution of synthesized multi-view images is restricted to 256×256. (3) ImageDream [52] also fails to handle input image with a large elevation angle. We expect these limitations can be mitigated with better multi-view diffusion models such as Zero-1-to-3++ [44] in future works.

5 Conclusion

In this work, we present a large multi-view Gaussian model for high-resolution 3D content generation. Our model, distinct from previous methods reliant on NeRF and transformers, employs Gaussian splatting and U-Net to address the challenges of high memory requirements and low-resolution training. Additionally, we explore data augmentation for better robustness, and introduce a mesh extraction algorithm for the generated 3D Gaussians. Our approach achieves both high-resolution and high-efficiency for 3D objects generation, proving its versatility and applicability in various contexts.

Acknowledgements. This work is supported by the Sichuan Science and Technology Program (2023YFSY0008), National Natural Science Foundation of China (61632003, 61375022, 61403005), Grant SCITLAB-20017 of Intelligent Terminal Key Laboratory of SiChuan Province, Beijing Advanced Innovation Center for Intelligent Robots and Systems (2018IRS11), and PEK-SenseTime Joint Laboratory of Machine Vision. This project is also funded by the Ministry of Education, Singapore, under its MOE AcRF Tier 2 (MOE-T2EP20221-0012), NTU NAP, and under the RIE2020 Industry Alignment Fund - Industry Collaboration Projects (IAF-ICP) Funding Initiative, as well as cash and in-kind contribution from the industry partner(s).

References

1. Cao, Z., Hong, F., Wu, T., Pan, L., Liu, Z.: Large-vocabulary 3D diffusion model with transformer. arXiv preprint arXiv:2309.07920 (2023)
2. Chan, E.R., et al.: Efficient geometry-aware 3D generative adversarial networks. In: CVPR (2022)
3. Charatan, D., Li, S., Tagliasacchi, A., Sitzmann, V.: pixelSplat: 3D gaussian splats from image pairs for scalable generalizable 3D reconstruction. arXiv preprint arXiv:2312.12337 (2023)
4. Chen, G., Wang, W.: A survey on 3D gaussian splatting. arXiv preprint arXiv:2401.03890 (2024)
5. Chen, H., et al.: Single-stage diffusion nerf: a unified approach to 3D generation and reconstruction. arXiv preprint arXiv:2304.06714 (2023)
6. Chen, R., Chen, Y., Jiao, N., Jia, K.: Fantasia3D: disentangling geometry and appearance for high-quality text-to-3D content creation. arXiv preprint arXiv:2303.13873 (2023)
7. Chen, Y., et al.: Gaussianeditor: swift and controllable 3D editing with gaussian splatting. arXiv preprint arXiv:2311.14521 (2023)
8. Chen, Y., et al.: IT3D: improved text-to-3D generation with explicit view synthesis. arXiv preprint arXiv:2308.11473 (2023)
9. Chen, Z., Hong, F., Mei, H., Wang, G., Yang, L., Liu, Z.: Primdiffusion: volumetric primitives diffusion for 3D human generation. arXiv preprint arXiv:2312.04559 (2023)
10. Chen, Z., Wang, F., Liu, H.: Text-to-3D using gaussian splatting. arXiv preprint arXiv:2309.16585 (2023)
11. Deitke, M., et al.: Objaverse-xl: a universe of 10m+ 3D objects. arXiv preprint arXiv:2307.05663 (2023)

12. Deitke, M., et al.: Objaverse: a universe of annotated 3D objects. In: CVPR, pp. 13142–13153 (2023)
13. He, K., Zhang, X., Ren, S., Sun, J.: Deep residual learning for image recognition. In: CVPR, pp. 770–778 (2016)
14. Ho, J., Jain, A., Abbeel, P.: Denoising diffusion probabilistic models. In: NeurIPS, vol. 33, pp. 6840–6851 (2020)
15. Hong, Y., et al.: LRM: large reconstruction model for single image to 3D. arXiv preprint arXiv:2311.04400 (2023)
16. Jun, H., Nichol, A.: Shap-e: generating conditional 3D implicit functions. arXiv preprint arXiv:2305.02463 (2023)
17. Kerbl, B., Kopanas, G., Leimkühler, T., Drettakis, G.: 3D gaussian splatting for real-time radiance field rendering. ToG **42**(4), 1–14 (2023)
18. Laine, S., Hellsten, J., Karras, T., Seol, Y., Lehtinen, J., Aila, T.: Modular primitives for high-performance differentiable rendering. ToG **39**(6) (2020)
19. Li, J., et al.: Instant3d: fast text-to-3D with sparse-view generation and large reconstruction model. arXiv preprint arXiv:2311.06214 (2023)
20. Li, W., Chen, R., Chen, X., Tan, P.: Sweetdreamer: aligning geometric priors in 2D diffusion for consistent text-to-3D. arXiv preprint arXiv:2310.02596 (2023)
21. Li, Y., et al.: Focaldreamer: text-driven 3D editing via focal-fusion assembly. arXiv preprint arXiv:2308.10608 (2023)
22. Lin, C.H., et al.: Magic3d: high-resolution text-to-3D content creation. In: CVPR, pp. 300–309 (2023)
23. Ling, H., Kim, S.W., Torralba, A., Fidler, S., Kreis, K.: Align your gaussians: text-to-4D with dynamic 3D gaussians and composed diffusion models. arXiv preprint arXiv:2312.13763 (2023)
24. Liu, R., Wu, R., Van Hoorick, B., Tokmakov, P., Zakharov, S., Vondrick, C.: Zero-1-to-3: zero-shot one image to 3D object. arXiv preprint arXiv:2303.11328 (2023)
25. Liu, Y., et al.: Syncdreamer: generating multiview-consistent images from a single-view image. arXiv preprint arXiv:2309.03453 (2023)
26. Liu, Z., Feng, Y., Black, M.J., Nowrouzezahrai, D., Paull, L., Liu, W.: Meshdiffusion: score-based generative 3D mesh modeling. arXiv preprint arXiv:2303.08133 (2023)
27. Long, X., et al.: Wonder3d: single image to 3D using cross-domain diffusion. arXiv preprint arXiv:2310.15008 (2023)
28. Lorensen, W.E., Cline, H.E.: Marching cubes: a high resolution 3D surface construction algorithm. In: Seminal Graphics: Pioneering Efforts that Shaped the Field, pp. 347–353 (1998)
29. Loshchilov, I., Hutter, F.: Decoupled weight decay regularization. arXiv preprint arXiv:1711.05101 (2017)
30. Luo, T., Rockwell, C., Lee, H., Johnson, J.: Scalable 3D captioning with pretrained models. arXiv preprint arXiv:2306.07279 (2023)
31. Metzer, G., Richardson, E., Patashnik, O., Giryes, R., Cohen-Or, D.: Latent-nerf for shape-guided generation of 3D shapes and textures. arXiv preprint arXiv:2211.07600 (2022)
32. Mildenhall, B., Srinivasan, P.P., Tancik, M., Barron, J.T., Ramamoorthi, R., Ng, R.: Nerf: representing scenes as neural radiance fields for view synthesis. In: ECCV (2020)
33. Müller, N., Siddiqui, Y., Porzi, L., Bulo, S.R., Kontschieder, P., Nießner, M.: Diffrf: rendering-guided 3D radiance field diffusion. In: CVPR, pp. 4328–4338 (2023)
34. Müller, T., Evans, A., Schied, C., Keller, A.: Instant neural graphics primitives with a multiresolution hash encoding. ACM TOG **41**(4), 1–15 (2022)

35. Nichol, A., Jun, H., Dhariwal, P., Mishkin, P., Chen, M.: Point-e: a system for generating 3D point clouds from complex prompts. arXiv preprint arXiv:2212.08751 (2022)
36. Poole, B., Jain, A., Barron, J.T., Mildenhall, B.: Dreamfusion: text-to-3D using 2D diffusion. arXiv preprint arXiv:2209.14988 (2022)
37. Qian, G., et al.: Magic123: one image to high-quality 3D object generation using both 2D and 3D diffusion priors. arXiv preprint arXiv:2306.17843 (2023)
38. Qin, X., Zhang, Z., Huang, C., Dehghan, M., Zaiane, O.R., Jagersand, M.: U2-net: going deeper with nested u-structure for salient object detection. Pattern Recogn. **106**, 107404 (2020)
39. Ren, J., Pan, L., Tang, J., Zhang, C., Cao, A., Zeng, G., Liu, Z.: Dreamgaussian4d: generative 4D gaussian splatting. arXiv preprint arXiv:2312.17142 (2023)
40. Rombach, R., Blattmann, A., Lorenz, D., Esser, P., Ommer, B.: High-resolution image synthesis with latent diffusion models. In: CVPR, pp. 10684–10695 (2022)
41. Ronneberger, O., Fischer, P., Brox, T.: U-Net: convolutional networks for biomedical image segmentation. In: Navab, N., Hornegger, J., Wells, W.M., Frangi, A.F. (eds.) MICCAI 2015. LNCS, vol. 9351, pp. 234–241. Springer, Cham (2015). https://doi.org/10.1007/978-3-319-24574-4_28
42. Saharia, C., et al.: Photorealistic text-to-image diffusion models with deep language understanding. In: NeurIPS, vol. 35, pp. 36479–36494 (2022)
43. Shen, T., Gao, J., Yin, K., Liu, M.Y., Fidler, S.: Deep marching tetrahedra: a hybrid representation for high-resolution 3D shape synthesis. In: NeurIPS (2021)
44. Shi, R., et al.: Zero123++: a single image to consistent multi-view diffusion base model (2023)
45. Shi, Y., Wang, P., Ye, J., Long, M., Li, K., Yang, X.: Mvdream: multi-view diffusion for 3D generation. arXiv preprint arXiv:2308.16512 (2023)
46. Song, J., Meng, C., Ermon, S.: Denoising diffusion implicit models. arXiv preprint arXiv:2010.02502 (2020)
47. Szymanowicz, S., Rupprecht, C., Vedaldi, A.: Splatter image: ultra-fast single-view 3D reconstruction. arXiv (2023)
48. Tang, J., Ren, J., Zhou, H., Liu, Z., Zeng, G.: Dreamgaussian: generative gaussian splatting for efficient 3D content creation. arXiv preprint arXiv:2309.16653 (2023)
49. Tang, J., et al.: Delicate textured mesh recovery from nerf via adaptive surface refinement. arXiv preprint arXiv:2303.02091 (2022)
50. Tsalicoglou, C., Manhardt, F., Tonioni, A., Niemeyer, M., Tombari, F.: Textmesh: generation of realistic 3d meshes from text prompts. arXiv preprint arXiv:2304.12439 (2023)
51. Vaswani, A., et al.: Attention is all you need. In: NeurIPS, vol. 30 (2017)
52. Wang, P., Shi, Y.: Imagedream: image-prompt multi-view diffusion for 3D generation. arXiv preprint arXiv:2312.02201 (2023)
53. Wang, P., et al.: PF-LRM: pose-free large reconstruction model for joint pose and shape prediction. arXiv preprint arXiv:2311.12024 (2023)
54. Wang, T., et al.: Rodin: a generative model for sculpting 3D digital avatars using diffusion. In: CVPR, pp. 4563–4573 (2023)
55. Wang, Z., et al.: Prolificdreamer: high-fidelity and diverse text-to-3D generation with variational score distillation. arXiv preprint arXiv:2305.16213 (2023)
56. Weng, Z., et al.: Single-view 3D human digitalization with large reconstruction models. arXiv preprint arXiv:2401.12175 (2024)
57. Xu, D., et al.: AGG: amortized generative 3D gaussians for single image to 3D. arXiv preprint arXiv:2401.04099 (2024)

58. Xu, Y., et al.: Dmv3d: denoising multi-view diffusion using 3D large reconstruction model. arXiv preprint arXiv:2311.09217 (2023)
59. Yariv, L., Puny, O., Neverova, N., Gafni, O., Lipman, Y.: Mosaic-SDF for 3D generative models. arXiv preprint arXiv:2312.09222 (2023)
60. Yi, T., et al.: Gaussiandreamer: fast generation from text to 3D gaussian splatting with point cloud priors. arXiv preprint arXiv:2310.08529 (2023)
61. Zhang, R., Isola, P., Efros, A.A., Shechtman, E., Wang, O.: The unreasonable effectiveness of deep features as a perceptual metric. In: CVPR (2018)
62. Zhao, Z., et al.: Michelangelo: conditional 3D shape generation based on shape-image-text aligned latent representation. arXiv preprint arXiv:2306.17115 (2023)
63. Zou, Z.X., et al.: Triplane meets gaussian splatting: fast and generalizable single-view 3D reconstruction with transformers. arXiv preprint arXiv:2312.09147 (2023)

Mahalanobis Distance-Based Multi-view Optimal Transport for Multi-view Crowd Localization

Qi Zhang[1], Kaiyi Zhang[1,2], Antoni B. Chan[3], and Hui Huang[1](✉)

[1] Visual Computing Research Center, College of Computer Science and Software Engineering, Shenzhen University, Shenzhen, China
qi.zhang.opt@gmail.com, zhangkaiyi2022@email.szu.edu.cn,
hhzhiyan@gmail.com
[2] Guangdong Laboratory of Artificial Intelligence and Digital Economy (SZ), Shenzhen, China
[3] Department of Computer Science, City University of Hong Kong, Hong Kong SAR, China
abchan@cityu.edu.hk

Abstract. Multi-view crowd localization predicts the ground locations of all people in the scene. Typical methods usually estimate the crowd density maps on the ground plane first, and then obtain the crowd locations. However, existing methods' performances are limited by the ambiguity of the density maps in crowded areas, where local peaks can be smoothed away. To mitigate the weakness of density map supervision, optimal transport-based point supervision methods have been proposed in the single-image crowd localization tasks, but have not been explored for multi-view crowd localization yet. Thus, in this paper, we propose a novel Mahalanobis distance-based multi-view optimal transport (M-MVOT) loss specifically designed for multi-view crowd localization. First, we replace the Euclidean-based transport cost with the Mahalanobis distance, which defines elliptical iso-contours in the cost function whose long-axis and short-axis directions are guided by the view ray direction. Second, the object-to-camera distance in each view is used to adjust the optimal transport cost of each location further, where the wrong predictions far away from the camera are more heavily penalized. Finally, we propose a strategy to consider all the input camera views in the model loss (M-MVOT) by computing the optimal transport cost for each ground-truth point based on its closest camera. Experiments demonstrate the advantage of the proposed method over density map-based or common Euclidean distance-based optimal transport loss on several multi-view crowd localization datasets.

Keywords: Multi-view · Optimal transport · Crowd localization

Supplementary Information The online version contains supplementary material available at https://doi.org/10.1007/978-3-031-73235-5_2.

1 Introduction

Multi-view crowd localization [10,11,26,31] has been proposed to predict the people's locations on the ground of the scene, which could be used for applications like crowd analysis, autonomous driving, public traffic management, *etc.*. It fuses multi-camera information via feature extraction and projection of each

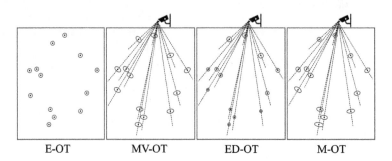

Fig. 1. The comparison of different optimal transport (OT) losses on the ground plane. The original OT uses Euclidean distance cost (E-OT) and treats all deviations from the ground-truth equally. MV-OT uses the view ray direction to the camera to change the cost function, ED-OT considers the camera distance in E-OT, while M-OT considers both the influence of the view ray direction and the distance to the camera.

camera view into a common ground plane, followed by multi-view fusion and decoding. Current methods mainly rely on fixed-size Gaussian-kernel density maps as supervision to train the multi-view crowd localization models. The people locations are further extracted after a local non-maximum suppression (NMS) as a post-processing step on the estimated crowd density maps, where the pixel-wise mean square error (MSE) loss between the predicted and ground-truth density maps is adopted to train the whole model. However, these methods' localization performance is limited by the ambiguity of the density maps in crowded areas, where there may not be clear density peaks over each person for localization.

To mitigate the problems of Gaussian density map supervision in the single-image crowd localization task, point-supervision methods based on optimal transport (OT) [23,34] have been proposed and achieved significant gains in localization performance compared to Gaussian density map methods trained with MSE loss. OT loss directly uses point annotations as supervision and generates more compact density maps [34]. However, point supervision for multi-view crowd localization methods has not been explored yet.

In this paper, we explore using point supervision for multi-view crowd localization and propose a novel Mahalanobis distance-based multi-view optimal transport (M-MVOT) loss. In the M-MVOT loss, the transport cost is defined using the Mahalanobis distance, which adjusts the cost based on the view-ray direction between the ground-truth point and the camera, and the distance to the camera. Specifically, the Mahalanobis distance defines elliptical iso-contours

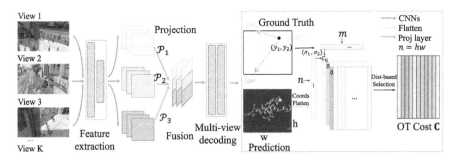

Fig. 2. The model architecture and the proposed Mahalanobis distance-based multi-view optimal transport (M-MVOT) loss for multi-view crowd localization. The model consists of feature extraction, projection, and multi-view fusion and decoding. In the proposed M-MVOT, each point's transport cost **C** is calculated via the Mahalanobis distance instead of the common Euclidean distance **under the closest camera**, which is **directed by the view ray and adjusted by the object-to-camera distance**.

of the cost function for a ground-truth annotation, where the ellipse's long- and short-axes are guided by the view-ray direction of the camera, and the axes extents are guided by the distance of the point to the camera. The projection step of the multi-view crowd localization framework causes streaking artifacts of the features along the view ray on the ground plane, which could smear the predicted density map making localization less accurate. In order to counteract this effect, we penalize predictions that deviate from the ground-truth position more along the view ray. This can be implemented by assigning larger weights in the Mahalanobis distance along the view ray direction. In addition, the object-to-camera distance in each view is used to further adjust the cost (denoted as M-OT), where the locations far away from the camera are more heavily penalized in order to overcome the more severe projection errors there, compared to locations near to the camera.

Finally, to consider all the input camera views in the model loss (M-MVOT), we integrate the transport costs over multiple views, by setting each point's optimal transport cost to be calculated with the closest camera. In Fig. 1, we compare different optimal transport loss together, where E-OT uses Euclidean distance and considers each point equally regardless of camera locations, MV-OT uses Mahalanobis distance based on view ray direction, ED-OT considers the camera distance in E-OT, and M-OT considers both the view ray direction and camera distance. M-MVOT considers all camera views together based on M-OT (see Fig. 3 right). Figure 2 shows the brief architecture of the multi-view crowd localization model and the proposed M-MVOT loss. The model consists of single-view feature extraction, projection, and multi-view fusion and decoding steps. In the proposed M-MVOT loss, the cost for transporting pixels from the predicted density map to the ground-truth points is measured with the Mahalanobis distance to obtain the transport cost **C** for the optimal transport problem. The main contributions of this work are as follows.

- To the best of our knowledge, this is the first study on using point supervision optimal transport (OT) for training multi-view crowd localization.
- We propose a new OT loss specifically designed for multi-view crowd localization, which uses view-ray direction and object-to-camera distance to adjust the Mahalanobis distance transport cost and create a multi-view version of OT based on calculating the cost under the closest camera.
- The experiments validate that the proposed M-MVOT loss achieves better localization performance than density map-based MSE and Euclidean distance-based OT losses.

2 Related Work

2.1 Multi-view Crowd Localization

Traditional Methods. Early traditional multi-view crowd localization methods first detect each person in the images and then associate the camera view results via correspondence matching with hand-crafted features [8,27,33,36]. W. Ge et al. [9] presented a Bayesian people detection and counting approach by sampling from a posterior distribution over crowd configurations. Traditional methods rely on hand-crafted features and background subtraction preprocessing, which limit their performance and application scenarios under heavy occlusions.

Deep Learning Methods. With the development of deep learning frameworks and more multi-view crowd datasets, deep learning-based multi-view crowd localization methods [3,5,31] have seen great progress. The early deep-learning methods did not consider the feature alignment between camera views [5]. MVMS [42] proposed an end-to-end multi-view fusion framework with a feature projection layer for the wide-area crowd-counting task. MVDet [11] also used feature perspective transformations to fuse multi-view information for the multi-view crowd localization. These are the first methods utilizing end-to-end frameworks for multi-view crowd localization tasks, which include single-view feature extraction, projection, and multi-view fusion and decoding. In more recent work, SHOT [31] utilized multi-height projection with an attention-based soft selection module and replaced the perspective transformation with a homography transformation layer in the model for better localization performance. MVDeTr [10] proposed the multi-view deformable transformer network MVDeTr for multi-view fusion by extending the deformable transformer framework in [44]. CVCS [43] has studied multi-view crowd counting methods under the cross-view cross-scene setting, and proposed a large synthetic multi-view crowd dataset CVCS, which could also be used for multi-view crowd localization tasks. 3DROM [26] proposed a data augmentation method to relieve the model overfitting issue by generating random 3D cylinder occlusion masks in the 3D space. 3DROM achieved state-of-the-art performance on 2 small single-scene datasets MultiviewX and Wildtrack, which have fixed cameras for training and testing. However, it is unclear how well 3DROM performs on larger cross-view cross-scene datasets, such as the CVCS dataset [43].

Overall, the existing multi-view crowd localization methods [10,11,26,31] are trained with fixed-size Gaussian density maps as supervision. However, using Gaussian density maps causes issues for localization in dense crowd regions since when the Gaussian kernels significantly overlap, the local peaks will be smoothed out and the locations unrecoverable. Note that reducing the Gaussian width could potentially solve this problem, but leads to a more difficult learning problem. Considering that point-based supervision, such as optimal transport loss, achieved better performance on single-image crowd localization task, in this paper, we explore the multi-view optimal transport loss for the multi-view crowd localization task.

2.2 Single-Image Crowd Localization

Single-image crowd localization aims to estimate people's locations in crowd images with points, which are usually on each person's head. The task is similar to people detection, which is based on bounding boxes, but crowd localization only uses points to indicate people's locations. Most single-image counting methods are based on Gaussian density map supervision [2,6,7,13,14,16,21,24,28–30,39–41], and their density map predictions could also be used for crowd localization. However, due to the smoothness of the predicted Gaussian density maps in crowded regions, their crowd localization performance is quite limited.

Thus, point supervision methods [1,19,20,22,32,38], blob segmentation [15] and inverse distance maps [17] have been proposed, many of which show better localization performance, especially the **optimal transport**-based point-supervision methods [23,34,35]. [22] proposed Bayesian loss, which constructed a density contribution probability model from the point annotations. TopoCount [1] introduced a topological constraint to reason about the spatial arrangement of dots by a persistence loss based on the theory of persistent homology. Z. Ma [23] minimized the unbalanced optimal transport distance between the ground-truth points and predicted density maps, and applied it to crowd counting and localization. Concurrently, GL [34] explored the unbalanced optimal transport (UOT) for crowd counting and proposed a perspective-guided transport cost function to better handle perspective transformations. [34] also proved that the MSE loss with Gaussian ground-truth density maps and the Bayesian loss in [22] are special suboptimal cases of the unbalanced optimal transport loss. P2PNet [32] proposed a pure point framework in a one-to-one matching manner using the Hungarian algorithm. Most previous works perform localization from the density map by first thresholding the density map and then finding local peaks. W. Lin [18] proposed the optimal transport minimization (OT-M) algorithm for crowd localization from density maps, which is an iterative algorithm that finds the set of locations that have minimal OT loss with the density map.

Despite OT's success in single-view crowd localization, point supervision with optimal transport for multi-view crowd localization is an unexplored area. The multi-view setting introduces new difficulties, such as variations in the camera view, projection errors due to poor camera calibration, and artifacts introduced when projecting features from the camera view to the 2D ground plane, which

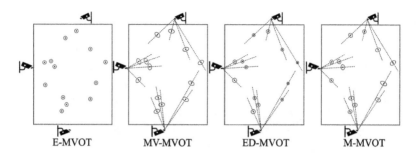

Fig. 3. The comparison of different versions for the multi-view optimal transport, where the cost for each ground-truth point is calculated using the closest camera. E-MVOT considers each point equally regardless of the camera views, which is the same as E-OT for a single camera view. MV-MVOT replaces the Euclidean distance with the Mahalanobis distance and the direction is guided by the view ray direction. ED-MVOT introduces the object-to-camera influence in the optimal transport cost of E-OT. M-MVOT considers both the view-ray direction guidance and the object-to-camera distance adjustment in the optimal transport based on M-OT.

could hinder model training and localization ability. *To address these issues, our proposed loss considers the camera view direction and distance when computing the transport cost, which is implemented via a Mahalanobis distance function.*

3 Multi-view Optimal Transport

In this section, we first review optimal transport based on Euclidean distance cost (E-OT) for single-image crowd localization. Then, we introduce the proposed Mahalanobis distance optimal transport (M-OT), where the cost can be guided either by the view ray direction of each GT location (MV-OT), or by the object-to-camera distance of each location (ED-OT), or both (M-OT). Finally, the proposed M-OT is applied under multi-views, where each prediction point's optimal transport cost is calculated under the closest camera view via M-OT, denoted as M-MVOT.

3.1 Euclidean Distance Optimal Transport

We follow the loss formulation in [34], which is based on an unbalanced optimal transport formulation with Euclidean distance cost (E-OT). Denote $\mathcal{X} = \{(a_i, \mathbf{x}_i)\}_{i=1}^n$ as the predicted density map, where a_i is pixel $\mathbf{x}_i$'s predicted density, $\mathbf{a} = [a_i]_i$ is the density of all pixels, and n is the number of pixels. We denote $\mathcal{Y} = \{(b_j, \mathbf{y}_j)\}_{j=1}^m$ as the ground-truth dot map, where $b_j \in \{0, 1\}$ indicates there is a person at location $\mathbf{y}_j$, $\mathbf{b} = [b_j]_j = \mathbf{1}_m$, and m is the ground-truth number of people. The entropic-regularized unbalanced optimal transport loss is as follows.

$$\mathcal{L}_\mathbf{C}^\tau = \min_{\mathbf{P} \in \mathcal{R}_+^{\mathbf{n} \times \mathbf{m}}} \langle \mathbf{C}, \mathbf{P} \rangle - \epsilon \sum_{ij} P_{ij} \log P_{ij}$$
$$+ \tau \|\mathbf{P}\mathbf{1_m} - \mathbf{a}\|_2^2 + \tau \|\mathbf{P^T}\mathbf{1_n} - \mathbf{b}\|_1, \quad (1)$$

where $\mathbf{P} \in \mathcal{R}_+^{\mathbf{n} \times \mathbf{m}}$ is the transport plan matrix, which assigns each density $\mathbf{x}_i$ of $\mathcal{X}$ to location $\mathbf{y}_j$ of $\mathcal{Y}$ for measuring the cost. $\mathbf{C} \in \mathcal{R}_+^{\mathbf{n} \times \mathbf{m}}$ is the transport cost matrix, where C_{ij} measures the cost of moving the predicted density at $\mathbf{x}_i$ to ground-truth dot location $\mathbf{y}_j$. The cost function is defined as the exponential Euclidean distance between the predicted density of $\mathbf{x}_i$ and the ground-truth dot $\mathbf{y}_j$:

$$C_{ij} = c(\mathbf{x}_i, \mathbf{y}_j) = \exp(\|\mathbf{x}_i - \mathbf{y}_j\|). \quad (2)$$

The second term $H(P) = \sum_{ij} P_{ij} \log P_{ij}$ is the entropic regularization term. [34] sets $D_1(\mathbf{P}\mathbf{1_m}|\mathbf{a}) = \|\mathbf{P}\mathbf{1_m} - \mathbf{a}\|_2^2$, ensuring that all predicted density values have a corresponding annotation, and $D_2(\mathbf{P}^T\mathbf{1_n}|\mathbf{b}) = \|\mathbf{P^T}\mathbf{1_n} - \mathbf{b}\|_1$, ensuring that all ground-truth points are used in the transport plan. ϵ and τ are weighted factors to adjust the weight of the last three terms. The optimization of (1) can be computed via sinkorn iteration [25].

3.2 Mahalanobis Distance Optimal Transport

The Euclidean distance in the transport cost matrix $\mathbf{C}$ in (2) considers all directions of deviation from the ground-truth point equally. For multi-view crowd localization, we aim to apply the OT loss on the predicted density map on the ground plane. However, the projection step, which projects features from the camera view to the ground plane introduces streaking artifacts in the feature map due to missing information, i.e., the unknown height of the corresponding object in the 3D world, which may lead to streaking artifacts in the density map that hinder its localization ability. To counteract this effect, we use Mahalanobis distance to replace the Euclidean distance in (2), where the contours of the Mahalanobis distance are guided by the view ray direction at the location, denoted as MV-OT. Besides, the object-to-camera distance could also adjust the calculation of the cost matrix for different regions, denoted as ED-OT. And both the view ray direction and object-to-camera distance can be considered in the Mahalanobis distance optimal transport loss, denoted as M-OT.

View Ray-Directed Transport Cost (MV-OT). Using the Mahalanobis distance defines elliptical iso-contours in the transport cost function, where certain directions will incur larger costs than others. Here we set the ellipse shape to counteract the errors and streaking artifacts introduced by the projection step. Specifically, the short axis of the ellipse is set to be aligned with the **view ray** between the ground-truth point and the camera, and the long axis is set to be perpendicular to the **view ray** (MV-OT). In this way, the predictions that deviate from the ground-truth along the view ray from the ground-truth points are assigned with larger penalties (see "MV-OT" in Fig. 1, where more penalties are given along the view ray), in order to penalize prediction errors caused by projection distortion. Concretely, we first calculate the covariance matrix for each GT point:

$$\mathbf{S} = \mathbf{R}\mathbf{\Sigma}\mathbf{R}^{-1}, \mathbf{\Sigma} = \begin{bmatrix} \sigma_1^2 & 0 \\ 0 & \sigma_2^2 \end{bmatrix}, \mathbf{R} = \begin{bmatrix} \cos\beta & -\sin\beta \\ \sin\beta & \cos\beta \end{bmatrix}, \quad (3)$$

where $\mathbf{R}$ represents the rotation matrix from the original coordinate to the view-ray coordinate, and β is the rotation angle. $\mathbf{\Sigma}$ is the variance matrix with diagonal entries σ_1^2 and σ_2^2 corresponding to the variances along the view-ray direction and perpendicular to the view-ray direction, respectively. Here we set $\sigma_1^2 < \sigma_2^2$, and thus $1/\sigma_1^2 > 1/\sigma_2^2$, indicating deviating predictions along the view ray direction incur the larger cost and thus are more heavily penalized. Then, the Mahalanobis distance cost in MV-OT can be written as:

$$C_{ij} = c(\mathbf{x}_i, \mathbf{y}_j) = \exp(\sqrt{(\mathbf{x}_i - \mathbf{y}_j)^\mathrm{T} \mathbf{S}^{-1} (\mathbf{x}_i - \mathbf{y}_j)}). \quad (4)$$

Note when $\sigma_1^2 = \sigma_2^2 = 1$, Mahalanobis distance degenerates into the Euclidean distance. We choose $\sigma_1^2 = 1$ and $\sigma_2^2 = 1.2$ in the experiments for MV-OT.

Distance-Adjusted Transport Cost (ED-OT). The object-to-camera distance also influences the localization accuracy for the points, where faraway points tend to have larger prediction errors. Thus, we use the object-to-camera distance to adjust the cost calculation and give more penalties to faraway regions. We rewrite σ_1^2 and σ_2^2 as follows.

$$\sigma_1^2 = \sigma_2^2 = 1/\exp(\alpha * \mathrm{MinMaxNorm}(d_{cam})), \quad (5)$$

where d_{cam} denotes the distance from the ground-truth point to the camera, which can be obtained from the camera calibrations and point coordinates, and MinMaxNorm normalizes the object-to-camera distance into $[0, 1]$ through (d_{cam} - $\min(d_{cam}))/(\max(d_{cam})$ - $\min(d_{cam}))$. Here α is an adaptive factor to adjust the influence of distance to the cost function in (5) and (6). When d_{cam} is larger, σ_1^2 and σ_2^2 decrease, and thus more penalties are assigned. Since $\sigma_1^2 = \sigma_2^2$, the Euclidean distance is actually used in the cost, but the object-to-camera distance adjusts the Euclidean distance for different points, denoted as ED-OT.

View Ray-Directed and Distance-Adjusted Transport Cost (M-OT). To consider both the influence of the view ray direction and object-to-camera distance, we use the object-to-camera distance to adjust the long-axis and short-axis extents of the ellipse iso-contours in the Mahalanobis distance. Specifically, for the GT points that are far away from the camera (d_{cam} is larger), a larger variance is assigned perpendicular to the view ray (σ_2^2), which has the effect of penalizing deviations perpendicular to the view ray less, and penalizing deviations along the view ray (σ_1^2) more:

$$\sigma_1^2 = 1, \quad \sigma_2^2 = \exp(\alpha * \mathrm{MinMaxNorm}(d_{cam})). \quad (6)$$

By combining (3) and (6), we obtain the view ray-directed and distance-adjusted Mahalanobis distance cost function for OT, denoted as M-OT. Note that when $\alpha = 0$, then $\sigma_1^2 = \sigma_2^2 = 1$, and the proposed Mahalanobis distance cost of M-OT degenerates into E-OT in (2).

3.3 Mahalanobis Distance Multi-view Optimal Transport (M-MVOT)

We introduce the Mahalanobis distance optimal transport (M-OT) for single views in Sect. 3.2. We further introduce the Mahalanobis distance multi-view optimal transport (M-MVOT) by combining the proposed Mahalanobis distance optimal transport via a distance-based selection strategy. Specifically, the cost of a point is calculated using the closest camera view's M-OT defined as follows.

$$C_{ij} = \sum_{k=1}^{K} \mathbb{1}(d_{cam}^k) \exp(\sqrt{(\mathbf{x}_i - \mathbf{y}_j)^{\mathrm{T}} \mathbf{S}_k^{-1} (\mathbf{x}_i - \mathbf{y}_j)}), \qquad (7)$$

where $\mathbb{1}(d_{cam}^k) = \begin{cases} 1, & d_{cam}^k = \min_{p \in \{1, \cdots, K\}} d_{cam}^p, \\ 0, & \text{otherwise.} \end{cases}$ is binary and chooses the closest camera's M-OT value, K is the camera number, and d_{cam}^k means the distance from the ground-truth point to camera k.

Following this strategy, we extend our E-OT, MV-OT, ED-OT, and M-OT to multi-camera view settings for multi-view crowd localization tasks, denoted as E-MVOT, MV-MVOT, ED-MVOT and M-MVOT, respectively, where the closest camera is used to compute the cost function for a given ground-truth point. See details in Fig. 3.

4 Experiments and Results

4.1 Experiment Settings

Datasets. We evaluate the proposed multi-view optimal transport loss on 3 datasets in the experiments, including CVCS [43], Wildtrack [4] and MultiviewX [11]. **CVCS** is a synthetic multi-view crowd dataset, containing 31 scenes, where 23 are for training and the remaining 8 for testing. Each scene contains 100 multi-view frames with about 60–120 camera views. The size of the scenes in CVCS varies from 10 m × 20 m to 90 m × 80 m. Each scene contains 90–180 people. **Wildtrack** is a real-world single-scene multi-view dataset, whose scene size is 12 m × 36 m. There are 20 people per frame on average in the dataset. The dataset consists of 7 camera views and 400 synchronized multi-view frames, in which the first 360 frames are used for training and the 40 frames are for testing. **MultiviewX** is a synthetic single-scene multi-view dataset, whose scene size is 16 m × 25 m and consists of 6 camera views. The dataset includes 400 multi-view frames and the first 360 are used for training and 40 are for testing. MultiviewX has around 40 synthetic people per frame. The resolution of the images in the

3 datasets is 1920 × 1080. Note that CVCS is a *cross-scene* dataset, where the same scene and camera layout do not appear in both the training set and test set. In contrast, both Wildtrack and MultiviewX are *single-scene* datasets with the same camera layout for both training and testing. Thus, CVCS is a more challenging dataset, and can better demonstrate the generalization of multi-view localization methods to new scenes and camera layouts.

Comparison Methods. We compare the proposed M-MVOT loss based multi-view crowd localization model against the Euclidean distance version (E-MVOT) using the same architecture, as well as against existing methods MVDet [11], SHOT [31], MVDeTr [10], 3DROM [26]. We report the performance of previous methods on Wildtrack and MultiviewX according to the published results, and also apply them on the CVCS dataset for comparison. We also compare with RCNN [37], POM-CNN [8], DeepMCD [5], DeepOcc. [3], and Volumetric [12] on the Wildtrack and MultiviewX datasets.

Implementation Details. In the experiments, for CVCS, the input image resolution is 1280 × 720. In the training, 5 views are randomly selected for 5 times in each iteration per frame of each scene, and the same number of views (5) is randomly selected for 21 times per frame in the evaluation. For Wildtrack, the input image resolution is resized to 1280 × 720, and the scene map resolution is 120 × 360, where each pixel represents 0.1 m in the real world. For MultiviewX, the input image resolution is also resized to 1280 × 720, and the scene map resolution is 160 × 250, where each pixel is 0.1 m in the real world. The map distance threshold is 0.5 m in the real world as in [11] for all methods on the 3 datasets. On the CVCS dataset, the multi-view crowd localization model is implemented based on the existing multi-view crowd localization method MVDet [11]. On the Wildtrack and MultiviewX datasets, the implementation for the multi-view crowd localization is based on MVDeTr [10]. The ground-plane density map prediction losses in [10,11] are replaced with the proposed Mahalanobis distance multi-view optimal transport loss (M-MVOT). The models are trained with the proposed M-MVOT loss and the auxiliary 2D density map losses in [11] together. For Wildtrack and CVCS, τ is set to 1, and τ is set to 20 for MutiviewX. α is 1 on CVCS and 0.05 on both MultiviewX and Wildtrack. See more model implementation and training details in the supplemental.

Evaluation Metrics. We use 5 metrics to evaluate and compare the multi-view crowd localization methods: Multiple Object Detection Accuracy (MODA), Multiple Object Detection Precision (MODP), Precision (P), Recall (R), and F1_score (F1). Specifically, $MODA = 1 - (FP + FN)/(TP + FN)$ measures the detection accuracy. $MODP = (\sum(1 - d[d < t]/t))/TP$ measures the precision of detection, where d is the distance from a detected person point to its ground truth and t is the distance threshold. Meanwhile, $P = TP/(FP + TP)$, $R = TP/(TP + FN)$, and $F1 = 2P * R/(P + R)$. Here, TP, FP, and FN are the number of true positives, false positives, and false negatives, respectively. Overall, we use MODA and F1_score as the main metrics, since they comprehensively measure the performance of detecting the crowd both completely and precisely.

Table 1. The multi-view crowd localization results on the CVCS dataset. The distance threshold is 0.5 m in the world for all methods.

Method	MODA↑	MODP↑	Precision↑	Recall↑	F1_score↑
MVDet [11]	14.2	59.3	85.0	17.3	28.7
SHOT [31]	31.7	72.1	94.5	33.6	49.6
MVDeTr [10]	24.9	**79.6**	**98.1**	25.4	40.4
3DROM [26]	20.1	74.2	84.1	23.7	37.0
E-MVOT (ours)	43.1	74.3	85.6	51.8	64.5
M-MVOT (ours)	**43.5**	74.1	85.5	**52.3**	**64.9**

Table 2. Comparison of the multi-view people detection performance on single-scene datasets Wildtrack and MultiviewX. The proposed method is the best on MultiviewX and comparable on Wildtrack. The distance threshold is 0.5 m in the real world [11].

Dataset	MultiviewX					Wildtrack				
Method	MODA	MODP	P.	R.	F1.	MODA	MODP	P.	R.	F1.
RCNN [37]	18.7	46.4	63.5	43.9	51.9	11.3	18.4	68	43	52.7
POM-CNN [8]	-	-	-	-	-	23.2	30.5	75	55	63.5
DeepMCD [5]	70.0	73.0	85.7	83.3	84.5	67.8	64.2	85	82	83.5
DeepOcc. [3]	75.2	54.7	97.8	80.2	88.1	74.1	53.8	95	80	86.9
Volumetric [12]	84.2	80.3	97.5	86.4	91.6	88.6	73.8	95.3	93.2	94.2
MVDet [11]	83.9	79.6	96.8	86.7	91.5	88.2	75.7	94.7	93.6	94.1
SHOT [31]	88.3	82.0	96.6	91.5	94.0	90.2	76.5	96.1	94.0	95.0
MVDeTr [10]	93.7	**91.3**	**99.5**	94.2	97.8	91.5	**82.1**	**97.4**	94.0	95.7
3DROM [26]	95.0	84.9	99.0	96.1	97.5	**93.5**	75.9	97.2	96.2	**96.7**
E-MVOT (ours)	96.3	85.2	98.1	**98.1**	98.1	91.9	80.9	94.1	**98.0**	96.0
M-MVOT (ours)	**96.7**	86.1	98.8	97.9	**98.3**	92.1	81.3	94.5	97.8	96.1

4.2 Multi-view Crowd Localization Results

CVCS. In Table 1, we compare the proposed Mahalanobis distance multi-view optimal transport loss (M-MVOT) with other multi-view crowd localization methods and the Euclidean distance multi-view optimal transport loss (E-MVOT) on CVCS dataset. Table 1 shows that the proposed M-MVOT loss achieves better overall localization performance than existing multi-view crowd localization methods. Even though MVDeTr achieves the highest MODP, our proposed M-MVOT and E-MVOT are the best in terms of MODA and F1_score, which are significantly increased from previous methods due to M-MVOT and E-MVOT's higher recall rates. These results validate the advantages of our proposed M-MVOT loss, which considers the camera view variations in the task and can handle the localization ambiguities, as compared to Gaussian density map MSE loss. M-MVOT also outperforms E-MVOT, due to its ability to more flexi-

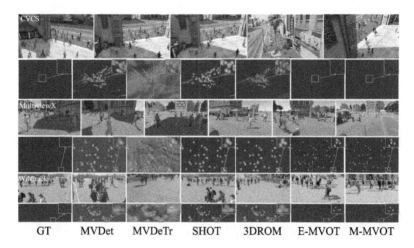

Fig. 4. The predicted crowd occupancy maps of different methods on the 3 datasets CVCS, MultiviewX, and Wildtrack (zoom in for better view). The value of crowd occupancy maps indicates the person probability of each location.

bly change the cost function based on the relationships between the camera view and ground-truth positions. Note that CVCS is a large multi-scene multi-view dataset, which is more challenging than other single-scene multi-view datasets.

Wildtrack and MultiviewX. In Table 2, we compare our M-MVOT/E-MVOT with other multi-view crowd localization methods on the MultiviewX (left) and Wildtrack datasets (right), respectively. On **MultiviewX**, the proposed M-MVOT and E-MVOT outperform the comparison methods, with the highest MODA and F1_score. M-MVOT also outperforms E-MVOT on the MultiviewX dataset (according to MODA and F1_score), which also demonstrates the advantage of the proposed M-MVOT over the common Euclidean distance optimal transport loss. On **Wildtrack**, 3DROM achieves the best results on the MODA and F1_score metrics, and M-MVOT and E-MVOT achieve the second-best results. The possible reason is that 3DROM is a data augmentation method that could deal with the overfitting issue on small datasets like Wildtrack. But M-MVOT is still better than E-MVOT, demonstrating the effectiveness of the view ray direction and camera distance guidance.

Visualization Results. We show the visualization result on the 3 datasets in Fig. 4. Generally, the proposed M-MVOT/E-MVOT shows better localization performance under crowded regions (in red boxes) and reduces the projection artifacts of comparison methods on faraway regions. Furthermore, as seen in the middle and the bottom of Fig. 4, M-MVOT predicts fewer false positives than E-MVOT on MultiviewX; M-MVOT also has fewer tail (streaking) artifacts than E-MVOT on Wildtrack, which proves the effectiveness of the proposed the view ray direction and distance guidance.

4.3 Ablation Study

Ablation Study on the MVOT Loss. We conducted an ablation study on the setting of the MVOT loss in Table 3 on the CVCS dataset, comparing the proposed M-MVOT loss with MSE, MVOT with Euclidean cost (E-MVOT), MVOT with Mahalanobis cost based on view ray guidance only (MV-MVOT), and camera distance only (ED-MVOT). From Table 3, M-MVOT achieves better performance than ED-MVOT, MV-MVOT, E-MVOT, and MSE loss, which demonstrates the advantage of the proposed distance guidance and view ray direction in the covariance matrix over other losses. Note that ED-MVOT got lower results, due to more penalties being added to faraway points in all directions, leading to fewer predictions (so it has higher precision). M-MVOT considers directions and adds relaxations to directions perpendicular to the view-ray, achieving better results. As shown in the red boxes of Fig. 5, M-MVOT predicts more accurate results than MV-MVOT and ED-MVOT, which demonstrates the effectiveness of using the view ray direction and object-to-camera distance in the covariance matrix of the MVOT loss.

Table 3. The ablation study on the MVOT loss for multi-view crowd localization on the CVCS dataset with other settings kept the same except for the training loss.

Loss	Dir.	Dis.	MODA	MODP	P	R	F1
MSE	✗	✗	14.2	59.3	85.0	17.3	28.7
E-MVOT	✗	✗	43.1	74.3	85.6	51.8	64.5
MV-MVOT	✓	✗	43.3	74.4	85.9	51.8	64.6
ED-MVOT	✗	✓	42.9	**75.0**	**87.7**	49.9	63.3
M-MVOT	✓	✓	**43.5**	74.1	85.5	**52.3**	**64.9**

Table 4. The ablation study on α and τ of M-MVOT loss on the MultiviewX dataset. The best results are achieved by $\alpha = 0.05$ and $\tau = 20$.

α	MODA	MODP	P.	R.	F1.	τ	MODA	MODP	P.	R.	F1.
0	96.3	85.2	98.1	**98.1**	98.1	1	95.2	**90.2**	99.1	96.1	97.6
0.05	**96.7**	**86.1**	98.8	97.9	**98.3**	10	96.1	86.5	**99.2**	96.9	98.0
0.2	**96.7**	84.3	98.6	**98.1**	**98.3**	20	**96.7**	86.1	98.8	**97.9**	**98.3**
0.6	96.3	84.5	98.2	**98.1**	98.1	30	95.2	83.8	97.9	97.3	97.6

Ablation Study on α and τ in M-MVOT. We conducted an ablation study on the α and τ of M-MVOT in Table 4 on the MultiviewX dataset. α adjusts the object-to-camera distance in the M-MVOT loss in (6). $\alpha = 0$ is equivalent to the

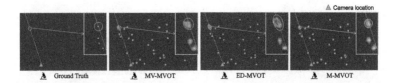

Fig. 5. Comparison of MV-MVOT, ED-MVOT, and M-MVOT. The blue triangle is the camera location. M-MVOT predicts fewer artifacts than MV-MVOT and ED-MVOT (see red boxes), demonstrating the effectiveness of distance adjustment and view-ray direction, respectively. (Color figure online)

Table 5. Comparisons with soft choices for fusing multi-cameras (Soft), a learnable covariance matrix (LCM) and multi-view fusion of single-view methods (Multi-UOT).

Comparison	MODA	MODP	P.	R.	F1.
Soft	96.5	86.0	98.7	97.8	98.2
LCM	95.1	81.5	96.7	**98.5**	97.6
Multi-UOT	23.9	72.0	75.2	35.7	48.4
M-MVOT	**96.7**	**86.1**	**98.8**	97.9	**98.3**

E-MVOT loss. When α increases, the effect of the distance change on the model is enlarged, which penalizes the faraway regions for more accurate localization performance. However, when α is too large, the penalties are too large, which may force the model to neglect the near regions. τ adjusts the effect of the last two regularization terms in (1) to ensure that all predicted density values have a corresponding annotation and all ground-truth points are used in the transport plan. When τ increases, MODA and Recall increase while MODP and Precision drop for detecting more points. But when τ is too large, the penalties of the last two terms are too large, harming the optimization of the main term. The best result is achieved at $\alpha = 0.05$, $\tau = 20$.

Ablation Study on the Multi-view Fusion Strategy. We conduct an ablation study on the multi-view fusion strategy, where soft weights learned from camera distance replace the binary choice strategy in Eq. (4). Table 5 shows that our proposed M-MVOT with binary multi-view fusion still outperforms the soft choices for fusing multi-cameras. The possible reason is that the binary camera choice forces the model to focus on the closest camera (the more reliable one) and reduces the complexity of the covariance matrix for model learning.

Comparison with Learnable Covariance Matrix (LCM). We compare our proposed M-MVOT, which uses manually designed covariance matrices, with a data-driven learnable covariance matrix (LCM) in Table 5, where the covariance matrices are estimated from the ground-plane features using two convolutional layers. Table 5 shows the results of using learnable covariance matrices (LCM), which is not as good as M-MVOT. The reason is that our M-MVOT considers the multi-views' unique domain knowledge in the covariance matrix and is more suitable for the multi-view task, although exploring adding priors to the learnable covariance matrix could be good future work.

Comparison with the Single-Image SOTA OT Method with a Simple Multi-view Fusion (Multi-UOT). We compare our proposed M-MVOT with the single-image SOTA OT method UOT [34] with a simple multi-view fusion. Specifically, we project and sum the 2D results of [34] on the ground and show results as Multi-UOT in Table 5. M-MVOT is much better than Multi-UOT and it demonstrates that single-image methods with simple multi-view fusion cannot fully utilize the multi-view information well, consistent with [11, 42].

5 Discussion and Conclusion

In this paper, we propose a novel Mahalanobis distance-based multi-view optimal transport (M-MVOT) loss, which is specially designed for multi-view crowd localization. As far as we know, this is the first study on point supervision for multi-view crowd localization, which advances the research on multi-view crowd localization tasks in the aspect of point supervision training losses. The proposed loss uses view-ray directed and object-to-camera distance-adjusted Mahalanobis distance in the cost calculation in order to counteract the distortions created during the projection step, and then it is applied to multi-camera views via calculating the cost under the closest camera. Overall, the experiments validate that the proposed M-MVOT achieves better localization performance, compared to density map-based mean squared error loss (MSE) and common Euclidean distance-based optimal transport loss. **Ethical considerations:** Multi-view crowd localization could be used for crowd analysis, autonomous driving, public traffic management, etc. For applications, we could encode the images as features, rely on synthetic datasets, or mask out human faces first instead of directly inputting images with human faces to the crowd localization models for higher privacy protection standards.

Acknowledgements. This work was supported in parts by NSFC (62202312, U21B2023, U2001206), Guangdong Basic and Applied Basic Research Foundation (2023B1515120026), DEGP Innovation Team (2022KCXTD025), City University of Hong Kong Strategic Research Grant (7005665), Shenzhen Science and Technology Program (KQTD 20210811090044003, RCJC20200714114435012), Guangdong Laboratory of Artificial Intelligence and Digital Economy (SZ), and Scientific Development Funds from Shenzhen University.

References

1. Abousamra, S., Hoai, M., Samaras, D., Chen, C.: Localization in the crowd with topological constraints. In: Proceedings of the AAAI Conference on Artificial Intelligence, vol. 35, pp. 872–881 (2021)
2. Bai, S., He, Z., Qiao, Y., Hu, H., Wu, W., Yan, J.: Adaptive dilated network with self-correction supervision for counting. In: Proceedings of the IEEE/CVF Conference on Computer Vision and Pattern Recognition, pp. 4594–4603 (2020)

3. Baqué, P., Fleuret, F., Fua, P.: Deep occlusion reasoning for multi-camera multi-target detection. In: Proceedings of the IEEE International Conference on Computer Vision, pp. 271–279 (2017)
4. Chavdarova, T., et al.: Wildtrack: a multi-camera HD dataset for dense unscripted pedestrian detection. In: Proceedings of the IEEE Conference on Computer Vision and Pattern Recognition, pp. 5030–5039 (2018)
5. Chavdarova, T., Fleuret, F.: Deep multi-camera people detection. In: 2017 16th IEEE International Conference on Machine Learning and Applications (ICMLA), pp. 848–853. IEEE (2017)
6. Cheng, Z.Q., Dai, Q., Li, H., Song, J., Wu, X., Hauptmann, A.G.: Rethinking spatial invariance of convolutional networks for object counting. In: Proceedings of the IEEE/CVF Conference on Computer Vision and Pattern Recognition, pp. 19638–19648 (2022)
7. Cheng, Z.Q., Li, J.X., Dai, Q., Wu, X., Hauptmann, A.G.: Learning spatial awareness to improve crowd counting. In: Proceedings of the IEEE/CVF International Conference on Computer Vision, pp. 6152–6161 (2019)
8. Fleuret, F., Berclaz, J., Lengagne, R., Fua, P.: Multicamera people tracking with a probabilistic occupancy map. IEEE Trans. Pattern Anal. Mach. Intell. **30**(2), 267–282 (2007)
9. Ge, W., Collins, R.T.: Crowd detection with a multiview sampler. In: ECCV, pp. 324–337 (2010)
10. Hou, Y., Zheng, L.: Multiview detection with shadow transformer (and view-coherent data augmentation). In: Proceedings of the 29th ACM International Conference on Multimedia, pp. 1673–1682 (2021)
11. Hou, Y., Zheng, L., Gould, S.: Multiview detection with feature perspective transformation. In: Vedaldi, A., Bischof, H., Brox, T., Frahm, J.-M. (eds.) ECCV 2020. LNCS, vol. 12352, pp. 1–18. Springer, Cham (2020). https://doi.org/10.1007/978-3-030-58571-6_1
12. Iskakov, K., Burkov, E., Lempitsky, V., Malkov, Y.: Learnable triangulation of human pose. In: ICCV (2019)
13. Jiang, X., et al.: Attention scaling for crowd counting. In: IEEE/CVF Conference on Computer Vision and Pattern Recognition (CVPR) (2020)
14. Kang, D., Chan, A.: Crowd counting by adaptively fusing predictions from an image pyramid. In: BMVC (2018)
15. Laradji, I.H., Rostamzadeh, N., Pinheiro, P.O., Vazquez, D., Schmidt, M.: Where are the blobs: counting by localization with point supervision. In: Proceedings of the European Conference on Computer Vision (ECCV), pp. 547–562 (2018)
16. Li, Y., Zhang, X., Chen, D.: Csrnet: dilated convolutional neural networks for understanding the highly congested scenes. In: CVPR, pp. 1091–1100 (2018)
17. Liang, D., Xu, W., Zhu, Y., Zhou, Y.: Focal inverse distance transform maps for crowd localization. IEEE Trans. Multimedia **25**, 6040–6052 (2022)
18. Lin, W., Chan, A.B.: Optimal transport minimization: crowd localization on density maps for semi-supervised counting. In: Proceedings of the IEEE/CVF Conference on Computer Vision and Pattern Recognition (CVPR), pp. 21663–21673 (2023)
19. Liu, C., Lu, H., Cao, Z., Liu, T.: Point-query quadtree for crowd counting, localization, and more. In: Proceedings of the IEEE/CVF International Conference on Computer Vision, pp. 1676–1685 (2023)
20. Liu, H., Zhao, Q., Ma, Y., Dai, F.: Bipartite matching for crowd counting with point supervision. In: IJCAI, pp. 860–866 (2021)

21. Liu, W., Salzmann, M., Fua, P.: Context-aware crowd counting. In: CVPR, pp. 5099–5108 (2019)
22. Ma, Z., Wei, X., Hong, X., Gong, Y.: Bayesian loss for crowd count estimation with point supervision. In: Proceedings of the IEEE/CVF International Conference on Computer Vision, pp. 6142–6151 (2019)
23. Ma, Z., Wei, X., Hong, X., Lin, H., Qiu, Y., Gong, Y.: Learning to count via unbalanced optimal transport. In: Proceedings of the AAAI Conference on Artificial Intelligence, vol. 35, pp. 2319–2327 (2021)
24. Oñoro-Rubio, D., López-Sastre, R.J.: Towards perspective-free object counting with deep learning. In: Leibe, B., Matas, J., Sebe, N., Welling, M. (eds.) ECCV 2016. LNCS, vol. 9911, pp. 615–629. Springer, Cham (2016). https://doi.org/10.1007/978-3-319-46478-7_38
25. Peyré, G., Cuturi, M., et al.: Computational optimal transport. Center for Research in Economics and Statistics Working Papers (2017-86) (2017)
26. Qiu, R., Xu, M., Yan, Y., Smith, J.S., Yang, X.: 3D random occlusion and multi-layer projection for deep multi-camera pedestrian localization. In: Avidan, S., Brostow, G., Cissé, M., Farinella, G.M., Hassner, T. (eds.) ECCV 2022, Part X. LNCS, vol. 13670, pp. 695–710. Springer, Cham (2022). https://doi.org/10.1007/978-3-031-20080-9_40
27. Sabzmeydani, P., Mori, G.: Detecting pedestrians by learning shapelet features. In: CVPR, pp. 1–8. IEEE (2007)
28. Sam, D.B., Surya, S., Babu, R.V.: Switching convolutional neural network for crowd counting. In: CVPR, pp. 4031–4039 (2017)
29. Shi, M., Yang, Z., et al.: Revisiting perspective information for efficient crowd counting. In: CVPR, pp. 7279–7288 (2019)
30. Sindagi, V.A., Patel, V.M.: Generating high-quality crowd density maps using contextual pyramid CNNs. In: ICCV, pp. 1879–1888 (2017)
31. Song, L., Wu, J., Yang, M., Zhang, Q., Li, Y., Yuan, J.: Stacked homography transformations for multi-view pedestrian detection. In: Proceedings of the IEEE/CVF International Conference on Computer Vision, pp. 6049–6057 (2021)
32. Song, Q., et al.: Rethinking counting and localization in crowds: a purely point-based framework. arXiv preprint arXiv:2107.12746 (2021)
33. Viola, P., Jones, M.J.: Robust real-time face detection. Int. J. Comput. Vision **57**(2), 137–154 (2004)
34. Wan, J., Liu, Z., Chan, A.B.: A generalized loss function for crowd counting and localization. In: Proceedings of the IEEE/CVF Conference on Computer Vision and Pattern Recognition, pp. 1974–1983 (2021)
35. Wang, B., Liu, H., Samaras, D., Nguyen, M.H.: Distribution matching for crowd counting. In: Advances in Neural Information Processing Systems, vol. 33, pp. 1595–1607 (2020)
36. Wu, B., Nevatia, R.: Detection and tracking of multiple, partially occluded humans by Bayesian combination of edgelet based part detectors. Int. J. Comput. Vision **75**(2), 247–266 (2007)
37. Xu, Y., Liu, X., Liu, Y., Zhu, S.C.: Multi-view people tracking via hierarchical trajectory composition. In: CVPR, pp. 4256–4265 (2016)
38. Yan, S., et al.: Multiview transformers for video recognition. In: Proceedings of the IEEE/CVF Conference on Computer Vision and Pattern Recognition, pp. 3333–3343 (2022)
39. Yan, Z., et al.: Perspective-guided convolution networks for crowd counting. In: Proceedings of the IEEE/CVF International Conference on Computer Vision, pp. 952–961 (2019)

40. Yang, Y., Li, G., Wu, Z., Su, L., Huang, Q., Sebe, N.: Reverse perspective network for perspective-aware object counting. In: Proceedings of the IEEE/CVF Conference on Computer Vision and Pattern Recognition, pp. 4374–4383 (2020)
41. Zhang, C., Li, H., et al.: Cross-scene crowd counting via deep convolutional neural networks. In: CVPR, pp. 833–841 (2015)
42. Zhang, Q., Chan, A.B.: Wide-area crowd counting via ground-plane density maps and multi-view fusion CNNs. In: CVPR, pp. 8297–8306 (2019)
43. Zhang, Q., Lin, W., Chan, A.B.: Cross-view cross-scene multi-view crowd counting. In: Proceedings of the IEEE/CVF Conference on Computer Vision and Pattern Recognition, pp. 557–567 (2021)
44. Zhu, X., Su, W., Lu, L., Li, B., Wang, X., Dai, J.: Deformable detr: deformable transformers for end-to-end object detection. arXiv preprint arXiv:2010.04159 (2020)

RAW-Adapter: Adapting Pre-trained Visual Model to Camera RAW Images

Ziteng Cui[1]($\boxtimes$) and Tatsuya Harada[1,2]

[1] The University of Tokyo, Bunkyo, Japan
{cui,harada}@mi.t.u-tokyo.ac.jp
[2] RIKEN AIP, Chuo-ku, Japan

Abstract. sRGB images are now the predominant choice for pre-training visual models in computer vision research, owing to their ease of acquisition and efficient storage. Meanwhile, the advantage of RAW images lies in their rich physical information under variable real-world challenging lighting conditions. For computer vision tasks directly based on camera RAW data, most existing studies adopt methods of integrating image signal processor (ISP) with backend networks, yet often overlook the interaction capabilities between the ISP stages and subsequent networks. Drawing inspiration from ongoing adapter research in NLP and CV areas, we introduce **RAW-Adapter**, a novel approach aimed at adapting sRGB pre-trained models to camera RAW data. RAW-Adapter comprises input-level adapters that employ learnable ISP stages to adjust RAW inputs, as well as model-level adapters to build connections between ISP stages and subsequent high-level networks. Additionally, RAW-Adapter is a general framework that could be used in various computer vision frameworks. Abundant experiments under different lighting conditions have shown our algorithm's state-of-the-art (SOTA) performance, demonstrating its effectiveness and efficiency across a range of real-world and synthetic datasets. Code is available at this url.

1 Introduction

In recent years, there has been a growing interest in revisiting vision tasks using unprocessed camera RAW images. How to leverage information-rich RAW image in computer vision tasks has become a current research hotspot in various sub-areas (*i.e.* denoising [2,76], view synthesis [46], object detection [27,48]). Compared to commonly used sRGB image, RAW image directly acquired by the camera sensor, encompasses abundant information unaffected or compressed by image signal processor (ISP), also offers physically meaningful information like noise distributions [47,68], owing to its linear correlation between image intensity and the radiant energy received by a camera. The acquisition of RAW data, facilitating enhanced detail capture and a higher dynamic range, imparts a unique advantage in addressing visual tasks under variable lighting conditions in the

Supplementary Information The online version contains supplementary material available at https://doi.org/10.1007/978-3-031-73235-5_3.

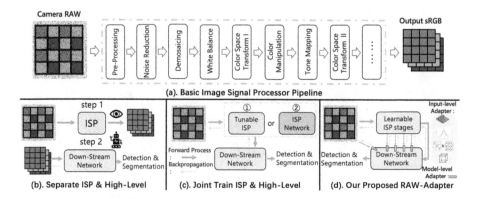

Fig. 1. (a). An overview of basic image signal processor (ISP) pipeline. (b). ISP and current visual model have different objectives. (c) Previous methods optimize ISP with down-stream visual model. (d) Our proposed RAW-Adapter.

real world. For instance, sunlight irradiance can reach as high as $1.3 \times 10^3 W/m^2$, while bright planet irradiance can be as low as $2.0 \times 10^{-6} W/m^2$ [32].

Meanwhile, sRGB has emerged as the primary choice for pre-training visual models in today's computer vision field, due to its scalability and ease of storage. Typically, sRGB images are derived from camera RAW data through the ISP pipeline. As shown in Fig. 1(a), the entire ISP pipeline consists of multiple modules to convert RAW images into vision-oriented sRGB images, each of these modules serves its own distinct purpose, with the majority being ill-posed and heavily dependent on prior information [5,16,26,35,42,56].

When adopt RAW data for computer vision tasks, the purpose of a manually designed ISP is to produce images that offer a superior visual experience [18,69], rather than optimizing for downstream visual tasks such as object detection or semantic segmentation (see Fig. 1(b)). Additionally, most companies' ISPs are black boxes, making it difficult to obtain information about the specific steps inside. Consequently, utilizing human vision-oriented ISP in certain conditions is sometimes even less satisfactory than directly using RAW [22,27,40,81].

To better take advantage of camera RAW data for various computer vision tasks. Researchers began to optimize the image signal processor (ISP) jointly with downstream networks. Since most ISP stages are non-differentiable and cannot be jointly backpropagation, there are two main lines of approaches to connect ISP stages with downstream networks (see Fig. 1(c)): ①. First-kind approaches maintain the modular design of the traditional ISP, involving the design of differentiable ISP modules through optimization algorithms [48,69,75], such as Hardware-in-the-loop [48] adopt covariance matrix adaptation evolution strategy (CMA-ES) [23]. ②. Second-kind approaches replaced the ISP part entirely with a neural network [18,55,72,73], such as Dirty-Pixel [18] adopt a stack of residual UNets [58] as pre-encoder, however, the additional neural network introduces a significant computational burden, especially when dealing with noise under various noise conditions, $\mathbb{P}_\mathbb{K}$ will predict the appropriate Gaussian kernel k for denoising to improve downstream visual tasks' performance, the predicted key parameters are the Gaussian kernel's major axis r_1, minor axis r_2 (see Fig. 4 left), here we set the kernel angle θ to 0 for simplification. Gaussian kernel k at pixel (x, y) would be:

$$k(x,y) = exp(-(b_0 x^2 + 2b_1 xy + b_2 y^2)), \tag{2}$$

where:

$$b_0 = \frac{cos(\theta)^2}{2r_1^2} + \frac{sin(\theta)^2}{2r_2^2}, b_1 = \frac{sin(2\theta)}{4r_1^2}((\frac{r_1}{r_2})^2 - 1), b_2 = \frac{sin(\theta)^2}{2r_1^2} + \frac{cos(\theta)^2}{2r_2^2}. \tag{3}$$

After gain ratio g and kernel k process on image $\mathbf{I}_1$, $\mathbb{P}_\mathbb{K}$ also predict a filter parameter σ (initial at 0) to keep the sharpness and recover details of generated image $\mathbf{I}_2$. Equation 4 shows the translation from $\mathbf{I}_1$ to $\mathbf{I}_2$, where ⊛ denotes kernel convolution and filter parameter σ is limited in a range of (0, 1) by a Sigmoid activation. For more details please refer to our supplementary part Sec. C.

$$\begin{aligned} \mathbb{P}_\mathbb{K}(\mathbf{I}_1, \mathtt{q}) &\to k\{r_1, r_2, \theta\}, g, \sigma, \\ \mathbf{I}_2' &= (g \cdot \mathbf{I}_1) \circledast k, \\ \mathbf{I}_2 &= \mathbf{I}_2' + (g \cdot \mathbf{I}_1 - \mathbf{I}_2') \cdot \sigma. \end{aligned} \tag{4}$$

White Balance & CCM Matrix: White balance (WB) mimics the color constancy of the human vision system (HVS) by aligning "white" color with the white object, resulting in the captured image reflecting the combination of light color and material reflectance [1,2,14]. In our work, we hope to find an adaptive white balance for different images under various lighting scenarios. Motivated by the design of Shades of Gray (SoG) [20] WB algorithm, where gray-world WB and Max-RGB WB can be regarded as a subcase, we've employed a learnable parameter ρ to replace the gray-world's L-1 average with an adaptive Minkowski distance average (see Eq. 5), the QAL block $\mathbb{P}_\mathbb{M}$ predicts Minkowski distance's hyper-parameter ρ, a demonstration of various ρ is shown in Fig. 4. After finding the suitable ρ, $\mathbf{I}_2$ would multiply to white balance matrix to generate $\mathbf{I}_3$.

Color Conversion Matrix (CCM) within the ISP is constrained by the specific camera model. Here we standardize the CCM as a single learnable 3 × 3 matrix, initialized as a unit diagonal matrix $\mathbf{E}_3$. The QAL block $\mathbb{P}_\mathbb{M}$ predicts 9 parameters here, which are then added to $\mathbf{E}_3$ to form the final 3 × 3 matrix $\mathbf{E}_{ccm}$. Then $\mathbf{I}_3$ would multiply to CCM $\mathbf{E}_{ccm}$ to generate $\mathbf{I}_4$, equation as follow:

$$\begin{aligned} \mathbb{P}_\mathbb{M}(\mathbf{I}_2, \mathtt{q}) &\to \rho, \mathbf{E}_{ccm}, \\ m_{i\in(r,g,b)} &= \sqrt[\rho]{avg(\mathbf{I}_2(i)^\rho)} / \sqrt[\rho]{avg((\mathbf{I}_2)^\rho)}, \\ \mathbf{I}_3 = \mathbf{I}_2 * \begin{bmatrix} m_r & & \\ & m_g & \\ & & m_b \end{bmatrix}&, \quad \mathbf{I}_4 = \mathbf{I}_3 * \mathbf{E}_{ccm}. \end{aligned} \tag{5}$$

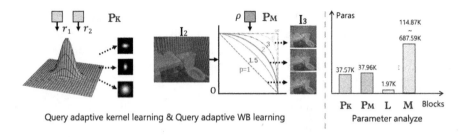

Fig. 4. Left, we use query adaptive learning (QAL) to predict key parameters is ISP process. Right, we show RAW-Adapter different blocks' parameter.

Color Manipulation Process: In ISP, color manipulation process is commonly achieved through lookup table (LUT), such as 1D and 3D LUTs [16]. Here, we consolidate color manipulation operations into a single 3D LUT, adjusting the color of image $\mathbf{I}_4$ to produce output image $\mathbf{I}_5$. Leveraging advancements in LUT techniques, we choose the latest neural implicit 3D LUT (NILUT) [11], for its speed efficiency and ability to facilitate end-to-end learning in our pipeline[1]. We denote NILUT [11] as $\mathbb{L}$, $\mathbb{L}$ maps the input pixel intensities R, G, B to a continuous coordinate space, followed by the utilization of implicit neural representation (INR) [61], which involves using a multi-layer perceptron (MLP) network to map the output pixel intensities to R', G', B':

$$\mathbf{I}_5(R', G', B') = \mathbb{L}(\mathbf{I}_4(R, G, B)). \tag{6}$$

Image $\mathbf{I}_5$ obtained through image-level adapters will be forwarded to the downstream network's backbone. Furthermore, $\mathbf{I}_5$'s features obtained in backbone will be fused with model-level adapters, which we will discuss in detail.

3.2 Model-Level Adapters

Input-level adapters guarantee the production of machine vision-oriented image $\mathbf{I}_5$ for high-level models. However, information at the ISP stages ($\mathbf{I}_1 \sim \mathbf{I}_4$) is almost overlooked. Inspired by adapter design in current NLP and CV area [8,28,33], we employ the prior information from the ISP stage as model-level adapters to guide subsequent models' perception. Additionally, model-level adapters promote a tighter integration between downstream tasks and the ISP stage. Dotted lines in Fig. 3(a) represent model-level adapters.

Model-level adapters $\mathbb{M}$ integrate the information from ISP stages to the network backbone. As shown in Fig. 3(a), we denote the stage $1 \sim 4$ as different stages in network backbone, image $\mathbf{I}_5$ would pass by backbone stages $1 \sim 4$ and then followed by detection or segmentation head. We utilize convolution layer c_i to extract features from $\mathbf{I}_1$ through $\mathbf{I}_4$, these extracted features are concatenated as $\mathbb{C}(\mathbf{I}_{1\sim4}) = \mathbb{C}(c_1(\mathbf{I}_1), c_2(\mathbf{I}_2), c_3(\mathbf{I}_3), c_4(\mathbf{I}_4))$. Subsequently, $\mathbb{C}(\mathbf{I}_{1\sim4})$ go through

---

[1] To save memory occupy, we change the channel number 128 in [11] to 32.

two residual blocks [25] and generate adapter f_1, f_1 would merge with stage 1's feature by a merge block, detail structure of merge block is shown in Fig. 3(c), which roughly includes concatenate process and a residual connect, finally merge block would output network feature for stage 2 and an additional adapter f_2. Then process would repeat in stage 2 and stage 3 of network backbone. We collectively refer to all structures related to model-level adapters as $\mathbb{M}$.

We show RAW-Adapter's different part parameter number in Fig. 4 right, including input-level adapters {$\mathbb{P}_\mathbb{K}$ (37.57K), $\mathbb{P}_\mathbb{M}$ (37.96K), $\mathbb{L}$ (1.97K)} and model level adapters $\mathbb{M}$ (114.87K $\sim$ 687.59K), model level adapters' parameter number depend on following network. Total parameter number of RAW-Adapter is around 0.2M to 0.8M, much smaller than following backbones ($\sim$25.6M in ResNet-50 [25], $\sim$197M in Swin-L [45]), also smaller than previous SOTA methods like SID [6] (11.99M) and Dirty-Pixel [18] (4.28M). Additionally, subsequent experiments will demonstrate that the performance of RAW-Adapter on backbone networks with fewer parameters is even superior to that of previous algorithms on backbone networks with higher parameters.

4 Experiments

4.1 Dataset and Experimental Setting

Dataset. We conducted experiments on object detection and semantic segmentation tasks, utilizing a combination of various synthetic and real-world RAW image datasets, an overview of the datasets can be found in Table 1.

For the object detection task, we take 2 open source real-world dataset PASCAL RAW [52] and LOD [27]. PASCAL RAW [52] is a normal-light condition dataset with 4259 RAW images, taken by a Nikon D3200 DSLR camera with 3 object classes. To verify the generalization capability of RAW-Adapter across various lighting conditions, we additionally synthesized low-light and overexposure datasets based on the PASCAL RAW dataset, named PASCAL RAW (dark) and PASCAL RAW (over-exp) respectively, the synthesized datasets are identical to the PASCAL RAW dataset in all aspects except for brightness levels. For PASCAL RAW (dark) and PASCAL RAW (over-exp) synthesis, the light intensity and environment irradiance exhibit a linear relationship on RAW images, thus we employed the synthesis method from the previous work [14,78]:

$$x_n \sim N(\mu = lx, \sigma^2 = \delta_r^2 + \delta_s lx)$$
$$y = lx + x_n, \tag{7}$$

where x denotes the original normal-light RAW image and y denotes degraded RAW image, $\delta_s = \sqrt{S}$ denotes shot noise while $\sqrt{S}$ is signal of the sensor, δ_r denotes read noise, l denotes the light intensity parameter which randomly chosen from [0.05, 0.4] in PASCAL RAW (dark), and randomly chosen from [1.5, 2.5] in PASCAL RAW (over-exp). Additionally, we follow PASCAL RAW [52]'s dataset split to separate the training set and test set.

Table 1. Dataset and framework setting in our experiments.

	PASCAL RAW	PASCAL RAW (dark/over-exp)	LOD	ADE20K RAW	ADE20K RAW (dark/over-exp)
Task	Object Detection			Semantic Segmentation	
Number	4259		2230	27,574	
Type	real-world	synthesis	real-world	synthesis	
Sensor	Nikon D3200 DSLR		Canon EOS 5D Mark IV	-	
Framework	RetinaNet [43] & Sparse-RCNN [64]			Segformer [70]	
Backbone	ResNet [25]			MIT [70]	
pre-train	ImageNet [44] pre-train weights				

Meanwhile, LOD [27] is a real-world dataset with 2230 low-light condition RAW images, taken by a Canon EOS 5D Mark IV camera with 8 object classes, we take 1800 images as training set and the other 430 images as test set.

For the semantic segmentation task, we utilized the widely-used sRGB dataset ADE20K [80] to generate RAW dataset. Leveraging state-of-the-art unprocessing methods InvISP [71], we synthesized RAW images corresponding to the ADE20K sRGB dataset. Using InvISP, we projected input sRGB images into RAW format, effectively creating an ADE20K RAW dataset. To simulate various lighting conditions, we employed the same synthesis method from PASCAL RAW (dark/over-exp) to generate low-light and over-exposure RAW images, name as ADE20K RAW (dark/over-exp), training & test split is same as ADE20K.

Implement Details. We build our framework based on the open-source computer vision toolbox `mmdetection` [7] and `mmsegmentation` [12], both object detection tasks and semantic segmentation tasks are initialed with ImageNet pre-train weights (see Table 1), and we apply the data augmentation pipeline in the default setting, mainly include random crop, random flip, and multi-scale test, etc. For the object detection task, we adopt the 2 mainstream object detectors: RetinaNet [43] and Sparse-RCNN [64] with ResNet [25] backbone. For the semantic segmentation task, we choose to use the mainstream segmentation framework Segformer [70] with MIT [70] backbone[2].

Comparison Methods. We conducted comparative experiments with the current state-of-the-art (SOTA) algorithms, including various open-sourced ISP methods [6,34,35,71,79] and joint-training method DirtyPixels [18], among these Karaimer et al. [35] is a traditional ISP method with various human manipulate steps, and InvISP [71] & Lite-ISP [79] & SID [6] & DNF [34] are current SOTA network-based ISP methods, where SID [6] and DNF [34] is especially for the low-light condition RAW data. For fairness compared with the above ISP methods, both the training and test RAW data are rendered using the respective compared ISP. DirtyPixels [18] is the current SOTA joint-training method which uses a stack of residual UNet [58] as a low-level processor and joint optimization

---

[2] For more experiments on other frameworks, please refer to supplementary materials.

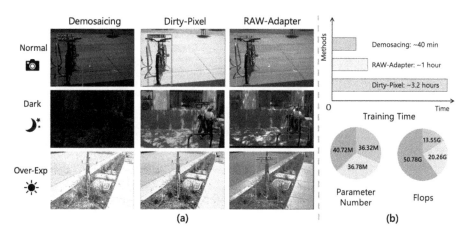

Fig. 5. (a). Detection performance on **PASCAL RAW** [52] (normal/dark/over-exp). (b). Efficiency comparison (blue: vanilla, yellow: RAW-Adapter, gray: Dirty-Pixel [18]).

with the following task-specific networks. For fairness, all comparison methods adopt the same data augmentation process and the same training setting, we will introduce detailed experiment results in the following section.

4.2 Object Detection Evaluation

For object detection task on PASCAL RAW [52] dataset, we adopt RetinaNet [43] with different size ResNet [25] backbones (ResNet-18, ResNet-50), all the models are trained on a single NVIDIA Tesla V100 GPU with SGD optimizer, the batch size is set to 4, training images are cropped into range of (400, 667) and training epochs are set to 50. Table 2 shows the detection results with (a). ResNet-18 and (b). ResNet-50 backbone, with comparisons of demosaiced RAW data ("Demosacing"), camera default ISP in [52], various current SOTA ISP solutions [6,34,35,79] and Dirty-Pixel [18], we can observe that sometimes ISP algorithms in normal and over-exposure conditions may even degraded the detection performance, and ISP in dark scenarios sometimes could improves detection performance, additionally manual design ISP solution [35] even out-perform the deep learning solutions. Previous joint-training method Dirty-Pixel [18] can improve detection performance. Overall, our RAW-Adapter method achieves the best performance and even outperforms some ISP algorithms utilizing ResNet-50 backbone when employing ResNet-18 backbone, which can significantly reduce computational load and model parameters.

Visualization of PASCAL RAW's detection results are shown in Fig. 5(a), we show the detection results under different lightness conditions, background are images generated from Dirty-Pixel [18] and RAW-Adapter (I_5). Our method could achieve satisfactory detection results across different lightness, while other methods face false alarm and miss detect. Efficiency comparison with Dirty-

Table 2. Comparison results on **PASCAL RAW** dataset [52], we take [43] with (a). ResNet-18 and (b). ResNet-50 backbone, **bold** denotes the best result.

(a) ResNet-18.

Method \ mAP	normal	over-exp	dark
Default ISP	88.3	-	-
Demosacing	87.7	87.7	80.3
Karaimer et al. [35]	86.0	85.6	81.9
Lite-ISP [79]	85.2	84.2	71.9
InvISP [71]	84.1	86.6	70.9
SID [6]	-	-	78.2
DNF [34]	-	-	81.1
Dirty-Pixel	88.6	88.0	80.8
RAW-Adapter	**88.7**	**88.7**	**82.5**

(b) ResNet-50.

Method \ mAP	normal	over-exp	dark
Default ISP	89.4	-	-
Demosacing	89.2	88.8	82.6
Karaimer et al. [35]	89.4	86.8	84.6
Lite-ISP [79]	89.3	85.1	73.5
InvISP [71]	89.1	87.3	74.7
SID [6]	-	-	81.5
DNF [34]	-	-	82.8
Dirty-Pixel	89.0	89.0	83.6
RAW-Adapter	**89.6**	**89.4**	**86.6**

Table 3. Comparison with ISP methods and Dirty-Pixel on **LOD** dataset [27]. We show the detection performance (mAP) ↑ of RetinaNet (R-Net) [43] and Sparse-RCNN (Sp-RCNN) [64], **bold** denotes the best result.

Methods	Demosaicing	Karaimer et al. [35]	InvISP [71]	SID [6]
mAP(R-Net)	58.5	54.4	56.9	49.1
mAP(Sp-RCNN)	57.7	52.2	49.4	43.1

Methods	DNF [34]	Dirty-Pixel [18]	RAW-Adapter (w/o $\mathbb{M}$)	RAW-Adapter
mAP(R-Net)	58.8	61.6	61.6	**62.2**
mAP(Sp-RCNN)	55.8	58.8	58.6	**59.2**

Pixel is shown in Fig. 5(b), our algorithm has achieved significant improvements in both training time acceleration and savings in Flops/Parameter Number.

Detection performance on LOD [27] is shown in Table 3, we adopt two detector RetinaNet [43] and Sparse-RCNN [64], both with ResNet-50 [25] backbone, training epochs are set to 35, Sparse-RCNN [64] are trained by 4 GPUs with Adam optimizer. With comparing of various ISP methods [6,34,35,79], Dirty-Pixel [18] and RAW-Adapter with only input-level adapters (w/o $\mathbb{M}$). Table 3 show that our algorithm can achieve the best performance on both two detectors.

4.3 Semantic Segmentation Evaluation

For semantic segmentation on ADE20K RAW dataset, we choose Segformer [70] as the segmentation framework with the different size MIT [70] backbones (MIT-B5, MIT-B3, MIT-B0), all the models are trained on 4 NVIDIA Tesla V100 GPU with Adam optimizer, the batch size is set to 4, training images are cropped

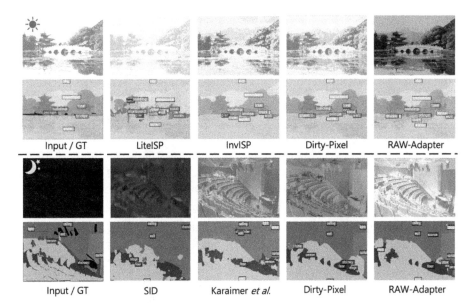

Fig. 6. Semantic segmentation results on over-exposure and low-light RAW data, compare with various ISP methods [6,35,71,79] and Dirty-Pixel [18].

into 512 × 512 and training iters are set to 80000. We make comparison with single demosacing, various ISP methods [6,34,35,71,79] and Dirty-Pixel [18], and RAW-Adapter with only input-level adapters (w/o $\mathbb{M}$).

Comparison results are shown in Table 4, where we compare both efficiency (parameters↓, inference time↓) and performance (mIOU↑), inference time is calculated on a single Tesla V100 GPU. We first show the results on vanilla demosaiced RAW data, then we show the results on RAW data processed by various ISP methods [6,34,35,71,79], followed by results of Dirty-Pixel [18], and our method without and with model-level adapter $\mathbb{M}$. For Table 4, we can find that directly employing ISP methods sometimes fails to enhance the performance of downstream segmentation tasks, particularly in low-light conditions. Joint-training method Dirty-Pixel [18] could improve down-stream performance and sometimes even degrade segmentation performance, but increase more network parameters. Overall, our proposed RAW-Adapter could effectively enhance performance while adding a slight parameter count and improving inference efficiency. RAW-Adapter on backbones with fewer parameters can even outperform previous methods on backbones with more parameters (i.e. The RAW-Adapter, employing the MIT-B3 backbone, achieves a dark condition mIOU of 40.52, surpassing other methods utilizing the MIT-B5 backbone despite having fewer than ∼ 36M parameters.).

The visualization of segmentation results are shown in Fig. 6, where row 1 and row 3 depict images rendered by different ISPs [6,35,71,79] and images generated from Dirty-Pixel [18] and RAW-Adapter ($\mathbf{I}_5$), row 2 and row 4 denote

Table 4. Comparison with previous methods on **ADE 20K RAW** (normal/over-exp/dark). **Bold** denotes the best result while underline denotes second best result.

	backbone	params(M) ↓	inference time(s) ↓	mIOU ↑ (normal)	mIOU ↑ (over-exp)	mIOU ↑ (dark)
Demosacing			0.105	45.26	42.08	30.97
Karaimer et al. [35]			0.525	46.42	43.71	39.79
InvISP [71]	MIT-B5	82.01	0.203	47.53	43.55	4.87
LiteISP [79]			0.261	43.22	42.01	5.52
DNF [34]			0.186	-	-	25.82
SID [6]			0.312	-	-	26.44
Dirty-Pixel [18]		86.29	0.159	47.76	<u>46.80</u>	40.02
	MIT-B3	48.92	0.098	47.19	45.33	38.93
	MIT-B0	8.00	<u>0.049</u>	35.43	32.10	27.19
RAW-Adapter (w/o M)	MIT-B5	82.09	0.148	<u>48.33</u>	46.48	<u>40.91</u>
	MIT-B3	44.72	0.086	47.01	45.00	39.34
	MIT-B0	**3.80**	**0.032**	35.09	32.82	27.29
	MIT-B5	82.53	0.167	**48.38**	**47.06**	**41.82**
RAW-Adapter	MIT-B3	45.16	0.102	47.59	45.66	40.52
	MIT-B0	<u>4.26</u>	0.053	35.67	33.44	27.78

our segmentation results appear superior compared to other methods. We were even pleased to discover that, although we do not add any human-vision-oriented loss function to constrain the model, RAW-Adapter still produces visually satisfactory image I_5 (see Fig. 6). This may be attributed to our preservation of traditional ISP stages in input-level adapters. Due to page limitations, we refer to more experimental results, visualization results, and ablation analysis of our method in the supplementary part.

5 Conclusion

In this paper, we introduce RAW-Adapter, an effective solution for adapting pre-trained sRGB models to camera RAW data. With input-level adapters and model-level adapters working in tandem, RAW-Adapter effectively forwards RAW images from various lighting conditions to downstream vision tasks, our method has achieved state-of-the-art (SOTA) results across multiple datasets.

For future research directions, we believe it is feasible to design a unified model capable of adapting to RAW visual tasks under different lighting conditions, without the need for retraining in each lighting scenario like RAW-Adapter. Additionally, designing multi-task decoders can enable the accommodation of various tasks, facilitating more effective integration of different visual tasks on RAW images with large-scale models.

Acknowledgements. This research is partially supported by JST Moonshot R&D Grant Number JPMJPS2011, CREST Grant Number JPMJCR2015 and Basic Research Grant (Super AI) of Institute for AI and Beyond of the University of Tokyo.

References

1. Afifi, M., Brown, M.S.: What else can fool deep learning? Addressing color constancy errors on deep neural network performance. In: 2019 IEEE/CVF International Conference on Computer Vision (ICCV), pp. 243–252 (2019). https://api.semanticscholar.org/CorpusID:207995696
2. Brooks, T., Mildenhall, B., Xue, T., Chen, J., Sharlet, D., Barron, J.T.: Unprocessing images for learned raw denoising. In: Proceedings of the IEEE/CVF Conference on Computer Vision and Pattern Recognition, pp. 11036–11045 (2019)
3. Buckler, M., Jayasuriya, S., Sampson, A.: Reconfiguring the imaging pipeline for computer vision. In: Proceedings of the IEEE International Conference on Computer Vision, pp. 975–984 (2017)
4. Carion, N., Massa, F., Synnaeve, G., Usunier, N., Kirillov, A., Zagoruyko, S.: End-to-end object detection with transformers. In: Vedaldi, A., Bischof, H., Brox, T., Frahm, J.-M. (eds.) ECCV 2020. LNCS, vol. 12346, pp. 213–229. Springer, Cham (2020). https://doi.org/10.1007/978-3-030-58452-8_13
5. Chatterjee, P., Joshi, N., Kang, S.B., Matsushita, Y.: Noise suppression in low-light images through joint denoising and demosaicing. In: CVPR 2011, pp. 321–328. IEEE (2011)
6. Chen, C., Chen, Q., Xu, J., Koltun, V.: Learning to see in the dark. In: Proceedings of the IEEE Conference on Computer Vision and Pattern Recognition, pp. 3291–3300 (2018)
7. Chen, K., Wang, J., et al.: Mmdetection: open mmlab detection toolbox and benchmark (2019)
8. Chen, Z., et al.: Vision transformer adapter for dense predictions. In: The Eleventh International Conference on Learning Representations (2023). https://openreview.net/forum?id=plKu2GByCNW
9. Cheng, B., Schwing, A.G., Kirillov, A.: Per-pixel classification is not all you need for semantic segmentation. In: NeurIPS (2021)
10. Conde, M.V., McDonagh, S., Maggioni, M., Leonardis, A., Pérez-Pellitero, E.: Model-based image signal processors via learnable dictionaries. In: Proceedings of the AAAI Conference on Artificial Intelligence, vol. 36, pp. 481–489 (2022)
11. Conde, M.V., Vazquez-Corral, J., Brown, M.S., Timofte, R.: Nilut: conditional neural implicit 3D lookup tables for image enhancement. In: Proceedings of the AAAI Conference on Artificial Intelligence (2024)
12. Contributors, M.: MMSegmentation: Openmmlab semantic segmentation toolbox and benchmark (2020). https://github.com/open-mmlab/mmsegmentation
13. Cui, Z., et al.: You only need 90k parameters to adapt light: a light weight transformer for image enhancement and exposure correction. In: BMVC, p. 238 (2022)
14. Cui, Z., Qi, G.J., Gu, L., You, S., Zhang, Z., Harada, T.: Multitask AET with orthogonal tangent regularity for dark object detection. In: Proceedings of the IEEE/CVF International Conference on Computer Vision (ICCV), pp. 2553–2562 (2021)

15. Cui, Z., et al.: Exploring resolution and degradation clues as self-supervised signal for low quality object detection. In: Avidan, S., Brostow, G., Cissé, M., Farinella, G.M., Hassner, T. (eds.) ECCV 2022. LNCS, vol. 13669, pp. 473–491. Springer, Cham (2022). https://doi.org/10.1007/978-3-031-20077-9_28
16. Delbracio, M., Kelly, D., Brown, M.S., Milanfar, P.: Mobile computational photography: a tour. Annu. Rev. Vis. Sci. **7**(1), 571–604 (2021)
17. Deng, J., Dong, W., Socher, R., Li, L.J., Li, K., Fei-Fei, L.: ImageNet: a large-scale hierarchical image database. In: CVPR 2009 (2009)
18. Diamond, S., Sitzmann, V., Julca-Aguilar, F., Boyd, S., Wetzstein, G., Heide, F.: Dirty pixels: towards end-to-end image processing and perception. ACM Trans. Graph. (SIGGRAPH) **40**(3), 1–15 (2021)
19. Dong, X., Yokoya, N.: Understanding dark scenes by contrasting multi-modal observations. In: Proceedings of the IEEE/CVF Winter Conference on Applications of Computer Vision (WACV), pp. 840–850 (2024)
20. Finlayson, G.D., Trezzi, E.: Shades of gray and colour constancy. In: Color and Imaging Conference, vol. 2004, pp. 37–41. Society for Imaging Science and Technology (2004)
21. Goyal, B., Lalonde, J.F., Li, Y., Gupta, M.: Robust scene inference under noise-blur dual corruptions. In: 2022 IEEE International Conference on Computational Photography (ICCP), pp. 1–12 (2022). https://api.semanticscholar.org/CorpusID:251040807
22. Guo, Y., Luo, F., Wu, X.: Learning degradation-independent representations for camera ISP pipelines. In: Proceedings of the IEEE/CVF Conference on Computer Vision and Pattern Recognition (CVPR), pp. 25774–25783 (2024)
23. Hansen, N., Ostermeier, A.: Adapting arbitrary normal mutation distributions in evolution strategies: the covariance matrix adaptation. In: Proceedings of IEEE International Conference on Evolutionary Computation, pp. 312–317 (1996). https://doi.org/10.1109/ICEC.1996.542381
24. Hasinoff, S.W., et al.: Burst photography for high dynamic range and low-light imaging on mobile cameras. ACM Trans. Graph. (Proc. SIGGRAPH Asia) **35**(6) (2016)
25. He, K., Zhang, X., Ren, S., Sun, J.: Deep residual learning for image recognition. In: Proceedings of the IEEE Conference on Computer Vision and Pattern Recognition, pp. 770–778 (2016)
26. Heide, F., et al.: Flexisp: a flexible camera image processing framework. ACM Trans. Graph. (ToG) **33**(6), 1–13 (2014)
27. Hong, Y., Wei, K., Chen, L., Fu, Y.: Crafting object detection in very low light. In: BMVC (2021)
28. Houlsby, N., et al.: Parameter-efficient transfer learning for NLP. In: International Conference on Machine Learning, pp. 2790–2799. PMLR (2019)
29. Hu, Y., He, H., Xu, C., Wang, B., Lin, S.: Exposure: a white-box photo post-processing framework. ACM Trans. Graph. (TOG) **37**(2), 26 (2018)
30. Ignatov, A., Van Gool, L., Timofte, R.: Replacing mobile camera ISP with a single deep learning model. arXiv preprint arXiv:2002.05509 (2020)
31. Iscen, A., Zhang, J., Lazebnik, S., Schmid, C.: Memory-efficient incremental learning through feature adaptation. In: Vedaldi, A., Bischof, H., Brox, T., Frahm, J.-M. (eds.) ECCV 2020. LNCS, vol. 12361, pp. 699–715. Springer, Cham (2020). https://doi.org/10.1007/978-3-030-58517-4_41

32. Jensen, H.W., Durand, F., Dorsey, J., Stark, M.M., Shirley, P., Premože, S.: A physically-based night sky model. In: Proceedings of the 28th Annual Conference on Computer Graphics and Interactive Techniques, SIGGRAPH 2001, pp. 399–408. Association for Computing Machinery, New York (2001). https://doi.org/10.1145/383259.383306
33. Jia, M., et al.: Visual prompt tuning. In: Avidan, S., Brostow, G., Cissé, M., Farinella, G.M., Hassner, T. (eds.) ECCV 2022. LNCS, vol. 13693, pp. 709–727. Springer, Cham (2022). https://doi.org/10.1007/978-3-031-19827-4_41
34. Jin, X., Han, L., Li, Z., Chai, Z., Guo, C., Li, C.: DNF: decouple and feedback network for seeing in the dark. In: ICCV (2023)
35. Karaimer, H.C., Brown, M.S.: A software platform for manipulating the camera imaging pipeline. In: Leibe, B., Matas, J., Sebe, N., Welling, M. (eds.) ECCV 2016. LNCS, vol. 9905, pp. 429–444. Springer, Cham (2016). https://doi.org/10.1007/978-3-319-46448-0_26
36. Ke, L., et al.: Segment anything in high quality. In: NeurIPS (2023)
37. Kim, W., Kim, G., Lee, J., Lee, S., Baek, S.H., Cho, S.: Paramisp: learned forward and inverse ISPS using camera parameters. In: Proceedings of the IEEE/CVF Conference on Computer Vision and Pattern Recognition, pp. 26067–26076 (2024)
38. Kirillov, A., et al.: Segment anything. In: Proceedings of the IEEE/CVF International Conference on Computer Vision (ICCV), pp. 4015–4026 (2023)
39. Koskinen, S., Yang, D., Kämäräinen, J.K.: Reverse imaging pipeline for raw RGB image augmentation. In: 2019 IEEE International Conference on Image Processing (ICIP), pp. 2896–2900 (2019). https://doi.org/10.1109/ICIP.2019.8804406
40. Li, Z., Lu, M., Zhang, X., Feng, X., Asif, M.S., Ma, Z.: Efficient visual computing with camera raw snapshots. TPAMI 1–18 (2024). https://doi.org/10.1109/TPAMI.2024.3359326
41. Li, Z., Jiang, H., Cao, M., Zheng, Y.: Polarized color image denoising. In: 2023 IEEE/CVF Conference on Computer Vision and Pattern Recognition (CVPR), pp. 9873–9882 (2023). https://doi.org/10.1109/CVPR52729.2023.00952
42. Liang, Z., Cai, J., Cao, Z., Zhang, L.: Cameranet: a two-stage framework for effective camera ISP learning. IEEE Trans. Image Process. **30**, 2248–2262 (2021). https://doi.org/10.1109/TIP.2021.3051486
43. Lin, T.Y., Goyal, P., Girshick, R., He, K., Dollár, P.: Focal loss for dense object detection. In: Proceedings of the IEEE International Conference on Computer Vision, pp. 2980–2988 (2017)
44. Lin, T.-Y., et al.: Microsoft COCO: common objects in context. In: Fleet, D., Pajdla, T., Schiele, B., Tuytelaars, T. (eds.) ECCV 2014. LNCS, vol. 8693, pp. 740–755. Springer, Cham (2014). https://doi.org/10.1007/978-3-319-10602-1_48
45. Liu, Z., et al.: Swin transformer: hierarchical vision transformer using shifted windows. In: Proceedings of the IEEE/CVF International Conference on Computer Vision (ICCV) (2021)
46. Mildenhall, B., Hedman, P., Martin-Brualla, R., Srinivasan, P.P., Barron, J.T.: Nerf in the dark: high dynamic range view synthesis from noisy raw images. In: Proceedings of the IEEE/CVF Conference on Computer Vision and Pattern Recognition, pp. 16190–16199 (2022)
47. Monakhova, K., Richter, S.R., Waller, L., Koltun, V.: Dancing under the stars: video denoising in starlight. In: Proceedings of the IEEE/CVF Conference on Computer Vision and Pattern Recognition (CVPR), pp. 16241–16251 (2022)

48. Mosleh, A., Sharma, A., Onzon, E., Mannan, F., Robidoux, N., Heide, F.: Hardware-in-the-loop end-to-end optimization of camera image processing pipelines. In: Proceedings of the IEEE/CVF Conference on Computer Vision and Pattern Recognition (CVPR) (2020)
49. Nam, S., Kim, S.J.: Modelling the scene dependent imaging in cameras with a deep neural network. In: 2017 IEEE International Conference on Computer Vision (ICCV), pp. 1726–1734 (2017). https://api.semanticscholar.org/CorpusID:23920472
50. Nguyen, R.M.H., Brown, M.S.: Raw image reconstruction using a self-contained SRGB-JPEG image with only 64 kb overhead. In: 2016 IEEE Conference on Computer Vision and Pattern Recognition (CVPR), pp. 1655–1663 (2016). https://doi.org/10.1109/CVPR.2016.183
51. Nishimura, J., Gerasimow, T., Sushma, R., Sutic, A., Wu, C.T., Michael, G.: Automatic ISP image quality tuning using nonlinear optimization. In: 2018 25th IEEE International Conference on Image Processing (ICIP), pp. 2471–2475. IEEE (2018)
52. Omid-Zohoor, A., Ta, D., Murmann, B.: Pascalraw: raw image database for object detection (2014)
53. Onzon, E., Mannan, F., Heide, F.: Neural auto-exposure for high-dynamic range object detection. In: 2021 IEEE/CVF Conference on Computer Vision and Pattern Recognition (CVPR), pp. 7706–7716 (2021). https://doi.org/10.1109/CVPR46437.2021.00762
54. Potlapalli, V., Zamir, S.W., Khan, S., Khan, F.: Promptir: prompting for all-in-one image restoration. In: Thirty-Seventh Conference on Neural Information Processing Systems (2023)
55. Qin, H., et al.: Attention-aware learning for hyperparameter prediction in image processing pipelines. In: Avidan, S., Brostow, G., Cissé, M., Farinella, G.M., Hassner, T. (eds.) ECCV 2022. LNCS, vol. 13679, pp. 271–287. Springer, Cham (2022). https://doi.org/10.1007/978-3-031-19800-7_16
56. Ramanath, R., Snyder, W.E., Yoo, Y., Drew, M.S.: Color image processing pipeline. IEEE Signal Process. Mag. **22**(1), 34–43 (2005)
57. Rebuffi, S.A., Bilen, H., Vedaldi, A.: Learning multiple visual domains with residual adapters. In: Advances in Neural Information Processing Systems, vol. 30 (2017)
58. Ronneberger, O., Fischer, P., Brox, T.: U-Net: convolutional networks for biomedical image segmentation. In: Navab, N., Hornegger, J., Wells, W.M., Frangi, A.F. (eds.) MICCAI 2015. LNCS, vol. 9351, pp. 234–241. Springer, Cham (2015). https://doi.org/10.1007/978-3-319-24574-4_28
59. Sayed, M., Brostow, G.: Improved handling of motion blur in online object detection. In: 2021 IEEE/CVF Conference on Computer Vision and Pattern Recognition (CVPR), pp. 1706–1716 (2021). https://doi.org/10.1109/CVPR46437.2021.00175
60. Schwartz, E., Giryes, R., Bronstein, A.M.: Deepisp: toward learning an end-to-end image processing pipeline. IEEE Trans. Image Process. **28**(2), 912–923 (2018)
61. Sitzmann, V., Martel, J.N., Bergman, A.W., Lindell, D.B., Wetzstein, G.: Implicit neural representations with periodic activation functions. In: Proceedings of NeurIPS (2020)
62. Souza, M., Heidrich, W.: MetaISP – exploiting global scene structure for accurate multi-device color rendition. In: Guthe, M., Grosch, T. (eds.) Vision, Modeling, and Visualization. The Eurographics Association (2023). https://doi.org/10.2312/vmv.20231236
63. Stickland, A.C., Murray, I.: Bert and pals: projected attention layers for efficient adaptation in multi-task learning. In: International Conference on Machine Learning, pp. 5986–5995. PMLR (2019)

64. Sun, P., et al.: SparseR-CNN: end-to-end object detection with learnable proposals. arXiv preprint arXiv:2011.12450 (2021)
65. Sung, Y.L., Cho, J., Bansal, M.: Vl-adapter: parameter-efficient transfer learning for vision-and-language tasks. In: Proceedings of the IEEE/CVF Conference on Computer Vision and Pattern Recognition, pp. 5227–5237 (2022)
66. Vaswani, A., et al.: Attention is all you need. In: Advances in Neural Information Processing Systems, vol. 30 (2017)
67. Wang, Y., et al.: Raw image reconstruction with learned compact metadata. In: Proceedings of the IEEE/CVF Conference on Computer Vision and Pattern Recognition (CVPR), pp. 18206–18215 (2023)
68. Wei, K., Fu, Y., Yang, J., Huang, H.: A physics-based noise formation model for extreme low-light raw denoising. In: Proceedings of the IEEE/CVF Conference on Computer Vision and Pattern Recognition (CVPR) (2020)
69. Wu, C.T., et al.: Visionisp: repurposing the image signal processor for computer vision applications. In: 2019 IEEE International Conference on Image Processing (ICIP). IEEE (2019). https://doi.org/10.1109/icip.2019.8803607
70. Xie, E., Wang, W., Yu, Z., Anandkumar, A., Alvarez, J.M., Luo, P.: Segformer: simple and efficient design for semantic segmentation with transformers. arXiv preprint arXiv:2105.15203 (2021)
71. Xing, Y., Qian, Z., Chen, Q.: Invertible image signal processing. In: Proceedings of the IEEE/CVF Conference on Computer Vision and Pattern Recognition, pp. 6287–6296 (2021)
72. Xu, R., et al.: Toward raw object detection: a new benchmark and a new model. In: Proceedings of the IEEE/CVF Conference on Computer Vision and Pattern Recognition, pp. 13384–13393 (2023)
73. Yoshimura, M., Otsuka, J., Irie, A., Ohashi, T.: Dynamicisp: dynamically controlled image signal processor for image recognition. In: Proceedings of the IEEE/CVF International Conference on Computer Vision (ICCV), pp. 12866–12876 (2023)
74. Yoshimura, M., Otsuka, J., Irie, A., Ohashi, T.: Rawgment: noise-accounted raw augmentation enables recognition in a wide variety of environments. In: Proceedings of the IEEE/CVF Conference on Computer Vision and Pattern Recognition (CVPR), pp. 14007–14017 (2023)
75. Yu, K., Li, Z., Peng, Y., Loy, C.C., Gu, J.: Reconfigisp: reconfigurable camera image processing pipeline. In: Proceedings of the IEEE/CVF International Conference on Computer Vision, pp. 4248–4257 (2021)
76. Zamir, S.W., et al.: Cycleisp: real image restoration via improved data synthesis. In: Proceedings of the IEEE/CVF Conference on Computer Vision and Pattern Recognition, pp. 2696–2705 (2020)
77. Zhang, R., et al.: Tip-adapter: training-free adaption of clip for few-shot classification. In: Avidan, S., Brostow, G., Cissé, M., Farinella, G.M., Hassner, T. (eds.) ECCV 2022. LNCS, vol. 13695, pp. 493–510. Springer, Cham (2022). https://doi.org/10.1007/978-3-031-19833-5_29
78. Zhang, Y., Zhang, Y., Zhang, Z., Zhang, M., Tian, R., Ding, M.: ISP-teacher: image signal process with disentanglement regularization for unsupervised domain adaptive dark object detection. In: Proceedings of the AAAI Conference on Artificial Intelligence, vol. 38, pp. 7387–7395 (2024)
79. Zhang, Z., Wang, H., Liu, M., Wang, R., Zuo, W., Zhang, J.: Learning raw-to-SRGB mappings with inaccurately aligned supervision. In: ICCV (2021)

80. Zhou, B., Zhao, H., Puig, X., Fidler, S., Barriuso, A., Torralba, A.: Scene parsing through ADE20K dataset. In: Proceedings of the IEEE Conference on Computer Vision and Pattern Recognition, pp. 633–641 (2017)
81. Zhou, W., Gao, S., Zhang, L., Lou, X.: Histogram of oriented gradients feature extraction from raw bayer pattern images. IEEE Trans. Circuits Syst. II Express Briefs **67**(5), 946–950 (2020). https://doi.org/10.1109/TCSII.2020.2980557

ns and Rule-Based Traffic

Kashyap Chitta[1,2(✉)], Daniel Dauner[1,2], and Andreas Geiger[1,2]

[1] University of Tübingen, Tübingen, Germany
kashyap.chitta@uni-tuebingen.de
[2] Tübingen AI Center, Tübingen, Germany
https://github.com/autonomousvision/sledge

Abstract. SLEDGE is the first generative simulator for vehicle motion planning trained on real-world driving logs. Its core component is a learned model that is able to generate agent bounding boxes and lane graphs. The model's outputs serve as an initial state for rule-based traffic simulation. The unique properties of the entities to be generated for SLEDGE, such as their connectivity and variable count per scene, render the naive application of most modern generative models to this task non-trivial. Therefore, together with a systematic study of existing lane graph representations, we introduce a novel raster-to-vector autoencoder. It encodes agents and the lane graph into distinct channels in a rasterized latent map. This facilitates both lane-conditioned agent generation and combined generation of lanes and agents with a Diffusion Transformer. Using generated entities in SLEDGE enables greater control over the simulation, e.g. upsampling turns or increasing traffic density. Further, SLEDGE can support 500 m long routes, a capability not found in existing data-driven simulators like nuPlan. It presents new challenges for planning algorithms, evidenced by failure rates of over 40% for PDM, the winner of the 2023 nuPlan challenge, when tested on hard routes and dense traffic generated by our model. Compared to nuPlan, SLEDGE requires 500× less storage to set up (<4 GB), making it a more accessible option and helping with democratizing future research in this field.

Keywords: Diffusion · Transformers · Simulation · Planning · Driving

1 Introduction

While recent breakthroughs in generative AI have revolutionized natural image synthesis [4,14], generative models are yet to find widespread adoption in

---

K. Chitta and D. Dauner—Equal contribution.

Supplementary Information The online version contains supplementary material available at https://doi.org/10.1007/978-3-031-73235-5_4.

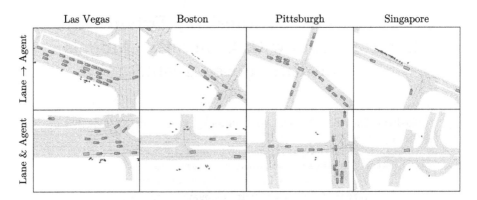

Fig. 1. SLEDGE. We show state snapshots of simulation environments generated by our approach in 4 cities, with the lanes, ego-vehicle, other vehicles, pedestrians, obstacles, and traffic lights. Our supplementary video visualizes clips with more examples.

autonomous driving. In contrast to the regular pixel grid of images, self-driving planners typically require abstract bird's eye view (BEV) representations as input which characterize the most important scene elements (e.g., lanes, traffic lights, static and dynamic objects) in a compact, vectorized format (see Fig. 2). These representations are a key component of data-driven simulators which are necessary for rigorous evaluation of planners [9,10,24,49]. However, learning a generative model on such irregular vectorized representations is challenging.

Consequently, many existing data-driven simulators [15,20] are initialized by simply replaying logs of abstract representations. They extract the local lane graph from a High Definition (HD) map and object bounding boxes from pre-recorded annotation logs. Planning algorithms can be tested in scenarios with restricted routes and durations (∼15 s long), to ensure that the environment during simulation is covered in the recording. Moreover, to provide sufficient diversity among routes for comprehensive testing, these simulators require huge databases, e.g., nuPlan [20] consists of 1300 h of driving logs which require over 2 TB of storage. Such high resource requirements heighten the barrier for entry into the field of vehicle motion planning.

In this paper, we study the generation of simulation-ready abstract representations of driving scenes. Using generative models as an alternative to log replays has the potential for significant compression [38]. However, the unique characteristics of abstract representations in driving scenes pose new challenges to modeling them, e.g., they require accurate topological connectivity, variable entity counts, and precise modeling of geometry (e.g., parallel lines). Due to these characteristics, generative models that operate on uniformly sized representations (e.g., images) are incompatible with our data-driven simulation task.

To tackle these complexities, we first perform a systematic study of the existing representations of lane graphs used in autonomous driving. We then propose a novel representation based on a raster-to-vector autoencoder. It represents a driving scene with a fixed-size BEV rasterized latent map (RLM). We learn to generate these RLMs with a Diffusion Transformer (DiT) [34]. Our model generates high-fidelity results enabling both lane-conditioned agent generation or joint lane and agent generation within a single flexible and scalable framework (Fig. 1). Independent to its use of generative models, SLEDGE also provides the previously missing functionality of simulating only agents within a certain radius of the ego-vehicle. By doing so, we can test on routes that are significantly longer than those currently used for evaluating planners. We find previously ignored failure modes of the state-of-the-art PDM-Closed planner [10], which is unable to complete 25–50% of our new 500 m long test routes despite having failure rates below 10% on existing benchmarks.

Contributions. (1) We formalize the task of abstract scene generation for autonomous driving with a challenging benchmark and corresponding metrics. (2) We perform a systematic exploration of modern generative models (with various architectures and representations) and propose a novel latent diffusion model for synthesizing abstract driving scenes that largely outperforms other baselines. (3) We present a simulation framework, SLEDGE, which is nearly 3 orders of magnitude more storage efficient than nuPlan yet enables more rigorous testing of planning algorithms with long-horizon simulations using rule-based traffic.

2 Related Work

Diffusion Models. Best known for their success in generative modeling of images [14,37] and video [3,4,47], diffusion models have recently found widespread adoption in diverse domains, including point clouds [29,32,48], floor plans [6,40], molecules [46], robot policies [7], traffic patterns [51], and many others. Scenario Diffusion [35], a pioneering approach in generating vehicles conditioned on HD maps with diffusion, is the closest existing approach to ours. This method uses latent diffusion with a raster-based vehicle decoder unlike our proposed transformer decoder head. Importantly, we offer significantly increased capabilities: generation of lane graphs, support for pedestrians, obstacles, and traffic lights, as well as long-horizon simulation environments with reactive agents.

Generating Lane Graphs. Lane graphs, the most important component of HD maps, are well-studied. They are often constructed through an offline mapping process often involving human annotators [13]. However, a surge of recent work on predicting lane graphs from sensor data [23,25] has sparked interest on generative modeling of these graphs. The first and only existing study on this task, HDMapGen [31], proposes an autoregressive approach for generating lane graphs node-by-node [8]. Our experiments show that this achieves reasonable results, but is unable to match the high quality and scalability offered by our model. Unlike HDMapGen, our approach jointly generates agents with lanes, and efficiently generates all elements in parallel. A concurrent project, DriveSceneGen [41], generates lanes and vehicles with image-space diffusion. Our approach covers agent types beyond vehicles, uses less heuristics, and is more efficient. Additionally, we show the successful integration of our model into a simulator.

Data-Driven Simulation. Developing an autonomous driving system necessitates rigorous testing which is costly and risky if conducted in the real world. Driving simulators are an alternative [12,20,45]. However, simulators face challenges in ensuring realism while initializing traffic scenes, simulating traffic. Instead, data-driven simulators address these challenges by replaying traffic scenes from real-world recordings [1,15,20,22]. The simulator can mine specific situations or even optimize the initial parameters for safety-critical scenarios [11,16,44]. Leveraging modern generative models, we take a step further than existing frameworks and learn the underlying distribution of the real-world data.

3 Method

Our goal is to design a driving scene synthesis framework that can be trained using real-world driving logs and incorporated into SLEDGE, our generative simulator with rule-based traffic. We base this framework on LDMs [37] as: (1) latent diffusion shows excellent training stability and scalability with compute. (2) One can easily construct a fixed-size latent space for diffusion that can be mapped to the variable sized set representation for simulation (Sect. 3.1) using detection-based transformer architectures [5,25]. Our LDM is trained in two stages: an autoencoder (Sect. 3.2) followed by a diffusion model (Sect. 3.3). We detail the simulation of scenes generated by the LDM in Sect. 3.4.

3.1 NuPlan Vector Representation

We combine sets of entities to represent scenes in the default nuPlan format.

Lanes. Our focus is on the generation of lanes, the central element of HD maps used in data-driven simulation. Each lane $\mathbf{L} \in \mathbb{R}^{20 \times 2}$ is geometrically represented by a polyline, i.e., a fixed set of 20 bird's eye view (BEV) points. These are bounded by two endpoints and form the lane centerline, connected along the driving direction. A lane may share endpoint(s) with predecessor and successor lanes. This information is encoded in an adjacency matrix $\mathbf{A} \in \mathbb{R}^{N \times N}$, where N is the number of lanes in a certain field of view (FOV). The set of all lane polylines $\mathcal{L}$ form the lane graph of the local map $\mathcal{M} = \{\mathcal{L}, \mathbf{A}\}$.

Traffic Lights. We then augment the lane graph with polylines representing traffic lights. These share the same 20 × 2 format as lanes, and come in two types (red and green). The set of red polylines ($\mathcal{R}$) contains the lane regions that are currently not traversable due to a red traffic light. The set of green polylines ($\mathcal{G}$), on the other hand, contains entities indicating lane segments where the road is safe to proceed along due to the presence of a green traffic signal.

Agents. Further, we expand the scene representation using oriented bounding boxes for the agents. Each bounding box is defined by a 2D center position, heading, 2D extent and optional speed. We consider three types of agent sets:

pedestrians ($\mathcal{P}$), vehicles ($\mathcal{V}$) and static objects ($\mathcal{O}$). Pedestrians and vehicle boxes are assigned a speed attribute, whereas static objects are not.

Ego Velocity. Finally, initializing a simulation requires the BEV ego velocity $\mathbf{v} \in \mathbb{R}^2$. Overall, we denote the scene state as $\mathcal{S} = \{\mathcal{M}, \mathcal{R}, \mathcal{G}, \mathcal{P}, \mathcal{V}, \mathcal{O}, \mathbf{v}\}$.

3.2 Raster-to-Vector Autoencoder

Each entity type in $\mathcal{S}$ is unique. To maintain overall scene consistency, we would like to model them with a single architecture, instead of creating several independent entity-specific generative models. Furthermore, most existing tools in the literature have been developed and optimized for 2D input domains [37]. To this end, we propose the raster-to-vector autoencoder (RVAE) which unifies all entity types in $\mathcal{S}$ into a compact, shared 2D representation well-suited for diffusion modeling. The autoencoder is inspired by work in object detection [5] and online mapping for autonomous driving [25].

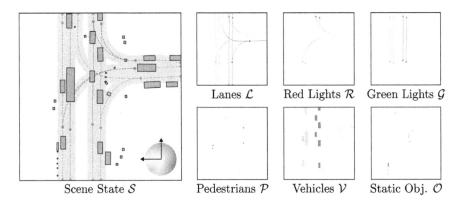

Fig. 2. Rasterized State Image (RSI). We encode $\mathcal{S}$ into a 12-channel image, with 2 channels per entity type. We visualize these encodings as optical flow fields.

Rasterization. We first define a function $\rho: \mathcal{S} \to \mathbf{I}$ that encodes the scene state into a rasterized state image (RSI) $\mathbf{I} \in \mathbb{R}^{W \times H \times 12}$. Our design of ρ is motivated by (and closely resembles) techniques used in motion planners [2,36]. As shown in Fig. 2, ρ maps the three polyline entity types ($\mathcal{L}, \mathcal{R}, \mathcal{G}$) and three bounding box entity types ($\mathcal{P}, \mathcal{V}, \mathcal{O}$) to image pixel locations, assigning 2 channels to each entity type which encode all their attributes. For polylines, we use a 2D directional vector $\Delta = [dx, dy]$, which points from any point to its successor, indicating the presence of a polyline traversing a specific pixel. A background value (i.e., $[0,0]$) is assigned to other regions. For a bounding box type entity, we rasterize it in BEV according to its position, extent, and heading. For dynamic bounding boxes ($\mathcal{P}, \mathcal{V}$), the values in the two channels within the box region represent the entity's 2D velocity. For static obstacles, we fill the rasterized region with

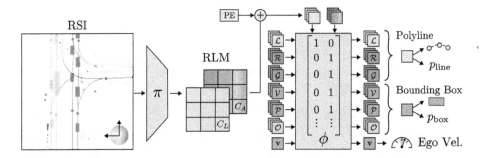

Fig. 3. Raster-to-Vector Autoencoder (RVAE). We represent scenes with a rasterized latent map (RLM) consisting of two channel groups. The 'Lanes' group is decoded into lane segments and the 'Agents' group into all other scene entities, via a transformer decoder with attention masking. The autoencoder is trained to predict polylines, bounding boxes, and the ego velocity in a simulation-compatible format.

the obstacle's orientation vector. We use a square field of view centered at and oriented as per the ego vehicle's pose for rasterization. The ego velocity **v** is encoded at the origin of **I** as an extra rasterized vehicle in $\mathcal{V}$.

Vectorization. This step is necessary to decode the unified RSI representation **I** back into the per-entity attributes (e.g., polylines, bounding boxes). We use a learned vectorization pipeline consisting of a raster encoder π and vector decoder head ϕ. The ResNet-50 [17] encoder takes the RSI and outputs a rasterized latent map (RLM) $\mathbf{M} = \pi(\mathbf{I})$ of shape $W' \times H' \times C$, where $H' = H/2^d$ and $W' = W/2^d$ for a downsampling factor $d \in \mathbb{N}$. C is a chosen channel dimension. In practice, the RLMs we use are compact, $H' = W' = 8$ and $C = 64$. Additionally, as shown in Fig. 3, we split the RLM's channels into 2 groups, $C = C_L + C_A$ for the lanes and agents respectively. For each group, we tokenize the latent vectors spatially with 1×1 patches, resulting in $W' \times H'$ lane tokens and $W' \times H'$ agent tokens. Following the DETR [5] paradigm, we use these tokens as keys and values for our transformer decoder ϕ. The decoder uses a fixed number of learnable queries of each entity type, which we cap to a maximum count per entity, based on statistics from our dataset. The final decoder layer is unique per entity type and outputs the attributes specific to that entity, e.g. a 20×2 set of point coordinates for a polyline, or a 6-dimensional descriptor (2D position, orientation, 2D extent, and speed) for a bounding box. It also outputs an existence attribute $p \in [0,1]$ for both polylines (p_{line}) and bounding boxes (p_{box}), which is used to handle variable counts of ground truth entities with a fixed number of queries.

Channel Group Masking. The motivation behind our design with two token groups is to enable agent generation conditioned on known lanes. To this end, the tokens for agents should contain no information about lanes. We implement a binary mask in the cross-attention mechanism of ϕ to achieve this. Specifically, queries for lanes $\mathcal{L}$ are prevented from attending to the keys and values of the agents tokens, and all other queries (i.e., $\mathcal{R}, \mathcal{G}, \mathcal{P}, \mathcal{V}, \mathcal{O}, \mathbf{v}$) cannot attend to the lanes tokens. Our experiments show the effectiveness of this approach.

Training. The autoencoder is optimized using both reconstruction and existence losses, and a KL divergence loss on the RLM. For reconstruction, we first match generated and ground truth entities using the Hungarian algorithm, as in [5]. We use a matching score based on the L1 error of the entity's position attributes. We then use the L1 error summed over all attributes and averaged over all matches as the training loss. For the existence variable, we use a binary cross entropy loss based on whether the query was matched to a ground truth entity.

3.3 Diffusion Transformer

We obtain RLMs $\mathbf{M}$ for each training example via the frozen, pretrained encoder π and use them to train a diffusion model δ with the DDPM algorithm [19].

Training. For each scene, we sample a noise scaling factor σ from a log-normal distribution and create a noisy sample $\hat{\mathbf{M}} = \mathbf{M} + \sigma \mathcal{E}$, with $\mathcal{E}$ sampled from a standard normal distribution of the same shape as $\mathbf{M}$. We model $\delta(\hat{\mathbf{M}}; \mathbf{c}, \sigma)$ as a Transformer [43], following DiT [34], where $\mathbf{c}$ is a conditioning vector. We choose $\mathbf{c}$ to be a one-hot label indicating the city to which the example belongs. This conditioning resolves ambiguities between locations (e.g. right- and left-hand driving in the US and Singapore). However, it is possible to adapt our framework to other conditioning, e.g., images, text descriptions, or learned clusters based on the autoencoder features. The DiT architecture is simple, scalable, and free from down- or upsampling operations, making it compatible with RLMs of any spatial resolution. It applies a series of self-attention blocks to the tokenized input $\hat{\mathbf{M}}$, with the conditioning on $\mathbf{c}$ and σ implemented using AdaLN-Zero [34]. We optimize the L2 reconstruction loss between $\mathcal{E}$ and $\delta(\hat{\mathbf{M}}; \mathbf{c}, \sigma)$.

Generation. During inference, we begin with an initial noisy sample $\hat{\mathbf{M}} \sim \mathcal{N}\left(0, \sigma_{\max}^2 \mathbf{I}\right)$, which undergoes iterative refinement from $\sigma = \sigma_{\max}$ to $\sigma = 0$ based on the reverse PDE defined via δ [19]. We then use the trained vector decoder ϕ from Sect. 3.2 to predict $\mathcal{L}, \mathcal{R}, \mathcal{G}, \mathcal{P}, \mathcal{V}, \mathcal{O}$, and $\mathbf{v}$. Only entities with an existence probability above a threshold τ are retained, and for overlapping bounding boxes, those with the highest existence probability are kept. The process of recovering the full scene state $\mathcal{S}$ further involves extracting the adjacency matrix $\mathbf{A}$ of the lane graph, which is not an explicit output of our model. We do this by simply matching lanes whose start and endpoints lie within a range of 1.5 m with orientations differing by less than 60°, which we found to be robust in practice given our highly accurate lane polyline predictions.

Conditional Generation via Inpainting. Diffusion models excel at inpainting, even without explicit training for this task [28]. Specifically, by executing the denoising process on only a subset of the tokens in the noisy sample $\hat{\mathbf{M}}$, with the remaining tokens extracted from a known and encoded scene, we can inpaint RLMs. We use this capability to perform two tasks: (1) lane conditioned

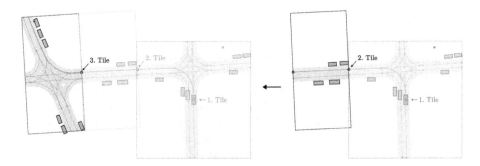

Fig. 4. Route Extrapolation by Inpainting. We show an example scenario generated by our DiT, where we iteratively sample poses along a route, warp the previous tile's RSI to this pose, and generate a new tile conditioned on the warped RSI.

agent generation and (2) route extrapolation. Lane conditioned agent generation involves encoding all lane tokens from a known map, and denoising all agent tokens. Route extrapolation, as illustrated in Fig. 4, involves encoding a subset of tokens within a known spatial region to denoise the unknown region. Specifically, we can iteratively sample poses along a generated route, warp the previously generated scene's RSI to the new pose with an affine transformation, and use a known region as conditioning for completing a newly created tile. We provide implementation details in the supplementary material.

3.4 SLEDGE Simulation Environments

Finally, we initialize a reactive simulation in SLEDGE using the generated initial scene state $\mathcal{S}$. In the following, we provide an overview of the steps involved.

Hard Routes and Traffic. To evaluate a planner in ambiguous situations, we must specify the driver intention, e.g. whether to turn left or right at an intersection. Existing replay-based simulators offer limited controllability over this, since they are unable to extract agents to simulate if the planner diverges significantly from the route followed by the human driver from the log recording. However, for generated scenes, we can extract multiple valid routes from the lane graph, e.g., we define 'hard' routes by selecting the route with the highest number of turns. In addition, our approach also provides a degree of control over traffic density. We define a 'hard' traffic setting by generating multiple valid traffic configurations along the desired route, and selecting the configuration with the largest number of generated agents. Our experiments show that these 'hard' settings provide new challenges to the state of the art for planning.

Behavior Simulation. We simulate non-ego vehicles in SLEDGE by projecting each to the center of a lane based on proximity and heading, from which it then laterally follows the lane centerlines. For longitudinal control, we use a simple policy called the Intelligent Driver Model (IDM) [42]. Upon choosing a connected lane sequence as a driving path, IDM calculates a longitudinal trajectory, iteratively adjusting acceleration based on current position, velocity, and distance to the leading vehicle. For pedestrians, we assume a constant velocity and heading while unrolling the simulation. Traffic lights are hard-coded to change states every 15 s. While these choices are simplistic, they are in line with the capabilities of today's best data-driven simulators [15,20]. SLEDGE is not incompatible by design with other types of policies for vehicle and pedestrian simulation, but we leave this exploration to future work.

Simulation Radius. By default, existing data-driven simulators like nuPlan simulate all the initialized agents at all timesteps. This severely limits the scalability of these simulators to long simulation horizons or large scenes. We propose a simple modification for SLEDGE wherein we only simulate agents at a distance below $\alpha = 64$ m to the ego vehicle at a given timestep, while holding the state of all other agents to be constant (Fig. 5). We demonstrate the scalability this provides by conducting experiments on simulations that are 10× longer than those in nuPlan, i.e., up to 150 s as opposed to nuPlan's 15 s.

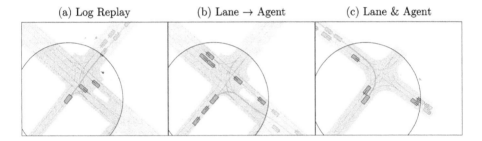

Fig. 5. Long Route Simulation. SLEDGE supports (a) replayed scenarios, (b) lane-conditioned agent generation, and (c) joint lane and agent generation. Importantly, we enable testing on arbitrarily long routes by dynamically simulating agents near the ego vehicle while keeping the state of distant agents fixed.

4 Experiments

We now present our experimental results. (1) We demonstrate the suitability of the RLM as a lane graph representation. (2) We benchmark a series of models for lane graph generation. (3) We showcase the effectiveness of SLEDGE environments for evaluating planners. For all experiments, we present concise descriptions of baselines, metrics and implementations in the main paper. Additional details can be found in the supplementary material.

Dataset. We use nuPlan [20], the largest publicly available dataset for vehicle motion planning. It comprises 1300 h of logs from 4 cities. We sample 450k train and 50k validation frames from these logs with sampling intervals of 30 s, 1 s, 2 s and 2 s between frames for Las Vegas, Boston, Pittsburgh and Singapore respectively, in order to obtain a balanced distribution while achieving high map coverage. Each city offers unique challenges for generative modeling, e.g. Las Vegas has large and dense intersections, while Singapore involves left-hand traffic. Each frame is limited to a 64 m × 64 m FOV centered at the ego vehicle.

Implementation. We consider a ResNet-50 for the raster encoder π and a transformer decoder with three layers for the vector decoder ϕ of the RVAE. For the generative tasks, we apply the DiT-L and DiT-XL (138M vs. 487M params) variants with a 1 × 1 patch size and DDPM noise scheduling as in [37].

4.1 Lane Graph Representations

In our first experiment, we evaluate various representations based on their ability to reconstruct the complete directed lane graph $\mathcal{M} = \{\mathcal{L}, \mathbf{A}\}$.

Table 1. 64 m × 64 m Lane Graph Reconstruction. We show the F1 score, lateral displacement and Chamfer distance for graphs extracted from each representation. We additionally include qualitative results (more in supplementary material). The RSI struggles with nearly overlapping segments at the beginning of forks (FNs) and overlaps at intersections (FPs). The RLM closes the gap towards the upper bound (vector).

Rep.	Fixed?	Split?	Size (KB)	GEO			TOPO		
				F1 ↑	Lat. ↓	Ch. ↓	F1 ↑	Lat. ↓	Ch. ↓
RSI	✓	✓	524.3	0.933	0.133	0.423	0.851	0.438	64.824
RLM	✓	✗	16.0	0.981	0.161	0.399	0.945	0.282	20.096
		✓	8.0	0.980	0.164	0.411	0.944	0.288	20.624
Vector	✗	✓	4.8	0.997	0.005	0.070	0.990	0.010	4.174

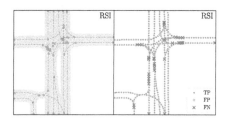
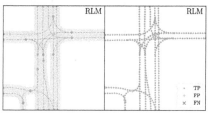

Baselines. We consider three baseline representations. **(1) RSI:** we use the skan library [33] to extract vector polylines from the 256 × 256 × 2 RSI representation of the lane graph. Skan is a highly optimized image processing pipeline for graph extraction (details and visualizations in supplementary). **(2) RLM w/o mask:** we train the RVAE without the channel group masking proposed in Sect. 3.2, which entangles information about agents and lanes into a single 8×8×64 latent map instead of an 8×8×32 tensor for the lane graph and an independent 8×8×32 tensor for agents. **(3) Vector:** As an upper bound, we additionally compute our metrics for the representation used as target labels for the autoencoder during training. This is a set of polylines shaped $N \times 20 \times 2$, where we cap N at 30 in our experiments. Note that the ground truth for evaluation in this experiment uses all polylines in the scene (which is sometimes greater than 30).

Metrics. Our metrics are adapted from the street map extraction literature [18]. We use three base metrics, which all operate on point sets sampled along graphs at a resolution of one point every 1.5 m. **(1) F1:** measures the harmonic mean of the precision and recall, which are estimated using Hungarian matching between point sets with a distance threshold of 1.5 m. Intuitively, this penalizes large structural errors, while ignoring small positional offsets. **(2) Lateral L2 (Lat.):** averaged over all true positive matched points, measures the lateral offset of each such point from its nearest ground truth lane centerline. In contrast to F1, it penalizes positional errors, while ignoring structurally incorrect and unmatched lanes. **(3) Chamfer:** averaged over all points in two point sets, this is the distance of each point to the closest in the other set. It requires both precise structure and details. These three base metrics are further applied in two settings. **(1) GEO** uses point sets sampled from the complete graph, making it independent of the predicted adjacency matrix $\mathbf{A}$. On the other hand, **(2) TOPO** uses $\mathbf{A}$ to extract sets of fully-connected sub-graphs corresponding to every tenth node of the graph (i.e., every 15 m). The base metrics are computed on these sub-graphs and averaged. Errors in $\mathbf{A}$ can lead to large missing sections of such sub-graphs, making TOPO suitable for evaluating connectivity.

Results. As shown in Table 1, the RSI is unable to match the RLM on all TOPO and both the F1 and Chamfer GEO metrics. Despite the significantly larger representation size of 524 KB per 64 m × 64 m scene, the reliance on heuristics to convert this representation back to its constituent entities serves as a major limiting factor for the RSI. Qualitatively, we observe that the key issue is dense groups of lines that nearly overlap at forks or cross over in intersections. For the 2 variants of RLM, we obtain similar reconstruction quality, largely closing the gap towards the upper bound vector representation. Importantly, the proposed channel group masking for the RLM (i.e., 'Split') disentangles the lane graph from the agents with no impact on the graph reconstruction fidelity. Finally, as we cap the maximum number of polylines to 30, we observe a very minor error rate (<1% drop in F1) in the upper bound vector representation, corresponding to the negligible fraction of scenes with over 30 lanes. However, the size of the vector representation varies significantly per scene, ranging from 2 to 30 lanes. The RLM achieves an ideal balance of high quality with a fixed size.

4.2 Lane Graph Generation

Next, we compare our proposed DiT to several generative models for lane graphs.

Baselines. We select four diverse baselines. **(1) VAE:** we train a convolutional VAE with a 2D decoder head to generate RSIs. **(2) RVAE:** we sample from the decoder ϕ of our proposed autoencoder. **(3) HDMapGen:** an autoregressive hierarchical graph neural network for lane graph generation [31], reimplemented

Table 2. 64 m × 64 m Lane Graph Generation. We show the route lengths, precision and recall (in %), and various Frechet distances between samples from the validation set and model outputs. We additionally include qualititve results (more examples in supplementary material). Latent diffusion combining DiT with an RLM obtains the best results. *Trained with ∼6× more compute than others, which already converge at the lower compute budget.

Arch.	Repr.	Route Length ↑	Prec. (CNN) ↑		Recall (CNN) ↑		Frechet (Urban Planning) ↓			
			ImNet	RVEnc	ImNet	RVEnc	Conne.	Densi.	Reach	Conve.
VAE	RSI	2.68 ± 3.66	26.58	0.00	8.70	0.16	9.45	0.99	2.86	13.06
RVAE	Vector	23.79 ± 9.96	10.41	4.56	16.28	8.14	15.63	12.57	3.08	17.72
HDMapGen	Vector	28.17 ± 14.81	19.10	7.48	17.17	12.45	7.02	3.03	2.49	18.10
DiT-L	RSI	24.78 ± 10.38	20.36	19.20	21.49	5.94	6.11	15.33	1.90	3.95
	RLM	32.51 ± 9.93	33.25	63.99	36.24	61.60	2.35	3.52	0.88	3.10
DiT-XL*	RLM	**35.37 ± 10.28**	**42.05**	**78.07**	**42.32**	**72.63**	**0.27**	**2.47**	**0.20**	**0.47**

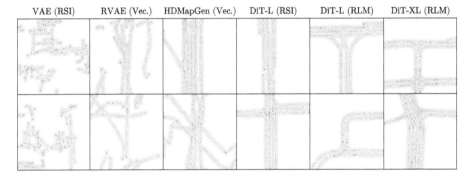

for nuPlan. **(4) DiT (RSI):** similar to the concurrent DriveSceneGen [41], this is an image diffusion model that generates the RSI.

Metrics. Lane graph generation does not have established evaluation protocols. Therefore, we use a comprehensive set of four metric types. **(1) Route Length** measures the mean and std of the longest valid ego vehicle route in the 64 m × 64 m FOV for 1k generated graphs. We assess sample quality with **(2) Precision** and coverage with **(3) Recall** based on the improved generative metrics from [21]. To obtain fixed-sized representations needed to estimate precision and recall with a nearest neighbour classifier, we rasterize the lane graphs and consider the penultimate feature vectors of a ResNet-50. We use two versions of these metrics with ResNets trained on ImageNet and the encoder π of our autoencoder. The remaining metrics are **(4) Frechet** distances taken from [8,31], based on graph features used in urban planning. They operate on nodes of the generated lane graphs with degree $\neq 2$, which are referred to in [31] as **key points**. *Connectivity:* this uses the degrees of all key points. *Density:* the number of key points in the 64 m × 64 m FOV. *Reach:* the number of valid paths found from key points to others. *Convenience:* Lengths of all valid paths from all key points. The Frechet metrics are scaled by suitable powers of 10 for readability. Precision, Recall, and Frechet distances are measured using 50k generated and ground truth graphs.

Results. Our results are shown in Table 2. All DiT variants generate more plausible layouts, with significantly better metrics than HDMapGen or the VAEs. For DiT-L, we observe higher coverage and visual fidelity when using the RLM representation instead of the RSI. In particular, the RLM-based models excel at creating coincident endpoints in intersections between connected lanes, which is crucial for smooth simulation. Scaling to DiT-XL provides further gains across all metrics, demonstrating the effectiveness of increasing the model capacity and compute budget. Since ImageNet features may be misaligned to BEV graphs, and the RVEnc features might favor RLM diffusion models, we focus on the urban planning Frechet metrics (in particular, Reach) in our subsequent experiments.

Scaling. We run a systematic analysis of scaling for our DiT with the RLM representation. We conduct a grid of experiments, considering **(1) 2 model sizes:** DiT-B and DiT-L, **(2) 3 dataset sizes:** 1×, 0.5×, and 0.25× our full dataset, and **(3) 3 compute budgets:** 24, 48 and 96 GPU hours. As shown in Fig. 6, the performance scales significantly with increased compute. While it also scales with more parameters, more data does not have a large impact. This is possibly because data diversity is more valuable than scale, and all of our 3 datasizes have similar diversity. This scaling behavior shows the potential of further improvements for SLEDGE with more training resources.

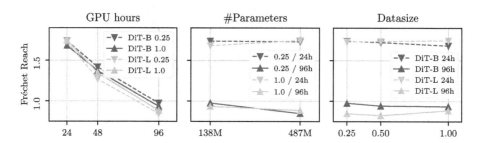

Fig. 6. Scaling. The DiT's performance scales significantly with increased compute. For our task, dataset size is less crucial, with all settings performing similarly.

4.3 SLEDGE Simulation of PDM-Closed

In our final experiment, we use the inpainting capabilities of DiT-XL in SLEDGE to demonstrate its utility for testing vehicle motion planners.

Tasks and Settings. We consider the two inpainting-based tasks described in Sect. 3.3: lane conditioned agent generation (**Lane → Agent**) and joint lane and agent generation via route extrapolation (**Lane & Agent**). For the Lane → Agent task, we consider 100 existing nuPlan logs and use the original route extracted from these logs as the 'easy' route. 'Hard' routes are generated from the same initial pose while maximizing the number of turns. For the **Lane & Agent** task, as we no longer rely on the nuPlan maps, we dynamically adjust the difficulty of generated routes during DiT inpainting. We generate 100 scenarios. For each, we perform iterative inpainting starting from an initial 64 m × 64 m area, and perform a depth-first search on the updated lane graph at every inpainting step. 'Hard' routes correspond to selecting the next point to inpaint from to be the endpoint reachable with the most new turns, and 'easy' routes are generated from endpoints with the fewest turns. We evaluate on route lengths of 100 m (with a simulation time of 30 s) and 500 m (150 s). We consider both 'easy' and 'hard' traffic densities: 'easy' is a single sample, and 'hard' is the scene with the largest number of agents among 8 DiT samples.

Metrics. We evaluate the **Planner Failure Rate (PFR)** in SLEDGE using PDM-Closed [10]. This is the winner of the 2023 nuPlan challenge, and the state of the art for motion planning in the short, 15 s scenarios possible with existing data-driven simulators. The planner 'fails' if it achieves less than 20% of the route's total progress, goes in the wrong driving direction for more than 6 m, goes off-road, or causes an at-fault collision. We also measure the **#Turns** and **#Agents**, which are proxies of the difficulty of the route and traffic.

Results. We show our results in Table 3. Importantly, we note that setting up an evaluation with SLEDGE only requires a 3 GB download of our DiT-L checkpoint for the Lane & Agent mode, and an additional 1 GB download of the nuPlan maps for the Lane → Agent mode, in contrast to the 2 TB of logs needed to set up nuPlan. When using easy routes and traffic, we observe similar PFRs for both replay-based and generative simulation despite this large compression factor. For replay, we observe over a 4× rise in PFR from 0.06 to 0.26 when extending the route from 100 m to 500 m. These are primarily due to PDM-Closed's inability to make lane changes or overtake slow vehicles, which are important planning behaviors that are not strongly penalized on current benchmarks like Val14 [10]. In all settings, switching to hard routes increases the number of turns, which is more challenging than straight driving and in turn deteriorates the planning

Table 3. Simulation of PDM-Closed in SLEDGE. Our simulator offers control over the route length, difficulty and traffic density. In several settings, we present new challenges for the existing state-of-the-art, leading to high failure rates of over 40%.

Task		→ Lane → Agent						Lane & Agent					
Length		→ 100 m			500 m			100 m			500 m		
Routes	Traffic	Turns	Agents	PFR	Turns	Agents	PFR	Turns	Agents	PFR	Turns	Agents	PFR
Replay		0.89	57.40	0.06	3.29	102.34	0.26	-	-	-	-	-	-
Easy	Easy	0.89	44.61	0.07	3.29	125.23	0.25	0.61	27.30	0.22	2.22	110.51	0.39
	Hard		56.44	0.11		167.47	0.39	0.57	39.11	0.20	2.30	173.91	0.44
Hard	Easy	1.18	44.66	0.14	4.20	128.79	0.28	1.23	27.14	0.29	3.66	107.11	0.45
	Hard		57.65	0.11		170.87	0.44	1.07	39.03	0.30	3.82	169.66	0.49

performance. In the most challenging settings with hard routes and traffic, the PFR increases to over 40%.

5 Conclusion

We present SLEDGE, a generative simulator for vehicle motion planning based on latent diffusion. We conduct several experiments to show that it is more realistic, compact, controllable, and diverse than other generative and replay-based approaches. Additionally, we establish several baselines and metrics for the generative simulation task. We hope our work can lay the foundation for accelerating progress in data-driven simulation and vehicle motion planning.

Limitations. Evaluating simulators is hard. We provide metrics for the lane graph generation sub-task, and preliminary experiments on testing rule-based planning, but using the simulator for other downstream tasks, such as reinforcement learning, will be important to showcase its full potential. For efficiency and robustness, we use a relatively small FOV and simulation radius, a simplistic lane representation consisting of only the centerline (assuming constant lane widths), and rule-based traffic behavior with IDM. These issues could be alleviated by further scaling of our model and extensions for learned motion behavior [41,50]. However, like other diffusion models, the compute requirements of our approach are already high. We see value in improving the efficiency of SLEDGE through relevant techniques for accelerating diffusion models [26,27,30,39].

Acknowledgments. This work was supported by the ERC Starting Grant LEGO-3D (850533), the DFG EXC number 2064/1 - project number 390727645, the German Federal Ministry of Education and Research: Tübingen AI Center, FKZ: 01IS18039A and the German Federal Ministry for Economic Affairs and Climate Action within the project NXT GEN AI METHODS. We thank the International Max Planck Research School for Intelligent Systems (IMPRS-IS) for supporting Kashyap Chitta and Daniel Dauner. We also thank Agniv Sharma for providing his reimplementation of HDMap-Gen, Bernhard Jaeger for proofreading, and the nuPlan team for open-sourcing their dataset and simulation tools to the community.

References

1. Althoff, M., Koschi, M., Manzinger, S.: Commonroad: composable benchmarks for motion planning on roads. In: Proceedings of IEEE Intelligent Vehicles Symposium (IV) (2017)
2. Bansal, M., Krizhevsky, A., Ogale, A.S.: Chauffeurnet: learning to drive by imitating the best and synthesizing the worst. In: Proceedings of Robotics: Science and Systems (RSS) (2019)
3. Blattmann, A., et al.: Stable Video Diffusion: Scaling Latent Video Diffusion Models to Large Datasets. arXiv:2311.15127 (2023)
4. Brooks, T., et al.: Video generation models as world simulators (2024). https://openai.com/research/video-generation-models-as-world-simulators

5. Carion, N., Massa, F., Synnaeve, G., Usunier, N., Kirillov, A., Zagoruyko, S.: End-to-end object detection with transformers. In: Vedaldi, A., Bischof, H., Brox, T., Frahm, J.-M. (eds.) ECCV 2020. LNCS, vol. 12346, pp. 213–229. Springer, Cham (2020). https://doi.org/10.1007/978-3-030-58452-8_13
6. Chen, J., Deng, R., Furukawa, Y.: Polydiffuse: polygonal shape reconstruction via guided set diffusion models. In: Advances in Neural Information Processing Systems (NeurIPS) (2023)
7. Chi, C., et al.: Diffusion policy: visuomotor policy learning via action diffusion. In: Proceedings of Robotics: Science and Systems (RSS) (2023)
8. Chu, H., et al.: Neural turtle graphics for modeling city road layouts. In: Proceedings of the IEEE International Conference on Computer Vision (ICCV) (2019)
9. NAVSIM Contributors: NAVSIM: data-driven non-reactive autonomous vehicle simulation (2024). https://github.com/autonomousvision/navsim
10. Dauner, D., Hallgarten, M., Geiger, A., Chitta, K.: Parting with misconceptions about learning-based vehicle motion planning. In: Proceedings of Conference on Robot Learning (CoRL) (2023)
11. Ding, W., Chen, B., Li, B., Eun, K.J., Zhao, D.: Multimodal safety-critical scenarios generation for decision-making algorithms evaluation. IEEE Robot. Autom. Lett. (RA-L) **6**(2), 1551–1558 (2021)
12. Dosovitskiy, A., Ros, G., Codevilla, F., Lopez, A., Koltun, V.: CARLA: an open urban driving simulator. In: Proceedings of Conference on Robot Learning (CoRL) (2017)
13. Elhousni, M., Lyu, Y., Zhang, Z., Huang, X.: Automatic Building and Labeling of HD Maps with Deep Learning. arXiv:2006.00644 (2020)
14. Esser, P., et al.: Scaling Rectified Flow Transformers for High-Resolution Image Synthesis. arXiv:2403.03206 (2024)
15. Gulino, C., et al.: Waymax: an accelerated, data-driven simulator for large-scale autonomous driving research. In: Advances in Neural Information Processing Systems (NIPS) Track on Datasets and Benchmarks (2023)
16. Hanselmann, N., Renz, K., Chitta, K., Bhattacharyya, A., Geiger, A.: King: generating safety-critical driving scenarios for robust imitation via kinematics gradients. In: Avidan, S., Brostow, G., Cissé, M., Farinella, G.M., Hassner, T. (eds.) ECCV 2022. LNCS, vol. 13698, pp. 335–352. Springer, Cham (2022). https://doi.org/10.1007/978-3-031-19839-7_20
17. He, K., Zhang, X., Ren, S., Sun, J.: Deep residual learning for image recognition. In: Proceedings of IEEE Conference on Computer Vision and Pattern Recognition (CVPR) (2016)
18. He, S., Balakrishnan, H.: Lane-level street map extraction from aerial imagery. In: Proceedings of the IEEE Winter Conference on Applications of Computer Vision (WACV) (2022)
19. Ho, J., Jain, A., Abbeel, P.: Denoising diffusion probabilistic models. arXiv:2006.11239 (2020)
20. Karnchanachari, N., et al.: Towards learning-based planning: the nuPlan benchmark for real-world autonomous driving. In: Proceedings of IEEE International Conference on Robotics and Automation (ICRA) (2024)
21. Kynkäänniemi, T., Karras, T., Laine, S., Lehtinen, J., Aila, T.: Improved precision and recall metric for assessing generative models. In: Advances in Neural Information Processing Systems (NeurIPS) (2019)
22. Li, Q., Peng, Z., Feng, L., Zhang, Q., Xue, Z., Zhou, B.: Metadrive: composing diverse driving scenarios for generalizable reinforcement learning. IEEE Trans. Pattern Anal. Mach. Intell. (PAMI) **45**(3), 3461–3475 (2022)

23. Li, T., et al.: LaneSegNet: map learning with lane segment perception for autonomous driving. In: Proceedings of the International Conference on Learning Representations (ICLR) (2024)
24. Li, Z., et al.: Is ego status all you need for open-loop end-to-end autonomous driving? arXiv:2312.03031 (2023)
25. Liao, B., et al.: MAPTR: structured modeling and learning for online vectorized HD map construction. In: Proceedings of the International Conference on Learning Representations (ICLR) (2023)
26. Lin, S., Wang, A., Yang, X.: SDXL-Lightning: Progressive Adversarial Diffusion Distillation. arXiv:2402.13929 (2024)
27. Liu, X., Gong, C., Liu, Q.: Flow Straight and Fast: Learning to Generate and Transfer Data with Rectified Flow. arXiv:2209.03003 (2022)
28. Lugmayr, A., Danelljan, M., Romero, A., Yu, F., Timofte, R., Van Gool, L.: RePaint: inpainting using denoising diffusion probabilistic models. In: Proceedings IEEE Conference on Computer Vision and Pattern Recognition (CVPR) (2022)
29. Luo, S., Hu, W.: Diffusion probabilistic models for 3D point cloud generation. In: Proceedings of IEEE Conference on Computer Vision and Pattern Recognition (CVPR) (2021)
30. Luo, S., Tan, Y., Huang, L., Li, J., Zhao, H.: Latent Consistency Models: Synthesizing High-Resolution Images with Few-Step Inference. arXiv:2310.04378 (2023)
31. Mi, L., et al.: Hdmapgen: a hierarchical graph generative model of high definition maps. In: Proceedings of IEEE Conference on Computer Vision and Pattern Recognition (CVPR) (2021)
32. Nichol, A., Jun, H., Dhariwal, P., Mishkin, P., Chen, M.: Point-E: A System for Generating 3D Point Clouds from Complex Prompts. arXiv:2212.08751 (2022)
33. Nunez-Iglesias, J., Blanch, A.J., Looker, O., Dixon, M.W.A., Tilley, L.: A new python library to analyse skeleton images confirms malaria parasite remodelling of the red blood cell membrane skeleton. PeerJ **6**, e4312 (2018)
34. Peebles, W., Xie, S.: Scalable diffusion models with transformers. In: Proceedings of the IEEE International Conference on Computer Vision (ICCV) (2023)
35. Pronovost, E., et al.: Scenario diffusion: controllable driving scenario generation with diffusion. In: Advances in Neural Information Processing Systems (NeurIPS) (2023)
36. Renz, K., Chitta, K., Mercea, O.B., Koepke, S., Akata, Z., Geiger, A.: Plant: explainable planning transformers via object-level representations. In: Proceedings of Conference on Robot Learning (CoRL) (2022)
37. Rombach, R., Blattmann, A., Lorenz, D., Esser, P., Ommer, B.: High-resolution image synthesis with latent diffusion models. In: Proceedings of IEEE Conference on Computer Vision and Pattern Recognition (CVPR) (2022)
38. Santurkar, S., Budden, D., Shavit, N.: Generative Compression. arXiv:1703.01467 (2017)
39. Sauer, A., Boesel, F., Dockhorn, T., Blattmann, A., Esser, P., Rombach, R.: Fast High-Resolution Image Synthesis with Latent Adversarial Diffusion Distillation. arXiv:2403.12015 (2024)
40. Shabani, M.A., Hosseini, S., Furukawa, Y.: Housediffusion: vector floorplan generation via a diffusion model with discrete and continuous denoising. In: Proceedings of IEEE Conference on Computer Vision and Pattern Recognition (CVPR) (2023)
41. Sun, S., et al.: DriveSceneGen: Generating Diverse and Realistic Driving Scenarios from Scratch. arXiv:2309.14685 (2023)
42. Treiber, M., Hennecke, A., Helbing, D.: Congested traffic states in empirical observations and microscopic simulations. Phys. Rev. E **62**(2), 1805 (2000)

43. Vaswani, A., et al.: Attention is all you need. In: Advances in Neural Information Processing Systems (NeurIPS), pp. 5998–6008 (2017)
44. Wang, J., et al.: Advsim: generating safety-critical scenarios for self-driving vehicles. In: Proceedings of IEEE Conference on Computer Vision and Pattern Recognition (CVPR) (2021)
45. Wymann, B., Dimitrakakisy, C., Sumnery, A., Espié, E., Guionneauz, C.: Torcs: the open racing car simulator (2015)
46. Xu, M., Powers, A., Dror, R., Ermon, S., Leskovec, J.: Geometric latent diffusion models for 3D molecule generation. In: Proceedings of the International Conference on Machine learning (ICML) (2023)
47. Yang, J., et al.: Generalized Predictive Model for Autonomous Driving. arXiv:2403.09630 (2024)
48. Zeng, X., et al.: Lion: latent point diffusion models for 3d shape generation. In: Advances in Neural Information Processing Systems (NeurIPS) (2022)
49. Zhai, J.T., et al.: Rethinking the open-loop evaluation of end-to-end autonomous driving in nuscenes. arXiv:2305.10430 (2023)
50. Zhong, Z., et al.: Language-guided traffic simulation via scene-level diffusion. In: Proceedings of Conference on Robot Learning (CoRL) (2023)
51. Zhong, Z., et al.: Guided conditional diffusion for controllable traffic simulation. In: Proceedings of IEEE International Conference on Robotics and Automation (ICRA) (2023)

AFreeCA: Annotation-Free Counting for All

Adriano D'Alessandro(✉)[iD], Ali Mahdavi-Amiri[iD], and Ghassan Hamarneh[iD]

Simon Fraser University, Burnaby, Canada
{acdaless,amahdavi,hamarneh}@sfu.ca

Abstract. Object counting methods typically rely on manually annotated datasets. The cost of creating such datasets has restricted the versatility of these networks to count objects from specific classes (such as humans or penguins), and counting objects from diverse categories remains a challenge. The availability of robust text-to-image latent diffusion models (LDMs) raises the question of whether these models can be utilized to generate counting datasets. However, LDMs struggle to create images with an exact number of objects based solely on text prompts but they can be used to offer a dependable *sorting* signal by adding and removing objects within an image. Leveraging this data, we initially introduce an unsupervised sorting methodology to learn object-related features that are subsequently refined and anchored for counting purposes using counting data generated by LDMs. Further, we present a density classifier-guided method for dividing an image into patches containing objects that can be reliably counted. Consequently, we can generate counting data for any type of object and count them in an unsupervised manner. Our approach outperforms unsupervised and few-shot alternatives and is not restricted to specific object classes for which counting data is available. Code available at: github.com/adrian-dalessandro/AFreeCA.

Keywords: Object Counting · Synthetic Data · Feature Learning

1 Introduction

Object counting is an important task in computer vision, with diverse applications such as crowd analysis [11,26,31], wildlife monitoring [1], and traffic analysis [9]. These tasks, historically reliant on fully-supervised methods, demand detailed annotations that are labor-intensive to collect. For instance, the annotation of 5,109 images for a crowd counting dataset necessitated 3,000 h, averaging 35 min per image [29]. Although semi-supervised approaches have sought to reduce this burden by requiring annotations for only a subset of the data, they

Supplementary Information The online version contains supplementary material available at https://doi.org/10.1007/978-3-031-73235-5_5.

Fig. 1. We propose a method which exploits synthetic counting data generated by Stable Diffusion. With this, we establish an *annotator-free* method that produces accurate count maps without location-based supervision for a wide range of object categories.

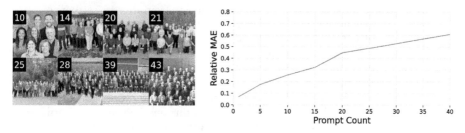

Fig. 2. *Left:* when given a prompt count of 20, Stable Diffusion outputs images with a similar but often incorrect object count. *Right:* as the prompt count increases, the relative error between the true underlying count and the prompt count increases.

still demand hundreds of hours of labor. In response, research into unsupervised counting methods has aimed to eliminate the annotation burden entirely. Despite these efforts, such methods remain challenging to develop and the existing methods are limited to crowd counting [2,17]. In this paper, using text-to-image latent diffusion models (LDMs), we propose an *unsupervised* counting method that can be applied to a *wide variety* of objects as in Fig. 1.

Recently, advancements in text-to-image LDMs, such as Stable Diffusion [23], have presented novel opportunities to address various challenges. These methods excel at producing synthetic images with a high degree of realism that accurately reflect the contents of text-based prompts. This capability enables the generation of labeled synthetic images automatically by integrating label information into the prompts. Recent works have highlighted the significant potential of this strategy for zero-shot recognition, few-shot recognition, and model pre-training [8,25]. However, these strategies have only focused on object recognition, raising the question of whether they can be applied to object counting.

A straightforward approach for extending the use of LDMs to object counting might involve prompting the network to produce images using explicit quantity labels, such as *"An image with 20 people"* with "20" used as the prompt count label as seen in Fig. 2 (Left). However, as previous work has also indicated, the

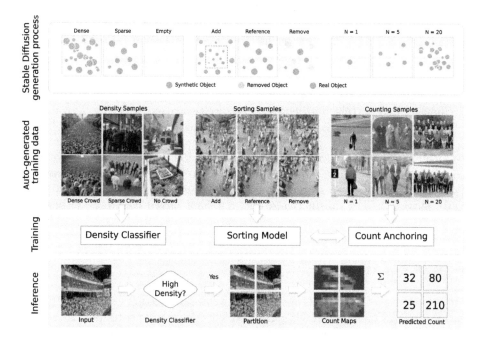

Fig. 3. Workflow. Our framework uses simple prompts to create synthetic data for training a sorting model, a density classifier, and a count anchoring network. These elements are combined into a model which can accurately count diverse object categories even within dense images by subdividing them into smaller, more manageable areas.

text encoders used in LDMs have a limited understanding of quantity [19]. To demonstrate this, we annotated 40 synthetic samples per-category across different crowd count categories and analyzed the discrepancy between prompted and actual counts, termed label noise. Figure 2 (Right) shows that the relative discrepancy increases with the magnitude of the prompt count label, suggesting that synthetic data is unreliable for images with high counts. To learn robust count-related features given this label noise we train a sorting network on a dataset comprised of image triplets, ordered by their known count ranking. These triplets are generated by synthetically adding to or removing objects from real images using LDMs, enabling the network to capture high-quality counting features. While these features do not directly correspond to specific count numbers, we anchor them using synthetic counting data. To accomplish this, we fine-tune only a linear layer on-top of the sorting network, ensuring the stability of the learned features [13, 27]. Further, given the diminished reliability of the synthetic data for high object counts, we adopt an approach at inference-time to divide dense regions into smaller patches. This partitioning strategy not only aligns the patch with the more reliable counting range of our network but it also helps in preserving high-frequency details by sourcing higher-resolution patches directly from the original images. Our method's workflow is detailed in Fig. 3. In summary, our contributions are listed in the following.

- To the best of our knowledge, we propose the *first unsupervised* counting method that can be used for a *variety of object categories*.
- We demonstrate that latent diffusion models are able to produce valuable counting and sorting data across diverse object categories. We introduce a method to leverage this data, initially focusing on learning to sort followed by a fine tuning strategy to learn counting.
- To maximize the utilization of image resolution and improve performance on highly dense images, we introduce a method that dynamically partitions dense images by leveraging guidance from a density classification network.
- We surpass SOTA unsupervised and zero-shot crowd counting methods on several crowd counting benchmark datasets. We also provide various ablation studies to justify our design.

2 Related Work

Fully-Supervised Counting. Object counting has advanced through density map learning, using Gaussian kernels convolved with dot map annotations [15,16,31]. There are a wide range of diverse approaches to leveraging these annotations. GauNet [4] employs locally connected Gaussian kernels for density map generation, introducing low-rank approximation with translation invariance. GLoss [28] reframes density map estimation as an unbalanced optimal transport problem, proposing a specialized loss function. And, AMSNet [10] uses neural architecture search and a multi-scale loss to incorporate structural information.

Unsupervised Counting. In our work, we define unsupervised counting as any method that eliminates the need for new annotations, thus reducing the annotation burden for object counting problems. This definition includes methods that utilize foundation models. CSS-CCNN [2] introduces an unsupervised approach using self-supervised pre-training with image rotation and Sinkhorn matching, requiring prior knowledge of maximum crowd count and power law distribution parameters. CrowdCLIP [17] leverages features from CLIP, a pre-trained language-image model, through multi-modal ranking with text prompts. They use a filtering strategy to select crowd-containing patches and match features to crowd count categories represented in text prompts.

Generative Models. Stable Diffusion (SD) [23] is a generative model for image synthesis based on latent diffusion models. SD generates images through a multi-step process. Initially, a real image is encoded via a variational autoencoder, yielding a compressed representation in a lower-dimensional latent space. Gaussian noise is then iteratively applied to these features. To restore meaningful features, SD employs the reverse diffusion process, an iterative denoising mechanism. Finally, the denoised features are decoded to synthesize a new image. SD can be conditioned on text embeddings, typically from a text-encoder like CLIP [21], enabling image generation guided by user-defined text prompts.

3 Methodology

To count objects in images from an *arbitrary* category, we propose using latent diffusion models (LDMs) to synthesize data. However, LDMs often misinterpret quantity in prompts, leading to discrepancies between the intended and actual object counts in generated images, known as *label noise*. This issue worsens with higher prompt counts, highlighting two key problems: the existence of a count threshold beyond which generated data becomes unreliable due to excessive label noise, and the diminished reliability of learned features for dense regions, increasing the risk of the model overfitting to noise. Our methodology addresses these problems by combining several strategies explained in the following.

First, we generate highly reliable synthetic sorting data by adding and removing objects from an image using Stable Diffusion (SD) [23] (Fig. 3). This process produces triplets of images ranked based on object count that can be used to learn robust counting features through training a *sorting network* [5,17,18]. Following this, we use SD to generate *noisy* synthetic counting data. This data is then used to fine-tune a counting network built upon our sorting network. By using the count data as anchors within the sorting features, we can preserve the reliability of the object quantity features obtained by the sorting network while establishing a correspondence between the features and the count.

As the data generated by SD tends to be more accurate for smaller counts, there could be a decline in performance when dealing with images containing a large number of objects. We extend the model's reach to a wider range of object counts by recognizing that any image can be partitioned into sub-regions with fewer objects. However, over-partitioning sparse regions can introduce noise due to boundary artifacts, double counting, etc. Thus, we need a method to identify dense regions within an image. As a result, we generate synthetic crowd density classification data and then fine-tune a density guidance network on top of our pre-trained sorting model to identify dense image regions. At the inference stage, this density guidance is instrumental in identifying image sections that might exceed the counting model's reliable counting range. These identified regions are then processed at a higher resolution (when available) to ensure accurate counting. By retrieving higher-resolution patches from the original image, we address issues related to feature loss caused by resizing images to a fixed size before feeding them into a network. Therefore, dividing an image into patches leads to more precise counting estimations.

3.1 Learning to Sort

Generating Synthetic Sorting Data. We generate a dataset of images sorted based on their object counts by modifying reference images to either add or remove objects, using a LDM for image-to-image synthesis or out-painting. Unlike previous approaches that focus on intra-image ranking for counting [18], our strategy can add objects and also remove objects while preserving the scene's perspective and original features. This broadens the diversity and range of object counts in the resulting image triplets. The process begins with a base dataset of real or generated reference images, represented as $\mathcal{D}^{\text{ref}} = \{x_i^{\text{ref}}\}^{N_{\text{ref}}}$, where x_i^{ref} denotes an image with an unknown number of objects c_i^{ref}, and N_{ref} is the total number of such images. To synthesize images with fewer objects, we apply Stable Diffusion (SD):

$$x_i^{\text{syn}-} = SD(x_i^{\text{ref}}, t_p, t_n) \tag{1}$$

where $x_i^{\text{syn}-}$ is the resultant image with fewer objects, guided by a text prompt t_p and, to prevent the addition of targeted objects, a negative prompt t_n. As an example, we set t_p to "an empty place" and t_n to "pedestrians, humans, people, crowds" for crowd counting problems. Due to the stochastic nature of the image generation process, $x_i^{\text{syn}-}$ contains an unknown number of objects c_i^{syn}. In the supplementary material, we empirically verify that the relationship, $c_i^{\text{ref}} \geq c_i^{\text{syn}}$, holds in 99% of cases.

For adding objects, we engage in a similar process using outpainting:

$$x_i^{\text{syn}+} = SD(x_i^{\text{ref}}, \mathcal{M}, t_p, t_n) \tag{2}$$

where $\mathcal{M}$ is a mask indicating where to outpaint and, thus, add new objects and is set to a thick band around the perimeter of the image x_i^{ref}, which is 1/3rd of the image size. This approach effectively increases the object count and scene density. We transform each image x_i^{ref} into four augmented versions with increased object counts $\{x_{ij}^{\text{syn}+}\}_{j=1}^4$ and four with decreased counts $\{x_{ik}^{\text{syn}-}\}_{k=1}^4$, from which the 16 ordered triplets $\{x_{ij}^{\text{syn}-} \geq x_i^{\text{ref}} \geq x_{ik}^{\text{syn}+}\}_{j,k=1}^4$ are generated.

Pre-training. For a given triplet of images $X = \{x^{\text{syn}-}, x^{ref}, x^{\text{syn}+}\}$ with rank labels $Y = \{0, 1, 2\}$, we generate a similarity matrix $[\mathcal{S}^y]_{i,j} = -|y_i - y_j|$. We encode X using network f_θ to produce feature vectors $Z = \{z^{syn-}, z^{ref}, z^{syn+}\}$; $z_i \in \mathbb{R}^{2048}$. We further utilize a sorting head v_Θ to produce a continuous valued output $\hat{y}_i$ from z_i as an estimate of the rank of x_i. We then calculate a second similarity matrix $[\mathcal{S}^z]_{i,j}$ using the cosine similarity between z_i and z_j. $\mathcal{S}^z$ and $\mathcal{S}^y$ encode the relational structure between examples and are used to align the feature space and label space with the ground truth ordering by minimizing the following losses (this is akin to RankSim [6] with the key difference that we use count sorting labels instead of ground truth values like they do):

$$\ell_{sort}^y = \sum_i^3 (\mathbf{rk}(S_i^y) - \mathbf{rk}(S_i^{\hat{y}}))^2 \qquad \ell_{sort}^z = \sum_i^3 (\mathbf{rk}(S_i^y) - \mathbf{rk}(S_i^z))^2, \tag{3}$$

where $S_i^{\hat{y}}$ is the predicted similarity matrix, S_i^* is i-th row of S^*, and $\mathbf{rk}$ is a non-differentiable ranking function. The total sorting loss is then defined as:

$$\mathcal{L}_{sort} = \ell_{sort}^y + \lambda \ell_{sort}^z, \tag{4}$$

where λ is set to 5.0. To optimize $\mathbf{rk}$, we follow the same strategy as in RankSim [6] and apply a blackbox combinatorial solver [20].

Fig. 4. Sorting Features. We calculate the channel-wise mean of the features produced by the sorting network to demonstrate where the network is active. The network appears to focus on the object of interest across a wide range of crowd densities.

3.2 Learning to Count from Synthetic Data

Our approach generates synthetic images with approximate object counts using SD for text-to-image synthesis:

$$\{x_i^s, c_i^p\} = SD(t_p, t_n), \tag{5}$$

where c_i^p is the object count from the prompt, and x_i^s are the generated images. This yields a dataset of synthetic examples $\mathcal{D}_{\text{cnt}}^s = \{x_i^s, c_i^p\}^{N_{\text{cnt}}}$. We use simple prompts to ensure accurate image generation, and to avoid complex prompt engineering. For crowd counting, prompts are straightforward, e.g., "20 people." Our experiments involve generating images for a wide range of counts from 1 to 1000, which we elaborate on in the supplementary material. We generate 150 images for each prompt category, plus 800 images with zero objects generated by random sampling scene category names from the Places365 [32] dataset for prompts and explicitly excluding objects using negative prompts. For training our counting network with the synthetic dataset $\mathcal{D}_{\text{cnt}}^s$ and the pre-trained sorting network f_θ, we adopt a strategy to preserve the integrity of f_θ features. Previous works have indicated that fine-tuning an entire network distorts the pre-trained features of that network and degrades performance on out-of-distribution data [13,27]. In Fig. 4, we observe that the feature maps produced by f_θ focus on the image regions where objects exist, which is a property that we would like to preserve. Given that we are training our counting network using synthetic data, and we are evaluating on real data, to avoid the potential for any feature distortion we only fine-tune a linear layer atop the pre-trained sorting network to anchor its features to actual count values.

To further ensure that our counting network is as accurate as possible, we automatically filter out outliers from the synthetic dataset using a feature space analysis approach, which is effective for mitigating the impacts of label noise [14, 30,33]. We adopt a simplified version of CleanNet [14], which calculates class prototypes as well as the similarity between samples and prototypes to detect noisy samples. We compute features z_i for each image x_i^s using f_θ and then create a prototype vector z_μ^c for each prompt count category c:

$$z_\mu^c = \frac{1}{N} \sum_{i=1}^{N} z_i^c. \tag{6}$$

We then filter out synthetic images whose features align closer with a different category's prototype than their own. Finally, we train a counting network, g_Φ optimizing a mean squared error:

$$\mathcal{L}_{count} = \frac{1}{N_{cnt}} \sum_{(x_i^s, c_i^p) \in \mathcal{D}_{cnt}^s} (g_\Phi \circ f_\theta(x_i^s) - c_i^p)^2. \tag{7}$$

3.3 Crowd Density Classification

Given that our counting networks is more reliable for smaller counts, we partition dense images into patches with lower counts to enhance estimation accuracy in each patch. This partitioning process is directed by a classifier that discerns between dense and sparse images. Given that dense image regions may be poorly represented by our counting model, a dataset is necessary for training such a classifier. We use stable diffusion to produce a classification dataset of synthetic images with various crowd densities $\mathcal{D}_{dense}^s = \{x_i^s, y_i^{den}\}^{N_{den}}$, where x_i^s is the synthetic image, y_i^{den} is the density category recorded in the prompt, and N_{den} is the size of the dataset. As highlighted in Fig. 3, we categorize densities into three groups: no crowd, sparse crowd, and dense crowd, embedding these categories directly into the generation prompts to create representative images for each. We re-purpose the zero-object examples from the earlier noisy count dataset for our "no crowd" category. We then use this synthetic dataset to train our classifier by fine-tuning a density classification layer, h_ϕ, on top of pre-trained sorting network. The loss function for this step is:

$$\mathcal{L}_{dense} = \frac{1}{N_{den}} \sum_{(x_i^s, y_i^{den}) \in \mathcal{D}_{den}^s} \ell_{CE}(h_\phi \circ f_\theta(x_i^s), y_i^{den}), \tag{8}$$

where ℓ_{CE} is the cross-entropy loss.

3.4 Density Classifier Guided Partitioning (DCGP)

We previously argued that by partitioning images into sub-region patches, each patch will have fewer objects and be more likely to fall within the accurate range of our counting model. However, images often contain groups of objects at a variety of sizes and crowd densities due to the perspective of the scene. Given this, we propose a technique called density classifier guided partitioning (DCGP), which treats image patches differently based on their estimated density.

For each image, we produce a count map to approximate the number of objects within the respective region (see Fig. 5, bottom). In cases where a region exhibits sparsity according to our density classifier, like the top-right quadrant of the count map depicted in Fig. 5, the original count maps can be directly employed to estimate the count within that region. Conversely, if a count map corresponds to a dense region, the spatial patch associated with that region, potentially at a higher resolution, is extracted from the input image. This patch is then forwarded to our counting network to obtain a more precise estimate,

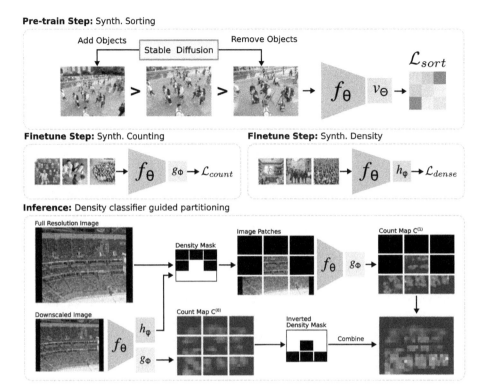

Fig. 5. Methodology. Our strategy involves three distinct steps supported by a synthetic training signal extracted from stable diffusion. The pre-training step sorts synthetic and real images to learn high quality object quantity features from the source distribution. The finetuning step utilizes the pre-trained features and synthetic data to train a counting head and a density classification head. Finally, the density classifier guides inference by partitioning dense images so that there are fewer objects per image.

considering the smaller number of objects within the patch, enabling the counting network to perform more effectively in estimation. This hybrid resolution counting estimation is eventually combined to find the final count.

Formally, to obtain a count map we adapt our pre-trained sorting network f_θ, omitting the global average pooling layer. This modified network produces feature maps $z \in \mathbb{R}^{H \times W \times 2048}$, where H is the height of the feature map and W is the width. We calculate the regional counts for each feature map element as follows:

$$c_{ij}^{(0)} = g_\Phi(z_{ij}/(H \cdot W)), \tag{9}$$

where $c_{ij}^{(0)}$ denotes the count for the (i, j)-th element, providing a comprehensive count map fir the image. We further classify each region's density using f_θ and a density classifier h_ϕ to generate a density map D, where D_{ij} represents the density class for the (i, j)-th region in z. This allows us to estimate which regions in the whole image are dense and require further processing.

Finally, we partition a high resolution version of a given image into a 3 × 3 grid of image patches. For sparse patches, we directly sum counts from $c^{(0)}$. For dense patches, we process them through the counting network for a precise patch count estimate $c^{(1)}$, summing these to achieve the image's final count.

Table 1. Crowd Counting Performance. We evaluate the performance of our method on the test set of crowd counting benchmark datasets. It is evident that our method outperforms other unsupervised and zero-shot techniques.

Method	Type	SHB		JHU		SHA		QNRF	
		MAE	MSE	MAE	MSE	MAE	MSE	MAE	MSE
ADSCNet [3]	Full supervised	6.4	11.3	–	–	55.4	97.7	71.3	132.5
GLoss [28]	Full supervised	7.3	11.7	59.9	259.5	61.3	95.4	84.3	147.5
GauNet [4]	Full supervised	6.2	9.9	58.2	245.1	54.8	89.1	81.6	153.7
CLIP-Count [12]	Zero-Shot	45.7	77.4	–	–	192.6	308.4	–	–
CSS-CCNN++ [2]	Unsupervised	–	–	197.9	611.9	195.6	293.2	414.0	652.1
CrowdCLIP [17]	Unsupervised	69.3	85.8	213.7	576.1	**146.1**	236.3	283.3	488.7
Ours	Unsupervised	**35.0**	**50.7**	**173.8**	**519.4**	152.7	**219.0**	**283.1**	**453.2**

Table 2. Object Counting Performance. Our method outperforms few-shot and zero-shot methods on the CARPK dataset.

Method	Type	CARPK	
		MAE	MSE
FamNet [22]	Few-shot	28.84	44.47
BMNet+ [24]	Few-shot	10.44	13.77
CLIP-Count [12]	Zero-shot	11.96	16.61
Ours	Unsupervised	**9.35**	**12.29**

4 Experiments and Results

4.1 Performance

Datasets. We benchmark on 4 crowd counting datasets: ShanghaiTechA (SHA) [31], ShanghaiTechB (SHB) [31], UCF-QNRF [11], and JHU-Crowd++ [26] with average crowd counts of 501, 123, 815, and 346 respectively. We further benchmark on a vehicle counting dataset [9] and a penguin counting dataset [1].

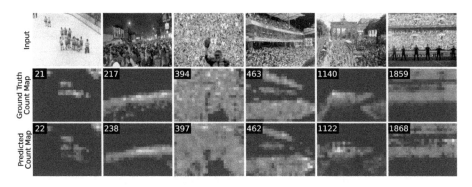

Fig. 6. Qualitative Crowd Counting. We compare the count maps produced by our model to the ground truth count maps. The count maps are annotated with the ground truth counts and predicted counts. Our method accurately localizes crowds across a range of crowd densities, without any location-based supervision.

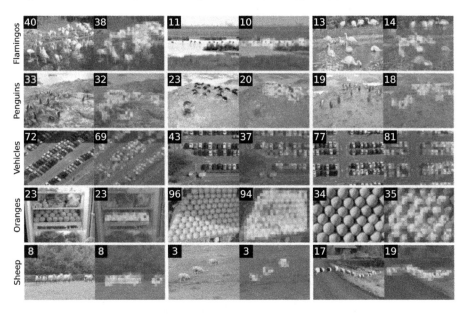

Fig. 7. Qualitative Object Counting. We evaluate our method across a diverse range of object categories and demonstrate its proficiency in accurately predicting object counts and also generating precise count maps for each evaluated category.

Main Results. Our results in Table 1 and Table 2 showcase our method's performance on both crowd and vehicle counting tasks, evaluated using mean absolute error (MAE) and mean squared error (MSE). We compare our approach to unsupervised, few-shot, and zero-shot methods, where authors have made those results available. Few-shot and zero-shot methods aim to create a general counting network for any category using a large, manually annotated dataset with a

Table 3. Partitioning Strategy. We explore the impact of the image partitioning rate for naive partitioning and density classifier guidance.

Strategy	Rate	SHB		JHU		SHA		QNRF	
		MAE	MSE	MAE	MSE	MAE	MSE	MAE	MSE
Fixed	1×1	42.1	68.1	203.6	649.6	196.6	321.4	392.5	707.1
Fixed	2×2	51.6	62.1	202.4	531.0	165.7	258.2	314.5	484.8
Fixed	3×3	47.1	57.1	219.5	501.8	186.9	278.8	425.6	559.9
Simple GP	$1 \times 1 \rightarrow 2 \times 2$	42.1	68.1	181.7	541.9	163.1	255.9	299.9	487.7
Simple GP	$1 \times 1 \rightarrow 3 \times 3$	42.1	68.1	173.9	**495.3**	155.7	225.4	309.5	480.2
DCGP	$1 \times 1 \rightarrow 2 \times 2$	38.5	61.0	182.8	562.9	156.9	230.8	324.0	541.7
DCGP	$1 \times 1 \rightarrow 3 \times 3$	**35.0**	**50.7**	**173.8**	519.4	**152.7**	**219.0**	**283.1**	**453.2**

Table 4. Pre-training Ablation Study. We explore the impact of different pre-training strategies on the final performance of the network

Pre-training	SHB		JHU		SHA		QNRF	
	MAE	MSE	MAE	MSE	MAE	MSE	MAE	MSE
ImageNet	68.1	104.7	304.9	654.9	316.2	410.0	423.1	655.9
Intra-Image Rank	52.8	76.3	256.4	610.4	186.6	291.2	380.0	694.9
ℓ^y_{sort} only	38.2	58.0	183.9	551.1	188.4	293.5	359.3	579.1
$\ell^y_{sort} + \ell^z_{sort}$ (Ours)	**35.0**	**50.7**	**173.8**	**519.4**	**152.7**	**219.0**	**283.1**	**453.2**

diverse set of categories. Few-shot methods rely on an exemplar, sampled from the target image, to define the category of interest, whereas zero-shot methods use a text prompt. Unlike these methods, which depend on extensive annotated datasets but produce a general counting network, our unsupervised approach learns to count objects for a single category without any manual annotations. Despite this difference, we include these methods in our comparison for two reasons: First, they share the common goal of removing or reducing the annotation burden. Second, there are no unsupervised counting methods which have provided evaluation on object categories beyond crowds.

Table 5. Counting Network Ablation Study. We explore the performance of different strategies when training the counting network with synthetic counting data.

Strategy	SHB		JHU		SHA		QNRF	
	MAE	MSE	MAE	MSE	MAE	MSE	MAE	MSE
$\mathcal{L}_{count}$ only train	122.9	154.6	318.5	785.7	408.8	531.8	711.2	1027.6
Full Network Finetune	43.9	74.6	292.4	749.0	340.8	462.5	664.8	965.2
Last Layer Finetune	**35.0**	**50.7**	**173.8**	**519.4**	**157.7**	**219.0**	**283.1**	**453.2**

Table 6. High Resolution Partitioning. We explore the performance implications of selecting partitions from a high-resolution image compared to a low resolution image.

Strategy	SHB		JHU		SHA		QNRF	
	MAE	MSE	MAE	MSE	MAE	MSE	MAE	MSE
Low Resolution	35.6	55.0	180.2	563.8	154.4	223.8	317.6	552.8
Full Resolution	35.0	50.7	173.8	519.4	152.7	219.0	283.1	453.2

With respect to crowd counting, we find that our method outperforms CLIP-Count [12] on all available datasets. It also outperforms both CrowdCLIP [17] and CSS-CCNN [2] on every dataset, with the sole exception of the SHA dataset in terms of MAE. Notably, our technique significantly improves on the JHU and SHB datasets, indicating that it particularly excels in scenarios with fewer objects. With respect to object counting, there are no unsupervised counting methods which have explored objects beyond pedestrians in crowds. However, we find that our unsupervised approach surpasses existing zero-shot and few-shot methods, as demonstrated on the CARPK dataset. Additionally, we assess our method on the penguins dataset [1], to establish a benchmark for future researchers, where we report a MAE of 5.1, and a MSE of 8.0.

In summary, our method not only outperforms existing unsupervised approaches in crowd counting but also outperforms zero-shot methods across both crowd and vehicle counting tasks. This suggests that our method provides superior performance for arbitrary categories.

Qualitative Results. We highlight qualitative results for crowd counting in Fig. 6 by comparing count maps estimated by our method to count maps calculated from ground truth dot maps. We demonstrate that our approach produces accurate count maps that localize crowds across a wide range of counts without *any* location-based supervision. We further extend our evaluation to arbitrary object categories, as seen in Fig. 7. For penguins and vehicles, we utilize sorting data generated from training set images of the penguins [1] and the CARPK [9] datasets, respectively, assessing performance on their test sets. For flamingos, oranges, and sheep, we create synthetic images with simple prompts (e.g., *"Many sheep. Photograph."*), combined with random scene names from the Places365 dataset [32]. We evaluate quality on a mix of images from the FSC147 dataset [22] and manually annotated public domain images, confirming our method's robustness in accurately localizing diverse object categories.

4.2 Ablation Study

Impact of Partition Strategy. In Table 3, we examine the effect of partitioning strategies on network performance. Initially, we explore a straightforward approach of dividing all images into an $M \times M$ grid at different rates of M, which we refer to as fixed partitioning. We then evaluate a simple guided partitioning strategy, where we use the density classifier to provide a single density estimate for the whole image. If the whole image is classified as dense, we use $M \times M$ partitioning, and 1×1 otherwise. Finally, we evaluate DCGP using different partitioning rates M, where we start with a 1×1 rate and partition with an $M \times M$ rate for image regions determined to be dense by h_ϕ. Notably, DCGP provides a significant performance boost compared to naively applying a 1×1 rate or $M \times M$ rate, indicating that our method is successfully addressing the challenges introduced by the synthetic label noise. Further, we determine that the optimal partitioning rate is 1×1 to 3×3, which we apply to all datasets.

Impact of Pre-training. Our approach introduces a pre-training strategy aimed at developing high-quality features for object quantity estimation, serving as a foundation for the subsequent training of our counting network. In Table 4, we explore different pre-training strategies to determine the impact of the feature learning strategy on the final performance of the network. We compare our method with networks pre-trained on ImageNet [7] and using an intra-image ranking signal [18], as well as examining the sorting signal alone. Our findings demonstrate that our pre-training technique significantly outperforms these alternatives, and that regularizing the feature space improves performance.

Counting Network Training. In Sect. 3.2 we argued that finetuning the sorting network with the noisy synthetic counting data can distort the learned features. In Table 5, we examine different strategies for producing the counting network. Initially, we train the network solely with the counting signal, which confirms that relying exclusively on noisy synthetic counting data yields inferior results. Further investigation into finetuning the entire pre-trained sorting network reveals that such an approach adversely affects performance, supporting our argument that we should favor linear probing to avoid feature distortion.

Impact of Image Resolution. At inference time, we resize all images to a fixed resolution. However, many images are of a higher resolution than this fixed size. When we partition an image, we sample the patches from the higher resolution image to preserve image details. In Table 6, we explore the impact of this strategy and find that it improves performance for all datasets. This effect is especially pronounced for datasets with very high resolution images like QNRF.

5 Conclusions and Future Work

In this paper, we provide an *unsupervised* counting method that can be applied to *multiple object categories*. We have tackled this challenging task by using synthetic counting data generated by latent diffusion models (LDMs), which provides a flexible strategy for effectively eliminating the annotation burden. However, LDMs face challenges in reliably understanding object quantity, which results in noisy annotations. To mitigate this issue, we employed LDMs to create two additional types of synthetic data: one by manipulating the number of objects within images, which yields an ordered image with a weak but very reliable object quantity label, and the other by generating synthetic images with varying object density categories, offering the ability to detect dense image regions. Given these components, we deploy a novel inference time strategy that automatically detects when images are dense and then partitions those images to reduce the total number of objects per sub-image. We demonstrate our method's superiority over the SOTA in unsupervised crowd counting and zero-shot crowd counting across multiple benchmark datasets. Our work not only significantly alleviates the annotation burden but outlines a new direction for unsupervised counting. Our approach, like others, has limitations. For instance, its performance falls short compared to supervised methods. Injecting limited reliable counting data into the counting network may enhance performance. Additionally, our method operates effectively on images generated by LDMs, but may not perform well on non-natural image datasets such as medical data. Addressing these challenges can be a future research direction.

Acknowledgments. This work was supported by the Natural Sciences and Engineering Research Council of Canada (NSERC) through the Discovery Grant program. We also extend our deepest gratitude to the Digital Research Alliance of Canada for the use of the Cedar supercomputing cluster, which was essential to our research. Finally, we thank the Computing Science Department at Simon Fraser University for their continued support and resources.

References

1. Arteta, C., Lempitsky, V., Zisserman, A.: Counting in the wild. In: European Conference on Computer Vision (2016)
2. Babu Sam, D., Agarwalla, A., Joseph, J., Sindagi, V.A., Babu, R.V., Patel, V.M.: Completely self-supervised crowd counting via distribution matching. In: Avidan, S., Brostow, G., Cissé, M., Farinella, G.M., Hassner, T. (eds.) Computer Vision - ECCV 2022, pp. 186–204. Springer, Cham (2022). https://doi.org/10.1007/978-3-031-19821-2_11
3. Bai, S., He, Z., Qiao, Y., Hu, H., Wu, W., Yan, J.: Adaptive dilated network with self-correction supervision for counting. In: Proceedings of the IEEE/CVF Conference on Computer Vision and Pattern Recognition (CVPR) (2020)
4. Cheng, Z.Q., Dai, Q., Li, H., Song, J., Wu, X., Hauptmann, A.G.: Rethinking spatial invariance of convolutional networks for object counting. In: Proceedings of the IEEE/CVF Conference on Computer Vision and Pattern Recognition, pp. 19638–19648 (2022)
5. D'Alessandro, A.C., Mahdavi-Amiri, A., Hamarneh, G.: Learning-to-count by learning-to-rank. In: 2023 20th Conference on Robots and Vision (CRV), pp. 105–112 (2023). https://doi.org/10.1109/CRV60082.2023.00021
6. Gong, Y., Mori, G., Tung, F.: RankSim: ranking similarity regularization for deep imbalanced regression. In: International Conference on Machine Learning (ICML) (2022)

7. He, K., Zhang, X., Ren, S., Sun, J.: Deep residual learning for image recognition. In: Proceedings of the IEEE Conference on Computer Vision and Pattern Recognition, pp. 770–778 (2016)
8. He, R., et al.: Is synthetic data from generative models ready for image recognition? In: The Eleventh International Conference on Learning Representations (ICLR) (2023). https://openreview.net/forum?id=nUmCcZ5RKF
9. Hsieh, M.R., Lin, Y.L., Hsu, W.H.: Drone-based object counting by spatially regularized regional proposal networks. In: The IEEE International Conference on Computer Vision (ICCV). IEEE (2017)
10. Hu, Y., Jiang, X., Liu, X., Zhang, B., Han, J., Cao, X., Doermann, D.: NAS-count: counting-by-density with neural architecture search. In: Vedaldi, A., Bischof, H., Brox, T., Frahm, J.-M. (eds.) ECCV 2020. LNCS, vol. 12367, pp. 747–766. Springer, Cham (2020). https://doi.org/10.1007/978-3-030-58542-6_45
11. Idrees, H., et al.: Composition loss for counting, density map estimation and localization in dense crowds. In: Proceedings of the European Conference on Computer Vision (ECCV), pp. 532–546 (2018)
12. Jiang, R., Liu, L., Chen, C.: Clip-count: towards text-guided zero-shot object counting. In: Proceedings of the 31st ACM International Conference on Multimedia, MM 2023, pp. 4535–4545. Association for Computing Machinery, New York (2023)
13. Kumar, A., Raghunathan, A., Jones, R.M., Ma, T., Liang, P.: Fine-tuning can distort pretrained features and underperform out-of-distribution. In: International Conference on Learning Representations (2022). https://openreview.net/forum?id=UYneFzXSJWh
14. Lee, K.H., He, X., Zhang, L., Yang, L.: Cleannet: transfer learning for scalable image classifier training with label noise. In: Proceedings of the IEEE Conference on Computer Vision and Pattern Recognition (CVPR) (2018)
15. Lempitsky, V., Zisserman, A.: Learning to count objects in images. Adv. Neural Inf. Process. Syst. **23** (2010)
16. Li, Y., Zhang, X., Chen, D.: CSRNET: dilated convolutional neural networks for understanding the highly congested scenes. In: Proceedings of the IEEE Conference on Computer Vision and Pattern Recognition, pp. 1091–1100 (2018)
17. Liang, D., Xie, J., Zou, Z., Ye, X., Xu, W., Bai, X.: Crowdclip: unsupervised crowd counting via vision-language model. In: Proceedings of the IEEE/CVF Conference on Computer Vision and Pattern Recognition (CVPR), pp. 2893–2903 (2023)
18. Liu, X., Van De Weijer, J., Bagdanov, A.D.: Leveraging unlabeled data for crowd counting by learning to rank. In: Proceedings of the IEEE Conference on Computer Vision and Pattern Recognition, pp. 7661–7669 (2018)
19. Paiss, R., et al.: Teaching clip to count to ten. In: Proceedings of the IEEE/CVF International Conference on Computer Vision (ICCV), pp. 3170–3180 (2023)
20. Pogančić, M.V., Paulus, A., Musil, V., Martius, G., Rolinek, M.: Differentiation of blackbox combinatorial solvers. In: International Conference on Learning Representations (2020). https://openreview.net/forum?id=BkevoJSYPB
21. Radford, A., et al.: Learning transferable visual models from natural language supervision. In: International Conference on Machine Learning, pp. 8748–8763. PMLR (2021)
22. Ranjan, V., Sharma, U., Nguyen, T., Hoai, M.: Learning to count everything. In: Proceedings of the IEEE/CVF Conference on Computer Vision and Pattern Recognition (CVPR) (2021)

23. Rombach, R., Blattmann, A., Lorenz, D., Esser, P., Ommer, B.: High-resolution image synthesis with latent diffusion models. In: Proceedings of the IEEE/CVF Conference on Computer Vision and Pattern Recognition, pp. 10684–10695 (2022)
24. Shi, M., Hao, L., Feng, C., Liu, C., Cao, Z.: Represent, compare, and learn: a similarity-aware framework for class-agnostic counting. In: Proceedings of IEEE/CVF Conference on Computer Vision and Pattern Recognition (CVPR) (2022)
25. Shipard, J., Wiliem, A., Thanh, K.N., Xiang, W., Fookes, C.: Diversity is definitely needed: improving model-agnostic zero-shot classification via stable diffusion. In: Proceedings of the IEEE/CVF Conference on Computer Vision and Pattern Recognition (CVPR) Workshops, pp. 769–778 (2023)
26. Sindagi, V.A., Yasarla, R., Patel, V.M.: Jhu-crowd++: large-scale crowd counting dataset and a benchmark method. IEEE Trans. Pattern Anal. Mach. Intell. **44**(5), 2594–2609 (2020)
27. Trivedi, P., Koutra, D., Thiagarajan, J.J.: A closer look at model adaptation using feature distortion and simplicity bias. In: The Eleventh International Conference on Learning Representations (2023). https://openreview.net/forum?id=wkg_b4-IwTZ
28. Wan, J., Liu, Z., Chan, A.B.: A generalized loss function for crowd counting and localization. In: Proceedings of the IEEE/CVF Conference on Computer Vision and Pattern Recognition, pp. 1974–1983 (2021)
29. Wang, Q., Gao, J., Lin, W., Li, X.: Nwpu-crowd: a large-scale benchmark for crowd counting and localization. IEEE Trans. Pattern Anal. Mach. Intell. (2020). https://doi.org/10.1109/TPAMI.2020.3013269
30. Wu, P., Zheng, S., Goswami, M., Metaxas, D., Chen, C.: A topological filter for learning with label noise. In: Larochelle, H., Ranzato, M., Hadsell, R., Balcan, M., Lin, H. (eds.) Advances in Neural Information Processing Systems, vol. 33, pp. 21382–21393. Curran Associates, Inc. (2020). https://proceedings.neurips.cc/paper_files/paper/2020/file/f4e3ce3e7b581ff32e40968298ba013d-Paper.pdf
31. Zhang, Y., Zhou, D., Chen, S., Gao, S., Ma, Y.: Single-image crowd counting via multi-column convolutional neural network. In: Proceedings of the IEEE Conference on Computer Vision and Pattern Recognition, pp. 589–597 (2016)
32. Zhou, B., Lapedriza, A., Khosla, A., Oliva, A., Torralba, A.: Places: a 10 million image database for scene recognition. IEEE Trans. Pattern Anal. Mach. Intell. **40**, 1452–1464 (2017)
33. Zhu, Z., Dong, Z., Liu, Y.: Detecting corrupted labels without training a model to predict. In: Proceedings of the 39th International Conference on Machine Learning. Proceedings of Machine Learning Research, vol. 162, pp. 27412–27427. PMLR (2022). https://proceedings.mlr.press/v162/zhu22a.html

Adversarially Robust Distillation by Reducing the Student-Teacher Variance Gap

Junhao Dong[1,2], Piotr Koniusz[3,4], Junxi Chen[5], and Yew-Soon Ong[1,2]

[1] Nanyang Technological University, Singapore, Singapore
{junhao003,asysong}@ntu.edu.sg
[2] Centre for Frontier AI Research, IHPC, A*STAR, Singapore, Singapore
[3] Australian National University, Canberra, Australia
[4] Data61 CSIRO, Canberra, Australia
piotr.koniusz@data61.csiro.au
[5] Sun Yat-sen University, Guangzhou, China
chenjx353@mail2.sysu.edu.cn

Abstract. Adversarial robustness generally relies on large-scale architectures and datasets, hindering resource-efficient deployment. For scalable solutions, adversarially robust knowledge distillation has emerged as a principle strategy, facilitating the transfer of robustness from a large-scale teacher model to a lightweight student model. However, existing works focus solely on sample-to-sample alignment of features or predictions between the teacher and student models, overlooking the vital role of their statistical alignment. Thus, we propose a novel adversarially robust knowledge distillation method that integrates the alignment of feature distributions between the teacher and student backbones under adversarial and clean sample sets. To motivate our idea, for an adversarially trained model (e.g., student or teacher), we show that the robust accuracy (evaluated on testing adversarial samples under an increasing perturbation radius) correlates negatively with the gap between the feature variance evaluated on testing adversarial samples and testing clean samples. Such a negative correlation exhibits a strong linear trend, suggesting that aligning the feature covariance of the student model toward the feature covariance of the teacher model should improve the adversarial robustness of the student model by reducing the variance gap. A similar trend is observed by reducing the variance gap between the gram matrices of the student and teacher models. Extensive evaluations highlight the state-of-the-art adversarial robustness and natural performance of our method across diverse datasets and distillation scenarios.

Supplementary Information The online version contains supplementary material available at https://doi.org/10.1007/978-3-031-73235-5_6.

1 Introduction

Deep Neural Networks (DNNs) [20,22,45] have been demonstrated to be susceptible to adversarial samples [49]. These malicious inputs, subtly altered with visually imperceptible perturbations, can disrupt deep learning-based systems [14,27]. Thus, establishing robustness against adversarial samples becomes vital for the practical applicability of DNNs in trustworthy real-world applications [11,30]. To this end, adversarial training has emerged as the most effective defense approach against adversarial samples [12,21,37]. However, achieving substantial adversarial robustness essentially necessitates the use of large-scale models or abundant training data [47,52]. To circumvent this and achieve robustness in resource-efficient scenarios, adversarially robust knowledge distillation [10,19,25,58,59] transfers adversarial robustness from a large-scale *teacher* model to a lightweight *student* model under some performance trade-off. Existing robust distillation approaches typically align feature representations or predictions between the teacher and student models. However, such a sample-to-sample feature alignment scheme differs from the feature-statistics alignment which often serves as a better mechanism for modeling properties of sets of samples. This oversight leads to a significant gap between the variance of features of adversarial samples and the variance of features of clean samples, consequently limiting the robust generalization ability of the student, as detailed in the motivation below.

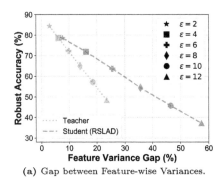
(a) Gap between Feature-wise Variances.

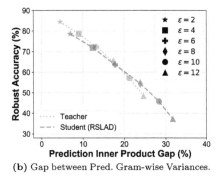
(b) Gap between Pred. Gram-wise Variances.

Fig. 1. Analysis on robust teacher and lightweight student (RSLAD [59]). Figure 1a: The robust accuracy (on test adversarial samples under an increasing perturbation radius $255 \cdot \epsilon$) correlates negatively with the gap between the feature variance evaluated on testing adversarial samples and testing clean samples. Figure 1b: A similar trend occurs for the gap between variances of prediction score-based gram matrices (inner product). The negative correlation with linear characteristics suggests that reducing the variance gaps of the student toward the variance gaps of the teacher should improve robustness.

Motivation. The feature variance has been linked with the generalization gap. For example, Huang et al. [26] observed that for functions $f \in \mathcal{F}$, where $\mathcal{F}$ is a finite class of functions, for each $\delta \in (0,1)$ with probability at least $1-\delta$, the gap between expected and empirical risks r and $\tilde{r}$ is bounded as:

$$|r(f,\mathcal{Q}) - \tilde{r}(f(\mathbf{X}),\mathbf{Y})| \leq \zeta + \sqrt{8(N-1)\sigma^2(f(\mathbf{X}))\iota/N},$$

where $\iota = \ln(3|\mathcal{F}|/\delta)$, $\zeta = 3.5\iota/N + 3/$, $|\mathcal{F}|$ is the cardinality of a class of functions $\mathcal{F}$, N is the number of samples, $\mathcal{Q}$ is the data distribution (e.g., $(\mathbf{x},y) \sim \mathcal{Q}$), whereas $f(\mathbf{X})$ and $\mathbf{Y}$ are features and labels for empirical samples (e.g., column vectors of $\mathbf{X}$), and $\sigma^2(\cdot)$ estimates their variance.

Considering the above bound, we suggest that matching the variance of features of adversarial samples with the variance of features of clean samples can naturally tighten the gap between the expected risk and the empirical risks of adversaries, i.e., $|r(f,\mathcal{Q}) - \tilde{r}(f(\mathbf{X}_{adv}),\mathbf{Y})|$ gets closer to $|r(f,\mathcal{Q}) - \tilde{r}(f(\mathbf{X}),\mathbf{Y})|$ when $\sigma^2(f(\mathbf{X}_{adv}))$ gets closer to $\sigma^2(f(\mathbf{X}))$. The bound in the equation is fulfilled simultaneously for both $\mathbf{X}$ and $\mathbf{X}_{adv}$ with probability at least $(1-\delta)^2$.

To support our claim, we study the robust accuracy of adversarial test samples w.r.t. the gap between the variance of features from adversarial test samples and that from clean samples (see Fig. 1a). Note that the larger the variance gap is, the worse the robust accuracy is, irrespective of whether the experiment is performed on the robust teacher or the student model (RSLAD [59]). Moreover, as RSLAD performs sample-to-sample distillation, its variance gap remains larger than that of the teacher across diverse perturbation radii, resulting in comparatively lower robustness. Figure 1b presents similar findings related to the variance gap of prediction score-based gram matrices (inner products).

Drawing on the insights from the above analyses, we hypothesize that bridging the gap in the robust accuracy of student and teacher can be achieved by aligning (i) the student's covariance to the teacher's covariance, (ii) the student's statistics for adversarial samples with the student's statistics on clean samples if the student is unable to distill the robust knowledge from the teacher, (iii) the student's prediction-based gram matrix with the teacher's gram matrix, as gram matrices capture the complementary information to covariance matrices.

Moreover, we introduce cost-effective parameter-level adversarial perturbations on the feature projection head alone to improve the student-teacher alignment of the multivariate Normal distributions (covariances). Looking at the spectrum of covariances, one may imagine that this mechanism helps find dominant directions in the eigenspace that are prone to misalignment, and the parameters of the feature projection head are minimized to overcome this effect.

Our key contributions can be summarized as follows:

i. We provide an intuitive motivation by analyzing the robust performance vs. the gap between variances of features of adversarial samples and clean samples. Such an exploration reveals a strong negative correlation with a linear trend, suggesting that aligning the covariance for adversarial features with the covariance for clean features helps improve the adversarial robustness.

ii. Contrary to prior robust distillation methods that typically conduct sample-to-sample feature alignment between the teacher and student models, or even align adversarial features of the student with the corresponding clean features of the teacher, we investigate the covariance alignment between the student and the teacher, as well as aligning the sub-covariance built from adversarial samples toward those from clean samples. Similarly, we investigate aligning score-based gram matrices for better robustness transfer.
iii. We provide comprehensive experiments across various datasets and scenarios, highlighting the superiority of our method, dubbed adverSarially robusT distillAtion by Reducing Student-teacHer varIance gaP (STARSHIP), over state-of-the-art adversarially robust knowledge distillation approaches.

2 Related Works

Adversarial Training. Goodfellow et al. [21] proposed to enhance robustness by adaptively incorporating adversaries [49] into the training data, which represents a "train-from-scratch" scheme. Existing works primarily focus on optimizing the input-loss landscape to mitigate the disruption impact of input-level noise on final predictions [13,16,37,43,55]. Wu et al. [51] introduced a double-perturbation mechanism that combines both input-level and parameter-level perturbations, flattening the parameter-loss landscape during adversarial training. Our method leverages parameter-level perturbations but in the context of exacerbating the student-teacher disagreement at the covariance level. To this end, the parameters of the feature projection head are optimized to overcome the introduced perturbations to attain robust alignment of student-teacher covariances, narrowing the variance gap between the student and teacher models. Despite the effectiveness of adversarial training, such methods heavily depend on large network architectures [52] and a massive amount of training data [15,47]. In contrast, we focus on obtaining the robustness w.r.t. lightweight models by robust distillation.

Vanilla Knowledge Distillation. Hinton et al. [23] proposed knowledge distillation with the goal of transferring the learned knowledge from a large-scale model (*teacher*) to a lightweight model (*student*) by leveraging ground-truth labels and the teacher's predictions. Subsequent knowledge distillation studies focus on the feature-level distillation [4,5,42,50], and the prediction-level distillation [2,6,28,39]. However, vanilla knowledge distillation has been identified as insufficient for robustness transfer from an adversarially trained teacher [19].

Adversarially Robust Knowledge Distillation. To bridge the robustness gap between large-scale and lightweight models, recent studies have investigated adversarially robust knowledge distillation [19,25,58,59]. Zi et al. [59] primarily resort to the soft labels of clean samples predicted by the teacher model to guide the student's predictions. Huang et al. [25] further integrated the gradient flow of the teacher model into the robust knowledge distillation to search for the worst-case matching point (adversarial sample). In contrast to these sample-to-sample alignment methods, our robust distillation method emphasizes a statistics-level robustness transfer that considers the properties of sample sets rather than individual samples. Thus, we align the covariance and gram matrices of the student toward those of the teacher for a more effective robust knowledge transfer (Fig. 2).

3 Proposed Method

Below, we introduce the background of our work and propose our robust knowledge distillation method, STARSHIP, motivated by observations in Fig. 1.

Background. Adversarially robust knowledge distillation generally requires an adversarially pre-trained teacher obtained via adversarial training from scratch. Let the network backbone and the classifier head be denoted as $f_\theta : \mathcal{X} \to \mathbb{R}^d$ and $g_{\theta'} : \mathbb{R}^d \to \mathbb{R}^c$, given network parameters θ and θ', respectively. Let d be the backbone feature dimension, and C be the number of categories. Adversarial training [37] on a dataset from distribution $\mathcal{D}$ solves the following min-max game:

$$\min_{\theta,\theta'} \mathbb{E}_{(\mathbf{x},y)\sim\mathcal{D}} \left[\max_{\|\delta\|_\infty < \epsilon} \mathcal{L}_{\text{CE}}\left(g_{\theta'}\left(f_\theta(\mathbf{x}+\delta)\right), y\right) \right], \quad (1)$$

where the perturbation δ forms the adversarial sample $\hat{\mathbf{x}}=\mathbf{x}+\delta$ restricted within radius ϵ of the original sample $\mathbf{x}$. The inner optimization maximizes the Cross-Entropy (CE) loss $\mathcal{L}_{\text{CE}}$ to find the worst-case adversarial sample posing misclassification, while the outer minimization optimizes the empirical risk on these

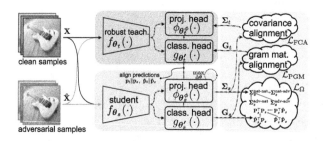

Fig. 2. Our STARSHIP pipeline. For brevity, assume we have a set of clean samples $\mathbf{X}$ and adversarial samples $\hat{\mathbf{X}}$. The student and the teacher backbones are equipped with the projection head and the classification head. Apart from aligning predictions which is a mere sample-to-sample strategy, in the green box we form covariance matrices $\mathbf{\Sigma}_t$ and $\mathbf{\Sigma}_s$, and align $\mathbf{\Sigma}_s$ toward $\mathbf{\Sigma}_t$ by $\mathcal{L}_{\text{FCA}}$. We also form prediction score based gram matrices $\mathbf{G}_t$ and $\mathbf{G}_s$, and align $\mathbf{G}_s$ toward $\mathbf{G}_t$ by $\mathcal{L}_{\text{PGM}}$. Moreover, $\mathcal{L}_\Omega$ aligns adversarial-adversarial, natural-adversarial and adversarial-natural sub-matrices of $\mathbf{\Sigma}_s$ and $\mathbf{G}_s$ with the natural-natural sub-matrices. $\Delta\theta$ denotes parameter perturbations.

adversaries. In the context of adversarially robust knowledge distillation, we generate adversarial samples by maximizing the prediction discrepancy between the teacher model $g_{\theta'_t}(f_{\theta_t}(\cdot))$ and the student model $g_{\theta'_s}(f_{\theta_s}(\cdot))$ as follows:

$$\hat{\mathbf{x}}^{t+1} = \Pi_{\mathbb{B}(\mathbf{x},\epsilon)}\left[\hat{\mathbf{x}}^t + \alpha \text{ sign}\left(\nabla_{\hat{\mathbf{x}}^t}\mathcal{L}_{\mathrm{KL}}\left(g_{\theta'_t}(f_{\theta_t}(\mathbf{x}))\,\middle\|\,g_{\theta'_s}(f_{\theta_s}(\hat{\mathbf{x}}^t))\right)\right)\right], \quad (2)$$

where $\mathcal{L}_{\mathrm{KL}}(\cdot\|\cdot)$ is the Kullback-Leibler (KL) divergence, α is the step size, $\hat{\mathbf{x}}^t$ is an adversarial sample at the tth iteration in an m-step generation, and $\hat{\mathbf{x}} = \hat{\mathbf{x}}^m$ is the final adversarial sample. As Zi et al. [59], we use soft labels of clean samples predicted by the teacher model to guide the adversarial sample generation.

In addition to adversary generation via Eq. (2), we also consider its adaptive variant following [25], which integrates the gradient flow of the teacher model into the adversary generation, albeit at a higher computational cost:

$$\hat{\mathbf{x}}_{\mathrm{ada}}^{t+1} = \Pi_{\mathbb{B}(\mathbf{x},\epsilon)}\left[\hat{\mathbf{x}}_{\mathrm{ada}}^t + \alpha \text{ sign}\left(\nabla_{\hat{\mathbf{x}}_{\mathrm{ada}}^t}\mathcal{L}_{\mathrm{KL}}\left(g_{\theta'_t}(f_{\theta_t}(\hat{\mathbf{x}}_{\mathrm{ada}}^t))\,\middle\|\,g_{\theta'_s}(f_{\theta_s}(\hat{\mathbf{x}}_{\mathrm{ada}}^t))\right)\right)\right], \quad (3)$$

where $\hat{\mathbf{x}}_{\mathrm{ada}}^t$ denotes the tth iteration during adaptive adversary generation. The primary distinction of Eq. (3) from Eq. (2) is the use of adaptive predictions from the teacher model for iterative adversarial samples, which can be regarded as adversaries with the worst-case prediction alignment. While Eq. (3) brings performance gains, it implicitly poses an increased computational cost.

3.1 Adversarially Robust Knowledge Distillation

In contrast to vanilla knowledge distillation, which aligns only clean samples, robust knowledge distillation additionally incorporates adversarial samples into knowledge transfer. Existing works primarily rely on fixed references (e.g., ground-truth labels or the teacher's predictions on a certain type of samples) to guide the distillation process [25,58,59]. In contrast, we treat adversarial samples not just as noise but as specialized augmentations that retain the visual semantic content of their original counterpart. For brevity, denote the teacher's predictions on clean samples as $\mathbf{p}_t = g_{\theta'_t}(f_{\theta_t}(\mathbf{x}))$ and on adversarial samples as $\hat{\mathbf{p}}_t = g_{\theta'_t}(f_{\theta_t}(\hat{\mathbf{x}}))$. For the student model, let $\mathbf{p}_s = g_{\theta'_s}(f_{\theta_s}(\mathbf{x}))$ and $\hat{\mathbf{p}}_s = g_{\theta'_s}(f_{\theta_s}(\hat{\mathbf{x}}))$ be the student's predictions on clean and adversarial samples. Then, the baseline loss used in our work, *Adversarially Robust Knowledge Distillation* (ARKD), is given as:

$$\mathcal{L}_{\mathrm{ARKD}} = \underbrace{(1-\beta)\,\mathcal{L}_{\mathrm{KL}}(\mathbf{p}_t\|\mathbf{p}_s)}_{\text{align clean predictions}} + \underbrace{\beta\,\mathcal{L}_{\mathrm{KL}}(\hat{\mathbf{p}}_t\|\hat{\mathbf{p}}_s)}_{\text{align adversarial predictions}}, \quad (4)$$

where $\beta \in [0,1]$ balances the focus of distillation between the clean samples and their adversarial counterparts. By adopting the ARKD loss, the student model can effectively probe responses of the adversarially pre-trained teacher model to learn its implicit robust behavior against adversarial samples.

3.2 Clean and Adversarial Feature Covariance Alignment

Following the motivation from Fig. 1 and our hypothesis that reducing the feature variance gap between the student and teacher models can facilitate distilling the robust behavior from the teacher model, we attach feature projection heads $\phi_{\boldsymbol{\theta}_s^\phi}(\cdot)$ and $\phi_{\boldsymbol{\theta}_t^\phi}(\cdot)$ with parameters $\boldsymbol{\theta}_s^\phi$ and $\boldsymbol{\theta}_t^\phi$ to the student and teacher models, respectively. Let $\boldsymbol{\phi}_t(\mathbf{x}; \boldsymbol{\theta}_t^\phi) \equiv \phi_{\boldsymbol{\theta}_t^\phi}(f_{\boldsymbol{\theta}_t}(\mathbf{x}))$ and $\boldsymbol{\phi}_t(\hat{\mathbf{x}}; \boldsymbol{\theta}_t^\phi) \equiv \phi_{\boldsymbol{\theta}_t^\phi}(f_{\boldsymbol{\theta}_t}(\hat{\mathbf{x}}))$ be the features obtained by combining the teacher projection head with the teacher backbone, given the clean sample $\mathbf{x}$ and its adversarial counterpart $\hat{\mathbf{x}}$.

Similarly, let $\boldsymbol{\phi}_s(\mathbf{x}; \boldsymbol{\theta}_s^\phi) \equiv \phi_{\boldsymbol{\theta}_s^\phi}(f_{\boldsymbol{\theta}_s}(\mathbf{x}))$ and $\boldsymbol{\phi}_s(\hat{\mathbf{x}}; \boldsymbol{\theta}_s^\phi) \equiv \phi_{\boldsymbol{\theta}_s^\phi}(f_{\boldsymbol{\theta}_s}(\hat{\mathbf{x}}))$ be the features obtained by combining the student projection head with the student backbone, given the clean sample $\mathbf{x}$ and its adversarial counterpart $\hat{\mathbf{x}}$.

Considering $\mathbf{X}$ and $\hat{\mathbf{X}}$ as the sets (*e.g.*, mini-batch) comprising clean and adversarial samples, respectively, organized as column vectors. Then, we simply denote the corresponding clean and adversarial features obtained from the teacher projection head as $\boldsymbol{\Phi}_t$ and $\hat{\boldsymbol{\Phi}}_t$, and from the student head as $\boldsymbol{\Phi}_s$ and $\hat{\boldsymbol{\Phi}}_s$. To highlight that these features depend on the parameters of projection heads, we denote $\boldsymbol{\Phi}_t|\boldsymbol{\theta}_t^\phi$ and $\hat{\boldsymbol{\Phi}}_t|\boldsymbol{\theta}_t^\phi$ for the teacher, and $\boldsymbol{\Phi}_s|\boldsymbol{\theta}_s^\phi$ and $\hat{\boldsymbol{\Phi}}_s|\boldsymbol{\theta}_s^\phi$ for the student.

We concatenate the feature sets $\boldsymbol{\Phi}_t$ and $\hat{\boldsymbol{\Phi}}_t$ along the channel dimension and compute covariance $\boldsymbol{\Sigma}_t|\hat{\boldsymbol{\Phi}}_t$ or $\boldsymbol{\Sigma}_t \in \mathbb{Q}^{2d \times 2d}$ for brevity. Symbol $\mathbb{Q}$ is the set of symmetric positive definite matrices. Similarly, we concatenate $\boldsymbol{\Phi}_s$ and $\hat{\boldsymbol{\Phi}}_s$ along the channel dimension and compute $\boldsymbol{\Sigma}_s|\hat{\boldsymbol{\Phi}}_s$, or equivalently, $\boldsymbol{\Sigma}_s \in \mathbb{Q}^{2d \times 2d}$. Note that the feature covariance matrices for the teacher and student models,

$$\boldsymbol{\Sigma}_t = \begin{bmatrix} \boldsymbol{\Sigma}_t^{\text{nat-nat}} & \boldsymbol{\Sigma}_t^{\text{nat-adv}} \\ \boldsymbol{\Sigma}_t^{\text{adv-nat}} & \boldsymbol{\Sigma}_t^{\text{adv-adv}} \end{bmatrix} \quad \text{and} \quad \boldsymbol{\Sigma}_s = \begin{bmatrix} \boldsymbol{\Sigma}_s^{\text{nat-nat}} & \boldsymbol{\Sigma}_s^{\text{nat-adv}} \\ \boldsymbol{\Sigma}_s^{\text{adv-nat}} & \boldsymbol{\Sigma}_s^{\text{adv-adv}} \end{bmatrix}, \tag{5}$$

contain four types of interactions, *i.e.*, *nat-nat*, *nat-adv*, *adv-nat*, and *adv-adv*, which capture clean-clean, clean-adversarial, adversarial-clean, and adversarial-adversarial feature interactions, respectively. Then, the *Feature Covariance Alignment* (FCA), as a function of the projection parameters $\boldsymbol{\theta}_t^\phi$ and $\boldsymbol{\theta}_s^\phi$, is given by:

$$\mathcal{L}_{\text{FCA}}(\boldsymbol{\theta}_t^\phi, \boldsymbol{\theta}_s^\phi) = \frac{1}{4} d^2(\boldsymbol{\Sigma}_t|\boldsymbol{\theta}_t^\phi, \boldsymbol{\Sigma}_s|\boldsymbol{\theta}_s^\phi), \tag{6}$$

where $d^2(\cdot, \cdot)$ can be any suitable distance metrics between the covariance matrices, *e.g.*, the Frobenius norm, the Power-Euclidean (Pow-E) metric [17,33,34], the fast spectral expectation of Max-pooling (MaxExp) [33,34], or a metric between Multivariate normal distributions, *e.g.*, the KL divergence (similar to [56,57]). The distance is scaled by 0.25, as $\boldsymbol{\Sigma}_t$ and $\boldsymbol{\Sigma}_s$ comprise four sub-matrices.

To improve the resilience of the learned feature-level statistical information and achieve better covariance alignment, we employ adversarial perturbations to optimize parameters of the projection heads, inspired by Wu *et al.* [51], who suggested that the robust generalization gap is bounded by the flatness of the parameter-loss landscape. We introduce parameter-level perturbations to induce a feature- and covariance-level disagreement between the teacher and student:

$$\min_{\theta_t^\phi, \theta_s^\phi} \max_{\Delta\theta \in \mathcal{V}_s} \left[(1-\beta) \left\| \boldsymbol{\Phi}_t | \theta_t^\phi - \boldsymbol{\Phi}_s | (\theta_s^\phi + \Delta\theta) \right\|_F^2 + \beta \left\| \hat{\boldsymbol{\Phi}}_t | \theta_t^\phi - \hat{\boldsymbol{\Phi}}_s | (\theta_s^\phi + \Delta\theta) \right\|_F^2 \right.$$
$$\left. + \mathcal{L}_{\text{FCA}}(\theta_t^\phi, \theta_s^\phi + \Delta\theta) \right], \tag{7}$$

where $\Delta\theta$ denotes the parameter-level perturbations, and $\mathcal{V}_s$ is the perturbation region set $\{\Delta\theta \in \mathcal{V}_s : \|\Delta\theta\|_F \leq \eta \|\theta_s^\phi\|_F\}$ for the perturbation intensity η. The parameter-level perturbation is then optimized using iterative gradient ascent.

3.3 Matching Gram Matrices for Prediction Scores

As evidenced by Fig. 1b, there exists a direct correlation between enhancing robust performance and reducing the variance gap across prediction score-based gram matrices derived from adversarial and clean samples. Inspired by this nearly linear negative correlation of robustness with the gap, we propose to align the prediction score-based student's gram matrix towards the teacher's counterpart. In analogy to Sect. 3.2, given clean and adversarial samples $\mathbf{X}$ and $\hat{\mathbf{X}}$ (represented as column vectors), we obtain prediction score matrices $\mathbf{P}_t$ and $\hat{\mathbf{P}}_t$ by stacking prediction score column vectors ($\mathbf{p}_t$ and $\hat{\mathbf{p}}_t$) obtained from the teacher's prediction head, as detailed in Sect. 3.1. By analogy, we also obtain prediction score matrices $\mathbf{P}_s$ and $\hat{\mathbf{P}}_s$ based on outputs of the student's prediction head for B samples. We form the gram matrices for the teacher and the student:

$$\mathbf{G}_t = \frac{1}{C} \begin{bmatrix} \mathbf{P}_t^\top \mathbf{P}_t & \mathbf{P}_t^\top \hat{\mathbf{P}}_t \\ \hat{\mathbf{P}}_t^\top \mathbf{P}_t & \hat{\mathbf{P}}_t^\top \hat{\mathbf{P}}_t \end{bmatrix} \quad \text{and} \quad \mathbf{G}_s = \frac{1}{C} \begin{bmatrix} \mathbf{P}_s^\top \mathbf{P}_s & \mathbf{P}_s^\top \hat{\mathbf{P}}_s \\ \hat{\mathbf{P}}_s^\top \mathbf{P}_s & \hat{\mathbf{P}}_s^\top \hat{\mathbf{P}}_s \end{bmatrix}. \tag{8}$$

Modeling sample-to-sample relations does not take the spectrum of gram matrices into account. Thus, we propose *Prediction-score Gram Matching* (PGM):

$$\mathcal{L}_{\text{PGM}} = \frac{1}{4} d^2(\mathbf{G}_t, \mathbf{G}_s), \tag{9}$$

which aligns the student's gram matrix with the teacher's gram matrix. Let $d^2(\cdot, \cdot)$ be any suitable distance measure (as detailed in Sect. 3.2) between symmetric positive definite matrices $\mathbf{G}_t \in \mathbb{Q}^{2B \times 2B}$, $\mathbf{G}_s \in \mathbb{Q}^{2B \times 2B}$, given B clean samples and their B adversarial counterparts in a batch of data.

3.4 Aligning Adversarial Statistics with Clean Statistics

Drawing insights from the standard adversarially robust distillation that aligns the student's representations of adversarial samples with the teacher's (or even student's) representations of clean samples, the analogous alignment can be performed between covariance sub-matrices or gram sub-matrices. Let a sub-matrix extractor $\psi_{ij}(\cdot)$ split a matrix into two halves along the rows and two halves along the columns. For $\boldsymbol{\Sigma}_s$ in Eq. (5), operations $\psi_{11}(\cdot), \psi_{12}(\cdot), \psi_{21}(\cdot)$ and $\psi_{22}(\cdot)$ return

$\Sigma_s^{\text{nat-nat}}$, $\Sigma_s^{\text{nat-adv}}$, $\Sigma_s^{\text{adv-nat}}$, and $\Sigma_s^{\text{adv-adv}}$, respectively, as defined in Eq. (5). $\psi_{ij}(\cdot)$ acts on $\mathbf{G}_s$ by analogy. Define a set of index pairs $\mathcal{I} = \{(1,2), (2,1), (2,2)\}$. The loss aligning sub-matrices of adversarial-adversarial and clean-adversarial statistical interactions toward clean-clean interactions is thus defined as:

$$\mathcal{L}_\Omega = \frac{1}{2} \sum_{(i,j)\in\mathcal{I}} \left\|\psi_{11}(\Sigma_s), \psi_{ij}(\Sigma_s)\right\|_F^2 + \left\|\psi_{11}(\mathbf{G}_s), \psi_{ij}(\mathbf{G}_s)\right\|_F^2. \tag{10}$$

The above alignment function leverages the Frobenius norm rather than non-Euclidean distances, as generally, sub-matrices extracted by $\psi_{12}(\cdot)$ and $\psi_{21}(\cdot)$ are non-symmetric indefinite (they become symmetric positive definite if they fully converge to $\psi_{11}(\cdot)$). Taking Σ_s as an example, the interpretation of the above loss is as follows: term $\psi_{12}(\cdot)$ (and $\psi_{21}(\cdot)$) captures correlations across feature channels of the clean samples and their adversarial counterparts.

Discussion. The loss functions $\mathcal{L}_{FCA}$ and $\mathcal{L}_{PGM}$ contain highly complementary statistics. Observe that even for perfectly aligned $\Sigma_s = \Sigma_t$, consider N zero-centered feature vectors $\phi_s^{(n)}$ and $\phi_t^{(n)}$, which were used to compute their covariance matrices, and note that the corresponding feature vectors are not aligned at all, i.e., $\sum_{n=1,\ldots,N} \phi_s^{(n)} {\phi_s^{(n)}}^\top = \sum_{n=\Pi(1,\ldots,N)} \phi_t^{(n)} {\phi_t^{(n)}}^\top, \forall \Pi$, where $\Pi(1,\ldots,N)$ is a random permutation of the sample indexes $\{1,\ldots,N\}$. Indeed, covariance is invariant to the sample order. In contrast, for two matched gram matrices $\mathbf{P}_s = \mathbf{P}_t$, we know that assuming classification scores lie on the ℓ_1 simplex, then $\sum_{c=1,\ldots,C} {\boldsymbol{P}_s^{(c,:)}}^\top \boldsymbol{P}_s^{(c,:)} = \sum_{c=\Pi(1,\ldots,C)} {\boldsymbol{P}_t^{(c,:)}}^\top \boldsymbol{P}_t^{(c,:)}, \forall \Pi$, where $\Pi(1,\ldots,C)$ is a random permutation of class indexes $\{1,\ldots,C\}$. Thus, $\mathcal{L}_{FCA}$ is invariant to the sample order, whereas $\mathcal{L}_{PGM}$ is invariant to the class order.

3.5 Objective Function

Below, we formalize the objective function of our STARSHIP by combining individual loss components introduced in the preceding sections as follows:

$$\mathcal{L} = \mathcal{L}_{\text{ARKD}} + \lambda_1 \mathcal{L}_{\text{FCA}} + \lambda_2 \mathcal{L}_{\text{PGM}} + \lambda_3 \mathcal{L}_\Omega, \tag{11}$$

where λ_1, λ_2, and λ_3 are loss weighting factors. Generally, $\lambda_1 = 1$ in all experiments, as $\mathcal{L}_{\text{FCA}}$ is a covariance-based alignment equivalent of the sample-to-sample alignment $\mathcal{L}_{\text{ARKD}}$ (which has the default weight of 1). Moreover, as $\mathcal{L}_\Omega$ operates on the student only by encouraging the "adversarial parts" of matrices Σ_s and $\mathbf{G}_s$ to align with their "non-adversarial parts", we set $\lambda_3 = 1$ in all experiments, and we only vary λ_2 to ensure a desired level of aligning the student's statistics with the teacher's statistics. We optimize the network parameters of the student model by minimizing $\mathcal{L}$. In inference, we use the distilled student.

Table 1. CIFAR-10 and CIFAR-100: Comparisons of our STARSHIP with other adversarially robust knowledge distillation methods when **distilled from a large-scale WRN-28 teacher model**. Adversarial perturbations are restricted within the ℓ_∞-norm radius $\epsilon = 8/255$. We report both clean and robust accuracies (%).

Type	Architecture	Method		CIFAR-10 Clean	PGD	CW	AA	CIFAR-100 Clean	PGD	CW	AA
Teacher	WRN-28	SCORE	[41]	88.61	64.95	61.79	61.03	63.64	35.46	32.14	31.13
Student	ResNet-18	ARD	[19]	84.35	53.21	51.58	49.40	58.20	30.87	27.97	26.02
		IAD	[58]	83.46	53.34	51.69	49.09	57.35	31.11	28.12	26.22
		RSLAD	[59]	84.42	54.52	53.46	51.36	57.97	32.57	29.28	27.52
		AKD	[38]	86.04	53.95	52.39	50.11	60.79	31.37	28.85	26.93
		AdaAD	[25]	86.38	56.23	54.16	52.36	61.26	32.24	29.19	27.46
		STARSHIP		86.47	57.45	55.50	53.78	61.54	**34.45**	**30.75**	**29.30**
		Ada-STARSHIP		**87.04**	**58.30**	**56.03**	**54.47**	**62.19**	33.52	30.24	28.28
	MNV2	ARD	[19]	82.10	52.82	50.80	48.44	57.26	30.77	27.96	25.79
		IAD	[58]	81.62	52.77	50.61	48.53	56.88	30.54	27.59	25.69
		RSLAD	[59]	84.33	53.98	52.37	50.38	59.38	31.32	28.38	26.54
		AKD	[38]	84.52	51.80	50.29	48.09	59.32	30.13	27.61	25.46
		AdaAD	[25]	85.45	53.16	51.25	49.49	59.97	31.31	28.69	26.49
		STARSHIP		85.71	55.95	53.79	51.85	60.25	**34.20**	**30.72**	**28.57**
		Ada-STARSHIP		**86.44**	**56.39**	**54.87**	**52.62**	**61.33**	32.89	29.57	27.69

4 Experiments

In this section, we provide our experimental settings and compare our STARSHIP method with other adversarially robust knowledge distillation approaches.

Datasets. We conduct all the experiments on three standard image classification datasets: CIFAR-10, CIFAR-100 [35], and ImageNet-100 [9]. Further details of these datasets can be found in Appendix A.1.

Implementation Details. Following previous works [25,58,59] & Robust-Bench [7], we adopt ResNet-18/34 [22], MobileNetV2 (MNV2) [46], and Wide-ResNet-28-10 (WRN-28) [53] as the teacher and student models. We also adopt the Vision Transformer (ViT) as the teacher architecture. Unless specified otherwise, we conduct adversarial sample generation based on the ℓ_∞-norm threat model with the perturbation radius $\epsilon = 8/255$. More experimental details are included in Appendix A.2. In all the experiments, we adopt the loss weighting factors $\lambda_1 = \lambda_3 = 1.0$. We determine the hyper-parameters $\lambda_2 = 2.0$ and $\beta = 0.8$ through cross-validation on CIFAR-10 and consistently apply them across all datasets without modifications. We provide hyper-parameter evaluations in Appendix E.

Table 2. CIFAR-10 and CIFAR-100: Comparisons of our STARSHIP with other adversarially robust knowledge distillation methods in the **self-distillation** setting. We report both clean and robust accuracies (%).

Type	Architecture	Method		CIFAR-10				CIFAR-100		
				Clean	PGD	CW	AA	Clean	PGD	CW	AA
Teacher	ResNet-18	TRADES	[55]	82.45	52.21	50.29	48.90	56.37	28.68	24.87	23.78
Student	ResNet-18	ARD	[19]	81.64	52.62	51.35	49.19	57.96	31.34	27.84	26.13
		IAD	[58]	80.66	52.63	52.21	48.90	56.45	31.87	28.00	26.66
		RSLAD	[59]	81.30	53.80	52.32	50.78	55.17	31.21	27.82	26.46
		AKD	[38]	82.30	52.84	51.39	49.71	56.37	30.02	26.48	25.44
		AdaAD	[25]	82.34	52.75	51.26	49.92	56.62	29.62	26.11	24.78
		STARSHIP		81.97	**55.72**	**54.06**	**52.42**	57.60	**32.19**	**28.19**	**27.05**
		Ada-STARSHIP		**82.62**	55.05	53.40	51.89	**58.09**	31.95	28.07	26.92
Teacher	MNV2	TRADES	[55]	81.04	50.87	48.46	47.15	54.11	27.28	23.39	22.36
Student	MNV2	ARD	[19]	81.25	53.02	50.69	48.85	55.64	30.93	27.47	26.05
		IAD	[58]	79.36	53.45	50.93	49.14	54.00	31.01	27.59	26.11
		RSLAD	[59]	80.01	53.35	51.04	49.74	53.52	29.95	26.66	25.47
		AKD	[38]	80.86	52.74	50.60	49.13	54.26	28.99	25.46	24.31
		AdaAD	[25]	80.48	50.80	48.14	46.98	53.97	28.31	24.69	23.51
		STARSHIP		80.97	**54.28**	51.58	**50.46**	56.21	**32.03**	**27.98**	**26.74**
		Ada-STARSHIP		**81.45**	54.17	**51.63**	50.28	**56.70**	31.79	27.85	26.36

4.1 Results

Robust Distillation from a Larger Teacher. We compare our STARSHIP method with the state-of-the-art adversarially robust knowledge distillation approaches in Table 1. We report the classification accuracies on clean samples and their adversarial counterparts generated by three standard adversarial attack approaches: 20-step Projected Gradient Descent (PGD) [37] with a fixed step size $\alpha = 2/255$, CW [3], and Auto-Attack (AA) [8]. For a fair comparison, all the results are obtained by robust knowledge distillation from the same WRN-28 teacher model. Table 1 shows that our STARSHIP and its adaptive variant achieve the best accuracy on both clean and adversarial samples. Notably, the incorporation of the adaptive adversary generation strategy in Ada-STARSHIP leads to further improvements in natural performance. The superior performance on both ResNet-18 [22] and MNV2 [46] showcases the versatility of our method.

Robust Self-distillation. In self-distillation setting, the teacher and student architectures are identical. Table 2 shows that our STARSHIP and its adaptive extension, Ada-STARSHIP, outperform other robust knowledge distillation methods across CIFAR-10/100. In addition, our method achieves comparable or even better performance on clean samples than the teacher model. We attribute such a gain to our $\mathcal{L}_\Omega$ (Eq. (10)), which facilitates an intrinsic self-distill within the student, $i.e.$, sub-matrices in this loss entirely operate on the student backbone. Given the teacher backbone in this experiment is not stronger than the student backbone, such an internal self-distillation enables the student model to autonomously refine its decision boundaries. More experimental results w.r.t. diverse settings of $\mathcal{L}_\Omega$ can be found in Appendix C.4.

Table 3. ImageNet-100: Robust accuracy (%) on the distilled student models using ResNet-18 and MNV2 backbones.

Type	Architecture	Method		Clean	PGD	AA
Teacher	ResNet-34	TRADES	[55]	72.66	40.88	34.70
Student	ResNet-18	ARD	[19]	65.41	38.34	30.94
		IAD	[58]	65.62	39.09	32.63
		RSLAD	[59]	66.60	39.12	32.18
		AKD	[38]	67.00	38.85	31.56
		AdaAD	[25]	68.64	38.99	32.24
		STARSHIP		69.68	**40.08**	**33.18**
		Ada-STARSHIP		**71.10**	39.86	32.74
	MNV2	ARD	[19]	65.90	37.28	30.20
		IAD	[58]	64.96	38.00	31.40
		RSLAD	[59]	65.82	37.86	31.66
		AKD	[38]	66.09	36.47	30.03
		AdaAD	[25]	67.26	37.16	30.55
		STARSHIP		68.20	**39.84**	**32.87**
		Ada-STARSHIP		**69.74**	39.21	32.58

Table 4. Robust accuracy (%) of models distilled from ViTs using ResNet-18 and MNV2 student backbones on CIFAR-10.

Type	Architecture	Method		Clean	PGD	AA
Teacher	ViT-B	AT-PRM	[40]	83.98	53.10	49.66
Student	ResNet-18	ARD	[19]	82.76	52.95	49.03
		IAD	[58]	82.27	53.42	49.48
		RSLAD	[59]	82.33	54.89	49.74
		AKD	[38]	82.86	53.44	49.26
		AdaAD	[25]	82.51	54.30	50.02
		STARSHIP		83.53	**55.89**	**52.07**
		Ada-STARSHIP		**83.60**	55.31	51.49
Teacher	DeiT-S	AT-PRM	[40]	81.59	53.45	49.20
Student	MNV2	ARD	[19]	80.41	54.12	49.62
		IAD	[58]	80.86	53.91	50.18
		RSLAD	[59]	81.62	53.08	49.02
		AKD	[38]	82.11	53.85	49.57
		AdaAD	[25]	83.45	**55.04**	51.48
		STARSHIP		**83.52**	54.92	**51.54**
		Ada-STARSHIP				

Table 5. Robust distillation (WRN-28 → ResNet-18) on CIFAR-10/100 with auxiliary synthetic training data. We report both clean and (Auto-Attack) robust accuracies (%) associated with the robustness gain Δ_{Rob} with additional data.

Type	Method		CIFAR-10			CIFAR-100		
			Clean	Robust	Δ_{Rob}	Clean	Robust	Δ_{Rob}
Teacher	SCORE	[41]	88.61	61.03	–	63.64	31.13	–
Student	ARD	[19]	83.93	51.04	+1.64	57.36	26.20	+0.18
	IAD	[58]	83.16	50.30	+1.21	56.13	26.54	+0.32
	RSLAD	[59]	83.79	52.42	+1.06	56.17	27.84	+0.32
	AKD	[38]	85.50	51.34	+1.23	58.95	27.26	+0.33
	AdaAD	[25]	85.93	53.08	+0.72	60.54	27.62	+0.16
	STARSHIP		85.67	54.49	+0.71	60.68	**29.43**	+0.13
	Ada-STARSHIP		**86.52**	**55.75**	+0.35	**63.61**	28.63	+0.35

Table 6. Extension of robust distillation (WRN-28 → ResNet-18/MNV2) with single-step adversary strategy (N-FGSM [29]) on CIFAR-10. We report accuracy (%) with the average training time per epoch.

Type	Architecture	Method		CIFAR-10				CIFAR-100			
				Clean	PGD	CW	AA	Clean	PGD	CW	AA
Teacher	WRN-28	SCORE	[41]	88.61	64.95	61.79	61.03	63.64	35.46	32.14	31.13
Student	ResNet-18	ARD	[19]	84.35	53.21	51.58	49.40	58.20	30.87	27.97	26.02
		IAD	[58]	83.46	53.34	51.69	49.09	57.35	31.11	28.12	26.22
		RSLAD	[59]	84.52	53.46	51.36	57.97	32.57	29.28	27.52	
		AKD	[38]	86.04	55.38	52.39	50.11	60.79	31.37	28.85	26.93
		AdaAD	[25]	86.38	56.23	54.16	52.36	61.26	32.24	29.19	27.46
		STARSHIP		86.47	57.45	55.50	53.78	61.54	**34.45**	**30.75**	**29.30**
		Ada-STARSHIP		**87.04**	**58.30**	**56.03**	**54.47**	**62.19**	33.52	30.24	28.28
	MNV2	ARD	[19]	82.10	52.82	50.80	48.44	57.26	30.77	27.96	25.79
		IAD	[58]	81.62	52.77	50.61	48.53	56.68	30.54	27.59	25.89
		RSLAD	[59]	84.33	53.96	52.37	50.38	59.38	31.32	28.38	26.54
		AKD	[38]	84.52	51.80	50.29	48.09	59.32	30.13	27.61	25.46
		AdaAD	[25]	85.45	53.16	51.25	49.49	59.97	31.73	28.69	26.49
		STARSHIP		85.71	55.95	53.79	51.85	60.25	**34.20**	**30.72**	**28.57**
		Ada-STARSHIP		**86.44**	**56.39**	**54.87**	**52.62**	**61.33**	32.89	29.57	27.69

Robust Distillation on ImageNet-100. Below, we extend our investigation to evaluate the generalization ability of our STARSHIP method in the context of larger-scale and practical images. As presented in Table 3, we perform robust knowledge distillation on ImageNet-100 [9] using different student backbones. We can observe that our STARSHIP method is superior to existing robustness transfer methods by retaining more comprehensive knowledge inherited from the teacher model in terms of clean accuracy and adversarial robustness.

Robust Distillation from a ViT-Based Teacher. ViT has emerged as a competitive alternative to the convolutional network, showcasing good adversarial robustness [1,40]. Thus, we investigate if the intrinsic robustness of ViTs can also be transferred to lightweight student backbones. Table 4 shows that our proposed STARSHIP consistently achieves superior robust accuracy that even surpasses the ViT-based teacher model. Such a robustness improvement also demonstrates the viability of our method in the context of inheriting robustness from ViT-based teachers without compromising natural performance.

Table 7. Ablation study (WRN-28 → ResNet-18) of three main components in our STARSHIP on CIFAR-10/100.

Table 8. Comparison of different distance metrics for our STARSHIP during robust knowledge distillation (WRN-28 → ResNet-18) on CIFAR-10/100.

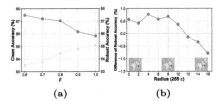

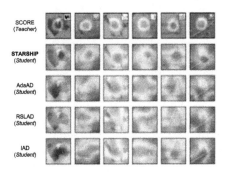

Fig. 3. Robust distillation on CIFAR-10 (WRN-28 → ResNet-18). Figure 3a: Trade-off between clean and (Auto-Attack) robust accuracies w.r.t. β that balances distillation between clean and adversarial samples. Figure 3b: Difference of robust accuracy between our Ada-STARSHIP and STARSHIP under diverse perturb. radii ϵ.

Fig. 4. Visualizations of attention of both the teacher and student models distilled via different robust distillation methods.

Robust Distillation with Additional Synthetic Data. Auxiliary data generated by Denoising Diffusion Probabilistic Models (DDPMs) [24] can further improve adversarial training. Nevertheless, there exists a gap in applying the auxiliary data in the context of robust knowledge distillation. To bridge this gap, we incorporate an additional 1M synthetic training data for CIFAR-10/100 when distilling from a large-scale teacher model. Table 5 shows that the auxiliary training data improves the adversarial robustness of distilled students compared to the original training in Table 1. STARSHIP achieves the most significant gain in robustness by incorporating auxiliary DDPM-generated data during robust knowledge distillation.

Single-Step Robust Distillation. Due to the multi-step adversary generation scheme, achieving adversarial robustness requires several times more computation resources than the standard training [18]. To mitigate such a computational burden, we explore the adversarially robust knowledge distillation with the single-step adversary generation strategy [29]. Table 6 shows that our STARSHIP can efficiently inherit non-trivial robustness from the teacher model based on single-step adversarial samples. Further details and more experiments can be found in Appendix B.2 and Appendix C.5, respectively.

5 Further Analyses

Ablation Studies. Below, we investigate modules of STARSHIP: (i) Feature Covariance Alignment (FCA) in Sect. 3.2, (ii) Prediction-score Gram Matching (PGM) in Eq. (9), and (iii) Statistical Sub-matrices Alignment (Ω) in Eq. (10). Table 7 shows results on CIFAR-10. Our baseline (first row) uses the standard prediction alignment in Eq. (4). Both FCA and PGM improve adversarial robustness. The statistical sub-matrices alignment (Ω) boosts the clean accuracy of student by promoting the prediction invariance between clean and adv. samples.

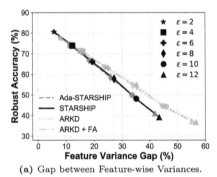
(a) Gap between Feature-wise Variances.

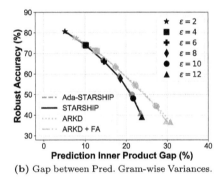
(b) Gap between Pred. Gram-wise Variances.

Fig. 5. The robust accuracy (%) of several student models w.r.t. their average gap (%) between feature-wise variances evaluated on the adversarial samples and clean samples (5a). In Fig. 5b, we show equivalent evaluations with the use of prediction-based gram-wise variances. The variance gaps are captured under several adversarial perturbation radii. Note that the experimental settings are the same as Fig. 1.

Clean vs. Robust Accuracy Trade-off. The trade-off between clean accuracy and adversarial robustness has been widely investigated in adversarial training [12,43,55]. Less is known about such a trade-off within the context of robust knowledge distillation. Thus, we study the effect of β, which balances the prediction alignment on clean and adversarial samples. Figure 3a shows an improved adversarial robustness associated with a decrease in clean performance as β gets larger. Furthermore, we explore the robustness of our STARSHIP and its adaptive variant (Ada-STARSHIP) against adversarial samples across diverse attack strengths in Fig. 3b. We adopt adversaries with $\epsilon = 8/255$ (signified by the dashed red line) for robust distillation. Compared with the vanilla STARSHIP, Ada-STARSHIP has a better performance on clean samples and their adversarial counterparts of lower attack strength ($\epsilon \leq 10$). In comparison, STARSHIP achieves better robustness against strong adversarial samples ($\epsilon > 10$). Such an intrinsic robustness trade-off makes our method applicable in diverse scenarios.

Visualization. As shown in Fig. 4, we provide visualizations of adversarial samples ($\epsilon = 8/255$) related to both the teacher and student models via Grad-CAM [48]. In comparison with other robust distillation methods, the student model distilled by our STARSHIP method shares similar attention regions to that of the teacher model. Such an observation further underscores the efficacy of our method in preserving the adversarially robust prediction alignment between the teacher and student. Additional visualization results are in Appendix D.

Discussion on Effectiveness of Our Method. Building on the insights from Fig. 1, we further analyze the robustness w.r.t. the variance gap between adversarial and clean representations under different attack strengths. Figure 5 evaluates our STARSHIP method and its adaptive variant. For comparison, we also introduce a Feature Alignment (FA) variant of our baseline method (ARKD) from Eq. (4). According to Fig. 5, relying on the feature or prediction alignment alone cannot reduce such a statistical gap, leading to relatively weaker robustness against adversarial samples. Unlike the sample-to-sample matching paradigm (*e.g.*, FA), our STARSHIP method introduces the alignment of second-order statistics [31,32,34,36,44,54] at both the feature and prediction levels, which introduces the multivariate Normal distribution alignment prior to the alignment process (*e.g.*, correlation alignment instead of mere matching of individual features or instances). Analysis of the parameter-level perturbations (Eq. (7)) is in Appendix C.6.

Performance comparison w.r.t. Different Distance Measures. Below, we explore the efficacy of several distance metrics on the alignment of covariance and gram matrices between the teacher and student distilled via our STARSHIP. As such matrices are symmetric positive (semi-)definite, distance $d^2(\cdot, \cdot)$ in Eq. (6) for feature covariance alignment and Eq. (9) for prediction-score gram matching can be achieved by non-Euclidean distances. Table 8 reports clean and robust accuracies of the distilled student model based on different distance metrics: (i) the Frobenius norm, (ii) the Power-Euclidean (Pow-E) metric [17,33,34], (iii) fast spectral expectation of Max-pooling (MaxExp) [33,34], and (iv) the KL divergence. Note that we conduct the alignment of multivariate Normal distributions (statistics) between the teacher and student models when adopting the KL divergence as the distance metric. We observe that the statistical alignment via MaxExp achieves the highest accuracy on both clean and adversarial examples, even surpassing the performance of distribution alignment via the KL divergence that adopts both the mean and covariance information. The main reason is that

MaxExp can balance (and partially equalize) the spectrum of the aligned statistics. It dampens the significance of leading eigenvalues and boosts the impact of non-leading eigenvalues. Thus, it encourages the student to distill all orthogonal variance vectors of the teacher covariance, not only those corresponding to the leading eigenvalues, preventing overfitting to leading principal directions.

6 Conclusions

Motivated by the implicit link between feature variance and model generalization, we have investigated the relation between the adversarial robustness and its correlation to the variance gap between feature variances of adversarial and clean samples. We have observed that a similar phenomenon holds for prediction-based gram matrices. We devised several alignment strategies leveraging second-order statistics to align the student's statistics with the teacher's statistics. We have also shown that a degree of self-distillation within the student model is beneficial for robustness. Leveraging second-order statistics to align the student toward the teacher helps capture well intricacies of the teacher's robust boundary through aligning correlations rather than sample-to-sample scores. Appendix F discusses limitations of our method.

Acknowledgments. This research is supported by National Research Foundation, Singapore and Infocomm Media Development Authority under its Trust Tech Funding Initiative, the Centre for Frontier Artificial Intelligence Research, Institute of High Performance Computing, A*Star, and the College of Computing and Data Science at Nanyang Technological University. Any opinions, findings, conclusions, or recommendations expressed in this material are those of the author(s) and do not reflect the views of National Research Foundation, Singapore, and Infocomm Media Development Authority. PK is supported by CSIRO's Science Digital.

References

1. Bai, Y., Mei, J., Yuille, A.L., Xie, C.: Are transformers more robust than cnns? Adv. Neural. Inf. Process. Syst. **34**, 26831–26843 (2021)
2. Beyer, L., Zhai, X., Royer, A., Markeeva, L., Anil, R., Kolesnikov, A.: Knowledge distillation: a good teacher is patient and consistent. In: Proceedings of the IEEE/CVF Conference on Computer Vision and Pattern Recognition, pp. 10925–10934 (2022)
3. Carlini, N., Wagner, D.: Towards evaluating the robustness of neural networks. In: 2017 IEEE Symposium on Security and Privacy (SP), pp. 39–57. IEEE (2017)
4. Chen, D., Mei, J.P., Zhang, H., Wang, C., Feng, Y., Chen, C.: Knowledge distillation with the reused teacher classifier. In: Proceedings of the IEEE/CVF Conference on Computer Vision and Pattern Recognition, pp. 11933–11942 (2022)
5. Chen, P., Liu, S., Zhao, H., Jia, J.: Distilling knowledge via knowledge review. In: Proceedings of the IEEE/CVF Conference on Computer Vision and Pattern Recognition, pp. 5008–5017 (2021)

6. Cho, J.H., Hariharan, B.: On the efficacy of knowledge distillation. In: Proceedings of the IEEE/CVF International Conference on Computer Vision, pp. 4794–4802 (2019)
7. Croce, F., et al.: Robustbench: a standardized adversarial robustness benchmark. In: Proceedings of the Neural Information Processing Systems Track on Datasets and Benchmarks (2021)
8. Croce, F., Hein, M.: Reliable evaluation of adversarial robustness with an ensemble of diverse parameter-free attacks. In: International Conference on Machine Learning, pp. 2206–2216. PMLR (2020)
9. Deng, J., Dong, W., Socher, R., Li, L.J., Li, K., Fei-Fei, L.: Imagenet: a large-scale hierarchical image database. In: 2009 IEEE Conference on Computer Vision and Pattern Recognition, pp. 248–255. IEEE (2009)
10. Dong, J., Koniusz, P., Chen, J., Wang, Z.J., Ong, Y.S.: Robust distillation via untargeted and targeted intermediate adversarial samples. In: Proceedings of the IEEE/CVF Conference on Computer Vision and Pattern Recognition, pp. 28432–28442 (2024)
11. Dong, J., Koniusz, P., Chen, J., Xie, X., Ong, Y.S.: Adversarially robust few-shot learning via parameter co-distillation of similarity and class concept learners. In: Proceedings of the IEEE/CVF Conference on Computer Vision and Pattern Recognition, pp. 28535–28544 (2024)
12. Dong, J., Moosavi-Dezfooli, S.M., Lai, J., Xie, X.: The enemy of my enemy is my friend: exploring inverse adversaries for improving adversarial training. In: Proceedings of the IEEE/CVF Conference on Computer Vision and Pattern Recognition (CVPR), pp. 24678–24687 (2023)
13. Dong, J., Wang, Y., Lai, J.H., Xie, X.: Improving adversarially robust few-shot image classification with generalizable representations. In: Proceedings of the IEEE/CVF Conference on Computer Vision and Pattern Recognition, pp. 9025–9034 (2022)
14. Dong, J., Wang, Y., Lai, J., Xie, X.: Restricted black-box adversarial attack against deepfake face swapping. IEEE Trans. Inf. Forensics Secur. **18**, 2596–2608 (2023). https://doi.org/10.1109/TIFS.2023.3266702
15. Dong, J., Wang, Y., Xie, X., Lai, J., Ong, Y.S.: Generalizable and discriminative representations for adversarially robust few-shot learning. IEEE Trans. Neural Netw. Learn. Syst., 1–14 (2024). https://doi.org/10.1109/TNNLS.2024.3379172
16. Dong, J., Yang, L., Wang, Y., Xie, X., Lai, J.: Towards intrinsic adversarial robustness through probabilistic training. IEEE Trans. Image Process. (2023)
17. Dryden, I.L., Koloydenko, A., Zhou, D.: Non-Euclidean statistics for covariance matrices, with applications to diffusion tensor imaging. Ann. Appl. Stat. **3**(3), 1102–1123 (2009)
18. Gao, R., et al.: Fast and reliable evaluation of adversarial robustness with minimum-margin attack. In: International Conference on Machine Learning, pp. 7144–7163. PMLR (2022)
19. Goldblum, M., Fowl, L., Feizi, S., Goldstein, T.: Adversarially robust distillation. In: Proceedings of the AAAI Conference on Artificial Intelligence, vol. 34, pp. 3996–4003 (2020)
20. Goodfellow, I., et al.: Generative adversarial nets. In: NIPS, pp. 2672–2680 (2014)
21. Goodfellow, I., Shlens, J., Szegedy, C.: Explaining and harnessing adversarial examples. In: ICLR (2015)
22. He, K., Zhang, X., Ren, S., Sun, J.: Deep residual learning for image recognition. In: Proceedings of the IEEE Conference on Computer Vision and Pattern Recognition (CVPR) (2016)

23. Hinton, G., Vinyals, O., Dean, J.: Distilling the knowledge in a neural network. arXiv preprint arXiv:1503.02531 (2015)
24. Ho, J., Jain, A., Abbeel, P.: Denoising diffusion probabilistic models. Adv. Neural. Inf. Process. Syst. **33**, 6840–6851 (2020)
25. Huang, B., Chen, M., Wang, Y., Lu, J., Cheng, M., Wang, W.: Boosting accuracy and robustness of student models via adaptive adversarial distillation. In: Proceedings of the IEEE/CVF Conference on Computer Vision and Pattern Recognition, pp. 24668–24677 (2023)
26. Huang, R., et al.: Feature variance regularization: a simple way to improve the generalizability of neural networks. Proceedings of the AAAI Conference on Artificial Intelligence, vol. 34, no. 04, pp. 4190–4197 (2020). https://doi.org/10.1609/aaai.v34i04.5840
27. Ilyas, A., Santurkar, S., Tsipras, D., Engstrom, L., Tran, B., Madry, A.: Adversarial examples are not bugs, they are features. In: Proceedings of the 33rd International Conference on Neural Information Processing Systems, pp. 125–136 (2019)
28. Jin, Y., Wang, J., Lin, D.: Multi-level logit distillation. In: Proceedings of the IEEE/CVF Conference on Computer Vision and Pattern Recognition, pp. 24276–24285 (2023)
29. de Jorge Aranda, P., et al.: Make some noise: reliable and efficient single-step adversarial training. Adv. Neural. Inf. Process. Syst. **35**, 12881–12893 (2022)
30. Kaur, D., Uslu, S., Rittichier, K.J., Durresi, A.: Trustworthy artificial intelligence: a review. ACM Comput. Surv. (CSUR) **55**(2), 1–38 (2022)
31. Koniusz, P., Cherian, A.: Sparse coding for third-order super-symmetric tensor descriptors with application to texture recognition. In: Proceedings of the IEEE Conference on Computer Vision and Pattern Recognition (CVPR) (2016)
32. Koniusz, P., Wang, L., Cherian, A.: Tensor representations for action recognition. IEEE Trans. Pattern Anal. Mach. Intell. **44**(2), 648–665 (2022). https://doi.org/10.1109/TPAMI.2021.3107160
33. Koniusz, P., Yan, F., Gosselin, P.H., Mikolajczyk, K.: Higher-order occurrence pooling on mid- and low-level features: visual concept detection. Technical report (2013). https://inria.hal.science/hal-00922524
34. Koniusz, P., Zhang, H.: Power normalizations in fine-grained image, few-shot image and graph classification. IEEE Trans. Pattern Anal. Mach. Intell. **44**(2), 591–609 (2022). https://doi.org/10.1109/TPAMI.2021.3107164
35. Krizhevsky, A., Hinton, G., et al.: Learning multiple layers of features from tiny images (2009)
36. Lin, T.Y., Maji, S., Koniusz, P.: Second-order democratic aggregation. In: Proceedings of the European Conference on Computer Vision (ECCV) (2018)
37. Madry, A., Makelov, A., Schmidt, L., Tsipras, D., Vladu, A.: Towards deep learning models resistant to adversarial attacks. In: 6th International Conference on Learning Representations, ICLR (2018)
38. Maroto, J., Ortiz-Jiménez, G., Frossard, P.: On the benefits of knowledge distillation for adversarial robustness. arXiv preprint arXiv:2203.07159 (2022)
39. Mirzadeh, S.I., Farajtabar, M., Li, A., Levine, N., Matsukawa, A., Ghasemzadeh, H.: Improved knowledge distillation via teacher assistant. In: Proceedings of the AAAI Conference on Artificial Intelligence, vol. 34, pp. 5191–5198 (2020)
40. Mo, Y., Wu, D., Wang, Y., Guo, Y., Wang, Y.: When adversarial training meets vision transformers: recipes from training to architecture. Adv. Neural. Inf. Process. Syst. **35**, 18599–18611 (2022)

41. Pang, T., Lin, M., Yang, X., Zhu, J., Yan, S.: Robustness and accuracy could be reconcilable by (proper) definition. In: International Conference on Machine Learning, pp. 17258–17277. PMLR (2022)
42. Park, W., Kim, D., Lu, Y., Cho, M.: Relational knowledge distillation. In: Proceedings of the IEEE/CVF Conference on Computer Vision and Pattern Recognition, pp. 3967–3976 (2019)
43. Rade, R., Moosavi-Dezfooli, S.: Reducing excessive margin to achieve a better accuracy vs. robustness trade-off. In: The Tenth International Conference on Learning Representations, ICLR (2022)
44. Rahman, S., Koniusz, P., Wang, L., Zhou, L., Moghadam, P., Sun, C.: Learning partial correlation based deep visual representation for image classification. In: Proceedings of the IEEE/CVF Conference on Computer Vision and Pattern Recognition (CVPR), pp. 6231–6240 (2023)
45. Ronneberger, O., Fischer, P., Brox, T.: U-net: convolutional networks for biomedical image segmentation. In: Navab, N., Hornegger, J., Wells, W.M., Frangi, A.F. (eds.) MICCAI 2015. LNCS, vol. 9351, pp. 234–241. Springer, Cham (2015). https://doi.org/10.1007/978-3-319-24574-4_28
46. Sandler, M., Howard, A., Zhu, M., Zhmoginov, A., Chen, L.C.: Mobilenetv2: inverted residuals and linear bottlenecks. In: Proceedings of the IEEE Conference on Computer Vision and Pattern Recognition, pp. 4510–4520 (2018)
47. Schmidt, L., Santurkar, S., Tsipras, D., Talwar, K., Madry, A.: Adversarially robust generalization requires more data. Adv. Neural Inf. Process. Syst. **31** (2018)
48. Selvaraju, R.R., Cogswell, M., Das, A., Vedantam, R., Parikh, D., Batra, D.: Gradcam: visual explanations from deep networks via gradient-based localization. In: Proceedings of the IEEE International Conference on Computer Vision, pp. 618–626 (2017)
49. Szegedy, C., et al.: Intriguing properties of neural networks. In: ICLR (2014)
50. Tian, Y., Krishnan, D., Isola, P.: Contrastive representation distillation. In: 8th International Conference on Learning Representations, ICLR (2020)
51. Wu, D., Xia, S.T., Wang, Y.: Adversarial weight perturbation helps robust generalization. Adv. Neural. Inf. Process. Syst. **33**, 2958–2969 (2020)
52. Xie, C., Yuille, A.L.: Intriguing properties of adversarial training at scale. In: 8th International Conference on Learning Representations, ICLR (2020)
53. Zagoruyko, S., Komodakis, N.: Wide residual networks. In: Proceedings of the British Machine Vision Conference (2016)
54. Zhang, H., Li, H., Koniusz, P.: Multi-level second-order few-shot learning. IEEE Trans. Multim. **25**, 2111–2126 (2023). https://doi.org/10.1109/TMM.2022.3142955
55. Zhang, H., Yu, Y., Jiao, J., Xing, E., El Ghaoui, L., Jordan, M.: Theoretically principled trade-off between robustness and accuracy. In: International Conference on Machine Learning, pp. 7472–7482. PMLR (2019)
56. Zhang, Y., Zhu, H., Chen, Y., Song, Z., Koniusz, P., King, I.: Mitigating the popularity bias of graph collaborative filtering: a dimensional collapse perspective. In: Oh, A., Naumann, T., Globerson, A., Saenko, K., Hardt, M., Levine, S. (eds.) Advances in Neural Information Processing Systems, vol. 36, pp. 67533–67550. Curran Associates, Inc. (2023)
57. Zhang, Y., et al.: Geometric view of soft decorrelation in self-supervised learning. In: Proceedings of the 30th ACM SIGKDD Conference on Knowledge Discovery and Data Mining. KDD 2024, Association for Computing Machinery, New York (2024)

58. Zhu, J., et al.: Reliable adversarial distillation with unreliable teachers. In: The Tenth International Conference on Learning Representations, ICLR (2022)
59. Zi, B., Zhao, S., Ma, X., Jiang, Y.G.: Revisiting adversarial robustness distillation: robust soft labels make student better. In: Proceedings of the IEEE/CVF International Conference on Computer Vision, pp. 16443–16452 (2021)

LN3DIFF: Scalable Latent Neural Fields Diffusion for Speedy 3D Generation

Yushi Lan[1](✉), Fangzhou Hong[1], Shuai Yang[2], Shangchen Zhou[1], Xuyi Meng[1], Bo Dai[3], Xingang Pan[1], and Chen Change Loy[1]

[1] S-Lab, Nanyang Technological University, Singapore, Singapore
yushi001@e.ntu.edu.sg, ccloy@ntu.edu.sg
[2] Wangxuan Institute of Computer Technology, Peking University, Beijing, China
[3] Shanghai Artificial Intelligence Laboratory, Shanghai, China

Abstract. The field of neural rendering has witnessed significant progress with advancements in generative models and differentiable rendering techniques. Though 2D diffusion has achieved success, a unified 3D diffusion pipeline remains unsettled. This paper introduces a novel framework called **LN3Diff** to address this gap and enable fast, high-quality, and generic conditional 3D generation. Our approach harnesses a 3D-aware architecture and variational autoencoder (VAE) to encode the input image(s) into a structured, compact, and 3D latent space. The latent is decoded by a transformer-based decoder into a high-capacity 3D neural field. Through training a diffusion model on this 3D-aware latent space, our method achieves superior performance on Objaverse, ShapeNet and FFHQ for conditional 3D generation. Moreover, it surpasses existing 3D diffusion methods in terms of inference speed, requiring no per-instance optimization. Video demos can be found on our project webpage: https://nirvanalan.github.io/projects/ln3diff.

Keywords: Generative Model · Reconstruction · Latent Diffusion Model

1 Introduction

The advancement of generative models [19,28] and differentiable rendering [83] has paved the way for a new research direction called neural rendering [83]. This field is continuously pushing the limits of view synthesis [51], editing [42], and particularly 3D object synthesis [6]. While 2D diffusion models [28,78] have outperformed GANs in image synthesis [14] in terms of quality [68], controllability [99], and scalability [72], a unified 3D diffusion pipeline has yet to be established.

3D object generation methods using diffusion models can be categorized into 2D-lifting and feed-forward 3D diffusion models. In 2D-lifting methods, score

---

Supplementary Information The online version contains supplementary material available at https://doi.org/10.1007/978-3-031-73235-5_7.

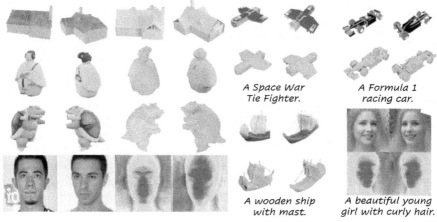

Fig. 1. We present LN3DIFF, which performs efficient 3D diffusion learning over a compact latent space. The resulting model enables both high-quality monocular 3D reconstruction and text-to-3D synthesis.

distillation sampling (SDS) [63,90] and Zero-123 [45,74] achieve 3D generation by leveraging pre-trained 2D diffusion models. However, SDS-based methods require costly per-instance optimization and are prone to the multi-face Janus problem [63]. Meanwhile, Zero-123 fails to enforce strict view consistency. On the other hand, feed-forward 3D diffusion models [9,33,44,52,89,97] enable fast 3D synthesis without per-instance optimization. However, these methods typically involve a two-stage pre-processing approach. First, during the data preparation stage, a shared decoder is learned over a large number of instances to ensure a shared latent space. This is followed by per-instance optimization to convert each 3D asset in the datasets into neural fields [93]. After this, the feed-forward diffusion model is trained on the prepared neural fields.

While the pipeline above is straightforward, it poses extra challenges to achieve high-quality 3D diffusion: **1) Scalability.** In the data preparation stage, existing methods face scalability issues due to using a shared, low-capacity MLP decoder for per-instance optimization. This approach is data inefficient, requiring over 50 views per instance [9,52] during training. Consequently, computation cost scales linearly with the dataset size, hindering scalability for large, diverse 3D datasets. **2) Efficiency.** Employing 3D-specific architectures [11,64,86] is computationally intensive and necessitates representation-specific designs [101]. Consequently, existing methods compress each 3D asset into neural fields [93] before training. However, this compression introduces high-dimensional 3D latent, increasing computational demands and training challenges. Limiting the neural field size [9] might mitigate these issues but at the cost of reconstruction quality. In addition, the auto-decoding paradigm can result in an unclean latent space [20,33,75], unsuitable for 3D diffusion training [68]. **3) Generalizability.** Existing 3D diffusion models primarily focus on unconditional generation over single classes, neglecting high-quality conditional 3D generation (*e.g.*, text-to-3D) across generic, category-free 3D datasets. Furthermore, projecting monocular input images into the diffusion latent space is crucial for conditional generation and editing [43,99], but this is challenging with the shared decoder designed for multi-view inputs.

In this study, we propose a novel framework called **L**atent **N**eural fields **3D** **Diff**usion (LN3DIFF) to address these challenges and enable fast, high-quality and generic conditional 3D generation. Our method involves training a variational autoencoder [39] (VAE) to compress input images into a lower-dimensional 3D-aware latent space, which is more expressive and flexible compared to pixel-space diffusion [14,28,31,78]. From this space, a 3D-aware transformer-based decoder gradually decodes the latent into a high-capacity 3D neural field. This autoencoding stage is trained amortized with differentiable rendering [83], incorporating novel view supervision for multi-view datasets [8,13] and adversarial supervision for monocular dataset [34]. Thanks to the high-capacity model design, our method is more *view efficient*, requiring only two views per instance during training. After training, we leverage the learned 3D latent space for conditional 3D diffusion learning, ensuring effective utilization of the trained model for high-quality 3D generation. The pre-trained encoder can amortize the data encoding over incoming data, thus streamlining operations and facilitating efficient 3D diffusion learning while remaining compatible with advances in 3D representations.

To enhance efficient information flow in the 3D space and promote coherent geometry reconstruction, we introduce a novel 3D-aware architecture tailored for fast and high-quality 3D reconstruction while maintaining a structured latent space. Specifically, we employ a convolutional tokenizer to encode the input image(s) into a *KL*-regularized 3D latent space, leveraging its superior perceptual compression ability [17]. We employ transformers [15,60] to enable flexible 3D-aware attention across 3D tokens in the latent space. Finally, we up-sample the 3D latent and apply differentiable rendering for image-space supervision, making our method a self-supervised 3D learner [77].

In summary, we contribute a 3D-representation-agnostic pipeline for building generic, high-quality 3D generative models. This pipeline provides opportunities to resolve a series of downstream 3D vision and graphics tasks. Specifically, we propose a novel 3D-aware reconstruction model that achieves high-quality 3D data encoding in an amortized manner. Learning in the compact latent space, our model demonstrates state-of-the-art 3D generation performance on the ShapeNet benchmark [8], surpassing both Generative Adversarial Network (GAN)-based and 3D diffusion-based approaches. Our method shows superior performance in monocular 3D reconstruction and conditional 3D generation on ShapeNet, FFHQ, and Objaverse datasets, with a fast inference speed, *e.g.*, 3× faster against existing latent-free 3D diffusion methods [1].

2 Related Work

3D-Aware GANs. GANs [19] have shown promising results in generating photorealistic images [3,35,36], inspiring researchers to explore 3D-aware generation [25,53,58]. Motivated by the recent success of neural rendering [49,51,59], researchers have introduced 3D inductive bias into the generation task [5,73], demonstrating impressive 3D-awareness synthesis through hybrid designs [6,21,29,55,57]. This has made 3D-aware generation applicable to a series of downstream applications [42,79,80,98]. However, GAN-based methods suffer from mode collapse [85] and struggle to model datasets with larger scale and diversity [14]. Besides, 3D reconstruction and editing with GANs require elaborately designed inversion algorithms [43].

3D-Aware Diffusion Models. The success of 2D diffusion models [28,78] has inspired their application to 3D generation. DreamFusion [32,63,90] adapted 2D models for 3D, but faces challenges like expensive optimization, mode collapse, and the Janus problem. Some methods propose learning the 3D prior in a 2D manner [7,45,46,84]. While these can produce photorealistic results, they lack view consistency and fail to fully capture the 3D structure. A canonical 3D diffusion pipeline involves a two-stage training process. First, an auto-decoder is pre-trained with multi-view images [16,33,44,52,75,89]. Then, 3D latent codes serve as the training corpus for diffusion. However, the auto-decoding stage leads to an unclean latent space and limited scalability. Moreover, the large latent codes, e.g., $256 \times 256 \times 96$ [89], hinder efficient diffusion learning [31].

Prior works, such as RenderDiffusion [1] and DMV3D [94], propose latent-free 3D diffusion by integrating rendering into diffusion sampling. However, this approach involves time-consuming volume rendering at each denoising step, significantly slowing down sampling. SSDNeRF [9] suggests a joint 3D reconstruction and diffusion approach but requires a complex training schedule and shows performance only in single-category unconditional generation. In contrast, our proposed LN3DIFF trains 3D diffusion on a compressed latent space without rendering operations. As shown in Sect. 4, our method outperforms others in 3D generation and monocular 3D reconstruction, achieving three times faster speed. Additionally, we demonstrate conditional 3D generation over diverse datasets, whereas RenderDiffusion and SSDNeRF focus on simpler classes. Other approaches, like 3DGen [23] and VolumeDiffusion [82], perform diffusion in the 3D latent space but heavily rely on 3D data (e.g., point clouds and voxels) and do not support monocular datasets like FFHQ [34]. Moreover, their methods are designed for U-Net, whereas our DiT-based architecture offers greater scalability.

Generalized 3D Reconstruction and View Synthesis. To bypass the per-scene optimization of NeRF, researchers have proposed learning a prior model through image-based rendering [10,30,70,88,95]. However, these methods are primarily designed for view synthesis and lack explicit 3D representations. LoLNeRF [67] learns a prior through auto-decoding but is limited to simple, category-specific settings. Moreover, these methods are intended for view synthesis and cannot generate new 3D objects. VQ3D [71] adapts the generalized reconstruction pipeline to 3D generative models. However, it uses a 2D architecture with autoregressive modeling over a 1D latent space, ignoring much of the inherent 3D structure. NeRF-VAE [40] directly models 3D likelihood with a

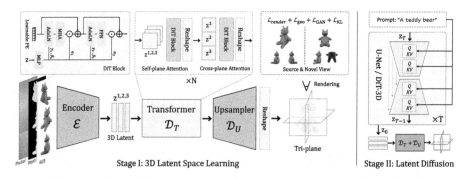

Fig. 2. Pipeline of LN3Diff. In the 3D latent space learning stage, a convolutional encoder $\mathcal{E}_\phi$ encodes a set of images $\mathcal{I}$ into the KL-regularized latent space. The encoded 3D latent is further decoded by a 3D-aware DiT transformer $\mathcal{D}_T$, in which we perform self-plane attention and cross-plane attention. The transformer-decoded latent is upsampled by a convolutional upsampler $\mathcal{D}_U$ towards a high-res tri-plane for rendering supervisions. In the next stage, we perform conditional diffusion learning over the compact latent space using either U-Net or DiT.

VAE posterior but is constrained to simple 3D scenes due to the limited capacity of VAE. Concurrently, LRM [10,30] has proposed a feedforward model for generalized monocular reconstruction. However, its latent space is not specifically designed for learning a generative model, which limits its effectiveness for 3D diffusion learning.

3 Scalable Latent Neural Fields Diffusion

This section introduces our latent 3D diffusion model, which learns efficient diffusion prior over the compressed latent space by a dedicated variational autoencoder. Specifically, the goal of training is to learn a variational encoder $\mathcal{E}_\phi$ that maps a set of posed 2D image(s) $\mathcal{I} = \{I_i, ..., I_V\}$, of a 3D object to a latent code $\mathbf{z}$, a denoiser $\epsilon_\theta(\mathbf{z}_t, t)$ to denoise the noisy latent code $\mathbf{z}_t$ given diffusion time step t, and a decoder $\mathcal{D}_\psi$ (including a Transformer $\mathcal{D}_T$ and an Upsampler $\mathcal{D}_U$) to map $\mathbf{z}_0$ to the 3D tri-plane $\widetilde{\mathcal{X}}$ corresponding to the input object.

Such design offers several advantages: (1) By explicitly separating the 3D data compression and diffusion stage, we avoid representation-specific 3D diffusion design [1,33,52,75,101] and achieve 3D representation/rendering-agnostic diffusion, which can be applied to any neural rendering techniques. (2) By leaving the high-dimensional 3D space, we reuse the well-studied Latent Diffusion Model (LDM) architecture [60,68,87] for computationally efficient learning and achieve better sampling performance with faster speed. (3) The trained 3D compression model in the first stage serves as an efficient and general-purpose 3D tokenizer, whose latent space can be easily re-used over downstream applications or extended to new datasets [12,96].

In the following subsections, we first discuss the compressive stage with a detailed framework design in Sect. 3.1. Based on that, we introduce the 3D diffusion generation stage in Sect. 3.2 and present the condition injection in Sect. 3.3. The method overview is shown in Fig. 2.

3.1 Perceptual 3D Latent Compression

As analyzed in Sect. 1, directly leveraging neural fields for diffusion training hinders model scalability and performance. Inspired by previous work [17,68], we propose to take multi-view image(s) as a proxy of the underlying 3D scene and compress the input image(s) into a compact 3D latent space. Though this paradigm is well-adopted in the image domain [17,68] with similar trials in specific 3D tasks [4,43,50,71], we, for the first time, demonstrate that a high-quality compression model is feasible, whose latent space serves as a compact proxy for efficient diffusion learning.

Encoder. Given a set of image(s) $\mathcal{I}$ of an 3D object where each image within the set $I \in \mathbb{R}^{H \times W \times 3}$ is an observation of an underlying 3D object from viewpoints $\mathcal{C} = \{\mathbf{c}_1, \ldots, \mathbf{c}_V\}$, LN3DIFF adopts a convolutional encoder $\mathcal{E}_\phi$ to encode the image set $\mathcal{I}$ into a latent representation $\mathbf{z} \sim \mathcal{E}_\phi(\mathcal{I})$. To inject camera condition, we concatenate Plücker coordinates $\mathbf{r}_i = (\mathbf{d}_i, \mathbf{p}_i \times \mathbf{d}_i) \in \mathbb{R}^6$ channel-wise as part of the input [76], where $\mathbf{d}_i$ is the normalized ray direction and $\mathbf{p}_i$ is the camera origin corresponding to the camera $\mathbf{c}_i$, and $\times$ denotes the cross product. For challenging datasets like Objaverse [13], we also concatenate the rendered depth map, making our input a dense 3D colored point cloud [92].

Unlike existing works [17,71] that operate on 1D order latent and ignore the internal structure, we choose to output 3D latent $\mathbf{z} \in \mathbb{R}^{h \times w \times d \times c}$ to facilitate 3D-aware operations, where $h = H/f, w = W/f$ are the spatial resolution with down-sample factor f, and d denotes the 3D dimension. Here we set $f = 8$ and $d = 3$ to make $\mathbf{z} \in \mathbb{R}^{h \times w \times 3 \times c}$ a tri-latent, which is similar to tri-plane [6,61] but in the compact 3D latent space. We further impose *KL-reg* [39] to encourage a well-structured latent space to facilitate diffusion training [68,87].

Decoder Transformer. The decoder aims to decode the compact 3D codes $\mathbf{z}$ for high-quality 3D reconstruction. Existing image-to-3D methods [4,6,20,43] adopt convolution as the building block, which lacks 3D-aware operations and impede information flow in the 3D space. Here, we adopt ViT [15,60] as the decoder backbone due to its flexibility and effectiveness. Inspired by Rodin [89], we made the following reformulation to the raw ViT decoder to encourage 3D inductive bias and avoid the mixing of uncorrelated 3D features: (1) *Self-plane Attention Block.* Given the input $\mathbf{z} \in \mathbb{R}^{l \times 3 \times c}$ where $l = h \times w$ is the sequence length, we treat each of the three latent planes as a data point and conduct self-attention within itself. This operation is efficient and encourages local feature aggregation. (2) *Cross-plane Attention Block.* To further encourage 3D inductive bias, we roll out $\mathbf{z}$ as a long sequence $l \times 3 \times c \rightarrow 3l \times c$ to conduct attention across planes, so that all tokens in the latent space could attend to each other.

In this way, we encourage global information flow for more coherent 3D reconstruction and generation. Compared to Rodin, our design is fully attention-based and naturally supports parallel computing without the expensive axis pooling aggregation.

Empirically, we observe that using DiT [60] block and injecting the latent $\mathbf{z}$ as conditions yields better performance compared to the ViT [15,56] block, which takes the latent $\mathbf{z}_0$ as the regular input. Specifically, the adaptive layer norm (adaLN) layer [60] fuses the input latent $\mathbf{z}$ with the learnable positional encoding for attention operations. Moreover, we interleave the two types of attention layers to make sure the overall parameters count consistent with the pre-defined DiT length, ensuring efficient training and inference. As all operations are defined in the token space, the decoder achieves efficient computation against Rodin [89] while promoting 3D priors.

Decoder Upsampler. After all the attention operations, we obtain the tokens from the last transformer layer $\tilde{\mathbf{z}}$ as the output. The context-rich tokens are reshaped back into spatial domain [24] and up-sampled by a convolutional decoder to the final tri-plane representation with shape $\hat{H} \times \hat{W} \times 3C$. Here, we adopt a lighter version of the convolutional decoder for efficient upsampling, where the three spatial latent of $\tilde{\mathbf{z}}$ are processed in parallel.

Learning a Perceptually Rich and Intact 3D Latent Space. Adversarial learning [19] has been widely applied in learning a compact and perceptually rich latent space [17,68]. In the 3D domain, the adversarial loss can also encourage correct 3D geometry when novel-view reconstruction supervisions are inapplicable [5,37,71], *e.g.*, the monocular dataset such as FFHQ. Inspired by previous research [37,71], we leverage adversarial loss to bypass this issue. Specifically, we impose an input-view discriminator to maintain perceptually-reasonable input view reconstruction, and an auxiliary novel-view discriminator to distinguish the rendered images between the input and novel views. We observe that if asking the novel-view discriminator to differentiate novel-view renderings and real images instead, the reconstruction model will suffer from *posterior collapse* [47], which outputs input-irrelevant but high-fidelity results to fool the novel-view discriminator. This phenomenon has also been observed by Kato *et al.* [37].

Training. After the decoder $\mathcal{D}_\psi$ decodes a high-resolution neural fields $\hat{\mathbf{z}}_0$ from the latent, we have $\hat{I} = \text{R}(\hat{\mathcal{X}}) = \text{R}(\mathcal{D}_\psi(\mathbf{z})) = \text{R}(\mathcal{D}_\psi(\mathcal{E}_\phi(\mathcal{I})))$, where R stands for differentiable rendering [83] and we take $\mathcal{D}_\psi(\mathbf{z}) = \text{R}(\mathcal{D}_\psi(\mathbf{z}))$ for brevity. Here, we choose $\hat{\mathcal{X}}$ as tri-plane [6,61] and R as volume rendering [51] for experiments. Note that our compression model is 3D representations/rendering agnostic and new neural rendering techniques [38] can be easily integrated by alternating the decoder architecture [81]. The final training objective reads as

$$\mathcal{L}(\phi, \psi) = \mathcal{L}_{\text{render}} + \lambda_{\text{geo}} \mathcal{L}_{\text{geo}} + \lambda_{\text{kl}} \mathcal{L}_{\text{KL}} + \lambda_{\text{GAN}} \mathcal{L}_{\text{GAN}}, \quad (1)$$

where $\mathcal{L}_{\text{render}}$ is a mixture of the L_1 and perceptual loss [100], $\mathcal{L}_{\text{reg}}$ encourages smooth geometry [91], $\mathcal{L}_{\text{KL}}$ is the *KL-reg* loss to regularize a structured latent space [68], and $\mathcal{L}_{\text{GAN}}$ improves perceptual quality and enforces correct geometry for monocular datasets. Note that $\mathcal{L}_{\text{render}}$ is applied to both input-view and randomly sampled novel-view images.

For category-specific datasets such as ShapeNet [8], we only supervise *one* novel view, which already yields good enough performance. For category-free datasets with diverse shape variations, *e.g.*, Objaverse [13], we supervise *four* novel views. Our method is more data-efficient against the existing state-of-the-art 3D diffusion method [9,52], which requires 50 views to converge. The implementation details are included in the supplementary material.

3.2 Latent Diffusion and Denoising

Latent Diffusion Models. LDM [68,87] is designed to acquire a prior distribution $p_\theta(\mathbf{z}_0)$ within the perceptual latent space, whose training data is the latent obtained online from the trained $\mathcal{E}_\phi$. Here, we use the score-based latent diffusion model [87], which is the continuous derivation of DDPM variational objective [28]. Specifically, the denoiser ϵ_θ parameterizes the score function score [78] as $\nabla_{\mathbf{z}_t} \log p(\mathbf{z}_t) := -\epsilon_\theta(\mathbf{z}_t, t)/\sigma_t$, with continuous time sequence t. By training to predict a denoised variant of the noisy input $\mathbf{z}_t$, ϵ_θ gradually learns to denoise from a standard Normal prior $\mathcal{N}(\mathbf{0}, \mathbf{I})$ by solving a reverse SDE [28].

Following LSGM [87], we formulate the learned prior at time t geometric mixing $\epsilon_\theta(\mathbf{z}_t, t) := \sigma_t(1 - \boldsymbol{\alpha}) \odot \mathbf{z}_t + \boldsymbol{\alpha} \odot \epsilon'_\theta(\mathbf{z}_t, t)$, where $\epsilon'_\theta(\mathbf{z}_t, t)$ is the denoiser output and $\alpha \in [0, 1]$ is a learnable scalar coefficient. Intuitively, this formulation can bring the denoiser input closer to a standard Normal distribution, which the reverse SDE can be solved faster. Similarly, Stable Diffusion [62,68] also scales the input latent by a factor to maintain a unit variance, which is pre-calculated on the billion-level dataset [72]. The training objective reads as

$$\mathcal{L}_{\text{diff}} = \mathbb{E}_{\mathcal{E}_\phi(I), \epsilon \sim \mathcal{N}(0,1), t}\left[\frac{w_t}{2}\|\epsilon - \epsilon_\theta(\mathbf{z}_t, t)\|_2^2\right], \qquad (2)$$

where $t \sim \mathcal{U}[0, 1]$ and w_t is an empirical time-dependent weighting function.

The denoiser ϵ_θ is realized by a time-dependent U-Net [69] or DiT [60]. During training, we obtain $\mathbf{z}_0$ online from the fixed $\mathcal{E}_\phi$, roll-out the tri-latent $h \times w \times 3 \times c \to h \times (3w) \times c$, and add time-dependent noise to get $\mathbf{z}_t$. Here, we choose the importance sampling schedule [87] with *velocity* [48] parameterization, which yields more stable behavior against ϵ parameterization for diffusion learning. After training, the denoised samples can be decoded to the 3D neural field (*i.e.*, tri-plane here) with a single forward pass through $\mathcal{D}_\psi$, on which neural rendering can be applied.

3.3 Conditioning Mechanisms

Compared with the existing approach that focuses on category-specific unconditional 3D diffusion model [1], we propose to inject CLIP embeddings [66] into the latent 3D diffusion model to support image/text-conditioned 3D generation. Given input condition $\mathbf{y}$, the diffusion model formulates the conditional distribution $p(\mathbf{z}|\mathbf{y})$ with $\epsilon_\theta(\mathbf{z}_t, t, \mathbf{y})$. The inputs $\mathbf{y}$ can be text captions for datasets like Objaverse, or images for general datasets like ShapeNet and FFHQ.

Text Conditioning. For datasets with text caption, we follow Stable Diffusion [68] to directly leverage the CLIP text encoder CLIP_T to encode the text caption as the conditions. All the output tokens 77×768 are used and injected to diffusion denoiser with cross attention blocks.

Image Conditioning. For datasets with images only, we encode the input I corresponding to the latent code $\mathbf{z}_0$ using CLIP image encoder CLIP_I and adopt the output embedding as the condition. To support both image and text conditions, we re-scale the image latent code with a pre-calculated factor to match the scale of the text latent.

Classifier-Free Guidance. We adopt *classifier-free guidance* [27] for latent conditioning to support conditional and unconditional generation. During diffusion model training, we randomly zero out the corresponding conditioning latent embeddings with 15% probability to jointly train the unconditional and conditional settings. During sampling, we perform a linear combination of the conditional and unconditional score estimates:

$$\hat{\epsilon}_\theta(\mathbf{z}_t, \tau_\theta(y)) = s\epsilon_\theta(\mathbf{z}_t, \tau_\theta(y)) + (1-s)\epsilon_\theta(\mathbf{z}_t), \tag{3}$$

where s is the guidance scale to control the mixing strength to balance sampling diversity and quality.

4 Experiments

Datasets. Following most previous work, we use ShapeNet [8] to benchmark the 3D generation performance. We use the Car, Chair, and Plane categories with 3514, 6700, and 4045 instances correspondingly. Each instance is randomly rendered from 50 views following a spherical uniform distribution. Moreover, to evaluate the performance over diverse high-quality 3D datasets, we also include the experiments over Objaverse [13], which is the largest 3D dataset with challenging categories and complicated geometry. We use the renderings provided by G-Objaverse [65] and choose a high-quality subset with around $176K$ 3D instances, where each consists of 40 random views.

Training Details. For ShapeNet and FFHQ training, we adopt a monocular input setting with $V = 1$ and target rendering size 128×128. For Objaverse, we adopt multi-view inputs with $V = 6$ and target rendering size 192×192. All training images are resized to $H = W = 256$ as input. The encoder $\mathcal{E}_\phi$ has down-sample factor $f = 8$ and the decoder upsampler $\mathcal{D}_U$ outputs tri-plane with size $\hat{H} = \hat{W} = 128$ and $C = 32$. To trade off rendering resolution and training batch size, we impose supervision over 80×80 randomly cropped patches. For adversarial loss, we use DINO [56] in vision-aided GAN [41] with non-saturating GAN loss [22] for discriminator training. For conditional diffusion training, we use the CLIP image embedding for ShapeNet and FFHQ, and CLIP text embedding from the official text caption for Objaverse. Both the autoencoding model and diffusion model are trained for $800K$ iterations, which take around 7 days with 8 A100 GPUs in total.

Metrics. Following prior work [9,18,52,75], we adopt both 2D and 3D metrics to benchmark the generation performance: Fréchet Inception Distance (FID@50K) [26] and Kernel Inception Distance (KID@50K) [2] to evaluate 2D renderings, as well as Coverage Score (COV) and Minimum Matching Distance (MMD) to benchmark 3D geometry. We calculate all metrics under 128×128 for fair comparisons across all baselines.

4.1 Evaluation

In this section, we compare our method with both state-of-the-art GAN-based methods: EG3D [6], GET3D [18] as well as recent diffusion-based methods: DiffRF [52], RenderDiffusion [1] and SSDNeRF [9]. Since LN3DIFF only leverages $v = 2$ for ShapeNet experiments, for SSDNeRF, we include both the official

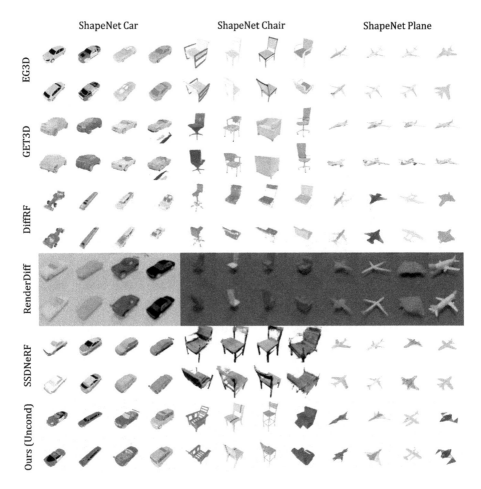

Fig. 3. ShapeNet Unconditional Generation. We show four samples for each method. Zoom in for the best view.

50-views trained SSDNeRF$_{V=50}$ version as well as the reproduced SSDNeRF$_{V=3}$ for fair comparison. We find SSDNeRF fails to converge with $V = 2$. We set the guidance scale in Eq. (3) to $s = 0$ for unconditional generation, and $s = 6.5$ for all conditional generation sampling.

Unconditional Generation on ShapeNet. To evaluate our methods against existing 3D generation methods, we include the quantitative and qualitative results for unconditional single-category 3D generation on ShapeNet in Table 1 and Fig. 3. We evaluate all baseline 3D diffusion methods with 250 DDIM steps and GAN-based baselines with $psi = 0.8$ to guarantee each sample is intact for COV/MMD evaluation. For FID/KID evaluation, we re-train the baselines and calculate the metrics using a fixed upper-sphere ellipsoid camera trajectory [77] across all datasets. For COV/MMD evaluation, we randomly sample 4096 points around the extracted sampled mesh and ground truth mesh surface and adopt Chamfer Distance for evaluation.

Table 1. Quantitative Comparison of Unconditional Generation on ShapeNet. The proposed LN3DIFF shows satisfactory performance on single-category generation.

Category	Method	FID@50K↓	KID@50K(%)↓	COV(%)↑	MMD(‰)↓
Car	EG3D [6]	33.33	1.4	35.32	3.95
	GET3D [18]	41.41	1.8	37.78	3.87
	DiffRF [52]	75.09	5.1	29.01	4.52
	RenderDiffusion [1]	46.5	4.1	–	–
	SSDNeRF$_{V=3}$ [9]	47.72	2.8	37.84	3.46
	SSDNeRF$^*_{V=50}$ [9]	45.37	2.1	**67.82**	2.50
	LN3DIFF(**Ours**)	**17.6**	**0.49**	43.12	**2.32**
Plane	EG3D [6]	14.47	0.54	18.12	4.50
	GET3D [18]	26.80	1.7	21.30	4.06
	DiffRF [52]	101.79	6.5	37.57	3.99
	RenderDiffusion [1]	43.5	5.9	–	–
	SSDNeRF$_{V=3}$ [9]	21.01	1.0	42.50	2.94
	LN3DIFF(**Ours**)	**8.84**	**0.36**	**43.40**	**2.71**
Chair	EG3D [6]	26.09	1.1	19.17	10.31
	GET3D [18]	35.33	1.5	28.07	9.10
	DiffRF [52]	99.37	4.9	17.05	14.97
	RenderDiffusion [1]	53.3	6.4	–	–
	SSDNeRF$_{V=3}$ [9]	65.04	3.0	**47.54**	6.71
	LN3DIFF(**Ours**)	**16.9**	**0.47**	47.1	**5.28**

As shown in Table 1, LN3DIFF achieves quantitatively better performance against all GAN-based baselines regarding rendering quality and 3D coverage. Figure 3 further demonstrates that GAN-based methods suffer greatly from mode collapse: in the ShapeNet Plane category, both EG3D and GET3D are limited to the white civil airplanes, which is fairly common in the dataset. Our methods can sample more diverse results with high-fidelity texture.

Compared against diffusion-based baselines, LN3DIFF also shows better visual quality with better quantitative metrics. SSDNeRF$_{V=50}$ shows a better coverage score, which benefits from leveraging more views during training. However, our method with $V = 2$ shows comparative performance against SSDNeRF$_{V=3}$ on the ShapeNet Chair and even better performance on the remaining datasets.

Conditional 3D Generation. Conditional 3D generation has the potential to streamline the 3D modeling process in both the gaming and film industries. As visualized in Fig. 4, we present our conditional generation performance on the ShapeNet dataset, where either text or image serves as the input prompt. Visually inspected, our method demonstrates promising performance in conditional generation, closely aligning the generated outputs with the input conditions. For the image-conditioned generation, our method yields semantically similar samples while maintaining diversity.

We also demonstrate the text-conditioned generation of Objaverse in Fig. 5 and Table 2. As shown, the diffusion model trained over LN3DIFF's latent space enables high-quality 3D generation over generic 3D datasets. This ability is unprecedented among existing 3D diffusion baselines and and marks a significant step toward highly controllable 3D generation. Qualitative comparisons against Shape-E and Point-E are provided in the supplementary materials.

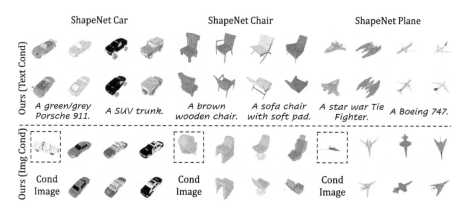

Fig. 4. ShapeNet Conditional Generation. We show conditional generation with both texts and image as inputs. Zoom in for the best view.

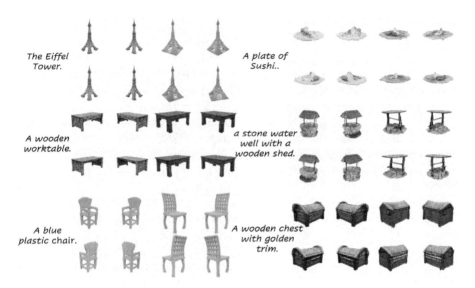

Fig. 5. Objaverse Conditional Generation Given Text Prompt. We show two samples for each prompt. Zoom in for the best view.

We compare our method against RenderDiffusion, the only 3D diffusion method that supports 3D generation over FFHQ. As shown in the lower part in Fig. 6, beyond view-consistent 3D generation, our method further supports conditional 3D generation at 128 × 128 resolution, while RenderDiffusion is limited to 64 × 64 resolution due to the expensive volume rendering integrated into diffusion training. Quantitatively, our method achieves an FID score of 36.6, compared to 59.3 by RenderDiffusion.

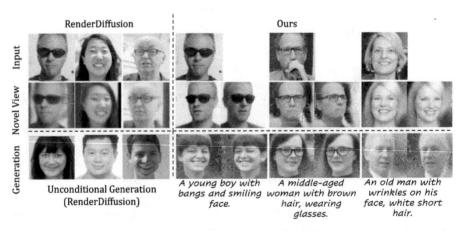

Fig. 6. FFHQ Monocular Reconstruction (upper half) and 3D Generation (lower half). For monocular reconstruction, we test our method with hold-out test set and visualize the input-view and novel-view. Compared to baseline, our method shows consistently better performance on both reconstruction and generation.

Monocular 3D Reconstruction Beyond the samples shown in Fig. 1, we also include monocular reconstruction results over FFHQ datasets in the upper half of Fig. 6 and compare against RenderDiffusion. As can be observed, our method demonstrates high fidelity and preserves semantic details even in self-occluded images. The novel view generated by RenderDiffusion appears blurry and misses semantic components that are not visible in the input view, such as the leg of the eyeglass.

Table 2. Quantitative Metrics on Text-to-3D. The proposed method outperforms Point-E and Shape-E on CLIP scores over two different backbones.

Method	ViT-B/32	ViT-L/14
Point-E [54]	26.35	21.40
Shape-E [33]	27.84	25.84
Ours	**29.12**	**27.80**

4.2 Ablation Study and Analysis

Reconstruction Arch Design. In Table 3, we benchmark each component of our auto-encoding architecture over a subset of Objaverse with $7K$ instances and record the PSNR at $100K$ iterations. Each component introduces consistent improvements with negligible parameter increases.

Table 3. Ablation of Reconstruction Arch Design. We ablate the design of our auto-encoding architecture. Each component contributes to a consistent gain in the reconstruction performance, indicating an improvement in the modeling capacity.

Design	PSNR@100K
2D Conv Baseline	17.46
+ ViT Block	18.92
ViT Block → DiT Block	20.61
+ Plucker Embedding	21.29
+ Cross-Plane Attention	21.70
+ Self-Plane Attention	**21.95**

Table 4. Diffusion Sampling Speed and Latent Size. We provide the sampling time per instance evaluated on 1 V100, along with the latent size. Our method achieves faster sampling speed while maintaining superior generation performance.

Method	V100-sec	Latent Size
Get3D/EG3D	<0.5	256
SSDNeRF	8.1	$128^2 \times 18$
RenderDiffusion	15.8	–
DiffRF	18.7	$32^3 \times 4$
LN3DIFF$_{\text{uncond}}$	**5.7**	$32^2 \times 12$
LN3DIFF$_{\text{cfg}}$	**7.5**	$32^2 \times 12$

Novel View Discriminator for Monocular Dataset. Novel view discriminator is crucial for monocular datasets like FFHQ. As shown in Fig. 7, without it, the VAE model fails to yield a plausible novel view.

Input Image   W/O Novel View Discriminator   W Novel View Discriminator

Fig. 7. Ablation of novel view discriminator.

Diffusion Sampling Speed and Latent Size. We report the sampling speed and latent space size comparison in Table 4. By performing on the compact latent space, our method achieves the fastest sampling while keeping the best generation performance. Though RenderDiffusion follows a latent-free design, its intermediate 3D neural field has a shape of $256^2 \times 96$ and hinders efficient diffusion training.

5 Conclusion and Discussions

In this work, we present a new paradigm of 3D generative model by learning the diffusion model over a compact 3D-aware latent space. A dedicated variational autoencoder encodes (multi-view) image(s) into a low-dim structured latent space, where conditional diffusion learning can be efficiently performed. We achieve state-of-the-art performance over ShapeNet and demonstrate our method over generic category-free Objaverse 3D datasets. Our work potentially facilitates numerous downstream applications in 3D vision and graphics tasks.

Limitations and Future Work. Our method comes with some limitations unresolved. VAE side, we observe that volume rendering is memory-consuming. Extending our decoder to more efficient 3D representations such as 3DGS [38] shall alleviate this issue. Besides, adding more real-world data such as MVImageNet [96] and more control conditions [99] is also worth exploring. Overall, our method is a step towards a general 3D diffusion model and can inspire future research in this direction.

Potential Negative Societal Impacts. The entity relational composition capabilities of LN3DIFF could be applied maliciously to real human figures. Additional potential impacts are discussed in the Supplementary File in depth.

Acknowledgement. This study is supported under the RIE2020 Industry Alignment Fund Industry Collaboration Projects (IAF-ICP) Funding Initiative, as well as cash and in-kind contributions from the industry partner(s). It is also supported by Singapore MOE AcRF Tier 2 (MOE-T2EP20221-0011).

References

1. Anciukevičius, T., et al.: RenderDiffusion: image diffusion for 3d reconstruction, inpainting and generation. In: CVPR (2023)
2. Bińkowski, M., Sutherland, D.J., Arbel, M., Gretton, A.: Demystifying MMD GANs. In: ICLR (2018)
3. Brock, A., Donahue, J., Simonyan, K.: Large scale GAN training for high fidelity natural image synthesis. In: ICLR (2019)
4. Cai, S., Obukhov, A., Dai, D., Van Gool, L.: Pix2NeRF: unsupervised conditional p-GAN for single image to neural radiance fields translation. In: CVPR (2022)
5. Chan, E., Monteiro, M., Kellnhofer, P., Wu, J., Wetzstein, G.: Pi-GAN: periodic implicit generative adversarial networks for 3D-aware image synthesis. In: CVPR (2021)
6. Chan, E.R., et al.: Efficient geometry-aware 3D generative adversarial networks. In: CVPR (2022)

7. Chan, E.R., et al.: GeNVS: generative novel view synthesis with 3D-aware diffusion models. arXiv (2023)
8. Chang, A.X., et al.: ShapeNet: an information-rich 3D model repository. arXiv preprint arXiv:1512.03012 (2015)
9. Chen, H., et al.: Single-stage diffusion nerf: a unified approach to 3D generation and reconstruction. In: ICCV (2023)
10. Chen, Y., Wang, T., Wu, T., Pan, X., Jia, K., Liu, Z.: Comboverse: compositional 3d assets creation using spatially-aware diffusion guidance. arXiv preprint arXiv:2403.12409 (2024)
11. Contributors, S.: SpConv: spatially sparse convolution library (2022). https://github.com/traveller59/spconv
12. Deitke, M., et al.: Objaverse-xl: a universe of 10m+ 3d objects. arXiv preprint arXiv:2307.05663 (2023)
13. Deitke, M., et al.: Objaverse: a universe of annotated 3D objects. arXiv preprint arXiv:2212.08051 (2022)
14. Dhariwal, P., Nichol, A.: Diffusion models beat gans on image synthesis. In: NeurIPS (2021)
15. Dosovitskiy, A., et al.: An image is worth 16×16 words: transformers for image recognition at scale. In: ICLR (2021)
16. Dupont, E., Kim, H., Eslami, S.M.A., Rezende, D.J., Rosenbaum, D.: From data to functa: your data point is a function and you can treat it like one. In: ICML (2022)
17. Esser, P., Rombach, R., Ommer, B.: Taming transformers for high-resolution image synthesis. In: CVPR (2021)
18. Gao, J., et al.: Get3D: a generative model of high quality 3D textured shapes learned from images. In: NeurIPS (2022)
19. Goodfellow, I.J., et al.: Generative adversarial nets. In: NeurIPS (2014)
20. Gu, J., Gao, Q., Zhai, S., Chen, B., Liu, L., Susskind, J.: Learning controllable 3D diffusion models from single-view images. arXiv preprint arXiv:2304.06700 (2023)
21. Gu, J., Liu, L., Wang, P., Theobalt, C.: StyleNeRF: a style-based 3D-aware generator for high-resolution image synthesis. In: ICLR (2021)
22. Gulrajani, I., Ahmed, F., Arjovsky, M., Dumoulin, V., Courville, A.C.: Improved training of wasserstein GANs. In: NeurIPS (2017)
23. Gupta, A., Xiong, W., Nie, Y., Jones, I., Oğuz, B.: 3dgen: triplane latent diffusion for textured mesh generation (2023). https://arxiv.org/abs/2303.05371
24. He, K., Chen, X., Xie, S., Li, Y., Doll'ar, P., Girshick, R.B.: Masked autoencoders are scalable vision learners. In: CVPR (2022)
25. Henzler, P., Mitra, N.J., Ritschel, T.: Escaping plato's cave: 3D shape from adversarial rendering. In: ICCV (2019)
26. Heusel, M., Ramsauer, H., Unterthiner, T., Nessler, B., Hochreiter, S.: GANs trained by a two time-scale update rule converge to a local nash equilibrium. In: NeurIPS (2017)
27. Ho, J.: Classifier-free diffusion guidance. In: NeurIPS (2021)
28. Ho, J., Jain, A., Abbeel, P.: Denoising diffusion probabilistic models. In: NeurIPS (2020)
29. Hong, F., Chen, Z., Lan, Y., Pan, L., Liu, Z.: EVA3D: compositional 3D human generation from 2d image collections. In: ICLR (2022)
30. Hong, Y., et al.: LRM: large reconstruction model for single image to 3D. In: ICLR (2024)
31. Hoogeboom, E., Heek, J., Salimans, T.: simple diffusion: end-to-end diffusion for high resolution images. In: ICML (2023)

32. Jain, A., Mildenhall, B., Barron, J.T., Abbeel, P., Poole, B.: Zero-shot text-guided object generation with dream fields. In: CVPR (2022)
33. Jun, H., Nichol, A.: Shap-E: Generating conditional 3D implicit functions. arXiv preprint arXiv:2305.02463 (2023)
34. Karras, T., Aila, T., Laine, S., Lehtinen, J.: Progressive growing of GANs for improved quality, stability, and variation. In: ICLR (2018)
35. Karras, T., Laine, S., Aila, T.: A style-based generator architecture for generative adversarial networks. In: CVPR (2019)
36. Karras, T., Laine, S., Aittala, M., Hellsten, J., Lehtinen, J., Aila, T.: Analyzing and improving the image quality of StyleGAN. In: CVPR (2020)
37. Kato, H., Harada, T.: Learning view priors for single-view 3D reconstruction. In: CVPR (2019)
38. Kerbl, B., Kopanas, G., Leimkühler, T., Drettakis, G.: 3D gaussian splatting for real-time radiance field rendering. ACM Trans. Graph. **42**(4), 1–14 (2023)
39. Kingma, D.P., Welling, M.: Auto-encoding variational bayes. arXiv (2013)
40. Kosiorek, A.R., et al.: NeRF-VAE: a geometry aware 3D scene generative model. In: ICML (2021)
41. Kumari, N., Zhang, R., Shechtman, E., Zhu, J.Y.: Ensembling off-the-shelf models for gan training. In: CVPR (2022)
42. Lan, Y., Loy, C.C., Dai, B.: DDF: correspondence distillation from nerf-based gan. IJCV **132**, 611–631 (2022)
43. Lan, Y., Meng, X., Yang, S., Loy, C.C., Dai, B.: E3dge: self-supervised geometry-aware encoder for style-based 3D gan inversion. In: CVPR (2023)
44. Lan, Y., et al.: Gaussian3Diff: 3D gaussian diffusion for 3D full head synthesis and editing. arXiv (2023)
45. Liu, R., Wu, R., Hoorick, B.V., Tokmakov, P., Zakharov, S., Vondrick, C.: Zero-1-to-3: zero-shot one image to 3D object (2023)
46. Long, X., et al.: Wonder3d: single image to 3d using cross-domain diffusion. In: CVPR (2024)
47. Lucas, J., Tucker, G., Grosse, R.B., Norouzi, M.: Understanding posterior collapse in generative latent variable models. In: ICLR (2019)
48. Meng, C., Gao, R., Kingma, D.P., Ermon, S., Ho, J., Salimans, T.: On distillation of guided diffusion models. In: CVPR, pp. 14297–14306 (2022)
49. Mescheder, L., Oechsle, M., Niemeyer, M., Nowozin, S., Geiger, A.: Occupancy networks: learning 3D reconstruction in function space. In: CVPR (2019)
50. Mi, L., Kundu, A., Ross, D., Dellaert, F., Snavely, N., Fathi, A.: im2nerf: image to neural radiance field in the wild. arXiv (2022)
51. Mildenhall, B., Srinivasan, P.P., Tancik, M., Barron, J.T., Ramamoorthi, R., Ng, R.: NeRF: representing scenes as neural radiance fields for view synthesis. In: Vedaldi, A., Bischof, H., Brox, T., Frahm, J.-M. (eds.) ECCV 2020. LNCS, vol. 12346, pp. 405–421. Springer, Cham (2020). https://doi.org/10.1007/978-3-030-58452-8_24
52. Müller, N., Siddiqui, Y., Porzi, L., Bulo, S.R., Kontschieder, P., Nießner, M.: DiffRF: rendering-guided 3D radiance field diffusion. In: CVPR (2023)
53. Nguyen-Phuoc, T., Li, C., Theis, L., Richardt, C., Yang, Y.: HoloGAN: unsupervised learning of 3D representations from natural images. In: ICCV (2019)
54. Nichol, A., Jun, H., Dhariwal, P., Mishkin, P., Chen, M.: Point-e: a system for generating 3d point clouds from complex prompts (2022)
55. Niemeyer, M., Geiger, A.: GIRAFFE: representing scenes as compositional generative neural feature fields. In: CVPR (2021)

56. Oquab, M., et al.: DINOv2: learning robust visual features without supervision (2023)
57. Or-El, R., Luo, X., Shan, M., Shechtman, E., Park, J.J., Kemelmacher-Shlizerman, I.: StyleSDF: high-resolution 3D-consistent image and geometry generation. In: CVPR (2021)
58. Pan, X., Dai, B., Liu, Z., Loy, C.C., Luo, P.: Do 2D GANs know 3D shape? unsupervised 3D Shape Reconstruction from 2D Image GANs. In: ICLR (2021)
59. Park, J.J., Florence, P., Straub, J., Newcombe, R., Lovegrove, S.: Deepsdf: learning continuous signed distance functions for shape representation. In: CVPR, pp. 165–174 (2019)
60. Peebles, W., Xie, S.: Scalable diffusion models with transformers. In: ICCV (2023)
61. Peng, S., Niemeyer, M., Mescheder, L., Pollefeys, M., Geiger, A.: Convolutional occupancy networks. In: Vedaldi, A., Bischof, H., Brox, T., Frahm, J.-M. (eds.) ECCV 2020. LNCS, vol. 12348, pp. 523–540. Springer, Cham (2020). https://doi.org/10.1007/978-3-030-58580-8_31
62. Podell, D., et al.: SDXL: improving latent diffusion models for high-resolution image synthesis. arXiv (2023)
63. Poole, B., Jain, A., Barron, J.T., Mildenhall, B.: DreamFusion: text-to-3D using 2D diffusion. In: ICLR (2022)
64. Qi, C., Su, H., Mo, K., Guibas, L.: PointNet: deep learning on point sets for 3D classification and segmentation. arXiv (2016)
65. Qiu, L., et al.: Richdreamer: a generalizable normal-depth diffusion model for detail richness in text-to-3d. arXiv preprint arXiv:2311.16918 (2023)
66. Radford, A., et al.: Learning transferable visual models from natural language supervision. In: ICML (2021)
67. Rebain, D., Matthews, M., Yi, K.M., Lagun, D., Tagliasacchi, A.: LOLNeRF: learn from one look. In: CVPR (2022)
68. Rombach, R., Blattmann, A., Lorenz, D., Esser, P., Ommer, B.: High-resolution image synthesis with latent diffusion models. In: CVPR (2022)
69. Ronneberger, O., Fischer, P., Brox, T.: U-Net: convolutional networks for biomedical image segmentation. In: Navab, N., Hornegger, J., Wells, W.M., Frangi, A.F. (eds.) MICCAI 2015. LNCS, vol. 9351, pp. 234–241. Springer, Cham (2015). https://doi.org/10.1007/978-3-319-24574-4_28
70. Sajjadi, M.S.M., et al.: Scene representation transformer: geometry-free novel view synthesis through set-latent scene representations. In: CVPR (2022)
71. Sargent, K., et al.: VQ3D: learning a 3D-aware generative model on imagenet. In: ICCV (2023)
72. Schuhmann, C., et al.: LAION-5B: an open large-scale dataset for training next generation image-text models. arXiv (2022)
73. Schwarz, K., Liao, Y., Niemeyer, M., Geiger, A.: GRAF: generative radiance fields for 3D-aware image synthesis. In: NeurIPS (2020)
74. Shi, R., et al.: Zero123++: a single image to consistent multi-view diffusion base model. arXiv (2023)
75. Shue, J., Chan, E., Po, R., Ankner, Z., Wu, J., Wetzstein, G.: 3d neural field generation using triplane diffusion. In: CVPR (2022)
76. Sitzmann, V., Rezchikov, S., Freeman, W.T., Tenenbaum, J.B., Durand, F.: Light field networks: neural scene representations with single-evaluation rendering. In: NeurIPS (2021)
77. Sitzmann, V., Zollhöfer, M., Wetzstein, G.: Scene representation networks: continuous 3D-structure-aware neural scene representations. In: NeurIPS (2019)

78. Song, Y., Sohl-Dickstein, J., Kingma, D.P., Kumar, A., Ermon, S., Poole, B.: Score-based generative modeling through stochastic differential equations. In: ICLR (2021)
79. Sun, J., Wang, X., Shi, Y., Wang, L., Wang, J., Liu, Y.: Ide-3d: interactive disentangled editing for high-resolution 3D-aware portrait synthesis. ACM Trans. Graph. (TOG) **41**(6), 1–10 (2022)
80. Sun, J., et al.: FENeRF: face editing in neural radiance fields. arXiv (2021)
81. Szymanowicz, S., Rupprecht, C., Vedaldi, A.: Splatter image: ultra-fast single-view 3D reconstruction. arXiv (2023)
82. Tang, Z., et al.: Volumediffusion: flexible text-to-3d generation with efficient volumetric encoder (2023)
83. Tewari, A., et al.: Advances in neural rendering. Comput. Graph. Forum **41** (2021)
84. Tewari, A., et al.: Diffusion with forward models: solving stochastic inverse problems without direct supervision. In: NeurIPS (2023)
85. Thanh-Tung, H., Tran, T.: Catastrophic forgetting and mode collapse in gans. In: IJCNN, pp. 1–10 (2020)
86. Thomas, H., Qi, C.R., Deschaud, J.E., Marcotegui, B., Goulette, F., Guibas, L.J.: KPConv: flexible and deformable convolution for point clouds. In: ICCV (2019)
87. Vahdat, A., Kreis, K., Kautz, J.: Score-based generative modeling in latent space. In: NeurIPS (2021)
88. Wang, Q., et al.: IBRNet: learning multi-view image-based rendering. In: CVPR (2021)
89. Wang, T., et al.: RODIN: a generative model for sculpting 3D digital avatars using diffusion. In: CVPR (2023)
90. Wang, Z., Lu, C., Wang, Y., Bao, F., Li, C., Su, H., Zhu, J.: Prolificdreamer: high-fidelity and diverse text-to-3D generation with variational score distillation. In: NeurIPS (2023)
91. Weng, C.Y., Srinivasan, P.P., Curless, B., Kemelmacher-Shlizerman, I.: PersonNeRF: personalized reconstruction from photo collections. In: CVPR, pp. 524–533 (2023)
92. Wu, C.Y., Johnson, J., Malik, J., Feichtenhofer, C., Gkioxari, G.: Multiview compressive coding for 3D reconstruction. arXiv preprint arXiv:2301.08247 (2023)
93. Xie, Y., et al.: Neural fields in visual computing and beyond. Comput. Graph. Forum **41** (2021)
94. Xu, Y., et al.: DMV3D: denoising multi-view diffusion using 3D large reconstruction model. In: ICLR (2024)
95. Yu, A., Ye, V., Tancik, M., Kanazawa, A.: PixelNeRF: neural radiance fields from one or few images. In: CVPR (2021)
96. Yu, X., et al.: MVImgNet: a large-scale dataset of multi-view images. In: CVPR (2023)
97. Zhang, B., Tang, J., Nießner, M., Wonka, P.: 3DShape2VecSet: a 3d shape representation for neural fields and generative diffusion models. ACM Trans. Graph. **42**(4) (2023). https://doi.org/10.1145/3592442
98. Zhang, J., et al.: Deformtoon3d: deformable 3D toonification from neural radiance fields. In: ICCV (2023)
99. Zhang, L., Rao, A., Agrawala, M.: Adding conditional control to text-to-image diffusion models. In: ICCV (2023)
100. Zhang, R., Isola, P., Efros, A.A., Shechtman, E., Wang, O.: The unreasonable effectiveness of deep features as a perceptual metric. In: CVPR (2018)
101. Zhou, L., Du, Y., Wu, J.: 3D shape generation and completion through point-voxel diffusion. In: ICCV (2021)

Hierarchical Temporal Context Learning for Camera-Based Semantic Scene Completion

Bohan Li[1,2], Jiajun Deng[3], Wenyao Zhang[1,2], Zhujin Liang[4], Dalong Du[4], Xin Jin[2(✉)], and Wenjun Zeng[1,2]

[1] Shanghai Jiao Tong University, Shanghai, China
{bohan_li,wy_zhang}@sjtu.edu.cn
[2] Ningbo Institute of Digital Twin, Eastern Institute of Technology, Ningbo, China
{jinxin,wenjunzengvp}@eitech.edu.cn
[3] The University of Adelaide, Adelaide, Australia
jiajun.deng@adelaide.edu.au
[4] PhiGent Robotics, Beijing, China
{zhujin.liang,dalong.du}@phigent.ai

Abstract. Camera-based 3D semantic scene completion (SSC) is pivotal for predicting complicated 3D layouts with limited 2D image observations. The existing mainstream solutions generally leverage temporal information by roughly stacking history frames to supplement the current frame, such straightforward temporal modeling inevitably diminishes valid clues and increases learning difficulty. To address this problem, we present **HTCL**, a novel **H**ierarchical **T**emporal **C**ontext **L**earning paradigm for improving camera-based semantic scene completion. The primary innovation of this work involves decomposing temporal context learning into two hierarchical steps: (a) cross-frame affinity measurement and (b) affinity-based dynamic refinement. Firstly, to separate critical relevant context from redundant information, we introduce the pattern affinity with scale-aware isolation and multiple independent learners for fine-grained contextual correspondence modeling. Subsequently, to dynamically compensate for incomplete observations, we adaptively refine the feature sampling locations based on initially identified locations with high affinity and their neighboring relevant regions. Our method ranks 1st on the SemanticKITTI benchmark and even surpasses LiDAR-based methods in terms of mIoU on the OpenOccupancy benchmark. Our code is available on https://github.com/Arlo0o/HTCL.

Keywords: Semantic Scene Completion · Temporal Context Learning

1 Introduction

The comprehension of holistic 3D scenes holds paramount importance in autonomous driving systems [16,22,46]. This capability directly impacts the

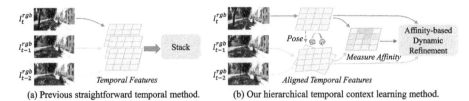

Fig. 1. Our hierarchical temporal context learning method versus previous straightforward temporal method (VoxFormer-T [22]) in semantic scene completion.

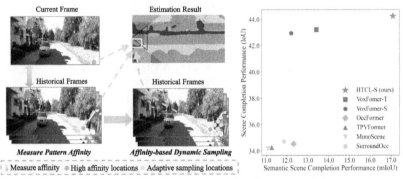

Fig. 2. (a) Overview pipeline of the proposed method, which measures contextual pattern affinity across temporal frames and dynamically samples relevant context. Our method shows promising performance in comprehending and completing semantic scenes even outside the camera's field of view, as indicated by the car highlighted with the yellow box. (b) Comparison with state-of-the-art camera-based semantic scene completion methods [7,16,22,46,52] on the SemanticKITTI test set. (Color figure online)

planning and obstacle avoidance functionalities of autonomous vehicles, thereby influencing their overall safety and efficiency. However, due to the limitations of real-world sensors such as restricted field of view and measurement noise, this task remains a challenging problem. 3D semantic scene completion (SSC) has been proposed to address the challenges by jointly inferring the geometry and semantics of the scenario from incomplete observations [7,14,16,22,34,35,47,51].

Given the inherent 3D nature, numerous semantic scene completion (SSC) solutions [14,34,35,47,51] rely on LiDAR as the primary technique for precise 3D location measurement. Although the LiDAR sensor provides accurate depth information, its deployment introduces a large overhead for both the cost and manual efforts. Therefore, it is necessary to explore an efficient approach for high-fidelity SSC with cost-effective devices. This motivation led to the investigation of camera-based solutions, which are characterized by superior deployment efficiency and the abundance of rich visual context [7,16,46,52].

The early attempts in camera-based SSC methods [7,36,39] commonly rely solely on the current frame, which can only provide very limited observation to recover the 3D geometry and semantics. To enrich contextual information, the pioneering work VoxFormer-T [22] proposes to make use of temporal coherence by stacking multiple history frames as supplements to the current frame. Nevertheless, as depicted in Fig. 1, this approach assumes that temporal features from different viewpoints are originally corresponded at the pixel level, thereby just using a straightforward aggregation. However, the shared semantic content undergoes uncertain positional changes across different perspectives. Therefore, directly integrating images from different timestamps may result in blurred predictive information. This ambiguity compromises the stability of semantic occupancy predictions and imposes difficulties on temporal modeling.

To alleviate this issue, we propose a new Hierarchical Temporal Context Learning (HTCL) paradigm, which improves the temporal feature aggregation for accurate 3D semantic scene completion. HTCL takes RGB images from different timestamps to hierarchically infer the 3D semantic occupancy for fine-grained scene comprehension. As depicted in Fig. 2 (a), our hierarchical temporal context modeling includes two sequential steps: (1) we explicitly measure the contextual pattern affinity between the current and historical frames, highlighting the most relevant patterns; (2) we adaptively refine the sampling locations based on the preliminary high-affinity locations and their nearby relevant context to dynamically compensate for incomplete observations. Specifically, we first leverage epipolar homo-warping to explicitly align the temporal invariant feature representations and establish temporal feature volumes to fully maintain fine-grained context. Then, we introduce *scale-aware isolation* and *incorporation of diverse independent learning* in the cross-frame affinity measurement to facilitate better affinity distribution modeling in SSC. Subsequently, to dynamically compensate for incomplete observations, we resort to investigating the critical locations with high affinities and the neighboring relevant context. Technically, a multi-level deformable 3D block conditioned with the affinity weights is involved to adaptively refine the sampling locations. Finally, a weighted voxel cross-attention is introduced to aggregate the reliable temporal content.

We conduct extensive experiments to validate the merits of our proposed HTCL. As shown in Fig. 2(b), our proposed method achieves significant superiority over existing camera-based methods in terms of geometry (IoU) and semantics (mIoU). Remarkably, our camera-based approach outperforms state-of-the-art VoxFormer-T [22] on the SemanticKITTI benchmark and even surpasses LiDAR-based methods on the OpenOccupancy benchmark in terms of mIoU. Our main contributions are summarized as follows:

- A temporal context learning paradigm with a hierarchical scheme to fully exploit dynamic and dependable 3D semantic scene completion.
- An affinity measurement strategy with scale-aware isolation and multiple independent learners for fine-grained contextual correspondence modeling.
- An affinity-based dynamic refinement schedule to reassemble the temporal content and adaptively compensate for incomplete observations.

- Our method achieves state-of-the-art performance among all camera-based SSC methods on the SemanticKITTI and OpenOccupancy benchmarks.

2 Related Work

2.1 3D Semantic Scene Completion

Semantic Scene Completion (SSC), also known as semantic occupancy prediction, is a dense 3D perception task that jointly addresses semantic segmentation and scene completion [2,6]. Numerous previous works leverage LiDAR as the primary input to take advantage of the 3D geometrical information [14,34,51]. Due to the cost-effectiveness and portability, camera-based 3D SSC is recently gaining increasing attention [2,6,7,9,16,20,22,34,35,38–40,47,49]. MonoScene [7] first proposes to infer geometry and semantics from a single RGB image with 2D-3D features projection. Inspired by this, a lot of the following works extend the domain of camera-based 3D scene perception [16,19,46,52]. TPVFormer [16] introduces a tri-perspective view to describe the fine-grained representation of a 3D scene. OccFormer [52] introduces a dual-path transformer network to process dense 3D features for semantic occupancy prediction. However, these methods attempt to describe the complicated 3D scene from single-timestep images, which is ineffective for such an inherently ill-posed problem due to incomplete visual cues. In this paper, we advocate delving into reliable temporal content to dynamically aggregate semantic context and compensate for incomplete observations.

2.2 Temporal Information Modeling in 3D Visual Perception

The utilization of temporal information is recently highlighted in temporal 3D object detection [17,21,24–26,29,44,50,50] and video depth estimation [5,27,45] to enhance prediction performance. Temporal 3D object detection solutions focus on coarse-grained predictions at region-level [24,26], while video depth estimation methods [5,27] aim to establish matching correspondence from sequential video frames. Consequently, such strategies are insufficient for semantic scene completion, where fine-grained features are essential for dense semantic perception. VoxFormer-T [22] builds the first temporal pipeline for camera-based SSC by simply stacking the features from different frames, while the temporal correspondence modeling for the dense perception task of SSC remains unexplored. In this paper, we propose to explicitly model the temporal context correlation with pattern affinity to aggregate reliable temporal content and compensate for incomplete observations.

3 Methodology

3.1 Preliminary

Given a set of input temporal RGB images $I_{set}^{rgb} = \{I_t^{rgb}, I_{t-1}^{rgb}, \cdots\}$, our objective is to simultaneously estimate the semantic and geometric properties of the 3D

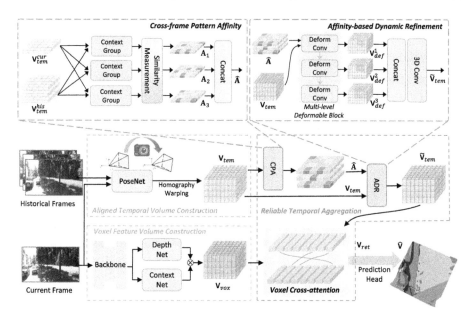

Fig. 3. Overall framework of our proposed method. Given temporal RGB images, the Aligned Temporal Volume is constructed with explicit epipolar homograph warping, while the Voxel Feature Volume is built by extending the LSS paradigm. Afterward, the Reliable Temporal Aggregation is introduced to dynamically aggregate reliable relevant temporal content for fine-grained semantic scene prediction.

scene. Notably, our focus is exclusively on current and historical image frames, omitting consideration of future frames [22] to formulate a more practical scheme for real-world applications. The scene is depicted as a voxel grid $\mathbf{V}$ with dimensions $\mathbb{R}^{H \times W \times Z}$, where H, W, Z denote the height, width and depth of the voxel grid, respectively. Each voxel within the grid is associated with a distinct semantic class represented by C, where C takes values from the set $\{c_0, c_1, \ldots, c_N\}$. The voxel can either correspond to empty space denoted as c_0 or to a particular semantic class from the set $\{c_1, c_2, \ldots, c_N\}$. Here, N represents the total count of available semantic classes. Our objective is to leverage the proposed framework, denoted as Θ, to learn a transformation:

$$\widehat{\mathbf{V}} = \Theta(I_t^{rgb}, I_{t-1}^{rgb}, \cdots), \tag{1}$$

where $\widehat{\mathbf{V}}$ denotes the estimated 3D semantic voxel grid, aiming to approximate the ground truth voxel grid $\mathbf{V}$.

3.2 Overall Framework

As depicted in Fig. 3, the overall framework of our proposed method mainly consists of three components: Aligned Temporal Volume Construction in the upper branch, Voxel Feature Volume Construction in the lower branch, and Reliable Temporal Aggregation for fine-grained SSC prediction.

Aligned Temporal Volume Construction. To construct the temporal feature volume $\mathbf{V}_{tem}$, we feed the current and historical frames into a lightweight PoseNet [5,15] to generate the temporal contextual volume $\mathbf{V}_{tem}$ with homography warping. Different from computing matching costs in typical temporal depth estimation solutions [5,27,45], we advocate to maintain the context features with $\mathbf{V}_{tem}$, further details are introduced in Sect. 3.3.

Voxel Feature Volume Construction. To construct the voxel feature volume $\mathbf{V}_{vox}$, a UNet backbone based on pre-trained EfficientNetB7 [41] is firstly employed to generate features with a spatial dimension of $\mathbb{R}^{H/4 \times W/4}$. Next, we extend the LSS [23,31] paradigm following recent studies [19,52] to build the voxel feature volume $\mathbf{V}_{vox}$ from the outer product of the contextual information and the depth distribution. To model the depth distribution, off-the-shelf monocular [3] or stereo [10] depth estimation networks are utilized with depth hypothesis planes of 192. By default, we employ the stereo depth estimation to form a stereo-based pipeline of *HTCL-S*. Moreover, we construct another monocular-based pipeline of *HTCL-M* with the monocular depth estimation, broadening the applicability to scenarios without stereo inputs.

Reliable Temporal Aggregation. We leverage the temporal volume $\mathbf{V}_{tem}$ to form the cross-frame affinity $\hat{\mathbf{A}}$, quantifying the contextual correspondence between current and historical features. Subsequently, we employ the cross-frame affinity to reassemble the temporal content and dynamically refine the sampling locations, yielding the reliable temporal volume $\widetilde{\mathbf{V}}_{tem}$. Further details on the Cross-frame Pattern Affinity (CPA) are introduced in Sect. 3.4, while the Affinity-based Dynamic Refinement (ADR) is presented in Sect. 3.5.

A Weighted Voxel Attention (WVA) is employed to aggregate reliable temporal content. Given $\mathbf{V}_{vox}$ and $\widetilde{\mathbf{V}}_{tem}$ as inputs, the query Q is generated from $\mathbf{V}_{vox}$, the key K and the value V are generated from $\widetilde{\mathbf{V}}_{tem}$. During the early training phase, unregulated temporal information could impair the learning of the voxel feature volume. To mitigate this, a flexible learning mechanism involving weighted voxel cross-attention is leveraged in the aggregation process:

$$\mathbf{V}_{ret} = \alpha \cdot \texttt{CrossAtt}(Q, K, V) + \mathbf{V}_{vox}, \qquad (2)$$

where the learnable coefficient α is initialized with 0 and gradually increases during training. $\mathbf{V}_{ret}$ represents the aggregated volume, which is fed into an SSC head with upsampling and a softmax layer for SSC prediction $\hat{\mathbf{V}}$ following [7,52].

3.3 Temporal Content Alignment

Given the fine-grained nature of the Semantic Scene Completion (SSC) task, constructing a dense temporal-aligned feature representation is crucial for accurate and robust perception. Rather than simply stacking the input images from different viewpoints [22], we propose to first align the temporal invariant content with explicit homography transformation.

As shown in Fig. 3, the current and historical frames are first fed into a lightweight PoseNet following video depth estimation [15,45] to generate the relative camera pose for photometric reprojection. Next, we leverage the current and historical frames to generate the current feature map F_t and historical feature maps $\{F_{t-1}, \cdots, F_{t-n}\}$. Following [5,45], we construct the warped historical features through homography warping with the relative camera pose and alternative depth hypothesis planes, which is formed as:

$$\text{Warp}(\mathbf{p}) = \mathbf{K}_i \cdot \left(\mathbf{R}_{0,i} \cdot \left(\mathbf{K}_0^{-1} \cdot \mathbf{p} \cdot d_j\right) + \mathbf{t}_{0,i}\right), \tag{3}$$

where $\{\mathbf{K}_i\}_{i=0}^{N-1}$ and $\{[\mathbf{R}_{0,i} \mid \mathbf{t}_{0,i}]\}_{i=1}^{N-1}$ denote camera intrinsic parameters and extrinsic parameters, respectively. d_j denotes j^{th} hypothesized depth of pixel $\mathbf{p}$ in F_t. Following that, we build a historical feature volume $\mathbf{V}_{tem}^{his}$ by aggregating all warped historical features in the canonical space. The historical feature volume contains geometric compatibility with different depth values between the current and historical frames. Next, we lift F_t along the depth dimension as [30,45] to generate the current feature volume $\mathbf{V}_{tem}^{cur}$. We concatenate $\mathbf{V}_{tem}^{his}$ and $\mathbf{V}_{tem}^{cur}$ following [45] to construct the temporal feature volume $\mathbf{V}_{tem}$:

$$\begin{aligned}\mathbf{V}_{tem} &= \text{Concat}\left\{(\mathbf{V}_{tem}^{cur}, \mathbf{V}_{tem}^{his}), \dim = \mathbb{C}\right\} \\ &= \text{Concat}\left\{\text{Lift}(F_t), \text{Warp}(F_{t-1}, \cdots, F_{t-n})\right\}.\end{aligned} \tag{4}$$

The temporal volume $\mathbf{V}_{tem}$ benefits the semantic scene modeling by explicitly aligning the contextual features across different timesteps. In the following section, we elaborate on fully exploiting reliable information according to contextual correspondence with $\mathbf{V}_{tem}$.

Why Feature Volume Instead of Cost Volume? The key distinction arises from the nature of camera-based Semantic Scene Completion, which is fundamentally not a matching task but rather a dense perception and reconstruction problem. Consequently, instead of directly computing matching costs within the temporal feature volume, our approach prioritizes the maintenance of fine-grained feature context. Moreover, to quantify the relevance of regional patterns within the temporal information, we construct auxiliary pattern affinity between the current and historical features.

3.4 Cross-Frame Pattern Affinity Measurement

Although the temporal volume is explicitly aligned, it mixes redundant context from different frames, making it insufficient to directly model the scene representations corresponding to the current frame. Therefore, we propose to construct Cross-frame Pattern Affinity (CPA) to measure the regional contextual correspondence between the historical feature volume $\mathbf{V}_{his}$ and current feature volume $\mathbf{V}_{cur}$.

Similarity Measurement. As a classic similarity metric, cosine similarity is commonly used in semantic analysis [12,13,33] and information retrieval [18,32] for correlation measurement. Given two vectors of α and β, the cosine similarity is calculated as:

$$\text{sim}(\alpha,\beta) = \cos(\alpha,\beta) = \frac{\alpha \cdot \beta}{||\alpha|| * ||\beta||}. \tag{5}$$

However, the original cosine similarity may yield high similarity scores with two dissimilar vectors [37]. This limitation is acknowledged and rectified through the scale-aware isolation [1], which takes into consideration different pattern scales. Nevertheless, these solutions tend to emphasize vector orientations and encounter challenges when assessing similarity within dense distributions. To overcome these drawbacks, ensemble learning techniques, as discussed in [48], leverage a diverse set of independent learners to address the aforementioned undesirable properties, thereby enhancing the effectiveness of dense similarity measurements.

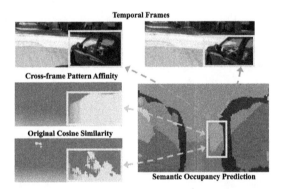

Fig. 4. Visualization of the heat maps from our proposed Cross-frame Pattern Affinity (CPA) and the original cosine similarity.

Given these concerns, we identify the criteria of an optimal similarity measurement strategy for fine-grained representations in SSC: *incorporation of diverse independent learning* and *scale-aware isolation*. In pursuit of this objective, we advocate employing the scale-aware isolated cosine similarity and taking multi-group context as inputs for affinity computation with dense distributions. Our strategy is implemented through two key steps:

- Incorporate different pattern scales from multi-group context to enable diverse independent similarity learning for fine-grained representations in SSC.
- Generate cosine similarities with scale-aware isolation and aggregate them for reliable pattern affinity measurement.

Multi-group Context Generation. To facilitate diverse independent similarity learning, 3D atrous convolutions with different dilation rates are employed to construct multi-group contextual features. Specifically, the historical feature volume $\mathbf{V}_{tem}^{his}$ is processed by a set of atrous convolutions to generate historical multi-group context $\mathbf{H}_i$ ($i \in (1, 2, 3)$):

$$\mathbf{H}_i = \text{GN}\left(\delta\left(\text{Atrous}_i(\mathbf{V}_{tem}^{his})\right)\right), \tag{6}$$

where GN and δ denote group normalization and GELU activation, respectively. The atrous convolutions are employed in parallel with dilation rates of 1, 2, and 4, respectively. Note that the current multi-group context $\mathbf{C}_i$ is generated in a symmetrical manner from the current feature volume $\mathbf{V}_{tem}^{cur}$:

$$\mathbf{C}_i = \text{GN}\left(\delta\left(\text{Atrous}_i(\mathbf{V}_{tem}^{cur})\right)\right). \tag{7}$$

Measure Pattern Affinity for Dense SSC. We upgrade Eq. 5 with two primary modifications to formulate the pattern affinity measurement for fine-grained contextual correspondence modeling in SSC. Firstly, we consider different pattern scales of the multi-group context and compute the pattern affinity $\mathbf{A}_i$ with each scale i. These independent group-scale affinity matrices are further aggregated along the channel dimension. Secondly, we subtract the respective averages during the affinity computation within each group scale to achieve scale-aware isolation. The formulas are represented as:

$$\mathbf{A}_i = \text{sim}(\mathbf{C}_i, \mathbf{H}_i) \tag{8}$$
$$= \frac{\sum_{j=0}^{C}(\mathbf{C}_i^j - \overline{\mathbf{C}}_i)(\mathbf{H}_i^j - \overline{\mathbf{H}}_i)}{\sqrt{\sum_{j=0}^{C}(\mathbf{C}_i^j - \overline{\mathbf{C}}_i)^2}\sqrt{\sum_{j=0}^{C}(\mathbf{H}_i^j - \overline{\mathbf{H}}_i)^2}},$$
$$\hat{\mathbf{A}} = \text{Concat}\left\{(\mathbf{A}_1, \mathbf{A}_2, \mathbf{A}_3), \dim = \mathbb{C}\right\}, \tag{9}$$

where the affinity matrices $\mathbf{A}_i$ of different group scales are concatenated along the channel dimension to get the cross-frame pattern affinity $\hat{\mathbf{A}}$. The input context matrices $\mathbf{C}_i$ and $\mathbf{H}_i$ are taken as high-dimension vectors with different group scales. $\overline{\mathbf{C}}_i$ and $\overline{\mathbf{H}}_i$ represent averaged context matrices of each group scale. As illustrated in Fig. 4, the affinity map from Cross-frame Pattern Affinity (CPA) effectively signifies contextual correspondence within the temporal content.

3.5 Affinity-Based Dynamic Refinement

Given our objective of completing and comprehending the 3D scene corresponding to the current frame, it is essential to assign greater weights to the most relevant locations. Simultaneously, investigating their neighboring relevant context is also critical to compensate for incomplete observations.

To this end, we propose to adaptively refine the feature sampling locations based on the obtained identified high-affinity locations and their neighboring relevant regions. We implement the above ideas with 3D deformable convolutions [11,42]. Specifically, we accomplish the dynamic refinement by introducing the affinity-based correspondence weights and deformable position offsets. In the context of a sampling grid window K_w, the formula is expressed as:

$$\mathbf{V}_{def} = \sum_{k=1}^{K_w} w_k \cdot \mathbf{V}_{tem}(\mathbf{p} + \mathbf{p}_k + \Delta \mathbf{p}_k) \cdot a_k, \tag{10}$$

where K_w represents the number of points in the sampling process. $\Delta \mathbf{p}_k$ denotes the additional offset in the sampling grid. w_k denotes the spatial feature weight and a_k represents the affinity weight from the cross-frame pattern affinity $\hat{\mathbf{A}}$.

To further reason about dynamic modeling through hierarchical context, we optimize the refinement process by considering contextual information from different feature levels. As illustrated in Fig. 3, we construct a multi-level deformable block with three cascade 3D deformable convolutions. The output features are aggregated to generate the reliable temporal volume $\widetilde{\mathbf{V}}_{tem}$:

$$\widetilde{\mathbf{V}}_{tem} = \mathbf{W} \left(\text{Concat} \left\{ (\mathbf{V}_{def}^1, \mathbf{V}_{def}^2, \mathbf{V}_{def}^3), \dim = \mathbb{C} \right\} \right). \tag{11}$$

where the multi-level deformable temporal volumes $\mathbf{V}_{def}^i$ ($i \in (1,2,3)$) are concatenated along the channel dimension and processed with a 3D convolution layer $\mathbf{W}$ for dimension reduction.

4 Experiment

4.1 Datasets and Metrics

SemanticKITTI. The SemanticKITTI [2] dataset comprises 22 outdoor scenes with LiDAR scans and stereo images. The ground truth is voxelized as 256×256×32 grids. Each voxel grid has a size of (0.2m, 0.2m, 0.2m) and is annotated with 21 semantic classes (19 semantics, 1 free and 1 unknown). Following [7,22], we divide the 22 outdoor scenes into 10 training scenes, 1 validation scene, and 1 test scene. Our proposed HTCL is evaluated on the SemanticKITTI with both temporal stereo images (HTCL-S) and monocular images (HTCL-M).

OpenOccupancy. The OpenOccupancy [43] dataset extends the nuScene [4] dataset by providing dense semantic occupancy annotations. The dataset holds 850 scenes of 34K keyframes with 360-degree LiDAR scans. We split the whole dataset into 28130 training frames and 6019 validation frames following [43]. Each frame holds 400K occupied voxels with 17 semantic labels. Note that we exclusively apply monocular-based HTCL-M on the OpenOccupancy dataset due to the unavailability of stereo images.

Evaluation Metrics. Following the previous works [7,22], we utilize the mean Intersection over Union (**mIoU**) as our primary metric to assess the performance of the semantic scene completion (SSC) task. Additionally, we report the Intersection over Union (**IoU**) to evaluate the performance of the class-agnostic scene completion (SC) task.

4.2 Experimental Setup

We follow the common practice [7,22,52] to initialize the encoder part of our UNet with the pretrained weight of EfficientNetB7 [41]. By default, our model takes the current frame and the previous 3 image frames as inputs. We implemented our model on PyTorch with a batch size of 4. The model is trained 24 epochs with the AdamW optimizer [28]. The learning rate is set to 1×10^{-4} with a weight decay of 0.01.

4.3 Performance

Quantitative Comparison. As reported in Table 1, we compare our HTCL with recent public methods on the SemanticKITTI dataset, including VoxFormer [22], OccFormer [52], SurroundOcc [46], TPVFormer [16] and MonoScene [7]. VoxFomer-T is a temporal baseline with current and historical 4 images as inputs. We can observe that our proposed method outperforms all the other methods significantly. Compared to VoxFomer-T, our method achieves a remarkable relative improvement in mIoU even with fewer historical inputs (3 vs. 4). We

Table 1. Quantitative results on the SemanticKITTI test set with the state-of-the-art SSC methods. The "S-T", "S" and "M" denote temporal stereo images, single-frame stereo images, and single-frame monocular images, respectively.

Methods	HTCL-S (ours)	VoxFormer-T	VoxFormer-S	OccFormer	SurroundOcc	TPVFormer	MonoScene
Input	S-T	S-T	S	M	M	M	M
IoU	**44.23**	43.21	42.95	34.53	34.72	34.25	34.16
mIoU	**17.09**	13.41	12.20	12.32	11.86	11.26	11.08
car	**27.30**	21.70	20.80	21.60	20.60	19.20	18.80
bicycle	1.80	**1.90**	1.00	1.50	1.60	1.00	0.50
motorcycle	**2.20**	1.60	0.70	1.70	1.20	0.50	0.70
truck	**5.70**	3.60	3.50	1.20	1.40	3.70	3.30
other-veh.	**5.40**	4.10	3.70	3.20	4.40	2.30	4.40
person	1.10	1.60	1.40	**2.20**	1.40	1.10	1.00
bicyclist	**3.10**	1.10	2.60	1.10	2.00	2.40	1.40
motorcyclist	**0.90**	0.00	0.20	0.20	0.10	0.30	0.40
road	**64.40**	54.10	53.90	55.90	56.90	55.10	54.70
parking	**33.80**	25.10	21.10	31.50	30.20	27.40	24.80
sidewalk	**34.80**	26.90	25.30	30.30	28.30	27.20	27.10
other.grd	**12.40**	7.30	5.60	6.50	6.80	6.50	5.70
building	**25.90**	23.50	19.80	15.70	15.20	14.80	14.40
fence	**21.10**	13.10	11.10	11.90	11.30	11.00	11.10
vegetation	**25.30**	24.40	22.40	16.80	14.90	13.90	14.90
trunk	**10.80**	8.10	7.50	3.90	3.40	2.60	2.40
terrain	**31.20**	24.20	21.30	21.30	19.30	20.40	19.50
pole	**9.00**	6.60	5.10	3.80	3.90	2.90	3.30
traf.sign	**8.30**	5.70	4.90	3.70	2.40	1.50	2.10

Table 2. Quantitative results on the OpenOccupancy validation set with the state-of-the-art SSC methods. The "L", "M", "M-D" and "M-T" denote LiDAR inputs, monocular images, monocular images with depth maps and temporal monocular images, respectively. The LiDAR points are projected and densified to generate the depth maps.

Methods	HTCL-M (ours)	JS3C-Net	LMSCNet	3DSketch	AICNet	TPVFormer	MonoScene
Input	M-T	L	L	M-D	M-D	M	M
IoU	21.4	**30.2**	27.3	25.6	23.8	15.3	18.4
mIoU	**14.1**	12.5	11.5	10.7	10.6	7.8	6.9
■ barrier	**14.8**	14.2	12.4	12.0	11.5	9.3	7.1
■ bicycle	**10.2**	3.4	4.2	5.1	4.0	4.1	3.9
■ bus	**14.8**	13.6	12.8	10.7	11.8	11.3	9.3
■ car	**18.9**	12.0	12.1	12.4	12.3	10.1	7.2
■ const. veh.	**7.6**	7.2	6.2	6.5	5.1	5.2	5.6
■ motorcycle	**11.3**	4.3	4.7	4.0	3.8	4.3	3.0
■ pedestrian	**12.3**	7.3	6.2	5.0	6.2	5.9	5.9
■ traffic cone	**9.6**	6.8	6.3	6.3	6.0	5.3	4.4
■ trailer	5.5	**9.2**	8.8	8.0	8.2	6.8	4.9
■ truck	**13.5**	9.1	7.2	7.2	7.5	6.5	4.2
■ drive. suf.	**32.5**	27.9	24.2	21.8	24.1	13.6	14.9
■ other flat	**21.7**	15.3	12.3	14.8	13.0	9.0	6.3
■ sidewalk	**20.7**	14.9	16.6	13.0	12.8	8.3	7.9
■ terrain	**17.7**	16.2	14.1	11.8	11.5	8.0	7.4
■ manmade	5.8	**14.0**	13.9	12.0	11.6	9.2	10.0
■ vegetation	8.5	**24.9**	22.2	21.2	20.2	8.2	7.6

also report quantitative results on the OpenOccupancy validation set in Table 2. To provide depth maps for AICNet [20] and 3DSketch [8], LiDAR points are projected and densified following OpenOccupancy [43]. Despite LiDAR's inherent advantage in terms of IoU due to more accurate 3D geometric measurement, our HTCL surpasses all the other methods (including LiDAR-based LMSCNet [35] and JS3C-Net [49]) in terms of mIoU, which demonstrates the effectiveness of our method for semantic scene completion.

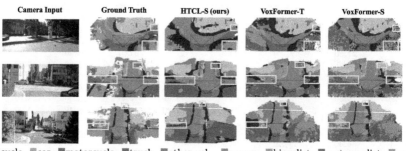

Fig. 5. Qualitative results on the SemanticKITTI validation set. Our HTCL-S captures more complete and accurate scenery layouts compared with VoxFormer.

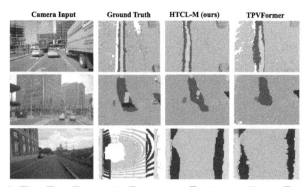

■bicycle ■bus ■car ■const.veh. ■motorcycle ■pedestrian ■truck ■sidewalk
■traffic cone ■trailer ■drive.suf. ■other flat ■terrain ■manmade ■vegetation

Fig. 6. Qualitative results on the OpenOccupancy validation set. Our HTCL-M generates more complete and comprehensive scenes compared with ground truth.

Qualitative Comparison. Figure 5 provides a qualitative comparison between our proposed method and VoxFormer on the SemanticKITTI validation set. We can observe that the real-world scenes are complex and the annotated ground truth is relatively sparse, which poses challenges for completely reconstructing the semantic scenes from limited visual cues. Compared with VoxFormer, our method captures a more complete and accurate scenery layout (e.g., crossroads in the second and third rows). Moreover, our method effectively hallucinates more proper scenery outside of the camera field of view (e.g., shadow areas in the first and second rows) and demonstrates significant superiority over moving objects (e.g., trucks in the second row). We also visualize the prediction results of our proposed method on the OpenOccupancy validation set as shown in Fig. 6. Our proposed method generates much denser and more realistic results compared with the ground truth.

Temporal Stereo Variants Evaluation. To ensure a fair and comprehensive comparison, we implement temporal stereo variants of the baselines as shown in Table 3. Following VoxFormer-T, we employ stacked temporal stereo images

Table 3. Evaluation results of temporal stereo variants. For MonoScene[‡], TPVFormer[‡] and OccFormer[‡], we employ stacked temporal stereo images as inputs following VoxFormer-T.

Methods	Input	mIoU (%) ↑	Time (s) ↓
MonoScene[‡] (2022)	Stereo-T	12.96	**0.281**
TPVFormer[‡] (2023)	Stereo-T	13.21	0.324
OccFormer[‡] (2023)	Stereo-T	13.57	0.348
VoxFormer-T (2023)	Stereo-T	13.35	0.307
HTCL-M (ours)	Mono-T	16.11	0.289
HTCL-S (ours)	Stereo-T	**17.13**	0.297

Table 4. Ablation study for different architectural components on the SemanticKITTI validation set. The full names of different components are in Sect. 4.4.

TCA		CPA		ADR		WVA		IoU (%)	mIoU (%)
Feature Volume	Cost Volume	Scale-aware Isolation	Multi-group	Affinity	Deformable	Coefficient	CrossAtt		
	✓	✓	✓	✓	✓	✓	✓	44.01	16.02
✓			✓	✓	✓	✓	✓	43.07	15.18
✓		✓		✓	✓	✓	✓	43.15	15.26
✓		✓	✓		✓	✓	✓	42.79	14.65
✓		✓	✓	✓		✓	✓	42.96	15.14
✓		✓	✓	✓	✓		✓	43.97	15.85
✓		✓	✓	✓	✓	✓		44.13	15.98
✓		✓	✓	✓	✓	✓	✓	**45.51**	**17.13**

as inputs to get variants of MonoScene[‡], TPVFormer[‡] and OccFormer[‡]. Note that VoxFormer-T originally employs 4 previous frames, while the other stereo variants employ 3 previous frames as ours. As shown in the table, our method efficiently achieves superior performance with the same temporal inputs.

4.4 Ablation Study

We conduct adequate ablations for our proposed method on the SemanticKITTI validation set. Specifically, we analyze the impact of different architectural components in Table 4 and study the influence of temporal inputs in Table 5.

Temporal Content Alignment (TCA). The ablation study for Temporal Content Alignment (TCA) is reported in the second row of Table 4. We can observe that replacing the cost volume with the feature volume yields obvious performance gains, improving the IoU and mIoU by 1.50 and 1.11, respectively. We attribute this enhancement to the fine-grained feature context preservation.

Cross-frame Pattern Affinity (CPA). The ablation of Cross-frame Pattern Affinity (CPA) is detailed in the third row of Table 4. As we can see, equipping the original cosine similarity with the scale-aware isolation and introducing the multi-group context generation lead to significant performance enhancement, improving the mIoU by 1.95 and 1.87, respectively.

Affinity-Based Dynamic Refinement (ADR). The ablation study for the Affinity-based Dynamic Refinement (ADR) is conducted by removing the affinity weights and replacing the deformable convolutions with normal convolutions, as detailed in the fourth row of Table 4. Leveraging the affinity information is effective in modeling contextual correspondence, as the procedure results in a noticeable performance gain of 2.72 IoU and 2.48 mIoU. Furthermore, dynamic refinement with the deformable convolutions offers efficient and flexible contextual modeling, improving IoU and mIoU by 2.55 and 1.99, respectively.

Weighted Voxel Attention (WVA). The ablation study about Weighted Voxel Attention (WVA) is shown in Fig. 7 and the fifth row of Table 4. We remove the learnable coefficient and replace the voxel cross-attention with naive concatenation for comparison. It is evident in Fig. 7 that the adoption of our proposed strategy leads to a more rapid and stable convergence of the entire model. Additionally, Table 4 shows noteworthy improvements through the incorporation of the learnable coefficient and the voxel cross-attention, enhancing the mIoU by 1.28 and 1.15, respectively.

Table 5. Effect of using a different number of temporal frames. These models are evaluated on the SemanticKITTI validation set.

Temporal Inputs					mIoU (%) ↑	Time (s) ↓
I_{t-1}^{rgb}	I_{t-2}^{rgb}	I_{t-3}^{rgb}	I_{t-4}^{rgb}	I_{t-5}^{rgb}		
✓					15.08	0.268
✓	✓				16.43	0.283
✓	✓	✓			17.13	0.297
✓	✓	✓	✓		17.31	0.311
✓	✓	✓	✓	✓	17.42	0.324

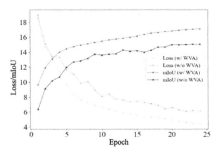

Fig. 7. The convergence and performance curves of WVA.

Temporal Inputs. We report the semantic scene completion performance and running time with temporal inputs of different frame numbers, which are detailed in Table 5. As we can observe, the effectiveness gain of more than 3 previous frames is relatively slight with more running time, thus we adopt 3 frames as the default setting to balance between efficiency and effectiveness.

5 Conclusion

In this paper, we introduce HTCL, an innovative hierarchical temporal in-context learning paradigm for semantic scene completion (SSC). To highlight the most relevant patterns, we introduce the pattern affinity to measure the contextual correspondence between the current and historical frames. Subsequently, to dynamically compensate for incomplete observations, we propose to adaptively refine the feature sampling locations based on the initially high-affinity locations and their neighboring relevant regions. Our method outperforms state-of-the-art camera-based methods and even surpasses LiDAR-based methods for semantic scene completion. We hope HTCL could inspire further research in camera-based temporal modeling for SSC and its applications in 3D visual perception.

Acknowledgments. This work was supported in part by NSFC 62302246 and ZJNSFC under Grant LQ23F010008, and supported by High Performance Computing Center at Eastern Institute of Technology, Ningbo, and Ningbo Institute of Digital Twin.

References

1. Anastasiu, D.C., Karypis, G.: L2ap: fast cosine similarity search with prefix l-2 norm bounds. In: International Conference on Data Engineering. IEEE (2014)
2. Behley, J., et al.: Semantickitti: a dataset for semantic scene understanding of lidar sequences. In: ICCV (2019)
3. Bhat, S.F., Alhashim, I., Wonka, P.: Adabins: depth estimation using adaptive bins. In: CVPR (2021)
4. Caesar, H., et al.: nuscenes: a multimodal dataset for autonomous driving. In: CVPR (2020)
5. Cai, C., Ji, P., Yan, Q., Xu, Y.: Riav-mvs: recurrent-indexing an asymmetric volume for multi-view stereo. In: CVPR (2023)
6. Cai, Y., Chen, X., Zhang, C., Lin, K.Y., Wang, X., Li, H.: Semantic scene completion via integrating instances and scene in-the-loop. In: CVPR (2021)
7. Cao, A.Q., de Charette, R.: Monoscene: monocular 3d semantic scene completion. In: CVPR (2022)
8. Chen, X., Lin, K.Y., Qian, C., Zeng, G., Li, H.: 3d sketch-aware semantic scene completion via semi-supervised structure prior. In: CVPR (2020)
9. Cheng, R., Agia, C., Ren, Y., Li, X., Bingbing, L.: S3cnet: a sparse semantic scene completion network for lidar point clouds. In: Conference on Robot Learning (2021)
10. Cheng, X., et al.: Hierarchical neural architecture search for deep stereo matching. In: NeurIPS, vol. 33 (2020)
11. Dai, J., Qi, H., Xiong, Y., Li, Y., Zhang, G., Hu, H., Wei, Y.: Deformable convolutional networks. In: ICCV (2017)
12. Evangelopoulos, N., Zhang, X., Prybutok, V.R.: Latent semantic analysis: five methodological recommendations. Eur. J. Inf. Syst. **21**(1), 70–86 (2012)
13. Evangelopoulos, N.E.: Latent semantic analysis. Wiley Interdisc, Rev. Cogn. Sci. **4**(6) (2013)
14. Garbade, M., Chen, Y.T., Sawatzky, J., Gall, J.: Two stream 3d semantic scene completion. In: CVPRW (2019)
15. Guizilini, V., Ambrus, R., Pillai, S., Raventos, A., Gaidon, A.: 3d packing for self-supervised monocular depth estimation. In: CVPR (2020)
16. Huang, Y., Zheng, W., Zhang, Y., Zhou, J., Lu, J.: Tri-perspective view for vision-based 3d semantic occupancy prediction. In: CVPR (2023)
17. Kopf, J., Rong, X., Huang, J.B.: Robust consistent video depth estimation. In: CVPR (2021)
18. Korenius, T., Laurikkala, J., Juhola, M.: On principal component analysis, cosine and euclidean measures in information retrieval. Inf. Sci. **177**(22), 4893–4905 (2007)
19. Li, B., et al.: Stereoscene: bev-assisted stereo matching empowers 3d semantic scene completion (2023)
20. Li, J., Han, K., Wang, P., Liu, Y., Yuan, X.: Anisotropic convolutional networks for 3d semantic scene completion. In: CVPR (2020)
21. Li, S., Luo, Y., Zhu, Y., Zhao, X., Li, Y., Shan, Y.: Enforcing temporal consistency in video depth estimation. In: ICCV (2021)
22. Li, Y., et al.: Voxformer: sparse voxel transformer for camera-based 3d semantic scene completion. In: CVPR (2023)
23. Li, Y., et al.: Bevdepth: acquisition of reliable depth for multi-view 3d object detection. In: AAAI (2023)

24. Lin, X., Lin, T., Pei, Z., Huang, L., Su, Z.: Sparse4d: multi-view 3d object detection with sparse spatial-temporal fusion. arXiv preprint arXiv:2211.10581 (2022)
25. Liu, Y., Wang, T., Zhang, X., Sun, J.: Petr: Position embedding transformation for multi-view 3d object detection. In: Avidan, S., Brostow, G., Cisse, M., Farinella, G.M., Hassner, T. (eds.) ECCV 2022, vol. 13687, pp. 531–548. Springer, Heidelberg (2022)
26. Liu, Y., et al.: Petrv2: a unified framework for 3d perception from multi-camera images. In: ICCV (2023)
27. Long, X., Liu, L., Li, W., Theobalt, C., Wang, W.: Multi-view depth estimation using epipolar spatio-temporal networks. In: CVPR (2021)
28. Loshchilov, I., Hutter, F.: Decoupled weight decay regularization. arXiv preprint arXiv:1711.05101 (2017)
29. Luo, X., Huang, J.B., Szeliski, R., Matzen, K., Kopf, J.: Consistent video depth estimation. ACM Trans. Graph. (ToG) **39** (2020)
30. Newcombe, R.A., Lovegrove, S.J., Davison, A.J.: Dtam: dense tracking and mapping in real-time. In: ICCV (2011)
31. Philion, J., Fidler, S.: Lift, splat, shoot: encoding images from arbitrary camera rigs by implicitly unprojecting to 3D. In: Vedaldi, A., Bischof, H., Brox, T., Frahm, J.-M. (eds.) ECCV 2020. LNCS, vol. 12359, pp. 194–210. Springer, Cham (2020). https://doi.org/10.1007/978-3-030-58568-6_12
32. Rahutomo, F., Kitasuka, T., Aritsugi, M.: Semantic cosine similarity. In: ICAST (2012)
33. Ramachandran, L., Gehringer, E.F.: Automated assessment of review quality using latent semantic analysis. In: ICALT. IEEE (2011)
34. Rist, C.B., Emmerichs, D., Enzweiler, M., Gavrila, D.M.: Semantic scene completion using local deep implicit functions on lidar data. IEEE Trans. Pattern Anal. Mach. Intell. **44**(10), 7205–7218 (2021)
35. Roldao, L., de Charette, R., Verroust-Blondet, A.: Lmscnet: lightweight multiscale 3d semantic completion. In: 3DV (2020)
36. Roldao, L., De Charette, R., Verroust-Blondet, A.: 3d semantic scene completion: a survey. Int. J. Comput. Vision **130**(8), 1978–2005 (2022)
37. Sarwar, B., Karypis, G., Konstan, J., Riedl, J.: Item-based collaborative filtering recommendation algorithms. In: WWW (2001)
38. Silberman, N., Hoiem, D., Kohli, P., Fergus, R.: Indoor segmentation and support inference from RGBD images. In: Fitzgibbon, A., Lazebnik, S., Perona, P., Sato, Y., Schmid, C. (eds.) ECCV 2012. LNCS, vol. 7576, pp. 746–760. Springer, Heidelberg (2012). https://doi.org/10.1007/978-3-642-33715-4_54
39. Song, S., Yu, F., Zeng, A., Chang, A.X., Savva, M., Funkhouser, T.: Semantic scene completion from a single depth image. In: CVPR (2017)
40. Straub, J., et al.: The replica dataset: a digital replica of indoor spaces. arXiv preprint arXiv:1906.05797 (2019)
41. Tan, M., Le, Q.: Efficientnet: rethinking model scaling for convolutional neural networks. In: ICML (2019)
42. Wang, F., Galliani, S., Vogel, C., Speciale, P., Pollefeys, M.: Patchmatchnet: learned multi-view patchmatch stereo. In: CVPR (2021)
43. Wang, X., et al.: Openoccupancy: a large scale benchmark for surrounding semantic occupancy perception. In: ICCV (2023)
44. Wang, Y., Wang, P., Yang, Z., Luo, C., Yang, Y., Xu, W.: Unos: unified unsupervised optical-flow and stereo-depth estimation by watching videos. In: CVPR (2019)

45. Watson, J., Mac Aodha, O., Prisacariu, V., Brostow, G., Firman, M.: The temporal opportunist: self-supervised multi-frame monocular depth. In: CVPR (2021)
46. Wei, Y., Zhao, L., Zheng, W., Zhu, Z., Zhou, J., Lu, J.: Surroundocc: multi-camera 3d occupancy prediction for autonomous driving. In: ICCV (2023)
47. Wu, S.C., Tateno, K., Navab, N., Tombari, F.: Scfusion: real-time incremental scene reconstruction with semantic completion. In: 3DV (2020)
48. Xia, P., Zhang, L., Li, F.: Learning similarity with cosine similarity ensemble. Inf. Sci. **307**, 39–52 (2015)
49. Yan, X., et al.: Sparse single sweep lidar point cloud segmentation via learning contextual shape priors from scene completion. In: AAAI (2021)
50. Zhang, H., Shen, C., Li, Y., Cao, Y., Liu, Y., Yan, Y.: Exploiting temporal consistency for real-time video depth estimation. In: ICCV (2019)
51. Zhang, J., Zhao, H., Yao, A., Chen, Y., Zhang, L., Liao, H.: Efficient semantic scene completion network with spatial group convolution. In: ECCV (2018)
52. Zhang, Y., Zhu, Z., Du, D.: Occformer: dual-path transformer for vision-based 3d semantic occupancy prediction. In: ICCV (2023)

Equi-GSPR: Equivariant SE(3) Graph Network Model for Sparse Point Cloud Registration

Xueyang Kang[1,2,3], Zhaoliang Luan[2,4], Kourosh Khoshelham[3], and Bing Wang[2(✉)]

[1] Faculty of Electrical Engineering, KU Leuven, Leuven, Belgium
alex.kang@kuleuven.com
[2] Spatial Intelligence Group, The Hong Kong Polytechnic University, Hung Hom, Hong Kong
bingwang@polyu.edu.hk
[3] Faculty of Engineering and IT, The University of Melbourne, Parkville, Australia
k.khoshelham@unimelb.edu.au
[4] IoTUS Lab, Queen Mary University of London, London, UK
z.luan@qmul.ac.uk

Abstract. Point cloud registration is a foundational task for 3D alignment and reconstruction applications. While both traditional and learning-based registration approaches have succeeded, leveraging the intrinsic symmetry of point cloud data, including rotation equivariance, has received insufficient attention. This prohibits the model from learning effectively, resulting in a requirement for more training data and increased model complexity. To address these challenges, we propose a graph neural network model embedded with a local Spherical Euclidean 3D equivariance property through **SE**(3) message passing based propagation. Our model is composed mainly of a descriptor module, equivariant graph layers, match similarity, and the final regression layers. Such modular design enables us to utilize sparsely sampled input points and initialize the descriptor by self-trained or pre-trained geometric feature descriptors easily. Experiments conducted on the 3DMatch and KITTI datasets exhibit the compelling and robust performance of our model compared to state-of-the-art approaches, while the model complexity remains relatively low at the same time.

Keywords: Equivariance · **SE**(3) · Graph Network Model · Point Cloud Registration · Feature Descriptor · Similarity

1 Introduction

The registration of point clouds typically involves formulating robust geometric feature descriptors and a subsequent complex matching process to predict feature

Supplementary Information The online version contains supplementary material available at https://doi.org/10.1007/978-3-031-73235-5_9.

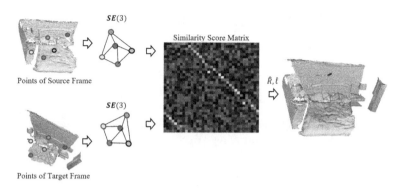

Fig. 1. The registration model converts the sparse point descriptors of the source and target frames into an equivariant graph feature representation, respectively. Then the **SE**(3) equivariant graph features are used for the similarity score calculation. The matched features are then decoded into the relative transform to align the two scans.

correspondences [8]. However, these correspondences established from raw point cloud often exhibit a high outlier-to-inlier ratio, leading to significant registration errors or complete failures. To enhance the robustness of registration processes, PointDSC [2] explicitly calculates local feature spatial consistency and evaluates pairwise 3D geometric feature descriptor similarity across two frames [10,46] to eliminate outliers from the alignment optimization process. Other approaches like Deep Global Registration (DGR) [9] treat correspondence prediction as a classification problem, utilizing concatenated coordinates of input point cloud pairs and employing a differentiable optimizer for pose refinement. Despite the effectiveness of these models on public datasets, their training requires accurate correspondence supervision, necessitating a complex point-to-point search process that is particularly vulnerable to numerous outliers.

Geometric feature descriptors, derived from keypoint neighborhoods of a specified range, often overlook the underlying geometric topology of the data, such as the global connectivity among points. This oversight results in feature descriptors lacking **SE**(3) rotation equivariance, thereby impeding efficient and robust learning of rotation-equivariant and invariant features. The recently introduced RoReg [44] model employs a rotation-guided detector to enhance rotation coherence matching and integrates it with RANSAC for pose estimation. However, it suffers from high computational demand and reduced processing speed. This highlights the need for more efficient rotation-equivariant model architectures to significantly enhance registration performance.

To address these challenges, we introduce a novel approach that leverages a graph convolution-based model to jointly learn **SE**(3) equivariant features, starting with feature descriptors extracted from sparsely sampled points across two frames. Our proposed **SE**(3) equivariant graph network model, aimed at sparse point cloud registration, is depicted in Fig. 1. Unlike Transformer and CNN-based models, our graph architecture captures both the topology and geometric features of point clouds, similar to other proposed geometric descriptors [40,53], facilitating the learning of fine-grained rigid rotation-equivariant feature representations for more robust and coherent point cloud registration through data symmetry. The primary contributions of our study are the following:

- Introduction of an equivariant graph model to facilitate neighbor feature aggregation and **SE**(3) equivariant coordinate embedding from either learned geometric descriptors for point cloud registration.
- Implementation of a novel matching approach within the implicit feature space, based on similarity evaluation and Low-Rank Feature Transformation (LRFT), eliminating the need for explicit point correspondence supervision and exhausting search.
- Development of a specific matrix rank-based regularizer to enable the model to automatically identify and mitigate the impact of correspondence outliers, enhancing the robustness of the registration process.

2 Related Work

The concept of equivariant properties can be embedded in the layers of Convolutional Neural Networks (CNNs) to depict **SO**(2) group characteristics, as initially proposed by Cohen [11], and later extended to encompass arbitrary continuous input [16]. Another area of study focuses on equivariance representation by employing steerable kernel filters [12,27,41,47–49] for equivariance learning. For more intricate tasks involving **SO**(3), techniques such as Vector Neurons [13] and Tensor Field Networks [42] can be viewed as implementations of the capsule network model [50], transitioning from scalar values to vectors. Following the introduction of the Transformer model, Lie-group-based Transformer models [18,19,26] have been developed to capture equivariance through attention mechanisms. Moreover, to address the intricate equivariance inherent in the input data, equivariant(n) graph neural networks [14,28,38] are utilized to learn equivariant features for dynamic and complex issues. As **SE**(3) features are maintained through message passing [5] within the graph model, these equivariant models exhibit considerable potential in addressing many longstanding challenges, such as predicting molecule structures [39] or quantum structures [21], as well as particle dynamic flow physics [4,29]. Some studies have explored the application of equivariant models in various point cloud tasks, including 3D detection [40], 3D point classification [53], point cloud-based place recognition [30], 3D shape point registration [7,55], and 3D shape reconstruction [6]. These applications underscore the learning efficiency gained from the leveraging of the intrinsic symmetries in input data.

Despite the prevalence of equivariant models in the microscopic realm [18,39], the application of such models for tasks like multi-view 3D reconstruction [23] or other intricate 3D challenges remains under-explored. A fundamental component of 3D reconstruction involves point cloud registration. Many conventional methods employ linear-algebra-based optimization techniques for iterative point cloud registration, such as point-to-plane registration [33,34], LOAM [45], and its variation F-LOAM [45]. In recent years, there has been a rise in deep learning models for registration purposes, relying on representative feature descriptors [10,46] or precise correspondence establishment between descriptors, like deep global registration [9]. To enhance registration accuracy, some studies focus on developing more resilient descriptors like rotation-equivariant descriptors [3,43,44] for subsequent correspondence matching or incorporating **SO**(2) rotation equivariance into the registration framework using cylindrical convolution, as in Spinnet [1]. Other research works concentrate on optimizing correspondence search explicitly, for instance, Stickypillars utilizes optimal transport for matching, and PointDSC [2] employs spectral matching to eliminate outliers from raw correspondences. In contrast to conventional models, works in [36,54] excel in rapid registration performance. Moreover, some end-to-end models improve overall performance from feature descriptor learning to feature association in a differentiable way, like 3D RegNet [32], or correspondence-free registration aimed at streamlining the 3D point cloud registration [56]. Notably, Banani et al. [15] introduce an unsupervised model for point cloud registration using differentiable rendering, while Predator [25] demonstrates registration capabilities even in applications with low input point cloud overlap and numerous outliers.

3 Method

Our registration process begins by extracting feature descriptors from downsampled point clouds. Equivariance is integrated into the features through equivariant graph convolution layers. Subsequently, the number of features in the pairwise graph is aggregated to a smaller number via Low-Rank based constraint. Finally, the similarity between the pairwise features of the two frames is calculated for relative transform prediction. A detailed illustration of the model is shown in Fig. 2. The input to our model consists of N points $\boldsymbol{X} = [\boldsymbol{x}_1, \ldots, \boldsymbol{x}_N] \in N \times \mathbb{R}^3$ from the source frame, and N points $\boldsymbol{Y} = [\boldsymbol{y}_1, \ldots, \boldsymbol{y}_N] \in N \times \mathbb{R}^3$ from the target frame, where $x_i \in \mathbb{R}^3$ and $y_j \in \mathbb{R}^3$ form a correspondence (i, j). It is important to note, for ease of subsequent similarity search, that the coordinates of points in each frame are rearranged in descending order based on the ray length $||\mathbf{r}(t)||^2$ from the point position to the sensor frame center $\mathbf{o_s}$. For numerical stability during training, the source scan is normalized to a canonical frame, and the target scan is transformed relative to the source frame, allowing the model to predict the relative transformation from source to target.

3.1 Feature Descriptor

We incorporate geometric details of nearby interest points into our graph model using feature descriptors. To extract these descriptors, we can reuse available pre-trained point-based descriptors or train a shallow Multi-Layer Perceptron (MLP) module with l_1 layers prior to the model in an end-to-end manner, inspired by PointNet++ [35]. The feature representation of point i in the next layer $l + 1$

is calculated by averaging the output of the mapping function $h(\cdot)$ applied to the relative positional coordinates of point i and neighboring point $k \in \mathcal{N}(i)$ (n points), along with the hidden feature $h_k^{l_1}$ from the prior layer l_1.

$$h_i^{l_1+1} = \frac{1}{n} \sum_{k \in \mathcal{N}(i)} f_h(h_k^{l_1}, x_k - x_i). \tag{1}$$

3.2 Equivariant Graph Network Model

By utilizing the equivariant graph representation (l_2 layers) as introduced by Satorras et al. [38], we can enhance the receptive field and representation for feature descriptors of interest points by incorporating **SE**(3) equivariant properties via graph feature aggregation. Our method leverages graph-based message passing techniques [22] to propagate **SE**(3) equi-features. For the construction of the graph $\mathcal{G} = (\mathcal{V}, \mathcal{E})$ with vertices $\mathcal{V}$ and $\mathcal{E}$ edges. The individual hidden point descriptors $h_i^{l_2} \in \mathbb{R}^{32}$ and the point coordinate embedding $x_i^{l_2} \in \mathbb{R}^3$ at layer l_2 are treated as the node and edge features, respectively. The graph convolutional layer updates the edge equi-message $m_{ik} \in \mathbb{R}^{3 \times 3}$, node hidden feature $h_i^{l_2} \in \mathbb{R}^{32}$,

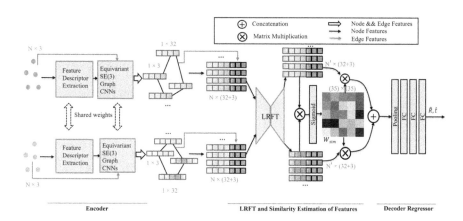

Fig. 2. The registration model consists of an encoder, a feature match block, and a decoder. Pointwise feature descriptors are extracted from the source and target scan points, passed through equivariant graph layers, and combined with coordinate embeddings to form a row-major order matrix. Next, the feature matrices from the source and target frames are compressed using MLPs-based Low-Rank Feature Transformation (LRFT). The aggregated features are used to create a similarity map through dot product of feature descriptors. In the decoder, features are weighted by similarity scores, then concatenated, and processed through pooling and fully connected layers to predict relative translation t_j^i and quaternion q_j^i.

and coordinate embedding $x_i^{l_2} \in \mathbb{R}^3$ at each equivariant layer.

$$m_{ik} = \phi_m(h_i^{l_2}, h_k^{l_2}, \left\|x_k^{l_2} - x_i^{l_2}\right\|^{\frac{1}{2}}), \tag{2}$$

$$x_i^{l_2+1} = x_i^{l_2} + C \sum_{k \in \mathcal{N}(i)} \exp(x_k^{l_2} - x_i^{l_2}) \phi_x(\text{proj}_{\mathcal{F}_{ik}} m_{ik}), \tag{3}$$

$$h_i^{l_2+1} = \phi_h(h_i^{l_2}, \sum_{k \in \mathcal{N}(i)} (\text{proj}_{\mathcal{F}_{ik}} m_{ik})), \tag{4}$$

where ϕ_m, ϕ_x, and ϕ_h represent 1D convolutional layers for the message, coordinate embedding, and hidden feature update, respectively. The normalizing factor C is applied to the exponentially weighted sum of mapped equi-message in Eq. (3). Additionally, a neighbouring search of $x_i^{l_2}$ within a specific radius is conducted to find the $\mathcal{N}(i)$ neighbouring feature descriptors for edge establishments, and this is used to prevent information overflow by confining the exchange of information within a local context, thereby reducing the complexity of the graph feature adjacency matrix from $O(n^2)$ to approximately $O(n)$. In Eq. (3), the projection of m_{ij} onto a locally equivariant frame ($\text{proj}_{\mathcal{F}_{ik}}(\cdot)$) helps to preserve the **SO**(3) feature invariance. The frame $\mathcal{F}_{ik}$ is constructed using pairwise coordinate embeddings as outlined in ClofNet [14],

$$\mathcal{F}_{ik} = (a_{ik}, b_{ik}, c_{ik}), \tag{5}$$

$$= (\frac{x_i^l - x_k^l}{\|x_i^l - x_k^l\|}, \frac{x_i^l \times x_k^l}{\|x_i^l \times x_k^l\|}, \frac{x_i^l - x_k^l}{\|x_i^l - x_k^l\|} \times \frac{x_i^l \times x_k^l}{\|x_i^l \times x_k^l\|}), \tag{6}$$

Consequently, the projection of m_{ik} into $\hat{m}_{ik}$ is formulated into the linear combination of axes of the local equi-frame scaled by the coefficients $(x_{ik}^a, x_{ik}^b, x_{ik}^b)$,

$$\text{proj}_{\mathcal{F}_{ik}} m_{ik} = \hat{m}_{ik} = x_{ik}^a a_{ik} + x_{ik}^b b_{ik} + x_{ik}^c c_{ik}. \tag{7}$$

The projection of edge message m_{ij} in Eq. (3) is performed in the local equi-frame (within bracket of Eq. (6)), to obtain projected message $\hat{m}_{ik}$, while the scalar coefficients in Eq. (7) remain **SO**(3) invariant. Consequently, the sum of equi-projected message in Eq. (4) is still a vector-based sum, maintaining the equivariance upon integration into the hidden layer ϕ_h.

3.3 Low-Rank Feature Transformation

Inspired by LoRA of language model [24], our approach diverges by not requiring pre-trained weights for fine-tuning. We employ two stacked linear forward layers with low-rank constraints in the middle of model (Fig. 3a) to map feature descriptors to aggregated descriptors. We name it as Low-Rank Feature Transformation (LRFT). This design enhances similarity match reliability and computational efficiency by performing matches on aggregated descriptors with integrated neighboring information. Specifically, The motivation for employing LRFT is twofold: 1) Theoretically, low-rank constraints in linear layers capture

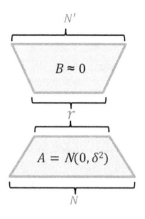 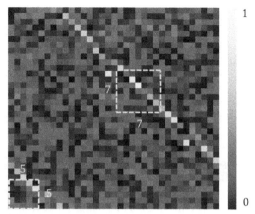

(a) Feature number reduction by Low-Rank Feature Transformation.

(b) Zoomed-in submatrix (35×35) from full similarity score matrix S ($N' \times N'$).

Fig. 3. Reducing the feature number through low-rank using MLP layers (a), and examining the similarity score matrix with submatrices for rank verification at bottom right (5×5) and center (7×7) of yellow dashed region as illustrated in subfigure (b). (Color figure online)

essential feature correlations within descriptors, as demonstrated by the matrix low-rank theorem (refer to Appendix Sec.1.2), leading to more reliable similarity matches; 2) Practically, our LRFT improves computational efficiency by aggregating feature descriptors prior to similarity matching, and It also enhances low-rank learning efficiency by training parameters during the forward pass, eliminating the need for pre-trained weights.

After the final layer ((l_2^*)th) of the equivariant graph module, the output graph features consist of both node and edge features. During this stage, we preserve each graph node feature $h_i^{l_2^*}, i \in N$ from the source frame and feature $h_j^{l_2^*}, j \in N$ from the target frame of the last graph layer. Next, the node feature is combined with the mean coordinate embeddings $\hat{x}_i^{l_2^*} = \frac{1}{n} \sum_{k \in \mathcal{N}(i)} x_k^{l_2^*}$, where $x_k^{l_2^*} \in \mathbb{R}^3$ is obtained from the edge embeddings (Eq. (3)) connected to the node feature $x_i^{l_2^*}$. Consequently, matrices for the source frame $\boldsymbol{H}_{src} \in \mathbb{R}^{N \times 35}$ and target frame $\boldsymbol{H}_{tar} \in \mathbb{R}^{N \times 35}$ are created by stacking of node features $(h_i^{l_2^*}, \hat{x}_i^{l_2^*})$ and coordinate embeddings along the column respectively. Before computing the feature similarity, the Low-Rank Feature Transformation (LRFT) technique is applied to compress the features into $\hat{\boldsymbol{H}}_{src} \in \mathbb{R}^{N' \times 35}$ and $\hat{\boldsymbol{H}}_{tar} \in \mathbb{R}^{N' \times 35}$ utilizing parameters of $\boldsymbol{A}$ and $\boldsymbol{B}$ mapping layers.

$$\hat{\boldsymbol{H}}_{src}, \hat{\boldsymbol{H}}_{tar} = (\boldsymbol{AB})^T (\boldsymbol{H}_{src}, \boldsymbol{H}_{tar}), \tag{8}$$

where $\boldsymbol{A}$ is a $N \times r$ matrix and $\boldsymbol{B}$ is $r \times N'$ matrix, and rank $r \ll \min(N, N')$, so that the number of node features are compressed into N' dimension after LRFT module mapping via multiplication AB, as depicted in the left of Fig. 3a. A is initialized from Gaussian distribution with standard deviation $\delta = \sqrt{r}$, while B is initialized with small constant close to zero.

The LRFT layers extract spatial context from neighboring feature descriptors through linear mapping, efficiently aggregating local information for decoder.

3.4 Similarity Calculation

Subsequently, we calculate the feature similarity score matrix (refer to Fig. 3b) for feature correspondence establishments using the compressed number of features after the LRFT module. This is achieved by computing the dot product $< \cdot >$ of features as an element. Prior to the multiplication, the respective element features h_i from $\hat{H}_{src} \in \mathbb{R}^{N' \times 35}$ and h_j from $\hat{H}_{tar} \in \mathbb{R}^{N' \times 35}$ are first normalized to $\hat{h}_i$ and $\hat{h}_j$.

$$S_{ij} = < \hat{h}_i \cdot \hat{h}_j >, \tag{9}$$

which forms a square similarity matrix $S \in \mathbb{R}^{N' \times N'}$, then subsequently normalized along each row to produce $\hat{S} \in \mathbb{R}^{N' \times N'}$. We compute the determinant of $\hat{S}$ by calculating trace to indicate whether the matrix rank has deficiency, which may arise from the presence of ambiguous correspondences. Accordingly, as per the match assignment rule, each row of the similarity matrix $\hat{S}$ should contain a singular value close to one, depicted as a light-colored square in Fig. 3b. Subsequently, the similarity matrix is employed to project the feature matrices $\hat{H}_{src}$ and $\hat{H}_{tar}$ through multiplication by $\hat{S}^T \hat{H}_{src} \in \mathbb{R}^{N' \times 35}$ and $\hat{S} \hat{H}_{tar} \in \mathbb{R}^{N' \times 35}$, respectively. The resulting matrices are concatenated to facilitate subsequent pooling and mapping through fully-connected layers. Furthermore, a regularizer is used to enforce the rank of $\hat{S}$ close to 35, ensuring that a submatrix $\hat{S}'$ with rank 35 out of $\hat{S}$ can be found,

$$\mathcal{L}_{Reg} = \left| (\mathbf{Trace}(\hat{S}^T \hat{S}))^{\frac{1}{2}} - 35 \right|. \tag{10}$$

Additionally, to eliminate outlier feature correspondences, each matched pair element $\hat{S}_{ij}$ undergoes verification through a submatrix full-rank check. This involves evaluating the determinant of a 7×7 submatrix centered at the feature element $\hat{S}_{ij}$ or a 5×5 submatrix at the border element of the $\hat{S}$ matrix (highlighted by the red dashed box in Fig. 3b). This verification process ensures local consistency in feature similarity matches, aiding in the identification of globally consistent and reliable match pattern search from the similarity matrix. Following verification, the final valid rank of the similarity matrix $\hat{S}$ is established as $r \leq 35$, with any invalid assignment row $\hat{S}_{i.}$ zeroed out to mask the corresponding feature for subsequent computations. The upper limit of 35 for rank r is derived from the ***Theorem***: $\mathbf{Rank}(AB) \leq \min(\mathbf{Rank}(A), \mathbf{Rank}(B))$. Please refer to the supplementary part for detailed proof the rank theorem.

3.5 Training Loss

The final layer predicts translation $\hat{t}$ and rotation matrix $\hat{R}$ in quaternion form. The ground-truth translation and rotation are denoted by t^* and R^* respectively.

$$\mathcal{L}_{total} = \mathcal{L}_{rot} + \mathcal{L}_{trans} + \beta \mathcal{L}_{Reg}, \tag{11}$$

β for regularizer is set to 0.05. The translation error (TE) and rotation error (RE) losses can be formulated as follows,

$$\mathcal{L}_{rot}(\hat{R}) = \arccos \frac{\text{Trace}(\hat{R}^T R^*) - 1}{2}, \tag{12}$$

$$\mathcal{L}_{trans}(\hat{t}) = \|\hat{t} - t^*\|^2. \tag{13}$$

The rotation error term $\mathcal{L}_{rot}$ and translation error term $\mathcal{L}_{trans}$ are measured in radians and meters, respectively. Given that the predicted transform is relative, from source to target frame, the scale of these transforms is typically in normal scale to avoid numerical stability issues in training.

4 Experiments

We evaluate the performance of the proposed model for point cloud registration in both indoor and outdoor environments. For indoor scenes, we utilize 3DMatch introduced by Zeng et al. [52]. The raw point clouds are uniformly downsampled to 1024 points through a voxel filter. For the outdoor evaluation, we select the KITTI dataset [20] with the same dataset split from the creators and follow the same downsampling process as in Choy et al., [10]. We report both the qualitative and quantitative results of the proposed model. Furthermore, we offer an in-depth analysis of the effect of varying parameter configurations and the contribution of each component to enhancing the model's performance. The computational efficiency of each model is presented in the metric table below.

Implementation Details. The key parameters of our model include the dimension of graph-relevant features and LRFT layers. Initially, the extracted feature descriptor dimension is 32 for subsequent graph learning. A critical aspect is the number of nearest neighbors for each node feature in the graph, set to 16 for constructing the graph using ball query for 3DMatch (ball radius at $0.3m$), while for KITTI, we employ kNN, selecting the nearest 16 points of query point to form a graph with 1024 nodes and 1024×16 edges. In graph learning, the node feature dimension is 32, and the edge embedding feature is 3, comprising coefficients projected onto the locally constructed coordinate frame as shown in Eq. (7). We use 4 equi-graph layers throughout the tests. The LRFT module consists of 3 parameters: input dimension N, internal rank r, and N'. Our model adopts a configuration of $1024/(32 + 3)/128$, where rank r is the sum of graph node feature dimension (32) and coordinate embedding dimension 3. The submatrix determinant check for similarity score matrix $\hat{S}$ is 5×5 along the borders and

7×7 within the matrix. A performance analysis comparing different parameter configurations is presented in the subsequent ablation section. All training and inference tasks are conducted on a single RTX 3090 GPU.

Evaluation Metrics. We employ the average Relative Error (RE) and Translation Error (TE) metrics from PointDSC [2] to assess the accuracy of predicted pose errors in successful registration. Additionally, we incorporate Registration Recall (RR) and *F1 score* as performance evaluation measures. To evaluate these metrics, we establish potential corresponding point pairs $(\boldsymbol{x}_i, \boldsymbol{y}_j) \in \Omega$ using input points from two frames, following the correspondence establishment approach outlined in PointDSC [2]. We apply the predicted transformation to the source frame point $\boldsymbol{x}_i$, recording a pairwise registration success only when the average Root Mean Square Error (RMSE) falls below a predefined threshold τ. The registration recall value δ is calculated as:

$$\delta = \sqrt{\frac{1}{\mathcal{N}(\Omega)} \sum_{(\boldsymbol{x}_i, \boldsymbol{y}_j) \in \Omega} \mathbb{1}[\|\hat{\boldsymbol{R}}\boldsymbol{x}_i + \hat{\boldsymbol{t}} - \boldsymbol{y}_j\|^2 < \tau]}, \quad (14)$$

where $\mathcal{N}(\Omega)$ denotes the total number of ground truth correspondences in set Ω. The symbol $\mathbb{1}$ functions as an indicator for condition satisfaction. Removing the conditional check within the indicator brackets on the equation's right side transforms it into a standard Root Mean Square Error (RMSE), $\sqrt{\frac{1}{\mathcal{N}(\Omega)} \sum_{(xi,yj) \in \Omega} |\hat{\boldsymbol{R}}xi + \hat{\boldsymbol{t}} - yj|^2}$. This RMSE metric is utilized in ablation experiments for parameter analysis. The *F1 score* is defined as $2 \cdot \frac{Precision \times Recall}{Precision + Recall}$.

Baseline Methods. 1) For the 3DMatch [52] benchmark, we compare our model with vanilla RANSAC implementations using various iterations and optimization refinements. We also include Go-ICP [51] and Super4PCS [31], which operate on raw points. Among learning-based methods, we compare with DGR [9] and PointDSC [2] combined with FCGF descriptors [10]. Additionally, we select D3Feat [3], SpinNet [1], and RoReg [44], which incorporate rotation invariance or equivariance. These learning methods do not support descriptor replacement, denoted by *. 2) For KITTI sequences [20], we implement the hand-crafted feature descriptor FPHF [37] due to performance saturation issues with FCGF [10] descriptors, as noted in PointDSC [2]. RoReg [44] is replaced with the registration model from the FCGF paper [10] (denoted as FCGF-Reg) due to public code limitations for KITTI dataset.

4.1 Indoor Fragments/Scans Registration

Point clouds are initially downsampled using a 5cm voxel size to generate 1024 sampled points. Registration success is evaluated using thresholds of 30cm for translational error (TE) and 15° for rotational error (RE). The correspondence distance threshold τ in Eq. (14) is set at 10cm. Comparative results between our proposed model and baseline approaches are presented in Table 1. Our model outperforms all comparison methods, despite slightly slower latency compared to RANSAC with 1k iterations. RoReg and SpinNet, ranking second and third, demonstrate minimal registration errors and maximal registration scores, highlighting the advantages of incorporating rotation features. While our model can integrate the FCGF descriptor, we present evaluation results using the PointNet++ learning descriptor for end-to-end training.

4.2 Outdoor Scenes Registration

The input point cloud from the KITTI sequences [20] is downsampled using a voxel size of 30 cm to generate 1024 sparse points for the experiments.

We set the registration thresholds at 60cm for Translation Error and 5° for Rotation Error. To measure Registration Recall (RR), we establish a threshold τ of 60 cm. Table 2 presents the quantitative results for comparison. Our proposed model demonstrates plausible performance compared to other methods, exhibiting minimal rotation and translation errors, and achieving the highest registration recall rate of 94.60% when compared to RoReg, the second-best model. However, RoReg has a remarkable weakness in its real-time performance, registering in over 30 min. Additionally, we visually display the registration sample outcomes of our proposed model on 3DMatch and KITTI below. For a more comprehensive visual comparison to baselines, please refer to the supplementary section.

Table 1. Evaluation results of registration methods on 3DMatch show non-learning-based approaches at the top and deep-learning registration models below. The learning-based feature descriptor FCGF [10] is tested with various learning baseline approaches. The symbol * indicates the original model implementation on 3D Match due to the lack of support for this feature descriptor replacement.

	RE (°) ↓	TE (cm) ↓	RR (%) ↑	F1 (%) ↑	Time (s) ↓
RANSAC-1k [17]	3.16	9.67	86.57	76.62	**0.08**
RANSAC-10k	2.69	8.25	90.70	80.76	0.58
RANSAC-100k	2.49	7.54	91.50	81.43	5.50
RANSAC-100k + refine	2.17	6.76	92.30	81.43	5.51
Go-ICP [51]	5.38	14.70	22.95	20.08	771.0
Super4PCS [31]	5.25	14.10	21.6	19.86	4.55
DGR [9] ↑	2.40	7.48	91.30	89.76	1.36
D3Feat* [3]	2.57	8.16	89.79	87.40	0.14
SpinNet* [1]	1.93	6.24	93.74	92.07	2.84
PointDSC [2]	2.06	6.55	93.28	89.35	0.09
RoReg* [44] ↓	1.84	6.28	93.70	91.60	2226
Ours	**1.67**	**5.68**	**94.60**	**94.35**	0.12

Table 2. Registration methods for evaluation on the KITTI dataset [20] involve testing the hand-crafted FPHF descriptor [37] in conjunction with different learning strategies. The symbol * indicates the lack of support for replacing the feature descriptor, as per the original implementations on KITTI.

	RE (°) ↓	TE (cm) ↓	RR (%) ↑	F1 (%) ↑	Time (s) ↓
RANSAC-1k [17]	2.51	38.23	11.89	14.13	0.20
RANSAC-10k	1.90	37.17	48.65	42.35	1.23
RANSAC-100k	1.32	25.88	74.37	73.13	13.7
RANSAC-100k + refine	1.28	18.42	77.20	74.07	15.65
Go-ICP [51]	5.62	42.15	9.63	12.93	802
Super4PCS [31]	4.83	32.27	21.04	23.72	6.29
FCGF-Reg* [10] ↓	1.95	18.51	70.86	68.90	**0.09**
DGR [9]	1.45	14.6	76.62	73.84	0.86
D3Feat* [3]	2.07	18.92	70.06	65.31	0.23
SpinNet* [1]	1.08	10.75	82.83	80.91	3.46
PointDSC* [2]	1.63	12.31	74.41	70.08	0.31
Ours	**0.92**	**8.74**	**83.83**	**85.09**	0.14

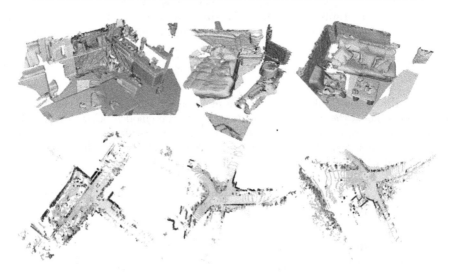

Fig. 4. The visual registration results of the proposed model on 3DMatch [52] and KITTI [20] are illustrated in the registration samples. Points from the target frame are represented in blue, whereas points converted from the source frame to the target frame by the predicted transform are visualized in yellow.(Color figure online)

To verify the proposed model performance under the different numbers of sampled input points, we also implement the sparse tests in Table 3, by comparing with FCGF registration, D3Feat with or without PointDSC combination, and SpinNet. Our model has a consistent performance on 3DMatch over the comparison methods. In addition, using more points can boost the proposed model registration performance with a marginal gain. Considering a good trade-off between accuracy and computational efficiency, 1024 is chosen as the input point number in our final implemented model, because there is only a small performance gap within 1% compared to the 2048 and 4096 number of points.

Table 3. RR results on 3DMatch with a different number of sampled points.

#Sampled Points	4096	2048	1024	512	256	Average
FCGF-Reg [17]	91.7	90.3	89.5	85.7	80.5	87.5
D3Feat [3]	91.9	90.4	89.8	86.0	82.5	88.1
D3Feat [3]+PointDSC [2]	92.1	92.5	90.8	87.4	83.6	89.3
SpinNet [1]	93.8	93.6	93.7	89.5	85.7	91.3
Ours	**95.3**	**94.8**	**94.6**	**91.3**	**88.5**	**92.9**

4.3 Ablation Study

Firstly, we present a table (Table 4) displaying different configurations and combinations of module blocks in the proposed model. Table analysis reveals significant accuracy enhancements with our model's learning-based descriptor over pre-trained FCGF and FPHF descriptors, as shown in rows 1 and 2, across multiple metrics. The descriptor learning layers (row 3) and equivariant graph CNN layers (row 4) are key to performance improvements. Replacing the equivariant with standard graph CNN layers (row 5) impairs rotation convergence, while omitting the LRFT module (row 6) marginally reduces performance. Ball query graph initialization (rows 8) outperforms KNN search (row 7) in efficiency and real-time outcomes on 3DMatch. Additionally, integrating rank regularizers (row 9) and sub-matrix rank verification (row 10) enhances model performance. Although equivariant layers introduce a slight computational delay (around 60ms), the overall performance gain is significant. Lastly, experiments of row 12-13 show that directly applying E-GCNN layers on the input or replacing the descriptor layers with equivariant structures such as SpinNet does not yield performance on par with our proposed method. Additionally, replacing E-GCNN layers with normal GCNN (row 11) results in significant performance degradation. These findings highlight the tailored effectiveness of our equi-approach.

For a more intuitive understanding of our model capability of learning equi-features, we provide t-SNE plots below for different encoder network outputs, showing that our Equi-Graph CNN produces features that are equivariant to rotated input. In comparison, features extracted by the conventional Graph CNN (GCNN) and SpinNet (**SO**(2) equivariant) as baselines do not exhibit rotation equivariance. To further elucidate the visual relationship between the rank of

Table 4. An ablation study was conducted on our model design, using 3DMatch [52] as the test dataset. The study involved comparing various descriptor combinations, exploring different combinations of layers for equi-feature learning, analyzing different methods of graph construction, and examining the impact of regularizers.

		RE (°) ↓	TE (cm) ↓	RR (%) ↑	F1 (%) ↑	Time (s) ↓
1.	FCGF Descriptor + Ours	1.62	6.24	93.87	94.28	0.15
2.	FPHF Descriptor + Ours	1.83	6.49	83.62	73.06	0.12
3.	w/o Feature Descriptor Layers	10.26	9.02	61.39	60.04	0.08
4.	w/o Equi-graph Layers	9.64	8.37	62.45	60.03	0.06
5.	Replacing by normal GCNN	8.32	5.94	68.52	67.54	0.10
6.	w/o LRFT layers	2.76	6.47	83.09	81.78	0.09
7.	KNN Graph Construction	1.72	**5.31**	92.37	93.74	0.16
8.	Ball Query Graph Construction	1.67	5.68	**94.60**	**94.35**	0.12
9.	w/o Rank Regularizer	6.41	7.92	76.45	78.09	0.10
10.	w/o Sub-matrix Rank Verification	2.58	6.93	87.76	88.36	0.07
11.	Descriptor Layers + GCNN	8.32	5.94	68.52	67.54	0.10
12.	w/o Descriptor Layers + Equi-GCNN	10.26	9.02	61.39	60.04	0.08
13.	SpinNet + Equi-GCNN	2.93	5.97	82.16	83.74	3.62
14.	Ours	1.67	5.68	94.60	94.35	0.12

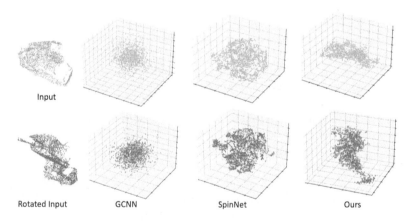

Fig. 5. The t-SNE comparisons of equi-features outputs.

the feature similarity score matrix $\hat{S}$ and point correspondences with or without equivariant features, please refer to the supplementary section.

Moreover, we provide the parameter configuration comparison of LRFT layers as well. It can be observed that varying the LRFT output feature number shows that increasing the middle-rank dimension enhances accuracy, yet beyond 200 ranks, error climbs a bit as depicted by the red curve. The relationship among

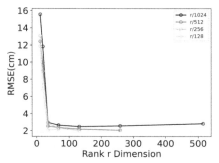 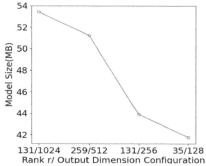

(a) RMSE results under the different rank r configurations, with four output dimension choices.

(b) Model size under the four best-chosen rank/output dimension configurations.

Fig. 6. The left is the RMSE results plot of four output feature sizes of the LRFT module under the various rank dimensions, and the model size of the best r/output dimension configuration of each curve in the left plot is presented at right.

the input feature number N, rank r, and the output number N' is defined as $r \ll N' < N$, indicating that N' should be at least twice the size of r, as evidenced by the termination position of curves in Fig. 6. Additionally, the optimal model size with the lowest RMSE is achieved with the 35/128 (r/N') LRFT configuration, corresponding to the yellow curve on the left.

5 Conclusion

We introduce an end-to-end model that leverages pre-trained feature descriptors or learns directly from raw scan points across two frames, incorporating equivariance embedding through graph layers, Low-Rank Feature Transformation and similarity score computation. Validation in both indoor and outdoor datasets confirms the superior performance of our proposed model. Ablation studies further substantiate the model design. Notably, the model's latency demonstrates its potential applicability in visual odometry. Future work could explore generalizing this framework to be input-order permutation invariant through graph attention layers or pooling, potentially integrating additional sensor modalities to address dynamic challenges.

Acknowledgements. This research work was jointly supported by the Young Scientists Fund of the National Natural Science Foundation of China (42301520), the Research Grants Council of Hong Kong (25206524), the Platform Project of UAV Research Centre (P0049516), the Seed Project of Smart Cities Research (P0051028), the Research Cultivation Program of ShenZhen University (2023JCT003), and the Pearl River Talent (2021JC02G046). Furthermore, we appreciate the University of Melbourne to provide the Graduate Research Scholarship for Xueyang to pursue his joint PhD program.

References

1. Ao, S., Hu, Q., Yang, B., Markham, A., Guo, Y.: Spinnet: learning a general surface descriptor for 3d point cloud registration. In: Proceedings of the IEEE/CVF Conference on Computer Vision and Pattern Recognition, pp. 11753–11762 (2021)
2. Bai, X., et al.: Pointdsc: robust point cloud registration using deep spatial consistency. In: Proceedings of the IEEE/CVF Conference on Computer Vision and Pattern Recognition, pp. 15859–15869 (2021)
3. Bai, X., Luo, Z., Zhou, L., Fu, H., Quan, L., Tai, C.L.: D3feat: joint learning of dense detection and description of 3d local features. In: Proceedings of the IEEE/CVF Conference on Computer Vision and Pattern Recognition, pp. 6359–6367 (2020)
4. Bogatskiy, A., Anderson, B., Offermann, J., Roussi, M., Miller, D., Kondor, R.: Lorentz group equivariant neural network for particle physics. In: International Conference on Machine Learning, pp. 992–1002. PMLR (2020)
5. Brandstetter, J., Hesselink, R., van der Pol, E., Bekkers, E.J., Welling, M.: Geometric and physical quantities improve e (3) equivariant message passing. arXiv preprint arXiv:2110.02905 (2021)
6. Chatzipantazis, E., Pertigkiozoglou, S., Dobriban, E., Daniilidis, K.: SE(3)-equivariant attention networks for shape reconstruction in function space. arXiv preprint arXiv:2204.02394 (2022)
7. Chen, H., Liu, S., Chen, W., Li, H., Hill, R.: Equivariant point network for 3d point cloud analysis. In: Proceedings of the IEEE/CVF Conference on Computer Vision and Pattern Recognition, pp. 14514–14523 (2021)
8. Cheng, Y., Huang, Z., Quan, S., Cao, X., Zhang, S., Yang, J.: Sampling locally, hypothesis globally: accurate 3d point cloud registration with a ransac variant. Visual Intell. **1**(1), 20 (2023)
9. Choy, C., Dong, W., Koltun, V.: Deep global registration. In: Proceedings of the IEEE/CVF Conference on Computer Vision and Pattern Recognition, pp. 2514–2523 (2020)
10. Choy, C., Park, J., Koltun, V.: Fully convolutional geometric features. In: Proceedings of the IEEE/CVF International Conference on Computer Vision, pp. 8958–8966 (2019)
11. Cohen, T., Welling, M.: Group equivariant convolutional networks. In: International Conference on Machine Learning, pp. 2990–2999. PMLR (2016)
12. Cohen, T.S., Geiger, M., Köhler, J., Welling, M.: Spherical CNNs. arXiv preprint arXiv:1801.10130 (2018)
13. Deng, C., Litany, O., Duan, Y., Poulenard, A., Tagliasacchi, A., Guibas, L.J.: Vector neurons: a general framework for so (3)-equivariant networks. In: Proceedings of the IEEE/CVF International Conference on Computer Vision, pp. 12200–12209 (2021)
14. Du, W., et al.: SE(3) equivariant graph neural networks with complete local frames. In: International Conference on Machine Learning, pp. 5583–5608. PMLR (2022)
15. El Banani, M., Gao, L., Johnson, J.: Unsupervisedr&r: unsupervised point cloud registration via differentiable rendering. In: Proceedings of the IEEE/CVF Conference on Computer Vision and Pattern Recognition, pp. 7129–7139 (2021)
16. Finzi, M., Stanton, S., Izmailov, P., Wilson, A.G.: Generalizing convolutional neural networks for equivariance to lie groups on arbitrary continuous data (2020)
17. Fischler, M.A., Bolles, R.C.: Random sample consensus: a paradigm for model fitting with applications to image analysis and automated cartography. Commun. ACM **24**(6), 381–395 (1981)

18. Fuchs, F., Worrall, D., Fischer, V., Welling, M.: SE(3)-transformers: 3d roto-translation equivariant attention networks. Adv. Neural. Inf. Process. Syst. **33**, 1970–1981 (2020)
19. Fuchs, F.B., Wagstaff, E., Dauparas, J., Posner, I.: Iterative SE(3)-transformers. In: Geometric Science of Information: 5th International Conference, GSI 2021, Paris, 21–23 July 2021, Proceedings 5, pp. 585–595. Springer (2021)
20. Geiger, A., Lenz, P., Urtasun, R.: Are we ready for autonomous driving? the Kitti vision benchmark suite. In: 2012 IEEE Conference on Computer Vision and Pattern Recognition, pp. 3354–3361. IEEE (2012)
21. Gilmer, J., Schoenholz, S.S., Riley, P.F., Vinyals, O., Dahl, G.E.: Neural message passing for quantum chemistry. In: International Conference on Machine Learning, pp. 1263–1272. PMLR (2017)
22. Gilmer, J., Schoenholz, S.S., Riley, P.F., Vinyals, O., Dahl, G.E.: Message passing neural networks. In: Schütt, K.T., Chmiela, S., von Lilienfeld, O.A., Tkatchenko, A., Tsuda, K., Müller, K.-R. (eds.) Machine Learning Meets Quantum Physics. LNP, vol. 968, pp. 199–214. Springer, Cham (2020). https://doi.org/10.1007/978-3-030-40245-7_10
23. Gojcic, Z., Zhou, C., Wegner, J.D., Guibas, L.J., Birdal, T.: Learning multiview 3D point cloud registration. In: Proceedings of the IEEE/CVF Conference on Computer Vision and Pattern Recognition, pp. 1759–1769 (2020)
24. Hu, E.J., et al.: Lora: low-rank adaptation of large language models. arXiv preprint arXiv:2106.09685 (2021)
25. Huang, S., Gojcic, Z., Usvyatsov, M., Wieser, A., Schindler, K.: Predator: registration of 3d point clouds with low overlap. In: Proceedings of the IEEE/CVF Conference on Computer Vision and Pattern Recognition (CVPR), pp. 4267–4276 (2021)
26. Hutchinson, M., Lan, C.L., Zaidi, S., Dupont, E., Teh, Y.W., Kim, H.: Lietransformer: equivariant self-attention for lie groups (2020)
27. Jenner, E., Weiler, M.: Steerable partial differential operators for equivariant neural networks. In: International Conference on Learning Representations (2022). https://openreview.net/forum?id=N9W24a4zU
28. Keriven, N., Peyré, G.: Universal invariant and equivariant graph neural networks. Adv. Neural Inf. Process. Syst. **32** (2019)
29. Köhler, J., Klein, L., Noé, F.: Equivariant flows: exact likelihood generative learning for symmetric densities. In: International Conference on Machine Learning, pp. 5361–5370. PMLR (2020)
30. Lin, C.E., Song, J., Zhang, R., Zhu, M., Ghaffari, M.: SE(3)-equivariant point cloud-based place recognition. In: Conference on Robot Learning, pp. 1520–1530. PMLR (2023)
31. Mellado, N., Aiger, D., Mitra, N.J.: Super 4pcs fast global pointcloud registration via smart indexing. In: Computer Graphics Forum, vol. 33, pp. 205–215. Wiley Online Library (2014)
32. Pais, G.D., Ramalingam, S., Govindu, V.M., Nascimento, J.C., Chellappa, R., Miraldo, P.: 3dregnet: a deep neural network for 3d point registration. In: Proceedings of the IEEE/CVF Conference on Computer Vision and Pattern Recognition, pp. 7193–7203 (2020)
33. Park, J., Zhou, Q.Y., Koltun, V.: Colored point cloud registration revisited. In: Proceedings of the IEEE International Conference on Computer Vision, pp. 143–152 (2017)
34. Park, S.Y., Subbarao, M.: An accurate and fast point-to-plane registration technique. Pattern Recogn. Lett. **24**(16), 2967–2976 (2003)

35. Qi, C.R., Yi, L., Su, H., Guibas, L.J.: Pointnet++: deep hierarchical feature learning on point sets in a metric space. Adv. Neural Inf. Process. Syst. **30** (2017)
36. Qin, Z., Yu, H., Wang, C., Guo, Y., Peng, Y., Xu, K.: Geometric transformer for fast and robust point cloud registration. In: Proceedings of the IEEE/CVF Conference on Computer Vision and Pattern Recognition, pp. 11143–11152 (2022)
37. Rusu, R.B., Blodow, N., Beetz, M.: Fast point feature histograms (FPFH) for 3D registration. In: 2009 IEEE International Conference on Robotics and Automation, pp. 3212–3217. IEEE (2009)
38. Satorras, V.G., Hoogeboom, E., Welling, M.: E(n) equivariant graph neural networks (2021)
39. Schütt, K.T., Sauceda, H.E., Kindermans, P.J., Tkatchenko, A., Müller, K.R.: Schnet–a deep learning architecture for molecules and materials. J. Chem. Phys. **148**(24) (2018)
40. Shi, W., Rajkumar, R.: Point-gnn: graph neural network for 3d object detection in a point cloud. In: Proceedings of the IEEE/CVF Conference on Computer Vision and Pattern Recognition, pp. 1711–1719 (2020)
41. Sosnovik, I., Szmaja, M., Smeulders, A.: Scale-tooltool steerable networks. arXiv preprint arXiv:1910.11093 (2019)
42. Thomas, N., et al.:: Tensor field networks: rotation-and translation-equivariant neural networks for 3D point clouds. arXiv preprint arXiv:1802.08219 (2018)
43. Wang, H., Liu, Y., Dong, Z., Wang, W.: You only hypothesize once: Point cloud registration with rotation-equivariant descriptors. In: Proceedings of the 30th ACM International Conference on Multimedia, pp. 1630–1641 (2022)
44. Wang, H., et al.: Roreg: pairwise point cloud registration with oriented descriptors and local rotations. IEEE Trans. Pattern Anal. Mach. Intell. (2023)
45. Wang, H., Wang, C., Chen, C.L., Xie, L.: F-loam: fast lidar odometry and mapping. In: 2021 IEEE/RSJ International Conference on Intelligent Robots and Systems (IROS), pp. 4390–4396. IEEE (2021)
46. Wang, Y., Solomon, J.M.: Deep closest point: learning representations for point cloud registration. In: Proceedings of the IEEE/CVF International Conference on Computer Vision, pp. 3523–3532 (2019)
47. Weiler, M., Cesa, G.: General E(2)-equivariant steerable CNNs. In: Conference on Neural Information Processing Systems (NeurIPS) (2019)
48. Weiler, M., Geiger, M., Welling, M., Boomsma, W., Cohen, T.S.: 3d steerable CNNs: learning rotationally equivariant features in volumetric data. Adv. Neural Inf. Process. Syst. **31** (2018)
49. Weiler, M., Hamprecht, F.A., Storath, M.: Learning steerable filters for rotation equivariant CNNs. In: Proceedings of the IEEE Conference on Computer Vision and Pattern Recognition, pp. 849–858 (2018)
50. Xinyi, Z., Chen, L.: Capsule graph neural network. In: International Conference on Learning Representations (2018)
51. Yang, J., Li, H., Campbell, D., Jia, Y.: Go-ICP: a globally optimal solution to 3D ICP point-set registration. IEEE Trans. Pattern Anal. Mach. Intell. **38**(11), 2241–2254 (2015)
52. Zeng, A., Song, S., Nießner, M., Fisher, M., Xiao, J., Funkhouser, T.: 3dmatch: learning local geometric descriptors from rgb-d reconstructions. In: Proceedings of the IEEE Conference on Computer Vision and Pattern Recognition, pp. 1802–1811 (2017)
53. Zhang, Y., Rabbat, M.: A graph-cnn for 3d point cloud classification. In: 2018 IEEE International Conference on Acoustics, Speech and Signal Processing (ICASSP), pp. 6279–6283. IEEE (2018)

54. Zhou, Q.Y., Park, J., Koltun, V.: Fast global registration. In: Computer Vision–ECCV 2016: 14th European Conference, Amsterdam, 11–14 October 2016, Proceedings, Part II 14, pp. 766–782. Springer (2016)
55. Zhu, M., Ghaffari, M., Clark, W.A., Peng, H.: E2pn: efficient SE(3)-equivariant point network. In: Proceedings of the IEEE/CVF Conference on Computer Vision and Pattern Recognition, pp. 1223–1232 (2023)
56. Zhu, M., Ghaffari, M., Peng, H.: Correspondence-free point cloud registration with so (3)-equivariant implicit shape representations. In: Conference on Robot Learning, pp. 1412–1422. PMLR (2022)

GTP-4o: Modality-Prompted Heterogeneous Graph Learning for Omni-Modal Biomedical Representation

Chenxin Li[1], Xinyu Liu[1], Cheng Wang[1], Yifan Liu[1], Weihao Yu[1], Jing Shao[2], and Yixuan Yuan[1(✉)]

[1] The Chinese University of Hong Kong, Ma Liu Shui, Hong Kong
chenxinli@link.cuhk.edu.hk, yxyuan@ee.cuhk.edu.hk
[2] Shanghai AI Laboratory, Shanghai, China

Abstract. Recent advances in learning multi-modal representation have witnessed the success in biomedical domains. While established techniques enable handling multi-modal information, the challenges are posed when extended to various clinical modalities and practical modality-missing setting due to the inherent modality gaps. To tackle these, we propose an innovative Modality-prompted He̱t̲e̲rogeneous G̲raph for O̲mni-modal Learning (GTP-4o), which embeds the numerous disparate clinical modalities into a unified representation, completes the deficient embedding of missing modality and reformulates the cross-modal learning with a graph-based aggregation. Specially, we establish a heterogeneous graph embedding to explicitly capture the diverse semantic properties on both the modality-specific features (nodes) and the cross-modal relations (edges). Then, we design a modality-prompted completion that enables completing the inadequate graph representation of missing modality through a graph prompting mechanism, which generates hallucination graphic topologies to steer the missing embedding towards the intact representation. Through the completed graph, we meticulously develop a knowledge-guided hierarchical cross-modal aggregation consisting of a global meta-path neighbouring to uncover the potential heterogeneous neighbors along the pathways driven by domain knowledge, and a local multi-relation aggregation module for the comprehensive cross-modal interaction across various heterogeneous relations. We assess the efficacy of our methodology on rigorous benchmarking experiments against prior state-of-the-arts. In a nutshell, GTP-4o presents an initial foray into the intriguing realm of embedding, relating and perceiving the heterogeneous patterns from various clinical modalities holistically via a graph theory. Project page: https://gtp4-o.github.io/.

Keywords: Biomedical Data · Multimodal Learning · Graph Networks

---

X. Liu and C. Wang—Equal second-author contribution.

© The Author(s), under exclusive license to Springer Nature Switzerland AG 2025
A. Leonardis et al. (Eds.): ECCV 2024, LNCS 15062, pp. 168–187, 2025.
https://doi.org/10.1007/978-3-031-73235-5_10

1 Introduction

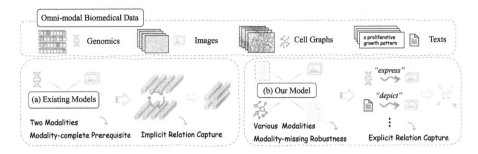

Fig. 1. Methodology Comparison. Unlike (a) prior methods, (b) our framework enables learning unified omni-modal representation from various clinical modalities with modality missing and explicit capture of the cross-modal relations through the established heterogeneous graph representation.

Each modality has its own perspective to reflect the specific data characteristics [30,33,35,75,87]. Integrating multi-modal data empowers the models with various insights into the conditions of subjects at the macroscopic, microscopic, and molecular levels, and allows for an accurate and comprehensive disease diagnosis [40,44,52,68,69,77]. For instance, multimodal fusion of various imaging techniques has significantly improved gastrointestinal lesion detection and characterization in endoscopic scenes [28,32,49,56]. Similarly, incorporating genomic information with pathological images can improve the prediction accuracy of cancer grading [5,6,16,72,76]. A relevant task, survival prediction, which aims to predict the time interval to a significant event such as death or disease relapse, can also benefit from such multi-modal inclusion [7]. Besides, the cell graphs constructed by the cell nuclei segmentation of pathological images, are shown to provide more fine-grained microscopic information [71]. Recent advances in visual language models also sparks the works in learning from biomedical images and texts [80], whereby the diagnostic texts usually encapsulates abstract semantic information [11,13]. These progress presents potential for extending the capacity boundary of biomedical multi-modal models to omni-modal representation to handle a broader range of clinical modalities.

Established multimodal methods typically follow the principle that first extracts uni-modal features, then learns cross-modal relations in paired multi-modal data [42,60,73,80]. Early researches design meticulous fusion techniques for multimodal information and wish to maximize the benefits of each modality [19,41]. Due to the imbalanced learning process with inherent modal disparity and heterogeneity [22,81,82], recent efforts pivot towards improving collaborative learning of multiple modalities by balancing and adjusting learning of each modality [15,59,74]. Through deriving the modality-relevant weighting factors, these methods dynamically modulates the learning and fusing of multimodal information on features [74], gradients [59], attentions [15,50], etc.

Despite the success in alleviating the modality gap, the challenge remains severe when applied to (especially a broad range of) biomedical modalities, primarily due to the two featured challenges. The first challenge lies in the large semantic heterogeneity exhibiting on biological modalities. A straightforward example is that an *"dog"* in natural images shares similar object-related semantics its sound, while the semantic relation and local correspondence between genomic profiles and pathological images is highly ambiguous [5,6,43,53,72]. Through prior methods employ optimal transport [72] or cross-modal attention [7] to capture fine-grained correlation across genes and images, they still overlook the heterogeneity in a high-order space, i.e., relations across modalities. Every two modalities have their own relation with specific semantics and attributes. As shown in Fig. 1, the relation across images and genomics is semantically related to *"express"*, while that across images and texts could be abstracted as *"depict"*. Therefore, these observations inspire us to introduce a unified non-Euclidean representation that explicitly captures the heterogeneous attributes on both modal features and cross-modal relations.

Secondly, in clinical practice, it is common to encounter partial absence in some modalities due to privacy and ethical considerations. The limitations in data collection technology and the concerns surrounding bioinformatics security make it more challenging to access all the data modalities. However, most multimodal methods have a common assumption on the data completeness [60,73,80]. Once a modality is missing regardless of training or testing, the multimodal fusion becomes unreachable, which leads to sub-optimal performance [26]. Therefore, we are committed to designing algorithms to adaptively complete the feature space messed up by the missing of modality such that all the representation from the missing and existing modality could be handled in a unified fashion.

To address the aforementioned challenges, we propose a modality-prompted heterogeneous graph framework for omni-modal learning (GTP-4o) that allows unifying representations under various biomedical modalities with potential modality missing. Specially, we establish a heterogeneous graph embedding [38, 39,51] to explicitly capture the heterogeneous attributes on both modal features and cross-modal relations. Then, we design a modality-prompted completion that completes the deficient graph embedding of missing modality through a novel graph prompting module, which generates hallucination nodes to steer the embedding towards the original complete space. Through the completed graph, we meticulously develop a knowledge-guided hierarchical aggregation that includes a knowledge-derived global meta-path neighbouring to capture the potential heterogeneous neighbors, and a local multi-relation aggregation for the comprehensive interaction of modal information across various heterogeneous relations. GTP-4o presents the first exploration in learning unified representations from various heterogeneous clinical modalities including genomics, pathological images, cell graphs, and diagnostic texts. Our contributions are as follows:

– This paper introduces the new problem of learning unified multimodal representations from various diverse clinical modalities, and presents the first effort

to embed and relate heterogeneous multimodal features through a graph representation and aggregation.
- We propose a modality-prompted completion module to complete the corrupted graph embedding of the missing modality by a graph prompting strategy, which generates hallucination nodes to steer the missing embedding towards the complete representation.
- We present a knowledge-guided hierarchical cross-modal aggregation, employing a global meta-path neighbouring to capture heterogeneous neighbors, and a local multi-relation aggregation module for information interaction across various heterogeneous relations.
- Extensive experiments on comprehensive benchmarks of disease diagnosis including pathological glioma grading and survival outcome prediction exhibits the efficacy of our method against prior state-of-the-arts.

2 Related Work

Biomedical Multimodal Learning. Utilizing multimodal data has gained significant attention for accurate and comprehensive imaging analysis [9,36,55,87] and diagnosis [2,10,64,76]. For instance, the comprehensive features from pathological images [57,89], genomics [46,48,78], are employed in joint for an accurate cancer-related diagnosis, e.g. glioma grading [14] and survival analysis [7]. Meanwhile, with the eGTP-4once of Visual Language Models (VLMs), much efforts have been devoted to enhancing the recognition and analysis ability of vision models by further incorporating the textual information from clinical text reports [80,83]. Inspired by the trends, this paper introduces the new problem of learning unified features from various disparate clinical modalities, including genomics [7,46], pathological images [57], cell graphs [71] and text descriptions [83].

Handling the modality heterogeneity is critical when integrating multimodal information [47]. Early works focus on studying early or late fusion methods [5,19,21,61,66], which integrates the predictions from individually separated models for the final decision. However, these methods suffer from either neglecting intra-modality dynamics [47] or failing to fully relating cross-modal information. Recent progress in intermediate fusion [6,25] has shown promise, which learns uni-modal features and capture cross-modal interactions at the same time, by leveraging the power of cross-modal attention [6,54,72,86]. However, they try to model all potential cross-modal relations with the learned attentions. Different from them, we present to explicitly capture the heterogeneity of modal features and cross-modal relations resorting to a heterogeneous graph space.

Graph Representation in Pathology. Graph representation has shown its promise in the field of pathology analysis [3,4,67]. Following previous efforts of Multiple Instance Learning (MIL) that split the high-resolution whole slide pathological images (aka., WSIs) into a bag of instances and pre-define the connective local areas in the Euclidean space, recent graph-based methods [3,4,17,85] models the interactions among instances flexibility via the graphs

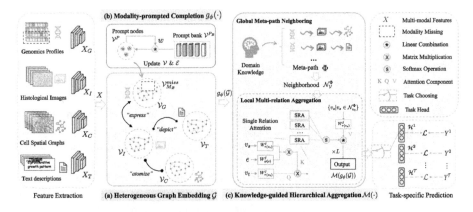

Fig. 2. Pipeline Overview of GTP-4o. We instantiate the omni-modal biomedical features (Sect. 3.1), and embed them onto (a) the heterogeneous graph space (Sect. 3.2). Then, we introduce (b) the modality-prompted completion via graph prompting to complete the missing embedding (Sect. 3.3). After that, we design (c) the knowledge-guided hierarchical aggregation from a global meta-neighbouring to uncover the heterogeneous neighbourhoods and a local multi-relation aggregation to interact features across various heterogeneous relations (Sect. 3.4).

topology. For instance, PatchGCN [4] models pathological images with homogeneous graphs, and regress survival data with a graph convolutional neural network (GCN) [23]. GTNMIL is designed as a graph-based MIL using graph transformer networks [85]. Recent methods [3,17] extend the prior practice to handling WSIs with heterogeneous graphs, introducing heterogeneity in each patch by different resolution levels [17], or semantic representations via pretext tasks [3]. However, these methods only considers heterogeneity in the image modality, while the more challenging multimodal scenario is left to study. To fill this gap, this paper explores the graph representation in a way more complex setting, i.e., learning from various disparate clinical modalities with significant heterogeneity.

3 Method

Overview. After performing data processing and feature extraction (Sect. 3.1), the omni-modal embedding for a patient subject could be represent by a 4-tuples of four modalities, including genomics (G), pathological images (I), cell spatial graphs (C) and diagnostic texts (T), $X = \{X_G, X_I, X_C, X_T\}$, with different number of instances in each modality, while the common dimension d. Then, we establish the heterogeneous graph representation $\mathcal{G}$ by transforming modal features to the graph space (Sect. 3.2). After that, the modality-prompted completion is performed, which employs a graph prompting $g_\phi(\cdot)$ to transform the incomplete graphic embedding to a prompted and completed representation $g_\phi(\mathcal{G})$ (Sect. 3.3). Afterwards, we conduct a knowledge-guided hierarchical aggregation that is parameterized by $\mathcal{M}$, including a global neighbouring via knowledge-derived meta-paths Φ, and a local multi-relation aggregation along various heterogeneous relations (Sect. 3.4). The final aggregated features $\mathcal{M} \circ g_\phi(\mathcal{G})$ end up with forwarding a task-specific head $\mathcal{H}^{\mathcal{T}}$ to obtain the specific prediction for task $\mathcal{T}$, based on which we we optimize the network parameters $\mathcal{M}, \mathcal{H}^{\mathcal{T}}$ and prompt parameters g_ϕ w.r.t. the task loss $\mathcal{L}$. (Sect. 3.5).

3.1 Data Processing and Feature Extraction

Genomic Profiles. Following [5], the genomic profiles that we use includes Copy Number Variation (CNV), bulk RNA-Seq expressions and mutation status [5]. We merge the mutation status and CNV and feed it and RNA-Seq as separate groups of gnomic data into Self-Normalizing Neural Network (SNN) [24] to get the embedding $X_G \in \mathbb{R}^{N_G \times d}$, where N_G equals to the number of genomic groups.

Pathological Images. Following [6,63], we divide WSIs into a series of non-overlapping patches, employ a ImageNet pretrained ResNet-50 to extract the features from each patch, and feed them into a projection layer to obtain $X_I \in \mathbb{R}^{N_I \times d}$, where N_I is the number of patches.

Cell Spatial Graphs. Cell graph representations explicitly capture selected fine-grained features of cells [58]. We utilize the procedure in PathomicFusion [5] to segment cells for each slide, curate graph topology. and use a graph convolutional network (GCN) [23] backbone to obtain the aggregated graph embedding, as $X_C \in \mathbb{R}^{N_c \times d}$, where N_C equals to the number of curated cell graphs.

Text Descriptions. As no actual medical reports are provided in the used materials, we employ an open-source multimodal Large Language Model (LLM), MiniGPT-4 [88] to generate the customized text descriptions for each pathological image[1]. The prompt is customized as follows.

Prompt: *"This is a pathology slide with glioma cells. Write a caption for this slide based on the following properties:* ❶ *size and shape of cells,* ❷ *color of cells,* ❸ *growth pattern and cellularity of cells,* ❹ *uclear atypia and pleomorphism of cells,* ❺ *necrosis of cells,* ❻ *microvascular proliferation of cells,* ❼ *mitotic activity of cells."*

We obtain several sentences for each slide, like: *"Some cells show signs of necrosis, with dark spots in the cytoplasm"*. We take each sentence as a modal instance, and employs a MedBERT [62] to obtain the embedding $X_T \in \mathbb{R}^{N_T \times d}$, where N_T equals to the number of curated sentences.

---

[1] We find that another foundation VLM, BLIP-2 [37] showing effective on curating captions for 3D representation [27,31,73] does not work well for pathological images.

3.2 Heterogeneous Graph Embedding

With the obtained modal features $X = \{X_G, X_I, X_C, X_T\}$ with the same dimension d, the heterogeneous graph embedding could be established by a feature to graph transformation. Formally, a heterogeneous graph space is formulated by $\mathcal{G} = \{\mathcal{V}, \mathcal{E}, \mathcal{A}, \mathcal{R}\}$, where $\mathcal{V}$ and $\mathcal{E}$ represents the set of entities (i.e., vertices or nodes) and relations (i.e., edges) that has been established in the theory of a classic directed graph. The further introduced $\mathcal{A}$ and $\mathcal{R}$ represent the attribute set of nodes and edges, respectively, by which we can explicitly define the heterogeneous properties for the features of modal entities and cross-modal relations. A function $\tau(v) = a \in \mathcal{A}$, is defined to map each node v to an attribute in the set $\mathcal{A}$, according to its modality, As a result, the attribute set of nodes can be formulated as $\mathcal{A} = \{G, I, C, T\}$. Furthermore, the edges $e \in \mathcal{E}$ in the heterogeneous embedding represent the relations from the source nodes $v_s \in \mathcal{V}_s$ to the target nodes $v_t \in \mathcal{V}_t$[2], therefore the attribute of an edge $e : v_s \to v_t$ is determined is determined by the attribute of the source node v_s and target node v_t as well as their actual semantic relations. Thus, a function $\varphi(e) = r \in \mathcal{R}$ that maps each edge $e \in \mathcal{E}$ to a specific attribute $r \in \mathcal{R}$ is introduced. We formulate this attribute set of relations $\mathcal{R}$ by the prior knowledge of biomedical modalities, $\mathcal{R} = \{\text{"express"}, \text{"depict"}, \text{"atomize"}, \text{"intra} - \text{modal"}, \}$. This set represents semantic relations of *"express"* between genomics and images, *"depict"* between images and texts, and *"atomize"* between images and cell graphs. We also model all relations between *"intra-modal"* instances. To obtain initial input graph embedding $\mathcal{V}$, we perform a non-linear projection on the modal features X. Edge embeddings are computed as cosine correlations between head and tail nodes.

3.3 Modality-Prompted Completion

The modality-prompted completion aims to adapt the deficient embedding of missing modality by updating its with some prompted entities that could be learnt. Formally, we introduce a graph prompt operation, which could be parameterized by g_ϕ, transforming a input graph representation $\mathcal{G}$ into $g_\phi(\mathcal{G})$. Hopefully, it is learnt to transform the missing graph embedding back to its original complete status.

General Prompting. Given the missing modality $M_\varnothing \in \{G, I, C, T\}$, some specific subjects have all the instances of that modality missed, such that the representation of them at $M_\varnothing$ is ruined, as $\mathcal{V}_{M_\varnothing} = \{\varnothing\}$. There are also some patient subjects not affected by the missing, still with the complete data and the representation $\mathcal{V}_{M_\varnothing}$ maintained. We sample hallucination nodes $v^P \in \mathcal{V}^P$, where $\mathcal{V}^P \in \mathbb{R}^{N_P \times d}$, as a basic prompt scheme for graph completion. We extract the representation prior of the missing modality $M_\varnothing$ by collecting the modality-specific feature from all subjects, except for the subjects with the incomplete data at the missing modality, $\mathcal{V}_{M_\varnothing}$ The subjects with the incomplete data at

---

[2] For simplicity, index s and t is omitted when cross-modal relations are not involved.

the missing modality, i.e., $\mathcal{V}_{M_\varnothing} = \{\varnothing\}$ Then we initialize the features of the N_P prompt entities by a Gaussian sampling from the extract modality prior, motivated by the intuition that the same modality among different subjects share a basically similar distribution. After initialized, the set of prompt nodes $\mathcal{V}^P$, could be optimized effectively along with the model training.

Entity-Dependent Prompting. The introduced prompted entities are agnostic to the context, bringing the risk of yielding sub-optimal results. To encode the entity-dependent contextual information, we further introduce a prompt bank that contains a set of prompt components, $\mathcal{V}^{P_B} \in \mathbb{R}^{N_B \times d}$, where N_B is the number of prompt components. We take the these components as a series of base prompt, the weights $\boldsymbol{w}$ of which could be obtained in an entity-independent fashion. That is, we pass each input node v^P through a channel-downscaling linear layer to obtain a compact feature vector, followed by a softmax operation, thus yielding the weights $w \in \mathbb{R}^{N_B}$,

$$w = \texttt{Softmax}\left(W_{d \times N_B}(v^P)\right), \qquad (1)$$

where $W_{d \times N_P}$ denotes the linear projection layer that transforms the feature dimension from d to N_P. Then we use these weights to modulate the prompt components for each query entity v^P, and sum the general prompts and the entity-dependent prompts that perceives the graphic context,

$$v^P \leftarrow v^P + \sum_{i=1}^{N_B} w_i \cdot v_i^{P_B}, \qquad (2)$$

where $v_i^{P_B}$ denotes i-th component in the prompt bank. By doing so, the prompted graph embedding could be described with the formulation of graph prompt function $g_\phi(\cdot)$, by which the nodes $\mathcal{V}$ and edges $\mathcal{E}$ of a graph $\mathcal{G}$ to prompt would be transformed as,

$$\mathcal{V} = \{\mathcal{V}_{/M_\varnothing}, \mathcal{V}^P\}, \ \mathcal{E} = \{\mathcal{E}_{/M_\varnothing}, \underset{\forall \varphi(e) \in \mathcal{R}}{\texttt{EdgeUpdate}}(\mathcal{V}_{/M_\varnothing}, \mathcal{V}^P)\} \qquad (3)$$

where $\mathcal{V}_{/M_\varnothing}$ denotes the node embedding at all the modalities except for $M_\varnothing$ and $\mathcal{E}_{/M_\varnothing}$ denotes the edge space when removing all the nodes at the modality $M_\varnothing$. $\underset{\forall \varphi(e) \in \mathcal{R}}{\texttt{EdgeUpdate}}(\cdot, \cdot)$ defines the operation that updates the features of edge e between two sets of nodes if the relation can be retrieved in the attribute space of edge $\varphi(e) \in \mathcal{R}$. Effectively, the graph embedding of the missing modality $M_\varnothing$ are adapted through inserted with the prompt nodes as well as uncovered with some ruined relations.

3.4 Knowledge-Guided Hierarchical Aggregation

With the completed graph $g_\phi(\mathcal{G})$, the knowledge-guided hierarchical aggregation module effectively embeds the knowledge prior into a series of meta-paths, thereby we can search for the global cross-modal heterogeneous neighboring.

With the found neighbouring, the local multi-relation aggregation module is performed across various heterogeneous edges, and the overall hierarchical aggregation [34,84] module can be parameterized by a network function $\mathcal{M}(\cdot)$.

Global Meta-path Neighbouring. The aggregation of graph information highly depends on the established neighboring rules [65,79], and we design novel meta-paths as global information pathways, allowing for interaction of two heterogeneous entities. Given the entities $\mathcal{V}$ and meta-paths Φ in the heterogeneous graph, the neighbors derived from meta-paths for all the entities are uniquely identified [79], as $\mathcal{N}_\mathcal{V}^\Phi$. Hence, our insight is to embed the domain knowledge into the formulation of Φ by considering the semantic relations across the clinical modalities. Recall that the edge attribute space $\mathcal{R}$ is explicitly defined by the biological relations among modalities, we derive that, with the exception of "intra-modal" relations, i.e., the entities at the same modality, all heterogeneous nodes can only interact with the nodes whose attributes are semantically related to themselves in a single-hop propagation [45,79]. Following this principle, we adopt a random walking strategy [12] to search for the optimal meta-paths from all potential candidates. Specially, we randomly start from an entity of one modality, then iterate over all the heterogeneous nodes with valid semantic relations in the attribute space, i.e., $\varphi(e_c) \in \mathcal{R}$. Repeating the iterations, we show that an appropriate customization of meta-paths could be $\Phi = \{G \xrightarrow{\text{"express"}} I \xrightarrow{\text{"atomize"}} C,\ G \xrightarrow{\text{"express"}} I \xrightarrow{\text{"depict"}} T,\ C \xrightarrow{\text{"atomize"}} I \xrightarrow{\text{"express"}} G,\ T \xrightarrow{\text{"depict"}} I \xrightarrow{\text{"express"}} G\}$. Following [79], all meta-paths Φ are formulated with the maximum lengths within two hops. In practical usage, when querying the neighbourhood $\mathcal{N}_v^\Phi$ for an entity v, we first project it onto the entity attribute space $\mathcal{A}$ by $\tau(v)$, and then iterate over all meta-paths in Φ with a given number of hops H. After that, all reached entities attributes are collected, and the entities of those collected attributes are taken as the neighbours along the meta-paths,

$$\mathcal{N}_v^\Phi = \left\{ v' \mid \tau(v') \in \{ \underset{i \in [1,|\Phi|]}{\|} \underset{H}{\text{Reach}}(\Phi_i) \} \right\}, \quad (4)$$

where $\underset{H}{\text{Reach}}$ denotes the operation that collects all the reached attributes by walking along a meta-path Φ_i with H hops. $\|_{i \in [1,|\Phi|]}$ is the concatenation operator for all the resulted elements.

Local Multi-relation Aggregation. With the derived entity-wise $\mathcal{N}_v^\Phi$ neighbours, we perform the information propagation for each target node $v_t \in \mathcal{V}_t$ as a local feature aggregation from all its neighbored source nodes $\mathcal{V}_s$. To model node-wise interaction [3,18], we introduce a Multi-Head Attention (MHA) mechanism that models the target node features as **Query** and source node features as **Key** and **Value**. We embed the target node v_t and source node v_s by different linear projection layers $W_{\tau(v_s)}^j$ and $W_{\tau(v_s)}^j$, with each attention head j,

$$v_s^{K,j} = W_{\tau(v_s)}^j \cdot v_s^{(l-1)}, \quad v_t^{Q,j} = W_{\tau(v_t)}^j \cdot v_t^{(l-1)},$$
$$v_s^{V,j} = W_{\tau(v_s)}^j \cdot v_s^{(l-1)}, \quad (5)$$

where $v_*^{(l-1)}$ represents the input node feature for node $v \in \mathcal{V}$ from the $(l-1)$-th layer. The projection layers are capable of mapping node features from different node attributes to an embedding space that is invariant across node attributes. The features of edges from the $(l-1)$-th layer $e_{v_s \to v_t}^{(l-1)}$ are also projected by a linear projection layer $W_{\varphi(e)}$, serving as a **Key** feature, $e_{v_s \to v_t}^{K} = W_{\varphi(e)} \cdot e_{v_s \to v_t}^{(l-1)}$. Once node embeddings are projected, we calculate the dot-product between the query and key vectors. Besides, we multiply the linearly transformed edge embedding with the similarity score to integrate the edge features into graph $\mathcal{G}$,

$$\text{SHA}(e, j) = \left(v_s^{K,j} \cdot e_{v_s \to v_t}^{K} \cdot v_t^{Q,j} \right) / \sqrt{d}, \tag{6}$$

where d denotes the dimension of node embeddings, $\text{SHA}(e, j)$ represents the attention score of edge e by the Single-Head Attention at j-head. We concatenate the scores obtained from each head and apply a softmax to them,

$$\text{SRA}(e) = \underset{\forall v_s \in \mathcal{N}_{v_t}^{\Phi}}{\text{Softmax}} \left(\underset{j \in [1,h]}{||} \text{SHA}(e, j) \right), \tag{7}$$

where $\text{SRA}(e)$ represents Single-Relation Attention, providing the final attention score of the edges aggregating all the heads $j \in [1, h]$. $\mathcal{N}_{v_t}^{\Phi}$ is the set of the neighbours to the target node v_t. Then, we perform target-specific aggregation to update the feature of each target node by averaging its neighboring node features. For each target node v_t, we conduct a softmax operation on all the attention vectors from its neighboring nodes and then aggregate the information of all neighboring source nodes of v_t together. The updated node features $v_t^{(l)}$ for $\mathcal{G}^{(l)}$ can be represented as,

$$v_t^{(l)} = \bigoplus_{\forall v_s \in \mathcal{N}_{v_t}^{\Phi}} \left(\underset{j \in [1,h]}{||} \left(v_s^{V,j} \cdot \text{SRA}(e) \right) \right), \tag{8}$$

where $\oplus$ is an aggregation operator, e.g., mean aggregation. The updated graph $\mathcal{G}^{(l)}$ is returned as the output of the l-th layer. Such operation is scalable by using L layers of aggregation. We further introduce modality-specific pooling for all nodes within the modality to obtain the prototype features for all modalities. Then the graph-level feature can be determined by a mean readout layer [3,79].

3.5 Overall Optimization

The aggregated multimodal representation could be obtained by $\mathcal{M} \circ g_\phi(\mathcal{G})$ from the heterogeneous graph $\mathcal{G}(\cdot)$, modality-wise graph prompting $g_\phi(\cdot)$ and knowledge-guided hierarchical aggregation $\mathcal{M}(\cdot)$. While the diverse diagnostic tasks including the glioma grading (a classification task) and the survival prediction (a integration prediction task) that may differ in the formulation, it is shown that they could be transformed to a uniform supervised learning fashion after some manipulations of task head [5]. Formally, the task-specific task head $\mathcal{H}^\mathcal{T}$ for the task $\mathcal{T}$ is introduced, with the task label denoted by $y^\mathcal{T}$,

$$\min_{\mathcal{M},\mathcal{H},\phi} \mathbb{E}_\mathcal{G} \; \mathcal{L}\left(\mathcal{H}^\mathcal{T} \circ \mathcal{M} \circ \phi(\mathcal{G}), y^\mathcal{T}\right). \tag{9}$$

where $\mathcal{L}$ denotes the loss function, which could be implemented by a NLL (negative log-likelihood) loss.

4 Experiments

4.1 Datasets and Settings

Datasets. We evaluate our method using data from The Cancer Genome Atlas (TCGA) [1], a public database that includes genomic and clinical data from thousands of cancer patients. We select the datasets of Glioblastoma & Lower Grade Glioma (GBMLGG) and Kidney Renal Clear Cell Carcinoma (KIRC). For TCGA-GBMLGG, following [5], we use ROIs from diagnostic slides and apply sparse stain normalization [5] to match all images to a standard H&E histology image, creating a total of 1505 images for 769 patients, with WHO grading labels from G2 to G4. We curate 80 CNA and 240 RNA-Seq genomic features for each patient. Note that there are 40% of the patients with inherently actual missing RNA-Seq data. For the KIRC dataset, we use manually extracted 512×512 ROIs from diagnostic whole slide images for 417 patients in CCRCC, yielding 1251 images total that are similarly normalized with stain normalization. We pair these images with 117 CNV and 240 RNA-Seq genomic features. There are grading labels by Fuhrman Grading from G1 to G4.

Evaluation. For each cancer dataset, we perform 5-fold cross-validation and report the average test performance. Different metrics are leveraged for specific evaluation tasks, pathological glioma grading with Area Under the Curve (AUC) and Accuracy (ACC), and survival outcome prediction with concordance index (C-Index). Due to inherent missing of partly genetic modality in GBMLGG, we deploy the framework directly without modifying the data. While for KIRC with the complete modality data, we simulate the same situations in GBMLGG, randomly dropping RNA-Seq data with 40% subjects over the whole dataset. To ensure the consistent missing cases at training and test, we constrain the proportion of incomplete subjects in the training and test splits to be equal when producing the five-fold validation. In order to explore the missing issues under more modalities and missing ratios, we also perform experiments under simulated missing settings in Sect. 4.3.

Implementation. The framework is optimized by the Adam optimizer, with a learning rate of 1×10^{-3} and a weight decay of 1×10^{-5} for the graph aggregation $\mathcal{M}$ and task head $\mathcal{H}$, over 150 epochs with early stopping. We adopt a smaller learning rate of 2×10^{-4} specially for optimizing prompt nodes $\mathcal{V}_P$ and the prompt bank components $\mathcal{V}^{P_B}$ in graph prompt function $g_\phi(\cdot)$. All the multimodal instances of a patient subject are jointly fed to the networks to get the final

Data augmentations are performed on the training graphs, which involve randomly dropping edges and nodes, and adding Gaussian noise to the node and edge features [29,70]. The dropout ratio of each dropout layer is selected as 0.2. Regarding the hyper-parameters, we have the number of prompt nodes and prompt bank components in Eq. 2 with $N_P = 5$ and $N_B = 5$, and the dimension of input graph representation as $d = 512$.

Table 1. Performance Comparison. We report results on four TCGA benchmarks, using various modality combinations of Genomics, Images, Cell graphs and Texts.

Methods	Modality				Glioma Grading (AUC/ACC)				Survival Pred. (C-Idx)	
	G	I	C	T	GBM	LGG	KI	RC	GBMLGG	KIRC
SNN [24]	✓	✗	✗	✗	0.8527	0.6583	0.8100	0.7790	0.7974	0.6639
SNNTrans [63]	✓	✗	✗	✗	0.8678	0.6725	0.8084	0.7755	0.7970	0.6671
AttMIL [20]	✗	✓	✗	✗	0.9063	0.7533	0.8252	0.7803	0.7908	0.6850
TransMIL [63]	✗	✓	✗	✗	0.9149	0.7683	0.8295	0.7899	0.8017	0.6876
PatchGCN [4]	✗	✓	✗	✗	0.8802	0.7429	0.8288	0.7896	0.7806	0.6795
GTNMIL [85]	✗	✓	✗	✗	0.9225	0.7966	0.8323	0.7980	0.8162	0.6953
HEAT [3]	✗	✓	✗	✗	0.9289	0.8057	0.8300	0.7961	0.8223	0.7059
Pathomic [5]	✓	✓	✗	✗	0.9172	0.7618	0.8295	0.7899	0.8101	0.7152
Porpoise [8]	✓	✓	✗	✗	0.9199	0.7789	0.8278	0.7800	0.8179	0.7179
MCAT [6]	✓	✓	✗	✗	0.9288	0.7929	0.8352	0.7957	0.8274	0.7235
TransFusion [86]	✓	✓	✗	✗	0.9209	0.7815	0.8299	0.7910	0.8251	0.7230
GTP-4o (Ours)	✓	✓	✗	✗	0.9256	0.8036	0.8349	0.7985	0.8296	0.7273
Pathomic [5]	✓	✓	✓	✗	0.9195	0.7674	0.8280	0.7889	0.8199	0.7211
TransFusion [86]	✓	✓	✓	✗	0.9225	0.7952	0.8318	0.7973	0.8283	0.7260
GTP-4o (Ours)	✓	✓	✓	✗	0.9336	0.8068	0.8331	0.8021	0.8329	0.7315
TransFusion [86]	✓	✓	✓	✓	0.9245	0.7986	0.8325	0.7990	0.8296	0.7289
GTP-4o (Ours)	✓	✓	✓	✓	0.9389	0.8126	0.8416	0.8068	0.8351	0.7336

4.2 Comparison with State-of-the-Arts

We compare the proposed method against several SOTA methods in Table 1. For fair comparison, we apply identical settings for all experiments, and use the official code of compared works to deploy on our tasks when necessary.

Unimodal Models. Existing methods to analyze genomic data and pathological images are introduced. For genomic data, we employ SNN [24] for survival outcome prediction in the TCGA [5,6], and SNNTrans [24,63] that incorporates SNN as the feature extractor and TransMIL [63] for a global aggregation. For pathological images, we report the results of the SOTA MIL methods including

the transformer-based models: AttnMIL [20], TransMIL [63], and the graph-based models PatchGCN [4], GTNMIL [85], HEAT [3]. It appears that using multimodal data consistently improves the performance under various metrics.
Multimodal Models. We compare the SOTA multimodal methods including Pathomic [5], Porpoise [8] and MCAT [6], which only focus on extracting complementary multimodal information from the genomics and pathological images. It appears that there is a gain in using multimodal complementary information for various diagnostic tasks. Furthermore, as extending to more modalities is still unexplored, we compare our GTP-4o with other baselines by extend existing work Pathomic [5] with the cell graph modality, and also compare with a simple baseline TransFusion [86] which concentrates the intra-modal representations learned by uni-modal models [63]. From the table, the proposed GTP-4o exhibits the obvious improvement under most of different biomedical modalities.

4.3 Further Results

Table 2. Ablation of GTP-4o Variants. Results on TCGA-GBMLGG benchmarks over tasks of Glioma Grading (GG.) and Survival Prediction (SP.) are reported.

Components	Variants	GG. (AUC)	GG. (AUC)	SP. (C-Idx)
Graph Representation	No *Heterogeneous Embedding*	0.9232	0.8030	0.8168
	No *Heterogeneous Relation*	0.9259	0.8048	0.8201
Modality Completion	No *Completion (Zero-init Missing)*	0.9087	0.7875	0.7946
	No *Completion (Drop Missing)*	0.9288	0.8061	0.8233
	No *Prompt Bank*	0.9275	0.8081	0.8280
Hierarchical Aggregation	No *Aggregation (Plain Mean)*	0.9329	0.8067	0.8311
	No *Knowledge Guidance*	0.9350	0.8071	0.8342
Full Model	The Proposed GTP-4o	0.9389	0.8126	0.8416

Ablation Studies. The effect of removing each component of GTP-4o is presented in Table 2. *No Heterogeneous Embedding* removes all the heterogeneous properties in the embedding such that it degrades to a simple homogeneous graph. *No Heterogeneous Relation* removes the heterogeneous properties of edges while maintaining the diverse attributes among the node features. *No Completion (Zero-init Missing)* handles the missing modality without using the proposed graph prompt completion while sets the features of the missing modality to zero values. *No Completion (Drop Missing)* directly drops the modality data in all patients if it occurs missing for some patients. *No Aggregation (Plain Mean)* removes the knowledge-guided aggregation while performs the plain mean aggregation among the k-NN heterogeneous neighbours ($k = 15$). *No Knowledge Guidance* removes the knowledge guidance for aggregation while uses the random meta-paths. Our ablation study results confirm the pivotal roles of our designs in the overall performance and effectiveness of the model.

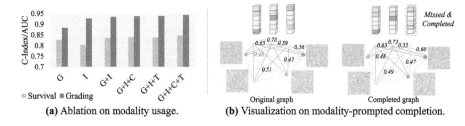

Fig. 3. (a) **Analysis of Modality Usage.** We provide the results of GTP-4o by using either Genes, Images, Cell graphs, Texts, or their combinations, on benchmarks of survival prediction (C-Index) and glioma grading (AUC). (b) **Analysis of Modality-prompted Completion.** We compare the relation pattern (similarity) in the original graph and the graph that is first removed specific instances then completed.

Impact of Modality Usage. Figure 3(a) shows the impact of using various combinations of modalities by GTP-4o. For the case of single modality, it is observed that each modality has its advantage for a specific task, as the relative performance of using only genes and only images is opposite for the tasks of glioma grading and survival analysis. We can also see that when more modalities like cell graphs and text descriptions are introduced, the performance of the model is improved on both two tasks. This suggests that GTP-4o is not only capable of generalizing to the various combinations of medical modality usage, but also able to deliver superior performance in terms of AUC (for glioma grading) and C-Index (for survival prediction).

Impact of Modality Missing and Completion. To validate graph prompting's effectiveness, we compared graphs built from original full instances and completed graphs with arbitrary missing instances for a non-missing case (TCGA-02-0006) in Fig. 3(b). The completed graph shows similar relation patterns to the real one, suggesting biological validity of our proposed completion method. We further explored various missing settings by simulating missing data in image and non-RNA genomics modalities on TCGA-GBMLGG benchmarks. Figure 4 illustrates GTP-4o's performance compared to the baseline without graph-prompted completion under different missing ratios. Results confirm the proposed completion method's effectiveness across various modality missing scenarios.

Limitation and Future Works. The current deployment is limited by the fact that no real-world clinical text reports are available for the datasets, thus we have to generate synthetic text descriptions by LLMs, probably bringing some data noise. Another limitation is that some additional modalities such as tabular data, are not considered in this paper, which could serve as future works.

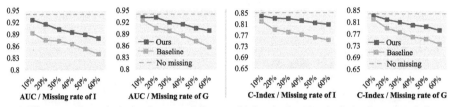

Fig. 4. Analysis of Modality Missing. We study the results of **(a)** glioma grading and **(b)** survival prediction with the various missing ratios of Images and Genes. We compare the full framework of Ours and the version without our completion (baseline).

5 Conclusion

Increasing biomedical multimodal data provides not only opportunities for accurate and comprehensive diagnosis but also challenges for learning against the modality heterogeneity as well as the missingness issues. This study presents GTP-4o, which signifies a pioneering exploration into learning unified representations from various clinical modalities via the graph theory, exhibiting the robustness to heterogeneous modalities. Unlike prior methods, GTP-4o explores capturing explicit relations via a heterogeneous graph embedding. A novel graph prompting is proposed to complete deficient graph representations of missing modalities, and a hierarchical multimodal aggregation employs a global meta-path prior to guide the local aggregation across various heterogeneous relations. Extensive experiments demonstrate the efficacy of GTP-4o on disease diagnosis.

Acknowledgments. This work was supported by Research Grants Council (RGC) General Research Fund 14204321.

References

1. https://gdc.cancer.gov
2. Ali, S., Li, J., Pei, Y., Khurram, R., Rehman, K.U., Mahmood, T.: A comprehensive survey on brain tumor diagnosis using deep learning and emerging hybrid techniques with multi-modal mr image. Archiv. Comput. Methods Eng. **29**(7), 4871–4896 (2022)
3. Chan, T.H., Cendra, F.J., Ma, L., Yin, G., Yu, L.: Histopathology whole slide image analysis with heterogeneous graph representation learning. In: CVPR, pp. 15661–15670 (2023)
4. Chen, R.J., et al.: Whole slide images are 2D point clouds: context-aware survival prediction using patch-based graph convolutional networks. In: de Bruijne, M., et al. (eds.) MICCAI 2021. LNCS, vol. 12908, pp. 339–349. Springer, Cham (2021). https://doi.org/10.1007/978-3-030-87237-3_33
5. Chen, R.J., et al.: Pathomic fusion: an integrated framework for fusing histopathology and genomic features for cancer diagnosis and prognosis. IEEE Trans. Med. Imaging **41**(4), 757–770 (2022). https://doi.org/10.1109/TMI.2020.3021387

6. Chen, R.J., et al.: Multimodal co-attention transformer for survival prediction in gigapixel whole slide images. In: Proceedings of the IEEE/CVF International Conference on Computer Vision, pp. 4015–4025 (2021)
7. Chen, R.J., et al.: Multimodal co-attention transformer for survival prediction in gigapixel whole slide images. In: Proceedings of the IEEE/CVF International Conference on Computer Vision, pp. 4015–4025 (2021)
8. Chen, R.J., et al.: Pan-cancer integrative histology-genomic analysis via multimodal deep learning. Cancer Cell **40**(8), 865–878 (2022)
9. Chen, Y., Liu, C., Huang, W., Cheng, S., Arcucci, R., Xiong, Z.: Generative text-guided 3d vision-language pretraining for unified medical image segmentation. arXiv preprint arXiv:2306.04811 (2023)
10. Chen, Y., Liu, C., Liu, X., Arcucci, R., Xiong, Z.: Bimcv-r: a landmark dataset for 3d CT text-image retrieval. arXiv preprint arXiv:2403.15992 (2024)
11. Chen, Z., Li, W., Xing, X., Yuan, Y.: Medical federated learning with joint graph purification for noisy label learning. MIA (2023)
12. Codling, E.A., Plank, M.J., Benhamou, S.: Random walk models in biology. J. R. Soc. Interface **5**(25), 813–834 (2008)
13. Ding, Z., Dong, Q., Xu, H., Li, C., Ding, X., Huang, Y.: Unsupervised anomaly segmentation for brain lesions using dual semantic-manifold reconstruction. In: Tanveer, M., Agarwal, S., Ozawa, S., Ekbal, A., Jatowt, A. (eds.) ICONIP 2022, Part III, pp. 133–144. Springer, Cham (2023). https://doi.org/10.1007/978-3-031-30111-7_12
14. Doyle, S., Hwang, M., Shah, K., Madabhushi, A., Feldman, M., Tomaszewski, J.: Automated grading of prostate cancer using architectural and textural image features. In: 2007 4th IEEE International Symposium on Biomedical Imaging: From Nano to Macro, pp. 1284–1287. IEEE (2007)
15. Gao, P., et al.: Dynamic fusion with intra-and inter-modality attention flow for visual question answering. In: CVPR, pp. 6639–6648 (2019)
16. He, Z., Li, W., Zhang, T., Yuan, Y.: H 2 gm: a hierarchical hypergraph matching framework for brain landmark alignment. In: MICCAI, pp. 548–558 (2023)
17. Hou, W., et al.: H2-mil: exploring hierarchical representation with heterogeneous multiple instance learning for whole slide image analysis. Proc. AAAI Conf. Artif. Intell. **36**, 933–941 (2022)
18. Hu, Z., Dong, Y., Wang, K., Sun, Y.: Heterogeneous graph transformer. In: Proceedings of the Web Conference 2020, pp. 2704–2710 (2020)
19. Huang, S.C., Pareek, A., Seyyedi, S., Banerjee, I., Lungren, M.P.: Fusion of medical imaging and electronic health records using deep learning: a systematic review and implementation guidelines. NPJ Digit. Med. **3**(1), 136 (2020)
20. Ilse, M., Tomczak, J., Welling, M.: Attention-based deep multiple instance learning. In: International Conference on Machine Learning, pp. 2127–2136. PMLR (2018)
21. Joo, S., et al.: Multimodal deep learning models for the prediction of pathologic response to neoadjuvant chemotherapy in breast cancer. Sci. Rep. **11**(1), 18800 (2021)
22. Kim, S., Lee, N., Lee, J., Hyun, D., Park, C.: Heterogeneous graph learning for multi-modal medical data analysis. Proc. AAAI Conf. Artif. Intell. **37**, 5141–5150 (2023)
23. Kipf, T.N., Welling, M.: Semi-supervised classification with graph convolutional networks. In: ICLR (2017)
24. Klambauer, G., Unterthiner, T., Mayr, A., Hochreiter, S.: Self-normalizing neural networks. Adv. Neural Inf. Process. Syst. **30** (2017)

25. Kumar, A., Fulham, M., Feng, D., Kim, J.: Co-learning feature fusion maps from PET-CT images of lung cancer. IEEE Trans. Med. Imaging **39**(1), 204–217 (2019)
26. Lee, Y.L., Tsai, Y.H., Chiu, W.C., Lee, C.Y.: Multimodal prompting with missing modalities for visual recognition. In: CVPR, pp. 14943–14952 (2023)
27. Li, C., Feng, B.Y., Fan, Z., Pan, P., Wang, Z.: Steganerf: embedding invisible information within neural radiance fields. In: Proceedings of the IEEE/CVF International Conference on Computer Vision, pp. 441–453 (2023)
28. Li, C., et al.: Endosparse: real-time sparse view synthesis of endoscopic scenes using gaussian splatting. arXiv preprint arXiv:2407.01029 (2024)
29. Li, C., et al.: Knowledge condensation distillation. In: European Conference on Computer Vision, pp. 19–35. Springer, Cham (2022)
30. Li, C., et al.: Domain generalization on medical imaging classification using episodic training with task augmentation. Comput. Biol. Med. **141**, 105144 (2022)
31. Li, C., et al.: Gaussianstego: a generalizable stenography pipeline for generative 3d gaussians splatting. arXiv preprint arXiv:2407.01301 (2024)
32. Li, C., et al.: Endora: video generation models as endoscopy simulators. arXiv preprint arXiv:2403.11050 (2024)
33. Li, C., Liu, X., Li, W., Wang, C., Liu, H., Yuan, Y.: U-kan makes strong backbone for medical image segmentation and generation. arXiv preprint arXiv:2406.02918 (2024)
34. Li, C., et al.: Hierarchical deep network with uncertainty-aware semi-supervised learning for vessel segmentation. Neural Comput. Appl. 1–14 (2022)
35. Li, C., Zhang, Y., Li, J., Huang, Y., Ding, X.: Unsupervised anomaly segmentation using image-semantic cycle translation. arXiv preprint arXiv:2103.09094 (2021)
36. Li, C., Zhang, Y., Liang, Z., Ma, W., Huang, Y., Ding, X.: Consistent posterior distributions under vessel-mixing: a regularization for cross-domain retinal artery/vein classification. In: 2021 IEEE International Conference on Image Processing (ICIP), pp. 61–65. IEEE (2021)
37. Li, J., Li, D., Savarese, S., Hoi, S.: Blip-2: bootstrapping language-image pretraining with frozen image encoders and large language models. arXiv preprint arXiv:2301.12597 (2023)
38. Li, W., Chen, Z., Li, B., Zhang, D., Yuan, Y.: HTD: heterogeneous task decoupling for two-stage object detection. TIP (2021)
39. Li, W., Guo, X., Yuan, Y.: Novel scenes & classes: towards adaptive open-set object detection. In: ICCV, pp. 15780–15790 (2023)
40. Li, W., Liu, J., Han, B., Yuan, Y.: Adjustment and alignment for unbiased open set domain adaptation. In: CVPR, pp. 24110–24119 (2023)
41. Li, W., Liu, X., Yao, X., Yuan, Y.: Scan: cross domain object detection with semantic conditioned adaptation. In: AAAI, pp. 1421–1428 (2022)
42. Li, W., Liu, X., Yuan, Y.: Sigma: semantic-complete graph matching for domain adaptive object detection. In: CVPR (2022)
43. Li, W., Liu, X., Yuan, Y.: Sigma++: improved semantic-complete graph matching for domain adaptive object detection. In: TPAMI (2023)
44. Li, X., Jia, M., Islam, M.T., Yu, L., Xing, L.: Self-supervised feature learning via exploiting multi-modal data for retinal disease diagnosis. IEEE Trans. Med. Imaging **39**(12), 4023–4033 (2020)
45. Liang, Z., et al.: Unsupervised large-scale social network alignment via cross network embedding. In: Proceedings of the 30th ACM International Conference on Information and Knowledge Management, pp. 1008–1017 (2021)
46. Liberzon, A., et al.: The molecular signatures database hallmark gene set collection. Cell Syst. **1**(6), 417–425 (2015)

47. Lipkova, J., et al.: Artificial intelligence for multimodal data integration in oncology. Cancer Cell **40**(10), 1095–1110 (2022)
48. Liu, D., Yang, X., Wu, X.: Tumor immune microenvironment characterization identifies prognosis and immunotherapy-related gene signatures in melanoma. Front. Immunol. **12**, 663495 (2021)
49. Liu, H., Liu, Y., Li, C., Li, W., Yuan, Y.: LGS: a light-weight 4d gaussian splatting for efficient surgical scene reconstruction. arXiv preprint arXiv:2406.16073 (2024)
50. Liu, X., et al.: Stereo vision meta-lens-assisted driving vision. ACS Photonics (2024)
51. Liu, X., Li, W., Yang, Q., Li, B., Yuan, Y.: Towards robust adaptive object detection under noisy annotations. In: CVPR, pp. 14207–14216 (2022)
52. Liu, X., Li, W., Yuan, Y.: Intervention and interaction federated abnormality detection with noisy clients. In: Wang, L., Dou, Q., Fletcher, P. T., Speidel, S., Li, S. (eds.) MICCAI 2022, Part VIII, pp. 309–319. Springer, Cham (2022). https://doi.org/10.1007/978-3-031-16452-1_30
53. Liu, X., Li, W., Yuan, Y.: Decoupled unbiased teacher for source-free domain adaptive medical object detection. IEEE Trans. Neural Netw. Learn. Syst. **35**(6), 7287–7298 (2024)
54. Liu, X., Peng, H., Zheng, N., Yang, Y., Hu, H., Yuan, Y.: Efficientvit: memory efficient vision transformer with cascaded group attention. In: Proceedings of the IEEE/CVF Conference on Computer Vision and Pattern Recognition, pp. 14420–14430 (2023)
55. Liu, X., Yuan, Y.: A source-free domain adaptive polyp detection framework with style diversification flow. IEEE Trans. Med. Imaging **41**(7), 1897–1908 (2022)
56. Liu, Y., Li, C., Yang, C., Yuan, Y.: Endogaussian: Gaussian splatting for deformable surgical scene reconstruction. arXiv preprint arXiv:2401.12561 (2024)
57. Lu, M.Y., Williamson, D.F., Chen, T.Y., Chen, R.J., Barbieri, M., Mahmood, F.: Data-efficient and weakly supervised computational pathology on whole-slide images. Nat. Biomed. Eng. **5**(6), 555–570 (2021)
58. Marusyk, A., Almendro, V., Polyak, K.: Intra-tumour heterogeneity: a looking glass for cancer? Nat. Rev. Cancer **12**(5), 323–334 (2012)
59. Peng, X., Wei, Y., Deng, A., Wang, D., Hu, D.: Balanced multimodal learning via on-the-fly gradient modulation. In: CVPR, pp. 8238–8247 (2022)
60. Radford, A., et al.: Learning transferable visual models from natural language supervision. In: International Conference on Machine Learning, pp. 8748–8763. PMLR (2021)
61. Ramachandram, D., Taylor, G.W.: Deep multimodal learning: a survey on recent advances and trends. IEEE Signal Process. Mag. **34**(6), 96–108 (2017)
62. Rasmy, L., Xiang, Y., Xie, Z., Tao, C., Zhi, D.: Med-bert: pretrained contextualized embeddings on large-scale structured electronic health records for disease prediction. NPJ Digit. Med. **4**(1), 86 (2021)
63. Shao, Z., et al.: Transmil: transformer based correlated multiple instance learning for whole slide image classification. Adv. Neural. Inf. Process. Syst. **34**, 2136–2147 (2021)
64. Sun, L., et al.: Few-shot medical image segmentation using a global correlation network with discriminative embedding. Comput. Biol. Med. **140**, 105067 (2022)
65. Sun, Y., Han, J., Yan, X., Yu, P.S., Wu, T.: Pathsim: meta path-based top-k similarity search in heterogeneous information networks. VLDB **4**(11), 992–1003 (2011)

66. Wang, Z., Li, R., Wang, M., Li, A.: Gpdbn: deep bilinear network integrating both genomic data and pathological images for breast cancer prognosis prediction. Bioinformatics **37**(18), 2963–2970 (2021)
67. Wang, Z., et al.: Online disease diagnosis with inductive heterogeneous graph convolutional networks. In: Proceedings of the Web Conference 2021, pp. 3349–3358 (2021)
68. Wuyang, L., Chen, Y., Jie, L., Xinyu, L., Xiaoqing, G., Yixuan, Y.: Joint polyp detection and segmentation with heterogeneous endoscopic data. In: 3rd International Workshop and Challenge on Computer Vision in Endoscopy (EndoCV 2021): Co-located with the 17th IEEE International Symposium on Biomedical Imaging (ISBI 2021), pp. 69–79. CEUR-WS Team (2021)
69. Xu, H., Li, C., Zhang, L., Ding, Z., Lu, T., Hu, H.: Immunotherapy efficacy prediction through a feature re-calibrated 2.5 d neural network. Comput. Methods Prog. Biomed. **249**, 108135 (2024)
70. Xu, H., Zhang, Y., Sun, L., Li, C., Huang, Y., Ding, X.: AFSC: adaptive Fourier space compression for anomaly detection. arXiv preprint arXiv:2204.07963 (2022)
71. Xu, R., Li, Y., Wang, C., Xu, S., Meng, W., Zhang, X.: Instance segmentation of biological images using graph convolutional network. Eng. Appl. Artif. Intell. **110**, 104739 (2022)
72. Xu, Y., Chen, H.: Multimodal optimal transport-based co-attention transformer with global structure consistency for survival prediction. In: Proceedings of the IEEE/CVF International Conference on Computer Vision (ICCV), pp. 21241–21251 (2023)
73. Xue, L., et al.: Ulip: learning a unified representation of language, images, and point clouds for 3d understanding. In: CVPR, pp. 1179–1189 (2023)
74. Xue, Z., Marculescu, R.: Dynamic multimodal fusion. In: CVPR, pp. 2574–2583 (2023)
75. Yang, Q., Guo, X., Chen, Z., Woo, P.Y., Yuan, Y.: D2-net: dual disentanglement network for brain tumor segmentation with missing modalities. IEEE Trans. Med. Imaging **41**(10), 2953–2964 (2022)
76. Yang, Q., Li, W., Li, B., Yuan, Y.: MRM: masked relation modeling for medical image pre-training with genetics. In: ICCV, pp. 21452–21462 (2023)
77. Yang, Q., Yuan, Y.: Learning dynamic convolutions for multi-modal 3D MRI brain tumor segmentation. In: Crimi, A., Bakas, S. (eds.) BrainLes 2020. LNCS, vol. 12659, pp. 441–451. Springer, Cham (2021). https://doi.org/10.1007/978-3-030-72087-2_39
78. Zeng, Y., et al.: Exploration of the immune cell infiltration-related gene signature in the prognosis of melanoma. Aging (Albany, NY) **13**(3), 3459 (2021)
79. Zhang, C., Song, D., Huang, C., Swami, A., Chawla, N.V.: Heterogeneous graph neural network. In: Proceedings of the 25th ACM SIGKDD International Conference on Knowledge Discovery and Data Mining, pp. 793–803 (2019)
80. Zhang, S., et al.: Large-scale domain-specific pretraining for biomedical vision-language processing. arXiv preprint arXiv:2303.00915 (2023)
81. Zhang, Y., et al.: Modality-aware mutual learning for multi-modal medical image segmentation. In: de Bruijne, M., et al. (eds.) MICCAI 2021. LNCS, vol. 12901, pp. 589–599. Springer, Cham (2021). https://doi.org/10.1007/978-3-030-87193-2_56
82. Zhang, Y., Fang, Q., Qian, S., Xu, C.: Multi-modal multi-relational feature aggregation network for medical knowledge representation learning. In: Proceedings of the 28th ACM International Conference on Multimedia, pp. 3956–3965 (2020)

83. Zhang, Y., Jiang, H., Miura, Y., Manning, C.D., Langlotz, C.P.: Contrastive learning of medical visual representations from paired images and text. In: Machine Learning for Healthcare Conference, pp. 2–25. PMLR (2022)
84. Zhang, Y., et al.: Generator versus segmentor: pseudo-healthy synthesis. In: de Bruijne, M., et al. (eds.) MICCAI 2021. LNCS, vol. 12906, pp. 150–160. Springer, Cham (2021). https://doi.org/10.1007/978-3-030-87231-1_15
85. Zheng, Y., et al.: A graph-transformer for whole slide image classification. IEEE Trans. Med. Imaging **41**(11), 3003–3015 (2022)
86. Zhou, F., Chen, H.: Cross-modal translation and alignment for survival analysis. In: Proceedings of the IEEE/CVF International Conference on Computer Vision, pp. 21485–21494 (2023)
87. Zhou, T., Ruan, S., Canu, S.: A review: deep learning for medical image segmentation using multi-modality fusion. Array **3**, 100004 (2019)
88. Zhu, D., Chen, J., Shen, X., Li, X., Elhoseiny, M.: Minigpt-4: enhancing vision-language understanding with advanced large language models. arXiv preprint arXiv:2304.10592 (2023)
89. Zhu, X., Yao, J., Zhu, F., Huang, J.: WSISA: making survival prediction from whole slide histopathological images. In: Proceedings of the IEEE Conference on Computer Vision and Pattern Recognition, pp. 7234–7242 (2017)

PromptCCD: Learning Gaussian Mixture Prompt Pool for Continual Category Discovery

Fernando Julio Cendra[1], Bingchen Zhao[2], and Kai Han[1]

[1] The University of Hong Kong, Pok Fu Lam, Hong Kong
kaihanx@hku.hk
[2] the University of Edinburgh, Edinburgh, UK

Abstract. We tackle the problem of Continual Category Discovery (CCD), which aims to automatically discover novel categories in a continuous stream of unlabeled data while mitigating the challenge of catastrophic forgetting—an open problem that persists even in conventional, fully supervised continual learning. To address this challenge, we propose PromptCCD, a simple yet effective framework that utilizes a Gaussian Mixture Model (GMM) as a prompting method for CCD. At the core of PromptCCD lies the Gaussian Mixture Prompting (GMP) module, which acts as a dynamic pool that updates over time to facilitate representation learning and prevent forgetting during category discovery. Moreover, GMP enables on-the-fly estimation of category numbers, allowing PromptCCD to discover categories in unlabeled data without prior knowledge of the category numbers. We extend the standard evaluation metric for Generalized Category Discovery (GCD) to CCD and benchmark state-of-the-art methods on diverse public datasets. PromptCCD significantly outperforms existing methods, demonstrating its effectiveness. Project page: https://visual-ai.github.io/promptccd.

Keywords: Continual Category Discovery · Prompt Learning

1 Introduction

Deep learning models have achieved impressive performance in numerous computer vision tasks. However, the majority of these tasks have traditionally been conducted in a closed-world setting, where the models handle known categories and predefined scenarios. The crux lies in developing systems that can effectively and efficiently operate in the real, open-world we inhabit.

Category discovery, initially studied as Novel Class Discovery (NCD) [17] and subsequently extended to Generalized Category Discovery (GCD) [46],

---

Supplementary Information The online version contains supplementary material available at https://doi.org/10.1007/978-3-031-73235-5_11.

has recently emerged as an important open-world research problem, attracting increasing attention and efforts. NCD tackles the challenge of automatically discovering unseen categories in unlabelled data by leveraging the labelled data from seen categories, bridging the gap between known and novel categories. Meanwhile, GCD extends this challenge by allowing the unknown data to come from both labelled and novel categories. However, existing efforts on NCD and GCD mainly consider only static datasets. Our world, however, is inherently dynamic, necessitating intelligent systems that are not only able to discover novel categories but also can retain past knowledge while accommodating new information. Therefore, it is desired to develop methods that can discover novel classes from the unlabelled images over time. This task, called Continual Category Discovery (CCD) [55], extends the challenging open-world category discovery problem in a continual learning scenario (see Fig. 1). There are two major challenges in CCD. The first challenge is *catastrophic forgetting*, a well-known issue in continual learning settings [9]. Traditional techniques for mitigating forgetting, such as rehearsal-based [38], distillation-based [31], architecture-based [30], and prompting-based methods [49,50], assume fully labelled data at each stage, which is incompatible with the CCD framework where the goal is to work with unlabelled data streams. The second challenge is *discovering novel visual concepts*. While GCD is a related task, it operates under the assumption of static sets containing both labelled and unlabelled data. However, this assumption does not hold for the CCD task, where all data in the continuous stream during the discovery stage are unlabelled.

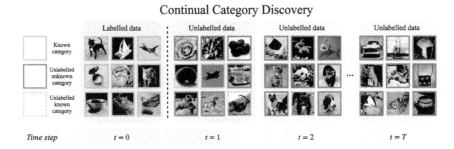

Fig. 1. Overview of the Continual Category Discovery task. In the initial stage, the model learns from labelled data, while in the subsequent stages, the model learns from a continuous data stream containing unlabelled instances from known and novel classes.

Recently, vision foundation models such as [6,37] have achieved remarkable progress and shown promise in various vision tasks, from image classification and object detection to more complex tasks like scene understanding. Given the capabilities of these foundation models, especially the self-supervised ones, we are interested in unleashing the potential of such models for dynamic environments by repurposing them to effectively tackle the challenging CCD problem. To this end, we propose PromptCCD framework. This framework empowers the model to leverage any prompt pool for solving CCD. Specifically, within our framework, we introduce a *plug-and-play* Gaussian Mixture Prompting (GMP) module. This module utilizes a Gaussian mixture prompt pool to model the data distribution at each discovery stage dynamically. By enhancing the feature representation with our adaptive queried Gaussian mixture prompts, our method excels at identifying novel visual categories across successive stages. Simultaneously, these prompts enable the model to seamlessly adapt to emerging data while preserving its performance on previously discovered categories, thus mitigating catastrophic forgetting. In addition to outperforming existing CCD solutions, our framework provides the unique advantage of enabling *on-the-fly* estimation of the category number in the unlabelled data, which is often assumed to be predetermined in prior works [55]. In this paper, we make the following contributions: (1) We propose a prompt learning framework for Continual Category Discovery (CCD), named *PromptCCD*. It can effectively repurpose the self-supervised vision foundation model for the challenging task of CCD, only introducing a small amount of extra learnable prompt parameters, and thus possessing strong scalability for practical use. (2) Within the proposed framework, we introduce *Gaussian Mixture Prompting* (GMP) module, a novel prompt learning technique that leverages Gaussian mixture components to enhance the representation learning and effectively address the issue of catastrophic forgetting when dealing with previously learned data. Notably, GMP's prompt serves a dual role, *i.e.*, as a task prompt, guiding the model during training, and as a class prototype, which is essential for CCD as label information is absent during class discovery. Moreover, GMP can be seamlessly integrated with other methods, enhancing their overall performance. An additional distinctive feature of GMP lies in its ability to estimate categories *on-the-fly*, making it well-suited for handling open-world tasks. (3) Finally, to evaluate the performance of the model for CCD, we extend the standard GCD metric to a new metric, called *continual ACC (cACC)*. Extensive experiments on both generic and fine-grained datasets demonstrate that our method significantly outperforms state-of-the-art CCD methods.

2 Related Work

Semi-supervised Learning aims at learning a classifier using both labelled and unlabelled data [7,36]. Most works assume that the unlabelled data contains instances from the *same* categories in the labelled data [36]. Pseudo-labeling [40], consistency regularization [2,28,44,45], and non-parametric classification [1] are among the popular methods. Some recent works do not assume the categories in the unlabelled and labelled set to be the same, such as [20,43,53], yet their focus is still on improving the performance of the categories from the labelled set.

Continual Learning aims to train models that can learn to perform on a sequence of tasks, with the restriction of the model can only see the data for the current task it is trained on [9]. Catastrophic forgetting [35] is a phenomenon that when the model is trained on a new task, it will quickly forget the knowledge on the task it has been trained on before, resulting in a catastrophic reduction of performance on the old tasks. There exists a rich literature on designing methods that enable the model to both learn to do the new task and maintain the knowledge of old tasks [3,4,14,30,31,38,50]. However, these works all assume that the incoming tasks have all labels provided. In contrast, CCD assumes that the new data is fully unlabelled and can have category overlap with previous tasks.

Novel/Generalized Category Discovery addresses the problem where there are novel categories in the unlabelled data and the goal is to automatically categorize the unlabelled samples, leveraging the labelled samples from the seen categories. Novel Category Discovery (NCD), formalized by DTC [17], assumes no overlap between the unlabelled and labelled data. Several successful NCD methods have emerged, showing promising performance through ranking statistics [15,16,21,56], data augmentation [59], and specialized objective function [13,22]. The problem is later extended to Generalized Category Discovery (GCD) [46] by considering that the unlabelled data may contain samples from both known and novel categories. [46] finetunes a pretrained model using both self-supervised [8] and supervised contrastive losses [24] and subsequently obtains the label assignment using a semi-supervised k-means algorithm. SimGCD [51] introduces a strong parametric baseline based on [46] for GCD, obtaining strong performance. Other GCD methods focus on fine-grained categories [11], automatic category estimation [18,58], and prompt learning [48,54].

Continual Category Discovery is a challenging but relatively under-explored problem. NCDwF [23] studies NCD under the continual learning setting, where the model first learns from labelled data and subsequently focuses on novel category discovery solely from unlabelled data. NCDwF shows that feature distillation and mutual information-based regularizers are effective for this problem. Concurrent to NCDwF, FRoST [41] introduces a replay-based method that stores feature prototypes from labelled data during the discovery phase. MSc-iNCD [32] leverages pretrained self-supervised learning models to address this problem. Grow & Merge [55] studies GCD under the continual learning setting, where the model has access to the labelled data in the initial stage and the unlabelled data in sequences in the subsequent stages. This method utilizes a growing phase to detect novel categories and a merging phase to distil knowledge from both novel and previously learned categories into a single model. Other methods addressing the GCD problem under the continual learning setting include PA-CGCD [25], which prevents forgetting using a proxy-anchor-based method, and MetaGCD [52], which balances class discovery and prevents forgetting using a meta-learning framework. Another method, IGCD [57], studies GCD under the continual learning setting in a slightly different way with an emphasis on the iNaturalist dataset for plant and animal species discovery. In each stage, IGCD takes a partially labelled set of images as input, rather than a set of fully unlabelled data like [25,52,55]. In this paper, we consider Continual Category Discovery (CCD) as the setting studied in [25,52,55]. In CCD, the model receives the labelled set at the initial stage and is tasked to discover categories from the unlabelled data in the subsequent stages.

3 Method

Problem Statement. The dataset in CCD contains both labelled data D^l and unlabelled data D^u. The labelled data $D^l = \{(x_i, y_i)\}_{i=1}^{N}$ contains tuples of the input $x_i \in \mathcal{X}$ and its corresponding labels $y_i \in \mathcal{Y}$ and it is only used in the initial stage for the model to learn useful features for the category discovery. In the following T discovery stages, at each stage, we receive a part of the unlabelled data $D_t^u \subset D^u$ that can be used to train the model. The unlabelled data D_t^u at each stage does not contain the labels, and it contains both known categories from previous stages and also novel categories. The goal of CCD is to train a model $\mathcal{H}_\theta : \mathcal{X} \to \mathcal{Z}$ parameterized by θ that first learns from labelled D^l and then in the following T discovery stages, learns from unlabelled data D_t^u such that $\mathcal{H}_\theta$ can be used to discover novel classes and assign class labels to all unlabelled instances utilizing representative feature without forgetting previous knowledge.

3.1 PromptCCD-B (*Baseline*): Learning Prompt Pool for CCD

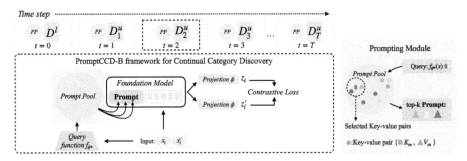

Fig. 2. Our baseline CCD framework adopts a prompt-based continual learning technique by utilizing a prompt pool module to adapt the vision foundation model for CCD.

Prompt learning [49,50] has been shown effective for supervised continual learning. With properly designed prompts, the necessity of extensive modification for the model when handling the growing data stream can be greatly reduced. However, these methods can not be directly applied to the CCD task, as they assume the data stream to be fully annotated, which is not the case for CCD.

To address this gap, we propose a novel baseline prompt learning framework for CCD denoted as PromptCCD-B, taking inspiration from [49,50] which learn a pool of prompts to adopt a large-scale pretrained model on ImageNet-21K [39] (in a supervised manner) for supervised continual learning. This baseline is designed to learn a shared pool of prompts that can effectively adapt the self-supervised foundation model to tackle the CCD challenge. Specifically, the model extracts a feature from a query example using a frozen pretrained model, and the feature will be used to retrieve the top-k most relevant prompts from the fixed-size M prompts in the shared pool. These prompts are then used to guide the representation learning process by prepending them with the input embeddings, optimised with contrastive learning at each learning stage.

The overall framework of our baseline is shown in Fig. 2. Given a model $\mathcal{H}_\theta : \{\phi, f_\theta\}$, where ϕ is a projection head, and $f_\theta = \{f_e, f_b\}$ is the transformer-based feature backbone which consists of input embedding layer f_e and self-attention blocks f_b. An input image $x \in \mathbb{R}^{H \times W \times 3}$ where H, W represent the height and width of the image, is first split into L tokens (patches) such that $x_q \in \mathbb{R}^{L \times (h \times w \times 3)}$ where h, w represent the height and width of the image patches. These patches are then projected by the input embedding layer $x_e = f_e(x_q) \in \mathbb{R}^{L \times z}$. A learnable prompt pool with M prompts is denoted as $\mathbb{V} = \{(K_m, V_m)\}_{m=1}^{M}$ where $K_m \in \mathbb{R}^z$ and $V_m \in \mathbb{R}^{L_{pp} \times z}$ are the key-value learnable pairs and L_{pp} is the prompt pool's token length. We define a query function f_{θ^*} (*non-trainable*) to map the input image x to the feature space. The query process on the prompt pool operates in a key-value fashion. For a given query $f_{\theta^*}(x)$, we find the top-k most similar keys in the prompt pool and retrieve the associated value by:

$$\mathcal{V}_{\text{top-k}} = \{V_i | K_i \in \mathcal{T}_{\mathbb{V}}^{k}(f_{\theta^*}(x))\}, \qquad (1)$$

where $\mathcal{T}_{\mathbb{V}}^{k}$ is a set of the top-k similar keys in $\mathbb{V}$. These retrieved prompts are then prepended to the patch embeddings to aid the learning process $x_{total} = [\mathcal{V}_{\text{top-k}}; x_e]$. The baseline method is trained with contrastive learning, let $\{x_i, x_i'\}$ be two randomly augmented views of the same image x_i. We obtain their representations as $z_i = \phi(f_\theta(x_i))$ and $z_i' = \phi(f_\theta(x_i'))$. To optimize the prompt pool, we pull the selected keys closer to the corresponding query features by making use of a cosine distance loss:

$$\mathcal{L}_i^{\cos} = \sum_{K_m \in \mathcal{T}_{\mathbb{V}}^{k}(f_{\theta^*}(x))} \gamma(f_{\theta^*}(x_i), K_m), \qquad (2)$$

where γ is the cosine distance function. Finally, when the training of stage t is finished, we transfer the current prompt pool $\mathbb{V}$ to the next stage.

Model Optimization. To optimize the model's representation, we follow the GCD literature to adopt the contrastive loss:

$$\mathcal{L}_i^{\text{rep}} = -\frac{1}{|\mathbb{N}(i)|} \sum_{p \in \mathbb{N}(i)} \log \frac{\exp(z_i \cdot z_p / \tau)}{\sum_n \mathbb{1}_{[n \neq i]} \exp(z_i \cdot z_n / \tau)}, \qquad (3)$$

where $\mathbb{1}_{[n \neq i]}$ is an indicator function such that it equals to 1 *iff* $n \neq i$, and τ is the temperature value. If x_i is a labelled image, $\mathbb{N}(i)$ corresponds to images with the same label y in the mini-batch B. While if x_i is an unlabelled image, $\mathbb{N}(i)$ contains only the index of the other augmented view x'_i of the image, i.e., $z_p = z'_i$. For the baseline model optimization, at the initial stage, i.e., $t = 0$, we have our initial labelled set D^l, and each image may have more than one positive sample; while at the subsequent stages, i.e., $t > 0$, we only have access to the unlabelled data D^u_t, and each image has only one positive sample, i.e., its another augmented view. The prompt parameters are also simultaneously optimized during the model optimization process.

Limitations of Baseline. Our baseline can achieve reasonably good performance, as can be seen in Sect. 4. It has several limitations. First, our baseline lacks an explicit mechanism to prevent forgetting. Without label information to guide it, the model may inadvertently bias its representation learning towards the current unlabelled data during fine-tuning, resulting in representation bias and forgetting. Another limitation arises from the fixed size of the prompt pool in our baseline framework. We rely on a predefined prompt pool size, which restricts the model's scalability. Consequently, the prompt pool's parameters may hinder the model's ability to discover a growing number of new categories. Lastly, our baseline framework lacks an efficient mechanism to estimate the number of categories dynamically, which is a crucial challenge for category discovery as per the GCD literature, but remains an open challenge under-explored in CCD.

3.2 PromptCCD: Learning Gaussian Mixture Prompt Pool for CCD

To address the aforementioned limitations in our baseline framework PromptCCD-B, here, we propose a novel Gaussian Mixture Prompting (GMP) module, which learns a parameter-efficient Gaussian Mixture Model (GMM) as the prompt pool, leading to a new framework, called PromptCCD (see Fig. 3).

Gaussian Mixtures Prompting (GMP) Module. The GMM is formulated as:

$$p(z) = \sum_{c=1}^{C} \pi_c \mathcal{N}(z|\mu_c, \Sigma_c) \quad \text{s.t.} \quad \sum_{c=1}^{C} \pi_c = 1, \quad (4)$$

where C is the number of Gaussian components, π_i is the learnable mixture weight, μ_i is the mean of each components, and Σ_i is the covariance of each component. Given the feature $\hat{z}_i = f_\theta(x_i)$ corresponding to the [CLS] token in the backbone, we calculate the log probability density value of each of the mixture components with the queried feature $\hat{z}_i$ and obtain a set of log-likelihood values for different GMM components. Finally, we find the top-k components in GMM with the highest log-likelihood values and retrieve the associated components' means:

$$\mu_{\text{top-k}} = \{\mu_c | c \in \mathcal{T}^k_{\text{GMM}}(p(\hat{z}_i))\}, \quad (5)$$

where $\mathcal{T}^k_{\text{GMM}}$ is a set of the top-k component(s) c. Similar to our baseline framework, PromptCCD-B, a set of embeddings $x_{total} = [\mu_{\text{top-k}}; x_e]$ is formed by prepending the selected prompts with the patch embeddings. We then

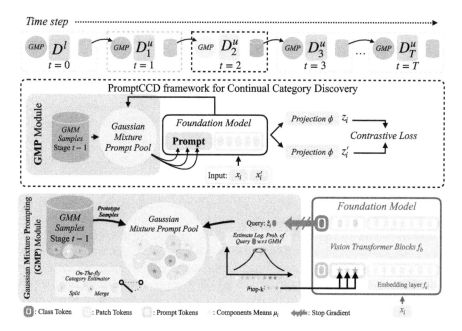

Fig. 3. Overview of our proposed PromptCCD framework and Gaussian Mixture Prompting (GMP) module. PromptCCD continually discovers new categories while retaining previously discovered ones by learning a dynamic GMP pool to adapt the vision foundation model for CCD. Specifically, we address CCD by making use of GMP modules to estimate the probability of input $\hat{z}_i$ by calculating the log-likelihood and use the top-k mean of components μ_i as prompts to guide the foundation model. Lastly, to retain previously learned prompts, we generate prototype samples from the fitted GMM at time step $t-1$ and fit the current GMM with these samples at time step t.

apply the same contrastive learning objective as in Sect. 3.1 to optimize our PromptCCD framework. The Gaussian mixture prompt pool serves as the core component that supports continuous category discovery across stages. After the training at stage t is done, we use the fitted GMM to sample a set of samples $\mathcal{Z}_t^s$ with S samples for each component c in the GMM. These samples are used to prevent forgetting previously learned knowledge, we achieve this by using these samples $\mathcal{Z}_t^s$ to fit the GMM of the next stage $t+1$. A pseudocode of the training procedure is provided in Sect. S1, supplementary material.

Our GMP possesses several unique strengths over the existing prompting techniques for supervised continual learning [49,50]. First, GMP's prompt serves a dual role, namely (1) as a task prompt to instruct the model and (2) as class prototypes to act as parametric replay sample distribution for discovered classes. The second role, which is unique and important for CCD/GCD, not only allows the model to draw unlimited replay samples to facilitate the representation tuning and class discovery in the next time step but also allows the model to transfer knowledge of previously discovered and novel categories and incorporate this information when making the decision to discover a novel category. Second, our GMP module enables easy adjustment of parameters and efficient dynamic expansion across stages. This allows our model to enjoy great scalability which is especially important when handling a growing number of categories. Finally, the GMM-based design of our GMP module allows us to equip it with an automatic split-and-merge mechanism, allowing our model to estimate the unknown number of categories in the unlabelled data stream.

3.3 PromptCCD-U: Unknown Number of Classes in Unlabelled Data

When the class numbers in the unlabelled data are unknown, one way to approach this problem is to estimate it offline using the non-parametric clustering method introduced in [46] at each time step. In CCD, considering the continual learning nature of the problem, it would be more plausible to estimate the class numbers on-the-fly without introducing extra models or an offline process. Inspired by GPC [58], which introduces a GMM-based category number estimation method for GCD, by automatically splitting and merging clusters during learning through assessing the cluster's compactness and separability using a Markov chain Monte Carlo (MCMC) algorithm. We incorporate this key idea into our CCD framework, making use of our learned prompt pool in our GMP module, further enabling the capability of our framework for automatic category number estimation. We denote this extended variant of our framework as PromptCCD-U.

Specifically, consider stage $t = 1$ of CCD. We first extract the features for all the unlabelled samples D_t^u and also use our GMM in our GMP fitted in the previous stage $t = 0$ to generate a set of pseudo features (as a replay for previously learned classes from the labelled data D^l). Let the combined features be $\mathcal{Z}$. We then fit them into the GMM in our GMP. As the class number in D_t^u is unknown, we start the fitting by setting an initial class number of the known class number in D^l and incorporate a *split-and-merge* mechanism as in [58] to allow for the dynamic adjustment of the GMM. Particularly, for each of the Gaussian components of the GMM and we further decompose it into two sub-components, i.e., $\mu_{c,1}, \mu_{c,2}$ and $\Sigma_{c,1}, \Sigma_{c,2}$. We then calculate the Hastings ratio which measures the compactness and separability of the clusters during the fitting iteration. The Hastings ratio for splitting a cluster is defined as:

$$H_s = \frac{\Gamma(N_{c,1})h(\mathcal{Z}_{c,1})\Gamma(N_{c,2})h(\mathcal{Z}_{c,2})}{\Gamma(N_c)h(\mathcal{Z}_c)}, \tag{6}$$

where Γ is the factorial function, h is the marginal likelihood function of the observed data $\mathcal{Z}$, $\mathcal{Z}_{c,1}$ denotes the data points assigned to the subcluster $\{c, 1\}$, and $N_{c,1}$ is the number of data points in the subcluster $\{c, 1\}$. Note that H_s is in the range of $(0, +\infty)$, thus we will use $p_s = min(1, H_s)$ as a valid probability for performing the splitting operation. When the fitting of the GMM is converged, the number of the resulting GMM components is then the class number of all classes seen so far. The number of new classes in D_t^u can be obtained by simply subtracting the previously learned class number.

4 Experiments

In this section, we describe our experimental setups in Sect. 4.1. Next, we present our main experimental results in Sect. 4.2. Finally, in Sect. 4.3 we analyze the effectiveness of our model's components and design choices.

4.1 Experimental Setups

Datasets. We conduct our experiments on various benchmark datasets, namely CIFAR100 (C100) [27], ImageNet-100 (IN-100) [42], TinyImageNet (Tiny) [29], Caltech-UCSD Birds-200-2011 (CUB) [47], FGVC-Aircraft [34], Stanford-Cars (SCars) [26], and Caltech-101 (C-101) [12]. Statistics of the benchmark datasets are shown in Table 1. CCD task consists of several stages. We set the number of stages to 4 with data splits presented in Table 2 following [55].

Table 1. Statistics of the CCD benchmark datasets following the splits in Table 2.

Stages	C100 [27]		IN-100 [42]		Tiny [29]		C-101 [12]		Aircraft [34]		SCars [26]		CUB [47]	
	C	#	C	#	C	#	C	#	C	#	C	#	C	#
Stage 0 (D^l)	70	30.45K	70	77.46K	140	60.90K	71	4.70K	70	1.98K	130	4.62K	140	3.65K
Stage 1 (D_1^u)	80	5.95K	80	15.14K	160	11.90K	81	0.73K	80	0.37K	152	0.98K	160	0.71K
Stage 2 (D_2^u)	90	6.55K	90	16.66K	180	13.10K	91	0.65K	90	0.43K	174	1.13K	180	0.79K
Stage 3 (D_3^u)	100	7.05K	100	17.94K	200	14.10K	101	1.12K	100	0.55K	196	1.38K	200	0.85K

Table 2. Data splits.

Class splits	D^l	D_1^u	D_2^u	D_3^u
$\{y_i \mid y_i \leq 0.7 * \|\mathcal{Y}\|\}$	87%	7%	3%	3%
$\{y_i \mid 0.7 * \|\mathcal{Y}\| < y_i \leq 0.8 * \|\mathcal{Y}\|\}$	0%	70%	20%	10%
$\{y_i \mid 0.8 * \|\mathcal{Y}\| < y_i \leq 0.9 * \|\mathcal{Y}\|\}$	0%	0%	90%	10%
$\{y_i \mid 0.9 * \|\mathcal{Y}\| < y_i \leq \|\mathcal{Y}\|\}$	0%	0%	0%	100%

Implementation Details. We use ViT-B/16 backbone [10] pretrained with DINO [6,37] for all experiments. Please note that [49,50] utilized a pretrained model with supervision, which is suitable for the standard supervised continual learning task. However, it is not well-suited to use such pretrained models for CCD task due to label information leakage. During training, only the final block of the vision transformer is finetuned for 200 epochs with a batch size of 128, using SGD optimizer and cosine decay learning rate scheduler with an initial learning rate of 0.1 and minimum learning rate of 0.0001, and weight decay of 0.00005. For the GMP module, we optimize the GMM every 30 epochs and start the prompt learning when the epoch is greater than 30. We set top-k to be 5, and the number of GMM samples to 100. We pick the final model by selecting the best performing model on 'Old' ACC using the validation set (evaluated every 10 epochs). All input images are resized to 224 × 224 and augmented to match the DINO pretrained model settings. For our method, we finetune the last transformer block of the model f_b and the projection head ϕ (Sect. S8 for details) using the loss introduced in Sect. 3. For other compared methods, we carefully chose the right hyper-parameters following their original papers. Finally, we dynamically estimate the class number using the method described in Sect. 3.3, following a procedure similar to [58]. We build our framework with PyTorch on a single NVIDIA RTX 3090 GPU.

Evaluation Metric. The model is finetuned at each stage. At test time, the classification token [CLS] features are used for clustering. For the clustering algorithm and label assignment, we use semi-supervised k-means (SS-k-means) [46] on the unlabelled sets D_t^u and measure the accuracy given the ground truth y_i and the clustering prediction $\hat{y}_i$ such that:

Algorithm 1 *Continual ACC (cACC) evaluation metric*
Input: Models $\{f_\theta^t \mid t = 1, \ldots, T\}$ and datasets $\{D^l, D^u\}$.
Output: $cACC$ value.
Require: SS-k-MEANS(Model, Labelled set, Unlabelled set).
Require: Initialize set $\mathbb{A}^L \leftarrow D^l$.
1: **for** $t \in \{1, \cdots, T\}$ **do**
2: $\quad ACC_t, D_t^{u^*} \leftarrow$ SS-k-MEANS($f_\theta^t, \mathbb{A}^L, D_t^u$)
3: $\quad \mathbb{A}^L \leftarrow \mathbb{A}^L \cup D_t^{u^*}$ // *append $D_t^{u^*}$ (w/ assigned labels) to $\mathbb{A}^L$*
4: $ACCs \leftarrow \{ACC_t \mid t = 1, \ldots, T\}$
5: $cACC \leftarrow$ AVERAGE($ACCs$)
6: **return** $cACC$

$$ACC = \max_{g \in \mathcal{G}(\mathcal{Y}_U)} \frac{1}{|D_t^u|} \sum_{i=1}^{|D_t^u|} \mathbb{1}\{y_i = g(\hat{y}_i)\}, \tag{7}$$

where $\mathcal{G}(\mathcal{Y}_U)$ represents a set of all permutations of class labels in the unlabelled set D_t^u. For the evaluation across stages in CCD, based on the standard clustering accuracy ACC for GCD, we introduce a new metric, called *continual ACC* (*cACC*), for the continual setting considering the sequential data stream. Commonly, in GCD, the ACC values are evaluated for 'All', 'Old', and 'New' splits of the dataset. In CCD, for one time step t, 'All' indicates the overall accuracy on the entire set D_t^u. 'Old' and 'New' indicate the accuracy from instances of unlabelled data from D_t^{uo} and D_t^{un} respectively. The evaluation protocol of cACC is summarized in Algorithm 1. Instead of relying solely on the labelled data D^l to guide the SS-k-means clustering algorithm, $cACC$ incorporates labelled data from $\{D^l, D_1^{u^*}, \ldots, D_{t-1}^{u^*}\}$, where $D_i^{u^*}$ represents data with assigned labels from previously unlabelled data D_i^u. High-quality label assignments facilitate the subsequent category discovery while low-quality label assignments accumulate errors for the subsequent category discovery.

Comparison with Other Methods. We compare our method with the other representative CCD methods: 1) Grow & Merge (G&M) [55]; 2) MetaGCD [52]; 3) PA-CGCD [25]; and re-implement GCD methods for CCD task, including 4) ORCA [5]; 5) GCD [46]; 6) SimGCD [51]. As G&M's encoder is based on ResNet18 network [19], we re-implement their dynamic branch mechanism with the ViT backbone and observe improved performance for their method compared to their original results (see Sect. S5). We also re-implement GCD and SimGCD for CCD settings by incorporating a replay-based method. At each stage, the model saves samples for discovered classes and mixes them with incoming streamed images. Lastly, we adopt L2P's [50] and DualPrompt's [49] prompt pool modules, following their original prompt pool hyperparameter choices, and integrate them with PromptCCD-B framework as our baselines.

4.2 Main Results

We evaluate our method in two scenarios: when the class number C, is known (Tables 3, 4, and 5) in each unlabelled set at different stages, and when C is unknown (Table 6). We report the $cACC$ by averaging the results across all stages. We also provide the breakdown results for each stage in Sect. S2. In addition, we also report results using other metrics in Sect. S4 and Sect. S6.

Table 3. The $cACC$ results of our method with different prompt pool designs for CCD on generic and fine-grained benchmark datasets where C is *known* in each unlabelled set. The experiments are conducted five times with different random seeds.

Method	Prompt Pool	CIFAR100			ImageNet-100		
		All	Old	New	All	Old	New
PromptCCD-B (Ours)	L2P [50]	51.59 ± 6.3	67.27 ± 8.7	46.14 ± 6.1	66.14 ± 2.3	81.05 ± 1.5	61.36 ± 3.2
PromptCCD-B (Ours)	DP [49]	59.60 ± 1.2	78.93 ± 1.3	54.14 ± 1.6	70.64 ± 1.3	83.46 ± 0.4	67.24 ± 1.8
PromptCCD (Ours)	GMP (Ours)	**63.97** ± 1.4	76.67 ± 2.6	60.01 ± 1.7	**75.38** ± 0.7	81.16 ± 0.7	73.71 ± 0.8

Method	Prompt Pool	TinyImageNet			CUB		
		All	Old	New	All	Old	New
PromptCCD-B (Ours)	L2P [50]	56.66 ± 0.4	66.05 ± 0.8	53.69 ± 0.4	51.31 ± 1.0	72.43 ± 1.0	44.27 ± 1.4
PromptCCD-B (Ours)	DP [49]	58.61 ± 1.5	66.61 ± 0.6	55.84 ± 1.7	56.30 ± 1.1	78.64 ± 1.7	48.91 ± 1.1
PromptCCD (Ours)	GMP (Ours)	**61.15** ± 1.0	66.29 ± 2.0	58.83 ± 1.0	**56.65** ± 1.0	79.88 ± 2.5	48.96 ± 0.8

Table 4. Comparison with other methods for CCD leveraging pretrained DINO and DINOv2 models on generic datasets with the *known* C in each unlabelled set.

Method	Pretrained Model	CIFAR100			ImageNet-100			TinyImageNet			Caltech-101		
		All	Old	New	All	Old	New	All	Old	New	All	Old	New
ORCA [5]	DINO	60.91	66.61	58.33	40.29	45.85	35.40	54.71	63.13	51.93	76.77	82.80	73.20
GCD [46]	DINO	58.18	72.27	52.83	69.41	81.56	65.65	55.20	65.87	51.61	78.27	86.60	72.92
SimGCD [51]	DINO	25.56	38.76	20.43	31.38	40.47	27.44	33.40	29.11	34.74	33.65	37.53	31.62
GCD w/replay	DINO	49.93	73.15	41.47	72.04	83.75	69.01	56.33	67.54	52.60	76.51	86.14	72.48
SimGCD w/replay	DINO	40.13	66.72	30.91	47.53	67.86	39.18	37.45	58.15	30.36	49.38	52.72	47.99
Grow & Merge [55]	DINO	57.43	63.68	55.31	67.84	75.10	66.60	52.14	59.68	49.96	75.75	83.66	71.59
MetaGCD [52]	DINO	55.49	69.38	48.98	66.41	80.54	60.65	55.26	66.12	50.79	80.75	89.02	75.86
PA-CGCD [25]	DINO	58.25	87.11	49.04	64.79	91.15	57.83	51.13	74.95	43.52	77.96	94.75	69.66
PromptCCD w/GMP (Ours)	DINO	**64.17**	75.57	60.34	**76.16**	81.76	74.35	**61.84**	66.54	60.26	**82.44**	89.08	79.72
GCD [46]	DINOv2	65.35	77.06	60.46	71.58	83.02	68.05	59.05	77.44	53.41	83.00	88.65	79.80
MetaGCD [52]	DINOv2	52.10	79.64	43.13	70.20	82.62	64.66	56.15	74.69	49.37	83.05	88.08	80.89
PA-CGCD [25]	DINOv2	54.36	79.19	45.65	74.82	88.20	72.02	52.10	68.07	46.32	83.06	94.07	77.55
PromptCCD w/GMP (Ours)	DINOv2	**69.73**	78.01	66.16	**76.28**	82.61	74.53	**68.20**	75.56	65.23	**83.86**	87.93	81.42

Variants of PromptCCD. In our comparison with our baseline PromptCCD-B, as shown in Table 3, our PromptCCD w/GMP demonstrates superior performance for CCD. Specifically, our model outperforms the baselines across all 'All' accuracy, while the baselines experience performance degradation in later stages (see Sect. S2). We attribute this decline to the baselines' non-scalable prompt pool parameters, which restrict their ability to *instruct* the model as the parameter count grows. In contrast, our scalable prompting technique leverages Gaussian mixture models to construct a flexible pool of prompts. Additionally, we ensure knowledge retention by sampling learned mixture components for fitting subsequent GMM.

Comparison with Known Class Numbers. The CCD benchmark results for generic and fine-grained datasets are presented in Table 4 and 5. Remarkably,

Table 5. Comparison with other methods for CCD leveraging pretrained DINO and DINOv2 models on fine-grained datasets with the *known* C in each unlabelled set.

Method	Pretrained Model	Aircraft			Stanford Cars			CUB		
		All	Old	New	All	Old	New	All	Old	New
ORCA [5]	DINO	30.77	25.71	32.44	20.79	33.40	17.60	41.73	66.19	34.14
GCD [46]	DINO	47.37	61.43	42.53	39.21	58.29	33.45	54.98	75.47	48.15
SimGCD [51]	DINO	29.03	35.72	25.61	21.01	40.93	16.48	39.89	59.25	33.75
GCD w/replay	DINO	45.63	62.38	39.89	39.87	58.18	33.89	54.66	74.64	47.81
SimGCD w/replay	DINO	37.44	61.43	28.96	22.76	49.04	16.65	42.08	72.65	31.92
Grow & Merge [55]	DINO	31.06	33.33	30.78	21.90	35.29	18.17	38.87	65.00	30.29
MetaGCD [52]	DINO	44.63	59.05	39.39	35.98	56.97	29.96	44.59	74.40	35.40
PA-CGCD [25]	DINO	48.24	73.09	40.60	43.88	80.43	33.54	52.48	77.26	44.74
PromptCCD w/GMP (Ours)	DINO	**52.64**	60.48	50.23	**44.07**	66.36	36.83	**55.45**	75.48	48.56
GCD [46]	DINOv2	57.87	63.80	55.39	58.52	71.65	53.80	66.70	83.33	60.81
MetaGCD [52]	DINOv2	54.90	64.29	52.08	57.16	71.87	52.01	62.19	82.50	55.13
PA-CGCD [25]	DINOv2	58.15	77.62	51.08	64.91	89.64	57.84	66.88	92.62	58.48
PromptCCD w/GMP (Ours)	DINOv2	**62.71**	68.33	60.82	**65.08**	76.60	60.75	**67.81**	81.55	62.81

our PromptCCD *w*/GMP outperforms other approaches across all datasets in terms of overall accuracy *'All'*. Notably, our approach maintains the balance between the *'Old'* and *'New'* accuracy compared to existing methods and achieving improved performance. As our model is based on GCD [46], we show that simply integrating our Gaussian mixture prompt module enables effective adaptation to the CCD settings. This highlights the robustness and versatility of our approach in handling CCD.

Table 6. Comparison with other methods for CCD leveraging the pretrained DINO model when the class number C in each unlabelled set is *unknown*.

	Est. method	CIFAR100			ImageNet-100			TinyImageNet			CUB		
Category discovery at stage --→		1	2	3	1	2	3	1	2	3	1	2	3
Estimated category C	GPC	77	78	81	73	73	83	158	164	168	163	172	175
Ground truth category C	–	80	90	100	80	90	100	160	180	200	160	180	200
Methods		All	Old	New	All	Old	New	All	Old	New	All	Old	New
GCD [46]	GPC	57.20	68.23	52.87	56.60	77.75	48.28	53.68	64.71	50.00	50.02	71.31	42.82
Grow & Merge [55]	GPC	56.59	60.59	54.02	52.83	72.34	45.46	53.44	56.49	51.94	36.88	63.81	28.10
MetaGCD [52]	GPC	53.20	66.22	47.68	51.40	75.73	41.53	57.92	62.82	54.64	44.56	70.24	36.57
PA-CGCD [25]	GPC	56.46	82.86	47.68	48.63	84.51	36.98	50.40	70.17	43.87	50.61	74.53	42.60
PromptCCD-U w/GMP (Ours)	GPC	**63.20**	71.19	59.90	**63.14**	76.68	56.87	**60.81**	71.07	56.81	**51.20**	73.33	43.51

Comparison with Unknown Class Numbers. To show the performance comparison for each model in a more realistic setting where C is unknown, we also report the benchmark results in Table 6, where we show 5 representative methods, *i.e.*, [25,46,52,55], and ours. Our method consistently outperforms all other methods by a large margin across the board, demonstrating the superior performance of our approach in the more realistic case when the class number is unknown.

Qualitative Analysis. Lastly, to visualize the feature representation generated by our method, we use t-SNE algorithm [33] to visualize the high-dimensional features of $\{D^l, D^u_t\}$ on each stage. For the sake of comparison, we also provide the visualization for the feature representation generated by Grow & Merge [55]. The qualitative visualization can be seen in Fig. 4; nodes of the same colour indicate that the instances belong to the same category. Moreover, for stage $t > 0$, we only highlight the feature's node belonging to unknown novel categories. It is observed that across stages, our cluster features are more discriminative.

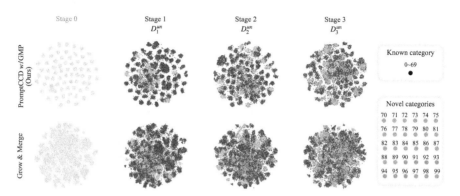

Fig. 4. t-SNE visualization of CIFAR100 with features from our model PromptCCD w/GMP and Grow & Merge on each stage.

4.3 Model Component Analysis

Top-k vs Random Prompts. In Table 8, to validate the effectiveness of using top-k prompts, we compare the results by using top-k and random-k prompts. We observe that using random-k prompts hurts the performance, as evidenced by that the performance using random-k is worse than without using any prompts. In contrast, our top-k strategy leads to significantly improved performance, especially for the *'New' ACC*. This observation suggests that prompting with top-k class prototypes indeed aids in the discovery of novel classes.

The Choices of Numbers for top-k and GMM Samples in GMP. To investigate the effectiveness of our GMP module, we analyzed each component in our prompt module and present the results in Table 7. The results show a

Table 7. Ablation study on different components of our GMP on C100 and CUB.

top-k	GMM	C100 Avg. ACC			CUB Avg. ACC		
Prompts	Samples	All	Old	New	All	Old	New
0	0	58.18	72.27	52.83	54.98	75.47	48.15
5	0	61.48	74.68	57.55	53.54	74.28	46.47
5	20	62.21	75.71	57.90	54.37	74.88	46.41
5	200	61.00	72.46	57.08	51.67	73.33	44.08
2	100	61.39	73.04	57.64	53.36	73.45	46.04
5	100	**64.17**	75.57	60.34	**55.45**	75.48	48.56
10	100	61.03	72.91	56.97	52.76	71.67	46.02

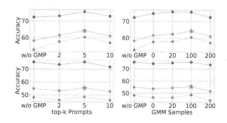

Fig. 5. Performance curves depicted from Table 7 ablation results.

clear advantage of adopting the GMP into our framework. The number of top-k prompts and the number of GMM samples are identified as important factors. The optimal configuration is top-5 for prompt selection, and 100 samples for sampling (Fig. 5), which appears to be a good trade-off.

Improving other CCD Methods with Our GMP. Thanks to the great flexibility of our GMP, it can serve as a *plug-and-play* module and be seamlessly integrated with other methods. Table 9 presents the results of integrating GMP with G&M and MetaGCD methods, showcasing significant improvements and highlighting the effectiveness of our GMP module. Nevertheless, our PromptCCD w/GMP substantially outperforms these enhanced CCD methods.

Table 8. Study on the effectiveness of top-k prompts compared with randomly picked prompts from GMM.

PromptCCD	CIFAR100 Avg. ACC			ImageNet-100 Avg. ACC		
	All	Old	New	All	Old	New
w/o GMP	58.18	72.27	52.83	69.41	81.56	65.65
GMP (random-k)	59.98	73.81 +1.54	55.68 +2.85	68.30	80.09 -1.47	63.50 -2.15
GMP (top-k) (Ours)	**64.17**	**75.57** +3.30	**60.34** +7.51	**76.16**	**81.76** +0.20	**74.35** +8.70
PromptCCD	TinyImageNet Avg. ACC			CUB Avg. ACC		
	All	Old	New	All	Old	New
w/o GMP	55.20	65.87	51.61	54.98	75.47	48.15
GMP (random-k)	55.69	63.95 -1.92	52.52 +0.91	51.46	73.10 -2.37	43.90 -4.25
GMP (top-k) (Ours)	**61.84**	**66.54** +0.67	**60.26** +8.65	**55.45**	**75.48** +0.01	**48.56** +0.41

Table 9. Study on the effectiveness of GMP module on different models. Here, we show the performances on CIFAR100.

Method	Prompt	All	Old	New
G&M	w/o GMP	57.43	63.68	55.31
G&M	w/GMP	**61.14**	**64.94**	**59.10**
MetaGCD	w/o GMP	55.49	69.38	48.98
MetaGCD	w/GMP	**58.99**	**69.69**	**54.29**
PromptCCD	w/o GMP	58.18	72.27	52.83
PromptCCD	w/GMP	**64.17**	**75.57**	**60.34**

Further Study on the Baseline Prompts. We also explore the impact of prompt pool size for PromptCCD-B $w/\{L2P, DP\}$ as indicated in Table 10, by experimenting with different prompt sizes. We find that varying the number of prompts does not significantly affect performance, even when the size aligns with the total number of classes in the CUB. In all cases, the results are significantly worse than the results obtained by our GMP above. This reveals the limitations of our baseline prompt method, particularly its reliance on fixed-size prompts and lack of scalability. Moreover, this also highlights the importance and effectiveness of our GMP design, the superior effectiveness of which can not be achieved by simply changing the prompt pool size of the baseline L2P and DP prompt pool size.

Table 10. Study on the PromptCCD-B $w/\{L2P, DP\}$ pool size on CUB.

	w/ L2P			w/ DP		
Pool Size	All	Old	New	All	Old	New
5	48.69	70.12	41.51	55.54	77.78	48.21
10	50.57	73.22	43.28	55.21	77.24	48.04
20	49.32	70.60	42.29	55.41	76.31	48.18
40	48.59	69.26	41.43	53.97	77.38	46.26
100	51.84	73.09	44.39	55.12	77.26	47.82
200	49.40	71.79	41.94	54.54	79.05	46.29

Please refer to the supplementary material for additional details and results.

5 Conclusion

In this paper, we have introduced PromptCCD, a simple yet effective framework for Continual Category Discovery (CCD) that tackles the challenge of discovering novel categories in the continuous stream of unlabelled data without catastrophic forgetting. By introducing the GMP module, PromptCCD dynamically updates the data representation and prevents forgetting during category discovery. Additionally, GMP enables on-the-fly estimation of category numbers, eliminating the need for prior knowledge of the category numbers. Our extensive evaluations on diverse datasets, along with our extended evaluation metric $cACC$, show that PromptCCD outperforms existing methods, highlighting its effectiveness in CCD.

Acknowledgments. This work is supported by the Hong Kong Research Grants Council - General Research Fund (Grant No.: 17211024).

References

1. Assran, M., et al.: Semi-supervised learning of visual features by non-parametrically predicting view assignments with support samples. In: ICCV (2021)
2. Berthelot, D., Carlini, N., Goodfellow, I., Papernot, N., Oliver, A., Raffel, C.: Mixmatch: a holistic approach to semi-supervised learning. In: NeurIPS (2019)
3. Boschini, M., Bonicelli, L., Buzzega, P., Porrello, A., Calderara, S.: Class-incremental continual learning into the eXtended DER-Verse. IEEE Trans. Pattern Anal. Mach. Intell. **45**(5), 5497–5512 (2023)
4. Buzzega, P., Boschini, M., Porrello, A., Abati, D., Calderara, S.: Dark experience for general continual learning: a strong, simple baseline. In: NeurIPS (2020)
5. Cao, K., Brbić, M., Leskovec, J.: Open-world semi-supervised learning. In: ICLR (2022)
6. Caron, M., et al.: Emerging properties in self-supervised vision transformers. In: ICCV (2021)
7. Chapelle, O., Schölkopf, B., Zien, A.: Semi-supervised Learning. MIT Press (2006)
8. Chen, T., Kornblith, S., Norouzi, M., Hinton, G.: A simple framework for contrastive learning of visual representations. In: ICML (2020)
9. De Lange, M., et al.: A continual learning survey: defying forgetting in classification tasks. IEEE TPAMI (2021)
10. Dosovitskiy, A., et al.: An image is worth 16x16 words: transformers for image recognition at scale. In: ICLR (2021)
11. Fei, Y., Zhao, Z., Yang, S., Zhao, B.: Xcon: learning with experts for fine-grained category discovery. In: BMVC (2022)
12. Fei-Fei, L., Fergus, R., Perona, P.: One-shot learning of object categories. IEEE TPAMI (2006)
13. Fini, E., Sangineto, E., Lathuilière, S., Zhong, Z., Nabi, M., Ricci, E.: A unified objective for novel class discovery. In: ICCV (2021)
14. Graves, A., et al.: Hybrid computing using a neural network with dynamic external memory. Nature (2016)
15. Han, K., Rebuffi, S.A., Ehrhardt, S., Vedaldi, A., Zisserman, A.: Automatically discovering and learning new visual categories with ranking statistics. In: ICLR (2020)
16. Han, K., Rebuffi, S.-A., Ehrhardt, S., Vedaldi, A., Zisserman, A.: AutoNovel: automatically discovering and learning novel visual categories. IEEE Trans. Pattern Anal. Mach. Intell. **44**(10), 6767–6781 (2022)
17. Han, K., Vedaldi, A., Zisserman, A.: Learning to discover novel visual categories via deep transfer clustering. In: ICCV (2019)

18. Hao, S., Han, K., Wong, K.Y.K.: CIPR: an efficient framework with cross-instance positive relations for generalized category discovery. In: TMLR (2024)
19. He, K., Zhang, X., Ren, S., Sun, J.: Deep residual learning for image recognition. In: CVPR (2016)
20. Huang, J., et al.: Trash to treasure: harvesting OOD data with cross-modal matching for open-set semi-supervised learning. In: ICCV (2021)
21. Jia, X., Han, K., Zhu, Y., Green, B.: Joint representation learning and novel category discovery on single- and multi-modal data. In: ICCV (2021)
22. Joseph, K.J., et al.: Spacing loss for discovering novel categories. In: CVPR Workshop (2022)
23. Joseph, K.J., et al.: Novel class discovery without forgetting. In: Avidan, S., Brostow, G., Cissé, M., Farinella, G.M., Hassner, T. (eds.) ECCV 2022, Part XXIV, pp. 570–586. Springer, Cham (2022). https://doi.org/10.1007/978-3-031-20053-3_33
24. Khosla, P., et al.: Supervised contrastive learning. In: NeurIPS (2020)
25. Kim, H., Suh, S., Kim, D., Jeong, D., Cho, H., Kim, J.: Proxy anchor-based unsupervised learning for continuous generalized category discovery. In: ICCV (2023)
26. Krause, J., Stark, M., Deng, J., Fei-Fei, L.: 3D object representations for fine-grained categorization. In: ICCV Workshop (2013)
27. Krizhevsky, A., Hinton, G.: Learning multiple layers of features from tiny images. Master's thesis, Department of Computer Science, University of Toronto (2009)
28. Laine, S., Aila, T.: Temporal ensembling for semi-supervised learning. In: ICLR (2017)
29. Le, Y., Yang, X.: Tiny imagenet visual recognition challenge. In: CS 231N (2015)
30. Li, X., Zhou, Y., Wu, T., Socher, R., Xiong, C.: Learn to grow: a continual structure learning framework for overcoming catastrophic forgetting. In: ICML (2019)
31. Li, Z., Hoiem, D.: Learning without forgetting. In: Leibe, B., Matas, J., Sebe, N., Welling, M. (eds.) ECCV 2016. LNCS, vol. 9908, pp. 614–629. Springer, Cham (2016). https://doi.org/10.1007/978-3-319-46493-0_37
32. Liu, M., Roy, S., Zhong, Z., Sebe, N., Ricci, E.: Large-scale pre-trained models are surprisingly strong in incremental novel class discovery. In: ICPR (2024)
33. Van der Maaten, L., Hinton, G.: Visualizing data using t-sne. JMLR (2008)
34. Maji, S., Kannala, J., Rahtu, E., Blaschko, M., Vedaldi, A.: Fine-grained visual classification of aircraft. arXiv preprint arXiv:1306.5151 (2013)
35. McCloskey, M., Cohen, N.J.: Catastrophic interference in connectionist networks: the sequential learning problem. In: Psychology of Learning and Motivation. Elsevier (1989)
36. Oliver, A., Odena, A., Raffel, C.A., Cubuk, E.D., Goodfellow, I.: Realistic evaluation of deep semi-supervised learning algorithms. In: NeurIPS (2018)
37. Oquab, M., et al.: Dinov2: learning robust visual features without supervision. In: TMLR (2023)
38. Rebuffi, S.A., Kolesnikov, A., Sperl, G., Lampert, C.H.: icarl: Incremental classifier and representation learning. In: CVPR (2017)
39. Ridnik, T., Ben-Baruch, E., Noy, A., Zelnik-Manor, L.: Imagenet-21k pretraining for the masses. In: NeurIPS (2021)
40. Rizve, M.N., Duarte, K., Rawat, Y.S., Shah, M.: In defense of pseudo-labeling: an uncertainty-aware pseudo-label selection framework for semi-supervised learning. In: ICLR (2021)
41. Roy, S., Liu, M., Zhong, Z., Sebe, N., Ricci, E.: Class-incremental novel class discovery. In: ECCV (2022)
42. Russakovsky, O., et al.: Imagenet large scale visual recognition challenge. In: IJCV (2015)

43. Saito, K., Kim, D., Saenko, K.: Openmatch: open-set consistency regularization for semi-supervised learning with outliers. In: NeurIPS (2021)
44. Sohn, K., et al.: Fixmatch: simplifying semi-supervised learning with consistency and confidence. In: NeurIPS (2020)
45. Tarvainen, A., Valpola, H.: Mean teachers are better role models: weight-averaged consistency targets improve semi-supervised deep learning results. In: NeurIPS (2017)
46. Vaze, S., Han, K., Vedaldi, A., Zisserman, A.: Generalized category discovery. In: CVPR (2022)
47. Wah, C., Branson, S., Welinder, P., Perona, P., Belongie, S.: The Caltech-UCSD Birds-200-2011 Dataset. California Institute of Technology (2011)
48. Wang, H., Vaze, S., Han, K.: Sptnet: an efficient alternative framework for generalized category discovery with spatial prompt tuning. In: ICLR (2024)
49. Wang, Z., et al.: DualPrompt: complementary prompting for rehearsal-free continual learning. In: Avidan, S., Brostow, G., Cissé, M., Farinella, G.M., Hassner, T. (eds.) ECCV 2022, Part XXVI, pp. 631–648. Springer, Cham (2022). https://doi.org/10.1007/978-3-031-19809-0_36
50. Wang, Z., et al.: Learning to prompt for continual learning. In: CVPR (2022)
51. Wen, X., Zhao, B., Qi, X.: Parametric classification for generalized category discovery: a baseline study. In: ICCV (2023)
52. Wu, Y., Chi, Z., Wang, Y., Feng, S.: Metagcd: learning to continually learn in generalized category discovery. In: ICCV (2023)
53. Yu, Q., Ikami, D., Irie, G., Aizawa, K.: Multi-task curriculum framework for open-set semi-supervised learning. In: ECCV (2020)
54. Zhang, S., Khan, S., Shen, Z., Naseer, M., Chen, G., Khan, F.S.: Promptcal: contrastive affinity learning via auxiliary prompts for generalized novel category discovery. In: CVPR (2023)
55. Zhang, X., et al.: Grow and merge: a unified framework for continuous categories discovery. In: NeurIPS (2022)
56. Zhao, B., Han, K.: Novel visual category discovery with dual ranking statistics and mutual knowledge distillation. In: NeurIPS (2021)
57. Zhao, B., Mac Aodha, O.: Incremental generalized category discovery. In: ICCV (2023)
58. Zhao, B., Wen, X., Han, K.: Learning semi-supervised gaussian mixture models for generalized category discovery. In: ICCV (2023)
59. Zhong, Z., Zhu, L., Luo, Z., Li, S., Yang, Y., Sebe, N.: Openmix: reviving known knowledge for discovering novel visual categories in an open world. In: CVPR (2021)

Sapiens: Foundation for Human Vision Models

Rawal Khirodkar[✉], Timur Bagautdinov, Julieta Martinez, Su Zhaoen, Austin James, Peter Selednik, Stuart Anderson, and Shunsuke Saito

Codec Avatars Lab, Meta, Menlo Park, USA
rawalkhirodkar@gmail.com

Abstract. We present *Sapiens*, a family of models for four fundamental human-centric vision tasks – 2D pose estimation, body-part segmentation, depth estimation, and surface normal prediction. Our models natively support 1K high-resolution inference and are extremely easy to adapt for individual tasks by simply fine-tuning foundation models pretrained on over 300 million in-the-wild human images. We observe that, given the same computational budget, self-supervised pretraining on a curated dataset of human images significantly boosts the performance for a diverse set of human-centric tasks. The resulting models exhibit remarkable generalization to in-the-wild data, even when labeled data is scarce or entirely synthetic. Our simple model design also brings scalability – model performance across tasks significantly improves as we scale the number of parameters from 0.3 to 2 billion. Sapiens consistently surpasses existing complex baselines across various human-centric benchmarks. Specifically, we achieve significant improvements over the prior state-of-the-art on Humans-5K (pose) by 7.6 mAP, Humans-2K (part-seg) by 17.1 mIoU, Hi4D (depth) by 22.4% relative RMSE, and THuman2 (normal) by 53.5% relative angular error.

"*Sapiens*-pertaining to, or resembling modern humans."

Keywords: Human-centric · Large-scale pretraining · In-the-wild

1 Introduction

Recent years have witnessed remarkable strides towards generating photorealistic humans in 2D [17,28,50,118] and 3D [69,89,102,109]. The success of these methods is greatly attributed to the robust estimation of various assets such as 2D keypoints [14,67], fine-grained body-part segmentation [119], depth [113], and surface normals [89,108]. However, robust and accurate estimation of these assets is still an active research area, and complicated systems to boost performance for individual tasks often hinder wider adoption. Moreover, obtaining accurate ground-truth annotation in-the-wild is notoriously difficult to scale. Our goal is to provide a unified framework and models to infer these assets in-the-wild to unlock a wide range of human-centric applications for everybody.

Fig. 1. *Sapiens* models are finetuned for four human tasks - 2D pose estimation, body-part segmentation, depth prediction and normal prediction. Our models generalize across a variety of in-the-wild face, upper-body, full-body and multi-person images.

We argue that such human-centric models should satisfy three criteria: generalization, broad applicability, and high fidelity. Generalization ensures robustness to unseen conditions, enabling the model to perform consistently across varied environments. Broad applicability indicates the versatility of the model, making it suitable for a wide range of tasks with minimal modifications. High fidelity denotes the ability of the model to produce precise, high-resolution outputs, essential for faithful human generation tasks such as 2D to 3D lifting. This paper details the development of models that embody these attributes, collectively referred to as *Sapiens*.

Following the insights from [34,79,91], leveraging large datasets and scalable model architectures is key for generalization. For broader applicability, we adopt the pretrain-then-finetune approach, enabling post-pretraining adaptation to specific tasks with minimal adjustments. This approach raises a critical question: *What type of data is most effective for pretraining?* Given computational limits, should the emphasis be on collecting as many human images as possible, or is it preferable to pretrain on a less curated set to better reflect real-world variability? Existing methods often overlook the pretraining data distribution in the context of downstream tasks. To study the influence of pretraining data distribution on human-specific tasks, we collect the Humans-300M dataset, featuring 300 million diverse human images. These unlabelled images are used to pretrain a family of vision transformers [27] from scratch, with parameter counts ranging from 300M to 2B.

Among various self-supervision methods for learning general-purpose visual features from large datasets [5,19,34,45,46,121], we choose the masked-autoencoder (MAE) approach [45] for its simplicity and efficiency in pretraining. MAE, having a single-pass inference model compared to contrastive or multi-inference strategies, allows processing a larger volume of images with the same computational resources. For higher-fidelity, in contrast to prior methods, we increase the native input resolution of our pretraining to 1024 pixels, resulting in a $\sim 4\times$ increase in FLOPs compared to the largest existing vision backbone [91]. Each model is pretrained on 1.2 trillion tokens. Table 1 outlines a comparison with earlier approaches. For finetuning on human-centric tasks [15,101,113,119], we use a consistent encoder-decoder architecture. The encoder is initialized with weights from pretraining, while the decoder, a lightweight and task-specific head, is initialized randomly. Both components are then finetuned in an end-to-end manner. We focus on four key tasks - 2D pose estimation, body-part segmentation, depth, and normal estimation, as shown in Fig. 1.

Consistently with prior studies [56,122], we affirm the critical impact of label quality on the model's in-the-wild performance. Public benchmarks [23,41,55] often contain noisy labels, providing inconsistent supervisory signals during model fine-tuning. At the same time, it is important to utilize fine-grained and precise annotations to align closely with our primary goal of 3D human digitization. To this end, we propose a substantially denser set of 2D whole body keypoints for pose estimation and a detailed class vocabulary for body part segmentation, surpassing the scope of previous datasets (please refer to Fig. 1). Specifically, we introduce a comprehensive collection of 308 keypoints encompassing the body, hands, feet, surface, and face. Additionally, we expand the segmentation class vocabulary to 28 classes, covering body parts such as the hair, tongue, teeth, upper/lower lip, and torso. To guarantee the quality and consistency of annotations and a high degree of automation, we utilize a multi-view capture setup to collect pose and segmentation annotations. We also utilize human-centric synthetic data for depth and normal estimation, leveraging 600 detailed scans from RenderPeople [82] to generate high-resolution depth maps and surface normals.

We show that the combination of domain-specific large-scale pretraining with limited, yet high-quality annotations leads to robust in-the-wild generalization. Overall, our method demonstrates an effective strategy for developing highly precise discriminative models capable of performing in real-world scenarios without the need for collecting a costly and diverse set of annotations.

Table 1. Comparison of state-of-the-art pretrained vision models. Sapiens adopts a higher resolution backbone on a large dataset of in-the-wild human images.

Method	Dataset	#Params	GFLOPs	Image size	Domain
DINO [16]	ImageNet1k	86 M	17.6	224	General
iBOT [121]	ImageNet21k	307 M	61.6	224	General
DINOv2 [79]	LVD-142M	1 B	291.0	224	General
ViT-6.5B [91]	IG-3B	6.5 B	1657.0	224	General
AIM [34]	DFN-2B	6.5 B	1657.0	224	General
Sapiens (Ours)	Humans-300M	2 B	8709.0	1024	Human

Our contributions are summarized as follows.

- We introduce *Sapiens*, a family of vision transformers pretrained on a large-scale, curated dataset of diverse human images.
- This study shows that simple data curation and large-scale pretraining significantly boost the model's performance with the same computational budget.
- Our models, fine-tuned with high-quality annotations or even synthetic data, demonstrate robust in-the-wild generalization.
- The first 1K high-resolution model that natively supports high-fidelity inference essential for human-centric tasks, achieving state-of-the-art performance on benchmarks for 2D pose estimation, body-part segmentation, depth, and normal estimation.

2 Related Work

Our work explores the limits of training large architectures on a large number of in-the-wild human images. We build on prior work from different areas: pretraining at scale, human vision tasks, and large vision transformers.

Pretraining at Scale. The remarkable success of large-scale pretraining [26,95] followed by task-specific finetuning for language modeling [2,13,53,96,99,100] has established this approach as a standard practice. Similarly, computer vision methods [1,4,33,34,42,79,83,85,87,120] are progressively embracing extensive data scales for pretraining. The emergence of large datasets, such as LAION-5B [90], Instagram-3.5B [77], JFT-300M [92], LVD-142M [79], Visual Genome [60], and YFCC100M [97], has enabled the exploration of a data corpus well beyond the scope of traditional benchmarks [61,67,86]. Salient work in this domain includes DINOv2 [79], MAWS [91], and AIM [34]. DINOv2 achieves state-of-the-art performance in generating self-supervised features by scaling the contrastive iBot [121] method on the LDV-142M dataset [79]. MAWS [91] studies the scaling of masked-autoencoders (MAE) [45] on billion images. AIM [34] explores the scalability of autoregressive visual pretraining similar to BERT [26] for vision transformers [27]. In contrast to these methods which mainly focus on general image pretraining or zero-shot image classification, we take a distinctly human-centric approach: our models leverage a vast collection of human images for pretraining, subsequently fine-tuning for a range of human-related tasks.

Human Vision Tasks. The pursuit of large-scale 3D human digitization [8,44,64,74] remains a pivotal goal in computer vision [12]. Significant progress has been made within controlled or studio environments [3,59,63,69,70,76,89], yet challenges persist in extending these methods to unconstrained environments [29]. To address these challenges, developing versatile models capable of multiple fundamental tasks such as keypoint estimation [21,36,47,51,57,78,80, 93,106], body-part segmentation [35,40,41,75,104,105], depth estimation [9,10, 32,43,52,66,84,113], and surface normal prediction [6,7,31,39,62,88,101,108] from images in natural settings is crucial. In this work, we aim to develop models for these essential human vision tasks which generalize to in-the-wild settings.

Scaling Architectures. Currently, the largest publicly-accessible language models contain upwards of 100B parameters [49], while the more commonly used language models [94,100] contain around 7B parameters. In contrast, Vision Transformers (ViT) [27], despite sharing a similar architecture, have not been scaled to this extent successfully. While there are notable endeavors in this direction, including the development of a dense ViT-4B [20] trained on both text and images, and the formulation of techniques for the stable training of a ViT-22B [25], commonly utilized vision backbones still range between 300M to 600M parameters [24,38,48,68] and are primarily pretrained at an image resolution of about 224 pixels. Similarly, existing transformer-based image generation models, such as DiT [81] use less than 700M parameters, and operate on a highly compressed latent space. To address this gap, we introduce Sapiens - a collection of large, high-resolution ViT models that are pretrained natively at a 1024 pixel image resolution on millions of human images.

3 Method

3.1 Humans-300M Dataset

We utilize a large proprietary dataset for pretraining of approximately 1 billion in-the-wild images, focusing exclusively on human images. The preprocessing involves discarding images with watermarks, text, artistic depictions, or unnatural elements. Subsequently, we use an off-the-shelf person bounding-box detector [103] to filter images, retaining those with a detection score above 0.9 and bounding box dimensions exceeding 300 pixels. Figure 2 provides an overview of

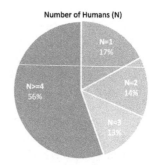

Fig. 2. Overview of number of humans per image in the Humans-300M dataset.

the distribution of the number of people per image in our dataset, noting that over 248 million images contain multiple subjects.

3.2 Pretraining

We follow the masked-autoencoder [45] (MAE) approach for pretraining. Our model is trained to reconstruct the original human image given its partial observation. Like all autoencoders, our model has an encoder that maps the visible image to a latent representation and a decoder that reconstructs the original image from this latent representation. Our pretraining dataset consists of both single and multi-human images; each image is resized to a fixed size with a square aspect ratio. Similar to ViT [27], we divide an image into regular non-overlapping patches with a fixed patch size. A subset of these patches is randomly selected and masked, leaving the rest visible. The proportion of masked patches to visible ones is defined as the masking ratio, which remains fixed throughout training. We refer to MAE [45] for more details. Figure 3 (*Top*) shows the reconstruction of our pretrained model on unseen human images. Our models exhibit generalization across a variety of image characteristics including scales, crops, the age and ethnicity of subjects, and number of subjects. Each patch token in our model accounts for 0.02% of the image area compared to 0.4% in standard ViTs, a 16× reduction - this provides a fine-grained inter-token reasoning for our models. Figure 3 (*Bottom*) shows that even with an increased mask ratio of 95%, our model achieves a plausible reconstruction of human anatomy on held-out samples.

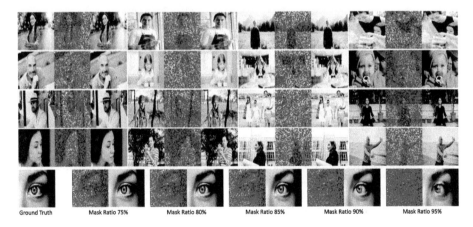

Fig. 3. In-the-wild generalization of *Sapiens* on unseen images. *Top*: Each triplet illustrates the ground truth (left), the masked image (center), and the MAE reconstruction (right), with a masking ratio of 75%, a patch size of 16, and an image size of 1024. *Bottom*: Varying the mask ratio between [0.75, 0.95] during inference reveals a minimal reduction in quality, underscoring the model's robust understanding of human images.

3.3 2D Pose Estimation

We follow the top-down 2D pose estimation paradigm, which aims to detect the locations of K keypoints from an input image $\mathbf{I} \in \mathbb{R}^{H \times W \times 3}$. Most methods pose this problem as heatmap prediction, where each of K heatmaps represents the probability of the corresponding keypoint being at any spatial location. Similar to [111], we define a pose estimation transformer, $\mathcal{P}$, for keypoint detection. The bounding box at training and inference is scaled to $H \times W$ and is provided as an input to $\mathcal{P}$. Let $\mathbf{y} \in \mathbb{R}^{H \times W \times K}$ denote the K heatmaps corresponding to the ground truth keypoints for a given input $\mathbf{I}$. The pose estimator transforms input $\mathbf{I}$ to a set of predicted heatmaps, $\hat{\mathbf{y}} \in \mathbb{R}^{H \times W \times K}$, such that $\hat{\mathbf{y}} = \mathcal{P}(\mathbf{I})$. $\mathcal{P}$ is trained to minimize the mean squared loss $\mathcal{L}_{\text{pose}} = \text{MSE}(\mathbf{y}, \hat{\mathbf{y}})$. During finetuning, the encoder of $\mathcal{P}$ is initialized with the weights from pretaining, and the decoder is initialized randomly. The aspect ratio $H : W$ is set to be $4 : 3$, with the pretrained positional embedding being interpolated accordingly [58]. We use lightweight decoders with deconvolution and convolution operations.

We finetune both the encoder and the decoder in $\mathcal{P}$ across multiple skeleton formats, including $K = 17$ [67], $K = 133$ [55] and a new highly-detailed skeleton format, with $K = 308$, as shown in Fig. 4 (*Left*). Compared to existing formats with at most 68 facial keypoints, our annotations consist of 243 facial keypoints, including representative points around the eyes, eyebrows, lips, nose, and ears. This design is tailored to meticulously capture the nuanced details of facial expressions in the real world. With these keypoints, we manually annotated 1 million images at 4K resolution from an indoor capture setup.

Fig. 4. Ground-truth annotations for 2D pose estimation and body-part segmentation.

3.4 Body-Part Segmentation

Commonly referred to as human parsing, body-part segmentation aims to classify pixels in the input image $\mathbf{I}$ into C classes. Most methods [41] transform this problem to estimating per-pixel class probabilities to create a probability map $\hat{\mathbf{p}} \in \mathbb{R}^{H \times W \times C}$ such that $\hat{\mathbf{p}} = \mathcal{S}(\mathbf{I})$, where $\mathcal{S}$ is the segmentation model. As outlined previously, we adopt the same encoder-decoder architecture and initialization scheme for $\mathcal{S}$. $\mathcal{S}$ is finetuned to minimize the weighted cross-entropy loss between the actual $\mathbf{p}$ and predicted $\hat{\mathbf{p}}$ probability maps, $\mathcal{L}_{\text{seg}} = \texttt{WeightedCE}(\mathbf{p}, \hat{\mathbf{p}})$.

We finetune $\mathcal{S}$ across two part-segmentation vocabularies: a standard set with $C = 20$ [41] and a new larger vocabulary with $C = 28$, as illustrated in Fig. 4 (*Right*). Our proposed vocabulary goes beyond previous datasets in important ways. It distinguishes between the upper and lower halves of limbs and incorporates more detailed classifications such as upper/lower lips, teeth, and tongue. To this end, we manually annotate $100K$ images at $4K$ resolution with this vocabulary.

3.5 Depth Estimation

For depth estimation, we adopt the architecture used for segmentation, with the modification that the decoder output channel is set to 1 for regression. We denote the ground-truth depth map of image $\mathbf{I}$ by $\mathbf{d} \in \mathbb{R}^{H \times W}$, the depth estimator by $\mathcal{D}$, where $\hat{\mathbf{d}} = \mathcal{D}(\mathbf{I})$, and M as the number of human pixels in the image. For the relative depth estimation, we normalize $\mathbf{d}$ to the range $[0, 1]$ using max and min depths in the image. The $\mathcal{L}_{\text{depth}}$ loss [32] for $\mathcal{D}$ is defined as follows:

$$\Delta \mathbf{d} = \log(\mathbf{d}) - \log(\hat{\mathbf{d}}), \tag{1}$$

$$\overline{\Delta \mathbf{d}} = \frac{1}{M} \sum_{i=1}^{M} \Delta \mathbf{d}_i, \qquad \overline{(\Delta \mathbf{d})^2} = \frac{1}{M} \sum_{i=1}^{M} (\Delta \mathbf{d}_i)^2, \tag{2}$$

$$\mathcal{L}_{\text{depth}} = \sqrt{\overline{(\Delta \mathbf{d})^2} - \frac{1}{2}(\overline{\Delta \mathbf{d}})^2}. \tag{3}$$

We render $500{,}000$ synthetic images using 600 high-resolution photogrammetry human scans as shown in Fig. 5 to obtain a robust monocular depth estimation model with high-fidelity. A random background is selected from a 100 HDRI environment map collection. We place a virtual camera within the scene, randomly adjusting its focal length, rotation, and translation to capture images and their associated ground-truth depth maps at $4K$ resolution.

Fig. 5. Ground-truth synthetic annotations for depth and surface normal estimation.

3.6 Surface Normal Estimation

Similar to previous tasks, we set the decoder output channels of the normal estimator $\mathcal{N}$ to be 3, corresponding to the xyz components of the normal vector at each pixel. The generated synthetic data is also used as supervision for surface normal estimation. Let **n** be the ground-truth normal map for image **I** and $\hat{\mathbf{n}} = \mathcal{N}(\mathbf{I})$. Similar to depth, the loss $\mathcal{L}_{\text{normal}}$ is only computed for human pixels in the image and is defined as follows:

$$\mathcal{L}_{\text{normal}} = ||\mathbf{n} - \hat{\mathbf{n}}||_1 + (1 - \mathbf{n} \cdot \hat{\mathbf{n}}) \tag{4}$$

4 Experiments

In this section, we initially provide an overview of the implementation details. Subsequently, we conduct comprehensive benchmarking across four tasks: pose estimation, part segmentation, depth estimation, and normal estimation.

4.1 Implementation Details

Our largest model, Sapiens-2B, is pretrained using 1024 A100 GPUs for 18 days using PyTorch. We use the AdamW [73] optimizer for all our experiments. The learning schedule includes a brief linear warm-up, followed by cosine annealing [72] for pretraining and linear decay [65] for finetuning. All models are pretrained from scratch at a resolution of 1024 × 1024 with a patch size of 16. For finetuning, the input image is resized to a 4:3 ratio, *i.e.* 1024 × 768. We use standard augmentations like cropping, scaling, flipping, and photometric distortions. A random background from non-human COCO [67] images is added for

Table 2. Sapiens encoder specifications for pretraining on Human-300M dataset.

Model	#Params	FLOPs	Hidden size	Layers	Heads	Batch size
Sapiens-0.3 B	0.336 B	1.242 T	1024	24	16	98,304
Sapiens-0.6 B	0.664 B	2.583 T	1280	32	16	65,536
Sapiens-1 B	1.169 B	4.647 T	1536	40	24	40,960
Sapiens-2 B	2.163 B	8.709 T	1920	48	32	20,480

segmentation, depth, and normal prediction tasks. Importantly, we use differential learning rates [115] to preserve generalization *i.e.* lower learning rates for initial layers and progressively higher rates for subsequent layers. The layer-wise learning rate decay is set to 0.85 with a weight decay of 0.1 for the encoder. We detail the design specifications of Sapiens in Table 2. Following [34,100], we prioritize scaling models by width rather than depth. Note that the Sapiens-0.3B model, while architecturally similar to the traditional ViT-Large, consists of twentyfold more FLOPs due to its higher resolution.

4.2 2D Pose Estimation

We finetune Sapiens for face, body, feet, and hand ($K = 308$) pose estimation on our high-fidelity annotations. For training, we use the train set with $1M$ images and for evaluation, we use the test set, named Humans-5K, with $5K$ images. Our evaluation is top-down [111] *i.e.* we use an off-the-shelf detector [37] for bounding-box and conduct single human pose inference. Table 3 shows a comparison of our models with existing methods for whole-body pose estimation. We evaluate all methods on 114 common keypoints between our 308 keypoint vocabulary and the 133 keypoint vocabulary from COCO-WholeBody [55]. Sapiens-0.6B surpasses the current state-of-the-art, DWPose-l [114] by +2.8 AP. Contrary to DWPose [114], which utilizes a complex student-teacher framework with feature distillation tailored for the task, Sapiens adopts a general encoder-decoder architecture with large human-centric pretraining.

Interestingly, even with the same parameter count, our models demonstrate superior performance compared to their counterparts. For instance, Sapiens-0.3B exceeds VitPose+-L by +5.6 AP, and Sapiens-0.6B outperforms VitPose+-H by +7.9 AP. Within the Sapiens family, our results indicate a direct correlation between model size and performance. Sapiens-2B sets a state-of-the-art with 61.1 AP, a significant improvement of **+7.6 AP** to the prior art. Despite fine-tuning with annotations from a indoor capture studio, Sapiens demonstrate robust generalization to real-world, as shown in Fig. 6.

Table 3. Pose estimation results on Humans-5K test set. Flip test is used.

Model	Input Size	Body		Foot		Face		Hand		Whole-body	
		AP	AR	AP	AR	AP	AR	AP	AR	AP	AR
DeepPose [98]	384 × 288	32.1	43.5	25.3	41.2	37.8	53.9	15.7	31.6	23.9	37.2
SimpleBaseline [106]	384 × 288	52.3	60.1	49.8	62.5	59.6	67.3	41.4	51.8	44.6	53.7
HRNet [93]	384 × 288	55.8	62.6	45.2	55.4	58.9	64.5	39.3	47.6	45.7	53.9
ZoomNAS [110]	384 × 288	59.7	66.3	48.1	57.9	74.5	79.2	49.8	60.6	52.1	60.7
ViTPose+-L [112]	256 × 192	61.0	66.8	62.4	68.2	50.1	55.7	41.5	47.3	47.8	53.6
ViTPose+-H [112]	256 × 192	61.6	67.4	63.2	69.0	50.7	56.3	42.0	47.8	48.3	54.1
RTMPose-x [54]	384 × 288	57.1	63.7	55.3	66.8	74.4	78.5	46.3	55.0	51.9	59.6
DWPose-m [114]	256 × 192	54.2	61.4	49.9	63.0	68.5	74.2	40.1	50.0	47.7	55.8
DWPose-l [114]	384 × 288	57.9	64.2	56.5	67.4	74.3	78.4	49.3	57.4	53.1	60.6
Sapiens-0.3B (Ours)	1024 × 768	58.1	64.5	56.8	67.7	74.5	78.6	49.6	57.7	53.4 (+0.3)	60.9 (+0.3)
Sapiens-0.6B (Ours)	1024 × 768	59.8	65.5	64.7	72.3	75.2	79.0	52.1	60.3	56.2 (+2.8)	62.4 (+2.1)
Sapiens-1B (Ours)	1024 × 768	62.9	68.2	68.3	75.1	76.4	79.7	55.9	63.4	59.4 (+5.9)	65.3 (+5.1)
Sapiens-2B (Ours)	1024 × 768	**64.7**	**69.9**	**69.4**	**76.2**	**76.9**	**79.9**	**57.1**	**64.4**	**61.1**(+7.6)	**67.1**(+7.0)

Fig. 6. Pose estimation with Sapiens-1B for 308 keypoints on in-the-wild images.

4.3 Body-Part Segmentation

We fine-tune and evaluate our annotations with a segmentation vocabulary of 28 classes. Our train set consists of $100K$ images, and the test set, Humans-2K, consists of $2K$ images. We compare Sapiens with existing body-part segmentation methods fine-tuned on our train set. Importantly, we use suggested pretrained checkpoints by each method as initialization. Similar to pose, we observe generalization to segmentation as shown in Table 4.

Table 4. We report mIoU and mAcc on Humans-2K test set. Methods with * are trained by us.

Model	mIoU (%)	mAcc (%)
FCN* [71]	48.2	57.6
SegFormer* [107]	53.5	62.9
Mask2Former* [22]	58.7	68.3
DeepLabV3+* [18]	64.1	74.8
Sapiens-0.3 B (Ours)	76.7	86.1
Sapiens-0.6 B (Ours)	77.8	86.3
Sapiens-1 B (Ours)	79.9	89.1
Sapiens-2 B (Ours)	**81.2**	**89.4**

Interestingly, our smallest model, Sapiens-0.3B outperforms existing state-of-the-art segmentation methods like Mask2Former [22] and DeepLabV3+ [18] by 12.6 mIoU due to its higher resolution and large human-centric pretraining. Furthermore, increasing the model size improves segmentation performance.

Sapiens-2B achieves the best performance of 81.2 mIoU and 89.4 mAcc on the test set. Figure 7 shows the qualitative results of our models.

4.4 Depth Estimation

We evaluate our models on THuman2.0 [117] and Hi4D [116] datasets for depth estimation. THuman2.0 consists of 526 high-quality human scans, from which we derive three sets of images for testing: a) face, b) upper body, and c) full body using a virtual camera. THuman2.0 with 1578 images thus enables the evaluation of our models' performance on single-human images across multiple scales. Conversely, the Hi4D dataset focuses on multi-human scenarios, with each sequence showcasing two subjects engaged in activities involving human-human interactions. We select sequences from pair 28, pair 32, and pair 37, featuring 6 unique subjects from camera 4, totaling 1195 multi-human real images for testing. We follow the relative-depth evaluation protocols established by MiDaS-v3.1 [11], reporting standard metrics such as AbsRel and δ_1. In addition, we also report RMSE as our primary metric since δ_1 does not effectively reflect performance in human scenes characterized by subtle depth variations.

Table 5 compares our models with existing state-of-the-art monocular depth estimators. Sapiens-2B, finetuned solely on synthetic data, remarkably outperforms prior art across all single-human scales and multi-human scenarios. We observe a 20% RMSE reduction compared to the top-performing Depth-Anything model on Hi4D images. It is important to highlight that while baseline models are trained on a variety of scenes, Sapiens specializes in human-centric

Fig. 7. Human segmentation with Sapiens-1B for 28 categories on in-the-wild images.

Table 5. Comparison of Sapiens for monocular depth estimation on human images.

Method	TH2.0-Face			TH2.0-UprBody			TH2.0-FullBody			Hi4D		
	RMSE ↓	AbsRel ↓	δ_1 ↑	RMSE	AbsRel	δ_1	RMSE	AbsRel	δ_1	RMSE	AbsRel	δ_1
MiDaS-L [11]	0.114	0.097	0.925	0.398	0.271	0.868	0.701	0.689	0.782	0.261	0.082	0.975
MiDaS-Swin2 [11]	0.050	0.036	0.995	0.122	0.081	0.948	0.292	0.171	0.862	0.209	0.063	0.997
DepthAny-B [113]	0.039	0.026	0.999	0.048	0.028	0.999	0.061	0.030	0.999	0.143	**0.034**	0.997
DepthAny-L [113]	0.039	0.027	0.999	0.048	0.027	0.999	0.060	0.030	0.999	0.147	0.035	0.997
Sapiens-0.3B (Ours)	0.012	0.008	1.000	0.015	0.009	1.000	0.021	0.010	1.000	0.148	0.046	1.000
Sapiens-0.6B (Ours)	0.011	0.008	1.000	0.015	0.009	1.000	0.021	0.010	1.000	0.142	0.044	1.000
Sapiens-1B (Ours)	0.009	0.006	1.000	0.012	0.007	1.000	0.019	0.009	1.000	0.125	0.039	1.000
Sapiens-2B (Ours)	**0.008**	**0.005**	**1.000**	**0.010**	**0.006**	**1.000**	**0.016**	**0.008**	**1.000**	**0.114**	0.036	**1.000**

depth estimation. Figure 8 presents a qualitative comparison of depth estimation between Sapiens-1B and DepthAnything-L. To ensure a fair comparison, the predicted depth is renormalized using the human mask in the baseline visualizations.

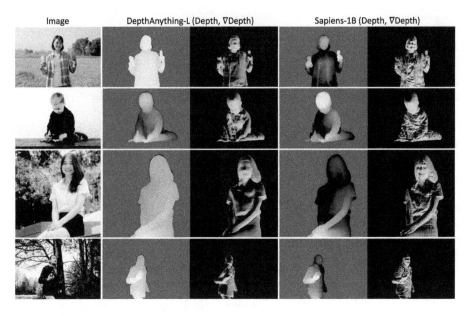

Fig. 8. We compare our depth prediction with DepthAnything [113]. To showcase the consistency of predicted depth, we also visualize the ∇depth as pseudo surface normals.

4.5 Surface Normal Estimation

The datasets for surface normal evaluation are identical to those used for depth estimation. Following [30], we report the mean and median angular error, along with the percentage of pixels within $t°$ error for $t \in \{11.25°, 22.5°, 30°\}$. Table 6 compares our models with existing human-specific surface normal estimators. All our models outperform existing methods by a significant margin. Sapiens-2B achieves a mean error of around $12°$ on the THuman2.0 (single-human) and Hi4D (multi-human) datasets. We qualitatively compare Sapiens-1B with PIFuHD [89] and ECON [108] for surface normal estimation in Fig. 9. Note that PIFuHD [89] is trained with the identical set of 3D scans as ours, and ECON [108] is trained with 4000 scans that are a super set of our 3D scan data.

Table 6. Comparison of Sapiens for surface normal estimation on human images.

Method	THuman2.0 [117]					Hi4D [116]				
	Angular Error°		% Within $t°$			Angular Error°		% Within $t°$		
	Mean	Median	11.25°	22.5°	30°	Mean	Median	11.25°	22.5°	30°
PIFuHD [89]	30.51	27.13	15.81	42.97	58.86	22.39	19.26	22.98	60.14	77.02
HDNet [52]	34.82	30.60	17.44	39.26	54.51	28.60	26.85	19.08	57.93	70.14
ICON [109]	28.74	25.52	22.81	47.83	63.73	20.18	17.52	26.81	66.34	82.73
ECON [108]	25.45	23.67	32.95	55.86	69.03	18.46	16.47	29.35	68.12	84.88
Sapiens-0.3 B	13.02	10.33	57.37	86.20	92.7	15.04	12.22	47.07	81.49	90.70
Sapiens-0.6 B	12.86	10.23	57.85	86.68	93.30	14.06	11.47	50.59	84.37	92.54
Sapiens-1 B	12.11	9.40	61.97	88.03	93.84	12.18	**9.59**	**60.36**	88.62	94.44
Sapiens-2 B	**11.84**	**9.16**	**63.16**	**88.60**	**94.18**	12.14	9.62	60.22	**89.08**	**94.74**

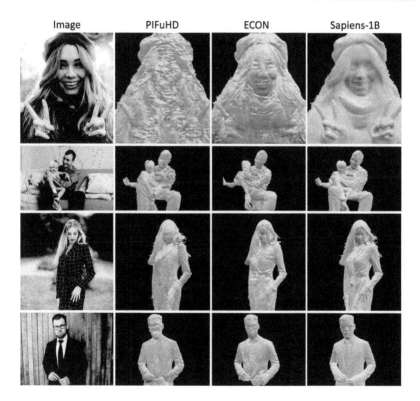

Fig. 9. Qualitative comparison of Sapiens-1B with PIFuHD [89] and ECON [108] for monocular surface normal estimation on challenging in-the-wild images.

4.6 Discussion

Importance of Pretraining Data Source. The feature quality is closely linked to the pretraining data quality. We assess the importance of pretraining on various data sources for human-centric tasks by pretraining Sapiens-0.3B on each dataset under identical training schedules and number of iterations. We fine-tune the model on each task and select early checkpoints for evaluation, reasoning that early-stage fine-tuning better reflects the model's generalization capability. We investigate the impact of pretraining at scale on general images (which may include humans) versus exclusively human images using Sapiens. We randomly select 100 million and 300 million general images from our 1 billion image corpus to create the General-100M and General-300M datasets, respectively. Table 7 showcases the comparison of pretraining outcomes. We report mAP for pose on Humans-5K, mIoU for segmentation on Humans-2K, RMSE for depth on THuman2.0, and mean angular error in degrees for normal estimation on Hi4D. Aligned with findings from [112], our results show that pretraining with Human300M leads to superior performance across all metrics, highlighting the benefits of human-centric pretraining within a fixed computational budget.

We also study the effect of number of unique human images seen during pretraining with normal estimation performance. We report % within 30°. Again, we maintain identical conditions for Sapiens-0.3B pretraining and finetuning. Figure 10 shows a steady improvement in performance as the pretraining data size increases without saturation. In summary, the diversity of human images during pretraining directly correlates with improved generalization to down-stream tasks.

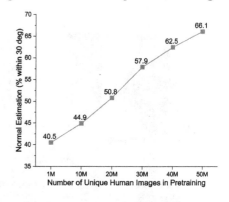

Fig. 10. Sapiens-0.3B's normal estimation performance with unique human images seen during pretraining.

Zero-Shot Generalization. Our models exhibit broad generalization to a variety of settings. For instance, in segmentation, Sapiens are finetuned on single-human images with limited subject diversity, minimal background variation, and solely third-person views (see Fig. 4). Nevertheless, our large-scale pretraining enables generalization across number of subjects, varying ages, and egocentric views, as shown in Fig. 11. These observations similarly hold for other tasks.

Table 7. Comparison of Sapiens-0.3B pretrained on various data sources. A domain-specific pretraining yields superior results compared to general data sources.

Pretraining Source	#Images	Pose (↑)	Seg(↑)	Depth(↓)	Normal(↓)
Random Initialization	–	30.2	40.3	0.720	35.4
General-100M	100M	35.7	50.1	0.351	27.5
General-300M	300M	37.3	52.8	0.347	26.8
Humans-100M	100M	43.6	61.2	0.316	24.0
Humans-300M (Full)	300M	**47.0**	**66.5**	**0.288**	**21.8**

| Multi-Human Generalization | Age Generalization | Viewpoint Generalization |

Fig. 11. *Sapiens* achieve broad generalization via large human-centric pretraining.

Applications. Accurately predicting pose, body-part segmentation, depth and surface-normals for in-the-wild human images enables a wide range of applications, such as 3D human scans and the generation of controllable images. We adapt ControlNet [118] to integrate keypoints for faces and hands predicted by Sapiens, and train this modified model on a subset of the Humans-300M dataset, as shown in Fig. 12. This grants more precise control over image generation, producing anatomically consistent human images.

| Image | Predicted Pose | Pose-conditioned Human Image Synthesis |

Fig. 12. Image synthesis using ControlNet [118] trained on Sapiens pose predictions.

Limitations. While our models generally perform well, they are not perfect. Human images with complex/rare poses, crowding, and severe occlusion are challenging (see supplemental for details). Although aggressive data augmentation and a detect-and-crop strategy could mitigate these issues, we envision our models as a tool for acquiring large-scale, real-world supervision with human-in-the-loop to develop the next generations of human vision models.

5 Conclusion

Sapiens represents a significant step toward elevating human-centric vision models into the realm of foundation models. Our models demonstrate strong generalization capabilities on a variety of human-centric tasks. We attribute the state-of-the-art performance of our models to: (i) large-scale pretraining on a large curated dataset, which is specifically tailored to understanding humans, (ii) scaled high-resolution and high-capacity vision transformer backbones, and (iii) high-quality annotations on augmented studio and synthetic data. We believe that these models can become a key building block for a multitude of downstream tasks, and provide access to high-quality vision backbones to a significantly wider part of the community. A potential direction for future work would be extending *Sapiens* to 3D and multi-modal datasets.

References

1. Abnar, S., Dehghani, M., Neyshabur, B., Sedghi, H.: Exploring the limits of large scale pre-training. arXiv preprint arXiv:2110.02095 (2021)
2. Achiam, J., et al.: Gpt-4 technical report. arXiv preprint arXiv:2303.08774 (2023)
3. Alldieck, T., Zanfir, M., Sminchisescu, C.: Photorealistic monocular 3d reconstruction of humans wearing clothing. In: Proceedings of the IEEE/CVF Conference on Computer Vision and Pattern Recognition, pp. 1506–1515 (2022)
4. Bai, Y., et al.: Sequential modeling enables scalable learning for large vision models. arXiv preprint arXiv:2312.00785 (2023)
5. Bao, H., Dong, L., Piao, S., Wei, F.: Beit: bert pre-training of image transformers. arXiv preprint arXiv:2106.08254 (2021)
6. Barron, J.T., Malik, J.: Intrinsic scene properties from a single rgb-d image. In: Proceedings of the IEEE Conference on Computer Vision and Pattern Recognition, pp. 17–24 (2013)
7. Barron, J.T., Malik, J.: Shape, illumination, and reflectance from shading. IEEE Trans. Pattern Anal. Mach. Intell. **37**(8), 1670–1687 (2014)
8. Bartol, K., Bojanić, D., Petković, T., Pribanić, T.: A review of body measurement using 3d scanning. IEEE Access **9**, 67281–67301 (2021)
9. Bhat, S.F., Alhashim, I., Wonka, P.: Adabins: depth estimation using adaptive bins. In: Proceedings of the IEEE/CVF Conference on Computer Vision and Pattern Recognition, pp. 4009–4018 (2021)
10. Bhat, S.F., Birkl, R., Wofk, D., Wonka, P., Müller, M.: Zoedepth: zero-shot transfer by combining relative and metric depth. arXiv preprint arXiv:2302.12288 (2023)
11. Birkl, R., Wofk, D., Müller, M.: Midas v3. 1–a model zoo for robust monocular relative depth estimation. arXiv preprint arXiv:2307.14460 (2023)
12. Bojic, L.: Metaverse through the prism of power and addiction: what will happen when the virtual world becomes more attractive than reality? Eur. J. Futures Res. **10**(1), 1–24 (2022)
13. Brown, T., et al.: Language models are few-shot learners. Adv. Neural. Inf. Process. Syst. **33**, 1877–1901 (2020)
14. Cao, Z., Simon, T., Wei, S., Sheikh, Y.: Realtime multi-person 2d pose estimation using part affinity fields. arXiv preprint arXiv:1611.08050 (2016)
15. Cao, Z., Hidalgo, G., Simon, T., Wei, S.E., Sheikh, Y.: Openpose: real-time multi-person 2d pose estimation using part affinity fields. arXiv preprint arXiv:1812.08008 (2018)
16. Caron, M., et al.: Emerging properties in self-supervised vision transformers. In: Proceedings of the IEEE/CVF International Conference on Computer Vision, pp. 9650–9660 (2021)
17. Chan, C., Ginosar, S., Zhou, T., Efros, A.A.: Everybody dance now. In: Proceedings of the IEEE/CVF International Conference on Computer Vision, pp. 5933–5942 (2019)

18. Chen, L.C., Zhu, Y., Papandreou, G., Schroff, F., Adam, H.: Encoder-decoder with atrous separable convolution for semantic image segmentation. In: Proceedings of the European Conference on Computer Vision (ECCV), pp. 801–818 (2018)
19. Chen, T., Kornblith, S., Norouzi, M., Hinton, G.: A simple framework for contrastive learning of visual representations. In: International Conference on Machine Learning, pp. 1597–1607. PMLR (2020)
20. Chen, X., et al.: Pali: a jointly-scaled multilingual language-image model. arXiv preprint arXiv:2209.06794 (2022)
21. Chen, Y., Wang, Z., Peng, Y., Zhang, Z., Yu, G., Sun, J.: Cascaded pyramid network for multi-person pose estimation. In: Proceedings of the IEEE Conference on Computer Vision and Pattern Recognition, pp. 7103–7112 (2018)
22. Cheng, B., Misra, I., Schwing, A., Kirillov, A., Girdhar, R.: Masked-attention mask transformer for universal image segmentation. arXiv preprint arXiv:2112.01527 (2021)
23. Couprie, C., Farabet, C., Najman, L., LeCun, Y.: Indoor semantic segmentation using depth information. arXiv preprint arXiv:1301.3572 (2013)
24. Dai, Z., Liu, H., Le, Q.V., Tan, M.: Coatnet: marrying convolution and attention for all data sizes. Adv. Neural. Inf. Process. Syst. **34**, 3965–3977 (2021)
25. Dehghani, M., et al.: Scaling vision transformers to 22 billion parameters. In: International Conference on Machine Learning, pp. 7480–7512. PMLR (2023)
26. Devlin, J., Chang, M.W., Lee, K., Toutanova, K.: Bert: pre-training of deep bidirectional transformers for language understanding. arXiv preprint arXiv:1810.04805 (2018)
27. Dosovitskiy, A., et al.: An image is worth 16x16 words: transformers for image recognition at scale. arXiv preprint arXiv:2010.11929 (2020)
28. Drobyshev, N., Chelishev, J., Khakhulin, T., Ivakhnenko, A., Lempitsky, V., Zakharov, E.: Megaportraits: one-shot megapixel neural head avatars. In: Proceedings of the 30th ACM International Conference on Multimedia, pp. 2663–2671 (2022)
29. Du, Y., Kips, R., Pumarola, A., Starke, S., Thabet, A., Sanakoyeu, A.: Avatars grow legs: generating smooth human motion from sparse tracking inputs with diffusion model. In: Proceedings of the IEEE/CVF Conference on Computer Vision and Pattern Recognition, pp. 481–490 (2023)
30. Eftekhar, A., Sax, A., Malik, J., Zamir, A.: Omnidata: a scalable pipeline for making multi-task mid-level vision datasets from 3d scans. In: Proceedings of the IEEE/CVF International Conference on Computer Visionm, pp. 10786–10796 (2021)
31. Eigen, D., Fergus, R.: Predicting depth, surface normals and semantic labels with a common multi-scale convolutional architecture. In: Proceedings of the IEEE International Conference on Computer Vision, pp. 2650–2658 (2015)
32. Eigen, D., Puhrsch, C., Fergus, R.: Depth map prediction from a single image using a multi-scale deep network. Adv. Neural Inf. Process. Syst. **27** (2014)
33. El-Nouby, A., Izacard, G., Touvron, H., Laptev, I., Jegou, H., Grave, E.: Are large-scale datasets necessary for self-supervised pre-training? arXiv preprint arXiv:2112.10740 (2021)
34. El-Nouby, A., et al.: Scalable pre-training of large autoregressive image models. arXiv preprint arXiv:2401.08541 (2024)
35. Fang, H.S., Lu, G., Fang, X., Xie, J., Tai, Y.W., Lu, C.: Weakly and semi supervised human body part parsing via pose-guided knowledge transfer. arXiv preprint arXiv:1805.04310 (2018)

36. Fang, H.S., Xie, S., Tai, Y.W., Lu, C.: RMPE: regional multi-person pose estimation. In: Proceedings of the IEEE International Conference on Computer Vision, pp. 2334–2343 (2017)
37. Fang, Y., Sun, Q., Wang, X., Huang, T., Wang, X., Cao, Y.: Eva-02: a visual representation for neon genesis. arXiv preprint arXiv:2303.11331 (2023)
38. Fang, Y., et al.: Eva: exploring the limits of masked visual representation learning at scale. In: Proceedings of the IEEE/CVF Conference on Computer Vision and Pattern Recognition, pp. 19358–19369 (2023)
39. Fouhey, D.F., Gupta, A., Hebert, M.: Data-driven 3d primitives for single image understanding. In: Proceedings of the IEEE International Conference on Computer Vision, pp. 3392–3399 (2013)
40. Gong, K., Liang, X., Li, Y., Chen, Y., Yang, M., Lin, L.: Instance-level human parsing via part grouping network. In: Proceedings of the European Conference on Computer Vision (ECCV), pp. 770–785 (2018)
41. Gong, K., Liang, X., Zhang, D., Shen, X., Lin, L.: Look into person: self-supervised structure-sensitive learning and a new benchmark for human parsing. In: Proceedings of the IEEE Conference on Computer Vision and Pattern Recognition, pp. 932–940 (2017)
42. Goyal, P., et al.: Self-supervised pretraining of visual features in the wild. arXiv preprint arXiv:2103.01988 (2021)
43. Guizilini, V., Vasiljevic, I., Chen, D., Ambrus, R., Gaidon, A.: Towards zero-shot scale-aware monocular depth estimation. In: Proceedings of the IEEE/CVF International Conference on Computer Vision, pp. 9233–9243 (2023)
44. Halstead, M.A., Barsky, B.A., Klein, S.A., Mandell, R.B.: Reconstructing curved surfaces from specular reflection patterns using spline surface fitting of normals. In: Proceedings of the 23rd Annual Conference on Computer Graphics and Interactive Techniques, pp. 335–342 (1996)
45. He, K., Chen, X., Xie, S., Li, Y., Dollár, P., Girshick, R.: Masked autoencoders are scalable vision learners. In: Proceedings of the IEEE/CVF Conference on Computer Vision and Pattern Recognition, pp. 16000–16009 (2022)
46. He, K., Fan, H., Wu, Y., Xie, S., Girshick, R.: Momentum contrast for unsupervised visual representation learning. In: Proceedings of the IEEE/CVF Conference on Computer Vision and Pattern Recognition, pp. 9729–9738 (2020)
47. He, K., Gkioxari, G., Dollár, P., Girshick, R.B.: Mask r-cnn. arXiv preprint arXiv:1703.06870 (2017)
48. He, K., Zhang, X., Ren, S., Sun, J.: Deep residual learning for image recognition. In: Proceedings of the IEEE Conference on Computer Vision and Pattern Recognition, pp. 770–778 (2016)
49. Hoffmann, J., et al.: Training compute-optimal large language models. arXiv preprint arXiv:2203.15556 (2022)
50. Hu, L., Gao, X., Zhang, P., Sun, K., Zhang, B., Bo, L.: Animate anyone: consistent and controllable image-to-video synthesis for character animation. arXiv preprint arXiv:2311.17117 (2023)
51. Huang, S., Gong, M., Tao, D.: A coarse-fine network for keypoint localization. In: Proceedings of the IEEE International Conference on Computer Vision, pp. 3028–3037 (2017)
52. Jafarian, Y., Park, H.S.: Learning high fidelity depths of dressed humans by watching social media dance videos. In: Proceedings of the IEEE/CVF Conference on Computer Vision and Pattern Recognition, pp. 12753–12762 (2021)
53. Jiang, A.Q., et al.: Mistral 7b. arXiv preprint arXiv:2310.06825 (2023)

54. Jiang, T., et al.: Rtmpose: real-time multi-person pose estimation based on mmpose. arXiv preprint arXiv:2303.07399 (2023)
55. Jin, S., et al.: Whole-body human pose estimation in the wild. In: Vedaldi, A., Bischof, H., Brox, T., Frahm, J.-M. (eds.) ECCV 2020. LNCS, vol. 12354, pp. 196–214. Springer, Cham (2020). https://doi.org/10.1007/978-3-030-58545-7_12
56. Kato, N., Li, T., Nishino, K., Uchida, Y.: Improving multi-person pose estimation using label correction. arXiv preprint arXiv:1811.03331 (2018)
57. Khirodkar, R., Chari, V., Agrawal, A., Tyagi, A.: Multi-instance pose networks: rethinking top-down pose estimation. In: Proceedings of the IEEE/CVF International Conference on Computer Vision, pp. 3122–3131 (2021)
58. Kirillov, A., et al.: Segment anything. arXiv preprint arXiv:2304.02643 (2023)
59. Kocabas, M., Chang, J.H.R., Gabriel, J., Tuzel, O., Ranjan, A.: Hugs: human Gaussian splats. arXiv preprint arXiv:2311.17910 (2023)
60. Krishna, R., et al.: Visual genome: connecting language and vision using crowdsourced dense image annotations. Int. J. Comput. Vision **123**, 32–73 (2017)
61. Krizhevsky, A., Hinton, G., et al.: Learning multiple layers of features from tiny images. arXiv preprint (2009)
62. Ladický, L., Zeisl, B., Pollefeys, M.: Discriminatively trained dense surface normal estimation. In: ECCV 2014, Part V 13, pp. 468–484. Springer (2014)
63. Lawrence, J., et al.: Project starline: a high-fidelity telepresence system. arXiv preprint (2021)
64. Levoy, M., et al.: The digital michelangelo project: 3d scanning of large statues. In: Proceedings of the 27th Annual Conference on Computer Graphics and Interactive Techniques, pp. 131–144 (2000)
65. Lewkowycz, A.: How to decay your learning rate. arXiv preprint arXiv:2103.12682 (2021)
66. Li, Z., Wang, X., Liu, X., Jiang, J.: Binsformer: revisiting adaptive bins for monocular depth estimation. arXiv preprint arXiv:2204.00987 (2022)
67. Lin, T.-Y., et al.: Microsoft COCO: common objects in context. In: Fleet, D., Pajdla, T., Schiele, B., Tuytelaars, T. (eds.) ECCV 2014. LNCS, vol. 8693, pp. 740–755. Springer, Cham (2014). https://doi.org/10.1007/978-3-319-10602-1_48
68. Liu, Z., et al.: Swin transformer v2: scaling up capacity and resolution. In: Proceedings of the IEEE/CVF Conference on Computer Vision and Pattern Recognition, pp. 12009–12019 (2022)
69. Lombardi, S., Simon, T., Saragih, J., Schwartz, G., Lehrmann, A., Sheikh, Y.: Neural volumes: learning dynamic renderable volumes from images. arXiv preprint arXiv:1906.07751 (2019)
70. Lombardi, S., Simon, T., Schwartz, G., Zollhoefer, M., Sheikh, Y., Saragih, J.: Mixture of volumetric primitives for efficient neural rendering. ACM Trans. Graph. **40**(4), 1–13 (2021)
71. Long, J., Shelhamer, E., Darrell, T., Berkeley, U.: Fully convolutional networks for semantic segmentation. arXiv preprint arXiv:1411.4038 (2014)
72. Loshchilov, I., Hutter, F.: SGDR: stochastic gradient descent with warm restarts. arXiv preprint arXiv:1608.03983 (2016)
73. Loshchilov, I., Hutter, F.: Decoupled weight decay regularization. arXiv preprint arXiv:1711.05101 (2017)
74. Lowe, D.G.: Three-dimensional object recognition from single two-dimensional images. Artif. Intell. **31**(3), 355–395 (1987)
75. Luo, Y., Zheng, Z., Zheng, L., Guan, T., Yu, J., Yang, Y.: Macro-micro adversarial network for human parsing. In: Proceedings of the European Conference on Computer Vision (ECCV), pp. 418–434 (2018)

76. Ma, S., et al.:: Pixel codec avatars. In: Proceedings of the IEEE/CVF Conference on Computer Vision and Pattern Recognition, pp. 64–73 (2021)
77. Mahajan, D., et al.: Exploring the limits of weakly supervised pretraining. In: Proceedings of the European Conference on Computer Vision (ECCV), pp. 181–196 (2018)
78. Newell, A., Yang, K., Deng, J.: Stacked hourglass networks for human pose estimation. In: European Conference on Computer Vision, pp. 483–499. Springer (2016)
79. Oquab, M., et al.: Dinov2: learning robust visual features without supervision. arXiv preprint arXiv:2304.07193 (2023)
80. Papandreou, G., et al.: Towards accurate multi-person pose estimation in the wild. In: Proceedings of the IEEE Conference on Computer Vision and Pattern Recognition, pp. 4903–4911 (2017)
81. Peebles, W., Xie, S.: Scalable diffusion models with transformers. In: Proceedings of the IEEE/CVF International Conference on Computer Vision, pp. 4195–4205 (2023)
82. Render people: 3d people for architectural visualization — renderpeople, https://renderpeople.com/3d-people. Accessed 22 Feb 2024
83. Radford, A., et al.: Learning transferable visual models from natural language supervision. In: International Conference on Machine Learning, pp. 8748–8763. PMLR (2021)
84. Ranftl, R., Lasinger, K., Hafner, D., Schindler, K., Koltun, V.: Towards robust monocular depth estimation: mixing datasets for zero-shot cross-dataset transfer. IEEE Trans. Pattern Anal. Mach. Intell. **44**(3), 1623–1637 (2020)
85. Rombach, R., Blattmann, A., Lorenz, D., Esser, P., Ommer, B.: High-resolution image synthesis with latent diffusion models. In: Proceedings of the IEEE/CVF Conference on Computer Vision and Pattern Recognition, pp. 10684–10695 (2022)
86. Russakovsky, O., et al.: Imagenet large scale visual recognition challenge. Int. J. Comput. Vision **115**, 211–252 (2015)
87. Saharia, C., et al.: Photorealistic text-to-image diffusion models with deep language understanding. Adv. Neural. Inf. Process. Syst. **35**, 36479–36494 (2022)
88. Saito, S., Huang, Z., Natsume, R., Morishima, S., Kanazawa, A., Li, H.: Pifu: pixel-aligned implicit function for high-resolution clothed human digitization. In: Proceedings of the IEEE/CVF International Conference on Computer Vision, pp. 2304–2314 (2019)
89. Saito, S., Simon, T., Saragih, J., Joo, H.: Pifuhd: multi-level pixel-aligned implicit function for high-resolution 3d human digitization. In: Proceedings of the IEEE/CVF Conference on Computer Vision and Pattern Recognition, pp. 84–93 (2020)
90. Schuhmann, C., et al.: Laion-5b: an open large-scale dataset for training next generation image-text models. Adv. Neural. Inf. Process. Syst. **35**, 25278–25294 (2022)
91. Singh, M., et al.: The effectiveness of MAE pre-pretraining for billion-scale pretraining. arXiv preprint arXiv:2303.13496 (2023)
92. Sun, C., Shrivastava, A., Singh, S., Gupta, A.: Revisiting unreasonable effectiveness of data in deep learning era. In: Proceedings of the IEEE International Conference on Computer Vision, pp. 843–852 (2017)
93. Sun, K., Xiao, B., Liu, D., Wang, J.: Deep high-resolution representation learning for human pose estimation. In: Proceedings of the IEEE Conference on Computer Vision and Pattern Recognition, pp. 5693–5703 (2019)

94. Taori, R., et al.: Alpaca: a strong, replicable instruction-following model. Stanford Center for Research on Foundation Models **3**(6), 7 (2023). https://crfmstanford.edu/2023/03/13/alpaca.html
95. Tay, Y., et al.: Scale efficiently: insights from pre-training and fine-tuning transformers. arXiv preprint arXiv:2109.10686 (2021)
96. Team, G., et al.: Gemini: a family of highly capable multimodal models. arXiv preprint arXiv:2312.11805 (2023)
97. Thomee, B., et al.: Yfcc100m: the new data in multimedia research. Commun. ACM **59**(2), 64–73 (2016)
98. Toshev, A., Szegedy, C.: Deeppose: human pose estimation via deep neural networks. In: Proceedings of the IEEE Conference on Computer Vision and Pattern Recognition, pp. 1653–1660 (2014)
99. Touvron, H., et al.: Llama: open and efficient foundation language models. arXiv preprint arXiv:2302.13971 (2023)
100. Touvron, H., et al.: Llama 2: open foundation and fine-tuned chat models. arXiv preprint arXiv:2307.09288 (2023)
101. Wang, X., Fouhey, D., Gupta, A.: Designing deep networks for surface normal estimation. In: Proceedings of the IEEE Conference on Computer Vision and Pattern Recognition, pp. 539–547 (2015)
102. Weng, C.Y., Curless, B., Srinivasan, P.P., Barron, J.T., Kemelmacher-Shlizerman, I.: Humannerf: free-viewpoint rendering of moving people from monocular video. In: Proceedings of the IEEE/CVF Conference on Computer Vision and Pattern Recognition, pp. 16210–16220 (2022)
103. Wu, Y., Kirillov, A., Massa, F., Lo, W.Y., Girshick, R.: Detectron2 (2019). https://github.com/facebookresearch/detectron2
104. Xia, F., Wang, P., Chen, L.C., Yuille, A.L.: Zoom better to see clearer: human and object parsing with hierarchical auto-zoom net. In: ECCV 2016, Part V 14. pp. 648–663. Springer (2016)
105. Xia, F., Wang, P., Chen, X., Yuille, A.L.: Joint multi-person pose estimation and semantic part segmentation. In: Proceedings of the IEEE Conference on Computer Vision and Pattern Recognition, pp. 6769–6778 (2017)
106. Xiao, B., Wu, H., Wei, Y.: Simple baselines for human pose estimation and tracking. In: Proceedings of the European Conference on Computer Vision (ECCV), pp. 466–481 (2018)
107. Xie, E., Wang, W., Yu, Z., Anandkumar, A., Alvarez, J.M., Luo, P.: Segformer: simple and efficient design for semantic segmentation with transformers. Adv. Neural. Inf. Process. Syst. **34**, 12077–12090 (2021)
108. Xiu, Y., Yang, J., Cao, X., Tzionas, D., Black, M.J.: Econ: explicit clothed humans optimized via normal integration. arXiv preprint arXiv:2212.07422 (2022)
109. Xiu, Y., Yang, J., Tzionas, D., Black, M.J.: Icon: implicit clothed humans obtained from normals. In: 2022 IEEE/CVF Conference on Computer Vision and Pattern Recognition (CVPR), pp. 13286–13296. IEEE (2022)
110. Xu, L., et al.: Zoomnas: searching for whole-body human pose estimation in the wild. IEEE Trans. Pattern Anal. Mach. Intell. **45**(4), 5296–5313 (2022)
111. Xu, Y., Zhang, J., Zhang, Q., Tao, D.: Vitpose: simple vision transformer baselines for human pose estimation. Adv. Neural. Inf. Process. Syst. **35**, 38571–38584 (2022)
112. Xu, Y., Zhang, J., Zhang, Q., Tao, D.: Vitpose+: vision transformer foundation model for generic body pose estimation. arXiv preprint arXiv:2212.04246 (2022)

113. Yang, L., Kang, B., Huang, Z., Xu, X., Feng, J., Zhao, H.: Depth anything: unleashing the power of large-scale unlabeled data. arXiv preprint arXiv:2401.10891 (2024)
114. Yang, Z., Zeng, A., Yuan, C., Li, Y.: Effective whole-body pose estimation with two-stages distillation. In: Proceedings of the IEEE/CVF International Conference on Computer Vision, pp. 4210–4220 (2023)
115. Yang, Z., Dai, Z., Yang, Y., Carbonell, J., Salakhutdinov, R.R., Le, Q.V.: Xlnet: generalized autoregressive pretraining for language understanding. Adv. Neural Inf. Process. Syst. **32** (2019)
116. Yin, Y., Guo, C., Kaufmann, M., Zarate, J.J., Song, J., Hilliges, O.: Hi4d: 4d instance segmentation of close human interaction. In: Proceedings of the IEEE/CVF Conference on Computer Vision and Pattern Recognition, pp. 17016–17027 (2023)
117. Yu, T., Zheng, Z., Guo, K., Liu, P., Dai, Q., Liu, Y.: Function4d: real-time human volumetric capture from very sparse consumer RGBD sensors. In: Proceedings of the IEEE/CVF Conference on Computer Vision and Pattern Recognition, pp. 5746–5756 (2021)
118. Zhang, L., Rao, A., Agrawala, M.: Adding conditional control to text-to-image diffusion models. In: Proceedings of the IEEE/CVF International Conference on Computer Vision, pp. 3836–3847 (2023)
119. Zhang, S.H., et al.: Pose2seg: detection free human instance segmentation. In: Proceedings of the IEEE Conference on Computer Vision and Pattern Recognition, pp. 889–898 (2019)
120. Zhang, X., et al.: Gpt-4v (ision) as a generalist evaluator for vision-language tasks. arXiv preprint arXiv:2311.01361 (2023)
121. Zhou, J., et al.: ibot: image bert pre-training with online tokenizer. arXiv preprint arXiv:2111.07832 (2021)
122. Zlateski, A., Jaroensri, R., Sharma, P., Durand, F.: On the importance of label quality for semantic segmentation. In: Proceedings of the IEEE Conference on Computer Vision and Pattern Recognition, pp. 1479–1487 (2018)

Linearly Controllable GAN: Unsupervised Feature Categorization and Decomposition for Image Generation and Manipulation

Sehyung Lee[✉], Mijung Kim, Yeongnam Chae, and Björn Stenger

Rakuten Institute of Technology, Rakuten Group, Inc., Tokyo, Japan
{sehyung.lee,mijung.a.kim,yeongnam.chae,bjorn.stenger}@rakuten.com

Abstract. This paper introduces an approach to linearly controllable generative adversarial networks (LC-GAN) driven by unsupervised learning. Departing from traditional methods relying on supervision signals or post-processing for latent feature disentanglement, our proposed technique enables unsupervised learning using only image data through contrastive feature categorization and spectral regularization. In our framework, the discriminator constructs geometry- and appearance-related feature spaces using a combination of image augmentation and contrastive representation learning. Leveraging these feature spaces, the generator autonomously categorizes input latent codes into geometry- and appearance-related features. Subsequently, the categorized features undergo projection into a subspace via our proposed spectral regularization, with each component controlling a distinct aspect of the generated image. Beyond providing fine-grained control over the generative model, our approach achieves state-of-the-art image generation quality on benchmark datasets, including FFHQ, CelebA-HQ, and AFHQ-V2.

Keywords: Controllable GAN · Unsupervised learning · Generative model

1 Introduction

Generative Adversarial Networks (GAN) have emerged as powerful tools for image generation, with the StyleGAN series standing out as prominent models in this domain [16–18]. These models leverage latent codes to shape the generative process; however, interpreting these codes remains a challenging task. Numerous studies have delved into the StyleGAN latent space to enhance controllability. Yet, many controllable StyleGAN-based models necessitate pre-trained classifiers [32], supervision information [31], or reliance on 3D morphable face models

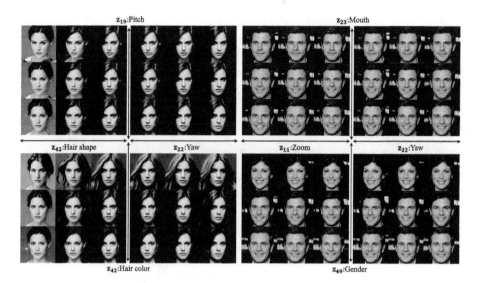

Fig. 1. Illustrative outcomes from LC-GAN trained on CelebA-HQ dataset. The method autonomously learns significant semantics and relates them to input latent codes in a fully unsupervised manner. Leveraging linear control over latent codes, the model generates diverse and customizable outputs with fine-grained adjustments.

(3DMM) [7,34]. Such dependencies can limit generalizability and introduce additional labeling efforts when applied to novel datasets.

An alternative approach involves examining the trained model parameters using subspace projection techniques like Principal Component Analysis (PCA) or integrating additional regularization terms during neural network training, such as Hessian penalty [25] and OroJaR [36], to encourage feature disentanglement. However, subspace analysis-based methods [10,29,30] often require preprocessing steps for projecting model parameters during image generation and post-processing steps for identifying interpretable directions after model training. This multi-step process can be cumbersome and time-consuming and potentially lead to a distortion of the original distribution of the training data. Moreover, orthogonal regularizer-based methods employ a stochastic estimator to approximate the first- or second-order finite difference, which necessarily requires many samples for accurate estimation. This increased demand for samples poses computational challenges, especially in scenarios with very high-dimensional latent codes, potentially affecting the scalability of these regularization techniques and leading to instability in GAN training. Another drawback of approaches such as [11,25,33,36], jointly optimizing orthogonal constraints with GAN, is the potential generation of less diverse and natural images, reflected in much higher Fréchet Inception Distance (FID) scores.

To overcome these limitations, we introduce Linearly Controllable GAN (LC-GAN), a novel approach that employs end-to-end learning for high-quality image synthesis, eliminating the need for pre-trained classifiers, supervision information, or stochastic estimations of first-/second-order finite differences. Our method begins by decomposing the input noise vector, sampled from a Gaussian distribution, into distinct geometry and appearance codes. Initially, the discriminator is trained to construct geometry and appearance embedding feature spaces by clustering training and augmented images through a combination of image augmentation and contrastive feature clustering. Subsequently, the generator utilizes these embedding feature spaces to ensure the consistency of its conditionally generated images by partially resampling the noise, facilitating geometry or appearance feature changes. Additionally, the features, initially separated into geometry and appearance codes, undergo further decomposition into more fine-grained subcategories through a feature selection mechanism. This process directly maps semantic changes in the generated images to changes in the input noise vector, enabling each subcategory to linearly and independently activate semantically meaningful features. This fine-grained control provides the ability to make specific changes, such as variations in appearance or viewpoint, while preserving other essential properties (Fig. 1).

In this work, we make the following contributions to the field of controllable image generation: (1) We introduce an unsupervised method that categorizes input latent codes into distinct geometry and appearance codes. (2) We present a mechanism for further decomposing features separated into geometry and appearance codes into more fine-grained subcategories. This approach not only enables control over semantically meaningful features during the image generation process but also enhances the interpretability of the input latent code. (3) We demonstrate higher quality of the generated images compared to existing state-of-the-art (SOTA) generative models. Our evaluation results show that LC-GAN outperforms other models in terms of both visual quality and feature controllability. We believe that these contributions advance the field of controllable image generation and provide a useful tool for various applications, such as image editing and synthesis. The source code can be found at https://github.com/rakutentech/lcgan.

2 Related Work

GAN [9] is widely used for image generation, but achieving precise control over specific image features remains challenging. To address this limitation, recent approaches have introduced additional data as supervision signals, including segmentation maps [31], image attributes [7,21,28], and text descriptions [22, 27,42]. The primary goal of these methods is to disentangle latent features, enabling more accurate control over the generated images. For instance, [31] uses semantic segmentation masks to guide the generation process, controlling object class and location in the resulting images.

Another approach to generating images involves the use of pre-trained models or synthetic data instead of supervision signals. For instance, [2,32] present GAN models that leverage pre-trained classifiers to verify if the generated images are correctly labeled into the target class. Attempts have also been made to learn 3D pose information from 2D GANs by disentangling pose in the latent space, but these efforts require additional 3D supervision, such as synthetic face datasets [21] or 3D morphable models [7,34,40].

On the other hand, several post-processing techniques [4,10,29,30,37] have been proposed to discover the semantic information encoded in the trained model's latent space through subspace projections. These methods show that latent codes can be disentangled without supervision, but the learned subspace still requires interpretation by visual inspection after training. However, these methods have some limitations, including the inconvenience of the post-processing and pre-processing steps required to identify the important directions and subspace projection. Moreover, this process may also result in some images being ignored due to being deemed unimportant through the subspace projections, potentially leading to a distortion of the original distribution of the training data.

Orthogonal regularizers such as HessianPenalty [25] and OroJaR [36] have been proposed as a means to encourage feature disentanglement and control in the training of neural networks. These regularizers operate by penalizing feature changes based on orthogonal constraints, which can facilitate the learning of more interpretable and independent features. The regularization term encourages the model to map input variations in specific directions, promoting the separation of different factors of variation in the latent space. Despite their potential benefits, orthogonal regularizers have limitations. Estimating first- or second-order finite differences, a key component of these methods, can incur a higher computational cost. This computational demand may impose constraints on training models with high-resolution images or large-scale architectures, limiting the applicability of these regularizers in such scenarios.

3 Linearly Controllable Generative Adversarial Network

The proposed method, LC-GAN, adopts an unsupervised approach to acquire an embedding feature space by integrating conditional feature clustering with real/fake image classification tasks. The discriminator network D outputs both real/fake classification scores and cluster assignments, facilitating the aggregation of features originating from diverse image augmentations. Throughout training, a contrastive loss is employed to encourage closer proximity within the same cluster and greater separation between different clusters, enabling the discriminator to learn a feature space that encapsulates desired image properties, thus enabling controllable image generation through input vector manipulation.

3.1 Embedding Feature Space for Feature Categorization

In this approach, the feature space is constructed by combining basic image augmentations with feature clustering techniques. The discriminator is extended with two projection heads, h^g and h^a, which map images into distinct embedding spaces: geometric change and appearance change. These feature spaces validate the consistency of generated images with intended feature changes. Figure 2

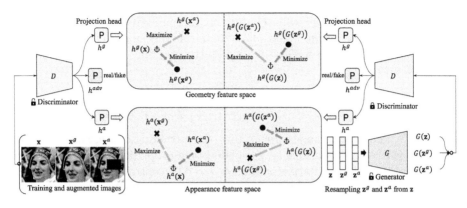

Fig. 2. Overview of geometry and appearance feature space construction and latent code categorization. The discriminator learns embedding feature spaces through contrastive feature clustering with samples generated from geometric and appearance augmentations. By clustering these features, the discriminator associates anchor and positive samples while separating negative samples in the feature space. The learned feature space is used to confirm the relationship between generated images $G(\mathbf{z})$ and re-generated images $G(\mathbf{z}^g)$ and $G(\mathbf{z}^a)$, created by partially resampling the noise of the drawn sample $\mathbf{z}$. This allows the generator to understand the desired feature changes with specific latent code resampling.

provides an overview of the training procedure. To construct individual feature spaces, a set of augmentations is applied, including random perspective transformation and random image erasing/color jittering. Augmented images simulating variations in geometry and appearance serve as positive and negative samples in corresponding and opposite projection heads to capture desired image changes. For instance, in the geometry feature space, positive samples are generated by applying the viewpoint change augmentation to an anchor image, while negative samples undergo appearance augmentation, introducing alterations unrelated to viewpoint changes. Subsequently, both the anchor image and augmented images undergo projection onto an L_2 normalized feature space using the corresponding projection head. The projection head associated with viewpoint change augmentation extracts features capturing changes in viewpoint. To enhance feature space learning, the discriminator employs contrastive feature clustering, aiming to bring positive samples of the same change type closer together in the feature space.

The contrastive loss for each training image is calculated using the following equation:

$$\mathcal{L}_{cl}^D = C(h^g(\mathbf{x}), h^g(\mathbf{x}^g), h^g(\mathbf{x}^a)) + C(h^a(\mathbf{x}), h^a(\mathbf{x}^a), h^a(\mathbf{x}^g)). \tag{1}$$

Here, h^g and h^a are the projection heads projecting the images into the geometry and appearance feature spaces, and $\mathbf{x}$, $\mathbf{x}^g$, and $\mathbf{x}^a$ are the training and augmented images by the geometry and appearance changes. The contrastive loss quantifies the dissimilarity between the feature representation of the anchor image and the positively augmented image, while considering the similarity between the feature representation of the anchor image and the negatively augmented image. The negative logarithm of the fraction is used as the loss function to be minimized during training.

$$C(\mathbf{f}, \mathbf{f}^+, \mathbf{f}^-) = -\log \frac{\exp(\mathbf{f}^T\mathbf{f}^+/\tau)}{\exp(\mathbf{f}^T\mathbf{f}^+/\tau) + \exp(\mathbf{f}^T\mathbf{f}^-/\tau)}, \quad (2)$$

where $\mathbf{f}$, $\mathbf{f}^+$, and $\mathbf{f}^-$ are the feature vectors of the training image, and positively and negatively augmented images, calculated by a projection head h of the discriminator, and the temperature parameter $\tau = 0.05$ controls the strength of the penalties applied to the positive and negative samples during the contrastive loss calculation.

The discriminator is trained using both contrastive loss and standard adversarial loss. Note that for the real *vs.* fake image classification, it does not utilize the augmented samples that are employed to construct the embedding feature space. The adversarial loss is computed as follows:

$$\mathcal{L}_{adv}^D = \mathop{\mathbb{E}}_{\mathbf{x} \sim p_{data}(\mathbf{x})} [\log D(\mathbf{x})] + \mathop{\mathbb{E}}_{\mathbf{z} \sim p_{\mathbf{z}}(\mathbf{z})} [\log(1 - D(G(\mathbf{z})))] \quad (3)$$

where p_{data} and $p_{\mathbf{z}}$ are the distributions of training data and noise samples, respectively, and $\mathbf{z}$ denotes random noise drawn from a Gaussian distribution with zero mean and unit variance, serving as the latent codes input to the generator network G. Additionally, we incorporate R_1 regularization [23] during discriminator training, defined as:

$$\mathcal{L}_{R_1}^D = \|\nabla D(\mathbf{x})\|^2. \quad (4)$$

This term encourages the discriminator to be more consistent in its predictions for real images, resulting in more stable training and better performance. The complete loss function for training the discriminator is the sum of the adversarial loss, the contrastive loss, and the regularization term: $\mathcal{L}^D = \mathcal{L}_{adv}^D + \lambda_{cl}\mathcal{L}_{cl}^D + \lambda_{R_1}\mathcal{L}_{R_1}^D$, where $\lambda_{cl} = 0.5$ and $\lambda_{R_1} = 10$ are experimentally set hyperparameters to balance the relative importance of different loss functions.

3.2 Linearly Controllable Generator

In Fig. 3, we present an overview of the generator architecture, characterized by two fundamental components: the mapping and synthesis networks. In this architecture, the generator strategically divides the input latent code $\mathbf{z}$ into two distinctive vectors. These vectors then undergo a linear mapping process independently through dedicated networks, resulting in intermediate latent codes. Subsequently, these codes play a crucial role within the image synthesis block to linearly control the desired output. Here, the geometry latent code takes the lead in orchestrating the creation of flowfields, dynamically shaping the feature maps generated based on the appearance latent code. The generator's training loss function is intricately composed of three essential terms: 1) a penalty for

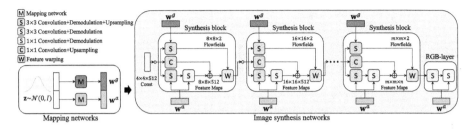

Fig. 3. Generator architecture. The generator comprises mapping networks and image synthesis networks. It begins with input vectors sampled from a Gaussian distribution, which are then divided into geometry and appearance latent codes, $\mathbf{w}^g$ and $\mathbf{w}^a$, by the corresponding mapping networks. These latent codes are subsequently used to produce images at an $m \times m$ resolution. The synthesis networks utilize these latent codes to create flowfields and generate feature maps, enabling precise control over the image synthesis process.

adversarial loss, 2) a emphasis on latent code categorization into geometry and appearance codes, and 3) decomposition of categorized features into more refined semantics.

Latent Code Categorization: Firstly, the adversarial loss, defined as

$$\mathcal{L}_{adv}^G = \mathop{\mathbb{E}}_{\mathbf{z} \sim p_\mathbf{z}(\mathbf{z})} \left[\log D(G(\mathbf{z})) \right], \tag{5}$$

penalizes the generator for failing to effectively deceive the discriminator. In addition, a conditional resampling is employed to enhance the generator's understanding of input vector alterations. As shown in Fig. 2, the generator's input $\mathbf{z}$ undergoes conditional resampling to yield $\mathbf{z}^g$ and $\mathbf{z}^a$, where the noise in the designated segment is resampled while retaining the other segments unchanged. This approach enables controlled modifications within specific feature spaces of the generated images. Notably, $\mathbf{z}^g$ is derived by resampling the segment associated with geometry augmentation, while $\mathbf{z}^a$ is obtained by resampling the segment correlated with appearance augmentation. Such conditioning establishes a direct relationship between the generated images and the alterations in the latent code. The generated images $G(\mathbf{z})$, $G(\mathbf{z}^g)$, and $G(\mathbf{z}^a)$ are evaluated as following

$$\mathcal{L}_{cl}^G = C(h^g(G(\mathbf{z})), h^g(G(\mathbf{z}^g)), h^g(G(\mathbf{z}^a))) + C(h^a(G(\mathbf{z})), h^a(G(\mathbf{z}^a)), h^a(G(\mathbf{z}^g))). \tag{6}$$

This loss function validates whether the generated output contains the desired property changes by conditionally modifying the sampled latent vector based on the constructed embedding feature spaces of the discriminator.

Spectral Regularization for Feature Selection and Mapping: Drawing upon the formulation in Eq. 6, the generator gains an understanding of the semantics encoded within the latent code segment. To enable more fine-grained control by disentangling the appearance and geometry latent codes into finer semantics in a fully unsupervised manner, we introduce a spectral regularization approach leveraging covariance matrix parameterization.

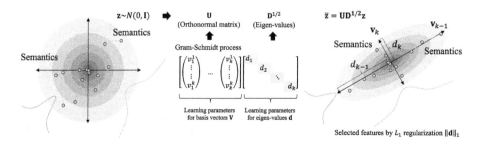

Fig. 4. Spectral regularization. To enable fine-grained control, the geometry and appearance features, categorized through contrastive feature learning, undergo further analysis and decomposition via a feature selection mechanism. This involves applying L_1 regularization to the eigen-values of the covariance matrix, allowing for the automatic selection of important features and achieving more semantic control over the generative process.

Our approach focuses on learning an anisotropic Gaussian distribution by parameterizing the covariance matrix. By doing so, we enable the model to automatically select important features while minimizing the number of representation dimensions through the application of L_1 regularization on its eigen-values. We draw input noise from a Gaussian distribution $\mathcal{N}(0, \Sigma)$ with a learnable covariance matrix Σ. This adaptive covariance matrix, being symmetric, can be decomposed into $\Sigma = \mathbf{UDU}^T$ according to the eigen-decomposition, where $\mathbf{U}$ consists of orthonormal vectors and $\mathbf{D}$ is a diagonal matrix containing the eigen-values of the covariance matrix.

To learn these matrices, as shown in Fig. 4, we initialize learning parameters determining the basis vectors and lengths of the axes, denoted as $\mathbf{V} = \{\mathbf{v}_1, ..., \mathbf{v}_k\}$, where $\mathbf{v}_k = \{v_k^1, ..., v_k^k\}$, and $\mathbf{d} = \{d_1, ..., d_k\}$ at the first layer of the mapping network. These parameters are then transformed into square and diagonal matrices, $\mathbf{V} \in \mathbb{R}^{k \times k}$ and $\mathbf{D}^{1/2} \in \mathbb{R}^{k \times k}$. Subsequently, $\mathbf{V}$ is converted into $\mathbf{U}$ using the Gram-Schmidt process. The transformed vectors are input into mapping networks consisting of 12 fully connected layers, generating the geometry and appearance latent codes. It is noteworthy that no non-linear activation functions are applied in the mapping networks to maintain the linearity of the Gaussian distribution.

The spectral regularizer is defined as the L_1 regularization of the eigen-values of the covariance matrix, aiming to suppress behavior associated with small eigen-values:

$$\mathcal{L}_s^G = ||\mathbf{d}^g||_1 + ||\mathbf{d}^a||_1. \qquad (7)$$

Here, $||\mathbf{d}^g||_1$ and $||\mathbf{d}^a||_1$ represent the L_1 norms of the vectors corresponding to the mapping networks for geometry and appearance latent codes. Applying L_1 regularization to the eigen-values of the covariance matrix encourages some eigen-values to become zero, promoting sparsity in the representation. This can lead to simpler and more interpretable representation learning in each geometry and appearance latent code by reducing the number of non-zero eigen-values, and consequently, the important features are autonomously selected.

The entire training loss function for the generator is calculated as the combination of the explained three terms: $\mathcal{L}^G = \mathcal{L}^G_{adv} + \lambda_{cl}\mathcal{L}^G_{cl} + \lambda_s\mathcal{L}^G_s$. Here, $\lambda_s = 1e^{-7}$ is experimentally set to control the importance of the sparsity term in the overall loss function. This value is determined through empirical testing and tuning to achieve the desired balance between adversarial, contrastive, and spectral regularization losses during training.

4 Experiments

Datasets: We trained our model using three public datasets: FFHQ (FF) [17], CelebA-HQ (CA) [14], and AFHQ-V2 (AF) [5]. FF contains 70K high-quality human faces with variations in pose, expression, and lighting. CA includes 30K celebrity faces with diverse appearances, and AF comprises 15,803 animal faces with a wide range of visual characteristics. To adapt to different image sizes, we used eight, seven, and six residual blocks for 1024^2, 512^2, and 256^2 resolutions, respectively. The datasets were resized as follows: AF images to 512^2 and 256^2, and FFHQ and CelebA-HQ images to all three resolutions.

Training and Implementation Details: The model was trained using the PyTorch framework, with Albumentations used to apply image augmentations [3]. The Adam optimizer [20] was used with a learning rate of 0.002 for the 256^2 and 512^2 resolution datasets, and 0.001 for the 1024^2 resolution dataset, with $\beta_1 = 0.0$ and $\beta_2 = 0.99$. A batch size of $B = 32$ was used, and the model was trained for 450K, 700K, and 900K epochs for the AF, CA, and FF datasets. To improve the quality of generated images, the exponential moving average of the generator parameters [39] was employed with a decay rate of 0.9999, starting from the 5K-th training iteration. Additionally, FreezeD [24], which freezes the weight parameters of the discriminator layers until a spatial resolution of 64^2, was applied from the 150K-th iteration for the AF dataset, and from the 300K-th and 500K-th iteration for the CA and FF datasets. The projection heads of the discriminator consist of three fully connected layers, producing a 256-dimensional output vector. The length of the input latent vector $\mathbf{z}$ is 128, which is divided into two 64-dimensional vectors. Each latent code is then transformed into the geometry and appearance latent codes, $\mathbf{w}^g$ and $\mathbf{w}^a$, using mapping networks composed of subspace transformation and fully connected layers. To accelerate training, we employed a lazy regularization strategy, applying contrastive loss and spectral regularization every other iteration, and only adversarial loss otherwise. This approach sped up training while maintaining image quality. We trained the networks on 256^2, 512^2, and 1024^2 resolution images using four, four, and eight NVIDIA-H100 GPUs, respectively, with training times of approximately 8, 17, and 19 h for 100K iterations.

Table 1. FID evaluation results on FFHQ (FF), CelebA-HQ (CA), and AFHQ-V2 (AF) datasets at different image resolutions (256^2:L, 512^2:M, and 1024^2:H).

Method	FF-L	CA-L	AF-L	FF-M	CA-M	AF-M	FF-H	CA-H	controllability
StyleGAN2 [15]	–	–	–	–	–	4.62	2.70	–	✗
StyleGAN3 [16]	–	–	–	–	–	4.40	3.07	–	✗
SWAGAN-Bi [8]	5.22	–	–	–	–	–	4.06	–	✗
StyleNAT [35]	2.05	–	–	–	–	–	4.17	–	✗
MSG-GAN [13]	–	–	–	–	–	–	5.80	6.37	✗
CIPS [1]	4.38	–	–	6.18	–	–	10.07	–	✗
StyleSwin [41]	2.81	3.25	–	–	–	–	5.07	4.43	✗
HiT-B [43]	2.95	3.39	–	–	–	–	6.37	8.83	✗
WaveDiff [26]	–	5.94	–	–	6.40	–	–	–	✗
DDGAN [38]	–	7.64	–	–	8.43	–	–	–	✗
SeFa [30]	6.87	6.43	9.48	4.21	4.52	5.98	9.36	5.83	✓
EigenGAN [11]	13.90	14.05	–	13.99	11.07	–	18.51	16.31	✓
LC-GAN	3.65	3.72	5.77	3.36	3.49	4.74	3.32	3.46	✓

Fig. 5. Examples of images generated by the trained models. From left to right, the image sets show generated images by models trained on the FFHQ, CelebA-HQ, and AFHQ-V2 datasets at different image resolutions.

4.1 High-quality Image Generation: Comparison with SOTAs

We compared the performance of our method with SOTA image generation methods using the FF, CA, and AF datasets. The quality of the generated images was evaluated using the FID metric [12], which measures the similarity between the distribution of real and generated images in the feature space of an Inception-v3 network. Lower FID scores indicate better image quality. To ensure a fair comparison, we followed the testing protocols used in other papers, such as [1,8,13,15,16,26,35,38,41,43].

For the FF dataset, we randomly sampled 50K images from the training dataset and generated 50K images using our trained generator, then calculated the FID between them. For the CA and AF datasets, we calculated FIDs between 30K and 15,803 generated images, respectively, and the entire training images. The compared algorithms were chosen among the SOTA methods that followed

Fig. 6. Demonstration of semantic control in generated images. The image pairs from top to bottom are generated using LC-GAN, SeFa, and EigenGAN, where the image pairs were created by controlling the same semantic aspect (yaw change).

the same FID measurement protocols. Additionally, we trained SeFa [30] and EigenGAN [11] on all training datasets to ensure a comprehensive comparative analysis. However, attempts to apply EigenGAN on the AF dataset with different settings were unsuccessful due to mode collapse.

Table 1 provides the quantitative evaluation results, demonstrating the superiority of our proposed method in terms of FID scores. Particularly noteworthy are the excellent FID scores for FF and CA, comparable to StyleGAN3 and StyleSWIN, which prioritize high-quality and high-resolution image generation without emphasizing control ability. Figure 5 presents a selection of representative images generated by our LC-GAN, illustrating its capability to produce realistic and diverse images.

4.2 Ablation Study of Image Property Controllability

Evaluation Criteria: To evaluate the effectiveness of our proposed method in controlling image properties, we conducted an experiment to generate pairs of images. One image was generated randomly, while the other was generated by resampling one of the latent codes. Our goal was to ensure that the model was properly trained and that the generated image pairs had similar semantics, while the resampled image properties were updated. We used the Arcface-ResNet100 [6] to calculate the feature distance between the pairs of generated images. This measure indicates the semantic similarity between human faces, with a decision boundary trained to distinguish between the same and different identities using a threshold of 1. The feature distance between images of the same person should be less than 1, while the feature distance between images of different people should be greater than 1. In addition, we used MediaPipe's face mesh model [19] to obtain facial landmarks, enabling us to visually assess the geometric consistency of the generated faces across different identities. We calculated the distance between corresponding landmarks of the image pairs to evaluate the consistency of the generated faces. Overall, this experiment allowed us to demonstrate the effectiveness of our proposed method in controlling image properties while maintaining appearance similarity and geometric consistency.

Identity-Preserving Image Generation: In our experiments, we focused on generating human face images while preserving their identity. We generated 1,000 pairs of images using generators trained on the CA and FF datasets by resampling only the viewpoint component. Additionally, we used SeFa and EigenGAN to generate other sets of image pairs, manually exploring the control direction associated with the viewpoint in the latent space. To evaluate the quality of the generated image pairs, we plotted the distance distributions in Fig. 7. The distances for our method were consistently lower compared to those obtained using SeFa and EigenGAN. To further assess identity similarity, we tested Arcface on randomly selected image pairs from the training dataset, and no false positive results were observed. This demonstrated the precise measurement of identity similarity by Arcface. Notably, Arcface consistently identified a significant majority of the image pairs generated by our method as depicting the same person, highlighting the robustness of our approach in preserving identity information. In contrast, the image pairs generated by SeFa and EigenGAN exhibited larger distance values, primarily due to variations introduced in both facial pose and appearance through principal direction changes. This impact was particularly notable in the CA-H and FF-H tests, where even slight changes in high-resolution images had a significant effect on the results. The results demonstrate the superior accuracy of our approach in maintaining identity consistency in the generated image pairs (Fig. 6 and Table 2).

Table 2. **Perceptual distance** between image pairs using Arcface-ResNet100.

Method	CA-L	FF-L	CA-M	FF-M	CA-H	FF-H
Random pairs	1.41	1.41	1.41	1.41	1.41	1.41
SeFa [30]	0.91	0.85	1.01	0.94	1.09	1.13
EigenGAN [11]	0.99	0.98	1.03	0.91	0.97	0.99
LC-GAN	0.71	0.64	0.78	0.74	0.71	0.61

Viewpoint-Preserving Image Generation: In this experiment, our goal was to generate images that preserve viewpoints by controlling the appearance, resulting in 1,000 pairs of images. To assess the effectiveness of our method, we measured the distance between facial landmarks for each image pair. The results, summarized in Table 3, compare the facial landmark consistency of our approach with random pairs, SeFa, and EigenGAN across datasets with varying resolutions. Our approach consistently outperformed SeFa and EigenGAN, achieving the lowest facial landmark distances across all datasets, demonstrating superior preservation of viewpoints in generated images. SeFa and EigenGAN often compromised viewpoint preservation by slightly altering facial poses when regenerating images with appearance changes. Specifically, LC-GAN excels in learning

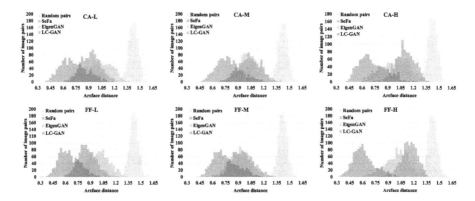

Fig. 7. Image pair similarity using Arcface feature distances. The results show that image pairs generated by our method have lower mean distances.

Fig. 8. Demonstration of semantic control in generated images. Image pairs from top to bottom are generated using LC-GAN, SeFa, and EigenGAN, illustrating controlled changes in the same semantic aspect (identity), along with face landmarks estimated by MediaPipe's face mesh model.

features related to controlling mouth movements and subtle facial expressions independently of identity changes. In contrast, SeFa and EigenGAN often exhibit entanglement of these features with changes in identity. Visual examples illustrating the effectiveness of our method in preserving viewpoints are available in Fig. 8.

Effectiveness of Contrastive Loss and Spectral Regularization: To validate the effectiveness of contrastive loss (CL) and spectral regularization (SR), we conducted ablation experiments by training the model with and without these terms. Focusing on the CA-L and FF-L datasets, we found that removing both terms led to entangled features solely within the appearance latent code. The top images in Fig. 9 demonstrate the lack of disentanglement when both CL and SR are removed. In contrast, applying CL without SR resulted in more uniformly distributed features across dimensions. When controlling a target attribute, it

Table 3. Landmark distance between image pairs using MediaPipe's face mesh.

Method	CA-L	FF-L	CA-M	FF-M	CA-H	FF-H
Random pairs	7.63	8.70	15.17	17.08	30.15	34.71
SeFa [30]	2.41	2.64	5.63	6.49	10.42	11.34
EigenGAN [11]	2.97	2.61	5.86	4.93	13.46	13.99
LC-GAN	1.65	1.97	3.66	4.19	5.83	9.09

Fig. 9. Comparison of ablated models controlling a dimension for yaw changes: without CL and SR (top), with CL only (middle), and full model (bottom).

Fig. 10. Exploring the latent space: Our approach facilitates the exploration of semantically meaningful features controlling various image properties.

required editing multiple dimensions to achieve similar changes to the full model, as the target features were spread across more dimensions. The middle images in Fig. 9 illustrate the increased need for manipulating multiple dimensions to achieve similar variation compared to the full model. Regarding FID scores, no significant differences were observed, although there was a slight degradation when each component was applied individually: 3.54 and 3.66 for CA-L, and 3.30 and 3.45 for FF-L, without CL and SR, and only with CL, respectively.

Visual Analysis: Our exploration of the learned semantics revealed that the proposed method excels in capturing both global features and intricate, specific attributes. Notably, the model can independently manipulate detailed behaviors such as zoom-in/out and mouth movements, as shown in Fig. 10. The generated results highlight the model's enhanced controllability over these attributes. For a more comprehensive exploration of the diverse set of discovered semantic attributes, we recommend referring to the provided source code.

5 Conclusion and Limitations

We presented LC-GAN, a novel GAN framework for controllable image generation. By integrating unsupervised disentanglement techniques with the Style-GAN architecture, LC-GAN produces high-quality images with control over image properties. Extensive experiments on multiple datasets, including FF, CA, and AF at different resolutions, demonstrated that LC-GAN outperforms SOTA models in terms of FID scores and controllability in image synthesis. However, LC-GAN has some limitations. The training process is time-consuming and requires larger GPU memory, especially when incorporating contrastive loss. Additionally, achieving precise metric-level control, such as a 20-degree rotation, can be challenging due to the unsupervised nature of the method. Future research should focus on addressing these limitations and improving the efficiency and precision of controllable image generation methods.

References

1. Anokhin, I., Demochkin, K., Khakhulin, T., Sterkin, G., Lempitsky, V., Korzhenkov, D.: Image generators with conditionally-independent pixel synthesis. In: Proceedings of the IEEE/CVF Conference on Computer Vision and Pattern Recognition, pp. 14278–14287 (2021)
2. Axel Sauer, A.G.: Counterfactual generative networks. In: International Conference on Learning Representations (2021)
3. Buslaev, A., Iglovikov, V.I., Khvedchenya, E., Parinov, A., Druzhinin, M., Kalinin, A.A.: Albumentations: fast and flexible image augmentations. Information **11**(2), 125 (2020)
4. Choi, J., Lee, J., Yoon, C., Park, J.H., Hwang, G., Kang, M.: Do not escape from the manifold: Discovering the local coordinates on the latent space of gans. arXiv preprint arXiv:2106.06959 (2021)
5. Choi, Y., Uh, Y., Yoo, J., Ha, J.W.: Stargan v2: diverse image synthesis for multiple domains. In: Proceedings of the IEEE Conference on Computer Vision and Pattern Recognition (2020)
6. Deng, J., Guo, J., Xue, N., Zafeiriou, S.: Arcface: additive angular margin loss for deep face recognition. In: Proceedings of the IEEE/CVF Conference on Computer Vision and Pattern Recognition, pp. 4690–4699 (2019)
7. Deng, Y., Yang, J., Chen, D., Wen, F., Tong, X.: Disentangled and controllable face image generation via 3d imitative-contrastive learning. In: Proceedings of the IEEE/CVF Conference on Computer Vision and Pattern Recognition, pp. 5154–5163 (2020)
8. Gal, R., Hochberg, D.C., Bermano, A., Cohen-Or, D.: Swagan: a style-based wavelet-driven generative model. ACM Trans. Graph. **40**(4), 1–11 (2021)
9. Goodfellow, I., et al.: Generative adversarial nets. Adv. Neural Inf. Process. Syst. **27** (2014)

10. Härkönen, E., Hertzmann, A., Lehtinen, J., Paris, S.: Ganspace: discovering interpretable GAN controls. Adv. Neural. Inf. Process. Syst. **33**, 9841–9850 (2020)
11. He, Z., Kan, M., Shan, S.: Eigengan: layer-wise eigen-learning for GANs. In: Proceedings of the IEEE/CVF International Conference on Computer Vision, pp. 14408–14417 (2021)
12. Heusel, M., Ramsauer, H., Unterthiner, T., Nessler, B., Hochreiter, S.: GANs trained by a two time-scale update rule converge to a local nash equilibrium. Adv. Neural Inf. Process. Syst. **30** (2017)
13. Karnewar, A., Wang, O.: MSG-GAN: multi-scale gradients for generative adversarial networks. In: Proceedings of the IEEE/CVF Conference on Computer Vision and Pattern Recognition, pp. 7799–7808 (2020)
14. Karras, T., Aila, T., Laine, S., Lehtinen, J.: Progressive growing of GANs for improved quality, stability, and variation. In: International Conference on Learning Representations (2018). https://openreview.net/forum?id=Hk99zCeAb
15. Karras, T., Aittala, M., Hellsten, J., Laine, S., Lehtinen, J., Aila, T.: Training generative adversarial networks with limited data. Adv. Neural. Inf. Process. Syst. **33**, 12104–12114 (2020)
16. Karras, T., et al.: Alias-free generative adversarial networks. Adv. Neural. Inf. Process. Syst. **34**, 852–863 (2021)
17. Karras, T., Laine, S., Aila, T.: A style-based generator architecture for generative adversarial networks. In: Proceedings of the IEEE/CVF Conference on Computer Vision and Pattern Recognition, pp. 4401–4410 (2019)
18. Karras, T., Laine, S., Aittala, M., Hellsten, J., Lehtinen, J., Aila, T.: Analyzing and improving the image quality of stylegan. In: Proceedings of the IEEE/CVF Conference on Computer Vision and Pattern Recognition, pp. 8110–8119 (2020)
19. Kartynnik, Y., Avatski, A., Grishchenko, I., Grundmann, M.: Real-time facial surface geometry from monocular video on mobile GPUS. arXiv preprint arXiv:1907.06724 (2019)
20. Kingma, D.P., Ba, J.: Adam: a method for stochastic optimization. In: International Conference on Learning Representations (2015)
21. Kowalski, M., Garbin, S.J., Estellers, V., Baltrušaitis, T., Johnson, M., Shotton, J.: CONFIG: controllable neural face image generation. In: Vedaldi, A., Bischof, H., Brox, T., Frahm, J.-M. (eds.) ECCV 2020. LNCS, vol. 12356, pp. 299–315. Springer, Cham (2020). https://doi.org/10.1007/978-3-030-58621-8_18
22. Li, B., Qi, X., Lukasiewicz, T., Torr, P.: Controllable text-to-image generation. Adv. Neural Inf. Process. Syst. **32** (2019)
23. Mescheder, L., Geiger, A., Nowozin, S.: Which training methods for GANs do actually converge? In: International Conference on Machine Learning, pp. 3481–3490. PMLR (2018)
24. Mo, S., Cho, M., Shin, J.: Freeze the discriminator: a simple baseline for fine-tuning GANs. In: CVPR AI for Content Creation Workshop (2020)
25. Peebles, W., Peebles, J., Zhu, J.-Y., Efros, A., Torralba, A.: The Hessian penalty: a weak prior for unsupervised disentanglement. In: Vedaldi, A., Bischof, H., Brox, T., Frahm, J.-M. (eds.) ECCV 2020. LNCS, vol. 12351, pp. 581–597. Springer, Cham (2020). https://doi.org/10.1007/978-3-030-58539-6_35
26. Phung, H., Dao, Q., Tran, A.: Wavelet diffusion models are fast and scalable image generators. In: Proceedings of the IEEE/CVF Conference on Computer Vision and Pattern Recognition (CVPR) (2023)
27. Qiao, T., Zhang, J., Xu, D., Tao, D.: Mirrorgan: learning text-to-image generation by redescription. In: Proceedings of the IEEE/CVF Conference on Computer Vision and Pattern Recognition, pp. 1505–1514 (2019)

28. Qiu, H., Yu, B., Gong, D., Li, Z., Liu, W., Tao, D.: Synface: face recognition with synthetic data. In: Proceedings of the IEEE/CVF International Conference on Computer Vision, pp. 10880–10890 (2021)
29. Shen, Y., Gu, J., Tang, X., Zhou, B.: Interpreting the latent space of GANs for semantic face editing. In: Proceedings of the IEEE/CVF Conference on Computer Vision and Pattern Recognition, pp. 9243–9252 (2020)
30. Shen, Y., Zhou, B.: Closed-form factorization of latent semantics in GANs. In: Proceedings of the IEEE/CVF Conference on Computer Vision and Pattern Recognition, pp. 1532–1540 (2021)
31. Shi, Y., Yang, X., Wan, Y., Shen, X.: Semanticstylegan: learning compositional generative priors for controllable image synthesis and editing. In: Proceedings of the IEEE/CVF Conference on Computer Vision and Pattern Recognition, pp. 11254–11264 (2022)
32. Shoshan, A., Bhonker, N., Kviatkovsky, I., Medioni, G.: Gan-control: explicitly controllable GANs. In: Proceedings of the IEEE/CVF International Conference on Computer Vision, pp. 14083–14093 (2021)
33. Song, Y., Sebe, N., Wang, W.: Orthogonal SVD covariance conditioning and latent disentanglement. IEEE Trans. Pattern Anal. Mach. Intell. (2022)
34. Tewari, A., et al.: Stylerig: rigging stylegan for 3d control over portrait images. In: Proceedings of the IEEE/CVF Conference on Computer Vision and Pattern Recognition, pp. 6142–6151 (2020)
35. Walton, S., Hassani, A., Xu, X., Wang, Z., Shi, H.: Stylenat: giving each head a new perspective. arXiv preprint arXiv:2211.05770 (2022)
36. Wei, Y., et al.: Orthogonal Jacobian regularization for unsupervised disentanglement in image generation. In: Proceedings of the IEEE/CVF International Conference on Computer Vision, pp. 6721–6730 (2021)
37. Wu, Z., Lischinski, D., Shechtman, E.: Stylespace analysis: disentangled controls for stylegan image generation. In: Proceedings of the IEEE/CVF Conference on Computer Vision and Pattern Recognition, pp. 12863–12872 (2021)
38. Xiao, Z., Kreis, K., Vahdat, A.: Tackling the generative learning trilemma with denoising diffusion GANs. In: International Conference on Learning Representations (2022). https://openreview.net/forum?id=JprM0p-q0Co
39. Yaz, Y., et al.: The unusual effectiveness of averaging in GAN training. In: International Conference on Learning Representations (2018)
40. Yin, X., Yu, X., Sohn, K., Liu, X., Chandraker, M.: Towards large-pose face frontalization in the wild. In: Proceedings of the IEEE International Conference on Computer Vision, pp. 3990–3999 (2017)
41. Zhang, B., et al.: Styleswin: transformer-based GAN for high-resolution image generation. In: Proceedings of the IEEE/CVF Conference on Computer Vision and Pattern Recognition, pp. 11304–11314 (2022)
42. Zhang, H., et al.: Stackgan: text to photo-realistic image synthesis with stacked generative adversarial networks. In: Proceedings of the IEEE International Conference on Computer Vision, pp. 5907–5915 (2017)
43. Zhao, L., Zhang, Z., Chen, T., Metaxas, D., Zhang, H.: Improved transformer for high-resolution GANs. Adv. Neural. Inf. Process. Syst. **34**, 18367–18380 (2021)

Generating Human Interaction Motions in Scenes with Text Control

Hongwei Yi[1,2(✉)], Justus Thies[2,3], Michael J. Black[2], Xue Bin Peng[1,4], and Davis Rempe[1]

[1] NVIDIA, Toronto, Canada
hongwei.yi@tuebingen.mpg.de
[2] Max Planck Institute for Intelligent Systems, Tubingen, Germany
[3] Technical University of Darmstadt, Darmstadt, Germany
[4] Simon Fraser University, Burnaby, Canada

Abstract. We present TeSMo, a text-controlled scene-aware motion generation method based on denoising diffusion models. Previous text-to-motion methods focus on characters in isolation without considering scenes due to the limited availability of datasets that include motion, text descriptions, and interactive scenes. Our approach begins with pre-training a scene-agnostic text-to-motion diffusion model, emphasizing goal-reaching constraints on large-scale motion-capture datasets. We then enhance this model with a scene-aware component, fine-tuned using data augmented with detailed scene information, including ground plane and object shapes. To facilitate training, we embed annotated navigation and interaction motions within scenes. The proposed method produces realistic and diverse human-object interactions, such as navigation and sitting, in different scenes with various object shapes, orientations, initial body positions, and poses. Extensive experiments demonstrate that our approach surpasses prior techniques in terms of the plausibility of human-scene interactions and the realism and variety of the generated motions. Code and data are available at https://research.nvidia.com/labs/toronto-ai/tesmo.

Keywords: Scene-Aware Human Motion Generation · Text-to-Motion

1 Introduction

Generating realistic human movements that can interact with 3D scenes is crucial for many applications, ranging from gaming to embodied AI. For example, character animators for games and films need to author motions that successfully navigate through cluttered scenes and realistically interact with target objects, while still maintaining artistic control over the style of the movement. One natural way to control style is through text, e.g., "skip happily to the chair and sit

---

Supplementary Information The online version contains supplementary material available at https://doi.org/10.1007/978-3-031-73235-5_14.

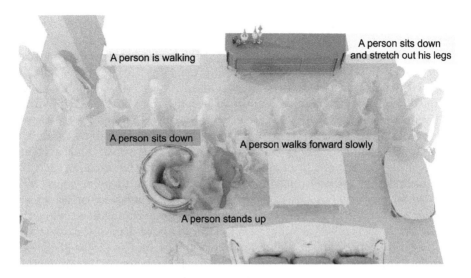

Fig. 1. We present TeSMo, a method for generating diverse and plausible human-scene interactions from text input. Given a 3D scene, TeSMo generates scene-aware motions, such as walking in free space and sitting on a chair. Our model can be easily controlled using textual descriptions, start positions, and goal positions.

down". Recently, diffusion models have shown remarkable capabilities in generating human motion from user inputs. Text prompts [36,47] let users control style, while methods incorporating spatial constraints enable more fine-grained control, such as specifying desired joint positions and trajectories [18,33,41]. However, these works have predominantly focused on characters in isolation, without considering environmental context or object interactions.

In this work, we aim to incorporate scene-awareness into user-controllable human motion generation models. However, learning to generate motions involving scene interactions is challenging, even without text prompts. Unlike large-scale motion capture datasets that depict humans in isolation [26], datasets with paired examples of 3D human motion and scene/object geometry are limited. Prior work uses small paired datasets without text annotations to train VAEs [9,34,49] or diffusion models [16,30] that generate human scene interactions with limited scope and diversity. Reinforcement learning methods are able to learn interaction motions from limited supervision [11,22,52], and can generate behaviors that are not present in the training motion dataset. However, designing reward functions that lead to natural movements for a diverse range of interactions is difficult and tedious.

To address these challenges, we introduce a method for Text-conditioned Scene-aware Motion generation, called TeSMo. As shown in Fig. 1, our method generates realistic motions that navigate around obstacles and interact with objects, while being conditioned on a text prompt to enable stylistic diversity. Our key idea is to combine the power of general, but scene-agnostic, text-to-motion diffusion models with paired human-scene data that captures realistic interactions. First, we pre-train a text-conditioned diffusion model [36] on a diverse motion dataset with no objects (e.g., HumanML3D [7]), allowing it to learn a realistic motion prior and the correlation with text. We then fine-tune the model with an augmented scene-aware component that takes scene information as input, thereby refining motion outputs to be consistent with the environment.

Given a target object with which to interact and a text prompt describing the desired motion, we decompose the problem of generating a suitable motion in a scene into two components, *navigation* (e.g., approaching a chair while avoiding obstacles) and *interaction* (e.g., sitting on the chair). Both stages leverage diffusion models that are pre-trained on scene-agnostic data, then fine-tuned with an added scene-aware branch. The *navigation* model generates a pelvis trajectory that reaches a goal pose near the interaction object. During fine-tuning, the scene-aware branch takes, as input, a top-down 2D floor map of the scene and is trained on our new dataset containing locomotion sequences [26] in 3D indoor rooms [6]. The generated pelvis trajectory is then lifted to a full-body motion using motion in-painting [33]. Next, the *interaction* model generates a full-body motion conditioned on a goal pelvis pose and a detailed 3D representation of the target object. To further improve generalization to novel objects, the model is fine-tuned using augmented data that re-targets interactions [9] to a variety of object shapes while maintaining realistic human-object contacts.

Experiments demonstrate that our navigation approach outperforms prior work in terms of goal reaching and obstacle avoidance, while producing full-body motions on par with scene-agnostic diffusion models [18,41]. Meanwhile, our interaction model generates motions with fewer object penetrations than the state-of-the-art approach [52], being preferred 71.9% of the time in a perceptual study. The central contribution of this work includes: (**1**) a novel approach to enable scene-aware and text-conditioned motion generation by fine-tuning an augmented model on top of a pre-trained text-to-motion diffusion model, (**2**) a method, TeSMo, that leverages this approach for navigation and interaction components to generate high-quality motions in a scene from text, (**3**) data augmentation strategies for placing navigation and interaction motions with text annotations realistically in scenes to enable scene-aware fine-tuning.

2 Related Work

2.1 Scene-Aware Motion Generation

Motion synthesis in computer graphics has a rich history, encompassing areas such as locomotion [1,19,23,50], human-scene/object interaction [21,35], and dynamic object interaction [3,24,25]. We refer readers to an extensive survey [53] for an overview and focus on scene-aware motion generation in this section.

A particular challenge in modeling scene-aware motion is the lack of paired, high-quality human-scene datasets. One line of work [37,38] employs a two-stage method that first predicts the root path, followed by the full-body motion based on the scene and predicted path. However, these methods suffer from low-quality motion generation, attributed to the noise in the training datasets captured from monocular RGB-D videos [10]. Neural State Machine (NSM) [34] proposes the use of phase labeling [15] and local expert networks [5,17,45] to generate high-quality object interactions, such as sitting and carrying, after training on a small human-object mocap dataset. Nonetheless, it struggles with recognizing walkable regions in 3D scenes, often failing to avoid obstacles. Therefore, later work in this vein requires using the A* algorithm for collision-free path planning [9]. These and related approaches [48,49] are moreover limited by the diversity of the small human-scene interaction datasets with no text annotations.

Various approaches ameliorate the data issue by creating synthetic data with captured [42,43] or generated [20] motions placed in scenes heuristically. HUMANISE [39] does this for text-conditioned scene interactions, but rely entirely on short synthetic sequences for training, where the realism is limited by the data generation heuristics used. The reinforcement learning (RL) approach DIMOS [52] learns autoregressive policies to reach goal poses in a scene without requiring paired human-scene data for training, but still relies on A* and is constrained by the accuracy of goal pose generation [51]. RL with physical simulators [2,11,28,40] has been used to produce physically plausible movements but faces challenges in generalizing across varied scenes and objects.

Unlike most prior works, our approach is text-conditioned and leverages a mix of both scene-agnostic and paired human-scene data. Pre-training is done with a diverse scene-agnostic dataset, while scene-aware fine-tuning uses motion data with scene context. For training, we adopt both synthetic data creation with real motions and data augmentation of real-world human-object interactions [9].

2.2 Diffusion-Based Motion Generation

Recently, diffusion models have demonstrated the ability to generate high-quality human motions, especially when conditioned on a text prompt [29,36,47]. In addition to text, several diffusion models add spatial controllability. Some works [33,36] adopt image inpainting techniques to incorporate dense trajectories of spatial joint constraints into generated motions. OmniControl [41] and GMD [18] allow control with sparse signals and a pre-defined root path, respectively.

A few diffusion works handle interactions with objects or scenes. TRACE [32] generates 2D trajectories for pedestrians based on a rasterized street map. SceneDiffuser [16] conditions generation on a full scanned scene point cloud, but motion quality is limited due to noisy training data [10]. Another approach [30] tackles single-object interactions through hierarchical generation of milestone poses followed by dense motion, but it lacks text control. A concurrent line of work enables text conditioning for single-object interactions [4,27], but they focus on humans manipulating dynamic objects rather than interactions in full scenes.

We leverage a pre-trained text-to-motion diffusion model [36] and a fine-tuned scene-aware branch to enable both text controllability and scene-awareness with diffusion. We break motion generation into navigation and interaction with static

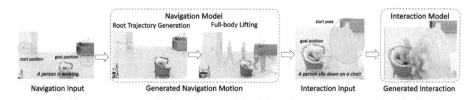

Fig. 2. Pipeline overview: given the start position (green arrow), goal position (red arrow), 3D scene, and text description, the navigation root trajectory is first generated and then the full-body motion is completed through in-painting. Subsequently, the interaction is generated from a start pose (i.e., the end pose from navigation), goal position, and the target object, enabling the generation of object-specific motion. (Color figure online)

objects by conditioning on 2D floor maps and 3D geometry, respectively, and create specialized human-scene data to enable diversity and quality.

3 Text-Conditioned Scene-Aware Motion Generation

3.1 Overview

Given a 3D scene and a target interaction object, our goal is to generate a plausible human-scene interaction, where the motion style can be controlled by a user-specified text prompt. Our approach decomposes this task into two components, *navigation* and *interaction*, as illustrated in Fig. 2. Both components are diffusion models that leverage a fine-tuning routine to enable scene-awareness without losing user controllability, as introduced in Sect. 3.2. To interact with an object, the character must first navigate to a location in the scene near the object, which is easily calculated heuristically or specified by the user, if desired. As described in Sect. 3.3, we design a hierarchical *navigation* model, which generates a root trajectory starting from an initial location that moves to the goal location while navigating around obstacles in the scene. The generated root trajectory is then lifted into a full-body motion using in-painting techniques [33,41]. Since the navigation model gets close to the object in the first stage, for generating the actual object *interaction*, we can focus on scenarios where the character is already near the object. This allows a one-stage motion generation model that directly predicts the full-body motion from the starting pose (*i.e.*, the last pose of navigation), a goal pelvis pose, and the object (as detailed in Sect. 3.4).

3.2 Background: Controllable Human Motion Diffusion Models

Motion Diffusion Models. Diffusion models have been successfully used to generate both top-down trajectories [32] and full-body motions [36,47]. These models generate motions by iteratively denoising a temporal sequence of N poses (e.g., root positions or full-body joint positions/angles) $\mathbf{x} = \begin{bmatrix} \mathbf{x}^1, \ldots, \mathbf{x}^N \end{bmatrix}$. During training, the model learns to reverse a forward diffusion process, which starts from a clean motion $\mathbf{x}_0 \sim q(\mathbf{x}_0)$, sampled from the training data, and after T diffusion steps is approximately Gaussian $\mathbf{x}_T \sim \mathcal{N}(\mathbf{0}, \mathbf{I})$. Then at each step t of motion denoising, the reverse process is defined as:

$$p_\phi(\mathbf{x}_{t-1}|\mathbf{x}_t, \mathbf{c}) = \mathcal{N}\left(\mathbf{x}_{t-1}; \boldsymbol{\mu}_\phi(\mathbf{x}_t, \mathbf{c}, t), \beta_t \mathbf{I}\right) \tag{1}$$

where $\mathbf{c}$ is some conditioning signal (e.g., a text prompt), and β_t depends on a pre-defined variance schedule. The denoising model $\boldsymbol{\mu}_\phi$ with parameters ϕ predicts the denoised motion $\hat{\mathbf{x}}_0$ from a noisy input motion $\mathbf{x}_t$ [13]. The model is trained by sampling a motion $\mathbf{x}_0$ from the dataset, adding random noise, and supervising the denoiser with a reconstruction loss $\|\mathbf{x}_0 - \hat{\mathbf{x}}_0\|^2$.

Augmented controllability. In the image domain, general pre-trained diffusion models are specialized for new tasks using an augmented ControlNet [46] branch, which takes in a new conditioning signal and is fine-tuned on top of the frozen base diffusion model. OmniControl [41] adopts this idea to the human motion domain. For motion diffusion models with a transformer encoder architecture, they propose an augmented transformer branch that takes in kinematic joint constraints (e.g., pelvis or other joint positions) and at each layer connects back to the base model through a linear layer that is initialized to all zeros.

As described in Sects. 3.3 and 3.4, our key insight is to use an augmented control branch to enable scene awareness. We first train a strong scene-agnostic motion diffusion model to generate realistic motion from a text prompt, and then fine-tune an augmented branch that takes scene information as input (e.g., a 2D floor map or 3D geometry). This new branch adapts generated motion to be scene-compliant, while still maintaining realism and text controllability.

Test-Time Guidance. At test time, diffusion models can be controlled to meet specific objectives through guidance. We directly apply guidance to the clean motion prediction from the model $\hat{\mathbf{x}}_0$ [14,32]. At each denoising step, the predicted $\hat{\mathbf{x}}_0$ is perturbed with the gradient of an analytic objective function $\mathcal{J}$ as $\tilde{\mathbf{x}}_0 = \hat{\mathbf{x}}_0 - \alpha \nabla_{\mathbf{x}_t} \mathcal{J}(\hat{\mathbf{x}}_0)$ where α controls the strength of the guidance and $\mathbf{x}_t$ is the noisy input motion at step t. The predicted mean $\boldsymbol{\mu}_\phi$ is then calculated with the updated motion prediction $\tilde{\mathbf{x}}_0$ as in [14,32]. As detailed later, we define guidance objectives for avoiding collisions and reaching goals.

3.3 Navigation Motion Generation

The goal of the navigation stage is for the character to reach a goal location near the target object using realistic locomotion behaviors that can be controlled by the user via text. We design a hierarchical method that first generates a dense root trajectory with a diffusion model, then leverages a powerful in-painting model [33] to generate a full-body motion for the predicted trajectory. This approach facilitates accurate goal-reaching with the root-only model, while allowing diverse text control through the in-painting model.

Root Trajectory Generation. Our root trajectory diffusion model, shown in Fig. 3(a), operates on motions where each pose is specified by $\mathbf{x}^n =$

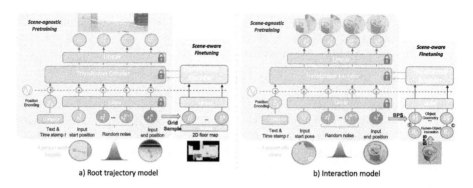

Fig. 3. Network architecture of the (**a**) root trajectory model and (**b**) interaction motion model. Initially, the base transformer encoder is trained on scene-agnostic motion data using start pose, target pose, and text as input. Subsequently, a scene-aware component is fine-tuned, which incorporates the 2D floor map (a) or 3D object (b).

$[x, y, z, \cos\theta, \sin\theta]_n$, with (x, y, z) being the pelvis position and θ the pelvis rotation, both of which are represented in the coordinate frame of the *first* pose in the sequence. The model is conditioned on a text prompt along with starting and ending goal positions and orientations. In contrast to the representation from prior work [7], which uses relative pelvis velocity and rotation, our representation using absolute coordinates facilitates constraining the outputs of the model with goal poses. Inspired by motion in-painting models [33,36], given a start pose **s** and end goal pose **g**, at each denoising step, we mask out the input $\mathbf{x}_t$ such that $\mathbf{x}_t^1 = \mathbf{s}$ and $\mathbf{x}_t^N = \mathbf{g}$, thereby providing clean goal poses directly to the model. To achieve this, a binary mask $\mathbf{m} = [\mathbf{m}^1, \ldots, \mathbf{m}^N]$ with the same dimensionality as $\mathbf{x}_t$ is defined, where $\mathbf{m}^1$ and $\mathbf{m}^N$ are a vector of 1's and all other $\mathbf{m}^n$ are 0's. During training, overwriting occurs with $\tilde{\mathbf{x}}_t = \mathbf{m} * \mathbf{x}_0 + (\mathbf{1} - \mathbf{m}) * \mathbf{x}_t$ where $*$ indicates element-wise multiplication and $\mathbf{x}_0$ is a ground truth root trajectory. We then concatenate the mask with the overwritten motion $[\tilde{\mathbf{x}}_t; \mathbf{m}]$ and use this as input to the model to indicate which frames have been overwritten.

At test time, goal-reaching is improved using a guidance objective $\mathcal{J}_g = (\hat{\mathbf{x}}_0^N - \mathbf{g})^2$ that measures the error between the end pelvis position and orientation of the predicted clean trajectory $\hat{\mathbf{x}}_0^N$ and the final goal pose.

Incorporating Scene Representation. The model as described so far is trained on a locomotion subset of the HumanML3D dataset [7] to enable generating realistic, text-conditioned root trajectories. However, it will be entirely unaware of the given 3D scene. To take the scene into account and avoid degenerating the text-following and goal-reaching performance, we augment the base diffusion model with a control branch that takes a representation of the scene as input. This scene-aware branch is a separate transformer encoder that is fine-tuned on top of the frozen base model. As input, we extract the walkable regions from the 3D geometry of the scene and project them to a bird's-eye view, yielding a 2D floor map $\mathcal{M}$. Following [32], a Resnet-18 [12] encodes the map $\mathcal{M}$ as feature grid, and at denoising step t, each 2D projected pelvis position $(x, z) \in \boldsymbol{x}_t^n$ is queried in the feature grid $\mathcal{M}$ to get the corresponding feature $\mathbf{f}_t^n$. The resulting sequence of features $\mathbf{f}_t = \left[\mathbf{f}_t^1, \ldots, \mathbf{f}_t^N\right]$, along with the text prompt and noisy motion $\mathbf{x}_t$, become the input to the separated transformer branch.

At test time, a collision guidance objective further encourages scene compliance. This is defined as $\mathcal{J}_c = \text{SDF}(\hat{\mathbf{x}}_0, \mathcal{M})$ where SDF calculates the 2D transform distance map from the 2D floor map, then queries the 2D distance value at each time step of the root trajectory. Positive distances, indicating pelvis positions outside the walkable region, are averaged to get the final loss.

Scene-Aware Training and Data. To train the scene-aware branch, it is important to have a dataset featuring realistic motions navigating through scenes with corresponding text prompts. For this purpose, we create the **Loco-3D-FRONT** dataset by integrating locomotion sequences from HumanML3D into diverse 3D environments from 3D-FRONT [6]. Each motion is placed within a different scene with randomized initial translation and orientation, following the methodology outlined in [43], as depicted in Fig. 4(a). Additionally, we apply left-right mirroring to both the motion and its interactive 3D scenes to augment the dataset [7]. This results in a dataset of approximately 9,500 walking motions, each motion accompanied by textual descriptions and 10 plausible 3D scenes on average, resulting in 95k locomotion-scene training pairs.

Added Control with Trajectory Blending. Our root trajectory diffusion model generates scene-aware motions and, unlike many prior works [9,52], does not require a navigation mesh to compute A* [8] paths to follow. However, a user may want a character to take the shortest path to an object by following the A* path, or to control the general shape of the path by drawing a 2D route themselves. To enable this, we propose to fuse an input 2D trajectory $\mathbf{p} \in \mathbb{R}^{N \times 2}$ with our model's predicted clean trajectory at every denoising step. At step t, we extract the 2D (x, z) components $\hat{\mathbf{p}}_0$ from the predicted root trajectory $\hat{\mathbf{x}}_0$ and interpolate them with the input trajectory $\tilde{\mathbf{p}}_0 = s * \hat{\mathbf{p}}_0 + (1 - s) * \mathbf{p}$ where s is the blending scale that controls how closely the generated trajectory matches the input. We then overwrite the 2D components of $\hat{\mathbf{x}}_0$ with $\tilde{\mathbf{p}}_0$ and continue denoising. This blending procedure ensures outputs roughly follow the desired path but still maintain realism inherent to the trained diffusion model.

Lifting to Full-Body Poses. To lift the generated pelvis trajectory to a full-body motion, we leverage the existing text-to-motion in-painting method PriorMDM [33], which takes a dense 2D root trajectory as input. By using this strong model that is pre-trained for text-to-motion, we can effectively generate natural and scene-aware full-body motion, while offering diverse stylistic control through text.

3.4 Object-Driven Interaction Motion Generation

After navigation, the character has reached a location near the target object and next should execute a desired interaction motion. Due to the fine-grained relationship between the body and object geometry during interactions, we propose a single diffusion model to directly generate full-body motion, unlike the two-stage navigation approach from Sect. 3.3.

Interaction Motion Generation. The interaction motion model operates on a sequence of full-body poses and is shown in Fig. 3(b). Our pose representation extends that of HumanML3D [7] to add the absolute pelvis position and heading $(x, y, z, \cos\theta, \sin\theta)$, similar to our navigation model. Each pose in the motion is $\mathbf{x}^n = \left[x, y, z, \sin\theta, \cos\theta, \dot{r}^a, \dot{r}^x, \dot{r}^z, r^y, \mathbf{j}^p, \mathbf{j}^v, \mathbf{j}^r, \mathbf{c}^f\right]_n \in \mathbb{R}^{268}$ with $\dot{r}^a$ root angular velocity, $(\dot{r}^x, \dot{r}^z)$ root linear velocity, r^y root height, $\mathbf{c}^f$ foot contacts, and $\mathbf{j}^p$, $\mathbf{j}^v$, $\mathbf{j}^r$ the local joint positions, velocities, and rotations, respectively.

The model is conditioned on a text prompt along with a starting full-body pose (*i.e.*, the final pose of the navigation stage) and a final goal pelvis position and orientation. The goal pelvis pose can usually be computed heuristically, but may also be provided by the user or predicted by another network [9]. The same masking procedure described in Sect. 3.3 is used to pass the start and end goals as input to the model. At test time, we also use the same goal-reaching guidance to improve the accuracy of hitting the final pelvis pose.

Object Representation. The base interaction diffusion model is first trained on a dataset of interaction motions from HumanML3D and SAMP [9] without any objects, which helps develop a strong prior on interaction movements driven by text prompts. Similar to navigation, we then augment the base model with a new object-aware transformer encoder and fine-tune this encoder separately.

For the input to this branch at each denoising step t, we leverage Basis Point Sets (BPS) [31] to calculate two key features: object geometry and the human-object relationship. First, a sphere with a radius of 1.0m is defined around the object's center, and 1024 points are randomly sampled inside this sphere to form the BPS. The distance between each point in the BPS and the object's surface is then calculated, capturing the object's geometric features and stored as $\mathbf{B}_O \in \mathbb{R}^{1024}$. Next, for each body pose $\mathbf{x}_t^n$ at timestep n in the noisy input sequence, we calculate the minimum distance from each BPS point to any body joint, giving $\mathbf{B}^n \in \mathbb{R}^{1024}$. The resulting sequence of features $\mathbf{B}_H = \left[\mathbf{B}^1, \ldots, \mathbf{B}^N\right]$ represents the human-object relationship throughout the entire motion. Finally, the object and human-object interaction features are concatenated with the original pose representation at each timestep $[\mathbf{x}_t^n; \mathbf{B}^n; \mathbf{B}_O]$ and fed to an MLP to generate a merged representation, which serves as the input to the scene-aware branch.

At test time, a collision objective is used to discourage penetrations between human and object. This is very similar to the collision loss described in Sect. 3.3, but the SDF volume is computed for the 3D object and body vertices that are inside the object are penalized. Please see the supplementary material for details.

Scene-Aware Training and Data. To train the scene-aware branch, we utilize the SAMP dataset [9], which captures motions and objects simultaneously. Specifically, we focus on "sitting" and "stand-up" interactions extracted from 80 sitting

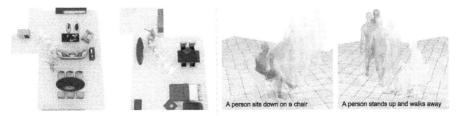

a) Loco-3D-Front: locomotion in different rooms   b) Interaction with different objects and text description

Fig. 4. (**a**) Loco-3D-FRONT contains locomotion placed in 3D-FRONT [6] scenes without collisions. (**b**) We augment SAMP [9] by randomly selecting chairs from 3D-FRONT to match the motions and annotating a text description for each sub-sequence.

motion sequences in the SAMP dataset involving chairs of varying heights, as shown in Fig. 4(b). To diversify the object geometry, we randomly select objects from 3D-FRONT [6] to match the contact vertices on human poses in the original SAMP motion sequences. This matching is achieved using the contact loss and collision loss techniques outlined in MOVER [44].

The original SAMP motions are often lengthy (∼100 s) and lack paired textual descriptions. For instance, a "sit" motion sequence involves walking to an object, sitting down, standing up, and moving away. To effectively learn individual skills, we extract sub-sequences containing specific interactions that begin or end with a sitting pose, such as "walk then sit", "stand up then sit", "stand up from sitting", and "walk from sitting." Furthermore, we annotate textual descriptions for each sub-sequence, which often incorporate the style of sitting poses, such as "a person walks and sits down on a chair while crossing their arms." Applying left-right data augmentation to motion and objects results in approximately 200 sub-sequences for each motion sequence, each paired with corresponding text descriptions and featuring various objects.

4 Experimental Evaluation

4.1 Evaluation Data and Metrics

Navigation. Navigation performance is assessed using the test set of Loco-3D-FRONT, comprising roughly 1000 sequences. Our metrics evaluate the generated root trajectory and the full-body motion after in-painting separately. For the root trajectory, we measure goal-reaching accuracy for the 2D (horizontal xz) root **position** (m), **orientation** (rad), and root **height** (m). The **collision ratio** evaluates the consistency of root motions with the environment. For the full-body motion after in-painting, we use **FID**, **R-precision**, and **diversity** from prior work [7], along with **foot skating** ratio [18], which evaluates the physical plausibility of motion-ground interaction. Please see the Sup. Mat. for details.

Interactions. To evaluate full-body human-object interactions, we use the established test split of the SAMP dataset [9], which contains motions related to

Table 1. Evaluation of navigation motion generation on the Loco-3D-FRONT test set. (**Left**) For generated pelvis trajectories, our approach achieves the best goal-reaching accuracy with low collision rate. (**Right**) After in-painting the full-body motion, our method maintains diverse and realistic motion that aligns with the given text prompt, competitive with diffusion-based scene-agnostic GMD and OmniControl.

Method	Root trajectory evaluation				Full-body motion evaluation			
	Goal-reaching error ↓							
	Pos.	Orient.	Height	Collision ↓	FID ↓	R-precision ↑	Diversity ↑	Foot skating ↓
Ground Truth	-	-	-	-	0.010	0.672	7.553	0.000
GMD [18]	0.374	1.231	-	-	13.160	0.114	4.488	0.181
OmniContol [41]	1.226	1.018	1.159	-	22.930	**0.458**	**7.128**	0.094
TRACE [32]	0.205	0.152	0.010	0.055	22.669	0.144	6.501	0.058
Ours (1-stage train)	0.197	0.132	0.013	**0.028**	22.372	0.152	6.347	0.062
Ours	**0.169**	**0.119**	**0.008**	0.031	**20.465**	0.376	6.415	**0.056**

sitting. Same as navigation, we analyze goal-reaching accuracy through position, orientation, and height errors. Furthermore, we assess physical plausibility by computing average **penetration values** and **penetration ratios** between the generated motion and interaction objects. We also perform a **user study** to compare methods. Please see more details in the Sup. Mat.

4.2 Comparisons

Navigation. We conduct a comparative analysis of our method with previous scene-aware and scene-agnostic motion generation approaches in Table 1. Every method is conditioned on a text prompt along with start/end goal poses. The TRACE baseline and our method TeSMo also receive the 2D floor map as input.

We first compare to GMD [18] and OmniControl [41], previous scene-agnostic text-to-motion diffusion models trained on HumanML3D to follow a diverse range of kinematic motion constraints. GMD utilizes the horizontal pelvis positions (x, z) of both the start and end goals to generate a dense root trajectory and subsequently the full-body motion. OmniControl takes as input the horizontal pelvis positions (x, z) along with the height y to directly generate full-body motion in a single stage. Our navigation model achieves better goal-reaching accuracy, e.g., 16.9 cm for root position, since it is trained specifically for the goal-reaching locomotion task. More importantly, the right half of Table 1 shows that our method's full-body motion after in-painting is comparable in realism, text-following, and diversity, while achieving the best foot skating results. This demonstrates our approach adds scene-awareness to locomotion generation without compromising realism or text control.

To justify our two-branch model architecture, we adapt TRACE [32], a recent root trajectory generation model designed to take a 2D map of the environment as input. The adapted TRACE architecture is very similar to our model in Fig. 3(a), but instead of using a separate scene-aware branch, the base transformer directly takes the encoded 2D floor map features as input. This results

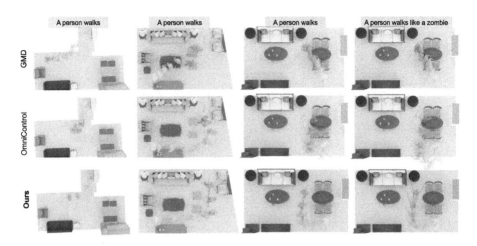

Fig. 5. Navigation generation performance. The start pose is the green arrow, and the goal pose is the red arrow. Our method more accurately reaches the goal and avoids obstacles while style is controlled by a text prompt. (Color figure online)

Table 2. Evaluation of human-object interaction motion generation on SAMP [9] sitting test set. Compared to DIMOS, our approach reaches the goal pose more accurately and exhibits fewer object penetrations, resulting in higher human preference.

Method	Goal-reaching error ↓			Object penetration ↓		User study
	Pos.	Height	Orient.	Value	Ratio	preference ↑
DIMOS [52]	0.2020	0.1283	0.4731	0.0193	0.1076	29.1%
Ours	**0.1445**	**0.0120**	**0.2410**	**0.0043**	**0.0611**	**71.9%**

in a single-branch architecture that must be trained from scratch, as opposed to our two-branch fine-tuning approach. Table 1 reveals that our method generates more plausible root trajectories with fewer collisions and more accurate goal-reaching. We also see that training our full two-branch architecture from scratch (*1-stage train* in Table 1), instead of using pre-training then fine-tuning, degrades both goal reaching and final full-body motion after in-painting.

A qualitative comparison of generated motions in different rooms is shown in Fig. 5. GMD tends to generate simple walking-straight trajectories. OmniControl and GMD do not reach the goal pose accurately and ignore the surroundings, leading to collisions with the environment. Our method TeSMo is able to generate diverse locomotion styles controlled by text in various scenes, achieving superior goal-reaching accuracy compared to other methods.

Interaction. Table 2 compares our approach to DIMOS [52], a state-of-the-art method to generate interactions trained with reinforcement learning. DIMOS requires a full-body final goal pose as input to the policy, unlike our approach which uses just the pelvis pose. Despite this, DIMOS struggles to reach the goal

Table 3. Test-time guidance evaluation. Adding guidance to reach goal poses and avoid collisions during inference improves performance. Lower is better for all metrics.

Guidance		Navigation		Interaction		
Goal Reach	Collision	Goal Pos.	Collision	Goal Pos.	Pen. Val.	Pen. Ratio
✗	✗	0.1568	0.0294	0.1445	0.0043	0.0611
✓	✗	**0.118**	0.0342	0.1453	0.0050	0.0554
✗	✓	0.1550	0.0013	0.1407	**0.0040**	**0.0414**
✓	✓	0.1241	**0.0012**	**0.1404**	0.0045	0.0494

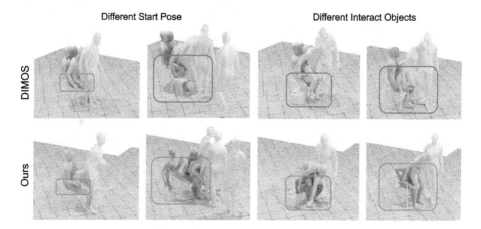

Fig. 6. Compared with DIMOS [52], our method generates more realistic human-object interactions with reduced floating and interpenetrations.

accurately, likely due to error accumulation during autoregressive rollout. Our method showcases fewer instances of interpenetration with interaction objects and the user study reveals a distinct preference for motions generated by our approach (preferred 71.9%) over those produced by DIMOS. Figure 6 compares the approaches qualitatively, where we see that more accurate goal-reaching reduces floating or penetrating the chair during sitting. Moreover, the interactions generated by DIMOS lack diversity, and cannot be conditioned on text.

4.3 Analysis of Capabilities

In Fig. 1, our method carries out a sequence of actions, enabling traversal and interaction with multiple objects within a scene. Figure 7 demonstrates additional key capabilities. In the top section, our method is controlled through a variety of text prompts. For interactions in particular, diverse text descriptions disambiguate between actions like sitting or standing up, and allow stylizing the sitting motion, e.g., with crossed arms. In the middle section, we enable user control over trajectories by adhering to a predefined A* path. By adjusting the

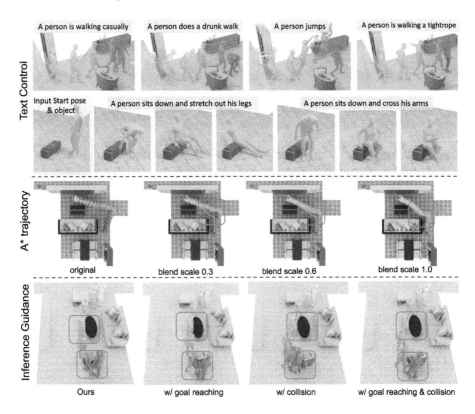

Fig. 7. TeSMo capabilities. (**Top**) Diverse text control; (**Middle**) Following A* path with adherence controlled by the blend scale; (**Bottom**) Test-time guidance encourages locomotion to reach the goal accurately without colliding with the environment.

blend scale, users can adjust how closely the generated trajectory follows A*. At the bottom of Fig. fig:capabilities, we harness guidance at test time to encourage motions to reach the goal while avoiding collisions and penetrations. As shown in Table 3, combining guidance losses gives improved results both for navigation and interactions.

5 Discussion and Future Work

We introduced TeSMo, a novel method for text-controlled scene-aware motion generation. While our navigation model ensures accurate goal-reaching and text-to-motion control, the two-stage process can disconnect the generated pelvis trajectory from the in-painted full-body poses. Exploring one-stage models for simultaneous pelvis and pose generation could streamline the process. Additionally, our 2D floor map approach limits handling intricate interactions, like stepping over a small stool.

Our approach aims at controllability, letting users specify text prompts or goal objects and locations. It may integrate with recent pipelines [40] that employ LLM planners to specify a sequence of actions and contact information that could be used to guide our motion generation. Looking ahead, we aim to broaden the spectrum of actions modeled by the system, to encompass activities such as lying down and touching. Furthermore, enabling interactions with dynamic objects will allow for more interactive and realistic scenarios.

Acknowledgments. Thanks Y. Huang and Y. Liu for technical support. Thanks M. Petrovich and N. Athanasiou for the fruitful discussion about text-to-motion synthesis. Thanks T. Niewiadomski, T. McConnell, and T. Alexiadis for running the user study.

Disclosure. https://files.is.tue.mpg.de/black/CoI_ECCV_2024.txt.

References

1. Agrawal, S., van de Panne, M.: Task-based locomotion. ACM Trans. Graph. **35**(4), 1–11 (2016). https://doi.org/10.1145/2897824.2925893
2. Chao, Y.W., Yang, J., Chen, W., Deng, J.: Learning to sit: synthesizing human-chair interactions via hierarchical control. In: Proceedings of the AAAI Conference on Artificial Intelligence, pp. 5887–5895 (2021)
3. Corona, E., Pumarola, A., Alenyà, G., Moreno-Noguer, F.: Context-aware human motion prediction. Cornell University - arXiv, Cornell University - arXiv (2019)
4. Diller, C., Dai, A.: CG-HOI: contact-guided 3D human-object interaction generation (2024)
5. Eigen, D., Ranzato, M., Sutskever, I.: Learning factored representations in a deep mixture of experts. arXiv Learning (2013)
6. Fu, H., et al.: 3D-front: 3D furnished rooms with layouts and semantics. In: Proceedings of the IEEE/CVF International Conference on Computer Vision, pp. 10933–10942 (2021)
7. Guo, C., et al.: Generating diverse and natural 3D human motions from text. In: Proceedings of the IEEE/CVF Conference on Computer Vision and Pattern Recognition (CVPR), pp. 5152–5161 (2022)
8. Hart, P.E., Nilsson, N.J., Raphael, B.: A formal basis for the heuristic determination of minimum cost paths. IEEE Trans. Syst. Sci. Cybern. **4**(2), 100–107 (1968)
9. Hassan, M., et al.: Stochastic scene-aware motion prediction. In: Proceedings of the IEEE/CVF International Conference on Computer Vision (ICCV), pp. 11374–11384 (2021)
10. Hassan, M., Choutas, V., Tzionas, D., Black, M.J.: Resolving 3D human pose ambiguities with 3D scene constraints. In: Proceedings of the IEEE/CVF International Conference on Computer Vision (ICCV) (2019)
11. Hassan, M., Guo, Y., Wang, T., Black, M., Fidler, S., Peng, X.B.: Synthesizing physical character-scene interactions. In: ACM SIGGRAPH 2023 Conference Proceedings. SIGGRAPH 2023. Association for Computing Machinery, New York (2023). https://doi.org/10.1145/3588432.3591525
12. He, K., Zhang, X., Ren, S., Sun, J.: Deep residual learning for image recognition. CoRR abs/1512.03385 (2015)

13. Ho, J., Jain, A., Abbeel, P.: Denoising diffusion probabilistic models. In: Larochelle, H., Ranzato, M., Hadsell, R., Balcan, M., Lin, H. (eds.) Advances in Neural Information Processing Systems, vol. 33, pp. 6840–6851. Curran Associates, Inc. (2020). https://proceedings.neurips.cc/paper/2020/file/4c5bcfec8584af0d967f1ab10179ca4b-Paper.pdf
14. Ho, J., Salimans, T., Gritsenko, A., Chan, W., Norouzi, M., Fleet, D.J.: Video diffusion models. arXiv preprint (2022)
15. Holden, D., Komura, T., Saito, J.: Phase-functioned neural networks for character control. ACM Trans. Graph. 1–13 (2017). https://doi.org/10.1145/3072959.3073663
16. Huang, S., et al.: Diffusion-based Generation, Optimization, and Planning in 3D Scenes. arXiv e-prints arXiv:2301.06015 (2023). https://doi.org/10.48550/arXiv.2301.06015
17. Jacobs, R.A., Jordan, M.I., Nowlan, S.J., Hinton, G.E.: Adaptive mixtures of local experts. Neural Comput. 79–87 (1991). https://doi.org/10.1162/neco.1991.3.1.79
18. Karunratanakul, K., Preechakul, K., Suwajanakorn, S., Tang, S.: Guided motion diffusion for controllable human motion synthesis. In: Proceedings of the IEEE/CVF International Conference on Computer Vision, pp. 2151–2162 (2023)
19. Kovar, L., Gleicher, M., Pighin, F.: Motion graphs. In: Seminal Graphics Papers: Pushing the Boundaries, vol. 2, pp. 723–732 (2023)
20. Kulkarni, N., et al.: Nifty: neural object interaction fields for guided human motion synthesis (2023)
21. Lee, J., Chai, J., Reitsma, P.S.A., Hodgins, J.K., Pollard, N.S.: Interactive control of avatars animated with human motion data. In: Proceedings of the 29th Annual Conference on Computer Graphics and Interactive Techniques (2002). https://doi.org/10.1145/566570.566607
22. Lee, J., Joo, H.: Locomotion-action-manipulation: synthesizing human-scene interactions in complex 3D environments. In: International Conference on Computer Vision (ICCV) (2023)
23. Lee, K.H., Choi, M.G., Lee, J.: Motion patches. ACM Trans. Graph. 898–906 (2006). https://doi.org/10.1145/1141911.1141972
24. Li, J., Clegg, A., Mottaghi, R., Wu, J., Puig, X., Liu, C.K.: Controllable human-object interaction synthesis. arXiv preprint arXiv:2312.03913 (2023)
25. Li, J., Wu, J., Liu, C.K.: Object motion guided human motion synthesis. ACM Trans. Graph. (TOG) **42**(6), 1–11 (2023)
26. Mahmood, N., Ghorbani, N., Troje, N.F., Pons-Moll, G., Black, M.J.: AMASS: archive of motion capture as surface shapes. In: International Conference on Computer Vision, pp. 5442–5451 (2019)
27. Peng, X., Xie, Y., Wu, Z., Jampani, V., Sun, D., Jiang, H.: Hoi-diff: text-driven synthesis of 3D human-object interactions using diffusion models. arXiv preprint arXiv:2312.06553 (2023)
28. Peng, X.B., Guo, Y., Halper, L., Levine, S., Fidler, S.: ASE: large-scale reusable adversarial skill embeddings for physically simulated characters. ACM Trans. Graph. **41**(4) (2022)
29. Petrovich, M., et al.: STMC: multi-track timeline control for text-driven 3D human motion generation. arXiv:2401.08559 (2024)
30. Pi, H., Peng, S., Yang, M., Zhou, X., Bao, H.: Hierarchical generation of human-object interactions with diffusion probabilistic models. In: ICCV, pp. 15061–15073 (2023)

31. Prokudin, S., Lassner, C., Romero, J.: Efficient learning on point clouds with basis point sets. In: International Conference on Computer Vision (ICCV), pp. 4332–4341 (2019)
32. Rempe, D., et al.: Trace and pace: controllable pedestrian animation via guided trajectory diffusion. In: CVPR (2023)
33. Shafir, Y., Tevet, G., Kapon, R., Bermano, A.H.: Human motion diffusion as a generative prior. arXiv preprint arXiv:2303.01418 (2023)
34. Starke, S., Zhang, H., Komura, T., Saito, J.: Neural state machine for character-scene interactions. ACM Trans. Graph. **38**(6) (2019). https://doi.org/10.1145/3355089.3356505
35. Taheri, O., Choutas, V., Black, M.J., Tzionas, D.: GOAL: generating 4D whole-body motion for hand-object grasping. In: Conference on Computer Vision and Pattern Recognition (CVPR) (2022). https://goal.is.tue.mpg.de
36. Tevet, G., Raab, S., Gordon, B., Shafir, Y., Bermano, A.H., Cohen-Or, D.: Human motion diffusion model. arXiv preprint arXiv:2209.14916 (2022)
37. Wang, J., Xu, H., Xu, J., Liu, S., Wang, X.: Synthesizing long-term 3D human motion and interaction in 3D scenes. In: Proceedings of the IEEE/CVF Conference on Computer Vision and Pattern Recognition, pp. 9401–9411 (2021)
38. Wang, J., Yan, S., Dai, B., Lin, D.: Scene-aware generative network for human motion synthesis. In: Proceedings of the IEEE/CVF Conference on Computer Vision and Pattern Recognition, pp. 12206–12215 (2021)
39. Wang, Z., Chen, Y., Liu, T., Zhu, Y., Liang, W., Huang, S.: Humanise: language-conditioned human motion generation in 3D scenes. In: Advances in Neural Information Processing Systems (NeurIPS) (2022)
40. Xiao, Z., et al.: Unified human-scene interaction via prompted chain-of-contacts. In: International Conference on Learning Representations (ICLR) (2024)
41. Xie, Y., Jampani, V., Zhong, L., Sun, D., Jiang, H.: Omnicontrol: control any joint at any time for human motion generation. arXiv preprint arXiv:2310.08580 (2023)
42. Ye, S., et al.: Scene synthesis from human motion. In: SIGGRAPH Asia 2022 Conference Papers. SA 2022. Association for Computing Machinery, New York (2022). https://doi.org/10.1145/3550469.3555426
43. Yi, H., Huang, C.H.P., Tripathi, S., Hering, L., Thies, J., Black, M.J.: MIME: Human-aware 3D scene generation. In: Computer Vision and Pattern Recognition (CVPR) (2023)
44. Yi, H., et al.: Human-aware object placement for visual environment reconstruction. In: Computer Vision and Pattern Recognition (CVPR), pp. 3959–3970 (2022)
45. Yuksel, S.E., Wilson, J.N., Gader, P.D.: Twenty years of mixture of experts. IEEE Trans. Neural Netw. Learn. Syst. 1177–1193 (2012). https://doi.org/10.1109/tnnls.2012.2200299
46. Zhang, L., Rao, A., Agrawala, M.: Adding conditional control to text-to-image diffusion models. In: Proceedings of the IEEE/CVF International Conference on Computer Vision, pp. 3836–3847 (2023)
47. Zhang, M., Cai, Z., Pan, L., Hong, F., Guo, X., Yang, L., Liu, Z.: Motiondiffuse: text-driven human motion generation with diffusion model. arXiv preprint arXiv:2208.15001 (2022)
48. Zhang, W., Dabral, R., Leimkühler, T., Golyanik, V., Habermann, M., Theobalt, C.: Roam: robust and object-aware motion generation using neural pose descriptors. In: International Conference on 3D Vision (3DV) (2024)
49. Zhang, X., Bhatnagar, B.L., Starke, S., Guzov, V., Pons-Moll, G.: Couch: towards controllable human-chair interactions. In: Avidan, S., Brostow, G., Cissé, M.,

Farinella, G.M., Hassner, T. (eds.) ECCV 2022. LNCS, vol. 13665, pp. 518–535. Springer, Cham (2022). https://doi.org/10.1007/978-3-031-20065-6_30
50. Zhang, Y., Tang, S.: The wanderings of odysseus in 3D scenes. In: Proceedings of the IEEE/CVF Conference on Computer Vision and Pattern Recognition, pp. 20481–20491 (2022)
51. Zhao, K., Wang, S., Zhang, Y., Beeler, T., Tang, S.: Compositional human-scene interaction synthesis with semantic control. In: Avidan, S., Brostow, G., Cissé, M., Farinella, G.M., Hassner, T. (eds.) ECCV 2022. LNCS, vol. 13666, pp. 311–327. Springer, Cham (2022). https://doi.org/10.1007/978-3-031-20068-7_18
52. Zhao, K., Zhang, Y., Wang, S., Beeler, T., , Tang, S.: Synthesizing diverse human motions in 3D indoor scenes. In: International Conference on Computer Vision (ICCV) (2023)
53. Zhu, W., et al.: Human motion generation: a survey. IEEE Trans. Pattern Anal. Mach. Intell. (2023)

NOVUM: Neural Object Volumes for Robust Object Classification

Artur Jesslen[1](✉)[iD], Guofeng Zhang[2][iD], Angtian Wang[2][iD], Wufei Ma[2][iD], Alan Yuille[2][iD], and Adam Kortylewski[1,3][iD]

[1] University of Freiburg, Freiburg im Breisgau, Germany
jesslen@cs.uni-freiburg.de
[2] Johns Hopkins University, Baltimore, USA
[3] Max-Planck-Institute for Informatics, Saarbrücken, Germany

Abstract. Discriminative models for object classification typically learn image-based representations that do not capture the compositional and 3D nature of objects. In this work, we show that explicitly integrating 3D compositional object representations into deep networks for image classification leads to a largely enhanced generalization in out-of-distribution scenarios. In particular, we introduce a novel architecture, referred to as NOVUM, that consists of a feature extractor and a *neural object volume* for every target object class. Each neural object volume is a composition of 3D Gaussians that emit feature vectors. This compositional object representation allows for a highly robust and fast estimation of the object class by independently matching the features of the 3D Gaussians of each category to features extracted from an input image. Additionally, the object pose can be estimated via inverse rendering of the corresponding neural object volume. To enable the classification of objects, the neural features at each 3D Gaussian are trained discriminatively to be distinct from (i) the features of 3D Gaussians in other categories, (ii) features of other 3D Gaussians of the same object, and (iii) the background features. Our experiments show that NOVUM offers intriguing advantages over standard architectures due to the 3D compositional structure of the object representation, namely: (1) An exceptional robustness across a spectrum of real-world and synthetic out-of-distribution shifts and (2) an enhanced human interpretability compared to standard models, all while maintaining real-time inference and a competitive accuracy on in-distribution data. Code and model can be found at ⌂/GenIntel/NOVUM.

A. Jesslen and G. Zhang—Equal contribution.

Supplementary Information The online version contains supplementary material available at https://doi.org/10.1007/978-3-031-73235-5_15.

1 Introduction

Current deep learning architectures demonstrate advanced capabilities in visual recognition tasks, *e.g.*, object classification, detection, and pose estimation [6,13,22,23,32]. However, generalization to out-of-distribution (OOD) scenarios remains a fundamental challenge [14,20,25,49]. In contrast, human vision achieves a significantly better robustness under OOD scenarios, e.g., domain shift, and occlusions [2,21]. Some cognitive studies hypothesize that human vision relies on a compositional 3D representation of objects while perceiving the world through an analysis-by-synthesis process [27,47]. While object-centric [24,45,46] and compositional representations [7,18,20] show promise in improving sample efficiency and generalization of machine learning algorithms, so far these models are largely ignorant of the 3D nature of our world and mostly limited to simple synthetic domains. This raises the question: Can machines enhance generalization at real-world classification tasks through learning 3D object representations? In this work, we embed a 3D compositional object representation explicitly into the neural network architecture for object classification. We take inspiration from recent advances in neural rendering [19] and prior works on pose estimation [17,37] to design a 3D-aware neural network for image classification. In particular, we propose NOVUM, which is composed of a feature extractor and *neural object volumes* for every object category (Fig. 1). Each neural object volume is a spatial composition of Gaussian ellipsoidal kernels that emit feature vectors. During inference, an image is classified by matching the Gaussian features of each category to the input feature map. In this way, object classification is realized using the individual Gaussian features only and without requiring the 3D spatial geometry, hence facilitating a fast and robust classification inference. Each Gaussian contributes to the prediction by focusing individually on different evidences (different parts of objects in our case). When combined together, the lack of evidence from part of the Gaussians can be compensated by the rest, leading to high robustness against outliers or occlusions. Moreover, the inherent 3D representation in our model can also estimate the 3D object pose via inverse rendering of the feature volume using Gaussian splatting. We use 3D pose annotations of the objects to train the neural features to be distinct from features of other categories' neural object volumes, as well as spatially apart features of the same neural object volume, and the background clutter. Intuitively, the feature extractor learns to classify every pixel in the image as being either part of the object or as background context.

We evaluate NOVUM on a variety of datasets that contain 3D pose annotations and OOD shifts, such as real-world OOD shifts on the OOD-CV dataset [48], and synthetic OOD shifts on the corrupted PASCAL3D+ [14] and occluded PASCAL3D+ [37]. Our experiments show that NOVUM is exceptionally robust compared to other state-of-the-art architectures (both CNNs and Transformers) at object classification while performing on par with in-distribution data in terms of accuracy and inference speed. Moreover, the 3D pose predictions obtained via inverse rendering are competitive to baseline models which are limited to perform 3D pose estimation only. Finally, we show that the Gaussian

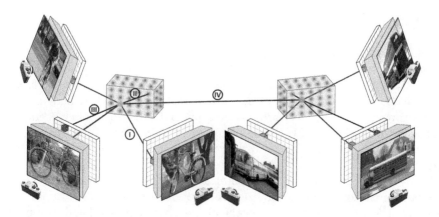

Fig. 1. Schematic overview of how NOVUM is trained. The model consists of a shared backbone (yellow) and one neural object volume for each object class (grey), which are represented as 3D Gaussians on a cuboid shape. During training, the backbone first computes feature maps of the training images. Given the class label and the 3D object pose, the backbone is trained in a contrastive manner using four types of losses: (I) To make features of the same Gaussian similar across instances (green), while at the same time making the features distinct (red) from (II) features of Gaussians from the same object, (III) background features, and (IV) features of Gaussians from other objects. (Color figure online)

matching result provides intuitive human interpretable information about the model prediction, by showing where the model perceives corresponding object parts.

In conclusion, NOVUM introduces a 3D volume representation for classification, achieving robustness through compositionality and 3D-awareness, while offering enhanced human interpretability by visualizing individual Gaussian kernel matches, and still enabling real-time inference.

2 Related Work

Robust Image Classification. Image classification is a classical task in computer vision. Multiple influential architectures such as ResNet [13], some transformers [23,36] were designed for this task. However, these models have primarily targeted in-distribution data, leaving a significant gap when faced with out-of-distribution (OOD) data such as common synthetic corruptions [14] or real-world OOD data [49]. To bridge this OOD generalization gap, efforts have concentrated mainly on two fronts: data augmentation and architectural design. Data augmentation strategies involve leveraging learned augmentation policies [5], and data mixtures [15] to enhance the diversity of training samples, thereby fostering generalization by generating synthetic OOD data. On the other hand, architectural advances that incorporate general prior knowledge about the world, such as compositionality [7,18,20], or object-object centric representations [24,45,46], have also shown promising advances in terms of enhancing sample efficiency and generalization in neural networks. Our approach falls into the second category, as we introduce a 3D compositional representation into the architecture of neural networks for object classification. In contrast to standard discriminative methods relying on a single entangled feature representation of an image, our 3D and compositional representation leads to a largely enhanced robustness when faced with occlusions, corruptions, and real-world OOD scenarios.

Contrastive Learning. Studies have found that, in supervised settings, learning features that are discriminative during training does not guarantee that a model will generalize, instead, they can have inductive biases towards learning simple "shortcut" features and decision rules [16,28]. Hence, contrastive learning [11] is an interesting direction to prevent to learn these shortcuts since it is essentially a framework that learns similar/dissimilar representations by optimizing through the similarities of pairs in the representation space. Later the idea of using contrastive pair is extended to Triplet [31]. While traditional contrastive learning has focused on image-level or text-level representations, the extension to feature-level representations has gained lots of attention after InfoNCE [29]. Influential works in various fields include SimCLR [3], CLIP [30], CoKe [1]. In our paper, we adopt a three-level feature contrastive loss that encourages spatially distinct features, categorical specific features, and background specific features.

3D Neural Representations for Pose Estimation. Explicit 3D object representations can offer significant advantages in terms of generalization over purely image-based representations. Importantly, when using category-level 3D representations for vision tasks, it becomes imperative to not rely on detailed instance-specific features (*e.g.*, object colour or texture), but rather to enable the learning of general category-level features. As a result, employing features that remain invariant to such specificities emerges as a straightforward solution. Initially, [34] delved into computing correspondences between image features and a learned volumetric representation, leveraging HOG features and employing a Proposal-Validation process, which is slow. Later works including [17,37,38], have replaced HOG features with neural features capable of encoding richer information. This shift to neural features has facilitated the extension of render-and-compare methods, originally introduced at the pixel-level [4,41], to the feature level. This extension enhances the models' ability to generalize to rendering category-level instances. Conceptually, this method embodies an approximate analysis-by-synthesis approach, similar to the principles outlined in [10], which proves to be more robust against out-of-distribution (OOD) data in 3D pose estimation when contrasted with classical discriminative methods [26,35,50].

NOVUM builds on prior work on category-level pose estimation, such as neural mesh models [37,38]. NOVUM extends this line of work in multiple ways: (1) An architecture with a shared class-agnostic backbone, which enables us to introduce a class-contrastive loss for object classification. (2) A principled, efficient way of performing classification inference that exploits individual 3D Gaussians. (3) A comprehensive mathematical formulation that derives a vMF-based contrastive training loss. (4) A compositional 3D Gaussian representation [19] that can be matched robustly with volume rendering [40] even in strong OOD scenarios. We demonstrate the applicability of compositional 3D representations in classification tasks, hence pioneering a step towards highly advanced capabilities in terms of generalization, interpretability and multi-tasking in vision.

3 Method

In this section, we present a deep network with an integrated 3D and compositional volume representation of objects to achieve robust object classification (Sect. 3.1). We discuss how the model can be learned (Sect. 3.2) and how it can be applied to infer the object class as well as the 3D pose at test time (Sect. 3.3).

3.1 NOVUM: A Network Architecture with Neural Object Volumes

Motivation. Our aim is to achieve robustness at object classification by devising a neural network architecture that integrates a 3D compositional representation of objects. Specifically, our architecture explicitly uses an object-centric 3D representation to introduce compositionality on two levels of abstraction: (1) *Image-level compositionality*, *i.e.*, a representation that explicitly models an image as a composition of objects and background clutter. (2) *Object-level compositionality*, *i.e.*, a representation that explicitly models objects as a spatial composition of local elements. This compositional representation enhances model robustness by improving the models ability to classify objects even if only few local parts are recognizable (*e.g.*, due to occlusion or due to a novel object topology), or when the object is placed in an unusual context (*e.g.*, a bicycle underwater).

Neural Object Volumes. Building on recent advances in Gaussian splatting [19], we represent objects as a 3D density field via a spatial composition of K Gaussians that are placed on the surface geometry of each object category. Each Gaussian emits an associated feature vector, hence defining the volumetric object representation that we refer to as *neural object volume*. Each neural object volume represents an object category and is learned from feature maps extracted from 2D images by given an annotated 3D pose of the object (Fig. 1).

More formally, we define a neural volume density at spatial location $x \in \mathbb{R}^3$ as a mixture of three-dimensional Gaussian $\rho(x) = \sum_{k=1}^{K} \rho_k(x)$. Each Gaussian density is defined as $\rho_k(x) = \mathcal{N}(\mu_k, \Sigma_k)$, where $\mu_k \in \mathbb{R}^3$ is the 3D position of the k-th Gaussian and $\Sigma_k \in \mathbb{R}^{3\times 3}$ is its covariance matrix (defining the direction, shape and size of k-th kernel). In our setup, we do not assume detailed geometry of the object but simply arrange the Gaussians such that they form a cuboid volume with a pre-defined and diagonal covariance, such that the volume approximately encompasses the variable object instances in the corresponding object category. Each Gaussian is associated with a feature vector, denoted $C_k \in \mathbb{R}^D$. For each object category y, we learn a set of features $\mathcal{C}_y = \{C_k \in \mathbb{R}^D\}_{k=1}^{K}$. We define the set of Gaussian features from all object categories as $\mathcal{C} = \{\mathcal{C}_y\}_{y=1}^{Y}$, where Y is the total number of object categories. The neural object volume can be rendered into the feature space, using standard volume rendering:

$$\hat{C}_i(\alpha) = \int_{t_n}^{t_f} T(t) \sum_{k=1}^{K} \rho_k(\boldsymbol{r}_\alpha(t))\boldsymbol{C_k}\,\mathrm{d}t, \quad \text{where } T(t) = \exp\left(-\int_{t_n}^{t} \rho(\boldsymbol{r}_\alpha(s))\mathrm{d}s\right), \quad (1)$$

where the feature $\hat{C}_i(\alpha)$ at pixel position i in the rendered feature map is computed by aggregating the Gaussian features along the ray $\boldsymbol{r}_\alpha(t)$. The ray traverses from the camera center through the pixel i on the image plane with α denoting the camera view. Here, t ranges from the near plane t_n to the far plane t_f. The remainder of the image that is not covered by the rendered object volume is represented as background features $\mathcal{B} = \{\beta_n \in \mathbb{R}^D\}_{n=1}^{N_b}$ where N_b is a fixed hyperparameter, and $\mathcal{B}$ is shared among all object categories.

NOVUM. Our model architecture builds on a feature extractor Φ_w and a set of neural object volumes, one for each object category. The feature extractor computes a feature map $F = \Phi_w(I) \in \mathbb{R}^{D \times H \times W}$ from an input image I, where w denotes the parameters of the CNN backbone. The feature map F contains feature vectors $f_i \in \mathbb{R}^D$ at positions i on a 2D lattice.

Learning our model requires obtaining correspondences between a Gaussian k of a neural object volume and a location i in the feature map of a training image, and we obtain this correspondence by projecting Gaussian into the image feature map F. With given camera pose α, we use volume rendering to compute the contribution γ_{ik} of Gaussian feature C_k to image features f_i. We estimate a one-to-one correspondence between features and Gaussians by selecting the closest image feature f_i for each Gaussian, i.e., where γ_{ik} is the largest. Throughout the remaining paper, we denote $f_{k \to i}$ to indicate the extracted feature f_i at location i that Gaussian k with mean μ_k projects to. We relate the extracted image features to the Gaussians and background features by von-Mises-Fisher (vMF) probability distributions. In particular, we model the probability of generating the feature f_i from corresponding Gaussian C_k as $P(f_{k \to i}|C_k) = c_M(\kappa)e^{\kappa f_{k \to i} \cdot C_k}$, where C_k represents the mean of each vMF distribution ($\|f_{k \to i}\|= 1, \|C_k\|= 1$). We also model the probability of generating the feature f_i from background features as $P(f_i|\beta_n) = c_M(\kappa)e^{\kappa f_i \cdot \beta_n}$ for $\beta_n \in \mathcal{B}$. The concentration parameter κ, which determines the spread of the distribution and can be interpreted as an inverse temperature parameter, is defined as a global hyperparameter. Hence, the normalization constant $c_M(\kappa)$ is a constant and can be ignored during learning and inference.

3.2 Learning Discriminative 3D Volume Representations

Learning NOVUM is challenging because we not only need to maximize the likelihood functions $P(f_{k \to i}|C_k)$ and $P(f_i|\mathcal{B})$, but also learn a corresponding parameters w of the backbone. In particular, we maximize the probability that any extracted feature $f_{k \to i}$ was generated from $P(f_{k \to i}|C_k)$ instead of from any other alternatives. This motivates us to use contrastive learning where we compare the probability that an extracted feature $f_{k \to i}$ is generated by the correct

Gaussian C_k or from one of three alternative processes, namely, (i) from the Gaussians of other object classes, (ii) from non-neighboring Gaussians of the same object, and (iii) from the background features (see illustration in Fig. 1):

$$\frac{P(f_{k\to i}|C_k)}{\sum_{\substack{C_l \in \mathcal{C}_y \\ C_l \notin \mathcal{N}_k}} P(f_{k\to i}|C_l) + \omega_\beta \sum_{\beta_n \in \mathcal{B}} P(f_{k\to i}|\beta_n) + \omega_{\bar{y}} \sum_{C_m \in \mathcal{C}_{\bar{y}}} P(f_{k\to i}|C_m)}, \quad (2)$$

where $\mathcal{N}_k = \{C_r : \|\mu_k - \mu_r\| < \delta, k \neq r\}$ is the neighborhood of C_k and δ is a distance threshold controlling the size of neighborhood. y is the category of the image and $\bar{y}$ is a set of all other categories except y. $\omega_\beta = \frac{P(\beta_n)}{P(C_k)}$ is the ratio of the probability that an image feature corresponds to the background instead of the Gaussian k, and $\omega_{\bar{y}} = \frac{P(C_m)}{P(C_k)}$ is the ratio of the probability that an image feature corresponds to Gaussians of other categories instead of the Gaussian k. We compute the final loss $\mathcal{L}(\mathcal{C}, \mathcal{B}, w)$ by taking the logarithm and summing over all training examples – all sets of features $\{f_{k\to i}\}$ from the training set

$$-\sum_y \sum_{k=1}^{K} o_k \cdot \log \frac{e^{\kappa f_{k\to i} \cdot C_k}}{\sum_{\substack{C_l \in \mathcal{C}_y \\ C_l \notin \mathcal{N}_k}} e^{\kappa f_{k\to i} \cdot C_l} + \omega_\beta \sum_{\beta_n \in \mathcal{B}} e^{\kappa f_{k\to i} \cdot \beta_n} + \omega_{\bar{y}} \sum_{C_m \in \mathcal{C}_{\bar{y}}} e^{\kappa f_{k\to i} \cdot C_m}}, \quad (3)$$

where $o_k = 1$ if the Gaussian is visible and $o_k = 0$ otherwise.

Updating Gaussian and Background Features. The Gaussian and background features $\mathcal{C}$ and $\mathcal{B}$ are updated after every gradient-update of the feature extractor. Following [1,12], we use momentum update for the Gaussian features:

$$C_k \leftarrow C_k \cdot \sigma + f_{k\to i} \cdot (1-\sigma), \quad \|C_k\| = 1. \quad (4)$$

The background features are simply resampled from the newest batch of training images. In particular, we remove the oldest features in $\mathcal{B}$, i.e., $\mathcal{B} = \{\beta_n\}_{n=1}^{N} \setminus \{\beta_n\}_{n=1}^{T}$. Next, we randomly sample T new background features f_b from the feature map, where f_b is a feature that no Gaussian contributes to and add them into the background feature set $\mathcal{B}$ (i.e., $\mathcal{B} \leftarrow \mathcal{B} \cup \{f_b\}$). We note that σ and T are hyper-parameters of our model.

3.3 Inference of Object Category and 3D Pose

Object Classification via Feature Matching without Geometry. Our classification inference pipeline is illustrated in Fig. 2. We perform classification in a fast and robust manner via matching the extracted features to the learned Gaussian features and background features. In short, for each object category y, we compute the foreground likelihood $P(f_i|C_y)$ and the background likelihood $P(f_i|\mathcal{B})$ on all locations i in the feature map. In this process, we do not take into account the object geometry, which reduces the matching to a simple convolution operation, hence making it very fast. To classify an image, we compare the total likelihood scores of each class average over all locations i.

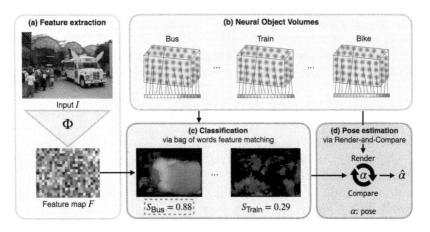

Fig. 2. Overview of the classification *inference pipeline*. NOVUM is composed of a backbone Φ and a set of neural object volumes represented as 3D Gaussians on a cuboid shape (green box) with colored associated features. During inference, an image is first processed by the backbone into a feature map F. The object class is predicted by *independently* matching the Gaussian features to the feature map (blue box). We color-code the detected Gaussians to highlight the interpretability of our method. Brightness shows the prediction confidence. Note that the model is only confident with the correct class even though the bus is an out-of-distribution sample. The 3D object pose can also be inferred via inverse rendering of the neural object volume (red box). (Color figure online)

In detail, we define a binary valued parameter $z_{i,k}$ such that $z_{i,k} = 1$ if the feature vector f_i matches best to any Gaussian feature $\{C_k\} \in \mathcal{C}_y$, and $z_{i,k} = 0$ if it matches best to a background feature. The object likelihood of the extracted feature map $F = \Phi_w(I)$ can then be computed as:

$$\prod_{f_i \in F} P(f_i|z_{i,k}, y) = \prod_{f_i \in F} P(f_i|C_k)^{z_{i,k}} \prod_{f_i \in F} \max_{\beta_n \in \mathcal{B}} P(f_i|\beta_n)^{1-z_{i,k}}. \tag{5}$$

As described in Sect. 3.1, the extracted features follow a vMF distribution. Thus the final prediction score of each object category y is:

$$S_y = \sum_{f_i \in F} \max\{\max_{C_k \in \mathcal{C}_y} f_i \cdot C_k, \max_{\beta_n \in \mathcal{B}} f_i \cdot \beta_n\}. \tag{6}$$

The final category prediction is $\hat{y} = \arg\max_{y \in Y}\{S_y\}$. Figure 2 (blue box) illustrates the matching process for different object classes by color coding the detected Gaussians. For the correct class, the Gaussians can be detected coherently even without taking geometry into account (as can be observed by the smooth color variation), while for wrong classes this is not the case. Our ability to visualize the predicted kernel correspondence demonstrates an advanced interpretability of the decision process compared to standard classifiers.

Volume Rendering for Pose Estimation. Given the predicted object category $\hat{y}$, we use the Gaussian features $\mathcal{C}_{\hat{y}}$ to estimate the camera pose α leveraging the 3D geometrical information of the neural object volumes. Following the vMF distribution, we optimize the pose α via feature reconstruction:

$$\mathcal{L}(\alpha) = \sum_{f_i \in FG} f_i \cdot \hat{C}_i(\alpha) + \sum_{f_b \in BG} \max_{\beta_n \in \mathcal{B}} f_b \cdot \beta_n, \qquad (7)$$

where FG is the set of foreground features that are covered by the rendered neural object, i.e., those features for which the aggregated volume density is bigger than a threshold $FG = \{f_i \in F, \sum_{k=1}^{K} \rho_k(\boldsymbol{r}_\alpha(t)) > \theta\}$. $BG = F \setminus FG$ is the set of features in the background. We estimate the pose by first finding the best initialization of the object pose α by computing the reconstruction loss (Eq. 7) for a set of pre-defined poses. Subsequently, we start gradient-based optimization using the initial pose that achieved the lowest loss to obtain the final pose prediction $\hat{\alpha}$.

4 Experiments

In this section, we evaluate NOVUM in terms of generalization on in-distribution and out-of-distribution data, 3D pose estimation and interpretability. We first discuss our experimental setup (Sect. 4.1), present baselines and results for classification (Sect. 4.2) and 3D pose estimation (Sect. 4.3). Additionally, we perform in-depth evaluations of interpretability and prediction consistency, and an ablation study (Sect. 4.4).

4.1 Setup

Datasets. We test NOVUM's robustness using four datasets where 3D pose annotations are available: PASCAL3D+ (P3D+) [43], occluded-P3D+ [39], corrupted-P3D+ [25], and Out-of-Distribution-CV (OOD-CV) [48]. PASCAL3D+ includes 12 object categories. Building on the P3D+ dataset, the occluded-P3D+ dataset is a test benchmark that evaluates robustness under multiple occlusion levels. It simulates realistic occlusion by superimposing occluders on top of the objects with three different levels: L1: 20%–40%, L2: 40%–60%, and L3: 60%–80%, where each level has corresponding percent of the objects and the background occluded. P3D+ dataset does not contain significant occlusion and is therefore referred to as occlusion level 0 (L0). The corrupted-P3D+ corresponds to P3D+ where we apply 12 types of corruptions [14,25] to each image of the original test images, and we choose a severity level of 4 out of 5. The OOD-CV dataset is a benchmark that includes real-world OOD examples of 10 object categories varying in terms of 5 nuisance factors: pose, shape, context, texture, and weather.

Implementation Details. Each neural object contains approximately $K = 1100$ Gaussians that are distributed uniformly on the cuboid. The shared feature extractor Φ is a ResNet50 [13] model with two upsampling layers and an input shape of 640 × 800. All features have a dimension of $D = 128$ and the size of the feature map F is $1/8^{th}$ of the input size. NOVUM is trained as described in Sect. 3.2, taking around 20 h on 4 RTX 3090 with 200 epochs. During training, we use $N = 2560$ background features. For each gradient step, we use $\sigma = 0.9$ for momentum update of the Gaussian features and sample $T = 5$ new background features to update $\mathcal{B}$. We set $\kappa = 1$ (see Appendix 9 for more details). We predict the object class as described in Sect. 3.3. The feature matching for classification takes around **0.01**s per sample on 1 RTX 3090, which is comparable to state-of-the-art feed-forward classification models. NOVUM can also infer the 3D object pose via inverse rendering by first evaluating the reconstruction loss on 144 initial poses (12 azimuth angles, 4 elevation angles, 3 in-plane rotations) and subsequently starting a gradient-based inverse rendering (Eq. 7) starting with lowest feature reconstruction loss as initialization. The pose inference pipeline takes around 0.21s per sample on 1 RTX 3090. We note that this inference might be further optimized *e.g.*, by caching intermediate matching results efficiently.

Evaluation. We evaluate our approach on two tasks: classification and 3D pose estimation. 3D pose estimation involves predicting azimuth, elevation, and rotations of an object with respect to a camera. Following [50], the pose estimation error is calculated between the predicted rotation matrix R_{pred} and the ground truth rotation matrix R_{gt} as $\Delta\left(R_{\text{pred}}, R_{\text{gt}}\right) = \left\|\log m\left(R_{\text{pred}}^T R_{\text{gt}}\right)\right\|_F / \sqrt{2}$. We measure accuracy using two thresholds $\frac{\pi}{18}$ and $\frac{\pi}{6}$.

Classification Baselines. We compare the performance of our approach to four competitive baseline architectures (*i.e.*, Resnet50, Swin-T, Convnext, and ViT-b-16) for the classification task. During training, these baselines are trained with a classification and a pose estimation head. Hence, 3D information is leveraged for these classification evaluations. For each baseline, we use a classification head for which the output is the number of classes in the dataset (*i.e.*, 12 for (occluded, corrupted)-P3D+; 10 for OOD-CV). We finetune each baselines during 200 epochs. In order to make baselines more robust, we apply standard data augmentation (i.e., scale, translation, rotation, and flipping) for each during training. More details about baselines can be found in the appendix.

3D-Pose Estimation. We compare the performance of our approach to five other baselines for the 3D pose estimation task. For Resnet50, Swin-T, Convnext, and ViT-b-16, we consider the pose estimation problem as a classification problem (following [50]) by using 42 intervals of ∼8.6° for each parameter that needs to be estimated (azimuth and elevation angle, and in-plane rotation). We fine-tune each baseline for 200 epochs. Similarly to classification, we apply standard data augmentation during training. We further evaluate against NeMo [37] that was explicitly designed for robust 3D pose estimation. We following the publicly available code and train a NeMo model for each class.

4.2 Robust Object Classification

We first evaluate the performance on IID data. As the L0 column of Table 1 shows, our approach achieves 99.5% for classification, which is comparable to other baselines. Furthermore, our approach manages to robustly classify images in various out-of-distribution scenarios. From Table 1, we can see that our representation allows to outperform all other traditional baselines with around 6% accuracy on average for different levels of occlusions and with up to 33% accuracy boost for images under five different types of nuisances in OOD-CV. For corrupted data, our approach also outperforms the baselines on average. In summary, NOVUM achieves **significant improvements in OOD generalization** while **maintaining state-of-the-art accuracy for IID** data for classification. Finally, it is worth noting that our approach is **more consistent** than all baselines. Our approach consistently outperforms baselines over all nuisances, which indicates the intrinsic robustness in our architecture.

Table 1. Classification accuracy results on P3D+, occluded-P3D+, OOD-CV and corrupted-P3D+ datasets. L0 corresponds to unoccluded images from Pascal3D+, and occlusion levels L1-L3 are from occluded-P3D+ dataset with occlusion ratios stated in Sect. 4.1. Our approach performs similarly in IID scenarios, while steadily outperforming all baselines in OOD scenarios. First is highlighted in **bold**, second is underlined. Higher is better. Full results in Appendix.

Dataset	P3D+	occluded-P3D+				OOD-CV	corrupted-P3D
Nuisance	L0	L1	L2	L3	Mean	Mean	Mean
Resnet50	99.3	93.8	77.8	45.2	72.3	51.4	78.7
Swin-T	99.4	93.6	77.5	46.2	72.4	64.2	78.9
Convnext	99.4	95.3	81.3	50.9	75.8	56.0	85.6
ViT-b-16	99.3	94.7	80.3	49.4	74.8	59.0	87.6
NOVUM	**99.5**	**97.2**	**88.3**	**59.2**	**81.6**	**85.2**	**91.3**

4.3 Robust 3D Pose Estimation

According to the results in Table 2, our approach outperforms all baselines significantly across IID and OOD scenarios, while also exceeding the performance of NeMo [37], the current state-of-the-art method for robust 3D pose estimation. This higher performance can be attributed to the better optimization of volume rendering which gives more stable gradients compared to mesh-based differentiable rendering.

4.4 Comprehensive Assessment of Our Representation

Interpretability. Our explicit volumetric representation can be leveraged to predict the object class and its pose in an image. Insightful information also lies

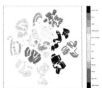  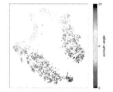

(a) **NOVUM**: t-SNE plot of learned feat. $\mathcal{C}$. (b) Nemo: t-SNE plot of learned features Θ. (c) **NOVUM**: t-SNE plot of all car samples. (d) Resnet50: t-SNE plot of all car samples.

Fig. 3. (a–b) t-SNE plots comparing (a) the learned features $\mathcal{C}$ of our approach and (b) the learned vertex features Θ of NeMo. As can be seen, our contrastive loss allows a much clearer distribution of the space while keeping Gaussian features from different classes far from each other (while the low-quality clustering observed in (b) may likely originates from the ImageNet pretraining). (c–d) t-SNE plots of the mean extracted feature for each car image of the test dataset. We observe a very clear organization of the samples according to the azimuth angle for (c) our approach while this organization is completely absent in (d) other feed-forward baselines (*e.g.*, Resnet50).

Table 2. 3D-Pose Estimation results for different datasets. A prediction is considered correct if the angular error is lower than threshold (*i.e.*, $\frac{\pi}{6}$, and $\frac{\pi}{18}$). Higher is better. Our approach shows it is capable of robust 3D pose estimation that performs similarly to the current state-of-the-art. Note that models marked with a '*' possess an explicit 3D representation.

Dataset	P3D+	occ-P3D+	cor-P3D+	OOD-CV	P3D+	occ-P3D+	cor-P3D+	OOD-CV
Threshold	$\pi/6$				$\pi/18$			
Resnet50	82.2	53.8	33.9	51.8	39.0	15.8	15.8	18.0
Swin-T	81.4	48.2	34.5	50.9	46.2	16.6	15.6	19.8
Convnext	82.4	49.3	37.1	50.7	38.9	14.1	24.1	19.9
ViT-b-16	82.0	50.8	38.0	48.0	38.0	15.0	21.3	21.5
NeMo*	86.1	62.2	48.0	51.6	**61.0**	31.8	43.4	21.9
NOVUM *	**88.2**	**63.1**	**49.1**	**52.9**	59.1	**32.7**	**43.4**	**22.8**

in the Gaussian feature matching between image features and the neural object volumes. Visualizing the matching results (see Fig. 4 and more results including videos in Appendix 10 and Figure S2), reveals which object parts are perceived by the model at any given location in the image. As illustrated in Fig. 4, the occluded parts of the objects are not colored, meaning that no Gaussian has been matched with these image features. Figures 3a and 3c moreover show that our learned features are disentangled and encode useful information in terms of object classes and 3D pose. When comparing Figs. 3a and 3b, we observe different cluster for each category while the low-quality clustering observed in the latter Figure may likely originates from the ImageNet pretraining. When comparing Figs. 3c and 3, we observe a consistent distribution of car instances depending of their pose in the former while in the latter, features don't have any explicit organization in terms of pose.

Table 3. Consistency of predictions for classification and 3D pose estimation. The last line shows the accuracy of those images with both object pose (threshold: $\frac{\pi}{6}$) and class correctly predicted. "cls." stands for classification.

Dataset	P3D+	occ-P3D+		
Task	L0	L1	L2	L3
classification	99.5	97.2	88.3	59.2
3D pose (th: $\frac{\pi}{6}$)	90.1	82.6	71.3	52.7
cls. & 3D pose	89.8	81.9	70.2	50.2

Table 4. Ablation studies of the background features and the shape of the representation on P3D+ and occluded-P3D+. We ablate both the background features and the object shape.

Components		P3D+	occ-P3D+		
$\mathcal{B}$	Shape	L0	L1	L2	L3
	single feature	93.2	90.3	80.4	44.0
✓	sphere	99.3	97.0	87.9	59.0
	cuboid	99.3	97.0	85.7	53.0
✓	cuboid	**99.5**	**97.2**	**88.3**	**59.2**

Consistency. Another valuable characteristics of our approach lies in the fact that it is consistent between the different tasks. In Table 3, we observe that the accuracy for samples that have correct class and pose prediction are limited by the pose estimation itself since they are fairly similar. In IID scenarios, the difference is only of 0.3%, while in OOD scenarios the difference is around 1% on average. We believe that this consistency comes from the common explicit

Fig. 4. Four qualitative results that were misclassified by ViT-b-16. We show for each: (left) the input image and (right) the extracted feature map and the predicted 3D pose overlaid. We color coded the features by encoding the color as a function of μ_k of the matched Gaussian C_k (as done in NOCS [42]). Hence, a smooth color gradient shows a high quality matching. In the extracted features, the brightness illustrates the confidence of matching with the Gaussian features.

volumetric representation that is shared for all tasks. Such a behavior would not be expected between different task-specific models that are trained separately.

Efficiency. For classification, NOVUM matches real-time performance as other CNN or transformer-based baselines, reaching an inference speed of 50FPS. Despite variations in parameter numbers among baselines (Swin-T: 28M, ViT-b-16: 86M, NOVUM: 83M), we find no correlation of the parameter count with OOD robustness. For pose estimation, compared to the render-and-compare-based method, NeMo [37], our model uses a significantly lower parameter count (NOVUM: 83M, NeMo: 996M, *i.e.*, an impressive 12x reduction) due to our class-agnostic backbone. We also observe that our model converges faster in the render-and-compare optimization compared to NeMo (NOVUM: 30 steps, NeMo: 300 steps), which can be attributed to our class-contrastive representation and the neural object volumes which yields better gradient during the optimization. More detailed comparisons can be found in Appendix 8.5.

Ablations. We ablate the geometry selection and background features during training. In NOVUM, we form a cuboid with Gaussians. An alternative could be to choose to (1) employ a finer-grained representation, (2) utilize a single Gaussian to represent the volume, or (3) adopt a spherical geometric representation instead of a cuboid. Opting for the first alternative would necessitate either establishing a deformable geometry or treating each sub-category as a distinct class, which is beyond the scope of this work. In Table 4 (more details in Appendix 8.6), we ablate the shape of our 3D representation. As expected, the choice of the representation's shape does not have a pronounced influence on performance, as was already underlined for meshes [37]. However, we observe a slight advantage in favor of the cuboid shape which may be attributed to the cuboid's closer approximation of the true shape. Additionally, we considered a *single feature vector* approach, where a single Gaussian per class is employed during training using contrastive learning. We observe a performance drop by up to 15%, which highlights the importance of the geometry during training. These findings corroborate the classification results of our method: by selectively omitting some geometric information (*i.e.*, the 3D structure), we can reach similar outcomes while significantly enhancing computational efficiency. Ultimately, we note that the background model $\mathcal{B}$ is beneficial during training since it promotes greater dispersion among neural features. This proves to be useful for inference in cases marked by occlusions but does not have visible effect in IID scenarios.

5 Limitations and Future Work

Despite the strong generalization performance, multi-task capabilities and interpretability, the proposed NOVUM model also has limitations. For example, the current necessity for pose annotation during training is suboptimal. However, we maintain an optimistic outlook regarding future advancements that may mitigate this requirement. Notably, recent studies [8,9,44] are investigating the demanding task of aligning objects with minimal examples, indicating a promising research trajectory. Currently, the geometry of the neural object volumes is fixed, which is suboptimal for object categories with large intra-class variability. Moving forward, exploring methods for flexible deformations would likely enhance the performance of the model, while also enabling the accurate segmentation of objects, thus enriching the scope of the proposed framework.

6 Conclusion

In this work, we introduced NOVUM, an architecture for object classification with an explicit compositional 3D object representation. We presented a framework for learning neural object volumes with associated features across various categories using annotated 3D object poses. Our experimental results shows that NOVUM achieves much higher robustness compared to other state-of-the-art architectures under OOD scenarios, *e.g.*, occlusion and corruption, with competitive performance in-distribution scenarios, and similar inference speed for image classification. Further, we demonstrate NOVUM can also estimate 3D pose accurately by following an inverse rendering approach, achieves an enhanced human interpretability and consistency of multi-task predictions. In conclusion, our results showcase NOVUM as a pioneering step towards highly advanced generalization capabilities in vision models.

Acknowledgements. Adam Kortylewski acknowledges support via his Emmy Noether Research Group funded by the German Research Foundation (DFG) under Grant No. 468670075. Alan Yuille acknowledges support from Army Research Laboratory award W911NF2320008 and Office of Naval Research N00014-21-1-2812.

References

1. Bai, Y., Wang, A., Kortylewski, A., Yuille, A.: Coke: localized contrastive learning for robust keypoint detection (2022)
2. Bengio, Y., Lecun, Y., Hinton, G.: Deep learning for AI. Commun. ACM **64**(7), 58–65 (2021)
3. Chen, T., Kornblith, S., Norouzi, M., Hinton, G.: A simple framework for contrastive learning of visual representations (2020)
4. Chen, X., Dong, Z., Song, J., Geiger, A., Hilliges, O.: Category level object pose estimation via neural analysis-by-synthesis. In: Vedaldi, A., Bischof, H., Brox, T., Frahm, J.-M. (eds.) ECCV 2020. LNCS, vol. 12371, pp. 139–156. Springer, Cham (2020). https://doi.org/10.1007/978-3-030-58574-7_9
5. Cubuk, E.D., Zoph, B., Mane, D., Vasudevan, V., Le, Q.V.: Autoaugment: learning augmentation policies from data. arXiv preprint arXiv:1805.09501 (2018)
6. Dosovitskiy, A., et al.: An image is worth 16x16 words: transformers for image recognition at scale. arXiv preprint arXiv:2010.11929 (2020)
7. George, D., et al.: A generative vision model that trains with high data efficiency and breaks text-based captchas. Science **358**(6368), eaag2612 (2017)
8. Goodwin, W., Havoutis, I., Posner, I.: You only look at one: category-level object representations for pose estimation from a single example. arXiv preprint arXiv:2305.12626 (2023)

9. Goodwin, W., Vaze, S., Havoutis, I., Posner, I.: Zero-shot category-level object pose estimation. In: Avidan, S., Brostow, G., Cissé, M., Farinella, G.M., Hassner, T. (eds.) ECCV 2022. LNCS, vol. 13699, pp. 516–532. Springer, Cham (2022). https://doi.org/10.1007/978-3-031-19842-7_30
10. Grenander, U.: A unified approach to pattern analysis. In: Advances in Computers, vol. 10, pp. 175–216. Elsevier (1970)
11. Hadsell, R., Chopra, S., LeCun, Y.: Dimensionality reduction by learning an invariant mapping. In: 2006 IEEE Computer Society Conference on Computer Vision and Pattern Recognition (CVPR 2006), vol. 2, pp. 1735–1742 (2006). https://doi.org/10.1109/CVPR.2006.100
12. He, K., Fan, H., Wu, Y., Xie, S., Girshick, R.: Momentum contrast for unsupervised visual representation learning. In: Proceedings of the IEEE/CVF Conference on Computer Vision and Pattern Recognition, pp. 9729–9738 (2020)
13. He, K., Zhang, X., Ren, S., Sun, J.: Deep residual learning for image recognition. In: Proceedings of the IEEE Conference on Computer Vision and Pattern Recognition, pp. 770–778 (2016)
14. Hendrycks, D., Dietterich, T.: Benchmarking neural network robustness to common corruptions and perturbations. In: International Conference on Learning Representations (2019)
15. Hendrycks, D., Mu, N., Cubuk, E.D., Zoph, B., Gilmer, J., Lakshminarayanan, B.: Augmix: a simple data processing method to improve robustness and uncertainty. arXiv preprint arXiv:1912.02781 (2019)
16. Hermann, K.L., Lampinen, A.K.: What shapes feature representations? Exploring datasets, architectures, and training (2020)
17. Iwase, S., Liu, X., Khirodkar, R., Yokota, R., Kitani, K.M.: Repose: fast 6D object pose refinement via deep texture rendering. In: Proceedings of the IEEE/CVF International Conference on Computer Vision (ICCV), pp. 3303–3312 (2021)
18. Jin, Y., Geman, S.: Context and hierarchy in a probabilistic image model. In: 2006 IEEE Computer Society Conference on Computer Vision and Pattern Recognition (CVPR 2006), vol. 2, pp. 2145–2152. IEEE (2006)
19. Kerbl, B., Kopanas, G., Leimkühler, T., Drettakis, G.: 3D gaussian splatting for real-time radiance field rendering. ACM Trans. Graph. **42**(4) (2023)
20. Kortylewski, A., He, J., Liu, Q., Yuille, A.L.: Compositional convolutional neural networks: a deep architecture with innate robustness to partial occlusion. In: Proceedings of the IEEE/CVF Conference on Computer Vision and Pattern Recognition (CVPR) (2020)
21. Kortylewski, A., Liu, Q., Wang, H., Zhang, Z., Yuille, A.: Combining compositional models and deep networks for robust object classification under occlusion. In: Proceedings of the IEEE/CVF Winter Conference on Applications of Computer Vision, pp. 1333–1341 (2020)
22. LeCun, Y., Bengio, Y., et al.: Convolutional networks for images, speech, and time series. In: The Handbook of Brain Theory and Neural Networks, vol. 3361, no. 10, p. 1995 (1995)
23. Liu, Z., et al.: Swin transformer: hierarchical vision transformer using shifted windows. In: Proceedings of the IEEE/CVF International Conference on Computer Vision, pp. 10012–10022 (2021)
24. Locatello, F., et al.: Object-centric learning with slot attention. In: Advances in Neural Information Processing Systems, vol. 33, pp. 11525–11538 (2020)
25. Michaelis, C., et al.: Benchmarking robustness in object detection: autonomous driving when winter is coming. arXiv preprint arXiv:1907.07484 (2019)

26. Mousavian, A., Anguelov, D., Flynn, J., Kosecka, J.: 3D bounding box estimation using deep learning and geometry. In: Proceedings of the IEEE Conference on Computer Vision and Pattern Recognition, pp. 7074–7082 (2017)
27. Neisser, U., et al.: Cognitive Psychology. Appleton-Century-Crofts (1967)
28. Nguyen, T., Raghu, M., Kornblith, S.: Do wide and deep networks learn the same things? Uncovering how neural network representations vary with width and depth (2021)
29. van den Oord, A., Li, Y., Vinyals, O.: Representation learning with contrastive predictive coding (2019)
30. Radford, A., et al.: Learning transferable visual models from natural language supervision (2021)
31. Schroff, F., Kalenichenko, D., Philbin, J.: Facenet: a unified embedding for face recognition and clustering. In: 2015 IEEE Conference on Computer Vision and Pattern Recognition (CVPR), pp. 815–823 (2015). https://doi.org/10.1109/CVPR.2015.7298682
32. Simonyan, K., Zisserman, A.: Very deep convolutional networks for large-scale image recognition. arXiv preprint arXiv:1409.1556 (2014)
33. Sra, S.: A short note on parameter approximation for von mises-fisher distributions: and a fast implementation of i_s (x). Comput. Stat. **27**, 177–190 (2012)
34. Stark, M., Goesele, M., Schiele, B.: Back to the future: learning shape models from 3D cad data. In: British Machine Vision Conference (BMVC), pp. 1–11 (2010). https://doi.org/10.5244/C.24.106
35. Tulsiani, S., Malik, J.: Viewpoints and keypoints. In: Proceedings of the IEEE Conference on Computer Vision and Pattern Recognition, pp. 1510–1519 (2015)
36. Vaswani, A., et al.: Attention is all you need. In: Advances in Neural Information Processing Systems, vol. 30 (2017)
37. Wang, A., Kortylewski, A., Yuille, A.: Nemo: neural mesh models of contrastive features for robust 3D pose estimation. In: International Conference on Learning Representations (2021)
38. Wang, A., Mei, S., Yuille, A.L., Kortylewski, A.: Neural view synthesis and matching for semi-supervised few-shot learning of 3D pose. In: Advances in Neural Information Processing Systems, vol. 34, pp. 7207–7219 (2021)
39. Wang, A., Sun, Y., Kortylewski, A., Yuille, A.L.: Robust object detection under occlusion with context-aware compositionalnets. In: Proceedings of the IEEE/CVF Conference on Computer Vision and Pattern Recognition, pp. 12645–12654 (2020)
40. Wang, A., Wang, P., Sun, J., Kortylewski, A., Yuille, A.: Voge: a differentiable volume renderer using gaussian ellipsoids for analysis-by-synthesis. In: The Eleventh International Conference on Learning Representations (2022)
41. Wang, H., Sridhar, S., Huang, J., Valentin, J., Song, S., Guibas, L.J.: Normalized object coordinate space for category-level 6D object pose and size estimation. In: Proceedings of the IEEE/CVF Conference on Computer Vision and Pattern Recognition, pp. 2642–2651 (2019)
42. Wang, H., Sridhar, S., Huang, J., Valentin, J., Song, S., Guibas, L.J.: Normalized object coordinate space for category-level 6D object pose and size estimation. In: Proceedings of the IEEE/CVF Conference on Computer Vision and Pattern Recognition (CVPR) (2019)
43. Xiang, Y., Mottaghi, R., Savarese, S.: Beyond pascal: a benchmark for 3D object detection in the wild. In: IEEE Winter Conference on Applications of Computer Vision, pp. 75–82. IEEE (2014)
44. Yang, H., Shi, J., Carlone, L.: TEASER: fast and certifiable point cloud registration. IEEE Trans. Robot. **37**(2), 314–333 (2020)

45. Yi, K., et al.: Clevrer: collision events for video representation and reasoning. arXiv preprint arXiv:1910.01442 (2019)
46. Yi, K., Wu, J., Gan, C., Torralba, A., Kohli, P., Tenenbaum, J.: Neural-symbolic VQA: disentangling reasoning from vision and language understanding. In: Advances in Neural Information Processing Systems, vol. 31 (2018)
47. Yuille, A., Kersten, D.: Vision as Bayesian inference: analysis by synthesis? Trends Cogn. Sci. **10**(7), 301–308 (2006)
48. Zhao, B., et al.: OOD-CV-v2: an extended benchmark for robustness to out-of-distribution shifts of individual nuisances in natural images. arXiv preprint arXiv:2304.10266 (2023)
49. Zhao, B., et al.: OOD-CV: a benchmark for robustness to individual nuisances in real-world out-of-distribution shifts. In: Proceedings of the European Conference on Computer Vision (ECCV) (2022)
50. Zhou, X., Karpur, A., Luo, L., Huang, Q.: Starmap for category-agnostic keypoint and viewpoint estimation. In: Proceedings of the European Conference on Computer Vision (ECCV), pp. 318–334 (2018)

Align Before Collaborate: Mitigating Feature Misalignment for Robust Multi-agent Perception

Kun Yang[1], Dingkang Yang[1], Ke Li[3], Dongling Xiao[3], Zedian Shao[4], Peng Sun[2], and Liang Song[1](✉)

[1] Academy for Engineering and Technology, Fudan University, Shanghai, China
{kunyang20,dkyang20,songl}@fudan.edu.cn
[2] Duke Kunshan University, Suzhou, China
[3] Tencent Youtu Lab, Shenzhen, China
[4] Duke University, Durham, USA

Abstract. Collaborative perception has received widespread attention recently since it enhances the perception ability of autonomous vehicles via inter-agent information sharing. However, the performance of existing systems is hindered by the unavoidable collaboration noises, which induce feature-level spatial misalignment over the collaborator-shared information. In this paper, we propose a model-agnostic and lightweight plugin to mitigate the feature-level misalignment issue, called dy<u>n</u>amic f<u>e</u>ature <u>a</u>lignmen<u>t</u> (NEAT). The merits of the NEAT plugin are threefold. First, we introduce an importance-guided query proposal to predict potential foreground regions with space-channel semantics and exclude environmental redundancies. On this basis, a deformable feature alignment is presented to explicitly align the collaborator-shared features through query-aware spatial associations, aggregating multi-grained visual clues with corrective mismatch properties. Ultimately, we perform a region cross-attention reinforcement to facilitate aligned representation diffusion and achieve global feature semantic enhancement. NEAT can be readily inserted into existing collaborative perception procedures and significantly improves the robustness of vanilla baselines against pose errors and transmission delay. Extensive experiments on four collaborative 3D object detection datasets under noisy settings confirm that NEAT provides consistent gains for most methods with distinct structures.

Keywords: Collaborative perception · Deformable feature alignment

1 Introduction

Perception is a fundamental capability of autonomous vehicles (AVs) to guarantee road safety in sophisticated driving scenarios [22]. Previous single-agent perception has been extensively explored in vision-oriented vehicular applications,

---

K. Yang and D. Yang—Equal contribution.

(a) Infrastructure side image   (b) Non-aligned point cloud fusion due to transmission delay   (c) Non-aligned point cloud fusion due to localization errors   (d) Aligned point cloud fusion with our NEAT aligner

Fig. 1. (a) shows the infrastructure side image where the ego vehicle is marked with the orange circle. (b) and (c) show the point cloud fusion mismatches due to transmission delay and pose errors, respectively. (d) Our NEAT plugin mitigates the feature spatial misalignment issue and produces the well-aligned point cloud fusion. (Color figure online)

including driver monitoring [39], object detection [24,42], and instance segmentation [21,52]. Nevertheless, the single-agent perception paradigm is generally vulnerable in realistic conditions due to limited sensor ranges [51] and viewpoint occlusions [50]. To this end, collaborative perception has been proposed as a promising solution to alleviate the inadequate observations of individual agents. Based on the Vehicle-to-Vehicle/Everything (V2V/X) communication, existing studies [4,7,8,13,27–29,34–37,45,47,49] effectively improve the perception capabilities of the ego agent through information exchange and perspective complementation among heterogeneous agents (*e.g.*, AVs and infrastructures), resulting in a more holistic and precise understanding of surrounding environments.

Collaborative perception systems are categorized as early [4,23], intermediate [13], and late [26,33] fusion, where the feature-level intermediate fusion is the most favored due to its superior trade-off between perception performance and bandwidth cost. However, existing approaches [27,34,37] design intermediate collaboration schemes assuming that the spatial transformations between agents are perfect, which are vulnerable to real-world collaboration noises, including pose errors and transmission delay. Specifically, these realistic noises induce spatial misalignment at the feature level, thereby obfuscating the object locations and degrading the collaboration performance. Figure 1 illustrates visual examples of the feature misalignment issues. We also conduct a toy comparison experiment on the OPV2V dataset [37] to evaluate the hazards posed by collaboration noises. From Fig. 2, the state-of-the-art (SOTA) methods CoBEVT [34], Where2comm [7], AttFuse [37], and F-Cooper [3] exhibit respectable detection precisions in the perfect setting regarding the Average Precision (AP)@IoU 0.7. Nevertheless, when confronted with the noisy setting in realistic scenarios, the potential feature misalignment results in unavoidable performance deterioration, even worse than the No Fusion baseline without collaboration. These findings confirm that the feature misalignment leads to severe performance bottlenecks in existing models. Accordingly, how to effectively mitigate these feature mismatches becomes the core of achieving robust collaborative perception.

Several solutions are proposed to mitigate feature-level misalignment in noisy conditions. For instance, some integrated frameworks leverage attention patterns

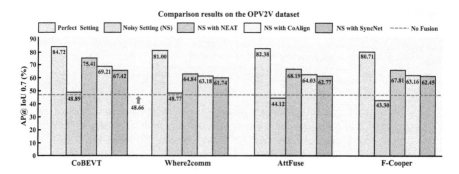

Fig. 2. We conduct a toy comparison experiment on the OPV2V dataset to evaluate the noise interference. The potential feature misalignment in the noisy setting cause the inevitable performance degradation of existing methods compared to the perfect setting. In this case, the NEAT plugin consistently improves the detection precision of the vanilla baselines. Under the same noisy conditions, NEAT brings more significant performance improvements for the baselines compared to SyncNet [11] and CoAlign [18].

with larger perceptive fields [36,41,45], multi-scale collaborator features [7,8,27], and historical information [28,31,48]. However, these approaches invariably lead to more complicated computations and larger bandwidth costs. Also, these model-specific solutions lack scalability and fail to be applied to other frameworks. In addition, various plugins enhance the robustness of existing approaches to specific noises, such as SyncNet [11] for transmission delay and CoAlign [18] for pose errors. Nevertheless, the history-based prediction module in SyncNet introduces significant computation and communication overheads, and the alignment module in CoAlign relies heavily on common visible objects between agents. Moreover, these plugins fail to cope with transmission delay and pose errors simultaneously since these two types of noises induce distinct mismatch patterns. From Fig. 1b, transmission delay causes temporal asynchrony and produces feature misalignment in the object motion directions, which are depicted by the direction-invariant red arrows. Moreover, pose errors bring about more irregular misalignment patterns due to random localization noises, as shown in Fig. 1c. To this end, we propose a lightweight plugin, NEAT, that can be readily integrated into existing frameworks and requires only the ego and collaborator features. As Fig. 2 shows, NEAT achieves more significant performance improvements than CoAlign [18] and SyncNet [11] under the same noisy conditions. Also, the extra parameter number brought from NEAT is only about 1MB (see Table 1).

In summary, we propose a *model-agnostic* and *lightweight* plugin, NEAT, to alleviate the feature misalignment dilemma and facilitate robust collaboration of multi-agent perception systems. Given ego and mismatched collaborator features, our NEAT plugin accomplishes dynamic feature alignment progressively through three tailored components. Specifically, we present the importance-guided query proposal (IQP) module to highlight potential foreground regions in collaborator features. IQP estimates the perceptually critical level of each location and introduces multi-scale views for robust query selection. Then, the deformable feature alignment (DFA) component is devised to integrate valuable visual clues for selected queries and facilitate local semantic enhancement in collaborator features. DFA introduces global context support for object-related token generation, then leverages the deformable cross-attention to construct query-aware spatial associations and aggregate semantically relevant representations. Ultimately, we propose the region cross-attention reinforcement (RCR) module to diffuse the locally enhanced features into the global representation, producing refined collaborator features with aligned properties for the subsequent feature fusion stage. Our contributions can be summarized as follows:

- The proposed NEAT model is the first dedicated plug-and-play design to address the feature misalignment issues caused by two types of collaboration noises, which can be readily integrated into most methods with diverse architectures and bring significant performance gains consistently.
- The proposed customized components progressively enhance perceptually critical semantics in collaborator-shared features and provide a universal solution for achieving the robust multi-agent perception system.
- Extensive experiments are conducted on multiple collaborative 3D object detection datasets. Comprehensive analysis in real-world and simulated scenarios under noisy settings shows the applicability and effectiveness of NEAT.

2 Related Work

2.1 Collaborative Perception

Collaborative perception aims to incorporate sensor observations from heterogeneous agents to improve the detection capability of the ego agent. Several intermediate collaboration efforts are presented to enhance robustness in noisy conditions, which can be categorized into integrated frameworks and scalable plugins. The former efforts employ framework-specific designs to mitigate noise. For instance, various works utilize attention patterns with larger receptive fields, including Swin Transformer in V2X-ViT [36], axial attention in CoBEVT [34], and deformable attention in SCOPE [45]. Moreover, FFNet [48] and CoBEVFlow [31] apply feature flow to predict the current frame to mitigate transmission delay. These prediction designs introduce significant computation and storage overheads and cause excessive cumulative errors. Multiscale solutions [7,8,17,28,30] employ multiscale collaborator features to overcome local noise interference. However, these methods invariably require sending multiple BEV features and depend on complicated feature fusion mechanisms. The latter efforts alleviate specific noises in a plug-and-play manner. Specifically, SyncNet [11] introduces historical information to predict features based on DiscoNet [13], and CoAlign [18] utilizes co-visible objects to calibrate the inter-agent transfer matrix. Nevertheless, these plugins lack scalability due to their dependence on specific conditions and noises, e.g., SyncNet's requirement for extensive stored collaborator features and CoAlign's dependence on co-visible objects. To summarize, the framework-specific methods fail to provide a unified solution for overcoming noise interference, and the existing plugins can only handle one specific noise. Accordingly, we propose the lightweight NEAT plugin to mitigate transmission delay and pose errors in a model-agnostic manner simultaneously.

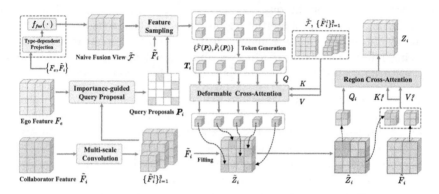

Fig. 3. NEAT plugin overview. Features F_e and $\tilde{F}_i$ are inputs. First, NEAT obtains the multi-scale views $\{\tilde{F}_i^l\}_{l=1}^3$ and utilizes the IQP module to produce the query proposals P_i. The naive fusion view $\tilde{\mathcal{F}}_i$ is built via type-dependent projection and $f_{fus}(\cdot)$ to provide global object-related semantics. Afterward, we sample features from $\{\tilde{\mathcal{F}}, \tilde{F}_i\}$ based on the selected 2D positions in P_i and generate the token embedding T_i. With deformable cross-attention and T_i, NEAT enhances the selected queries and obtains initial aligned feature $\tilde{Z}_i$ via the filling operation. Ultimately, we take a partitioned region as an example to introduce region cross-attention, which generates globally enhanced aligned feature Z_i for the subsequent feature fusion stage.

2.2 Vision Attention for Object Detection

Benefiting from the development of learning-based technologies [38,40,43,44,46], attention mechanisms are widely exploited in vision tasks due to their outstanding contextual modeling capabilities, such as DETR [2] and AutoAlign [5] for object detection. However, vanilla global-wise attention invariably brings heavy computation and storage burdens. Numerous efforts are proposed to alleviate this issue via efficient attention designs, including [1,6,12,14,19,25,32,53]. For instance, deformable DETR [32] leverages the advantages of sparse spatial sampling. AutoAlignV2 [6] builds cross-modal correlations and accelerates feature integration through a deformable attention model. For LiDAR-based object detection, Centerformer [53] presents two sparse attention schemes to facilitate the extraction of meaningful object representations, including window-aware cross-attention and deformable cross-attention. In this paper, we focus on the collaborative 3D object detection tasks in autonomous driving scenarios, which expose more challenges due to real-world collaboration noises. Consequently, we design a deformable attention-based feature alignment component to mitigate error-prone feature associations between collaborators and the ego agent, improving the robustness of existing collaborative perception systems against noises.

3 Methodology

In this section, we formulate the collaborative perception procedure to introduce the working context of our NEAT plugin.

Feature Extraction and Agent Communication. Consider a driving scene with N agents, where e-th agent is identified as the ego agent, and X_i denotes the local point cloud of i-th agent. Each agent utilizes a feature encoder $\Phi_{enc}(\cdot)$ to transform the point cloud X_i into Bird's Eye View (BEV) features as follows:

$$F_i = \Phi_{enc}(X_i) \in \mathbb{R}^{C \times H \times W}, \qquad (1)$$

where C, H, W stand for the channel, height, and width. To reduce the required bandwidth cost of F_i, the i-th collaborator adopts the communication mechanism $\Phi_{com}(\cdot)$ to obtain the compressed or sparse feature as $\tilde{F}_i = \Phi_{com}(F_i)$. Then, the i-th collaborator transmits feature $\tilde{F}_i$ and local pose information p_i to the ego agent for subsequent feature fusion stage.

Coordinate Transformation and Feature Fusion. Upon receiving the shared messages $\{\tilde{F}_i, p_i\}$ from the i-th collaborator, the ego agent obtain relative poses with $\{p_e, p_i\}$ and transform $\tilde{F}_i$ into the ego coordinate system via the coordinate transformation operator $\Gamma(\cdot)$ [36] as $\tilde{F}_i = \Gamma(\tilde{F}_i; p_e, p_i)$. After that, the ego agent produces the fused feature $\mathcal{F}_e$ for the final detection with the transformed features as follows:

$$\mathcal{F}_e = \Phi_{fus}(F_e, \{\tilde{F}_i\}_{i=1}^N) \in \mathbb{R}^{C \times H \times W}, \qquad (2)$$

where $\Phi_{fus}(\cdot)$ denotes the cross-agent feature fusion models, such as attention schemes [7,36,37] or graph models [28,29]. However, collaboration noises (*e.g.*, transmission delay and pose errors) cause feature-level spatial misalignment shown in Fig. 1 and hinder the feature fusion procedure, leading to the performance bottlenecks of existing frameworks under noisy conditions (see Fig. 2).

3.1 NEAT Overview

Unlike previous works [18,28,36,45,48] that superficially temper noise interference via integrated frameworks or plugins, we develop a *model-agnostic* and *lightweight* NEAT plugin to explicitly mitigate two types of collaboration noises and break performance bottlenecks of existing systems. As Fig. 3 shows, our NEAT plugin only requires the features F_e and $\tilde{F}_i$ as inputs. Then, three tailored components are progressively executed to produce the aligned feature Z_i for replacing $\tilde{F}_i$ in feature fusion. The summarized procedure includes: **(i)** $\tilde{F}_i$ is encoded into three scales with the same channel size using simple convolutions, and $\tilde{F}_i^l$ denotes the feature at l-th scale. With this basis, the importance-guided query proposal component produces the query proposals $\boldsymbol{P}_i \in \mathbb{R}^{N_p \times 2}$ with features $\{\tilde{F}_i^l\}_{l=1}^3$ and F_e, where N_p is the proposal number. **(ii)** The deformable feature alignment component first builds the naive fusion view $\tilde{\mathcal{F}}$ to provide global object-related information, then acquires the initial aligned feature $\tilde{Z}_i$ via critical semantics aggregation. **(iii)** The region cross-attention reinforcement component applies the features $\tilde{Z}_i$ and $\tilde{F}_i$ to perform feature refinement, generating the aligned feature Z_i for feature fusion. We will detail the implementation of the above three proposed components in Sects. 3.2, 3.3, and 3.4.

3.2 Importance-Guided Query Proposal

The query proposal phase aims to predict potential foreground regions, which will be queried in the feature alignment phase. We present an Importance-guided Query Proposal (IQP) component, which generates the importance map with space-channel semantics and applies multi-scale views to alleviate the noise interference for query selection. Unlike Where2comm [7], which relies on trained decoders for query selection, IQP employs space-channel saliency and multiscale solutions to accurately predict foreground regions under noisy conditions.

Importance Map Generation. To highlight object-related foreground regions within the features, we utilize a two-stage scheme to generate importance maps by exploring inter-pixel spatial interactions and intra-pixel channel semantics. First, a convolution-based decoder $f_{dec}(\cdot)$ is used to capture informative regions and yield the corresponding spatial confidence map M_i^s of $\tilde{F}_i$. Meanwhile, the channel-wise average pooling operation $\Psi_m(\cdot)$ integrates the visual semantics of different channels and produces the salient activation map M_i^c. The above schemes are formulated as:

$$M_i^s = \sigma(f_{dec}(\tilde{F}_i)), \ M_i^c = \sigma(\Psi_m(\tilde{F}_i)), \tag{3}$$

where $\sigma(\cdot)$ is the sigmoid activation. The importance map is obtained by these two maps as $M_i = M_i^s \odot M_i^c \in [0,1]^{H \times W}$, where $\odot$ denotes the element-wise multiplication. M_i reflects the perceptually critical level of each location within the collaborator feature $\tilde{F}_i$.

Query Proposal Selection. Since high-resolution perception views are sensitive to feature spatial misalignment [28], the single-scale importance maps inevitably induce erroneous foreground region predictions. To tackle this issue, we leverage multi-scale collaborator features to generate the importance maps $\{M_i^l\}_{l=1}^3$ with varying perception resolutions, which can correct initial importance estimations. Moreover, the corresponding importance map M_e of the ego feature F_e is also obtained to complement possible occlusions in collaborators' local observations. Given the above four importance maps, we synchronize their spatial dimensions by bilinear interpolation and obtain the corresponding pixel-wise average as $\tilde{M}_i$. Then, the heuristic selection approach $f_{sel}(\cdot; \delta)$ applies a predefined threshold δ to select the top $\delta\%$ of spatial locations from $\tilde{M}_i$ as query proposals $\bm{P}_i$, which is formulated as $\bm{P}_i = f_{sel}(\tilde{M}_i; \delta) \in \mathbb{R}^{N_p \times 2}$.

3.3 Deformable Feature Alignment

After the query selection phase, we introduce a Deformable Feature Alignment (DFA) component tailored for explicit feature alignment, which enhances the representation of query proposals with contextual visual clues and provides aligned properties. Considering the inability of static local attention [7,34] to cope with irregular feature misalignment and sparse BEV features, the deformable attention is adopted to build query-aware spatial associations and efficiently aggregate perceptually relevant semantics. DFA consists of the following two stages.

Object-Related Token Generation. Previous deformable operations [32,53] generate tokens with query features to learn sampling offsets and attention scores, which is not applicable to collaborative perception scenarios since single collaborator features fail to provide global context information. Accordingly, we propose to build a naive fusion view to provide global object-related semantics for token generation and mitigate feature misalignment in collaborators' local observations. Specifically, all features are first projected to the common space via type-dependent linear layers $\text{LN}_t(\cdot)$. The intuition is that agent discrepancies in sensor characteristics worsen feature mismatch, whereas the type-dependent projection can capture agent-specific attributes and bridge feature heterogeneity [36]. Then, we employ a general fusion operation $f_{fus}(\cdot)$ (e.g., convolution [27] and attention [37]) to generate the naive fusion view $\tilde{\mathcal{F}}$ as follows:

$$\tilde{\mathcal{F}} = f_{fus}(\text{LN}_t(F_e), \text{LN}_t(\tilde{F}_i)) \in \mathbb{R}^{C \times H \times W}. \quad (4)$$

As Fig. 3 shows, we extract the corresponding features of the query proposals $\boldsymbol{P}_i$ from $\tilde{\mathcal{F}}$ and $\tilde{F}_i$ and concatenate them along the channel dimension. The linear layer $\text{LN}(\cdot)$ fuses the concatenated features to augment the object-related information within the tokens. The generation process of object-related tokens $\boldsymbol{T}_i$ is formulated as follows:

$$\boldsymbol{T}_i = \text{LN}(\tilde{\mathcal{F}}(\boldsymbol{P}_i) \parallel \tilde{F}_i(\boldsymbol{P}_i)) \in \mathbb{R}^{N_p \times C}, \quad (5)$$

where $\parallel$ denotes the concatenation operation. $\boldsymbol{T}_i$ integrates the global semantics in the naive fusion view and the local context from the collaborators, facilitating subsequent attention learning to produce more robust aligned views.

Deformable Feature Aggregation. We first project $\boldsymbol{T}_i$ as positional encoding and obtain the token embedding as $\hat{\boldsymbol{T}}_i = \text{LN}(\boldsymbol{T}_i) + \boldsymbol{T}_i$. To aggregate comprehensive and multi-grained perceptual information, the multi-scale collaborator feature $\{\tilde{F}_i^l\}_{l=1}^3$ and naive fusion view $\tilde{\mathcal{F}}$ are utilized as attending features. Subsequently, the token embedding $\hat{\boldsymbol{T}}_i$ is passed into a linear layer to learn the sampling offset map $\Delta \boldsymbol{P}_i$ for each attending feature, providing the 2D spatial offsets $\{\Delta p_m \mid 1 \leq m \leq N_m\}$ for each query proposal p, where N_m denotes the sampled key point number. We leverage the learned offset maps to sample key points on the four attending features and extract these key points' features as $\{\tilde{F}_i^l(\boldsymbol{P}_i + \Delta \boldsymbol{P}_i)\}_{l=1}^3$ and $\tilde{\mathcal{F}}(\boldsymbol{P}_i + \Delta \boldsymbol{P}_i)$, providing alignment properties for queries and alleviating local misalignment. The attention scores of the sampled key points are obtained as $\phi(\text{LN}(\tilde{T}_i))$, where $\phi(\cdot)$ denotes the softmax function.

Given the sampled features and learned attention scores, we formulate the enhanced feature of each query proposal p as follows:

$$\text{DFA}(p) = \sum_{h=1}^{H} \boldsymbol{W}_h [\sum_{l=1}^{4} \sum_{m=1}^{N_m} \phi(\text{LN}(\tilde{T}_i(p))) \tilde{F}_i^l(p + \Delta p_m)], \qquad (6)$$

where h indexes the attention head, $\boldsymbol{W}_h$ is the learnable weight, and $\tilde{\mathcal{F}}(p + \Delta p_m)$ is rewritten as $\tilde{F}_i^4(p + \Delta p_m)$ only for formula simplification. From Fig. 3, to produce the aligned feature $\tilde{Z}_i$, the augmented feature $\text{DFA}(p)$ is filled into $\tilde{F}_i$ based on the 2D location of query proposal p. Previous deformable attention-based efforts [41,45] learn offsets and attention scores only with the local information of the ego agent, which inevitably leads to sub-optimal feature aggregation. In contrast, NEAT optimizes the offset and attention learning process through tokens $\tilde{Z}_i$ with global semantics. Also, NEAT significantly alleviates the computation overhead by reducing the sampled key point number N_m (from 15 to 5).

3.4 Region Cross-Attention Reinforcement

To generate the aligned feature Z_i through global representation refinement, we introduce a novel Region Cross-attention Reinforcement (RCR) component to diffuse locally-enhanced features in $\tilde{Z}_i$ to the surroundings and aggregate the relevant semantics from $\tilde{F}_i$. With the guidance of confidence priors, RCR filters out the irrelevant regions through region-wise correlation computation, reducing the attending range of global pixel-to-pixel attention adaptively. In contrast, global attention leads to excessive computation cost, while per-location attention [37] fails to achieve global feature refinement through a limited interaction range.

Specifically, the features $\tilde{Z}_i$ and $\tilde{F}_i$ are first concatenated as $Z_i^r = \tilde{Z}_i \| \tilde{F}_i \in \mathbb{R}^{C \times 2H \times W}$. Then, we partition $\tilde{Z}_i$ and Z_i^r into non-overlapping regions with size (S_h, S_w) and apply the 1×1 convolution $\omega_{1*1}(\cdot)$ to obtain the *query*, *key*, *value* embeddings as $Q_i = \omega_{1*1}(\tilde{Z}_i) \in \mathbb{R}^{hw \times S_h S_w \times C}$ and $K_i, V_i = \omega_{1*1}(Z_i^r) \in \mathbb{R}^{2hw \times S_h S_w \times C}$, where $h = \frac{H}{S_h}$ and $w = \frac{W}{S_w}$. The region-wise *query* and *key* embeddings are produced with the per-region average pooling operation $\Psi_a(\cdot)$:

$$Q_i^r = \Psi_a(Q_i \odot M_i^s), \ K_i^r = \Psi_a(K_i \odot M_i^r), \qquad (7)$$

where the confidence maps $M_i^s/M_i^r = \sigma(f_{dec}(\tilde{Z}_i/Z_i^r))$ are employed as prior information to introduce object-related semantics during region selection. To select semantically related regions, we compute the correlation matrix by matrix multiplication between Q_i^r and transposed K_i^r and apply the selection function $f_{sel}(\cdot; \delta_r)$ to select the top δ_r regions. After that, as Fig. 3 shows, the features of the selected regions are gathered to form the *key* and *value* embeddings $\{K_i^g, V_i^g\}$. The above process is summarized as follows:

$$K_i^g, V_i^g = \text{gather}(f_{sel}(Q_i^r(K_i^r)^T; \delta_r)) \in \mathbb{R}^{hw \times \delta_r S_h S_w \times C}. \qquad (8)$$

Ultimately, we adopt region cross-attention to reinforce the representation of each pixel within $\tilde{Z}_i$ and produce the aligned feature Z_i as follows:

$$Z_i = \phi(\frac{Q_i(K_i^g)^T}{\sqrt{C}})V_i^g \in \mathbb{R}^{hw \times S_h S_w \times C}. \tag{9}$$

We replace $\tilde{F}_i$ with the reshaped feature $Z_i \in \mathbb{R}^{C \times H \times W}$ and pass Z_i into the subsequent feature fusion stage. With the above RCR component, Z_i aggregates perceptually relevant semantics from the global context and mitigates the local spatial misalignment. Accordingly, Z_i provides more comprehensive and accurate spatial information and facilitates more robust feature fusion than feature $\tilde{F}_i$.

4 Experiments

4.1 Experimental Setup

Datasets and Evaluation Metrics. We conduct extensive experiments on four collaborative perception benchmark datasets, including V2XSet [36], OPV2V [37], OPV2V Culver City [37], and V2V4Real [35]. **V2XSet** is a large-scale simulation dataset supporting multi-agent V2X perception. This dataset contains 11,447 labeled point cloud frames, split into training/validation/testing sets with 6,694, 1,920, and 2,833 frames. The simulation dataset **OPV2V** provides raw sensor measurements from 2 to 7 autonomous vehicles in each scene. There are 10,914 frames of point clouds and RGB images, and the training, validation, and testing splits include 6,764, 1,981, and 2,169 frames. The testing dataset **OPV2V Culver City** contains approximately 600 point cloud frames with 3D annotations. **V2V4Real** is the first real-world dataset for V2V collaborative perception, where two self-driving vehicles are configured to record the surrounding environments with multi-modal sensors. To evaluate the object detection performance, we adopt the Average Precision (AP) at Intersection-over-Union (IoU) thresholds of 0.5 and 0.7 as experimental metrics following [10].

Implementation Details. The proposed models are implemented using Pytorch toolbox [20] and trained with Nvidia Tesla V100 GPUs by Adam optimizer [9]. We set the default noisy setting to simulate the realistic noise levels, where the localization and heading errors are sampled from Gaussian distributions with a standard deviation of 0.2 m and 0.2°, and the transmission delay is 100 ms. The batch sizes on the V2XSet, OPV2V, and V2V4Real datasets are $\{3, 3, 5\}$, and epoch numbers are $\{15, 15, 10\}$. The initial learning rate is 2e-3, decaying every ten epochs with a factor of 0.1. We implement the regression and classification decoders using two convolutional layers to produce predictions and use smooth absolute error loss and focal loss [15] for objective optimization. $f_{dec}(\cdot)$ reuses the parameters of the classification decoder and utilizes a maximum pooling operation to reduce the channel dimension. The selection thresholds δ on the V2XSet, OPV2V, and V2V4Real datasets are $\{25, 25, 35\}$. We employ the per-location attention [37] to implement the fusion method $f_{fus}(\cdot)$. The sampled key point number N_m is 5, and the attention head H is 8. The RCR component sets the region size (S_h, S_w) and threshold δ_r as $(4, 8)$ and 2, respectively. We provide the evaluation results on the testing sets of the four datasets.

Model Zoo. To evaluate the effectiveness of NEAT, we select five representative models with distinct structures as the comparison baselines. Specifically, **F-Cooper** [3] employs a heuristic max-out strategy to fuse collaborator features. **CORE** [27] designs a feature-level reconstruction task to preserve critical semantics and designs an attention-aware collaboration component. **AttFuse** [37] uses the per-location attention to learn the attention weights of corresponding pixels across different feature maps. **Where2comm** [7] presents a confidence-aware per-location attention design to facilitate pragmatic collaboration. **CoBEVT** [34] applies axial attention to aggregate meaningful object representations from local windows and sparse global tokens.

Table 1. Comparisons of space-time complexity and performance on the four datasets. The complexity results are reported in the average parameter number (Para.) and MACs across datasets. The detection results are reported in AP@0.5/0.7.

Models	Para. (M)	MACs (G)	V2XSet		OPV2V		V2V4Real		OPV2V Culver City	
			AP@0.5	AP@0.7	AP@0.5	AP@0.7	AP@0.5	AP@0.7	AP@0.5	AP@0.7
No Fusion	6.37	30.46	60.60	40.22	68.71	48.66	39.78	22.02	55.7	47.10
Late Fusion	6.37	31.19	54.93	30.74	76.24	56.14	49.41	22.05	60.18	39.23
When2com [16]	11.31	186.46	68.84	42.93	72.73	54.80	51.82	24.11	61.73	40.61
V2VNet [29]	14.45	435.62	79.11	49.28	77.58	62.27	57.16	26.73	71.46	51.50
DiscoNet [13]	9.63	147.85	79.83	54.16	80.66	63.82	55.94	26.61	70.62	50.36
V2X-ViT [36]	15.39	306.78	83.59	61.44	84.76	68.92	55.35	28.72	73.48	54.04
SCOPE [45]	19.71	355.45	84.26	62.17	85.02	71.13	56.28	28.04	72.27	53.66
FFNet [48]	13.74	251.02	78.76	55.49	83.26	65.58	57.02	28.55	70.51	52.83
CoBEVFlow [31]	15.56	295.27	82.94	58.35	85.18	69.24	59.37	30.63	74.06	54.37
F-Cooper [3]	8.21	160.25	71.47	46.92	79.28	58.46	53.19	25.93	68.68	48.02
F-Cooper + NEAT	9.50	165.41	**83.03**	**60.94**	**86.91**	**67.81**	**56.25**	**32.70**	**74.91**	**53.28**
CORE [27]	11.14	172.53	73.89	43.24	78.85	56.17	57.65	29.74	68.95	45.65
CORE + NEAT	12.51	178.08	**76.35**	**48.06**	**81.14**	**60.45**	**60.91**	**34.19**	**70.63**	**48.48**
AttFuse [37]	8.53	115.54	70.85	48.66	80.23	59.82	55.86	27.88	70.90	51.26
AttFuse + NEAT	9.85	120.98	**80.30**	**55.97**	**84.37**	**68.19**	**60.60**	**34.24**	**75.22**	**55.21**
Where2comm [7]	10.36	210.92	81.98	53.43	80.55	61.92	58.37	31.49	70.32	50.17
Where2comm + NEAT	11.64	216.67	**83.42**	**58.72**	**82.06**	**64.84**	**59.28**	**35.43**	**73.57**	**53.29**
CoBEVT [34]	12.03	238.14	81.12	54.30	81.32	63.19	58.20	28.72	71.54	52.33
CoBEVT + NEAT	13.37	244.75	**85.07**	**65.16**	**90.01**	**75.41**	**60.55**	**34.97**	**78.82**	**58.10**

4.2 Detection Performance Comparison

Table 1 presents the average space-time complexity and detection performance comparison between various baselines and the NEAT-based models under default noisy settings across the four datasets. The complexity is measured by the mainstream Parameter Number (M) and MACs (G). No Fusion represents the ego-centered perception without collaboration. Late Fusion refers to aggregating all agents' output boxes and producing the results by non-maximum suppression. Moreover, we consider representative and reproducible state-of-the-art (SOTA) methods, including When2com [16], V2VNet [29], V2X-ViT [36], DiscoNet [13], SCOPE [45], FFNet [48] and CoBEVFlow [31]. The key observations are listed below. **(i)** NEAT significantly improves the five selected models on all metrics with very slight complexity costs. For instance, the NEAT-based F-Cooper, CORE, AttFuse, Where2comm, and CoBEVT achieve an average gain of 8.85%, 4.10%, 6.50%, 3.81%, and 8.78% compared to their vanilla counterparts across the four datasets concerning AP@0.7. **(ii)** Compared to V2X-ViT [36], SCOPE [45] and FFNet [48] that implicitly mitigate the feature misalignment by integrated frameworks, the models equipped with NEAT achieve better performance. Specifically, the NEAT-based CoBEVT enables the method CoBEVT to achieve SOTA performance on all four datasets with a slight extra computation cost. The reasonable explanations include: (1) The importance-aware NEAT alleviates misalignment conditions by aggregating semantically relevant object representations, which is applicable to methods with diverse structures; (2) our plugin can flexibly address varying feature mismatches in real-world (*e.g.*, V2V4Real) and simulated scenes via global feature refinement.

4.3 Robustness to Localization Error

Evaluating the collaboration performance under diverse pose error levels is essential in assessing the system robustness. Figure 4 show the detection precision of the four models before and after equipping the NEAT plugin under varying pose noises. The localization error is sampled from Gaussian distributions with

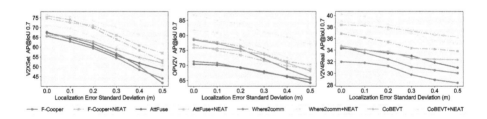

Fig. 4. Robustness assessment of the localization error on the three datasets.

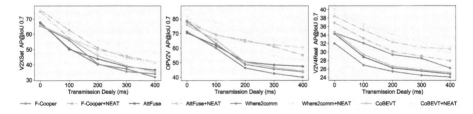

Fig. 5. Robustness assessment of the transmission delay on the three datasets.

a standard deviation of $\sigma_{xyz} \in [0, 0.5]$ m. **(i)** Intuitively, due to the flawed interagent spatial transformation caused by pose errors, the performance curves of all intermediate fusion methods show an evident downward trend with increasing error levels. In comparison, our NEAT plugin alleviates the performance degradation and significantly improves the vanilla baselines at all error levels. **(ii)** For instance, Where2comm [7] on V2XSet degrades to similar results as No Fusion under severe localization errors. In this case, NEAT mitigates the feature misalignment and enhances the corresponding performance favorably. These findings verify that NEAT can empower existing models with robust resistance to noise interference by aligning collaborator features.

4.4 Robustness to Transmission Delay

As Fig. 1b shows, temporal asynchrony caused by transmission delay also induces deleterious feature misalignment, resulting in subsequent two-side fusion errors among agents. Figure 5 provides the robustness analysis of vanilla baselines and NEAT-based models against varying delays. Inevitably, performance deterioration spreads across all vanilla methods as transmission delay continuously increases. In particular, the AP@0.7 result of F-Cooper [3] on OPV2V drops substantially, even lower than the baseline No Fusion (AP@0.7 = 48.66%) when the delay exceeds 300 ms. Conversely, NEAT provides significant gains for different models at all delay levels. That is, the NEAT-based models maintain high detection precision compared to the vanilla counterparts, even under a severe delay (*i.e.*, 400 ms). Our plugin's merit stems from the query-aware spatial associations to compensate for temporal asynchrony.

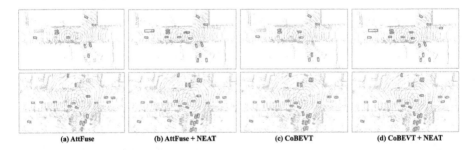

Fig. 6. Qualitative comparison results in real-world scenarios from the V2V4Real dataset [35]. Green and red boxes denote ground truths and detection results, respectively. Compared to the vanilla baselines, NEAT-based models significantly improve the detection results. (Color figure online)

4.5 Qualitative Evaluation

To intuitively evaluate the merits of NEAT, we select two challenging scenarios on V2V4Real to present the detection visualizations of vanilla baselines and

NEAT-based models. In Fig. 6, AttFuse [37] and CoBEVT [34] exhibit vulnerable performance in noisy settings due to massive missing predicted bounding boxes. Through the proposed plugin, the NEAT-based models produce more detection results that are well aligned with the ground truths. These observations confirm that NEAT improves the detection robustness of collaborative perception systems under noisy conditions.

4.6 Comparison of Plugin Methods

We compare the existing plugins SyncNet [11] and CoAlign [18] on V2XSet, OPV2V, and V2V4Real datasets to explore the effectiveness of NEAT. As shown in Table 2, four models are selected as baselines to observe the detection performance changes, including CoBEVT [34], AttFuse [37], F-Cooper [3], and CORE [27]. NEAT provides more significant and consistent performance gains across different baselines by jointly addressing the pose errors and transmission delay. In contrast, CoAlign and SyncNet offer sub-optimal solutions due to their noise-specific designs and inadequate improvements. In practice, our NEAT introduces extremely low spatio-temporal complexity overhead (*i.e.*, Para. = 1.31 M and MACs = 5.11 G) compared to the existing plugins, including CoAlign and SyncNet. Specifically, these two plugins bring average parameter numbers of 2.94/3.80M and MACs of 26.51/31.02G across datasets, respectively.

Table 2. Comparison results of plugin methods on the V2XSet, OPV2V, and V2V4Real datasets. We provide four models for different plugins as baselines for use in assembly, including CoBEVT, AttFuse, F-Cooper, and CORE. NEAT brings more significant gains to vanilla models on all metrics.

Models	V2XSet	OPV2V	V2V4Real
CoBEVT	81.12/54.30	81.32/63.19	58.20/28.72
CoBEVT + SyncNet	82.28/57.75	83.77/67.42	58.74/31.75
CoBEVT + CoAlign	83.15/60.20	85.65/69.21	59.05/31.22
CoBEVT + NEAT	**85.07/65.16**	**90.01/75.41**	**60.55/34.97**
AttFuse	70.85/48.66	80.23/59.82	55.86/27.88
AttFuse + SyncNet	75.07/52.34	81.92/62.77	58.14/30.79
AttFuse + CoAlign	76.82/51.90	82.53/64.03	57.32/31.37
AttFuse + NEAT	**80.30/55.97**	**84.37/68.19**	**60.60/34.24**
F-Cooper	71.47/46.92	79.28/58.46	53.19/25.93
F-Cooper + SyncNet	77.29/52.76	83.16/62.45	54.78/27.76
F-Cooper + CoAlign	76.73/53.42	82.83/63.16	55.02/29.64
F-Cooper + NEAT	**83.03/60.94**	**86.91/67.81**	**56.25/32.70**
CORE	73.89/43.24	78.85/56.17	57.65/29.74
CORE + SyncNet	75.23/45.53	79.82/56.96	58.37/31.32
CORE + CoAlign	74.41/45.77	80.07/57.54	58.56/31.41
CORE + NEAT	**76.35/48.06**	**81.14/60.45**	**60.91/34.19**

Table 3. Ablation study results of the proposed components on the V2XSet, OPV2V, and V2V4Real datasets. **FA**: deformable feature aggregation; **TG**: object-related token generation; **MG**: importance map generation; **PS**: query proposal selection; **RCR**: region cross-attention reinforcement.

CoBEVT [34] + NEAT					V2XSet	OPV2V	V2V4Real
FA	TG	MG	PS	RCR			
					81.12/54.30	81.32/63.19	58.20/28.72
✔					82.69/58.24	84.41/67.05	58.95/31.81
✔	✔				83.10/59.97	85.74/69.30	59.43/32.45
✔	✔	✔			83.56/62.08	87.26/72.16	59.87/33.50
✔	✔	✔	✔		83.94/63.25	88.33/73.58	60.02/34.33
✔	✔	✔	✔	✔	**85.07/65.16**	**90.01/75.41**	**60.55/34.97**

AttFuse [37] + NEAT					V2XSet	OPV2V	V2V4Real
FA	TG	MG	PS	RCR			
					70.85/48.66	80.23/59.82	55.86/27.88
✔					74.24/51.48	81.95/62.66	57.44/29.96
✔	✔				76.27/52.41	82.36/64.05	58.35/31.47
✔	✔	✔			77.35/54.02	82.95/65.51	59.23/32.64
✔	✔	✔	✔		78.64/54.63	83.14/66.38	59.82/33.22
✔	✔	✔	✔	✔	**80.30/55.97**	**84.37/68.19**	**60.60/34.24**

4.7 Ablation Studies

We select two representative models to perform thorough ablation studies on V2XSet, OPV2V, and V2V4Real to verify the necessity of the proposed components. As Table 3 shows, the vanilla baselines without any components in NEAT are employed as the performance reference. On this basis, we progressively add (1) FA, (2) TG, (3) MG, (4) PS, and (5) RCR and present the corresponding AP@0.5/0.7 results. **(i)** The continuously improved detection performance over the three datasets reveals that each component provides indispensable contributions. **(ii)** Particularly, deformable feature aggregation brings noteworthy gains by aggregating vital context semantics with corrective mismatch attributes, while object-related token generation facilitates attention learning by introducing global object-related semantics. **(iii)** Furthermore, region cross-attention reinforcement efficiently diffuses the locally enhanced representations with region-wise attention, achieving global feature refinement and significant improvements.

5 Conclusion

This paper proposes NEAT, a lightweight plugin, to address the feature misalignment issue induced by collaboration noises in multi-agent perception systems. NEAT accomplishes the explicit alignment of collaborator-shared features with customized components, ensuring subsequent high-quality collaboration and perception. Extensive experiments show that NEAT can be readily integrated into existing methods to provide a generic solution for improving system robustness.

Acknowledgement. This work is supported in part by the China Mobile Research Fund of the Chinese Ministry of Education under Grant KEH2310029 and in part by the Specific Research Fund of the innovation Platform for Academicians of Hainan Province under Grant YSPTZX202314. This work is also supported in part by the Shanghai Key Research Laboratory of NSAI, the Joint Laboratory on Networked AI Edge Computing, Fudan University-Changan, the National Natural Science Foundation of China (Grant No. 62250410368), and the Kunshan Government Research Fund 24KKSGR024.

References

1. Bai, X., et al.: Transfusion: robust lidar-camera fusion for 3D object detection with transformers. In: Proceedings of the IEEE/CVF Conference on Computer Vision and Pattern Recognition, pp. 1090–1099 (2022)
2. Carion, N., Massa, F., Synnaeve, G., Usunier, N., Kirillov, A., Zagoruyko, S.: End-to-end object detection with transformers. In: Vedaldi, A., Bischof, H., Brox, T., Frahm, J.-M. (eds.) ECCV 2020. LNCS, vol. 12346, pp. 213–229. Springer, Cham (2020). https://doi.org/10.1007/978-3-030-58452-8_13
3. Chen, Q., Ma, X., Tang, S., Guo, J., Yang, Q., Fu, S.: F-cooper: feature based cooperative perception for autonomous vehicle edge computing system using 3D point clouds. In: Proceedings of the 4th ACM/IEEE Symposium on Edge Computing, pp. 88–100 (2019)
4. Chen, Q., Tang, S., Yang, Q., Fu, S.: Cooper: cooperative perception for connected autonomous vehicles based on 3D point clouds. In: 2019 IEEE 39th International Conference on Distributed Computing Systems, pp. 514–524. IEEE (2019)

5. Chen, Z., et al.: Autoalign: pixel-instance feature aggregation for multi-modal 3D object detection. arXiv preprint arXiv:2201.06493 (2022)
6. Chen, Z., Li, Z., Zhang, S., Fang, L., Jiang, Q., Zhao, F.: Deformable feature aggregation for dynamic multi-modal 3D object detection. In: Avidan, S., Brostow, G., Cissé, M., Farinella, G.M., Hassner, T. (eds.) ECCV 2022. LNCS, vol. 13668, pp. 628–644. Springer, Cham (2022). https://doi.org/10.1007/978-3-031-20074-8_36
7. Hu, Y., Fang, S., Lei, Z., Zhong, Y., Chen, S.: Where2comm: communication-efficient collaborative perception via spatial confidence maps. In: Advances in Neural Information Processing Systems, vol. 35, pp. 4874–4886 (2022)
8. Hu, Y., Lu, Y., Xu, R., Xie, W., Chen, S., Wang, Y.: Collaboration helps camera overtake lidar in 3D detection. In: Proceedings of the IEEE/CVF Conference on Computer Vision and Pattern Recognition, pp. 9243–9252 (2023)
9. Kingma, D.P., Ba, J.: Adam: a method for stochastic optimization. In: International Conference on Learning Representations (2015)
10. Lang, A.H., Vora, S., Caesar, H., Zhou, L., Yang, J., Beijbom, O.: Pointpillars: fast encoders for object detection from point clouds. In: Proceedings of the IEEE/CVF Conference on Computer Vision and Pattern Recognition, pp. 12697–12705 (2019)
11. Lei, Z., Ren, S., Hu, Y., Zhang, W., Chen, S.: Latency-aware collaborative perception. In: Avidan, S., Brostow, G., Cissé, M., Farinella, G.M., Hassner, T. (eds.) ECCV 2022. LNCS, vol. 13692, pp. 316–332. Springer, Cham (2022). https://doi.org/10.1007/978-3-031-19824-3_19
12. Li, F., et al.: Lite DETR: an interleaved multi-scale encoder for efficient DETR. In: Proceedings of the IEEE/CVF Conference on Computer Vision and Pattern Recognition, pp. 18558–18567 (2023)
13. Li, Y., Ren, S., Wu, P., Chen, S., Feng, C., Zhang, W.: Learning distilled collaboration graph for multi-agent perception. In: Advances in Neural Information Processing Systems, vol. 34, pp. 29541–29552 (2021)
14. Li, Z., et al.: Bevformer: learning bird's-eye-view representation from multi-camera images via spatiotemporal transformers. In: Avidan, S., Brostow, G., Cissé, M., Farinella, G.M., Hassner, T. (eds.) ECCV 2022. LNCS, vol. 13669, pp. 1–18. Springer, Cham (2022). https://doi.org/10.1007/978-3-031-20077-9_1
15. Lin, T.Y., Goyal, P., Girshick, R., He, K., Dollár, P.: Focal loss for dense object detection. In: Proceedings of the IEEE International Conference on Computer Vision, pp. 2980–2988 (2017)
16. Liu, Y.C., Tian, J., Glaser, N., Kira, Z.: When2com: multi-agent perception via communication graph grouping. In: Proceedings of the IEEE/CVF Conference on Computer Vision and Pattern Recognition, pp. 4106–4115 (2020)
17. Lu, Y., Hu, Y., Zhong, Y., Wang, D., Chen, S., Wang, Y.: An extensible framework for open heterogeneous collaborative perception. arXiv preprint arXiv:2401.13964 (2024)
18. Lu, Y., et al.: Robust collaborative 3D object detection in presence of pose errors. In: 2023 IEEE International Conference on Robotics and Automation (ICRA), pp. 4812–4818. IEEE (2023)
19. Misra, I., Girdhar, R., Joulin, A.: An end-to-end transformer model for 3D object detection. In: Proceedings of the IEEE/CVF International Conference on Computer Vision, pp. 2906–2917 (2021)
20. Paszke, A., et al.: Pytorch: an imperative style, high-performance deep learning library. In: Advances in Neural Information Processing Systems, vol. 32 (2019)
21. Qi, C.R., Su, H., Mo, K., Guibas, L.J.: Pointnet: deep learning on point sets for 3D classification and segmentation. In: Proceedings of the IEEE/CVF Conference on Computer Vision and Pattern Recognition, pp. 652–660 (2017)

22. Ram, T., Chand, K.: Effect of drivers' risk perception and perception of driving tasks on road safety attitude. Transport. Res. F: Traffic Psychol. Behav. **42**, 162–176 (2016)
23. Rawashdeh, Z.Y., Wang, Z.: Collaborative automated driving: a machine learning-based method to enhance the accuracy of shared information. In: 2018 21st International Conference on Intelligent Transportation Systems, pp. 3961–3966. IEEE (2018)
24. Redmon, J., Divvala, S., Girshick, R., Farhadi, A.: You only look once: unified, real-time object detection. In: Proceedings of the IEEE/CVF Conference on Computer Vision and Pattern Recognition, pp. 779–788 (2016)
25. Sheng, H., et al.: Improving 3D object detection with channel-wise transformer. In: Proceedings of the IEEE/CVF International Conference on Computer Vision, pp. 2743–2752 (2021)
26. Shi, S., et al.: VIPS: real-time perception fusion for infrastructure-assisted autonomous driving. In: Proceedings of the 28th Annual International Conference on Mobile Computing and Networking, pp. 133–146 (2022)
27. Wang, B., Zhang, L., Wang, Z., Zhao, Y., Zhou, T.: Core: cooperative reconstruction for multi-agent perception. In: Proceedings of the IEEE/CVF International Conference on Computer Vision, pp. 8710–8720 (2023)
28. Wang, T., et al.: UMC: a unified bandwidth-efficient and multi-resolution based collaborative perception framework. In: Proceedings of the IEEE/CVF International Conference on Computer Vision, pp. 8187–8196 (2023)
29. Wang, T.-H., Manivasagam, S., Liang, M., Yang, B., Zeng, W., Urtasun, R.: V2VNet: vehicle-to-vehicle communication for joint perception and prediction. In: Vedaldi, A., Bischof, H., Brox, T., Frahm, J.-M. (eds.) ECCV 2020. LNCS, vol. 12347, pp. 605–621. Springer, Cham (2020). https://doi.org/10.1007/978-3-030-58536-5_36
30. Wang, Z., et al.: VIMI: vehicle-infrastructure multi-view intermediate fusion for camera-based 3D object detection. arXiv preprint arXiv:2303.10975 (2023)
31. Wei, S., et al.: Robust asynchronous collaborative 3D detection via bird's eye view flow. arXiv preprint arXiv:2309.16940 (2023)
32. Xizhou, Z., Weijie, S., Lewei, L., Bin, L., Xiaogang, W., Jifeng, D.: Deformable DETR: deformable transformers for end-to-end object detection. In: International Conference on Learning Representations (2021)
33. Xu, R., Chen, W., Xiang, H., Xia, X., Liu, L., Ma, J.: Model-agnostic multi-agent perception framework. In: 2023 IEEE International Conference on Robotics and Automation, pp. 1471–1478. IEEE (2023)
34. Xu, R., Tu, Z., Xiang, H., Shao, W., Zhou, B., Ma, J.: Cobevt: cooperative bird's eye view semantic segmentation with sparse transformers. In: Conference on Robot Learning (2022)
35. Xu, R., et al.: V2v4real: a real-world large-scale dataset for vehicle-to-vehicle cooperative perception. In: Proceedings of the IEEE/CVF Conference on Computer Vision and Pattern Recognition, pp. 13712–13722 (2023)
36. Xu, R., Xiang, H., Tu, Z., Xia, X., Yang, M.H., Ma, J.: V2X-ViT: vehicle-to-everything cooperative perception with vision transformer. In: Avidan, S., Brostow, G., Cissé, M., Farinella, G.M., Hassner, T. (eds.) ECCV 2022. LNCS, vol. 13699, pp. 107–124. Springer, Cham (2022). https://doi.org/10.1007/978-3-031-19842-7_7
37. Xu, R., Xiang, H., Xia, X., Han, X., Li, J., Ma, J.: Opv2v: an open benchmark dataset and fusion pipeline for perception with vehicle-to-vehicle communication. In: Proceedings of the International Conference on Robotics and Automation, pp. 2583–2589. IEEE (2022)

38. Yang, D., Huang, S., Kuang, H., Du, Y., Zhang, L.: Disentangled representation learning for multimodal emotion recognition. In: Proceedings of the 30th ACM International Conference on Multimedia (ACM MM), pp. 1642–1651 (2022)
39. Yang, D., et al.: Aide: a vision-driven multi-view, multi-modal, multi-tasking dataset for assistive driving perception. In: Proceedings of the IEEE/CVF International Conference on Computer Vision, pp. 20459–20470 (2023)
40. Yang, D., Yang, K., Li, M., Wang, S., Wang, S., Zhang, L.: Robust emotion recognition in context debiasing. In: Proceedings of the IEEE/CVF Conference on Computer Vision and Pattern Recognition (CVPR), pp. 12447–12457 (2024)
41. Yang, D., et al.: How2comm: communication-efficient and collaboration-pragmatic multi-agent perception. In: Advances in Neural Information Processing Systems, vol. 36 (2024)
42. Yang, K., et al.: A novel efficient multi-view traffic-related object detection framework. In: IEEE International Conference on Acoustics, Speech and Signal Processing, pp. 1–5 (2023)
43. Yang, K., Sun, P., Lin, J., Boukerche, A., Song, L.: A novel distributed task scheduling framework for supporting vehicular edge intelligence. In: 2022 IEEE 42nd International Conference on Distributed Computing Systems (ICDCS), pp. 972–982. IEEE (2022)
44. Yang, K., Sun, P., Yang, D., Lin, J., Boukerche, A., Song, L.: A novel hierarchical distributed vehicular edge computing framework for supporting intelligent driving. Ad Hoc Netw. **153**, 103343 (2024)
45. Yang, K., et al.: Spatio-temporal domain awareness for multi-agent collaborative perception. In: Proceedings of the IEEE/CVF International Conference on Computer Vision, pp. 23383–23392 (2023)
46. Yang, K., Yang, D., Zhang, J., Wang, H., Sun, P., Song, L.: What2comm: towards communication-efficient collaborative perception via feature decoupling. In: Proceedings of the 31th ACM International Conference on Multimedia (ACM MM), pp. 7686–7695 (2023)
47. Yu, H., et al.: Dair-v2x: a large-scale dataset for vehicle-infrastructure cooperative 3D object detection. In: Proceedings of the IEEE/CVF Conference on Computer Vision and Pattern Recognition, pp. 21361–21370 (2022)
48. Yu, H., Tang, Y., Xie, E., Mao, J., Luo, P., Nie, Z.: Flow-based feature fusion for vehicle-infrastructure cooperative 3D object detection. In: Advances in Neural Information Processing Systems, vol. 36 (2024)
49. Yu, H., et al.: V2x-seq: a large-scale sequential dataset for vehicle-infrastructure cooperative perception and forecasting. In: Proceedings of the IEEE/CVF Conference on Computer Vision and Pattern Recognition, pp. 5486–5495 (2023)
50. Yuan, X., Kortylewski, A., Sun, Y., Yuille, A.: Robust instance segmentation through reasoning about multi-object occlusion. In: Proceedings of the IEEE/CVF Conference on Computer Vision and Pattern Recognition, pp. 11141–11150 (2021)
51. Yuan, Z., Song, X., Bai, L., Wang, Z., Ouyang, W.: Temporal-channel transformer for 3D lidar-based video object detection for autonomous driving. IEEE Trans. Circuits Syst. Video Technol. **32**(4), 2068–2078 (2021)
52. Zhao, H., Shi, J., Qi, X., Wang, X., Jia, J.: Pyramid scene parsing network. In: Proceedings of the IEEE/CVF Conference on Computer Vision and Pattern Recognition, pp. 2881–2890 (2017)
53. Zhou, Z., Zhao, X., Wang, Y., Wang, P., Foroosh, H.: Centerformer: center-based transformer for 3D object detection. In: Avidan, S., Brostow, G., Cissé, M., Farinella, G.M., Hassner, T. (eds) ECCV 2022. LNCS, vol. 13698, pp. 496–513. Springer, Cham (2022). https://doi.org/10.1007/978-3-031-19839-7_29

HIMO: A New Benchmark for Full-Body Human Interacting with Multiple Objects

Xintao Lv[1], Liang Xu[1,2], Yichao Yan[1]([✉]), Xin Jin[2], Congsheng Xu[1], Shuwen Wu[1], Yifan Liu[1], Lincheng Li[3], Mengxiao Bi[3], Wenjun Zeng[2], and Xiaokang Yang[1]

[1] MoE Key Lab of Artificial Intelligence, AI Institute, Shanghai Jiao Tong University, Shanghai, China
yanyichao@sjtu.edu.cn
[2] Ningbo Institute of Digital Twin, Eastern Institute of Technology, Ningbo, China
[3] NetEase Fuxi AI Lab, Hangzhou, China
https://lvxintao.github.io/himo

Abstract. Generating human-object interactions (HOIs) is critical with the tremendous advances of digital avatars. Existing datasets are typically limited to humans interacting with a single object while neglecting the ubiquitous manipulation of multiple objects. Thus, we propose **HIMO**, a large-scale MoCap dataset of full-body human interacting with multiple objects, containing **3.3K** 4D HOI sequences and **4.08M** 3D HOI frames. We also annotate **HIMO** with detailed textual descriptions and temporal segments, benchmarking two novel tasks of HOI synthesis conditioned on either the whole text prompt or the segmented text prompts as fine-grained timeline control. To address these novel tasks, we propose a dual-branch conditional diffusion model with a mutual interaction module for HOI synthesis. Besides, an auto-regressive generation pipeline is also designed to obtain smooth transitions between HOI segments. Experimental results demonstrate the generalization ability to unseen object geometries and temporal compositions. Our data, codes, and models will be publicly available for research purposes.

Keywords: Human-Object Interaction · Multiple Objects · Text-driven HOI Synthesis · Temporal Segmentation

1 Introduction

Humans constantly interact with objects as daily routines. As a key component for human-centric vision tasks, the ability to synthesize human-object

---
X. Lv and L. Xu—Equal contribution.

Supplementary Information The online version contains supplementary material available at https://doi.org/10.1007/978-3-031-73235-5_17.

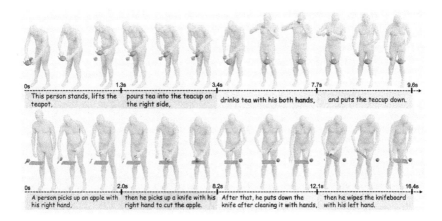

Fig. 1. Overview. HIMO is a large-scale dataset of full-body interacting with multiple objects, with fine-grained textual descriptions. The long interaction sequences and the corresponding texts are elaborately segmented temporally, facilitating the detailed timeline control of multi-step human-object interactions.

interactions (HOIs) is fundamental with numerous applications in video games, AR/VR, robotics, and embodied AI. However, most of the previous datasets and models [8,17,24,26,59] are limited to interacting with a single object, yet neglect the ubiquitous functionality combination of multiple objects. Intuitively, the multiple objects setting is more practical and allows for broader applications, such as manipulating multiple objects for robotics [44,70].

The scarcity of such datasets mainly results in the underdevelopment of the synthesis of human interacting with multiple objects, as listed in Table 1. GRAB [59] builds the dataset of 4D full-body grasping of daily objects ("4D" refers to 3D geometry and temporal streams, "full-body" emphasizes the body movements and dexterous finger motion), followed by several 4D HOI datasets with RGB modality [8,24], interacting with articulated [17] or sittable [26] objects. For text-driven HOI generation, [15,46] manually label the textual descriptions for the BEHAVE [8] and CHAIRS [26] datasets, and Li et al. [31,32] collect a dataset of human interacting with daily objects with textual annotations. Despite the significant development, all these datasets focus on single-object interactions. Thus, a dataset of full-body humans interacting with multiple objects incorporated with detailed textual descriptions is highly desired.

Capturing 4D HOIs is challenging due to the subtle finger motions, severe occlusions, and precise tracking of various objects. Compared with a single object, interacting with multiple objects is more complicated regarding the *spatial* movements between human-object and object-object, and the *temporal* schedule of several atomic HOI intervals. For example, the process of a person interacting with a teapot and teacup could be: "Lift the teapot" → "Pour tea to the teacup" → "Drink tea", as shown in Fig. 1. To facilitate the synthesis of **H**uman **I**nteracting with **M**ultiple **O**bjects, we build a large-scale dataset called

Table 1. Dataset comparisons. We compare the HIMO dataset with existing human-object interaction datasets. "Multi-object", "Segment" and "#" mean human interacting with multiple objects, temporal segmentation, and the number of.

Dataset	Modality				Scale			
	Hand	Multi-object	Text	Segment	#Subject	#Object	#Seq	#Frame
GRAB [59]	✓	✗	✗	✗	10	51	1,334	-
BEHAVE [8]	✗	✗	✗	✗	8	20	321	15K
InterCap [24]	✗	✗	✗	✗	10	10	223	67K
ARCTIC [17]	✓	✗	✗	✗	9	11	339	2.1M
CHAIRS [26]	✓	✗	✗	✗	46	81	1,390	-
Li et al. [32]	✓	✗	✓	✗	17	15	-	-
HIMO(Ours)	✓	✓	✓	✓	34	53	3,376	4.08M

HIMO. We adopt the optical MoCap system to obtain precise body movements and track the motion of objects attached by reflective markers. For the dexterous finger motions, we employ wearable inertial gloves to avoid occlusions. In total, **3.3K** 4D HOI sequences with **34** subjects performing the combinations of **53** daily objects are presented, resulting in **4.08M** 3D HOI frames. Various subjects, object combinations, and interaction patterns also ensure the diversity of **HIMO**.

To facilitate the study of text-driven HOI synthesis [12,15,20,62,73], we annotate the HOI sequences with fine-grained textual descriptions. Different from existing datasets, we additionally segment the long interaction sequences temporally and align the corresponding texts semantically, which allows for fine-grained timeline control, and a flexible schedule of multiple atomic HOI sequences. Revisiting the example of pouring water, after training based on our datasets, *novel* HOI compositions such as "Drink water" → "Lift the teapot" → "Add more tea to the teacup" could be synthesized benefited from our fine-grained temporal segments. Although temporal segmentation is widely adopted in video action domain [16,35,38,57], we are the first to introduce the annotations to 4D HOIs.

Our HIMO dataset enables two novel generative tasks: 1) Text-driven HOI synthesis with multiple objects (denoted as HIMO-Gen); 2) HIMO-Gen conditioned on segmented texts as timeline control (denoted as HIMO-SegGen). Concretely, instead of single text as the condition for HIMO-Gen, HIMO-SegGen requires a series of multiple consecutive texts as guidance. For HIMO-Gen, we generate the synchronized human motion and object motions conditioned on the language descriptions and the initial states of the human and the objects. Inspired by previous works on modeling human-human interaction generation [36,56,65,67] and human interacting with single object [31,32,46,64,68], it is straightforward to design a dual-branch conditional diffusion model to generate the human and object motion, respectively. However, this naive solution suffers from the following challenges. First, the generated human motion and object motions could be spatio-temporally misaligned. Second, implausible contacts between human-object and object-object are commonly encountered. To guarantee the coordination between human motion and object motions, we further design a mutual interaction module implemented as a stack of multiple mutual-attention layers. We also add an object-pairwise loss based on the relative distance between the geometry of objects to generate plausible object movements. Experiments show that our method can synthesize realistic and coordinated HOIs.

Furthermore, for the task of HIMO-SegGen, each segmented HOI clip paired with the corresponding text can be viewed as a sample to train the HIMO-Gen model, while the key of HIMO-SegGen lies in the smoothness between the generated clips. We introduce a simple yet effective scheme to auto-regressively generate one HOI clip at a time, conditioning the next clip generation with the last few frames of the previous clip. Experiment results show that conditioning on 10 frames achieves the best performance. Besides, we also show the generalization ability of our model to unseen objects and novel HOI compositions. Overall, our contributions can be summarized as follows:

- We collect a large-scale dataset of full-body human interacting with multiple daily objects called **HIMO**, with precise and diverse 4D-HOI sequences.
- We annotate the detailed textual descriptions for each long HOI sequence, together with the temporal segmentation labels, facilitating two novel downstream tasks of HIMO-Gen and HIMO-SegGen.
- We propose a dual-branch conditional diffusion model with a mutual interaction module to fuse the human and object features. We also propose a pipeline to auto-regressively generate the composite HOI sequences.

2 Related Work

2.1 Human-Object Interaction Datasets

Many human-object interactions together with the hand-object interaction datasets have been proposed to facilitate the development of modeling HOIs. Several works [10,21,22,25,39,40] focusing on hand-object interaction have been proposed. Though effective, modeling hand-object interactions falls significantly short of comprehending the nature of human-object interaction, since the interaction process involves more than just hand participation. Full-body interactions [8,17,24,26,42,59,69,74,76] may enhance our understanding of HOIs. GRAB [59], InterCap [24] and ARCTIC [17] broaden the range of 4D-HOIs to full-body interactions. However, all of these full-body datasets only involve one single object, while the combination of several different objects is common in our daily lives. The KIT Whole-body dataset [42] contains motion in which subjects manipulate several objects simultaneously. However, the human body is represented as a humanoid model instead of a more realistic SMPL-X model. We address the deficiency by introducing a full-body 4D-HOI dataset that involves full-body interaction with multiple objects. Comparisons with the existing HOI datasets are listed in Table 1.

2.2 Text-Driven Motion Generation

Text-driven human motion generation aims at generating human motions based on textual descriptions. Benefiting from large-scale motion-text dataset like KIT-ML [49], BABEL [52] and HumanML3D [20], numerous works [12,14,32,47, 62,66,73,80] have emerged. TEMOS [47] proposes a transformer-based VAE to encode motion and an additional DistillBert [55] to encode texts. MDM [62] adopts the diffusion model [23] for motion generation. MLDM [12] denoises the low-dimensional latent code of motion instead of the raw motion sequences. T2M-GPT [73] utilizes VQ-VAE [43] to compress motion sequences into a learned codebook and then generate human motion in an auto-regressive fashion. GMD [27] further adds keyframe conditions to synthesize controllable human motion. In this work, we base our model on the widely adopted MDM [62] framework.

2.3 Human-Object Interaction Generation

Generating plausible and realistic human-object interaction sequences has attracted much attention in recent years [11,13,15,19,29,32,33,46,58–61,64, 68,71,78]. GRAB [59] generates plausible hand poses grasping a given object. GOAL [58] and SAGA [64] extend it to full-body grasping motion generation. However, these methods convert the task into the generation of the final pose of a human interacting with the *static* object. Recent works like InterDiff [68] tackle the problem of synthesizing comprehensive full-body interactions with dynamic objects, which leverages a diffusion model [23] to predict future HOI with realistic contacts. IMoS [19] synthesizes vivid HOIs conditioned on an intent label, such as: "use", "lift" and "pass". Notice that none of the aforementioned methods can generate full-body HOI involving multiple objects.

2.4 Temporal Action Composition

Human motion composition [5,7,30,34,48,53,56,77] with text timeline control aims to generate arbitrary long human motion sequences with smooth and realistic transitions, which can be separated as two categories. First, several auto-regressive methods [5,30,34,53,77] are developed relying on the availability of motion transition annotations. These methods iteratively generate the subsequent motion based on the already generated motion. Second, a few diffusion-based methods [6,7,48,56,75] are proposed to modify the sampling process of diffusion models with overlaps of a predefined transition duration. In our setting of `HIMO-SegGen`, we also need to generate composite HOI sequences with smooth transitions. Thus, we adopt a simple yet effective auto-regressive pipeline to iteratively synthesize the HOI sequences.

3 The HIMO Dataset

We present HIMO, a large-scale 4D-HOI dataset of full-body human interacting with multiple objects, comprising accurate and diverse full-body human motion, object motions and mutual contacts. Next, we will introduce the data acquisition process in Sect. 3.1 and then describe the SMPL-X fitting process and the annotation of textual descriptions and the temporal segments in Sect. 3.2.

3.1 Data Acquisition

Hybrid Capture System. To obtain the accurate movement of the human body against occlusions, we follow [59] to adopt the optical MoCap scheme instead of multi-view RGB camera system [8,24], which guarantees much lower error [59,65]. We believe that high-quality moving sequences are more important than natural images for HOI motion data. An overview of our capturing setup can be seen in Fig. 2(a,c). We deploy Optitrack [3] MoCap system with 20 PrimeX-22 infrared cameras to capture the body motion, where each subject wearing MoCap suits with 41 reflective markers. Each camera can capture 2K frames at 120fps. To capture the dexterous finger movements, we select the inertial Noitom Perception Neuron Studio (PNS) gloves [2], which can precisely record the subtle finger poses in real-time. The inertial gloves can still work well under severe occlusions of human-object and object-object. We frequently re-calibrate the PNS gloves to ensure the capture quality. Spatially, the locating boards attached to the back of the hands provide the rotation information of the wrists, thus the human body movement can be integrated with the finger movements. Temporally, we utilize the Tentacle Sync device to provide timecodes for Optitrack and PNS gloves to synchronize them. Apart from the hybrid MoCap system for full-body motion capture, we also set up a Kinect camera to record the RGB videos from the front view with the resolution of 1080 × 768.

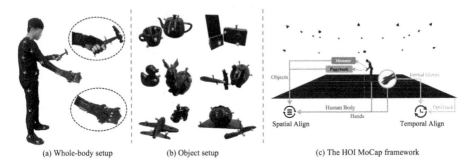

Fig. 2. Overview of the HIMO capture system. (a). We combine the Optitrack MoCap framework and the inertial gloves as a hybrid MoCap system. (b). Some examples of the 3D printed objects and the attached reflective markers. (c). The whole HOI MoCap framework, where the objects, the human body and hands of the subject are spatially aligned, and the Optitrack and inertial gloves are temporally synchronized.

Capturing Objects. We choose 53 common household objects in various scenes including the dining rooms, kitchens, living rooms and studies from ContactDB [9] and Sketchfab [4]. We further reshape them into appropriate sizes and

3D print them, so that the 3D object geometry is aligned with the real printed objects. To precisely track the object motion, we attach several (3–6) reflective markers on their surface with strong glue, as shown in Fig. 2(b), to track the rigid body composed of the attached markers. The coordinate system of the subject and the objects are naturally aligned. One critical rule of attaching the markers to objects is to mitigate the side effects of manipulating the objects. Empirical results show that 12.5 mm diameter spherical markers achieve more robust tracking performance than smaller markers. Since the Optitrack system tracks the 6-DoF poses of the centroid of the attached markers rather than the centroid of the object, thus we employ a post-calibration process to compensate for the bias between them. The object category definition is listed in the supplementary.

Recording Process. Each subject is asked to manipulate the pre-defined combinations of 2 or 3 objects, such as "pour tea from a *teapot* to a *teacup*" and "cut an *apple* on a *knifeboard* with a *knife*". Initially, the objects are randomly placed on a table with the height of 74 cm and the subject keeps the rest pose. For each sequence, the subject is required to involve all the provided objects into the manipulation and perform 3 times of each interaction task with different operating patterns for variability. Casual operations yet may not be relevant to the prescribed action are welcome to enhance the complexity of each interaction sequence, such as "shaking the goblet before drinking from it" or "cleaning the knifeboard before cutting food on it". Finally, 34 subjects participated in the collection of HIMO, resulting in **3,376** HOI sequences, **9.44** hours and **4.08M** frames, where 2,486 of them interacting with 2 objects and the other 890 sequences interacting with 3 objects. More details of the setting of the interaction categories can be found in the supplementary materials.

3.2 Data Postprocessing

Fitting SMPL-X Parameters. The expressive SMPL-X [45] parametric model is widely adopted by recent works [18,37,41,48,72] for its generality and flexible body part editing ability. We also adopt the SMPL-X model to represent the human body with finger articulations. Formally, the SMPL-X parameters consist of global orient $g \in \mathbb{R}^3$, body pose $\theta_b \in \mathbb{R}^{21 \times 3}$, finger poses $\theta_h \in \mathbb{R}^{30 \times 3}$, root translation $t \in \mathbb{R}^3$ and shape parameter $\beta \in \mathbb{R}^{10}$. An optimization algorithm is adopted to obtain the SMPL-X parameters for each HOI sequence based on our full-body MoCap data. We initialize the subjects' shape β based on their height and weight following [51]. The joint energy term $\mathbb{E}_j$ optimizes body joints to our MoCap data as:

$$\mathbb{E}_j = \sum_{n=0}^{N} \sum_{j=0}^{J} \|P_n^j - \hat{P}_n^j\|_2^2, \tag{1}$$

where N is the frame number of the sequence, J is the number of human joints, P_n^j and $\hat{P}_n^j$ represent our MoCap joint position and the joint position regressed

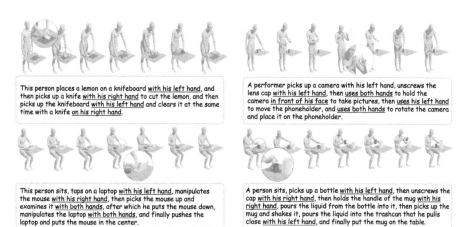

Fig. 3. More samples of the HIMO dataset. The subject is asked to interact with 2 or 3 objects based on their functionalities. The textual annotations comprise the detailed interaction order, the interaction pattern and the involved human body parts. Object names are depicted as the corresponding colors and the details are underlined.

from the SMPL-X model, respectively. The smoothing term $\mathbb{E}_s$ smooths the motion between frames and mitigates pose jittering as:

$$\mathbb{E}_s = \sum_{n=0}^{N-1} \sum_{j=0}^{J} \|\hat{P}_{n+1}^j - \hat{P}_n^j\|_2^2. \qquad (2)$$

Finally, a regularization term $\mathbb{E}_r$ is adopted to regularize the SMPL-X pose parameters from deviating as:

$$\mathbb{E}_r = \|\theta_b\|_2^2 + \|\theta_h\|_2^2. \qquad (3)$$

In total, the whole optimization objective can be summed as follows:

$$\mathbb{E} = \alpha \mathbb{E}_j + \lambda \mathbb{E}_s + \gamma \mathbb{E}_r, \qquad (4)$$

where $\alpha = 1, \lambda = 0.1, \gamma = 0.01$.

Textual Annotations. Text-driven motion generation thrives greatly thanks to the emergence of text-motion datasets [20,49,63]. To empower the research on text-driven HOI synthesis with multiple objects, i.e., HIMO-Gen, we elaborately annotate fine-grained textual descriptions of each motion sequence. Different from simple HOI manipulations with a single object and a single prompt, our HIMO dataset includes complex transitions among objects and temporal arrangements among different objects. Thus, the annotators are asked to describe the entire interaction procedure, emphasizing the detailed manipulation order, i.e., "First..., then..., finally...", the interaction pattern, i.e., "picking up", "rotating" and "pouring", and the involved human body parts, i.e., the left hand or right hand. We exploit an annotation tool based on [1] to enable the annotators to rotate and resize the view to observe the subtle HOIs patterns. More visualization results of the text-HOI pairs are presented in Fig. 3.

Temporal Segment Annotations. As aforementioned, the temporal segments of the long HOI sequences allow for fine-grained timeline control of decomposing the complex operation process into several simple ones. The granularity of the sequence splitting is small, as shown in Fig. 1. For example, the sequence of "pouring tea from the teapot to the teacup" can be split into "lifting the teapot", "pouring tea into the teacup" and "putting the teacup down". Similarly, we implement an annotation tool by [1] to better view the interaction details. The HOI sequence and its corresponding texts are simultaneously split into equal segments. Statistically, each sequence contains 2.60 segments on average.

4 Text-Driven HOI Synthesis

In this section, we benchmark two *novel* tasks of HIMO-Gen in Sect. 4.1 and HIMO-SegGen in Sect. 4.2. For HIMO-Gen, we propose a dual-branch diffusion-based framework to synthesize the motion of human and objects separately, with meticulously designed mutual interaction module and optimization objectives. For HIMO-SegGen, we apply an auto-regressive generation scheme.

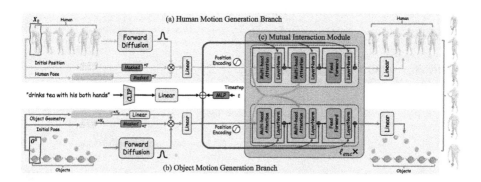

Fig. 4. Overview of the HOI synthesis framework. We propose a dual-branch conditional diffusion model to generate the human motion and object motions, respectively, conditioned on the textual description and the initial states of the human and objects. A mutual interaction module is also integrated for information fusion to generate coordinated HOI results.

4.1 The HIMO-Gen Framework

Data Representation. We denote the human motion as $\boldsymbol{H} \in \mathbb{R}^{T \times D_h}$ and the multiple object motions as $\boldsymbol{O} = \{\boldsymbol{O}_i\}_{i=0}^{N_o} \in \mathbb{R}^{T \times D_o}$, where T represents the frame number of the HOI sequence, N_o is the number of the objects, D_h and D_o represent the dimension of the human and object motion, respectively. We adopt the SMPL-X [45] parametric model to represent the human movements. The pose state of human at frame i is denoted as $\boldsymbol{H}^i$, which consists of the global joint positions $P^i \in \mathbb{R}^{52\times 3}$, global joint rotation $Q^i \in \mathbb{R}^{52\times 6}$ represented by the continuous 6D rotation format [79] and translation $t^i \in \mathbb{R}^3$. We also denote the object motion at frame i as $\boldsymbol{O}^i$ (We omit the subscript of the i-th object/geometry for better expression), which comprises of the relative rotation $\mathbf{R}^i \in \mathbb{R}^6$ with respect to the input object's frame and its global translation $\mathbf{T}^i \in \mathbb{R}^3$. To encode the object geometries $\boldsymbol{G} = \{\boldsymbol{G}_i\}_{i=0}^{N_o}$, we follow prior works [31,59] to adopt the Basis Point Set (BPS) representation [50]. We sample 1,024 points from the surface of the object meshes, and then for each sampled point, we calculate the directional vector with its nearest neighbor from the basis points set, resulting in $\boldsymbol{G} \in \mathbb{R}^{1024\times 3}$ for each object.

Problem Formulation. Given the textual description $\boldsymbol{L}$, the initial states of $\boldsymbol{H}^0$ and $\boldsymbol{O}^0$ as the *first* frame human motion and object motions and the object geometries $\boldsymbol{G}$. Our model aims to generate the spatio-temporally aligned human motion $\boldsymbol{H}$ together with the object motions $\boldsymbol{O}$ with plausible contacts.

Dual-Branch Diffusion Models. Our proposed framework is demonstrated in Fig. 4, which consists of two parallel diffusion-based branches for human and object motion generation and a mutual interaction module for feature fusion. Given a textual description $\boldsymbol{L}$ like "drinks tea with his both hands", we utilize the CLIP [54] as text encoder followed by a linear layer to obtain the semantic embeddings and then concatenate them with the features of the sampling timestep t extracted by an MLP process. Take the human motion generation branch as an example, the initial state $\boldsymbol{H}^0$ including the initial position and the human pose serves as the conditions. Following [31], we adopt the masked representation by zero padding $\boldsymbol{H}_0$ to T frames. The masked representations are then concatenated with the noised human motion representation $\boldsymbol{H}_t$, where t is the timestep of the forward diffusion process. For the object motion generation branch, the object geometry $\boldsymbol{G}$ is embedded by a linear layer and concatenated with the masked representation of the initial pose of objects and the noised representation of object motions. We separately feed the obtained human and object representations into a linear layer to obtain the motion embeddings and then pass them into the denoising mutual interaction module with positional encodings. After that, two linear layers are adopted to obtain the denoised motion representations of human and objects, respectively.

Mutual Interaction Module. To model the interaction between human and objects, we fuse the features of the two branches via a mutual interaction module as depicted in Fig. 4(c). The two branches share the same Transformer architecture with ℓ_{enc} blocks without sharing weights. Each Transformer block is comprised of two attention layers and one feed-forward layer. The first self-attention layer embeds the aggregated human/object features $\boldsymbol{H}^{(i)}$ and $\boldsymbol{O}^{(i)}$ into embeddings $\mathbf{e}_h^{(i)}$ or $\mathbf{e}_o^{(i)}$. The second attention layer functions as a *mutual attention*

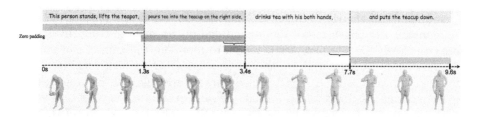

Fig. 5. The auto-regressive generation pipeline. The auto-regressive pipeline to iteratively generate the HOI synthesis results. We obtain the subsequent motion conditioned on the last few frames of the previously generated motion.

layer, where the key-value pair of the human branch Transformer is provided by the object hidden embedding $\mathbf{e}_o^{(i)}$, and vice versa, which can be formulated as:

$$\begin{aligned} \mathbf{H}^{(i+1)} &= FF(softmax(\frac{\mathbf{Q}_h \mathbf{K}_o^T}{\sqrt{C}}\mathbf{V}_o)); \mathbf{O}^{(i+1)} = FF(softmax(\frac{\mathbf{Q}_o \mathbf{K}_h^T}{\sqrt{C}}\mathbf{V}_h)), \\ \mathbf{Q}_h &= \mathbf{e}_h^{(i)}\mathbf{W}_h^{Q_h}, \mathbf{K}_h = \mathbf{H}^{(i)}\mathbf{W}_h^{K_h}, \mathbf{V}_h = \mathbf{H}^{(i)}\mathbf{W}_h^{V_h}, \\ \mathbf{Q}_o &= \mathbf{e}_o^{(i)}\mathbf{W}_o^{Q_o}, \mathbf{K}_o = \mathbf{O}^{(i)}\mathbf{W}_o^{K_o}, \mathbf{W}_o = \mathbf{O}^{(i)}\mathbf{W}_o^{V_o}, \end{aligned} \quad (5)$$

where the subscript h and o denote the human branch and the object branch, respectively, $\mathbf{W_h}$ and $\mathbf{W_o}$ denote the trainable parameters of the two branches.

Losses Formulation. To model the spatial relation between objects, we designed a novel object-pairwise loss. Our insight is that the relative distance between the interacted objects changes constantly with certain patterns, and explicitly adding constraints on the relative distance between the interacted objects is beneficial. For example, when cutting an apple with a knife, the knife gradually approaches and stays on the surface of the apple for a while, then moves away. We adopt the L2 loss as $\mathcal{L}_{dis}$ to keep the consistency between the generated results and the ground truth.

Besides, we also adopt the widely adopted geometric loss in the field of human motion, including joint position loss $\mathcal{L}_{pos}$ and the joint velocity loss $\mathcal{L}_{vel}$. To ensure the authenticity of our synthesized motion, we additionally utilized the interpenetration loss $\mathcal{L}_{pen}$ between human and objects. Here, we summarize our ultimate optimization objective as:

$$\mathcal{L} = \lambda_{vel}\mathcal{L}_{vel} + \lambda_{pos}\mathcal{L}_{pos} + \lambda_{pen}\mathcal{L}_{pen} + \lambda_{dis}\mathcal{L}_{dis}, \quad (6)$$

and we empirically set $\lambda_{vel} = \lambda_{pos} = \lambda_{pen} = 1$, and $\lambda_{dis} = 0.1$. More details of the formulations of the losses will be illustrated in the supplementary materials.

4.2 The HIMO-SegGen Framework

Problem Formulation. The only difference with HIMO-Gen lies in the input text prompts, where HIMO-Gen takes one single text description $\boldsymbol{L}$ as a condition, yet HIMO-SegGen is conditioned on a series of several atomic text prompts as $L = \{l_1, l_2, \cdots, l_m\}$. The key of HIMO-SegGen is to ensure smooth and realistic transitions between the generated HOI clips.

Generation Pipeline. Our generation pipeline is demonstrated in Fig. 5, which is simple yet effective. We segment the long HOI sequences into smaller motion-text pairs so that the natural transitions of consecutive HOIs are reserved. We first train the HIMO-Gen model to empower the ability to generate more fine-granularity HOIs. Different from the vanilla HIMO-Gen model which is only conditioned on the initial state of the first frame, we modify it to conditioned generation on the past *few* frames. During the inference time, we take the last few frames of the previous generation result as the condition for the next generation, to iteratively obtain the composite HOI results.

5 Experiments

In this section, we elaborate on the dataset and implementation details, the baseline methods and the evaluation metrics. Then we present extensive experimental results with ablation studies to show the effectiveness of our benchmark.

Dataset Details. We follow [20] to split our dataset into train, test, and validation sets with the ratio of 0.8, 0.15, and 0.05. The maximum motion length is set to 300 for HIMO-Gen and 100 for HIMO-SegGen. The maximum text length is

Table 2. Quantitative baseline comparisons on the **2-objects** partition of HIMO. $\pm$ indicates the 95% confidence interval and $\rightarrow$ means the closer the better. **Bold** indicates best result and underline indicates second best.

Method	R-Precision(Top 3)↑	FID ↓	MM-Dist↓	Diversity→	MModality↑
Real	$0.7988^{\pm 0.0081}$	$0.0176^{\pm 0.0065}$	$3.5659^{\pm 0.0109}$	$11.3973^{\pm 0.2577}$	–
IMoS [19]	$0.5013^{\pm 0.0120}$	$7.5890^{\pm 0.1121}$	$8.7402^{\pm 0.0310}$	$7.0033^{\pm 0.3205}$	$0.9920^{\pm 0.2004}$
MDM [62]	$0.6052^{\pm 0.0099}$	$6.8457^{\pm 0.3315}$	$8.0187^{\pm 0.0500}$	$\mathbf{11.3891^{\pm 0.2342}}$	$1.2880^{\pm 0.2110}$
priorMDM [56]	$0.5891^{\pm 0.0031}$	$7.8517^{\pm 0.2516}$	$7.2509^{\pm 0.0065}$	$12.5799^{\pm 0.1460}$	$1.5911^{\pm 0.1449}$
HIMO-Gen	$\underline{0.6369^{\pm 0.0032}}$	$\mathbf{1.4811^{\pm 0.0427}}$	$\underline{3.6491^{\pm 0.0101}}$	$\underline{11.6603^{\pm 0.2043}}$	$\underline{1.7863^{\pm 0.0570}}$
HIMO-SegGen	$\mathbf{0.6404^{\pm 0.0449}}$	$\underline{4.2004^{\pm 0.4729}}$	$\mathbf{3.6077^{\pm 0.0230}}$	$15.7317^{\pm 0.4090}$	$\mathbf{2.0495^{\pm 0.0030}}$

Table 3. Quantitative baseline comparisons on the **3-objects** partition of HIMO.

Method	R-Precision(Top 3)↑	FID ↓	MM-Dist↓	Diversity→	MModality↑
Real	$0.6988^{\pm 0.0054}$	$0.1811^{\pm 0.0442}$	$3.7696^{\pm 0.0279}$	$9.7674^{\pm 0.2180}$	–
IMoS [19]	$0.4662^{\pm 0.1010}$	$4.9902^{\pm 0.1772}$	$7.7702^{\pm 0.0585}$	$\underline{9.2310^{\pm 0.1130}}$	$0.6254^{\pm 0.1102}$
MDM [62]	$0.5025^{\pm 0.0136}$	$\mathbf{4.5713^{\pm 0.1100}}$	$6.3144^{\pm 0.0268}$	$8.8953^{\pm 0.2851}$	$\underline{0.7893^{\pm 0.5715}}$
priorMDM [56]	$\underline{0.5137^{\pm 0.0257}}$	$4.8210^{\pm 0.2030}$	$\underline{5.8900^{\pm 0.0232}}$	$\mathbf{9.3402^{\pm 0.0230}}$	$\mathbf{0.8120^{\pm 0.3280}}$
HIMO-Gen	$\mathbf{0.5350^{\pm 0.0185}}$	$\underline{4.7712^{\pm 0.1638}}$	$\mathbf{5.0866^{\pm 0.0412}}$	$8.9460^{\pm 0.1378}$	$0.7561^{\pm 0.0546}$

set to 40 and 15, respectively. To mitigate the influence of the order of several objects, we randomly shuffle the objects when loading the data.

Implementation Details. For both HIMO-Gen and HIMO-SegGen, the denoising network consists of $\ell_{enc} = 8$ layers of mutual interaction modules with 4 attention heads. We train the network using Adam [28] optimizer with a learning rate of 0.0001 and weight decay of 0.99. The training of HIMO-Gen and HIMO-SegGen takes about 9 and 12 h on a single A100 GPU with a batch size of 128.

Baseline Methods. We adopt the prominent text-to-motion methods MDM [62], priorMDM [56] and recent intent-driven HOI synthesis method IMoS [19] as our baselines. To be specific, we re-implement each model to support the condition input of object geometry and the initial states of human and objects. More details of the baselines are provided in the supplementary materials.

Evaluation Metrics. Following [20], we train a motion feature extractor and a text feature extractor in a contrastive manner for evaluation. Since we are evaluating the generation quality of HOIs, the motion features here include both human and object motion. Then we evaluate over the set of metrics proposed by [20]: *R-precision* and *MM-Dist* determine the relevance between the generated motion and the input prompts, *FID* measures the dissimilarity of the generated and ground-truth motion in latent feature space, *Diversity* measures the variability of the generated motions, and *MultiModality* shows the mean-variance of generated motion conditioned on a single text prompt.

5.1 Results and Analysis

Quantitative Comparisons. Quantitative results on the 2-objects and 3-objects partitions of HIMO are demonstrated in Table 2 and Table 3. In the 2-objects setting, our method HIMO-Gen outperforms the baseline methods on all metrics, which indicates our method has better generation quality over baseline methods. The quantitative results of HIMO-SegGen outperform HIMO-Gen in MM-Dist and MModality, which may be attributed to its fine-grained control of generated motion since HIMO-SegGen takes more concise descriptions as input conditions. In the 3-objects setting of HIMO, our method also outperforms baselines on the R-precision and MM-Dist metrics, which further shows the effectiveness of our method.

Qualitative Results. Qualitative results of our method are depicted in Fig. 6, which show that our method can synthesize plausible human-object interaction sequences. More generation results of HIMO-SegGen and visual comparisons will be presented in the supplementary materials.

Generalization Experiments. We find that our trained model can also apply to unseen object meshes. Besides, with our elaborate annotate temporal segments, our trained HIMO-SegGen model can generate *novel* HOI compositions. Due to the space limit, more results are listed in the supplementary materials.

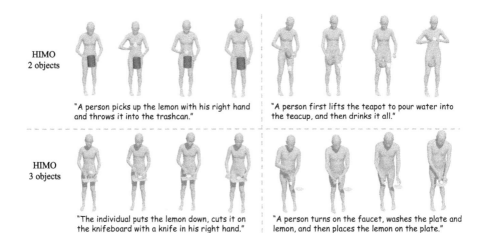

Fig. 6. Visualization results. Our HIMO-Gen framework can generate plausible and realistic sequences of human interacting with multiple objects.

5.2 Ablation Study

We conduct extensive ablation studies on HIMO-Gen and HIMO-SegGen, as depicted in Table 4 and Table 5. For HIMO-Gen, we experiment with three settings: 1) Replacing the **mutual interaction module** with the original transformer layer (w/o IM); 2) Removing the **object-pairwise loss** (w/o $\mathcal{L}_{dis}$); 3) Generating human motion conditioned on the generated objects motion as [32]. From Table 4, we can derive that the removal of the mutual interaction module results in a huge drop of R-precision, which verifies the effectiveness of the feature fusion between human and objects. The object-pairwise loss helps a lot in constraining the generated motions. Consecutive generation scheme brings huge performance drop to generate the human and object motions simultaneously. For HIMO-SegGen, we experiment on the conditioned past 1, 5, 10 and 20 frames. Given the results in Table 5, "10-frames" achieves the best performance in R-precision, FID and *MM-Dist*. More past frames ("20 frames") provide no

Table 4. Ablation studies of several model designs on **2-objects** partition of HIMO.

Method	R-Precision(Top 3)↑	FID ↓	MM-Dist↓	Diversity→	MModality↑
Real	$0.7988^{\pm 0.0081}$	$0.0176^{\pm 0.0065}$	$3.5659^{\pm 0.0109}$	$11.3973^{\pm 0.2577}$	—
Ours-w/o IM	$0.4710^{\pm 0.0509}$	$3.5113^{\pm 0.0029}$	$6.7135^{\pm 0.1029}$	$13.0256^{\pm 0.1207}$	$1.2086^{\pm 0.0492}$
Ours-w/o $\mathcal{L}_{dis}$	$\mathbf{0.6426^{\pm 0.0085}}$	$2.8694^{\pm 0.0488}$	$4.9342^{\pm 0.0191}$	$12.0789^{\pm 0.1153}$	$1.4394^{\pm 0.0549}$
Ours-consec.	$0.3902^{\pm 0.3046}$	$8.5024^{\pm 0.2310}$	$8.9982^{\pm 0.0120}$	$10.2788^{\pm 0.1113}$	$1.1209^{\pm 0.0890}$
HIMO-Gen	$\underline{0.6369^{\pm 0.0032}}$	$\mathbf{1.4811^{\pm 0.0427}}$	$\mathbf{3.6491^{\pm 0.0101}}$	$\mathbf{11.6603^{\pm 0.2043}}$	$\mathbf{1.7863^{\pm 0.0570}}$

Table 5. Ablation studies of the frame number of conditioned frames for HIMO-SegGen on **2-objects** partition of the HIMO dataset.

#Frame	R-Precision(Top 3)↑	FID ↓	MM-Dist↓	Diversity→	MModality↑
Real	$0.7988^{\pm 0.0081}$	$0.0176^{\pm 0.0065}$	$3.5659^{\pm 0.0109}$	$11.3973^{\pm 0.2577}$	–
1	$0.5502^{\pm 0.0230}$	$8.7320^{\pm 0.3208}$	$6.2094^{\pm 0.3553}$	$\mathbf{13.2999^{\pm 0.5034}}$	$0.8814^{\pm 0.0394}$
5	$0.5448^{\pm 0.0493}$	$7.3029^{\pm 0.2677}$	$\underline{5.5053^{\pm 0.0026}}$	$17.8045^{\pm 0.0956}$	$0.9500^{\pm 0.2374}$
10	$\mathbf{0.6404^{\pm 0.0449}}$	$\mathbf{4.2004^{\pm 0.4729}}$	$\mathbf{3.6077^{\pm 0.0230}}$	$15.7317^{\pm 0.4090}$	$\underline{2.0495^{\pm 0.0030}}$
20	$\underline{0.5508^{\pm 0.0891}}$	$\underline{5.6749^{\pm 0.1002}}$	$6.7708^{\pm 0.0355}$	$\underline{15.2112^{\pm 0.2210}}$	$\mathbf{2.3863^{\pm 0.3399}}$

additional performance gain, which may result from less attention to the predicted frames.

6 Conclusion

In this paper, we propose HIMO, a dataset of full-body human interacting with multiple household objects. We collect a large amount of 4D HOI sequences with precise and diverse human motions, object motions and mutual contacts. Based on that, we annotate the detailed textual descriptions for each HOI sequence to enable the text-driven HOI synthesis. We also equip it with elaborative temporal segments to decompose the long sequences into several atomic HOI steps, which facilitates the HOI synthesis driven by consecutive text prompts. Extensive experiments show plausible results, verifying that the HIMO dataset is promising for downstream HOI generation tasks.

Limitations: Our work has the following limitations: 1) **Large objects:** Our HIMO dataset is built mainly based on small household objects, without human interacting with large objects such as chairs and suitcases. 2) **RGB modality:** Though we set up a Kinect camera to capture the monocular RGB sequences, the human and object appearances are unnatural since humans are wearing tight MoCap suits and objects are attached with reflective markers. 3) **Facial expressions:** We neglect the facial expressions since our dataset mainly focuses on the manipulation of multiple objects, *i.e.*, the body movements, finger gestures and the object motions, which have little correlation with facial expressions.

Acknowledgements. This work was supported in part by NSFC (62201342, 62101325), and Shanghai Municipal Science and Technology Major Project (2021SHZDZX0102).

References

1. Aitviewer. https://eth-ait.github.io/aitviewer/
2. Noitom. https://noitom.com/

3. Optitrack. https://optitrack.com/
4. Sketchfab. https://sketchfab.com/feed
5. Athanasiou, N., Petrovich, M., Black, M.J., Varol, G.: Teach: temporal action composition for 3D humans. In: 2022 International Conference on 3D Vision (3DV), pp. 414–423. IEEE (2022)
6. Bar-Tal, O., Yariv, L., Lipman, Y., Dekel, T.: Multidiffusion: fusing diffusion paths for controlled image generation (2023)
7. Barquero, G., Escalera, S., Palmero, C.: Seamless human motion composition with blended positional encodings. arXiv preprint arXiv:2402.15509 (2024)
8. Bhatnagar, B.L., Xie, X., Petrov, I., Sminchisescu, C., Theobalt, C., Pons-Moll, G.: Behave: dataset and method for tracking human object interactions. In: CVPR. IEEE (2022)
9. Brahmbhatt, S., Ham, C., Kemp, C.C., Hays, J.: ContactDB: analyzing and predicting grasp contact via thermal imaging. In: CVPR (2019)
10. Brahmbhatt, S., Tang, C., Twigg, C.D., Kemp, C.C., Hays, J.: ContactPose: a dataset of grasps with object contact and hand pose. In: ECCV (2020)
11. Braun, J., Christen, S., Kocabas, M., Aksan, E., Hilliges, O.: Physically plausible full-body hand-object interaction synthesis. In: International Conference on 3D Vision (3DV) (2024)
12. Chen, X., et al.: Executing your commands via motion diffusion in latent space. In: CVPR, pp. 18000–18010 (2023)
13. Christen, S., Kocabas, M., Aksan, E., Hwangbo, J., Song, J., Hilliges, O.: D-grasp: physically plausible dynamic grasp synthesis for hand-object interactions. In: CVPR (2022)
14. Dabral, R., Mughal, M.H., Golyanik, V., Theobalt, C.: Mofusion: a framework for denoising-diffusion-based motion synthesis. In: CVPR (2023)
15. Diller, C., Dai, A.: CG-HOI: contact-guided 3D human-object interaction generation. arXiv preprint arXiv:2311.16097 (2023)
16. Ding, G., Sener, F., Yao, A.: Temporal action segmentation: an analysis of modern techniques. IEEE Trans. Pattern Anal. Mach. Intell. (2023)
17. Fan, Z., et al.: ARCTIC: a dataset for dexterous bimanual hand-object manipulation. In: CVPR (2023)
18. Feng, Y., Choutas, V., Bolkart, T., Tzionas, D., Black, M.J.: Collaborative regression of expressive bodies using moderation. In: 3DV (2021)
19. Ghosh, A., Dabral, R., Golyanik, V., Theobalt, C., Slusallek, P.: Imos: intent-driven full-body motion synthesis for human-object interactions. In: Eurographics (2023)
20. Guo, C., et al.: Generating diverse and natural 3D human motions from text. In: CVPR, pp. 5152–5161 (2022)
21. Hampali, S., Rad, M., Oberweger, M., Lepetit, V.: Honnotate: a method for 3D annotation of hand and object poses. In: CVPR (2020)
22. Hampali, S., Sarkar, S.D., Rad, M., Lepetit, V.: Keypoint transformer: solving joint identification in challenging hands and object interactions for accurate 3D pose estimation. In: CVPR (2022)
23. Ho, J., Jain, A., Abbeel, P.: Denoising diffusion probabilistic models (2020)
24. Huang, Y., Taheri, O., Black, M.J., Tzionas, D.: InterCap: joint markerless 3D tracking of humans and objects in interaction. In: Andres, B., Bernard, F., Cremers, D., Frintrop, S., Goldlücke, B., Ihrke, I. (eds.) DAGM GCPR 2022. LNCS, vol. 13485, pp. 281–299. Springer, Cham (2022). https://doi.org/10.1007/978-3-031-16788-1_18

25. Jian, J., Liu, X., Li, M., Hu, R., Liu, J.: Affordpose: a large-scale dataset of hand-object interactions with affordance-driven hand pose. In: ICCV, pp. 14713–14724 (2023)
26. Jiang, N., et al.: Full-body articulated human-object interaction. In: ICCV, pp. 9365–9376 (2023)
27. Karunratanakul, K., Preechakul, K., Suwajanakorn, S., Tang, S.: Guided motion diffusion for controllable human motion synthesis. In: ICCV, pp. 2151–2162 (2023)
28. Kingma, D.P., Ba, J.: Adam: a method for stochastic optimization (2017)
29. Kulkarni, N., et al.: Nifty: neural object interaction fields for guided human motion synthesis (2023)
30. Lee, T., Moon, G., Lee, K.M.: Multiact: long-term 3D human motion generation from multiple action labels. In: Proceedings of the AAAI Conference on Artificial Intelligence, vol. 37, pp. 1231–1239 (2023)
31. Li, J., Clegg, A., Mottaghi, R., Wu, J., Puig, X., Liu, C.K.: Controllable human-object interaction synthesis. arXiv preprint arXiv:2312.03913 (2023)
32. Li, J., Wu, J., Liu, C.K.: Object motion guided human motion synthesis. ACM Trans. Graph. (TOG) **42**(6), 1–11 (2023)
33. Li, Q., Wang, J., Loy, C.C., Dai, B.: Task-oriented human-object interactions generation with implicit neural representations (2023)
34. Li, S., Zhuang, S., Song, W., Zhang, X., Chen, H., Hao, A.: Sequential texts driven cohesive motions synthesis with natural transitions. In: Proceedings of the IEEE/CVF International Conference on Computer Vision, pp. 9498–9508 (2023)
35. Li, Y., Xue, Z., Xu, H.: OTAS: unsupervised boundary detection for object-centric temporal action segmentation. In: Proceedings of the IEEE/CVF Winter Conference on Applications of Computer Vision, pp. 6437–6446 (2024)
36. Liang, H., Zhang, W., Li, W., Yu, J., Xu, L.: Intergen: diffusion-based multi-human motion generation under complex interactions. arXiv preprint arXiv:2304.05684 (2023)
37. Lin, J., Zeng, A., Wang, H., Zhang, L., Li, Y.: One-stage 3D whole-body mesh recovery with component aware transformer. In: CVPR, pp. 21159–21168 (2023)
38. Liu, K., Li, Y., Liu, S., Tan, C., Shao, Z.: Reducing the label bias for timestamp supervised temporal action segmentation. In: Proceedings of the IEEE/CVF Conference on Computer Vision and Pattern Recognition, pp. 6503–6513 (2023)
39. Liu, Y., et al.: Taco: benchmarking generalizable bimanual tool-action-object understanding. arXiv preprint arXiv:2401.08399 (2024)
40. Liu, Y., et al.: HOI4D: a 4D egocentric dataset for category-level human-object interaction. In: CVPR, pp. 21013–21022 (2022)
41. Lu, S., et al.: Humantomato: text-aligned whole-body motion generation (2023)
42. Mandery, C., Terlemez, O., Do, M., Vahrenkamp, N., Asfour, T.: Unifying representations and large-scale whole-body motion databases for studying human motion. IEEE Trans. Rob. **32**(4), 796–809 (2016)
43. van den Oord, A., Vinyals, O., Kavukcuoglu, K.: Neural discrete representation learning (2018)
44. Pan, Z., Zeng, A., Li, Y., Yu, J., Hauser, K.: Algorithms and systems for manipulating multiple objects. IEEE Trans. Robot. **39**(1), 2–20 (2022)
45. Pavlakos, G., et al.: Expressive body capture: 3D hands, face, and body from a single image. In: CVPR, pp. 10975–10985 (2019)
46. Peng, X., Xie, Y., Wu, Z., Jampani, V., Sun, D., Jiang, H.: Hoi-diff: text-driven synthesis of 3D human-object interactions using diffusion models. arXiv preprint arXiv:2312.06553 (2023)

47. Petrovich, M., Black, M.J., Varol, G.: TEMOS: generating diverse human motions from textual descriptions. In: Avidan, S., Brostow, G., Cissé, M., Farinella, G.M., Hassner, T. (eds.) ECCV 2022. LNCS, vol. 13682, pp. 480–497. Springer, Cham (2022). https://doi.org/10.1007/978-3-031-20047-2_28
48. Petrovich, M., et al.: Multi-track timeline control for text-driven 3D human motion generation. arXiv preprint arXiv:2401.08559 (2024)
49. Plappert, M., Mandery, C., Asfour, T.: The kit motion-language dataset. Big Data **4**(4), 236–252 (2016)
50. Prokudin, S., Lassner, C., Romero, J.: Efficient learning on point clouds with basis point sets. In: ICCV, pp. 4332–4341 (2019)
51. Pujades, S., et al.: The virtual caliper: rapid creation of metrically accurate avatars from 3d measurements. IEEE Trans. Visual Comput. Graphics **25**(5), 1887–1897 (2019)
52. Punnakkal, A.R., Chandrasekaran, A., Athanasiou, N., Quiros-Ramirez, A., Black, M.J.: BABEL: bodies, action and behavior with english labels. In: CVPR, pp. 722–731. Computer Vision Foundation/IEEE (2021)
53. Qian, Y., Urbanek, J., Hauptmann, A.G., Won, J.: Breaking the limits of text-conditioned 3D motion synthesis with elaborative descriptions. In: Proceedings of the IEEE/CVF International Conference on Computer Vision, pp. 2306–2316 (2023)
54. Radford, A., et al.: Learning transferable visual models from natural language supervision (2021)
55. Sanh, V., Debut, L., Chaumond, J., Wolf, T.: Distilbert, a distilled version of bert: smaller, faster, cheaper and lighter (2020)
56. Shafir, Y., Tevet, G., Kapon, R., Bermano, A.H.: Human motion diffusion as a generative prior. arXiv preprint arXiv:2303.01418 (2023)
57. Shi, D., Zhong, Y., Cao, Q., Ma, L., Li, J., Tao, D.: Tridet: temporal action detection with relative boundary modeling. In: Proceedings of the IEEE/CVF Conference on Computer Vision and Pattern Recognition, pp. 18857–18866 (2023)
58. Taheri, O., Choutas, V., Black, M.J., Tzionas, D.: Goal: generating 4D whole-body motion for hand-object grasping. In: CVPR, pp. 13263–13273 (2022)
59. Taheri, O., Ghorbani, N., Black, M.J., Tzionas, D.: GRAB: a dataset of whole-body human grasping of objects. In: ECCV (2020). https://grab.is.tue.mpg.de
60. Taheri, O., et al.: Grip: generating interaction poses using latent consistency and spatial cues (2023)
61. Tendulkar, P., Surís, D., Vondrick, C.: Flex: full-body grasping without full-body grasps. In: CVPR (2023)
62. Tevet, G., Raab, S., Gordon, B., Shafir, Y., Cohen-or, D., Bermano, A.H.: Human motion diffusion model. In: The Eleventh International Conference on Learning Representations (2023). https://openreview.net/forum?id=SJ1kSyO2jwu
63. Wang, Z., Chen, Y., Liu, T., Zhu, Y., Liang, W., Huang, S.: Humanise: language-conditioned human motion generation in 3D scenes. In: NeurIPS (2022)
64. Wu, Y., et al.: Saga: stochastic whole-body grasping with contact. In: ECCV (2022)
65. Xu, L., et al.: Inter-x: towards versatile human-human interaction analysis. arXiv preprint arXiv:2312.16051 (2023)
66. Xu, L., et al.: Actformer: a GAN-based transformer towards general action-conditioned 3D human motion generation. In: ICCV, pp. 2228–2238 (2023)
67. Xu, L., et al.: Regennet: towards human action-reaction synthesis. In: CVPR, pp. 1759–1769 (2024)
68. Xu, S., Li, Z., Wang, Y.X., Gui, L.Y.: InterDiff: generating 3D human-object interactions with physics-informed diffusion. In: ICCV (2023)

69. Xu, X., Joo, H., Mori, G., Savva, M.: D3D-HOI: dynamic 3D human-object interactions from videos. arXiv preprint arXiv:2108.08420 (2021)
70. Yang, S., Zhang, W., Song, R., Cheng, J., Wang, H., Li, Y.: Watch and act: learning robotic manipulation from visual demonstration. IEEE Trans. Syst. Man Cybern. Syst. (2023)
71. Zhang, H., Ye, Y., Shiratori, T., Komura, T.: Manipnet: neural manipulation synthesis with a hand-object spatial representation. ACM Trans. Graph. **40**(4) (2021). https://doi.org/10.1145/3450626.3459830
72. Zhang, H., et al.: Pymaf-x: towards well-aligned full-body model regression from monocular images. PAMI **45**(10), 12287–12303 (2023)
73. Zhang, J., et al.: Generating human motion from textual descriptions with discrete representations. In: CVPR, pp. 14730–14740 (2023)
74. Zhang, J., et al.: Neuraldome: a neural modeling pipeline on multi-view human-object interactions. In: CVPR (2023)
75. Zhang, Q., Song, J., Huang, X., Chen, Y., Liu, M.Y.: Diffcollage: parallel generation of large content with diffusion models. arXiv preprint arXiv:2303.17076 (2023)
76. Zhang, X., Bhatnagar, B.L., Starke, S., Guzov, V., Pons-Moll, G.: Couch: towards controllable human-chair interactions (2022)
77. Zhang, Y., Black, M.J., Tang, S.: Perpetual motion: generating unbounded human motion. arXiv preprint arXiv:2007.13886 (2020)
78. Zheng, J., Zheng, Q., Fang, L., Liu, Y., Yi, L.: Cams: canonicalized manipulation spaces for category-level functional hand-object manipulation synthesis. In: CVPR, pp. 585–594 (2023)
79. Zhou, Y., Barnes, C., Lu, J., Yang, J., Li, H.: On the continuity of rotation representations in neural networks. In: CVPR, pp. 5745–5753. Computer Vision Foundation/IEEE (2019)
80. Zhou, Z., Wang, B.: UDE: a unified driving engine for human motion generation. In: CVPR, pp. 5632–5641 (2023)

SAIR: Learning Semantic-Aware Implicit Representation

Canyu Zhang[1], Xiaoguang Li[1], Qing Guo[2(✉)], and Song Wang[1]

[1] University of South Carolina, Columbia, USA
{canyu,xl22}@email.sc.edu, songwang@cec.sc.edu
[2] IHPC and CFAR, Agency for Science, Technology and Research (A*STAR),
Singapore, Singapore
tsingqguo@ieee.org

Abstract. Implicit representation of an image can map arbitrary coordinates in the continuous domain to their corresponding color values, presenting a powerful capability for image reconstruction. Nevertheless, existing implicit representation approaches only focus on building continuous appearance mapping, ignoring the continuities of the semantic information across pixels. Consequently, achieving the desired reconstruction results becomes challenging when the semantic information within input image is corrupted, such as when a large region is missing. To address the issue, we suggest learning *semantic-aware implicit representation (*SAIR*)*, that is, we make the implicit representation of each pixel rely on both its appearance and semantic information (*e.g.*, which object does the pixel belong to). To this end, we propose a framework with two modules: (1) a semantic implicit representation (SIR) for a corrupted image. Given an arbitrary coordinate in the continuous domain, we can obtain its respective text-aligned embedding indicating the object the pixel belongs. (2) an appearance implicit representation (AIR) based on the SIR. Given an arbitrary coordinate in the continuous domain, we can reconstruct its color whether or not the pixel is missed in the input. We validate the novel semantic-aware implicit representation method on the image inpainting task, and the extensive experiments demonstrate that our method surpasses state-of-the-art approaches by a significant margin.

Keywords: Image inpainting · Semantic-aware implicit representation · Visual language model

1 Introduction

Implicit neural representation has demonstrated surprising performance in the 2D image [5,10], video [4] and novel view [29,40,49] reconstruction. They achieve this by extracting appearance features from 2D images using an encoder and then associating continuous coordinates with corresponding appearance features

---

Canyu Zhang and Xiaoguang Li are co-first authors and contribute equally.
Code is available at https://github.com/tsingqguo/sair..

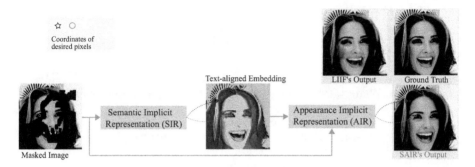

Fig. 1. Semantic-aware implicit representation (SAIR) is composed of semantic implicit representation (SIR) and appearance implicit representation (AIR). SIR processes the corrupted image and obtains its text-aligned embedding. AIR reconstructs the color.

through a neural network, ultimately translating them into the RGB color space. However, these methods tend to neglect the semantic meaning of individual pixels. Reconstructed images may contain noticeable artifacts or lose crucial semantic information, especially when dealing with degraded input images, such as those with large missing regions. As shown in Fig. 1, when the local appearance information is missing around the woman's eye, previous implicit representation methods like LIIF [5] fall short in accurately reconstructing the missing pixels.

To address this issue, we propose a novel approach called Semantic-Aware Implicit Representation (SAIR). In SAIR, each pixel's implicit representation depends on both its appearance and semantic context (e.g., the object it belongs to). This integration of continuous semantic mapping mitigates the limitations of only employing appearance implicit representation. Consequently, even in cases of degraded appearance information, the network can produce high-quality outputs with the aid of semantic information. This enhancement benefits various image processing tasks, including image generation, inpainting, editing, and semantic segmentation. As illustrated in Fig. 1, our method surpasses the existing implicit neural representation approaches that rely solely on appearance mapping on the image inpainting task. Remarkably, even when confronted with severely degraded input images, *e.g.*, a large region misses, our approach still can accurately fill in the missing pixels, yielding a natural and realistic result.

The proposed semantic-aware implicit representation involved two modules: (1) a semantic implicit representation (SIR) for a corrupted image whose large regions miss. Given an arbitrary coordinate in the continuous domain, the SIR can obtain its respective text-aligned embedding indicating the object the pixel belongs to. (2) an appearance implicit representation (AIR) based on the SIR. Given an arbitrary coordinate in the continuous domain, AIR can reconstruct its color whether or not the pixel is missed in the input. Specifically, to implement the SIR, we first use the modified CLIP [32] encoder to extract the text-aligned embedding from the input image. This specific modification (see Sect. 4.2) allows CLIP to output a spatial-aware embedding without introducing additional parameters and altering the feature space of CLIP. The text-aligned embedding can effectively reflect the pixel-level semantic information. However, this embedding has a smaller dimension compared to the input image. Moreover, when the input image is significantly degraded, the quality of the extracted embedding deteriorates considerably. To address those problems, we utilize the semantic implicit representation within the text-align embedding. This process not only expands the feature dimensions but also compensates for missing information when the input image is severely degraded.

To implement AIR, we utilize a separate implicit representation function that takes three inputs: the appearance embedding extracted from the input image using a CNN-based network, the enhanced text-aligned embedding by SIR (see Sect. 4.3), and the pixel coordinates which indicating the location information. This allows AIR to leverage both appearance and semantic information simultaneously. We validate the novel semantic-aware implicit representation (SAIR) method on the image inpainting task and conducted comprehensive experiments on the widely utilized CelebAHQ [26] and ADE20K [50] datasets. The extensive experiments demonstrate that our method surpasses state-of-the-art approaches by a significant margin.

In summary, our main contributions are listed as follows:

- We acknowledge the limitation of existing implicit representation methods that rely solely on building continuous appearance mapping, hindering their effectiveness in handling severely degraded images. To address this limitation, we introduce Semantic-Aware Implicit Representation (SAIR).
- We propose a novel framework to implement SAIR which involves two modules:(1) Semantic implicit representation (SIR) for enhancing semantic embedding, and (2) Appearance implicit representation (AIR), which builds upon SIR to leverage both semantic and appearance information simultaneously.
- Comprehensive experiments on the widely utilized CelebAHQ [26] and ADE20K [50] datasets demonstrate that our proposed method surpasses previous implicit representation approaches by a significant margin across four commonly used image quality evaluation metrics, *i.e.*, PSNR, SSIM, L1, and LPIPS.

2 Related Work

Implicit Neural Representation. Implicit neural functions find applications across a wide spectrum of domains, encompassing sound signals [36], 2D images [2,5,11,15], videos [4], and 3D shapes [8,12,43,44]. These functions offer a means to continuously parameterize signals, enabling the handling of diverse data types, such as point clouds in IM-NET [6] or video frames in NERV [3]. Implicit neural functions have demonstrated their ability to generate novel views, as exemplified by Nerf [29], which leverages an implicit neural field to synthesize new perspectives. Within the domain of image processing, methods like

LIIF [5] establish a connection between pixel features and RGB color, facilitating arbitrary-sized image super-resolution. LTE [15], a modification of LIIF, extends this concept by incorporating additional high-frequency information in Fourier space to address the limitations of a standalone MLP. Recently, However, these approaches lack explicit consideration of semantic information during training, which can result in potential inconsistencies at the semantic level.

Image Inpainting. Image inpainting techniques [1,7,16,17,27,31,39,45] are designed to restore corrupted image regions by leveraging information from non-missing portions, which can benefit many downstream tasks [18,19]. Established methods such as [22,30,34] employ edge information or smoothed images to guide the restoration process. Another noteworthy approach, as introduced by [25], relies on valid pixels to infer the missing ones. Furthermore, [9] incorporates an element-wise convolution block to reconstruct missing regions around the mask boundary while utilizing a generative network to address other missing areas. Extending upon these techniques, [20] advances the inpainting process by implementing element-wise filtering at both feature and image levels. Feature-level filtering is tailored for substantial missing regions, while image-level filtering refines local details. However, contemporary inpainting models face challenges when confronted with substantial missing regions, as reliable neighborhood features are often lacking. In such scenarios, text prompts prove invaluable as a robust guidance mechanism, enhancing the inpainting process.

Image-Text Cross-Model Learning. Cross-model networks have gained substantial attention across various image processing domains, including image semantic segmentation [28,41,52], image generation [21,33,38,53], and visual question answering (VQA) [42,48]. For instance, DF-GAN [38] represents a one-stage text-to-image backbone capable of directly synthesizing high-resolution images. In the realm of image segmentation, [28] leverages latent diffusion models (LDMs) to segment text-based real and AI-generated images. In VQA, [24] incorporates explicit object region information into the answering model. Furthermore, [46] harnesses text to assist the model in generating missing regions within images, thereby pushing the boundaries of image inpainting tasks. Additionally, language models like CLIP [32] have emerged to bridge the gap between image and semantic features. In this paper, we explore the influence of semantic information within the implicit neural function on the image inpainting task. Through the integration of semantic information, our objective is to endow the model with a more profound comprehension of the semantic meaning associated with specific image coordinates.

3 Preliminary: Local Image Implicit Representation

Given an image $\mathbf{I}$, an implicit representation for the image is to map coordinates in the continuous domain to corresponding color values; that is, we have

$$c_\mathbf{p} = \sum_{\mathbf{q} \in \mathcal{N}_\mathbf{p}} \omega_\mathbf{q} f_\theta(\mathbf{z}_\mathbf{q}^{\text{app}}, \text{dist}(\mathbf{p}, \mathbf{q})), \qquad (1)$$

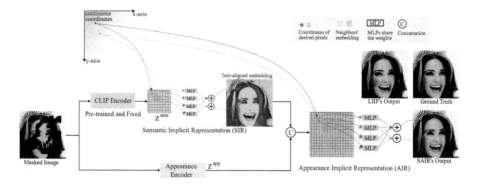

Fig. 2. Pipeline of our semantic-aware implicit representation (SAIR). The semantic implicit representation (SIR) is used to complete the missing semantic information. The appearance implicit representation (AIR) is used to complete missing details

where $\mathbf{p}$ is the continuous coordinates, the output $c_\mathbf{p}$ is the color of the pixel $\mathbf{p}$, $\mathcal{N}_\mathbf{p}$ contains all neighboring pixels of $\mathbf{p}$ within the image $\mathbf{I}$, $f_\theta(\cdot)$ is an MLP for coordinate-color mapping, $\omega_\mathbf{q}$ is the weight of q, and $\mathbf{z}_\mathbf{q}^{app}$ is the appearance feature of pixel $\mathbf{q}$. Note that, all pixels in $\mathcal{N}_\mathbf{p}$ are sampled from the input image $\mathbf{I}$ and their features $\{\mathbf{z}_\mathbf{q}^{app}\}$ are extracted through an encoder network for handling $\mathbf{I}$. Intuitively, the MLP is to transform the appearance embedding of a neighboring pixel to the color of the pixel $\mathbf{p}$ based on their spatial distance. Recent works have demonstrated that training above implicit representation via image quality loss (*e.g.*, L_1 loss) could remove noise or perform super-resolution [6,11,15]. However, when the neighboring pixels in $\mathcal{N}_\mathbf{p}$ miss, the implicit representation via Eq. 1 is affected. As shown in Fig. 3, the existing implicit representation approaches cannot properly reconstruct the pixels within missing regions.

4 Semantic-Aware Implicit Representation (SAIR)

4.1 Overview

To address the issue, we propose the semantic-aware implicit representation (SAIR), which contains two key modules, *i.e.*, semantic implicit representation (SIR) and appearance implicit representation (AIR). The first objective is to create a continuous semantic representation that enables us to fill in the missing semantic information within the input image. The second objective is to develop a continuous appearance representation that allows us to restore missing details.

Specifically, given an input image $\mathbf{I} \in \mathbb{R}^{H \times W \times 3}$ that may contain some missing regions indicated by a mask $\mathbf{M} \in \mathbb{R}^{H \times W}$, we aim to build the semantic implicit representation (SIR) to predict semantic embedding of an arbitrary given pixel whose coordinates could be non-integer values. The embedding could indicate the object the pixel belongs to. We formulate the process as

$$\mathbf{z}_\mathbf{p}^{sem} = \text{SIR}(\mathbf{I}, \mathbf{M}, \mathbf{p}), \tag{2}$$

where $\mathbf{z}_\mathbf{p}^{\text{sem}}$ denotes the semantic embedding of the pixel $\mathbf{p}$. Intuitively, we require the SIR to have three properties: ❶ The predicted semantic embedding should be well aligned with the extract category of the object the pixel belongs to. ❷ If the given coordinate (*i.e.*, $\mathbf{p}$) is within unlost regions but with non-integer values, SIR could estimate its semantic embedding accurately. This requires SIR to have the capability of interpolation. ❸ If the specified coordinate is within missing regions, SIR could complete the semantic embedding properly. We extend the local image implicit representation to the embedding level with text-aligned embeddings and propose the SIR in Sect. 4.2 to achieve the above three properties.

After getting the semantic embedding of the desired pixel, we further estimate the appearance (*e.g.*, color) of the pixel via the appearance implicit representation; that is, we have

$$c_\mathbf{p} = \text{AIR}(\mathbf{I}, \text{SIR}, \mathbf{p}), \tag{3}$$

where $c_\mathbf{p}$ denotes the color of the desired pixel $\mathbf{p}$. Intuitively, AIR is to predict the color of $\mathbf{p}$ according to the built semantic implicit representation (SIR) and input appearance. We detail the process in Sect. 4.3.

4.2 Semantic Implicit Representation (SIR)

We first use the modified CLIP model to extract the text-aligned embedding as the semantic embedding. Specifically, inspired by the recent work MaskCLIP [51], we remove the query and key embedding layers of the raw CLIP model and restructured the value-embedding and final linear layers into two separate 1×1 convolutional layers. This adjustment is made without introducing additional parameters or altering the feature space of CLIP, allowing the CLIP output a spatial-aware embedding tensor. Given the input image $\mathbf{I} \in \mathbb{R}^{H \times W \times 3}$, we feed it into the modified image encoder of CLIP and output a tensor $\mathbf{Z}^{\text{sem}} \in \mathbb{R}^{h \times w \times c}$ where h, w, and c are the height, width, and channel numbers. Note that $\mathbf{Z}^{\text{sem}}$ is not pixel-wise embedding with $h \ll H$ and $w \ll W$, which have much lower resolution than $\mathbf{I}$. MaskCLIP employs the naive resize operation to map the $\mathbf{Z}^{\text{sem}}$ to the same size as the input image, which cannot complete the missing semantic information. Instead, we propose to extend local image implicit representation to the text-aligned embedding and formulate the SIR as

$$\mathbf{z}_\mathbf{p}^{\text{sem}} = \text{SIR}(\mathbf{I}, \mathbf{M}, \mathbf{p}) = \sum_{\mathbf{q} \in \mathcal{N}_\mathbf{p}} \omega_\mathbf{q} f_\theta([\mathbf{z}_\mathbf{q}^{\text{sem}}, \mathbf{M}[\mathbf{q}]], \text{dist}(\mathbf{p}, \mathbf{q})), \tag{4}$$

where $\mathbf{z}_\mathbf{q}^{\text{sem}} = \mathbf{Z}^{\text{sem}}[\mathbf{q}]$ is the embedding of the $\mathbf{q}$ location at $\mathbf{Z}^{\text{sem}}$, and $\mathcal{N}_\mathbf{p}$ denotes the set of neighboring coordinates around $\mathbf{p}$. $\text{dist}(\mathbf{p}, \mathbf{q})$ measures the distance between $\mathbf{p}$ and $\mathbf{q}$. $f_\theta(\cdot)$ is a MLP with the θ being the weights. Intuitively, $f_\theta(\cdot)$ is to estimate the text-aligned embedding of the location $\mathbf{p}$ according to the known embedding of $\mathbf{q}$ and the spatial relationship between $\mathbf{p}$ and $\mathbf{q}$. Finally, all estimations based on different $\mathbf{q}$ are weightly combined through $\omega_\mathbf{q}$ that is also set as the area ratio of $\mathbf{p}$-$\mathbf{q}$-made rectangle in the whole neighboring area.

4.3 Appearance Implicit Representation (AIR)

With the built SIR, we aim to build the appearance implicit representation (AIR) that can estimate the colors of arbitrarily specified coordinates. In first step, we use a CNN-based appearance encoder to generate appearance feature $\mathbf{Z}^{app} = \text{AppEncoder}(\mathbf{I}, \mathbf{M})$, and $\mathbf{Z}^{app} \in \mathbb{R}^{H \times W \times C}$. Given a pixel's coordinates $\mathbf{p}$, we predict its color by

$$c_\mathbf{p} = \text{AIR}(\mathbf{I}, \mathbf{M}, \text{SIR}, \mathbf{p}) = \sum_{\mathbf{q} \in \mathcal{N}_\mathbf{p}} \omega_\mathbf{q} f_\beta([\mathbf{z}_\mathbf{q}^{app}, \text{SIR}(\mathbf{I}, \mathbf{M}, \mathbf{q})], \text{dist}(\mathbf{p}, \mathbf{q})), \quad (5)$$

where $\mathbf{z}_\mathbf{q}^{app} = \mathbf{Z}^{app}[\mathbf{q}]$ is the appearance embedding of $\mathbf{q}$-th pixel. The function f_β is a MLP with the β being its weights. Intuitively, we estimate the color of the $\mathbf{p}$-th pixel according to the appearance and semantic information of the neighboring pixels by jointly considering the spatial distance. For example, if a pixel $\mathbf{p}$ misses, the appearance feature of $\mathbf{p}$ (*i.e.*, $\mathbf{z}_\mathbf{p}^{app}$) is affected and tends to zero while the semantic information could be inferred from contexts. As shown in Fig. 2, even though the pixels around the left eye miss, we still know the missed pixels belong to the left eye category.

4.4 Implementation Details

Network Architecture. We utilize and modify the pre-trained ViT-B/16 image encoder of CLIP model to extract the semantic embedding. And we set the AppEncoder as a convolutional neural network and detail the architecture in Table 1, which is capable of generating features of the same size as the input image.

Our MLP modules $f_\alpha(\cdot)$ and $f_\beta(\cdot)$ are four-layer MLP with ReLU activation layers, and the hidden dimension is 256. **Loss functions.** During the training phase, we employ the L1 loss to measure the discrepancy between the predicted pixel color and the ground truth pixel color, which is utilized for calculating the reconstruction loss $\mathcal{L}_1$. To guarantee the feature after SIR remains in text-aligned feature space, we choose L1 loss to quantify the dissimilarity between the unmasked image's text-aligned feature $\mathbf{Z}_{unmask}^{sem} \in \mathbf{R}^{h \times w \times c}$, and the SIR reconstructed feature $\mathbf{Z}_{LR}^{reconstructed} \in \mathbf{R}^{h \times w \times c}$ without changing the resolution. The final loss function is $\mathcal{L} = \mathcal{L}_1 + \alpha \mathcal{L}_2$.

Table 1. Architecture of AppEncoder. $H \times W$ is the resolution of the input.

Output size	Operation
H × W	Conv(4, 64, 7, 1, 3), ReLU
H/2 × W/2	Conv(64, 128, 4, 2, 1), ReLU
H/4 × W/4	Conv(128, 256, 4, 2, 1), ReLU
Resnet × 8	
H/4 × W/4	Conv(256, 256, 3, 1, 1), ReLU
H/4 × W/4	Conv(256, 256, 3, 1, 1), ReLU
H/2 × W/2	Conv(256, 128, 4, 2, 1), ReLU
H × W	Conv(128, 64, 4, 2, 1), ReLU

Hyperparameters. We employ the Adam optimizer with parameters ($\beta_1 = 0.9$, $\beta_2 = 0.999$). The learning rate is set to 0.0001 and is halved every 100 epochs. Our models are trained for 200 epochs on two NVIDIA Tesla V100 GPUs, and the batch size is set to 16.

Table 2. Comparison results on the CelebAHQ dataset across varied mask ratios.

Method	0%–20%				20%–40%				40%–60%			
	PSNR↑	SSIM↑	L1↓	LPIPS↓	PSNR↑	SSIM↑	L1↓	LPIPS↓	PSNR↑	SSIM↑	L1↓	LPIPS↓
EdgeConnect	34.53	0.964	0.005	0.038	27.30	0.889	0.025	0.104	22.32	0.771	0.035	0.195
RFRNet	34.93	0.966	0.005	0.035	27.50	0.890	0.024	0.100	22.77	0.775	0.033	0.185
JPGNet	35.86	0.972	**0.004**	0.040	28.18	0.909	0.023	0.119	22.32	0.771	0.035	0.195
LAMA	36.04	0.973	0.008	0.024	29.14	0.932	0.020	0.029	22.94	0.854	0.033	0.152
MISF	36.32	0.976	0.012	0.019	29.85	0.932	0.021	0.055	23.91	0.868	0.042	0.133
LIIF	35.27	0.969	0.012	0.023	28.80	0.923	0.026	0.043	23.30	0.830	0.051	0.136
SAIR	**37.97**	**0.977**	0.010	**0.016**	**31.49**	**0.944**	**0.019**	**0.025**	**24.87**	0.870	**0.031**	**0.124**

5 Experimental Results

5.1 Setups

Datasets. We validate the effectiveness of proposed method through comprehensive experiments conducted on two widely used datasets: CelebAHQ [14] and ADE20K [50]. CelebAHQ is a large-scale dataset consisting of 30,000 high-resolution human face images, selected from the CelebA dataset [26]. These face images are categorized into 19 classes, and for our experiments, we use 25,000 images for training and 5,000 images for testing purposes. ADE20K, on the other hand, is a vast dataset comprising both outdoor and indoor scenes. It consists of 25,684 annotated training images, covering 150 semantic categories. We leverage this dataset to evaluate our method's performance on scene inpainting tasks. To create masked images for our experiments, we utilize the mask dataset [25] similar as the previous works [20]. This dataset offers over 9,000 irregular binary masks with varying mask ratios, spanning from 0% to 20%, 20% to 40%, and 40% to 60%. These masks are instrumental in generating realistic inpainting scenarios for evaluation purposes.

Baselines. We enhance our approach by incorporating semantic representations based on previous implicit neural function model LIIF [5]. By modifying image encoder and integrating semantic information, we obtain the semantic-aware implicit function, denoted as SAIR. For comparative analysis, we select state-of-the-art inpainting methods StructFlow [34], EdgeConnect [30], RFRNet [16], JPGNet [9], LAMA [37], MISF [20], and the implicit neural function without semantic information LIIF [5] as our baselines.

Evaluation Metrics. To assess the performance of all methods, we utilize four commonly employed image quality evaluation metrics: peak signal-to-noise ratio (PSNR), structural similarity index (SSIM), L1 loss, and learned perceptual image patch similarity (LPIPS) [47]. PSNR, SSIM, and L1 offer insights into the quality of the generated image, while LPIPS quantifies the perceptual distance between the restored image and the ground truth.

Table 3. Comparison results on the ADE20K dataset across varied mask ratios.

Method	0%–20%				20%–40%				40%–60%			
	PSNR↑	SSIM↑	L1↓	LPIPS↓	PSNR↑	SSIM↑	L1↓	LPIPS↓	PSNR↑	SSIM↑	L1↓	LPIPS↓
EdgeConnect	30.91	0.948	0.007	0.049	24.18	0.841	0.022	0.139	20.07	0.694	0.048	0.259
RFRNet	30.36	0.937	0.008	0.073	23.42	0.807	0.027	0.199	19.21	0.638	0.060	0.340
JPGNet	31.65	0.952	0.007	0.074	24.72	0.851	0.022	0.202	20.46	0.713	0.048	0.342
LAMA	31.07	0.956	0.009	0.036	24.15	0.859	0.025	0.116	20.15	0.713	0.048	0.257
MISF	31.45	0.954	0.006	**0.032**	24.97	0.859	**0.020**	0.117	20.59	0.717	0.044	0.233
LIIF	30.96	0.946	0.010	0.038	24.57	0.846	0.026	0.120	19.79	0.708	0.049	0.274
SAIR	**31.01**	**0.964**	**0.005**	0.034	**26.44**	**0.866**	0.023	**0.110**	**21.88**	**0.722**	**0.042**	**0.193**

5.2 Comparison Results

The results obtained on the CelebAHQ dataset are presented in Table 2, demonstrating a significant performance improvement achieved by incorporating semantic information into the models. For instance, SAIR outperforms MISF by 1.74 in PSNR for the 0–20% mask ratio. Moreover, SAIR surpasses LAMA by 2.35 in PSNR and 1.2% in SSIM for 20–40% ratio. In the 20–40% mask ratio, SAIR exhibits enhancements of 2.69 in PSNR and 7.1% in SSIM compared to LIIF. The results on the ADE20K dataset, as shown in Table 3, also reveal

Input   LIIF   LAMA   MISF   SAIR   GT

Fig. 3. Visual comparison with competitors: the first two cases are from the CelebAHQ dataset, while the last two are from the ADE20K dataset.

the effectiveness of incorporating semantic information. SAIR achieves a lowered LPIPS of 0.193 for the 40–60% mask ratio. And SAIR improves PSNR to 26.44 and SSIM to 86.6% in 20–40% ratio range. Notably, SAIR attains the best PSNR and SSIM performance for all mask ratios. These results demonstrate that semantic information aids in processing degraded images. Our approach overcomes the limitations imposed by noise in masked area appearance features by leveraging the guidance of semantic information.

Qualitative results from different models are presented in Fig. 3, showcasing significant enhancements achieved by our proposed method.

Figure 3 unmistakably illustrates that the implicit neural function models lacking semantic guidance tend to produce blurry reconstructions in the affected regions, often displaying a noticeable boundary between masked and unmasked areas. In contrast, models enriched with semantic information yield more visually coherent and pleasing results. As observed in the first row, it becomes apparent that traditional implicit neural functions like LIIF struggle to recover the 'eye' category when it is entirely masked. In such cases, the neighboring appearance features can only provide information about the 'face.' However, SAIR demonstrates its ability to reconstruct the 'eye' category effectively, benefitting from the restored semantic features. Furthermore, in the last row of Fig. 3, the original implicit neural function generates unexpected regions prominently.

Fig. 4. From left to right, we show the input masked image, masked semantic feature after CLIP encoder, and semantic feature after SIR.

In Fig. 4, we present a visual representation of the image features before and after the application of our SIR module. The pre-trained CLIP encoder cannot handle the masked regions ideally. And it becomes evident that our proposed SIR module effectively reconstructs the corrupted image feature.

To assess the impact of semantic information during the training process, we visually analyze the training progress of both LIIF and SAIR. The training loss curves depicted in Fig. 5 demonstrate that both models converge at a similar point. This observation suggests that the inclusion of semantic information can facilitate loss convergence without necessitating an extended training duration. Moreover, as seen in Fig. 5, the PSNR curve illustrates that the model enriched with semantic information consistently outperforms the original implicit representation model right from the outset.

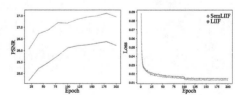

Fig. 5. PSNR vs Epoch and Training Loss vs Epoch.

5.3 Ablation Study

Study on Using Different Image Encoders. To demonstrate the compatibility of our semantic feature embedding with various image encoders, we conducted an ablation study in which we replaced our image encoder with the original LIIF encoder EDSR [23]. As indicated in Table 4, when compared to a model without the inclusion of semantic features (EDSR(wo)), the model that incorporates semantic features (EDSR(w)) also exhibited improvements, increasing the PSNR by 1.12 and the SSIM by 2.1%. These experiments provide compelling evidence that semantic information has the potential to enhance performance across different appearance feature spaces.

Table 4. Ablation study results on different image encoders and different implicit neural function models on CelebAHQ dataset.

Variant	All mask ratios	
	PSNR↑	SSIM↑
EDSR(wo)	30.26	0.892
EDSR(w)	**31.48**	**0.913**
LTE	30.60	0.931
SemLTE	**31.97**	**0.939**

Study on Using Different Implicit Neural Functions. In order to demonstrate the versatility of our semantic feature integration with various implicit neural functions, we conducted an ablation study using another implicit neural function known as LTE [15], which is specifically designed for image super-resolution tasks. In this study, we seamlessly incorporated semantic features into LTE, creating what we refer to as SemLTE. The resulting performance metrics are presented in Table 4, where SemLTE achieved significant improvements, elevating the PSNR to 31.97 and the SSIM to 93.9%. These outcomes affirm the adaptability of our proposed semantic implicit representation, showcasing its effectiveness when applied to different implicit neural functions.

Study on the Models with/without SIR Block. To further assess the effectiveness of SIR module, we conducted performance tests on the CLIP encoder, both with and without the SIR in the semantic segmentation task.

We use masked images as inputs to generate the segmentation results, which are compared with ground truth. In the setting 'without SIR', we initially employed the CLIP text encoder to produce category features **CLIP_T** $\in \mathbb{R}^{L \times C}$ for all categories in the dataset, where L represents the number of categories. Subsequently, we used **CLIP_T** to filter the semantic feature **CLIP_I**, yielding a pixel-wise segmentation map $S \in \mathbb{R}^{H \times W \times L}$. In the setting 'with SIR', we use the SIR block to reconstruct the CLIP semantic feature **CLIP_I**, the reconstructed feature is used for segmentation. The results presented in Table 5 indicate that the inclusion of the SIR block leads to a notable increase in mIoU by 0.28, demonstrating the effective capacity of the SIR model to reconstruct semantic features.

Table 5. Semantic segmentation results from the models with/without SIR on ADE20K dataset.

Variant	mIoU
CLIP Encoder	0.17
CLIP Encoder+ SIR	**0.45**

Study on not Filling the Semantic Feature (NFS). In the preceding section, we employed SIR to reconstruct the semantic feature of masked images. Here, we delve into an alternative scenario where we do not to fill in the masked semantic feature. In our experiments, we introduced masked semantic features into the implicit neural function alongside the appearance feature. However, as evident in the results presented in Table 6 under the label NFS, this approach yields suboptimal performance when compared to SAIR.

Specifically, it leads to a noticeable decrease of 2.04 in PSNR and a 2.1% reduction in SSIM. The presence of meaningless semantic information within the masked region exerts an adverse influence on the construction of the implicit representation. **Study on only using semantic feature to build implicit representation (OUS).** In this section, we explore the possibility of constructing a continuous representation using only semantic features, meaning that we exclusively input semantic information into the implicit neural function. The results is shown in Table 6 as the OUS. It's worth noting that the CLIP image encoder is trained to produce features that align with textual information. In essence, this experiment underscores the significance of integrating both semantic and image-level information to attain favorable outcomes in image generation tasks.

Table 6. Ablation study on *Not filling semantic feature* (NFS), *Only using semantic feature* (OUS), and SAM encoder on CelebAHQ dataset.

Variant	All mask ratios	
	PSNR↑	SSIM↑
NFS	30.32	0.923
OUS	31.11	0.929
SAM Encoder	31.72	0.935
SAIR	**32.36**	**0.944**

Using Other Semantic Embeddings. As an alternative to employing our Semantic Implicit Representation (SIR), we can also utilize existing models designed for semantic embeddings, such as the previously introduced semantic segmentation model SAM [13]. To demonstrate this, we replaced our CLIP encoder with the pre-trained SAM image encoder, and the results are presented in Table 6. Notably, it becomes evident that the CLIP encoder outperforms traditional semantic segmentation encoders in this context. This superiority is attributed to the CLIP encoder's capacity to capture rich textual information, further enhancing the inpainting task's performance.

Study on Using SAIR for Image Superresolution. Our approach extends beyond its primary application and is equally effective for image super-resolution tasks (mask ratio is 0). SIR block can significantly enhance the resolution of semantic maps to various higher levels. In comparison to existing method LIIF, our proposed technique demonstrates superior super-resolution capabilities, as evidenced by the PSNR/SSIM results presented in Table 7 across different upsample ratios on CelebAHQ dataset. All models are trained under × 4 setting. SAIR outperforms LIIF across all mask ratios, leveraging the wealth of semantic information embedded within the images.

Table 7. Ablation study results on using SAIR for image superresolution.

PSNR ↑ /SSIM ↑	Upscale ratio			
	× 2	× 4	× 6	× 8
LIIF	34.32/0.963	30.98/0.882	30.13/0.838	29.83/0.819
SAIR	**34.61/0.972**	**31.24**/0.881	**30.39/0.840**	**30.03/0.825**

Importance of Implicit Neural Function (INF). We design three variants of SAIR to validate the necessity of leveraging INR: ❶ We remove the INR from SIR by replacing Eq. 4 with a bilinear operation, i.e., $z_p^{sem} = SIR_{woINR}(I, q) = \sum_{q \in \mathcal{N}_p} W_q z_q^{sem}$, where $\{W_q\}$ is the set of bilinear weights. ❷ We remove the INR from AIR by replacing Eq. 5 with a convolutional layer and a bilinear operation, i.e., $c_p = AIR_{woINR}(I, M, SIR, p) = \sum_{q \in \mathcal{N}_p} W_q f_{cnn}([z_q^{app}, SIR(I, M, q)])$. Note that, SIR remains the same. ❸ We remove INR from both SIR and AIR by performing the above replacements at the same time. The experiments are conducted on CelebAHQ dataset. As shown in Table 8, once we remove INR from SIR or AIR, the performance decreases, demonstrating its superior capability compared to alternatives such as bilinear operations and CNN layers.

Table 8. Ablation study results on INF.

Variants	SIR w. INR	AIR w. INR	PSNR↑	SSIM↑
❶		✔	27.29	0.912
❷	✔		29.77	0.921
❸			26.09	0.893
SAIR	✔	✔	30.97	0.962

5.4 Discussion

Comparison with Diffusion Based Inpainting Methods. The diffusion model is a popular topic and has been widely used in image inpainting tasks. However, our proposed method surpasses diffusion-based approaches in three key aspects. First, inference speed. The inference speed of our method (0.043 s/image) is faster than that of the diffusion-based method, such as stable diffusion [35] (12 s/image). Second, arbitrary resolution. Our method can generate images at arbitrary resolutions during inference, a capability lacking in diffusion-based models. Third, high fidelity. Diffusion-based models prioritize naturalness over fidelity, resulting in generated results that deviate significantly from the ground truth. We compare our method with stable diffusion on the CelebAHQ dataset under the same setting, the results are PSNR ↑ 37.97 (ours) vs. 37.60 (stable diffusion) and L1 ↓ 0.01 (ours) vs. 0.032 (stable diffusion). The diffusion model tends to generate results with low fidelity.

6 Future Work

In this study, we have demonstrated the effectiveness of the Semantic-Aware Implicit Representation (SAIR) in the domain of image inpainting. While our proposed method has shown remarkable performance in this specific task, its broader applicability to other vision-related tasks has yet to be fully explored. As part of our future research endeavors, we plan to conduct additional experiments to assess the potential of our method in addressing various vision tasks beyond inpainting, such as segmentation and denoising.

7 Conclusion

In this paper, we tackle the limitations inherent in existing implicit representation techniques, which predominantly rely on appearance information and often falter when faced with severely degraded images. To address this challenge, we introduce a novel approach: the semantic-aware implicit representation (SAIR). By seamlessly using a semantic implicit representation (SIR) to handle the pixel-level semantic feature and a appearance implicit representation (AIR) to reconstruct the image color, our method effectively mitigates the impact of potentially degraded regions. To gauge the effectiveness of our approach, we conducted comprehensive experiments on two widely recognized datasets, CelebAHQ [26] and ADE20K [50]. The results unequivocally demonstrate that our method outperforms existing implicit representation and inpainting approaches by a substantial margin across four commonly employed image quality evaluation metrics. Our model's capacity to assist the implicit neural function in processing damaged images expands its utility and applicability, offering promising prospects for various image-related tasks.

Acknowledgments. This research is supported by the National Research Foundation, Singapore, and DSO National Laboratories under the AI Singapore Programme (AISG Award No: AISG2-GC-2023-008), Career Development Fund (CDF) of Agency for Science, Technology and Research (No.: C233312028), and National Research Foundation, Singapore and Infocomm Media Development Authority under its Trust Tech Funding Initiative (No. DTC-RGC-04).

References

1. Bar, A., Gandelsman, Y., Darrell, T., Globerson, A., Efros, A.: Visual prompting via image inpainting. In: NeurIPS, vol. 35, pp. 25005–25017 (2022)
2. Cao, Y., Li, T., Cao, X., Tsang, I., Liu, Y., Guo, Q.: IRAD: implicit representation-driven image resampling against adversarial attacks. In: ICLR (2024)
3. Chen, H., He, B., Wang, H., Ren, Y., Lim, S.N., Shrivastava, A.: Nerv: neural representations for videos. In: Ranzato, M., Beygelzimer, A., Dauphin, Y., Liang, P., Vaughan, J.W. (eds.) NeurIPS, vol. 34, pp. 21557–21568. Curran Associates, Inc. (2021)
4. Chen, J., et al.: LRR: language-driven resamplable continuous representation against adversarial tracking attacks. In: ICLR (2024)
5. Chen, Y., Liu, S., Wang, X.: Learning continuous image representation with local implicit image function. In: CVPR, pp. 8628–8638 (2021)
6. Chen, Z., Zhang, H.: Learning implicit fields for generative shape modeling. In: CVPR, pp. 5939–5948 (2019)

7. Feng, T., Feng, W., Li, W., Lin, D.: Cross-image context for single image inpainting. In: NeurIPS, vol. 35, pp. 1474–1487 (2022)
8. Grattarola, D., Vandergheynst, P.: Generalised implicit neural representations. arXiv preprint arXiv:2205.15674 (2022)
9. Guo, Q., Li, X., Juefei-Xu, F., Yu, H., Liu, Y., Wang, S.: Jpgnet: joint predictive filtering and generative network for image inpainting. In: ACM International Multimedia, pp. 386–394 (2021)
10. Guo, Z., Lan, C., Zhang, Z., Chen, Z., Lu, Y.: Versatile neural processes for learning implicit neural representations. arXiv preprint arXiv:2301.08883 (2023)
11. Ho, C.H., Vasconcelos, N.: Disco: adversarial defense with local implicit functions. arXiv preprint arXiv:2212.05630 (2022)
12. Hsu, J., Gu, J., Wu, G., Chiu, W., Yeung, S.: Capturing implicit hierarchical structure in 3D biomedical images with self-supervised hyperbolic representations. In: NeurIPS, vol. 34, pp. 5112–5123 (2021)
13. Kirillov, A., et al.: Segment anything. arXiv:2304.02643 (2023)
14. Lee, C.H., Liu, Z., Wu, L., Luo, P.: Maskgan: towards diverse and interactive facial image manipulation. In: CVPR (2020)
15. Lee, J., Jin, K.H.: Local texture estimator for implicit representation function. In: CVPR, pp. 1929–1938 (2022)
16. Li, J., Wang, N., Zhang, L., Du, B., Tao, D.: Recurrent feature reasoning for image inpainting. In: CVPR, pp. 7760–7768 (2020)
17. Li, W., Lin, Z., Zhou, K., Qi, L., Wang, Y., Jia, J.: Mat: mask-aware transformer for large hole image inpainting. In: CVPR, pp. 10758–10768 (2022)
18. Li, X., et al.: Leveraging inpainting for single-image shadow removal. In: Proceedings of the IEEE/CVF International Conference on Computer Vision, pp. 13055–13064 (2023)
19. Li, X., Guo, Q., Cai, P., Feng, W., Tsang, I., Wang, S.: Learning restoration is not enough: transfering identical mapping for single-image shadow removal. arXiv preprint arXiv:2305.10640 (2023)
20. Li, X., Guo, Q., Lin, D., Li, P., Feng, W., Wang, S.: Misf: multi-level interactive siamese filtering for high-fidelity image inpainting. In: CVPR, pp. 1869–1878 (2022)
21. Li, Z., Min, M.R., Li, K., Xu, C.: Stylet2i: toward compositional and high-fidelity text-to-image synthesis. In: CVPR, pp. 18197–18207 (2022)
22. Liao, L., Xiao, J., Wang, Z., Lin, C.W., Satoh, S.: Uncertainty-aware semantic guidance and estimation for image inpainting. IEEE J. Sel. Top. Signal Process. **15**(2), 310–323 (2020)
23. Lim, B., Son, S., Kim, H., Nah, S., Mu Lee, K.: Enhanced deep residual networks for single image super-resolution. In: CVPR Workshop, pp. 136–144 (2017)
24. Lin, Y., Xie, Y., Chen, D., Xu, Y., Zhu, C., Yuan, L.: Revive: regional visual representation matters in knowledge-based visual question answering. arXiv preprint arXiv:2206.01201 (2022)
25. Liu, G., Reda, F.A., Shih, K.J., Wang, T.C., Tao, A., Catanzaro, B.: Image inpainting for irregular holes using partial convolutions. In: ECCV, pp. 85–100 (2018)
26. Liu, Z., Luo, P., Wang, X., Tang, X.: Deep learning face attributes in the wild. In: ICCV (2015)
27. Lu, Y., et al.: Snowvision: segmenting, identifying, and discovering stamped curve patterns from fragments of pottery. Int. J. Comput. Vision **130**(11), 2707–2732 (2022)
28. Lüddecke, T., Ecker, A.: Image segmentation using text and image prompts. In: CVPR, pp. 7086–7096 (2022)

29. Mildenhall, B., Srinivasan, P.P., Tancik, M., Barron, J.T., Ramamoorthi, R., Ng, R.: Nerf: representing scenes as neural radiance fields for view synthesis. Commun. ACM **65**(1), 99–106 (2021)
30. Nazeri, K., Ng, E., Joseph, T., Qureshi, F., Ebrahimi, M.: Edgeconnect: structure guided image inpainting using edge prediction. In: ICCV Workshops (2019)
31. Ni, M., Li, X., Zuo, W.: NUWA-LIP: language-guided image inpainting with defect-free VQGAN. In: CVPR, pp. 14183–14192 (2023)
32. Radford, A., et al.: Learning transferable visual models from natural language supervision. In: International Conference on Machine Learning, pp. 8748–8763. PMLR (2021)
33. Ramesh, A., Dhariwal, P., Nichol, A., Chu, C., Chen, M.: Hierarchical text-conditional image generation with clip latents. arXiv preprint arXiv:2204.06125, vol. 1, no. 2, p. 3 (2022)
34. Ren, Y., Yu, X., Zhang, R., Li, T.H., Liu, S., Li, G.: Structureflow: image inpainting via structure-aware appearance flow. In: CVPR, pp. 181–190 (2019)
35. Rombach, R., Blattmann, A., Lorenz, D., Esser, P., Ommer, B.: High-resolution image synthesis with latent diffusion models (2021)
36. Su, K., Chen, M., Shlizerman, E.: Inras: implicit neural representation for audio scenes. In: NeurIPS, vol. 35, pp. 8144–8158 (2022)
37. Suvorov, R., et al.: Resolution-robust large mask inpainting with fourier convolutions. In: WACV (2022)
38. Tao, M., Tang, H., Wu, F., Jing, X.Y., Bao, B.K., Xu, C.: DF-GAN: a simple and effective baseline for text-to-image synthesis. In: CVPR, pp. 16515–16525 (2022)
39. Wang, Y., Tao, X., Qi, X., Shen, X., Jia, J.: Image inpainting via generative multi-column convolutional neural networks. In: NeurIPS, vol. 31 (2018)
40. Xie, Z., Zhang, J., Li, W., Zhang, F., Zhang, L.: S-nerf: neural radiance fields for street views. arXiv preprint arXiv:2303.00749 (2023)
41. Xu, J., et al.: Groupvit: semantic segmentation emerges from text supervision. In: CVPR, pp. 18134–18144 (2022)
42. Yang, Z., et al.: Tap: text-aware pre-training for text-VQA and text-caption. In: CVPR, pp. 8751–8761 (2021)
43. Yariv, L., Gu, J., Kasten, Y., Lipman, Y.: Volume rendering of neural implicit surfaces. In: NeurIPS, vol. 34, pp. 4805–4815 (2021)
44. Yin, F., Liu, W., Huang, Z., Cheng, P., Chen, T., Yu, G.: Coordinates are not lonely–codebook prior helps implicit neural 3D representations. arXiv preprint arXiv:2210.11170 (2022)
45. Zhang, C., et al.: Superinpaint: learning detail-enhanced attentional implicit representation for super-resolutional image inpainting. arXiv preprint arXiv:2307.14489 (2023)
46. Zhang, L., Chen, Q., Hu, B., Jiang, S.: Text-guided neural image inpainting. In: ACM Multimedia, pp. 1302–1310 (2020)
47. Zhang, R., Isola, P., Efros, A.A., Shechtman, E., Wang, O.: The unreasonable effectiveness of deep features as a perceptual metric. In: CVPR (2018)
48. Zhao, M., et al.: Towards video text visual question answering: benchmark and baseline. In: Thirty-Sixth Conference on Neural Information Processing Systems Datasets and Benchmarks Track (2022)
49. Zhenxing, M., Xu, D.: Switch-nerf: learning scene decomposition with mixture of experts for large-scale neural radiance fields. In: ICLR (2022)
50. Zhou, B., Zhao, H., Puig, X., Fidler, S., Barriuso, A., Torralba, A.: Scene parsing through ade20k dataset. In: CVPR, pp. 633–641 (2017)

51. Zhou, C., Loy, C.C., Dai, B.: Extract free dense labels from clip. In: Avidan, S., Brostow, G., Cissé, M., Farinella, G.M., Hassner, T. (eds.) ECCV 2022. LNCS, vol. 13688, pp. 696–712. Springer, Cham (2022). https://doi.org/10.1007/978-3-031-19815-1_40
52. Zhou, Z., Lei, Y., Zhang, B., Liu, L., Liu, Y.: Zegclip: towards adapting clip for zero-shot semantic segmentation. In: CVPR, pp. 11175–11185 (2023)
53. Zhu, Y., et al.: One model to edit them all: free-form text-driven image manipulation with semantic modulations. In: NeurIPS, vol. 35, pp. 25146–25159 (2022)

ColorMNet: A Memory-Based Deep Spatial-Temporal Feature Propagation Network for Video Colorization

Yixin Yang, Jiangxin Dong, Jinhui Tang, and Jinshan Pan(✉)

Nanjing University of Science and Technology, Nanjing, China
{yangyixin,jxdong,jinhuitang,jspan}@njust.edu.cn

Abstract. How to effectively explore spatial-temporal features is important for video colorization. Instead of stacking multiple frames along the temporal dimension or recurrently propagating estimated features that will accumulate errors or cannot explore information from far-apart frames, we develop a memory-based feature propagation module that can establish reliable connections with features from far-apart frames and alleviate the influence of inaccurately estimated features. To extract better features from each frame for the above-mentioned feature propagation, we explore the features from large-pretrained visual models to guide the feature estimation of each frame so that the estimated features can model complex scenarios. In addition, we note that adjacent frames usually contain similar contents. To explore this property for better spatial and temporal feature utilization, we develop a local attention module to aggregate the features from adjacent frames in a spatial-temporal neighborhood. We formulate our memory-based feature propagation module, large-pretrained visual model guided feature estimation module, and local attention module into an end-to-end trainable network (named ColorMNet) and show that it performs favorably against state-of-the-art methods on both the benchmark datasets and real-world scenarios. Our source codes and pre-trained models are available at: https://github.com/yyang181/colormnet.

Keywords: Exemplar-based video colorization · Deep convolutional neural network · Feature propagation

1 Introduction

Due to the technical limitations of old imaging devices, lots of videos captured in the last century are in black and white, making them less visually appealing on modern display devices. As most of these videos have historical values and are difficult to reproduce, it is of great need to colorize them.

Restoring high-quality colorized videos is challenging as it not only needs to handle the colorization of each frame but also requires exploring temporal information from the video sequences. Therefore, directly applying existing

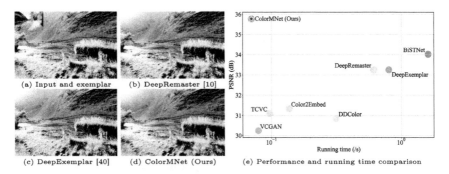

Fig. 1. Colorization results on a real-world video and model performance comparisons between our proposed ColorMNet and other methods on the DAVIS [26] dataset in terms of PSNR and running time. State-of-the-art methods [10,40] do not generate well-colorized images in (b) and (c). In contrast, by exploring the features from large-pretrained visual models to estimate robust spatial features for each frame, effectively propagating these features along the temporal dimension based on memory mechanisms for far-apart frames, and exploiting the video property that adjacent frames contain similar contents, our method accurately restores the colors on the grass and generates a realistic image in (d). (e) shows that the proposed ColorMNet performs favorably against state-of-the-art methods in terms of accuracy and running time. The size of the test images for measuring the running time is 960×536 pixels. (Color figure online)

image colorization methods [4,11,13,17,20,30,36,41,43] does not generate satisfactory colorized videos as minor perturbations in consecutive input video frames may lead to substantial differences in colorized video results. To overcome this, numerous methods model the temporal information from inter-frames by stacking multiple frames along the temporal dimension [1,10] or recurrently propagating features [18,32,34,40]. Although these approaches show better performance than the ones based on single image colorization, stacking multiple frames along the temporal dimension cannot effectively leverage spatial-temporal prior from adjacent frames and requires a large amount of GPU memory. In addition, recurrent-based feature propagation is not able to effectively explore long-range information, leading to unsatisfactory results for frames far apart.

To better explore long-range temporal information, several approaches [21,37] develop bidirectional recurrent-based feature propagation methods for video colorization. As the recurrent-based feature propagation treats the features of each frame equally, if the features are not estimated accurately, the errors will accumulate, thus affecting the final video colorization. Therefore, it is still challenging to effectively model temporal information from long-range frames.

In addition to the temporal information exploration, how to extract good features from each frame plays a significant role in video colorization. Existing methods [1,18,32,37,40,45] usually utilize a pretrained VGG [14] or ResNet-101 [7] to extract features from each frame. These methods are able to model local structures but are less effective for exploiting non-local and semantic structures, *e.g.*, complex scenes with multiple objects. To restore high-quality videos, it is of great interest to develop a better feature representation method that is able to characterize the non-local and semantic properties of each frame.

In this paper, we present a memory-based deep spatial-temporal feature propagation network for video colorization. Note that the robustness of the spatial features extracted from each frame is important, we first develop a large-pretrained visual model guided feature estimation (PVGFE) module, which is motivated by the success of large-pretrained visual models [23] in generating robust visual features to facilitate the spatial feature estimation. However, simply recurrently propagating the estimated spatial features or directly stacking them along the temporal dimension does not effectively explore temporal information for video colorization. Moreover, it requires a large amount of GPU memory capacity to store past frame representations when the spatial resolution is large and videos are long. To overcome these problems, we then propose a memory-based feature propagation (MFP) module that can not only adaptively explore and propagate useful features from far-apart frames but also reduce memory consumption. In addition, we note that adjacent frames of a video usually contain similar contents and thus develop a local attention (LA) module to better utilize the spatial and temporal features. Taken together, the memory-based deep spatial-temporal feature propagation network, called ColorMNet, is able to generate high-quality video colorization results (see Fig. 1(e)).

The main contributions are summarized as follows:

- We propose a large-pretrained visual model guided feature estimation module to model non-local and semantic structures of each frame for colorization.
- We develop a memory-based feature propagation module to adaptively explore temporal features from far-apart frames and reduce memory usage.
- We develop a local attention module to explore similar contents of adjacent frames for better video colorization.
- We formulate the proposed network into an end-to-end trainable framework and show that it performs favorably against state-of-the-art methods on both the benchmark datasets and real-world scenarios.

2 Related Work

User-Guided Image Colorization. Since the colorization problem is ill-posed, conventional image colorization methods usually adopt local user hints [2,8,20, 22,27,28,38,43,44] to make this problem well-posed. However, these methods do not fully exploit the property of video sequences and usually need to solve temporal consistency problems. In addition, their colorization performance for individual frame is usually far from satisfactory as estimating global and semantic features is challenging.

Automatic Video Colorization. Instead of using user-guided image methods, several approaches explore deep learning to solve video colorization. In [18,45], both the colorization performance and the temporal consistency are enhanced by recurrently propagating features from adjacent frames for video colorization. In [21], Liu et al.use bidirectional propagation to explore temporal features and introduces a self-regularization learning scheme to minimize the prediction difference obtained with different time steps. Although using feature propagation improves the temporal consistency, it is not a trivial task to estimate the spatial features from each frame due to the inherent complexity of scenes in videos. Moreover, these automatic video colorization methods [25,29,36] may work well on synthetic datasets but lack generality on diverse real-world scenarios.

Exemplar-Based Video Colorization. Exemplar-based methods aim to generate videos that are faithful to the exemplar with enhanced temporal consistency. In [40], Zhang et al.employ a recurrent-based feature propagation module to explore temporal features for latent frame restoration. In [10], Iizuka et al.propose stacking multiple frames along the temporal dimension to obtain better performance. To better explore spatial and temporal features, Chen et al. [1] adopt ResNets [7] instead of the commonly-used VGG [14] model for better spatial feature estimation and split video sequences into frame blocks for long-term spatiotemporal dependency. In [37], Yang et al.use bidirectional propagation to gradually propagate features and use optical flow models [31] to align adjacent frames. Recurrently propagating information from long-range frames or stacking multiple frames improves the colorization performance. However, if there are inaccurately estimated features of long-range frames, the errors will be accumulated, which thus affects video colorization. Additionally, when dealing with long videos, these methods often consume significant GPU memory and suffer from slow inference speeds, posing substantial challenges to video colorization.

3 ColorMNet

We denote the input grayscale frames as $\{X_i\}_{i=1}^{N}$ (where $X_i \in \mathbb{R}^{H \times W \times 1}$, $H \times W$ denotes the spatial resolution, N is the number of frames of the input video), and the given colorful exemplar as R (where $R \in \mathbb{R}^{H \times W \times 3}$). Our goal is to develop an effective and efficient video colorization method to restore colorful videos $\{I_i\}_{i=1}^{N}$ (where $I_i \in \mathbb{R}^{H \times W \times 3}$) with low GPU memory requirements. The proposed ColorMNet contains a large-pretrained visual model guided feature estimation (PVGFE) module to extract spatial features from each frame, a memory-based feature propagation (MFP) module that is able to adaptively explore the temporal features from far-apart frames, and a local attention (LA) module that is used to explore the similar contents from adjacent frames for better spatial and temporal feature utilization. Figure 2 illustrates the overview of the proposed method. In the following, we explain each module in detail.

3.1 PVGFE Module

To estimate robust spatial features, we explore the features learned from large-pretrained visual models as they are able to model the non-local and semantic information and are robust to numerous scenarios. Note that one of the large-pretrained visual models, i.e., DINOv2 [23], adopts ViT [5] as the main feature

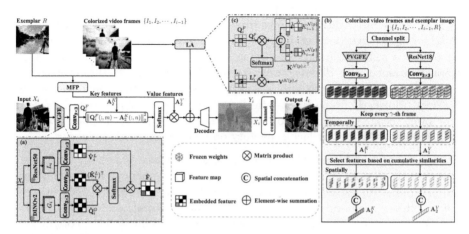

Fig. 2. An overview of the proposed ColorMNet. The core components of our method include: **(a)** large-pretrained visual model guided feature estimation (PVGFE) module, **(b)** memory-based feature propagation (MFP) module and **(c)** local attention (LA).

extractor and generates all-purpose visual features facilitating both image-level (classification) and pixel level (segmentation) tasks due to its robust feature representation ability. In this paper, we utilize the global features learned from DINOv2 to guide the local features learned from CNNs for better feature estimation of each frame.

Given the input grayscale frames $\{X_i\}_{i=1}^{N}$, we first extract features $\{G_i\}_{i=1}^{N}$ and $\{L_i\}_{i=1}^{N}$ by applying the pretrained DINOv2 and ResNet50 [7] to $\{X_i\}_{i=1}^{N}$, respectively. Then we use the cross attention [39] to fuse $\{G_i\}_{i=1}^{N}$ and $\{L_i\}_{i=1}^{N}$ for obtaining robust spatial features.

Specifically, we first extract the query feature Q_i^G from G_i, the key feature K_i^L and the value feature V_i^L from L_i by:

$$Q_i^G = \text{Conv}_{3\times3}(G_i), \tag{1a}$$

$$K_i^L = \text{Conv}_{3\times3}(L_i), \tag{1b}$$

$$V_i^L = \text{Conv}_{3\times3}(L_i), \tag{1c}$$

where $\{Q_i^G, K_i^L, V_i^L\} \in \mathbb{R}^{\hat{H}\times\hat{W}\times\hat{C}}$, $\hat{H}\times\hat{W}$ and $\hat{C}$ denote the spatial and channel dimensions, respectively; $\text{Conv}_{3\times3}(\cdot)$ denotes a convolution with the filter size of 3×3 pixels. We denote the matrix forms of Q_i^G, K_i^L, V_i^L as $\hat{\mathbf{Q}}_i^G$, $\hat{\mathbf{K}}_i^L$, $\hat{\mathbf{V}}_i^L$, and obtain the fused feature by:

$$\hat{\mathbf{F}}_i = \text{softmax}\left(\frac{\hat{\mathbf{Q}}_i^G(\hat{\mathbf{K}}_i^L)^\top}{\alpha}\right)\hat{\mathbf{V}}_i^L, \tag{2}$$

where $\hat{\mathbf{Q}}_i^G \in \mathbb{R}^{\hat{C}\times\hat{H}\hat{W}}$, $\hat{\mathbf{K}}_i^L \in \mathbb{R}^{\hat{C}\times\hat{H}\hat{W}}$ and $\hat{\mathbf{V}}_i^L \in \mathbb{R}^{\hat{C}\times\hat{H}\hat{W}}$ are obtained by reshaping tensors [39] from the original size $\mathbb{R}^{\hat{H}\times\hat{W}\times\hat{C}}$; softmax$(\cdot)$ denotes the softmax operation that is applied to each row of the matrix; α is a scaling factor.

In our implementation, we take the features of the last 4 layers of ViT-S/14 from DINOv2 and concatenate them together in channel dimension as the feature G_i. For the feature L_i, we take stage-4 features with stride 16 from the base ResNet50 [7]. Finally, the feature $\hat{\mathbf{F}}_i \in \mathbb{R}^{\hat{C} \times \hat{H}\hat{W}}$ is reshaped into $F_i \in \mathbb{R}^{\hat{H} \times \hat{W} \times \hat{C}}$ for the following processing.

3.2 MFP Module

Inspired by the efficient mechanisms of human brains in memorizing long-term information [3], i.e., paying more attention to frequently used information, we propose a memory-based feature propagation module to effectively and efficiently explore temporal features by establishing reliable connections with features from far-apart frames and alleviating the influence of inaccurately estimated features.

Assuming that we have predicted $i-1$ color frames $\{I_1, \cdots, I_{i-1}\}$ with their chrominance channels $\{Y_1, \cdots, Y_{i-1}\}$ and luminance channels $\{X_1, \cdots, X_{i-1}\}$ (i.e., the input grayscale frames), we aim to estimate the chrominance channel Y_i of the i-th color frame I_i based on $\{Y_1, \cdots, Y_{i-1}\}$ and the given exemplar R.

First, we extract features $\{F_1, F_2, \cdots, F_i\}$ and F_r from $\{X_1, X_2, \cdots, X_i\}$ and the luminance channel X_r of R using the proposed PVGFE module for global and semantic features in the spatial dimension. Meanwhile, we extract features $\{E_1, \cdots, E_{i-1}\}$ and E_r from $\{Y_1, \cdots, Y_{i-1}\}$ and the chrominance channel Y_r of R using a lightweight pretrained ResNet18 [7]. Then, we generate the embedded query, key, and value features by:

$$Q_i^F = \text{Conv}_{3\times 3}(F_i), \tag{3a}$$

$$\{K_j^F\}_{j=1}^{i-1} = \{\text{Conv}_{3\times 3}(F_j)\}_{j=1}^{i-1}, \tag{3b}$$

$$K_r^F = \text{Conv}_{3\times 3}(F_r), \tag{3c}$$

$$\{V_j^E\}_{j=1}^{i-1} = \{\text{Conv}_{3\times 3}(E_j)\}_{j=1}^{i-1}, \tag{3d}$$

$$V_r^E = \text{Conv}_{3\times 3}(E_r). \tag{3e}$$

As $\{K_1^F, \cdots, K_{i-1}^F\}$ and $\{V_1^E, \cdots, V_{i-1}^E\}$ contain valuable historical information of previously colorized video frames $\{I_1, \cdots, I_{i-1}\}$, they facilitate the establishment of long-range temporal correspondences between the contents of the current frame and those of the previously colorized video frames. However, memorizing all of the historical frames results in significant GPU memory consumption, especially as the number of colorized frames increases. As a trade-off between long-range correspondence and memory consumption, we keep every γ frame and discard the remaining ones that contain similar contents to obtain more temporally compact and representative features.

In particular, given that adjacent frames are mutually redundant, we first merge the temporal features by concatenating every γ frame, thereby establishing reliable connections with far-apart frames under constrained memory consumption, and obtain the aggregated key features $\mathbf{A}_1^K$ by:

$$\mathbf{A}_1^K = \text{Concat}(\mathbf{K}_\gamma^F, \mathbf{K}_{2\gamma}^F, \cdots, \mathbf{K}_{z\gamma}^F), \tag{4}$$

where $\{\mathbf{K}_\gamma^F, \mathbf{K}_{2\gamma}^F, \cdots, \mathbf{K}_{z\gamma}^F\}$ are the matrix forms of $\{K_\gamma^F, K_{2\gamma}^F, \cdots, K_{z\gamma}^F\}$, $z = \lfloor \frac{i-1}{\gamma} \rfloor$, $\lfloor \cdot \rfloor$ denotes the round down operation, and $\text{Concat}(\cdot)$ denotes the spatial dimension concatenation operation.

Note that errors are likely to accumulate when the features of some frames are not estimated accurately. In addition, the high dimension ($\hat{H}\hat{W}$) of $\mathbf{A}_1^K$ makes it impossible to handle lots of video frames with limited GPU memory capacity. To overcome these problems, we then aggregate $\mathbf{A}_1^K$ into a more spatially compact and less error-prone form by selecting better features with higher usage. However, the computation of the feature usage frequency relies on a sufficient number of colorized frames. Therefore, when the number of colorized frames is small (*i.e.*, $z < N_s$), we directly use $\mathbf{A}_2^K$ as the output of our proposed MFP module, which is defined as:

$$\mathbf{A}_2^K = \text{Concat}\left(\mathbf{K}_r^F, \mathbf{A}_1^K\right), \tag{5}$$

where $\mathbf{K}_r^F$ is the matrix form of K_r^F. When a sufficient number of frames is colorized (*i.e.*, $z = N_s$), we aggregate $\mathbf{A}_1^K$ by compressing the earlier N_e frames to reduce GPU memory consumption and alleviate the influence of inaccurately estimated features for better video colorization. After the aggregation, the total number of aggregated frames reduced from $z = N_s$ to $z = N_s - N_e$ and we use $\mathbf{A}_2^K$ in (5) as the output of the MFP module until z reaches N_s again. In the following, we explain the situation when $z = N_s$ in detail.

We first define the cumulative similarities $\{\mathbf{S}_\gamma, \cdots, \mathbf{S}_{z\gamma}\}$ for the key features $\{\mathbf{K}_\gamma^F, \cdots, \mathbf{K}_{z\gamma}^F\}$, which are aggregated in (4), as:

$$\mathbf{S}_\gamma = \sum_{j=\gamma+1}^{i-1} \sum_{m=1}^{\hat{H}\hat{W}} \left(\text{softmax}\left(-\mathbf{C}^j\right)\right), \mathbf{C}^j = (\mathbf{C}_{m,n}^j), \tag{6a}$$

$$\mathbf{C}_{m,n}^j = \left\|\mathbf{K}_j^F(:,m) - \mathbf{K}_\gamma^F(:,n)\right\|_2^2, \tag{6b}$$

where $\mathbf{S}_\gamma \in \mathbb{R}^{1 \times \hat{H}\hat{W}}$ and $\mathbf{C}^j \in \mathbb{R}^{\hat{H}\hat{W} \times \hat{H}\hat{W}}$, $\mathbf{K}_j^F(:,m)$ denotes the m-th feature vector in $\mathbf{K}_j^F$ and $\mathbf{K}_\gamma^F(:,n)$ denotes the n-th feature vector in $\mathbf{K}_\gamma^F$, $\{\mathbf{K}_j^F, \mathbf{K}_\gamma^F\} \in \mathbb{R}^{\hat{C}^k \times \hat{H}\hat{W}}$, $\|\cdot\|_2$ denotes the Euclidean distance calculation. Then, we normalize $\mathbf{S}_\gamma$ by dividing it by the number of frames in $\{\mathbf{K}_j^F\}_{j=\gamma+1}^{i-1}$ for fairness consideration and obtain $\mathbf{S}_\gamma^{'} = \frac{\mathbf{S}_\gamma}{i-1-\gamma}$, which can be utilized to estimate the probability that the feature $\mathbf{K}_\gamma^F$ is accurately predicted as $\mathbf{S}_\gamma^{'}$ describes how frequently the feature is used. We further define a top-M operation $\mathcal{T}_M(\cdot)$ to select the best M pixels of all features from the earlier N_e frames $\{\mathbf{K}_\gamma^F, \cdots, \mathbf{K}_{N_e\gamma}^F\}$, based on the highest M values in $\{\mathbf{S}_\gamma^{'}, \cdots, \mathbf{S}_{N_e\gamma}^{'}\}$ as:

$$\mathbf{K}^c = \text{Concat}\left(\mathbf{K}_\gamma^F, \cdots, \mathbf{K}_{N_e\gamma}^F\right), \tag{7a}$$

$$\mathbf{S}^c = \text{Concat}\left(\mathbf{S}_\gamma^{'}, \cdots, \mathbf{S}_{N_e\gamma}^{'}\right), \tag{7b}$$

$$\mathcal{T}_M\left(\mathbf{K}^c\right) = \text{Concat}\left(\mathbf{K}^c(:,1), \cdots, \mathbf{K}^c(:,M)\right), \tag{7c}$$

where $\mathbf{K}^c \in \mathbb{R}^{\hat{C}^k \times N_e \hat{H}\hat{W}}$, $\mathbf{S}^c \in \mathbb{R}^{1 \times N_e \hat{H}\hat{W}}$, and $\{\mathbf{K}^c(:,1), \cdots, \mathbf{K}^c(:,M)\}$ denote the M feature vectors in $\mathbf{K}^c$ that satisfy $\{\mathbf{S}^c(:,1), \cdots, \mathbf{S}^c(:,M)\}$ are the top-M values in $\mathbf{S}^c$. Then we obtain the aggregated key features $\mathbf{A}_2^K$ by:

$$\mathbf{A}_2^K = \text{Concat}\left(\mathcal{T}_M\left(\mathbf{K}^c\right), \mathbf{K}_r^F, \mathbf{K}_{(N_e+1)\gamma}^F, \cdots, \mathbf{K}_{z\gamma}^F\right), \quad (8)$$

where $\mathbf{A}_2^K \in \mathbb{R}^{\hat{C}^k \times T}, T = M + (1 + z - N_e)\hat{H}\hat{W}$. Similarly, we obtain the aggregated value features $\mathbf{A}_2^V \in \mathbb{R}^{\hat{C}^v \times T}$ by aggregating $\{V_r^E, \{V_j^E\}_{j=1}^{i-1}\}$.

To fully exploit the temporal information contained in $\mathbf{A}_2^K$ and $\mathbf{A}_2^V$, we use a commonly used L_2 similarity to find the features in $\mathbf{A}_2^K$ that are most similar to $\mathbf{Q}_i^F$, where $\mathbf{Q}_i^F \in \mathbb{R}^{\hat{C}^k \times \hat{H}\hat{W}}$ is the matrix form of Q_i^F, by:

$$\mathbf{W}_i = \text{softmax}\left(-\mathbf{D}^i\right), \mathbf{D}^i = (\mathbf{D}_{m,n}^i), \quad (9a)$$

$$\mathbf{D}_{m,n}^i = \left\|\mathbf{Q}_i^F(:,m) - \mathbf{A}_2^K(:,n)\right\|_2^2, \quad (9b)$$

where $\{\mathbf{W}_i, \mathbf{D}^i\} \in \mathbb{R}^{\hat{H}\hat{W} \times T}$, $\mathbf{Q}_i^F(:,m)$ denotes the m-th feature vector in $\mathbf{Q}_i^F$ and $\mathbf{A}_2^K(:,n)$ denotes the n-th feature vector in $\mathbf{A}_2^K$. Then, we reconstruct the i-th estimated feature $\mathbf{V}_i \in \mathbb{R}^{\hat{C}^v \times \hat{H}\hat{W}}$ of the chrominance channel by mapping the aggregated value features $\mathbf{A}_2^V$ based on $\mathbf{W}_i$ as:

$$\mathbf{V}_i = \mathbf{A}_2^V (\mathbf{W}_i)^\top. \quad (10)$$

Finally, we obtain the estimated feature V_i by reshaping $\mathbf{V}_i$ to its original size $\mathbb{R}^{\hat{H} \times \hat{W} \times \hat{C}^v}$. In the following, we enhance V_i by a local attention (LA) module.

3.3 LA Module

As adjacent frames contain similar contents that may be useful to complement the long-range information captured by MFP, we develop a local attention (LA) module to explore better spatial-temporal features.

We first use $Q_i^p \in \mathbb{R}^{1 \times 1 \times \hat{C}^k}$ to represent the feature Q_i^F in (3a) at the spatial location $p \in \mathbb{R}^{\hat{H} \times \hat{W}}$. Next, we formulate the past d key features in (3b) and past d value features in (3d) of the spatial-temporal neighborhood corresponding to Q_i^p as:

$$K^{\mathcal{N}(p),c} = \text{Concat}(K_{i-d}^{\mathcal{N}(p)}, \cdots, K_{i-1}^{\mathcal{N}(p)}), \quad (11a)$$

$$V^{\mathcal{N}(p),c} = \text{Concat}(V_{i-d}^{\mathcal{N}(p)}, \cdots, V_{i-1}^{\mathcal{N}(p)}), \quad (11b)$$

where $K_j^{\mathcal{N}(p)}$ and $V_j^{\mathcal{N}(p)}$ denote the j-th key feature and value feature corresponding to a $\lambda \times \lambda$ patch $\mathcal{N}(p)$ centered at p, $K^{\mathcal{N}(p),c} \in \mathbb{R}^{d \times \lambda \times \lambda \times \hat{C}^k}$ and $V^{\mathcal{N}(p),c} \in \mathbb{R}^{d \times \lambda \times \lambda \times \hat{C}^v}$. Then we apply the local attention to Q_i^p with $K^{\mathcal{N}(p),c}$ and $V^{\mathcal{N}(p),c}$ and obtain the output of LA module at location p as:

$$\mathbf{L}_i^p = \text{softmax}\left(\frac{Q_i^p(K^{\mathcal{N}(p),c})^\top}{\beta}\right) V^{\mathcal{N}(p),c}, \quad (12)$$

Table 1. Quantitative comparisons of the proposed method against state-of-the-art ones on the DAVIS [26] validation set (short frame length), the Videvo [16] validation set (medium frame length) and the NVCC2023 [12] validation set (long frame length). Our method achieves favorable performance in most of the metrics. Top 1_{st} and 2_{nd} results are marked in bold and underscore respectively. * denotes that we apply DVP [19] to the results of Color2Embed and DDColor as post-processing method. Note that for the LPIPS matrix on Videvo, we examine the performance of our method and BiSTNet in additional decimal places to determine the best one and the second best one as the performance of these two methods appears to be identical when displayed in the table with a limited number of decimal places. † denotes that two exemplars are used.

Methods	DDColor [13]	Color2Embed [44]	DDColor* [13]	Color2Embed* [44]	VCGAN [45]	TCVC [21]	DeepExemplar [40]	DeepRemaster [10]	BiSTNet† [37]	ColorMNet (Ours)
Categories			Image-based		Fully-automatic			Exemplar-based		
				Effectiveness evaluation						
DAVIS										
PSNR (dB)↑	30.84	31.33	30.67	31.05	30.24	31.10	33.24	33.25	<u>34.02</u>	**35.77**
FID↓	65.13	101.08	86.96	118.11	128.48	116.61	69.56	92.28	<u>44.69</u>	**38.39**
SSIM↑	0.926	0.951	0.936	0.943	0.924	0.955	0.950	0.961	<u>0.964</u>	**0.970**
LPIPS↓	0.085	0.076	0.086	0.086	0.100	0.080	0.062	0.060	<u>0.043</u>	**0.035**
FVD↓	687.33	864.11	1036.92	1037.21	557.02	1051.85	472.43	557.03	<u>333.56</u>	**180.84**
Colorfulness↑	**34.95**	20.49	27.13	20.27	15.32	20.96	21.08	19.15	26.23	<u>28.94</u>
Videvo										
PSNR (dB)↑	30.76	31.65	30.60	31.62	30.62	31.29	33.11	32.95	<u>34.12</u>	**34.35**
FID↓	45.08	66.73	50.80	73.02	97.86	80.74	54.93	68.15	<u>32.25</u>	**30.76**
SSIM↑	0.925	0.958	0.940	0.960	0.934	0.956	0.956	0.964	<u>0.968</u>	**0.972**
LPIPS↓	0.079	0.060	0.077	0.061	0.085	0.068	0.053	0.053	<u>0.036</u>	**0.036**
FVD↓	568.66	694.07	754.17	790.23	1104.69	989.14	479.38	503.11	<u>190.82</u>	**149.95**
Colorfulness↑	**31.70**	17.20	23.48	16.50	13.94	18.41	18.01	15.38	22.36	<u>24.83</u>
NVCC2023										
PSNR (dB)↑	29.95	30.90	30.16	30.66	29.77	30.45	32.03	32.25	<u>33.18</u>	**33.26**
FID↓	48.50	65.92	95.83	70.35	86.59	72.76	35.39	53.03	<u>25.55</u>	**20.16**
SSIM↑	0.888	0.933	0.911	0.927	0.866	0.935	0.930	<u>0.951</u>	0.949	**0.959**
LPIPS↓	0.114	0.091	0.099	0.098	0.130	0.089	0.073	0.071	<u>0.054</u>	**0.039**
FVD↓	704.16	776.82	857.21	866.37	988.39	898.79	301.32	426.91	<u>151.43</u>	**80.03**
Colorfulness↑	**45.37**	28.06	36.43	26.55	20.56	33.99	31.08	25.01	38.06	<u>39.30</u>
				Efficiency evaluation						
Running time (/s)	0.31	0.13	2.35	2.03	<u>0.08</u>	0.09	0.80	0.61	1.62	**0.07**
Memory (G)	/	/	/	/	5.63	<u>4.24</u>	19.0	16.6	34.9	**1.9**
Parameters (M)	227.9	176.54	227.9	176.54	96.57	198.38	<u>59.74</u>	54.28	158.17	123.61
				Temporal consistency evaluation						
CDC↓	0.007440	0.003961	0.003892	0.003947	0.005506	0.003871	0.003876	0.004285	<u>0.003870</u>	**0.003763**

where $\mathbf{Q}_i^p \in \mathbb{R}^{1 \times \hat{C}^k}$, $\mathbf{K}^{\mathcal{N}(p),c} \in \mathbb{R}^{d\lambda^2 \times \hat{C}^k}$ and $\mathbf{V}^{\mathcal{N}(p),c} \in \mathbb{R}^{d\lambda^2 \times \hat{C}^v}$ are matrices obtained by reshaping Q_i^p, $K^{\mathcal{N}(p),c}$ and $V^{\mathcal{N}(p),c}$ from their original size, β is the scaling factor. Finally, we obtain the feature $L_i \in \mathbb{R}^{\hat{H} \times \hat{W} \times \hat{C}^v}$ by reshaping $\mathbf{L}_i$ to its original size.

To restore the colors of the input frame X_i, we further adopt a simple but effective decoder. Specifically, we use a decoder $\mathcal{D}(\cdot)$ consisting of ResBlocks [7] followed by up-sampling interpolation layers to gradually refine the enhanced feature $V_i + L_i$ and obtain the predicted i-th chrominance channel Y_i as:

$$Y_i = \mathcal{D}(V_i + L_i). \tag{13}$$

4 Experimental Results

In this section, we first describe the experimental settings of the proposed ColorMNet. Then we evaluate the effectiveness of our approach against state-of-the-art methods. More experimental results are included in the supplemental material. The training code and test models will be available to the public.

4.1 Experimental Settings

Datasets. Following previous works [1,21,37], we use the datasets of DAVIS [26] and Videvo [16] for training and generate grayscale video frames using OpenCV library. For testing, we use three popular benchmark test datasets including the DAVIS validation set, the Videvo validation set and the official validation set of NTIRE 2023 Video Colorization Challenge [12] (NVCC2023 for short).

Exemplars. Following [1,10,37,40], we adopt a similar strategy to utilize the first frame of each video clip as an exemplar to colorize the video clip.

Evaluation Metrics. Following the experimental protocol of most existing colorization methods, we use peak signal-to-noise ratio (PSNR), structural similarity index measurement (SSIM) [35], the Fréchet Inception Distance (FID) [9], the learned perceptual image patch similarity (LPIPS) [42], the Fréchet Video Score (FVD) [33], and the colorfulness score (Colorfulness) [6] as evaluation metrics. These assessment matrices cover a spectrum of pixel-wise considerations, the distribution similarity between generated images as well as ground truth images, and the perception similarity.

Implementation Details. We train our model on a machine with one RTX A6000 GPU. We adopt the Adam optimizer [15] with default parameters using PyTorch [24] for 160,000 iterations. The batch size is set to 4. We adopt the CIE LAB color space for each frame in our experiments. The learning rate is set to a constant 2×10^{-5}. We empirically set $\gamma = 5$, $N_e = 5$, $N_s = 10$, $M = 128$ and $d = 1$. We employ a L_1 loss, computed as the mean absolute errors between the predicted images and the ground truths.

4.2 Comparisons with State-of-the-Art Methods

Quantitative Comparison. We benchmark our method against state-of-the-art ones on three datasets and report quantitative results. The competing methods include the automatic colorization techniques [21,45], single exemplar-based approaches [10,40], and a double exemplar-based method [37]. Furthermore, for enhanced self-containment, we incorporate comparisons with leading image colorization methods [13,44] and enhance the temporal consistency of these methods by applying DVP [19] as post-processing method. For [10,21,37,45], whose

Fig. 3. Qualitative comparisons on clip *parkour* from the validation set of DAVIS [26] dataset. (a)–(g) are the colorization results by DDColor [13], TCVC [21], VCGAN [45], DeepRemaster [10], DeepExemplar [40], BiSTNet[†] [37] and ColorMNet (Ours). (h) Ground truth. The evaluated methods do not generate realistic colorful images in (a)–(f). In contrast, our approach generates a well-colorized image in (g).

weights are trained on the same datasets as our method, we conduct tests using their official codes and weights provided by the authors. However, for [13,40,44], we train from scratch using the official codes on identical datasets as our method to ensure fair comparisons. In addition, we adhere to the commonly adopted protocol of using the same first frame ground truth image of each video clip as the exemplar for testing all exemplar-based methods [10,37,40,44], with the exception that we also include the last frame as an extra exemplar for testing the double exemplar-based method, BiSTNet [37].

Table 1 shows that the ColorMNet consistently generates competitive colorization results on the DAVIS [26] validation set, the Videvo [16] validation set, and the NVCC2023 [12] validation set, where our method performs better than the evaluated methods in terms of PSNR, SSIM, FID, LPIPS, and FVD, indicating that our method not only can generate both high-quality and high-fidelity colorization results, but also demonstrates the good generalization based on the favorable performance on the validation set in NVCC2023 (the training set in NVCC2023 is not included in the training phase).

Qualitative Evaluation. Figure 3(a) shows that the image-based method [13] generates results with non-uniform colors on the man's face and cloth. Automatic video colorization techniques [21,45] can not generate vivid colors (Fig. 3(b) and (c)). Exemplar-based methods [10,37,40] can not establish long-range correspondence and thus fail to restore the colors on the man's arms and cloth (Fig. 3(d)–(f)). In contrast, the proposed ColorMNet generates a vivid colorized image (Fig. 3(g)) by modeling both the spatial information from each frame and the temporal information from far-apart frames.

Evaluations on Real-World Videos. We further evaluate the proposed method on a real-world grayscale video, *Manhattan (1979)*. We obtain the exemplars (Fig. 4(b)) by searching the internet to find the most visually similar images to the input video frames. Figure 4(c) and (d) show that state-of-the-art meth-

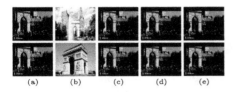

Fig. 4. Qualitative colorization comparisons on film *Manhattan (1979)*. (a) Input frame. (b) Exemplar images. (c)-(e) are the colorization results by DeepRemaster [10], DeepExemplar [40] and ColorMNet (Ours), respectively. Our ColorMNet generates error-free and realistic colors in (e) compared with (c) and (d).

Table 2. Quantitative evaluations on the effectiveness of the proposed PVGFE, MFP, and LA modules in our ColorMNet. Evaluated on the DAVIS [26] validation set. Our proposed ColorMNet achieves the best performance in terms of PSNR and SSIM among baseline methods.

Components	Feature extractor		Feature propagation			Locality		Metrics	
Methods	ResNet50	DINOv2	PVGFE	Stacking	Recurrent	MFP	LA	PSNR↑	SSIM↑
ColorMNet w/ ResNet50	✓					✓	✓	35.01	0.962
ColorMNet w/ DINOv2		✓				✓	✓	35.38	0.963
ColorMNet w/ Concatenation		✓				✓	✓	35.26	0.965
ColorMNet w/ Stacking		✓	✓	✓		✓	✓	33.94	0.961
ColorMNet w/ Recurrent		✓	✓		✓	✓	✓	35.26	0.966
ColorMNet w/o LA		✓	✓			✓		35.44	0.967
ColorMNet (Ours)		✓	✓			✓	✓	35.77	0.970

ods [10,40] do not colorize the objects (*e.g.*, the wall of the building, the trees and the sky) well. In contrast, our method generates better-colorized frames, where the colors look natural and realistic (Fig. 4(e)). In addition, our method demonstrates its robustness by consistently generating similar results even when provided with exemplar images that possess diverse colors and contents.

Efficiency Evaluation. Given that practical applications of video colorization often involve processing longer videos, where maximum GPU memory usage and inference speed are critical metrics, we further evaluate our method against three representative state-of-the-art exemplar-based video colorization approaches [10, 37,40]. Specifically, we record the maximum GPU memory consumption during the inference on a machine with an NVIDIA RTX A6000 GPU. The average running time is obtained using 300 test images with a 960 × 536 resolution.

Table 1 shows that the maximum GPU consumption of ColorMNet (ours) is only 11.2% of DeepRemaster [10], 10.0% of DeepExemplar [40] and 5.4% of BiSTNet [37]; the running time is at least 8× faster than the evaluated methods.

Temporal Consistency Evaluations. To examine whether the colorized videos generated by our method have a better temporal consistency property, we use the color distribution consistency index (CDC) [21] as the metric. Table 1 shows that our method has a lower CDC value when compared to exemplar-based methods [10,37,40] on the DAVIS [26] validation set, which indicates that our method is capable of generating videos with improved temporal consistency by exploring better temporal information.

5 Analysis and Discussion

To better understand how our method solves video colorization and demonstrate the effectiveness of its main components, we conduct a deeper analysis of the proposed approach. For the ablation studies in this section, we train our method and all alternative baselines on the training set of the DAVIS [26] dataset and the Videvo [16] dataset with 160,000 iterations for fair comparisons.

Fig. 5. Effectiveness of PVGFE for video colorization. (a) Input patch. (b)–(e) are the colorization results by ColorMNet$_{w/\ ResNet50}$, ColorMNet$_{w/\ DINOv2}$, ColorMNet$_{w/\ Concatenation}$ and ColorMNet (Ours), respectively. (f) Ground truth. Compared to the baselines, our approach yields a more natural colorized result in (e).

Effectiveness of PVGFE. The proposed PVGFE explores robust spatial features that can model both global semantic structures and local details for better video colorization. To demonstrate its effectiveness, we compare with baseline methods that respectively replace the PVGFE with the pretrained ResNet50 [7] (ColorMNet$_{w/\ ResNet50}$ for short), the pretrained DINOv2 [23] (ColorMNet$_{w/\ DINOv2}$ for short), and the concatenation of both pretrained ResNet50 and DINOv2 (ColorMNet$_{w/\ Concatenation}$ for short) in our implementation. Table 2 shows that our ColorMNet with the PVGFE outperforms all baseline methods. The qualitative comparisons in Fig. 5 show that the results obtained by baseline methods exhibit severe color distortions on the cloth of the player (Fig. 5(b)–(d)). In contrast, our proposed ColorMNet with the PVGFE generates a better-colorized frame in Fig. 5(e). Note that the PVGFE can adaptively enhance useful features while reducing the influence of useless information based on the similarities computed on input features by employing cross-attention (*i.e.*, (2)). However, a direct concatenation of features evenly without discrimination, *i.e.*, ColorMNet$_{w/\ Concatenation}$, is less effective for reducing the impact of useless features, thus degrading performance in Table 2 and Fig. 5(d).

Effectiveness of MFP. The proposed MFP propagates temporal features for better long-range correspondences. To investigate whether directly stacking multiple frames along the temporal dimension or recurrently propagating features can already generate competitive results, we compare with baseline methods that respectively replace the MFP with direct stacking of the features from all previous colorized frames and the exemplar image along the temporal dimension (ColorMNet$_{w/\ Stacking}$ for short) and the recurrent-based feature propagation [18,32,34,40] (ColorMNet$_{w/\ Recurrent}$ for short) in our implementation.

Table 2 shows that the PSNR value of our ColorMNet is at least 0.51dB higher than each baseline method, which illustrates the effectiveness of the proposed MFP in propagating features for video colorization. Figure 3(b) shows that the baseline that directly stacks frames is not able to generate a realistic image as spatial-temporal priors are not well-explored. The result obtained by the baseline with recurrent-based feature propagation contains significant color distortions on

(a)　　(b)　　(c)　　(d)　　(e)

Fig. 6. Effectiveness of the MFP module for video colorization. (a) Input frame and exemplar image. (b)–(d) are the colorization results by ColorMNet$_{w/\ Stacking}$, ColorMNet$_{w/\ Recurrent}$ and ColorMNet (Ours), respectively. (e) Ground truth. The baseline methods do not colorize the boy, the trees, and the sky well in (b) and (c). In contrast, our ColorMNet generates a colorized frame that contains fewer color distortions and more vivid colors in (d).

Table 3. Effectiveness of the MFP module for video colorization. (a) Input frame and exemplar image. (b)–(d) are the colorization results by ColorMNet$_{w/\ Stacking}$, ColorMNet$_{w/\ Recurrent}$ and ColorMNet (Ours), respectively. (e) Ground truth. The baseline methods do not colorize the boy, the trees, and the sky well in (b) and (c). In contrast, our ColorMNet generates a colorized frame that contains fewer color distortions and more vivid colors in (d).

Methods	Memory consumption (G)	Running time (/s)
ColorMNet$_{w/\ Stacking}$	24.1	1.04
ColorMNet (Ours)	1.9	0.07

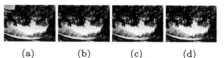

(a)　　(b)　　(c)　　(d)

Fig. 7. Effectiveness of the proposed LA module for video colorization. (a) Input frame and exemplar image. (b) and (c) are the colorization results by ColorMNet$_{w/o\ LA}$ and ColorMNet (Ours). (d) Ground truth. Our approach is able to generate better-colorized result in (c).

the boy (Fig. 3(c)), as the errors accumulate in the recurrent-based propagation steps. In contrast, the proposed ColorMNet using the MFP generates a vivid and error-free image in Fig. 3(d).

Efficiency of MFP. To examine the efficiency of the proposed MFP, we further evaluate the proposed ColorMNet against the baseline method with direct stacking (i.e., ColorMNet$_{w/\ Stacking}$) on the validation set of NVCC2023 [12] in terms of the maximum GPU memory consumption and the average running time. Table 3 shows that our ColorMNet only requires 7.8% of the maximum GPU consumption of the baseline method, but the average running time of our approach is nearly 14× faster than the baseline method, which indicates the efficiency of the proposed MFP.

Tables 2 and 3 show that our approach using the MFP achieves a favorable performance in terms of faster inference speed, lower GPU memory consumption, and better colorization results, which demonstrates the effectiveness and efficiency of the proposed MFP in video colorization.

Effectiveness of LA. To demonstrate the effect of the proposed LA, we further compare with a baseline method that removes the LA module (ColorMNet$_{w/o\ LA}$ for short) in our implementation. Table 2 shows that our ColorMNet using the

LA generates better results with higher PSNR and SSIM values than the baseline method. Figure 7(b) shows that the baseline method without the LA does not exploit the prior information among consecutive frames and thus cannot restore the colors on the sky and the leaves. However, our approach generates vivid and realistic colors in Fig. 7(c), which demonstrates that the proposed LA module is effective in capturing and leveraging better spatial-temporal features.

Limitations. We aim to enhance video colorization performance while reducing GPU memory usage. However, the proposed model size is relative large, which requires 123.61 Million parameters.

6 Conclusion

We present an effective memory-based deep spatial-temporal feature propagation network for video colorization. We develop a large-pretrained visual model guided feature estimation module to better explore robust spatial features. To establish reliable connections from far-apart frames, we propose a memory-based feature propagation module. We develop a local attention module to better utilize spatial-temporal priors. Both quantitative and qualitative experimental results show that our method performs favorably against state-of-the-art methods.

Acknowledgements. This work has been partly supported by the National Natural Science Foundation of China (Nos. U22B2049, 62272233, 62332010), the Fundamental Research Funds for the Central Universities (No. 30922010910), and the Postgraduate Research & Practice Innovation Program of Jiangsu Province (No. KYCX24_0680).

References

1. Chen, S., et al.: Exemplar-based video colorization with long-term spatiotemporal dependency. arXiv preprint arXiv:2303.15081 (2023)
2. Chen, X., Zou, D., Zhao, Q., Tan, P.: Manifold preserving edit propagation. ACM TOG **31**(6), 1–7 (2012)
3. Cheng, H.K., Schwing, A.G.: XMEM: long-term video object segmentation with an Atkinson-Shiffrin memory model. In: ECCV (2022)
4. Cheng, Z., Yang, Q., Sheng, B.: Deep colorization. In: ICCV (2015)
5. Dosovitskiy, A., et al.: An image is worth 16x16 words: transformers for image recognition at scale. In: ICLR (2021)
6. Hasler, D., Süsstrunk, S.: Measuring colorfulness in natural images. In: Human Vision and Electronic Imaging VIII (2023)
7. He, K., Zhang, X., Ren, S., Sun, J.: Deep residual learning for image recognition. In: CVPR (2016)
8. He, M., Chen, D., Liao, J., Sander, P.V., Yuan, L.: Deep exemplar-based colorization. ACM TOG **37**(4), 1–16 (2018)
9. Heusel, M., Ramsauer, H., Unterthiner, T., Nessler, B., Hochreiter, S.: GANs trained by a two time-scale update rule converge to a local NASH equilibrium. In: NeurIPS (2017)

10. Iizuka, S., Simo-Serra, E.: DeepRemaster: temporal source-reference attention networks for comprehensive video enhancement. ACM TOG **38**(6), 1–13 (2019)
11. Iizuka, S., Simo-Serra, E., Ishikawa, H.: Let there be color! joint end-to-end learning of global and local image priors for automatic image colorization with simultaneous classification. ACM TOG **35**(4), 1–11 (2016)
12. Kang, X., Lin, X., Zhang, K., et al.: NTIRE 2023 video colorization challenge. In: CVPRW (2023)
13. Kang, X., Yang, T., Ouyang, W., Ren, P., Li, L., Xie, X.: DDColor: towards photorealistic image colorization via dual decoders. In: ICCV (2023)
14. Karen Simonyan, A.Z.: Very deep convolutional networks for large-scale image recognition. In: ICLR (2015)
15. Kingma, D.P., Ba, J.: Adam: a method for stochastic optimization. In: ICLR (2015)
16. Lai, W.S., Huang, J.B., Wang, O., Shechtman, E., Yumer, E., Yang, M.H.: Learning blind video temporal consistency. In: ECCV (2018)
17. Larsson, G., Maire, M., Shakhnarovich, G.: Learning representations for automatic colorization. In: ECCV (2016)
18. Lei, C., Chen, Q.: Fully automatic video colorization with self-regularization and diversity. In: CVPR (2019)
19. Lei, C., Xing, Y., Chen, Q.: Blind video temporal consistency via deep video prior. In: NeurIPS (2020)
20. Levin, A., Lischinski, D., Weiss, Y.: Colorization using optimization. ACM TOG **23**(3), 689–694 (2004)
21. Liu, Y., et al.: Temporally consistent video colorization with deep feature propagation and self-regularization learning. arXiv preprint arXiv:2110.04562 (2021)
22. Luan, Q., Wen, F., Cohen-Or, D., Liang, L., Xu, Y., Shum, H.: Natural image colorization. In: ESRT (2007)
23. Oquab, M., Darcet, T., Moutakanni, T., et al.: Dinov2: Learning robust visual features without supervision. TMLR (2024)
24. Paszke, A., et al.: Pytorch: an imperative style, high-performance deep learning library. In: NeurIPS (2019)
25. Paul, S., Bhattacharya, S., Gupta, S.: Spatiotemporal colorization of video using 3D steerable pyramids. IEEE TCSVT **27**(8), 1605–1619 (2016)
26. Perazzi, F., et al.: A benchmark dataset and evaluation methodology for video object segmentation. In: CVPR (2016)
27. Qu, Y., Wong, T., Heng, P.: Manga colorization. ACM TOG **25**(3), 1214–1220 (2006)
28. Sangkloy, P., Lu, J., Fang, C., Yu, F., Hays, J.: Scribbler: controlling deep image synthesis with sketch and color. In: CVPR (2017)
29. Sheng, B., Sun, H., Magnor, M., Li, P.: Video colorization using parallel optimization in feature space. IEEE TCSVT **24**(3), 407–417 (2013)
30. Su, J.W., Chu, H.K., Huang, J.B.: Instance-aware image colorization. In: CVPR (2020)
31. Teed, Z., Deng, J.: Raft: recurrent all-pairs field transforms for optical flow. In: ECCV (2020)
32. Thasarathan, H., Nazeri, K., Ebrahimi, M.: Automatic temporally coherent video colorization. In: CRV (2019)
33. Unterthiner, T., van Steenkiste, S., Kurach, K., Marinier, R., Michalski, M., Gelly, S.: FVD: a new metric for video generation. In: ICLR (2019)
34. Wan, Z., Zhang, B., Chen, D., Liao, J.: Bringing old films back to life. In: CVPR (2022)

35. Wang, Z., Bovik, A.C., Sheikh, H.R., Simoncelli, E.P.: Image quality assessment: from error visibility to structural similarity. IEEE TIP **13**(4), 600–612 (2004)
36. Xu, Z., Wang, T., Fang, F., Sheng, Y., Zhang, G.: Stylization-based architecture for fast deep exemplar colorization. In: CVPR (2020)
37. Yang, Y., Peng, Z., Du, X., Tao, Z., Tang, J., Pan, J.: BistNet: semantic image prior guided bidirectional temporal feature fusion for deep exemplar-based video colorization. IEEE TPAMI 1–14 (2024)
38. Yatziv, L., Sapiro, G.: Fast image and video colorization using chrominance blending. IEEE TIP **15**(5), 1120–1129 (2006)
39. Zamir, S.W., Arora, A., Khan, S., Hayat, M., Khan, F.S., Yang, M.H.: Restormer: Efficient transformer for high-resolution image restoration. In: CVPR (2022)
40. Zhang, B., et al.: Deep exemplar-based video colorization. In: CVPR (2019)
41. Zhang, R., Isola, P., Efros, A.A.: Colorful image colorization. In: ECCV (2016)
42. Zhang, R., Isola, P., Efros, A.A., Shechtman, E., Wang, O.: The unreasonable effectiveness of deep features as a perceptual metric. In: CVPR (2018)
43. Zhang, R., et al.: Real-time user-guided image colorization with learned deep priors. ACM TOG **36**(4), 1–11 (2017)
44. Zhao, H., Wu, W., Liu, Y., He, D.: Color2embed: fast exemplar-based image colorization using color embeddings. arXiv preprint arXiv:2106.08017 (2021)
45. Zhao, Y., et al.: VCGAN: video colorization with hybrid generative adversarial network. IEEE TMM (2022)

UNIC: Universal Classification Models via Multi-teacher Distillation

Mert Bülent Sarıyıldız[✉], Philippe Weinzaepfel, Thomas Lucas, Diane Larlus, and Yannis Kalantidis

NAVER LABS Europe, Meylan, France
`bulent.sariyildiz@naverlabs.com`
`https://europe.naverlabs.com/unic`

Abstract. Pretrained models have become a commodity and offer strong results on a broad range of tasks. In this work, we focus on classification and seek to learn a unique encoder able to take from several complementary pretrained models. We aim at even *stronger generalization* across a variety of classification tasks. We propose to learn such an encoder via multi-teacher distillation. We first thoroughly analyze standard distillation when driven by multiple strong teachers with complementary strengths. Guided by this analysis, we gradually propose improvements to the basic distillation setup. Among those, we enrich the architecture of the encoder with a ladder of expendable projectors, which increases the impact of intermediate features during distillation, and we introduce teacher dropping, a regularization mechanism that better balances the teachers' influence. Our final distillation strategy leads to student models of the same capacity as any of the teachers, while retaining or improving upon the performance of the best teacher for each task.

Keywords: Multi-Teacher Distillation · Classification · Generalization

1 Introduction

Recent years have witnessed the rise of many pretrained models [6,55,75]. They often share the same architecture and sometimes even the same training data. They generalize to a broad range of tasks, but may particularly excel at specific visual recognition scenarios depending on the selected learning strategy. Self-supervised learning models [6–8] shine in transfer learning, *i.e.* generalization to novel classes, while models trained with masked modeling techniques [16,75] are often better suited to patch-level tasks. Meanwhile, supervised learning [12,27] is still best for specific classification tasks when labeled data is available during pretraining.

In this paper, our goal is to learn *a universal encoder* capable of strong generalization across a broad spectrum of classification tasks. More specifically, besides

---

Supplementary Information The online version contains supplementary material available at https://doi.org/10.1007/978-3-031-73235-5_20.

© The Author(s), under exclusive license to Springer Nature Switzerland AG 2025
A. Leonardis et al. (Eds.): ECCV 2024, LNCS 15062, pp. 353–371, 2025.
https://doi.org/10.1007/978-3-031-73235-5_20

ImageNet classification [46] – the dataset on which our teachers are trained and our students are distilled – we are further interested in the classification of novel classes, on new domains, as well as dense prediction tasks such as semantic segmentation or depth estimation. Our goal is to learn *a single encoder* that can be directly applied to all these tasks, out-of-the-box, without the need for any task-specific parameters besides a linear classifier per classification task.

Our approach uses multi-teacher distillation, drawing on the strengths of various specialized teachers to train an encoder that seeks to match or surpass the best teacher in each task. We conduct a comprehensive analysis of the distillation process from multiple teachers, evaluating our models on various tasks, including image-level classification on ImageNet-1K and 15 more transfer datasets, as well as patch-level classification tasks such as semantic segmentation and depth estimation. We leverage our findings to gradually devise a method that

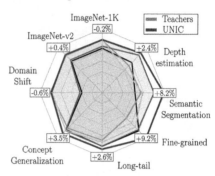

Fig. 1. Relative gains using our UNIC encoder distilled from four teachers (DINO, DeiT-III, iBOT, dBOT-ft), over the respective best teacher for each task using a *single encoder* and no task-specific parameters.

shows improved generalization across multiple tasks and axes. We modify the input of expandable projectors [7,8,48] (building what we call a *ladder of projectors*) so that they also act as information highways that propagate signal from intermediate layers to the distillation loss in a more direct manner. We analyze learning dynamics across teachers and further propose *teacher dropping*, an effective strategy for balancing the teachers' influence in multi-teacher distillation, resulting in significant gains for the tasks at which our distilled models were otherwise underperforming.

With all of our improvements added to the basic multi-teacher distillation setup, we are able to train models that exhibit strong generalization across a wide range of classification tasks on the image and patch levels, either retaining or improving the performance of the best teacher. As an example, we show in Fig. 1 that by distilling from four strong ViT-Base models trained on ImageNet (*i.e.* DINO [6], DeiT-III [56], iBOT [75], and dBOT-ft [29]) we are able to train a *universal encoder* excelling at all considered tasks. In our experimental study, we show that our findings further extend to the case of larger teachers like DINOv2 [35] and MetaCLIP [64] trained on arbitrary datasets. Finally, we study the way the distilled encoders utilize their weights: first, by quantifying performance drops after weights pruning, and second after reducing the dimension of the output feature space using PCA. These experiments show that distilled models have lower redundancy in both their weights and their features.

Contributions. To summarize, we conduct a thorough analysis of multi-teacher distillation for ViT encoders and use our findings to improve the distillation process and generalization power of the student. Among other simple but crucial modifications, we introduce improvements like ladder of projectors and teacher dropping regularization that enable us to learn models which retain or improve the performance of the best teachers across many diverse tasks. We refer to such models as **U**niversal **C**lassification models or **UNIC**. We finally perform extensive evaluations along multiple axes of generalization and study the ways the resulting models make use of their weights and feature space.

2 Related Work

Knowledge distillation (KD) was initially introduced as a model compression technique [5], where the goal is to train a smaller student model from the output of a teacher model [21]. While early work focused on predicting the final outputs of a classification model, the idea was rapidly extended to other forms of distillation, such as distilling intermediate representations [1,19,20,43,67,69,73]. These methods perform well but require careful layer selection and loss balancing [19]. In our work, instead of matching layer-wise representations between the student and teacher architectures, we add shortcut connections from intermediate layers of the student to the loss of each teacher.

Multi-teacher Knowledge Distillation. KD can naturally be extended to an ensemble of teachers so that student can benefit from their potential complementarity. While the final outputs of teachers trained for the same task can simply be averaged [2,13,21,69], multi-teacher distillation with teachers trained for different tasks is more challenging. UDON [70] first trains domain-specialist teachers which are subsequently distilled in a student model using adaptive data sampling for balancing the different domains. In [54], contrastive learning is used for ensemble distillation while [50] proposes a framework tailored for teachers trained with masked image modeling and contrastive learning. But such approaches are not straightforward to extend to teachers learned differently. Similarly, [65] combines self-supervised teachers from arbitrary heterogeneous pretext tasks. [11,14,45] focus on jointly utilizing pseudo- and true labels for multi-teacher distillation. Roth et al.[45] formulate multi-teacher distillation as continual learning and further propose a novel method for data partitioning based on confidence. Here we develop a more generic method for combining teachers, that is not limited to certain types of teachers or losses, and, unlike [28,45], does not require labeled data, nor classifiers associated with each teacher for obtaining pseudo-labels.

Loss Balancing is shown to be crucial in multi-task learning [9,22,24,72]. Similar strategies to automatically balance losses have also been proposed for multi-teacher distillation [13,30]. In [22], adaptive loss weights inversely proportional to the average of each loss are introduced, while [30] learns instance-level teacher importance weights using ground-truth labels. In [13], the random selection of one teacher per mini-batch is shown to help. Our experiments show that our proposed generalized teacher dropping strategy leads to better models compared to [13,22].

Distilling from A "Foundation Model" like CLIP [38] or DINOv2 [35] is an effective approach for tasks with limited training data [32,37,61]. Distilling from *multiple* foundation models allows for more versatile students. Recent works like AM-RADIO [41], SAM-CLIP [59], and Open Vocabulary SAM [71] combine the semantics captured by CLIP with the localization capabilities of models like DINOv2 [35] or SAM [25]. AM-RADIO [41] builds on the same base setup as our study, but employs no loss balancing. Another difference comes from the fact that their student encoder is only a part of the final model: AM-RADIO requires the teacher-specific projectors learned during distillation to also be used at test time, effectively increasing the parameters of the encoder with task-specific ones. Instead, our method performs well on multiple classification tasks *out-of-the-box*, without any additional parameters.

Combining Models Beyond Distillation. Other creative ways to combine multiple pretrained models have been proposed. Works like [33,39,40,53,62] explore different weight averaging strategies. They typically only combine models that differ by their hyper-parameter configuration. Aiming at generalization, [66] merges multiple ViTs, each specialized to a classification task, into a single encoder that solves all classification tasks jointly, via a gating network with multiple functions. Instead, our students are distilled from scratch, have a simple ViT architecture and tackle diverse classification tasks with simple linear probing.

Expendable projectors are extra modules that act as buffers between the final encoder output and the space where the loss is computed. They have been successfully used for both self-supervised [7,8] and supervised learning [48,60]. We extend this idea and add projectors during training to intermediate layers as well. Roth *et al.* [44] use several such projectors of varied dimensionality for metric learning, but do not use features from intermediate layers. Moreover, we use a specific set of projectors per teacher, similar to [2,41]. This way, projectors become *loss-specific*, *i.e.* they contribute to the loss for only one of the teachers.

3 Analyzing and Improving Multi-teacher Distillation

In this section we first present the multi-teacher distillation setup we use as a basis for our analysis (Sect. 3.1) and a summary of our evaluation protocol (Sect. 3.2). We then delve into challenges around multi-teacher distillation of ViT encoders (Sect. 3.3), and offer improvements to the basic setup to overcome them, like enhanced expendable teacher-specific projectors heads (Sect. 3.4) and strategies to more equally learn from all teachers (Sect. 3.5).

3.1 A Basic Distillation Setup

Our task is to distil M *teacher* models $\mathcal{T} = \{\mathcal{T}_1, \ldots, \mathcal{T}_M\}$ into a student model $\mathcal{S}$. An overview is shown in Fig. 2. Each teacher $t \in \mathcal{T}$ is a ViT [12] encoder

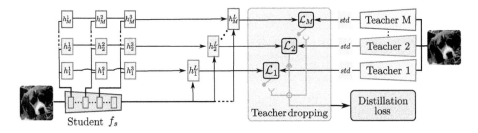

Fig. 2. Overview of our Multi-teacher Distillation Setup. The same input image is fed to each teacher and to student. We employ feature standardization at the output of all teachers (Sect. 3.3), a *ladder of expandable projectors* attached to student (Sect. 3.4) and *teacher dropping regularization* to balance teachers (Sect. 3.5). The latter enables us to adaptively select a subset of teachers to contribute to the loss simply using loss magnitudes. We use dedicated projectors for the CLS and patch tokens (Sect. 3.3).

that maps an image x to a set of d-dimensional feature vectors $y_{t,i} = f_t(x;i)$ for token i, which can either be one of the $H \times W$ patch tokens from $\mathcal{P}$ or the global CLS token c. We aim at learning the parameters f_s of the student $\mathcal{S}$, such that the output representations $z_i = f_s(x;i)$ excel at all the tasks that any of the teachers also shines at.

We append a *projector head* h_t per teacher to the student encoder's output which transforms each token into a teacher-specific representation $h_t(z_i)$. The loss for each teacher is then computed on $h_t(z_i)$, the output of the corresponding projector head. We consider these projector heads as *expendable, i.e.* they are removed after distillation and are not part of the student encoder. Their goal is to assist the learning process, taking inspiration from similar expendable projectors used in self-supervised [7] and supervised [48,60] representation learning. We set projector heads to be Multi-Layer Perceptrons (MLPs) with two linear layers, GeLU non-linearity and hidden dimension of $d_h = 4d$, where d is the feature dimension; we analyze projectors further in the next sections.

We use two common distillation losses: cosine and smooth-ℓ_1 (see supplementary material for details); the loss for token i from teacher t is given by:

$$\mathcal{L}_t(x;i) = 0.5 \times \left(\mathcal{L}^{cos}(h_t(z_i), y_{t,i}) + \mathcal{L}^{s\ell_1}(h_t(z_i), y_{t,i}) \right). \tag{1}$$

This loss is computed separately for the CLS and each of the patch tokens $\mathcal{P}$. To get the final loss, we sum losses from all teachers similar to [69], as well as over the CLS token c and the tokens of all patches:

$$\mathcal{L}(x) = \sum_{t \in \mathcal{T}} \left(\mathcal{L}_t(x;c) + \frac{1}{|\mathcal{P}|} \sum_{p \in \mathcal{P}} \mathcal{L}_t(x;p) \right), \tag{2}$$

where $|\mathcal{P}|$ is the number of patch tokens.

3.2 Protocol Summary

We first present a summary of the experimental protocol we use for the analysis in this section. Further details are presented in the supplementary material.

Datasets and Backbones. To better isolate the effects of different distillation components, we use the same training data and architectures for all teachers and students, *i.e.* the ImageNet-1K dataset [46] and ViT-Base [12], respectively. During distillation, we discard the labels of ImageNet and only use the images; no supervised loss is combined with the distillation losses presented above.

Table 1. Component analysis for distillation from two teachers. We report: image classification on 1) ImageNet-1K (IN-val) and 2) 15 transfer learning datasets (averaged), 3) semantic segmentation on ADE-20K, and 4) depth estimation on NYUd. Column legend: std: feature standardization, DP: dedicated projector heads for CLS/patch tokens, LP: ladder of projectors and *tdrop*: teacher dropping regularization.

	Model	std	DP	LP	tdrop	IN-val top-1 (↑)	Transfer top-1 (↑)	Segmentation mIoU (↑)	Depth RMSE (↓)
	Teacher models								
1.	DINO					77.7	72.4	30.4	0.570
2.	DeiT-III					83.6	68.5	32.3	0.589
3.	*best teacher*					83.6	72.4	32.3	0.570
	Multi-teacher distillation (DINO & DeiT-III teachers)								
4.	basic setup					78.7	73.1	33.9	0.560
5.	UNIC	✓				81.4	73.8	36.1	0.558
6.		✓	✓			82.2	74.1	36.9	0.551
7.		✓	✓	✓		82.7	74.2	37.4	0.546
8.		✓	✓	✓	✓	83.2	73.5	37.3	0.547

Teachers. We consider models learned using *self-supervised learning* (SSL), like DINO [6] or iBOT [75], and *supervised models* like DeiT-III [56] or fine-tuned dBoT [29], optimized for the classification task of ImageNet-1K. The former have proven extremely effective for generalization whereas the latter achieve state-of-the-art accuracy on the ImageNet-1K task. In this section, we present our analysis for $M = 2$ teachers, specifically DINO and Deit-III. More teachers and combinations are explored in Sect. 4 and in the supplementary material.

Tasks. We measure performance on many tasks, divided along the following axes: 1) Top-1 accuracy on the *training set classes* on the ImageNet-1K validation set [46] (IN-val); 2) *Transfer learning* performance on unseen classes; we report top-1 accuracy averaged over 15 diverse image classification datasets;[1]

---

[1] The 15 datasets are: 5 ImageNet-CoG levels [49] tailored for concept generalization, 8 small-scale fine-grained datasets (Aircraft, Cars196, DTD, EuroSAT, Flowers, Pets, Food101, SUN397) and two long-tail datasets (iNaturalist-2018 and 2019).

Dense prediction performance on 3) semantic segmentation and 4) depth estimation; we report mIoU on ADE-20k [74] and RMSE on NYUD [51], measured using a protocol that is essentially dense classification, *i.e.* using linear probes as in [35]. We learn linear probes for all tasks directly over encoder outputs z.

3.3 Analyzing Multi-teacher Distillation of ViT Tokens

In this section we analyze and revisit different aspects of distillation that are specific to ViT encoders, *e.g.* the use of CLS and patch tokens. The former is normally fed as input to image-level classifiers while patch tokens are important for dense prediction. In this section we study their statistics and explore how this affects design choices of the distillation setup. The top part of Table 1 compares the accuracy of the self-supervised DINO and supervised DeiT-III on the different evaluation axes. They show complementary strengths, *i.e.* they respectively perform well on transfer learning and the ImageNet-1K validation set (IN-val).

Equalizing Feature Statistics Across Tokens and Teachers. We start by analyzing the statistics of features extracted from the CLS and patch tokens of both teachers and show that this should be taken into account for multi-teacher distillation. We calculate such statistics and notice a number of discrepancies in their first and second moment values, both between CLS and patch tokens of a given teacher as well as across teachers. The norm and standard deviation for the CLS token features of DINO, for example, are double the ones for patch tokens of the same model, while the same statistics also differ across DeiT-III and DINO tokens (see supplementary material for more details).

To explore whether such statistical inconsistencies across features affect distillation, we add feature standardization on each teacher output, *i.e.* we normalize teacher features to zero mean and unit variance before computing the loss, which was shown to be useful in [19]. This not only equalizes any differences between CLS and patch tokens but also for tokens across teachers. For convenience and generality, we propose to learn such normalization statistics on-the-fly during distillation, using an exponential moving average. From Table 1 we see that the performance of models learned via distillation is consistently higher using feature standardization for both image- and patch-level tasks (rows 4 vs. 5).

▶ *Feature standardization improves multi-teacher distillation*

Projector Heads for CLS and Patch Tokens. Beside statistical differences, the CLS and patch token are also conceptually different: CLS is a global token expected to encode image-level semantics whereas the patch tokens encode local information. To better capture these specifics from CLS and patch tokens, we experiment with dedicated teacher-specific projector heads for each type of tokens. This comes at no added cost in practice, since we discard the projectors after distillation. We discuss expendable projectors further in Sect. 3.4. Comparing rows 5 and 6 in Table 1 we see that specializing the teacher-specific projector heads to either CLS or patch tokens leads to further gains.

▶ *Dedicated projectors for CLS/patches improve distillation performance*

Classification on ImageNet and Novel Classes. Results in Table 1 show that models learned via multi-teacher distillation lack in terms of ImageNet-1K performance compared to highly optimized models for that specific task, such as DeiT-III (82.2 vs. 83.6). One may suggest that this is due to the fact that we do not use labels during distillation. To test that, we also performed distillation using *only* the DeiT-III model as a teacher. In that case we were able to reach a top-1 accuracy of 83.1% on ImageNet. This is much higher than the 82.2% we get distilling jointly from multiple teachers and we therefore see that there is still space for improvement during distillation itself.

From Table 1 we also see that models learned via multi-teacher distillation greatly outperform DINO on transfer learning and classification of novel classes. This is also true for the recent iBOT [75] model, which also achieves state-of-the-art top-1 accuracy, *i.e.* 72.4% on average for transfer learning on our setup.

▶ *Multi-teacher distillation significantly improves generalization*

Multi-teacher Distillation for Dense Prediction. To assess the discriminative power of patch tokens individually, we consider two dense prediction tasks, semantic segmentation and depth prediction, after linear probing. Table 1 shows that even the basic multi-teacher distillation setup improves over the best teacher (row 4). More importantly, performance increases even further (row 6) using standardization and dedicated projectors for the CLS and patch tokens. The student encoder achieves *+4.6% higher mIoU* than the best teacher for segmentation. This result is even more impressive when compared to the performance of models that are targeting improved dense prediction. Our models, which are distilled from teachers trained with supervised and contrastive learning achieve dense prediction performance comparable to models known to excel at dense tasks, *i.e.* models trained via masked patch prediction like iBOT [75]: iBOT achieves 36.6% mIoU on ADE-20K, while our student reaches 36.9%.

▶ *Multi-teacher distillation improves the discriminative power of patch tokens*

Retaining Complementary Teacher Strengths. From the results in Table 1, we see that models learned with our multi-teacher distillation setup and simple modifications like feature standardization and dedicated projectors for CLS/patch tokens are starting to show strong generalization performance on a number of axes. We will use models distilled under this setup as the basis for the rest of our study. Such models seem to retain the complementary strengths of their teachers: They already outperform the best teacher on transfer learning and dense prediction tasks, while also enjoying decent performance on the ImageNet task.

▶ *Learning from multiple teachers can combine their strengths*

As we discuss above, there is however still room for improvement; we ideally want models to match or outperform the best teacher on all tasks. In the next sections, we analyze different aspects of our distillation setup and introduce further improvements towards that end.

3.4 A Ladder of Projectors for Distillation

The basic setup above uses expendable projector heads as a way of injecting teacher-specific parameters during distillation.[2] Such modules are appended at the end of the encoders and act as small "buffers" between the encoder output and the feature space considered by the loss. In this section, we propose to use more of these expendable modules in a complementary way: as *information highways* that propagate information from intermediate layers to the loss in a more direct manner. Intermediate layers have been used to improve distillation [15,30,69], typically by adding extra losses on top of those layers. However, this leads to a more challenging optimization. Besides, hyper-parameter tuning with many added losses is combinatorial, and it becomes cumbersome. These issues are far more prominent in the case of multiple teachers.

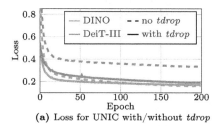
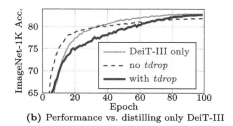

(a) Loss for UNIC with/without *tdrop* (b) Performance vs. distilling only DeiT-III

Fig. 3. Analyzing teacher dropping regularization (*tdrop*). (a) Loss for each of the two teachers during multi-teacher distillation, with and without *tdrop*. (b) ImageNet-1K top-1 accuracy when distilling from DINO & DeiT-III together, versus distilling only from DeiT-III, *i.e.* the teacher that excels at this task.

Instead of adding losses on intermediate representations, we propose to augment the existing expendable teacher-specific projector head to receive inputs from intermediate layers and append modules that connect all intermediate layer tokens directly to the teacher-specific projector head before the loss. We refer to such augmented projectors as a *ladder of projectors*. This architecture bares similarities to the adaptor architecture that is typically used for adapting a model to a new task [68]. In our case, however, the adaptor-like modules we append during distillation are *expendable*.

Specifically, we attach MLP projectors to intermediate layers and augment the input of the teacher-specific projectors h_t that until now operated only on the last layer of the student encoder. Let z^l denote the l-th layer output of the student encoder for $l = 1, \ldots, L$. The head for the ladder of projectors becomes:

$$h_t^{LP}(\{z^l : l \in L\}) = \sum_{l=1}^{L} h_t^l(z^l), \qquad (3)$$

---

[2] Projector heads are discarded after distillation and linear probes are learned over the encoder outputs z.

where h_t^l denotes the MLP projector head attached after layer $l \in L$. The architecture of h_t^l is identical to h_t, however, since we are adding multiple such projector heads, we significantly reduce the hidden dimension d_h^l and set $d_h^l = d$ when $l < L$. We explore architecture choices in the supplementary material.

From Table 1 we see that this *ladder of projectors* improves performance overall (row 8), especially for dense prediction. It seems that the dense connections lead to better prime patch tokens. Gains are also significant for supervised classification: ImageNet-1K accuracy is increased by +0.5%.

▶ *A ladder of projectors leads to improvements for both CLS and patch tokens*

3.5 Learning All Teachers Equally Well

The basic setup assumes that the final goal is for the distilled encoder to represent each teacher equally well. When distillation uses feature standardization across all teachers and simple losses like cosine and smooth-ℓ_1, there exists a straightforward way to compare how much each of the different teachers is learned: One may simply compare the *magnitudes* of the losses, that indicate how well we are approximating the feature space of each teacher.

Figure 3a displays the loss curves for multi-teacher distillation for UNIC models, using the setup presented in Sect. 3.3 (dashed lines). We see that the DINO teacher seems to be learned faster and better than DeiT-III.

▶ *Teachers do not equally contribute without further intervention*

It therefore comes as no surprise that our student lacks performance on ImageNet-1K, *i.e.* the task that DeiT-III excels at. But what if DINO was not even part of the distillation process? In Fig. 3b we show how ImageNet-1K accuracy changes during distillation using DINO & DeiT-III as teachers, and for the case of distilling *only* from DeiT-III. We see that our model learns faster using multiple teachers but converges to a lower accuracy: The student seems to exploit features from the additional teacher to ramp up performance faster, but fails to reach the accuracy of distilling DeiT-III alone (83.1%).

Figure 3 suggests that some form of loss balancing could be beneficial. Loss balancing is common in multi-task settings: In most cases it is done *manually* by adding hyperparameters that control each loss. Such an approach is however cumbersome for many teachers and losses like our case, something also discussed in [41]. It is important to avoid the combinatorial nature of manual tuning. Another way, would be to use some of the existing methods for loss balancing that are proposed for multi-task learning, *e.g.* methods like Adaloss [22]. We argue that the case of multi-teacher distillation over standardized features and simple regression losses is much simpler than multi-task learning when it comes to balancing the losses: The magnitudes of the losses are comparable and can be used for balancing and pacing the distillation process.

Teacher Dropping Regularization. We introduce a simple scheme for loss balancing that we name *teacher dropping*. Instead of designing some soft loss weighing algorithm, we take inspiration from methods like randomized dropout [52] and path dropping [23], and propose to "drop", *i.e.* zero-out the loss, for a subset of the teachers. Dropping teachers at random is however something that would not encourage loss equalization across teachers. Instead, we propose to directly use absolute magnitudes of the losses when selecting which teachers to drop, *i.e.* keeping the teacher whose loss magnitude is maximal and dropping any other teacher with some probability. This bares conceptual similarities to adaptive dropout [3], but our method is *non-parametric*, and simply exploits the fact that feature space losses on constrained representations are comparable.

We perform loss-based teacher dropping at the image level. At each iteration and for every image, we define a binary coefficient $\alpha_t = \{0, 1\}$ for each teacher t that is multiplied with the corresponding loss $\mathcal{L}_t$. This determines whether teacher t would be dropped or not for that image with probability p. To make sure there is always some signal to learn from, we choose to never drop the teacher with the maximum magnitude loss, *i.e.* the teacher that the current model approximates least well. All other teachers could be dropped with probability p. Specifically and for each image, the coefficient for teacher $t \in \mathcal{T}$ is given by:

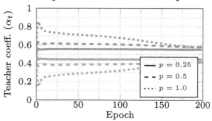

Fig. 4. Teacher coefficients α_t during distillation from DeiT and DINO.

$$\alpha_t = \begin{cases} 1 & \text{if } \mathcal{L}_t = \max_t \mathcal{L}_t, \\ (1 - \delta) & \text{if } \mathcal{L}_t \neq \max_i \mathcal{L}_i, \text{with } \delta \sim \text{Bernoulli}(p). \end{cases} \quad (4)$$

In all cases, the teacher that is least well approximated in the current iteration will always be used. We also experimented with patch-level teacher dropping but found no noticeable gains (see supplementary material).

Effect of Teacher Dropping during Distillation. We study the impact of teacher dropping during distillation in Fig. 3a: teacher dropping makes the loss magnitudes of the teachers much more similar as training progresses (solid lines). In Fig. 4 we plot how the teacher coefficients α_t vary during distillation; teacher utilization becomes more balanced and stabilizes after some epochs.

▶ *Teachers are distilled equally well with teacher dropping regularization*

How does Teacher Dropping Affect Performance? We compared teacher dropping regularization to manually balancing the teacher losses, random dropping [13], as well as to the recent Adaloss [22] loss balancing method. Starting from results in row 6 in Table 1, we found that none of these strategies is able to noticeably improve, let alone outperform results with teacher dropping (row 8). Specifically, Adaloss achieves 80.1/73.6/34.3/0.565 on the four tasks, respectively (see supplementary material for details). Besides performance, we believe the effectiveness and simplicity of the proposed teacher dropping is unparalleled.

We studied the impact of the teacher dropping probability p and found performances to be stable for different values. Yet, a higher p favours ImageNet performance, with a slight decrease on tasks where the student already outperforms the best teacher (see supplementary material).

From Table 1 (row 8) we see that teacher dropping boosts performance for ImageNet-1K, *i.e.* improves distillation on the task where our distilled models were lacking the most. When combining teacher dropping with a ladder of projectors, we are able to achieve 83.2%, our top performance on that task. This performance is only 0.4% lower than the highly optimized DeiT-III (row 3). What is more, we have also closed the observed gap between multi-teacher distillation and specialized distillation using DeiT-III alone. Teacher dropping significantly contributes to that end, *i.e.* increasing performance by 0.5% over our best model with ladder of projectors (rows 7 vs. 8).

▶ *Teacher dropping regularization is a simple and effective way to balance teachers, specifically designed for multi-teacher distillation*

3.6 Towards Universal Classification Models

Multi-teacher distillation using a ladder of projectors and teacher dropping regularization enables us to reach ImageNet classification performance comparable to the highly optimized DeiT-III, while simultaneously outperforming the best teacher on transfer learning performance on 15 datasets with mostly novel classes including long-tail ones, as well as on patch-level classification tasks like semantic segmentation and depth estimation. We contend this evidence demonstrates that our distilled models operate as more *universal* classification models. We will refer to models learned with our enhanced multi-teacher distillation setup as **UNIC** models (which stands for UNIversal Classification, pronounced *"unique"*).

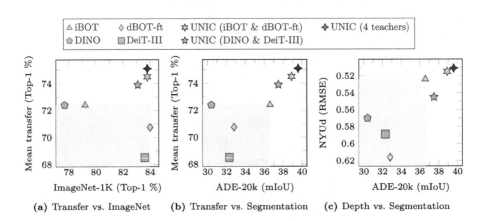

Fig. 5. Performance of different UNIC encoders on different pairs of tasks. We report performance for UNIC encoders distilled from DINO & DeiT-III, iBOT & dBOT-ft and all four teachers. We show results on ImageNet-1K **(a)**, over 15 transfer learning tasks **(a, b)**, semantic segmentation **(b, c)** and depth estimation **(c)**.

4 Experimental Study

Teachers. We report results distilling from two pairs of teachers (DeiT-III [56] & DINO [6] and iBOT [75] & dBOT-ft [29][3]), as well as using all four together. In all cases we use publicly available ViT-Base models trained on ImageNet-1K.

Extended Protocol. We use the protocol summarized in Sect. 3.2 and detailed in the supplementary material. We additionally report results on ImageNet-v2 [42], an alternative validation set for ImageNet, as well as two datasets for measuring performance under domain shift, *i.e.* ImageNet-R [18] and ImageNet-Sketch [58]. Besides reporting results for all 15 transfer datasets jointly, we further split the datasets into separate axes, *i.e.* for concept generalization [49], long-tail [57] and small-scale fine-grained recognition datasets (Aircraft [31], Cars196 [26], DTD [10], EuroSAT [17], Flowers [34], Pets [36], Food101 [4], SUN397 [63]).

In all cases we chose hyperparameters based on ImageNet-1K performance, the task which corresponds to the distillation data. See the supplementary material for further implementation and evaluation details. There, we further report results using the pre-existing classifiers in a plug-and-play manner, as well as for the case of distillation using synthetic data from the ImageNet-SD dataset [47].

Results. We summarize results for our best UNIC models from different teachers in Figs. 1 and 5. In Fig. 1 we show *relative* gains for a UNIC model trained from all four teachers, while in Fig. 5 we report results for models distilled from three different sets of teachers (DINO & DeiT-III, iBOT & dBOT-ft and all four teachers). A short summary of our most important observations follows.

1. **Stronger Teachers give Stronger Students.** From Fig. 5 we see that iBOT & dBOT-ft yield improved student models compared to DINO & DeiT-III.
2. **Adding more teachers seems to generally improve performance.** Distilling from all four teachers produces an even stronger student for most cases. This is also true when the additional teachers are not better than the existing ones: Besides ImageNet and transfer, adding DINO & DeiT-III to the ensemble also improves segmentation performance over iBOT.
3. **UNIC models excel at image-level classification.** UNIC from 4 teachers attains 83.8% and 80.3% top-1 accuracy on ImageNet-1K and ImageNet-v2, matching the top performance of the state-of-the-art dBOT-ft model (84% and 80%, respectively). Results are also strong on transfer learning, with UNIC achieving +2.7% higher top-1 on average than iBOT/DINO.
4. **Impressive gains on transfer to small fine-grained datasets.** UNIC achieves a +9.2% *relative gain* on average on 8 small-scale classification datasets, some for domains far outside the ImageNet training set used for all teachers and distillation (*i.e.* including satellite images and textures). Complementary teachers appear to be highly beneficial in this case.

---

[3] We use the dBOT model fine-tuned for ImageNet-1K classification.

5. **Strong gains for dense prediction with linear probing.** Strong gains are also observed on segmentation and depth estimation, for example on ADE-20K where UNIC achieves a +8.2% relative gain over iBOT. Although far from being the optimal protocol for the task, linear probing is best to evaluate the discriminative power of the patch tokens from the encoder.
6. **Retaining top teacher performance for domain shifts.** DeiT-III shows exceptionally high performance on ImageNet-R and Sketch (51.4% and 39.3% top-1 accuracy, respectively). Our best UNIC model retains this top performance, achieving 51.4% and 38.5%, respectively.

Distilling Arbitrary Models. We extend our study to larger teachers like MetaCLIP ViT-Huge/14 [64] and DINOv2 ViT-Giant/14 [35] trained on arbitrary datasets. We train a ViT-Large/14 student for 200 epochs at resolution 224, setting the teacher dropping probability to $p = 0.25$. In Table 2 we report results for k-NN and zero-shot classification on ImageNet-1K, as well as semantic segmentation on ADE-20K. UNIC* refers to a UNIC model without a ladder of projectors. These results offer some basic verification that our insights are also valid in this more generic distillation case: Our UNIC model outperforms DINOv2 on ImageNet-1K as well as the MetaCLIP on zero-shot classification.

Table 2. Distilling MetaCLIP and DINOv2

Model	k-NN	Z-shot	ADE-20K
Teacher Models			
MetaCLIP-H [64]	82.1	80.5	35.4
DINOv2-G [35]	83.4	–	**48.7**
AM-RADIO [41]	84.8	80.4	48.1
UNIC*-L	85.0	80.7	47.7
UNIC-L	**85.4**	**81.2**	47.1

Weight and Feature Space Utilization. In this section, we seek to better understand why multi-teacher distillation leads to overall stronger encoders. We do that by investigating the utilization of the encoder weights after pruning (Fig. 6a) and the feature space after dimensionality reduction (Fig. 6b). We report the change in accuracy on ImageNet-1K for our UNIC model and its teachers when we prune the weights or reduce the feature dimension before training linear probes. We prune encoder weights using ℓ_1-norm-based unstructured weight pruning, and perform dimensionality reduction using PCA with whitening.

From Fig. 6a, we see that the performance of UNIC drops more rapidly than any of the teachers as we increase the pruning ratio. This indicates that the encoder weights show improved synergy, working together more cohesively and efficiently to enhance the model's overall performance.

▶ *UNIC encoders utilize weights more effectively*

At the same time, in Fig. 6b, we see that our student preserves its base performance better than all teachers as we reduce the number of dimensions with PCA. It seems that the feature space of UNIC can be represented better with fewer principal components, possibly because of higher entanglement in the original feature space.

▶ *UNIC encoders are more resilient to dimensionality reduction*

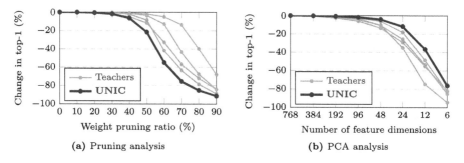

Fig. 6. Network utility analysis via ImageNet-1K linear probing for the four teachers and our student UNIC distilled from all of them. For each model, before training linear probes, we either **(a)** prune their weights or **(b)** reduce the dimension of their features via PCA. We report change in top-1 accuracy compared to their base performance. UNIC's encoder weights work together more cohesively **(a)**, and its feature space is more robust to dimensionality reduction **(b)**.

5 Conclusions

In this paper, we systematically analyze multi-teacher distillation and introduce improvements to the distillation process that significantly enhance the performance of student models across various benchmarks. More importantly, we show that it is possible to distil from multiple teachers with complementary strengths and learn models that match or improve the respective best teacher in both image- and patch-based classification tasks. In that regard, we view UNIC models as *universal* classification models, advancing the frontier of general representation learning without task-specific adaptation.

Acknowledgements. The authors would like to sincerely thank Myung-Ho Ju, Florent Perronnin, Rafael Sampaio de Rezende, Vassilina Nikoulina and Jean-Marc Andreoli for inspiring discussions and many thoughtful comments.

References

1. Ahn, S., Hu, S.X., Damianou, A., Lawrence, N.D., Dai, Z.: Variational information distillation for knowledge transfer. In: Proceedings of CVPR (2019)
2. Asif, U., Tang, J., Harrer, S.: Ensemble knowledge distillation for learning improved and efficient networks. In: Proceedings of ECAI (2020)
3. Ba, J., Frey, B.: Adaptive dropout for training deep neural networks. In: Proceedings of NeurIPS (2013)
4. Bossard, L., Guillaumin, M., Van Gool, L.: Food-101 – mining discriminative components with random forests. In: Fleet, D., Pajdla, T., Schiele, B., Tuytelaars, T. (eds.) ECCV 2014. LNCS, vol. 8694, pp. 446–461. Springer, Cham (2014). https://doi.org/10.1007/978-3-319-10599-4_29
5. Buciluǎ, C., Caruana, R., Niculescu-Mizil, A.: Model compression. In: Proceedings of SIGKDD (2006)

6. Caron, M., et al.: Emerging properties in self-supervised vision transformers. In: Proceedings of ICCV (2021)
7. Chen, T., Kornblith, S., Norouzi, M., Hinton, G.: A simple framework for contrastive learning of visual representations. In: Proceedings of ICML (2020)
8. Chen, X., He, K.: Exploring simple siamese representation learning. In: Proceedings of CVPR (2021)
9. Chen, Z., Badrinarayanan, V., Lee, C.Y., Rabinovich, A.: Gradnorm: gradient normalization for adaptive loss balancing in deep multitask networks. In: Proceedings of ICML (2018)
10. Cimpoi, M., Maji, S., Kokkinos, I., Mohamed, S., Vedaldi, A.: Describing textures in the wild. In: Proceedings of CVPR (2014)
11. Clark, K., Luong, M.T., Khandelwal, U., Manning, C.D., Le, Q.V.: Bam! born-again multi-task networks for natural language understanding. In: ACL (2019)
12. Dosovitskiy, A., et al.: An image is worth 16x16 words: Transformers for image recognition at scale. In: Proceedings of ICLR (2021)
13. Fukuda, T., Suzuki, M., Kurata, G., Thomas, S., Cui, J., Ramabhadran, B.: Efficient knowledge distillation from an ensemble of teachers. In: Interspeech (2017)
14. Ghiasi, G., Zoph, B., Cubuk, E.D., Le, Q.V., Lin, T.Y.: Multi-task self-training for learning general representations. In: Proceedings of CVPR (2021)
15. Hao, Z., et al.: Learning efficient vision transformers via fine-grained manifold distillation. In: Proceedings of NeurIPS (2022)
16. He, K., Chen, X., Xie, S., Li, Y., Dollár, P., Girshick, R.: Masked autoencoders are scalable vision learners. In: Proceedings of CVPR (2022)
17. Helber, P., Bischke, B., Dengel, A., Borth, D.: EuroSAT: a novel dataset and deep learning benchmark for land use and land cover classification. JSTAEORS (2019)
18. Hendrycks, D., et al.: The many faces of robustness: a critical analysis of out-of-distribution generalization. In: Proceedings of ICCV (2021)
19. Heo, B., Kim, J., Yun, S., Park, H., Kwak, N., Choi, J.Y.: A comprehensive overhaul of feature distillation. In: Proceedings of ICCV (2019)
20. Heo, B., Lee, M., Yun, S., Choi, J.Y.: Knowledge transfer via distillation of activation boundaries formed by hidden neurons. In: Proceedings of AAAI (2019)
21. Hinton, G., Vinyals, O., Dean, J.: Distilling the knowledge in a neural network. In: Proceedings of NeurIPS-W (2014)
22. Hu, H., Dey, D., Hebert, M., Bagnell, J.A.: Learning anytime predictions in neural networks via adaptive loss balancing. In: Proceedings of AAAI (2019)
23. Huang, G., Sun, Yu., Liu, Z., Sedra, D., Weinberger, K.Q.: Deep networks with stochastic depth. In: Leibe, B., Matas, J., Sebe, N., Welling, M. (eds.) ECCV 2016. LNCS, vol. 9908, pp. 646–661. Springer, Cham (2016). https://doi.org/10.1007/978-3-319-46493-0_39
24. Kendall, A., Gal, Y., Cipolla, R.: Multi-task learning using uncertainty to weigh losses for scene geometry and semantics. In: Proceedings of CVPR (2018)
25. Kirillov, A., et al.: Segment anything. arXiv:2304.02643 (2023)
26. Krause, J., Deng, J., Stark, M., Li, F.F.: Collecting a large-scale dataset of fine-grained cars. In: Proceedings of CVPR-W (2013)
27. Krizhevsky, A., Sutskever, I., Hinton, G.E.: ImageNet classification with deep convolutional neural networks. In: Proceedings of NeurIPS (2012)
28. Landgraf, S., Hillemann, M., Kapler, T., Ulrich, M.: Efficient multi-task uncertainties for joint semantic segmentation and monocular depth estimation. arXiv:2402.10580 (2024)
29. Liu, X., Zhou, J., Kong, T., Lin, X., Ji, R.: Exploring target representations for masked autoencoders. In: Proceedings of ICLR (2022)

30. Liu, Y., Zhang, W., Wang, J.: Adaptive multi-teacher multi-level knowledge distillation. Neurocomputing (2020)
31. Maji, S., Rahtu, E., Kannala, J., Blaschko, M., Vedaldi, A.: Fine-grained visual classification of aircraft. arXiv:1306.5151 (2013)
32. Marrie, J., Arbel, M., Mairal, J., Larlus, D.: On good practices for task-specific distillation of large pretrained models. arXiv:2402.11305 (2024)
33. Matena, M.S., Raffel, C.A.: Merging models with fisher-weighted averaging. In: Proc.eedings of NeurIPS (2022)
34. Nilsback, M.E., Zisserman, A.: Automated flower classification over a large number of classes. In: Proceedings of ICVGIP (2008)
35. Oquab, M., et al.: DINOv2: Learning robust visual features without supervision. TMLR (2024)
36. Parkhi, O.M., Vedaldi, A., Zisserman, A., Jawahar, C.V.: Cats and dogs. In: Proceedings of CVPR (2012)
37. Peng, Z., Dong, L., Bao, H., Wei, F., Ye, Q.: A unified view of masked image modeling. TMLR (2023)
38. Radford, A., et al.: Learning transferable visual models from natural language supervision. In: Proceedings of ICML (2021)
39. Ramé, A., Ahuja, K., Zhang, J., Cord, M., Bottou, L., Lopez-Paz, D.: Model ratatouille: Recycling diverse models for out-of-distribution generalization. In: Proceedings of ICML (2023)
40. Rame, A., Kirchmeyer, M., Rahier, T., Rakotomamonjy, A., Gallinari, P., Cord, M.: Diverse weight averaging for out-of-distribution generalization. In: Proceedings of NeurIPS (2022)
41. Ranzinger, M., Heinrich, G., Kautz, J., Molchanov, P.: AM-RADIO: Agglomerative model–reduce all domains into one. In: Proceedings of CVPR (2024)
42. Recht, B., Roelofs, R., Schmidt, L., Shankar, V.: Do ImageNet classifiers generalize to ImageNet? In: Proceedings of ICML (2019)
43. Romero, A., Ballas, N., Kahou, S.E., Chassang, A., Gatta, C., Bengio, Y.: Fitnets: hints for thin deep nets. In: Proceedings of ICLR (2015)
44. Roth, K., Milbich, T., Ommer, B., Cohen, J.P., Ghassemi, M.: Simultaneous similarity-based self-distillation for deep metric learning. In: Proceedings of ICML (2021)
45. Roth, K., Thede, L., Koepke, A.S., Vinyals, O., Henaff, O.J., Akata, Z.: Fantastic gains and where to find them: on the existence and prospect of general knowledge transfer between any pretrained model. In: Proceedings of ICLR (2024)
46. Russakovsky, O., et al.: ImageNet large scale visual recognition challenge. IJCV **115**(3) (2015)
47. Sariyildiz, M.B., Alahari, K., Larlus, D., Kalantidis, Y.: Fake it till you make it: Learning transferable representations from synthetic ImageNet clones. In: Proceedings of CVPR (2023)
48. Sariyildiz, M.B., Kalantidis, Y., Alahari, K., Larlus, D.: No reason for no supervision: Improved generalization in supervised models. In: Proceedings of ICLR (2023)
49. Sariyildiz, M.B., Kalantidis, Y., Larlus, D., Alahari, K.: Concept generalization in visual representation learning. In: Proceedings of ICCV (2021)
50. Shi, B., et al.: Hybrid distillation: Connecting masked autoencoders with contrastive learners. In: Proceedings of ICLR (2024)

51. Silberman, N., Hoiem, D., Kohli, P., Fergus, R.: Indoor segmentation and support inference from RGBD images. In: Fitzgibbon, A., Lazebnik, S., Perona, P., Sato, Y., Schmid, C. (eds.) ECCV 2012. LNCS, vol. 7576, pp. 746–760. Springer, Heidelberg (2012). https://doi.org/10.1007/978-3-642-33715-4_54
52. Srivastava, N., Hinton, G., Krizhevsky, A., Sutskever, I., Salakhutdinov, R.: Dropout: a simple way to prevent neural networks from overfitting. JMLR **15**(1) (2014)
53. Stoica, G., Bolya, D., Bjorner, J., Ramesh, P., Hearn, T., Hoffman, J.: Zipit! merging models from different tasks without training. In: Proceedings of ICLR (2024)
54. Tian, Y., Krishnan, D., Isola, P.: Contrastive representation distillation. In: Proceedings of ICLR (2020)
55. Touvron, H., Cord, M., Douze, M., Massa, F., Sablayrolles, A., Jégou, H.: Training data-efficient image transformers & distillation through attention. In: Proceedings of ICML (2021)
56. Touvron, H., Cord, M., Jegou, H.: DeiT III: Revenge of the ViT. In: Proc. ECCV (2022). https://doi.org/10.1007/978-3-031-20053-3_30
57. Van Horn, G., Met al.: The iNaturalist species classification and detection dataset. In: Proceedings of CVPR (2018)
58. Wang, H., Ge, S., Lipton, Z., Xing, E.P.: Learning robust global representations by penalizing local predictive power. In: Proceedings of NeurIPS (2019)
59. Wang, H., et al.: SAM-CLIP: merging vision foundation models towards semantic and spatial understanding. In: Proceedings of CVPR-W (2023)
60. Wang, Y., et al.: Revisiting the transferability of supervised pretraining: an MLP perspective. In: Proceedings of CVPR (2022)
61. Wei, L., Xie, L., Zhou, W., Li, H., Tian, Q.: Mvp: multimodality-guided visual pre-training. In: Proceedings of ECCV (2022)
62. Wortsman, M., et al.: Model soups: averaging weights of multiple fine-tuned models improves accuracy without increasing inference time. In: Proceedings of ICML (2022)
63. Xiao, J., Hays, J., Ehinger, K.A., Oliva, A., Torralba, A.: SUN database: large-scale scene recognition from abbey to zoo. In: Proceedings of CVPR (2010)
64. Xu, H., et al.: Demystifying clip data. In: Proceedings of ICLR (2024)
65. Yao, Y., Desai, N., Palaniswami, M.: MOMA: Distill from self-supervised teachers. arXiv:2302.02089 (2023)
66. Ye, P., et al.: Merging vision transformers from different tasks and domains. arXiv:2312.16240 (2023)
67. Yim, J., Joo, D., Bae, J., Kim, J.: A gift from knowledge distillation: fast optimization, network minimization and transfer learning. In: Proceedings of CVPR (2017)
68. Yin, D., Han, X., Li, B., Feng, H., Bai, J.: Parameter-efficient is not sufficient: Exploring parameter, memory, and time efficient adapter tuning for dense predictions. arXiv:2306.09729 (2023)
69. You, S., Xu, C., Xu, C., Tao, D.: Learning from multiple teacher networks. In: Proceedings of SIGKDD (2017)
70. Ypsilantis, N.A., Chen, K., Araujo, A., Chum, O.: Udon: Universal dynamic online distillation for generic image representations. arXiv: 2406.08332 (2024)
71. Yuan, H., Li, X., Zhou, C., Li, Y., Chen, K., Loy, C.C.: Open-vocabulary SAM: Segment and recognize twenty-thousand classes interactively. In: Proceedings of ECCV (2024)
72. Yun, H., Cho, H.: Achievement-based training progress balancing for multi-task learning. In: Proceedings of ICCV (2023)

73. Zagoruyko, S., Komodakis, N.: Paying more attention to attention: Improving the performance of convolutional neural networks via attention transfer. In: Proceedings of ICLR (2017)
74. Zhou, B., et al.: Semantic understanding of scenes through the ADE20k dataset. IJCV (2019)
75. Zhou, J., et al.: iBOT: image BERT pre-training with online tokenizer. In: Proceedings of ICLR (2022)

Instance-Dependent Noisy-Label Learning with Graphical Model Based Noise-Rate Estimation

Arpit Garg[1](✉)[iD], Cuong Nguyen[3][iD], Rafael Felix[1][iD], Thanh-Toan Do[2][iD], and Gustavo Carneiro[3][iD]

[1] Australian Institute for Machine Learning, University of Adelaide, Adelaide, SA, Australia
arpit.garg@adelaide.edu.au
[2] Department of Data Science and AI, Monash University, Melbourne, Australia
[3] Centre for Vision, Speech and Signal Processing, University of Surrey, Surrey, UK

Abstract. Deep learning faces a formidable challenge when handling noisy labels, as models tend to overfit samples affected by label noise. This challenge is further compounded by the presence of instance-dependent noise (IDN), a realistic form of label noise arising from ambiguous sample information. To address IDN, Label Noise Learning (LNL) incorporates a sample selection stage to differentiate clean and noisy-label samples. This stage uses an arbitrary criterion and a pre-defined curriculum that initially selects most samples as noisy and gradually decreases this selection rate during training. Such curriculum is sub-optimal since it does not consider the actual label noise rate in the training set. This paper addresses this issue with a new noise-rate estimation method that is easily integrated with most state-of-the-art (SOTA) LNL methods to produce a more effective curriculum. Synthetic and real-world benchmarks' results demonstrate that integrating our approach with SOTA LNL methods improves accuracy in most cases. (Code is available at https://github.com/arpit2412/NoiseRateLearning. Supported by the Engineering and Physical Sciences Research Council (EPSRC) through grant EP/Y018036/1 and the Australian Research Council (ARC) through grant FT190100525.)

Keywords: Noisy-labels · Instance-dependent noise · Label noise Learning

1 Introduction

Deep neural networks (DNNs) have demonstrated their effectiveness in various domains, including vision [47], language [42], and medicine [37]. However, its efficacy is largely dependent on high-quality training data, which can be resource-intensive to obtain [43]. Although cost-effective labelling techniques, such as data mining [11] and crowdsourcing [39], offer cheaper alternatives, they often compromise label quality [39]. Consequently, this can lead to erroneous labels in real-world datasets [21]. This is a relevant issue because even slight inaccuracies in the labels can significantly affect the performance of DNN due to their inherent memorisation capabilities [34,53]. This has prompted the development of algorithms resilient to noisy-label learning, with the aim of training models despite inaccuracies in training data labels. Although various strategies exist, our work uniquely proposes an approach centered on the estimation of the label noise rate from the training data, which is notably absent from current methodologies.

There are two types of label noise, namely instance-independent noise (IIN) [16] and instance-dependent noise (IDN) [44]. Such type of label noise typically influences the design principles of noisy-label learning algorithms. For example, IIN focuses on mislabellings that are independent of sample information [16], where estimating the underlying label transition matrix is a common way of handling this type of noise [51]. On the other hand, in the more realistic IDN, mislabelling is due to both sample information and true class labels [44], which generally require the combination of many label noise learning techniques, such as robust loss functions [28,56], and noisy-label sample selection [26,58]. Of the strategies mentioned above, sample selection approaches that classify training data into clean and noisy samples have produced competitive results on many benchmarks [9,11,15,21,26]. Such sample selection techniques require the definition of a classification criterion and a selection curriculum. Many studies on this topic focus on developing new sample selection criteria, such as the small-loss hypothesis [26], which states that noisy-label samples have higher loss values than clean-label samples, particularly at the early stage of training [1]. Another criterion type is the feature-based one. An example of such criterion is FINE [21] that discriminates clean and noisy-label samples via the distance to class-specific eigenvectors. In this technique, clean-label samples tend to lie closer to the class-specific dominant eigenvector of the latent representations than noisy-label samples. Another type of criterion is proposed in SSR [11], which introduces a selection criterion based on the K-nearest-neighbor (KNN) classification in the feature space. Furthermore, CC [58] uses a two-stage sampling procedure, including class-level feature clustering followed by a consistency score. An equally important problem in sample selection is the definition of the curriculum for selecting clean training samples, but it has received comparatively less attention.

The sample selection curriculum defines a threshold to be used with one of the criteria listed above to classify the training samples as clean or noisy at each training epoch [45]. For example, the threshold can be fixed to an arbitrary clustering score that separates clean and noisy samples [26], but this strategy does not account for the proportion of label noise in the training set, nor does it consider the dynamics of the selection of noisy-label samples during the training. The consideration of such dynamics has been studied in [16,49], which defined a curriculum of the noisy-label sampling rate $R(t)$ as a function of the training epoch $t \in \{1, \ldots, T\}$. The curriculum $R(t)$ defines a sampling rate close to 100%

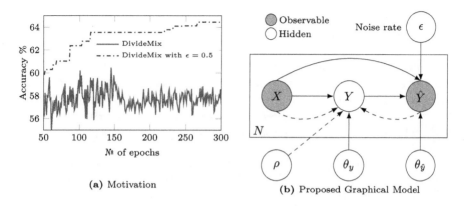

Fig. 1. (a) Comparison of test accuracy % (as a function of training epoch) between the original DivideMix [26] (solid, blue curve) and our modified DivideMix (dashed, red curve) that selects the clean and noisy data based on a fixed noise rate $R(t) = 1 - \epsilon = 50\%$ using the small-loss criterion on CIFAR100 [22] at 0.5 IDN [44]; (b) The proposed probabilistic graphical model that generates noisy-label $\hat{Y}$ conditioned on the image X, the latent clean-label Y and noise rate ϵ, where forward pass (solid lines) is parameterised by $\theta_y, \theta_{\hat{y}}$ and ϵ representing the generation step, and the backward pass (dashed lines) is parameterised by ρ. (Color figure online)

of the training set at the beginning of the training, which is then reduced to arbitrarily low rates at the end of the training. In practice, the function $R(t)$ is predefined [16] or learned by weighting a set of basis functions [49].

Although generally effective, these techniques do not consider the label noise rate estimated from the training set, making them vulnerable to over-fitting (if too many noisy-label samples are classified as clean) or under-fitting (if informative clean-label samples are classified as noisy). It can be argued that the estimation of the label transition matrix [6,7,48,51] aims to recover the noise rate affecting pairwise label transitions. However, label transition matrix techniques follow a quite different strategy compared to sample selection methods, where their main challenge is the general underconstrained aspect of the matrix estimation, making them sensitive to large noise rates and not scalable to a high number of classes [39]. Although several methods address label noise, none estimates the label noise rate directly from the training data to guide sample selection. Addressing this gap is the primary challenge of our study.

To underscore the importance of using the noise rate for sample selection during training, we experiment with CIFAR100 [22] at an IDN rate $\epsilon = 50\%$ [44] (noise rate specifications and other details are explained in Sect. 4). We use DivideMix [26] and replace its sample selection (based on an arbitrary clustering score [10,32,41]) by a thresholding process that classifies the $R(t) = 1 - \epsilon = 50\%$ largest loss samples as noisy, and the remaining ones as clean in all training epochs $t \in \{1, \ldots, T\}$. This sample selection is used for the semi-supervised learning of DivideMix [26]. As shown in Fig. 1a, the new sample selection approach based on the "manually provided" noise rate (dashed red curve) improves 6% in terms of prediction accuracy compared to the original DivideMix [26]

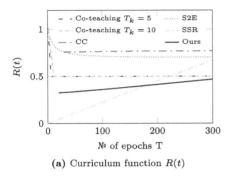 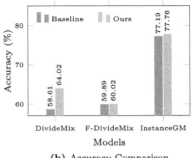

Fig. 2. (a) Visual comparison of different $R(t)$ on CIFAR100 [22] at 0.5 IDN [44]: *(i)* Co-teaching [16] with curriculum based on hyper-parameter $T_k \in \{5, 10\}$, where $R(t) = 1 - \tau \cdot \min(t/T_k, 1)$ with $\tau = \epsilon = 0.5$ being manually set; *(ii)* S2E [49], where $R(t)$ is estimated with a bi-level optimisation; *(iii)* SSR [11], where R(t) is based on a relabelling pattern, *(iv)* CC [58] following the small-loss hypothesis [26] and cosine similarity for R(t); and *(v)* ours. (b) Accuracy of noisy-label learning methods on CIFAR100 [22] at 0.5 IDN [44], including DivideMix [26], FINE [21] and InstanceGM [15], without (left, blue) and (right, orange) integration of our proposed graphical model for estimating the noise rate ϵ and sample selection rate based on ϵ. (Color figure online)

(solid blue curve) which relies on arbitrary thresholding. Similar conclusions can be achieved with other methods that apply sample selection strategies to address the noisy-label learning problem, as shown in the experiments.

In this paper, we introduce a new sample selection strategy centered on estimating the label noise rate of the training set. Our strategy is based on our novel noisy-label learning graphical model illustrated in Fig. 1b. This model can be seamlessly integrated with state-of-the-art (SOTA) noisy-label learning techniques, providing them with a precise noise rate estimate and subsequently refining the sample selection curriculum. In particular, our model's curriculum is not restrained by predefined $R(t)$ functions [16,49], but rather relies on a dynamically estimated noise rate sourced directly from the training set, as depicted in Fig. 2a. Our method dynamically estimates training set noise rates to mitigate overfitting and underfitting, seamlessly integrates with existing algorithms, and serves as a robust foundation for future noisy-label learning research. To summarise, our main contributions include:

- An innovative noisy-label learning graphical model, shown in Fig. 1b, that not only estimates but also leverages the noise rate from the training dataset to produce a refined sample selection curriculum for SOTA LNL methods.
- A simple and synergistic integration strategy of our novel graphical model with several SOTA noisy-label learning algorithms, such as DivideMix [26] and SSR [11], to improve their sample selection effectiveness, leading to an increased test accuracy, as shown in Fig. 2b.

Empirical evaluations show the role of our proposed sample selection methodology in improving the efficacy of leading noisy-label learning algorithms across various synthetic and real-world IDN benchmarks. Although our primary focus lies in addressing IDN problems, because of its more challenging and realistic nature, we have also applied our model to IIN problems, with detailed results in Table 6 and discussion in Appendix 3 of the supplementary material. With our innovative approach, we aim to improve the accuracy of SOTA LNL methods.

2 Related Work

It is well-known that DNNs suffer from overfitting when trained with noisy-labels [52], resulting in poor generalisation [28,55]. To mitigate the problems related to noisy-label learning, several techniques have been developed, including *noise-robust loss functions* [30], *noise-label sample selection* [21,26,40,58,59], *meta-learning* [57], and *re-labeling* followed by *sample selection* [11]. These techniques can handle many types of label noise (e.g., symmetric and asymmetric), but the most challenging type, i.e., IDN [39], tends to be successfully addressed with methods that have a training stage based on sample selection techniques [21,26,40,58,59]. These sample selection techniques separate clean and noisy-label samples, where clean samples are treated as labelled, and noisy samples are discarded [8,16,18,49] or treated as unlabeled samples [9,15,26,35,59] for semi-supervised learning [3]. Another solution consists of aligning clean and noisy samples with an information fusion method [19]. A major limitation of these approaches is the overly simplistic design of the curriculum to select clean and noisy-label samples during training. It is either pre-determined [16,18] or learned from a set of predetermined basis functions [49]. We argue that the design of a curriculum based on an estimated noise rate would benefit such methods, as motivated by Fig. 1a, but to the best of our knowledge, such estimation has not been explored by previous methods. The closest idea explored is based on the estimation of the type of noise instead of the noise rate [45]. It is possible to argue that the estimation of the label transition matrices [5,7,44,60], composed of the learned noise rate that affects the transition between pairs of labels, is a way to estimate the label noise rate. However, they portray results of comparatively lower accuracy for large real-world datasets or for IDN problems [39,44]. A possible reason for these poorer results is that label transition approaches suffer from identifiability issues [14], where any clean-label distribution assignment is acceptable as long as the distribution of observed labels can be reconstructed [13]. This makes the identification of the actual underlying clean-labels challenging. One solution is to consider the use of multiple annotations per sample to help analyse agreements and disagreements for improved identification of clean patterns [14,31]. However, the annotation of datasets with a single label per training sample is already challenging; the complexity is significantly larger to acquire multiple labels per sample.

An alternative technique to handle IDN problems is based on graphical models that represent the relationship between various observed and latent variables [15,45,50]. The approaches in [15,50] use graphical models that rely on a generative approach to produce noisy-labels from the respective image features and latent clean-labels. However, previous graphical models do not take into account the underlying noise rate parameter during modeling. Our work is the first graphical model approach to estimate the noise rate of the dataset. In addition, a close examination of existing literature reveals a trend: when traditional graphical models are integrated with SOTA noisy-label learning methods, the outcomes often fall short of expectations in terms of accuracy [2,15,50]. This observation underscores a strong point of our work. Our approach is uniquely tailored for easy integration with current SOTA noisy-label learning methods for sample selection, which, in turn, improves over the SOTA classification accuracy.

3 Method

In this section, we present our new graphical model that estimates the noise rate, which will be used in the sample selection process. Let $\mathcal{D} = \{(x_i, \hat{y}_i)\}_{i=1}^{N}$ be the noisy-label training set containing d-dimensional data vector $x_i \in \mathcal{X} \subseteq \mathbb{R}^d$ and its respective C-dimensional one-hot encoded observed (potentially corrupted) label $\hat{y}_i \in \hat{\mathcal{Y}} = \{\hat{y} : \hat{y} \in \{0,1\}^C \wedge \mathbf{1}_C^\top \hat{y} = 1\}$, where $\mathbf{1}_C$ is a vector of ones with C dimensions. The aim is to estimate the label noise rate ϵ, used for the generation of noisy-label training data from the observed training dataset $\mathcal{D}$ and integrate this label noise rate into the sample selection strategy.

3.1 Graphical Model

We portray the generation of noisy-label via the probabilistic graphical model shown in Fig. 1b. The observed random variables, denoted by shaded circles, are data X and the corresponding noisy-label $\hat{Y}$. We also have one latent variable, namely: the clean-label Y. Under our proposed modeling assumption, a noisy-label of a data instance can be generated as follows:

- sample an instance from $p(X)$, i.e. : $x \sim p(X)$
- sample a clean-label from the clean-label distribution:
  $y \sim \text{Categorical}(Y; f_{\theta_y}(x))$
- sample a noisy-label from the noisy-label distribution:
  $\hat{y} \sim \text{Categorical}(\hat{Y}; \epsilon \times f_{\theta_{\hat{y}}}(x, f_{\theta_y}(x)) + (1-\epsilon) \times y)$,

where Categorical(.) denotes a categorical distribution, $f_{\theta_y} : \mathcal{X} \to \Delta_{C-1}$ and $f_{\theta_{\hat{y}}} : \mathcal{X} \times \Delta_{C-1} \to \Delta_{C-1}$ denote two classifiers for the clean-label Y and noisy-label $\hat{Y}$, respectively, with $\Delta_{C-1} = \{s : s \in [0,1]^C \wedge \mathbf{1}_C^\top s = 1\}$ being the $(C-1)$-dimensional probability simplex. According to the data generation process shown in Fig. 1b, ϵ corresponds to $\mathbb{E}_{(x,\hat{y}) \sim p(X,\hat{Y})}[P(\hat{y} \neq y|x)]$, which is the label noise rate of the training dataset of interest. Our aim is to infer the parameters $\theta_y, \theta_{\hat{y}}$ and ϵ from a noisy-label dataset $\mathcal{D}$ by maximising the following log-likelihood:

$$\max_{\theta_y, \theta_{\hat{y}}, \epsilon} \mathbb{E}_{(x_i, \hat{y}_i) \sim \mathcal{D}}\left[\ln p(\hat{y}_i|x_i; \theta_y, \theta_{\hat{y}}, \epsilon)\right] = \max_{\theta_y, \theta_{\hat{y}}, \epsilon} \mathbb{E}_{(x_i, \hat{y}_i) \sim \mathcal{D}}\left[\ln \sum_{y_i} p(\hat{y}_i, y_i|x_i; \theta_y, \theta_{\hat{y}}, \epsilon)\right]. \quad (1)$$

Due to the presence of the clean-label y_i, it is difficult to directly evaluate the log-likelihood in Eq. (1). Therefore, we employ the *expectation - maximisation* (EM) algorithm [10] to maximise the log-likelihood. The main idea of the EM algorithm is to *(i)* construct a tight lower bound of the likelihood in Eq. (1) by estimating the posterior of the latent variable Y (known as *expectation step*) and *(ii)* maximise that lower bound (known as *maximisation step*). Formally, let $q(y_i|x, \hat{y}; \rho)$ be an arbitrary distribution over a clean-label y_i. The evidence lower bound (ELBO) of the log-likelihood in Eq. (1) can be obtained through Jensen's inequality and is presented as follows:

$$Q(\theta_y, \theta_{\hat{y}}, \epsilon, \rho) = \mathbb{E}_{(x_i, \hat{y}_i) \sim \mathcal{D}} \left[\mathbb{E}_{q(y_i|x_i; \hat{y}_i; \rho)} \left[\ln p(\hat{y}_i|x_i, y_i; \theta_y, \theta_{\hat{y}}, \epsilon) \right] \right.$$
$$\left. - \text{KL} \left[q(y_i|x_i, \hat{y}_i; \rho) \| p(y_i|x_i; \theta_y, \theta_{\hat{y}}, \epsilon) \right] \right]$$
$$= \mathbb{E}_{(x_i, \hat{y}_i) \sim \mathcal{D}} \left[\mathbb{E}_{q(y_i|x_i; \hat{y}_i; \rho)} \left[\ln p(\hat{y}_i|x_i, y_i; \theta_{\hat{y}}, \epsilon) \right.\right.$$
$$\left.\left. + \ln p(y_i|x_i; \theta_y) \right] + \mathbb{H} \left[q(y_i|x_i, \hat{y}_i; \rho) \right] \right], \quad (2)$$

where $\text{KL}[q\|p]$ is the Kullback - Leibler divergence between distributions q and p and $\mathbb{H}(q)$ is the entropy of the distribution q. The EM algorithm is then carried out iteratively by alternating the following two steps:

E step: we maximise the ELBO in (2) w.r.t. $q(y_i|x_i, \hat{y}_i; \rho)$. Theoretically, such an optimisation results in $\text{KL}\left[q(y_i|x_i, \hat{y}_i; \rho) \| p(y_i|x_i, \hat{y}_i)\right] = 0$ or $q(y_i|x_i, \hat{y}_i; \rho) = p(y_i|x_i, \hat{y}_i)$. This is equivalent to estimating the posterior of the clean-label y_i given noisy-label data $(x_i, \hat{y}_i)$. Obtaining the posterior $p(y_i|x_i, \hat{y}_i)$ is, however, intractable for most deep-learning models. To mitigate such an issue, we follow the *variational EM* approach [33] by employing an approximate posterior $q(y_i|x_i, \hat{y}_i; \rho^{(t)})$ that is the closest to the true posterior $p(y_i|x_i, \hat{y}_i)$, where:

$$\rho^{(t)} = \arg\max_{\rho} Q\left(\theta_y^{(t)}, \theta_{\hat{y}}^{(t)}, \epsilon^{(t)}, \rho\right), \quad (3)$$

with the superscript $^{(t)}$ denoting the parameters at the t-th iteration. Although this results in a non-tight lower bound of the log-likelihood in Eq. (1), it does increase the variational bound Q.

M step: we maximise the ELBO in Eq. (2) w.r.t. $\theta_y, \theta_{\hat{y}}$ and ϵ given $\rho^{(t)}$ obtained in the E step:

$$\theta_y^{(t+1)}, \theta_{\hat{y}}^{(t+1)}, \epsilon^{(t+1)} = \arg\max_{\theta_y, \theta_{\hat{y}}, \epsilon} Q\left(\theta_y, \theta_{\hat{y}}, \epsilon, \rho^{(t)}\right). \quad (4)$$

The estimated noise rate ϵ can then be integrated into certain noisy-label algorithms to train the models of interest as mentioned in Sect. 1. In addition, the inference of noise rate ϵ might be associated with the identifiability issue when estimating the clean-label Y [29], i.e., there exists multiple sets of ρ and θ_y, where each set can explain the observed noisy-label data equally well. Such issues are addressed in the following subsection.

3.2 Sample Selection

The identifiability issue when inferring the clean-label Y from noisy-label data $(X, \hat{Y})$ can be mitigated either by acquiring multiple noisy-labels [29] or introducing additional constraints, such as small loss hypothesis [16] or FINE [21]. Since requesting additional noisy-labels per training sample is not always possible, we follow the latter approach by imposing a constraint (scoring term), denoted as $L(\theta_y, \epsilon^{(t)})$, over θ_y in the M step via a sample selection approach based on the estimated noise rate $\epsilon^{(t)}$. Formally, we propose a new curriculum when selecting samples as follows:

Algorithm 1 Proposed noisy-label learning that relies on the estimation of noise rate ϵ to build a sample selection curriculum.

1: **procedure** NOISE RATE ESTIMATION AND INTEGRATION($\mathcal{D}, T, \lambda$)
2:      ▷ $\mathcal{D} = \{(x_i, \hat{y}_i)\}_{i=1}^{N}$: training set with noisy-label data   ◁
3:      ▷ T: number of epochs   ◁
4:      ▷ λ: a hyper-parameter   ◁
5:      ▷ Notation: $\Theta^{(t)} = (\theta_y^{(t)}, \theta_{\hat{y}}^{(t)}, \epsilon^{(t)})$ ◁
6:      Initialise $\theta_y^{(1)}, \theta_{\hat{y}}^{(1)}, \epsilon^{(1)}$ and $\rho^{(0)}$
7:      $\theta_y^{(1)} \leftarrow$ WARM UP($\mathcal{D}, \theta_y^{(1)}$)
8:      $t \leftarrow 0$
9:      **for** $n_{\text{epoch}} = 1 : T$ **do**
10:         **for** each mini-batch $\mathcal{S}$ in SHUFFLE($\mathcal{D}$) **do**
11:            $t \leftarrow t + 1$
12:            $\mathcal{S}_{\text{clean}}, \mathcal{S}_{\text{noisy}} \leftarrow$ SAMPLE SELECTION($\mathcal{S}, \theta_y^{(t)}, \epsilon^{(t)}$)   ▷ Eq. (7)
13:            ▷ Variational E-step as in Eq. (3)   ◁
14:            $\rho^{(t)} \leftarrow$ E STEP($\mathcal{S}, \theta_y^{(t)}, \theta_{\hat{y}}^{(t)}, \epsilon^{(t)}, \rho^{(t-1)}$)
15:            ▷ M-step as in Eq. (8)   ◁
16:            $\theta_y^{(t+1)}, \theta_{\hat{y}}^{(t+1)}, \epsilon^{(t+1)} \leftarrow$ M STEP($\mathcal{S}_{\text{clean}}, \mathcal{S}_{\text{noisy}}, \theta_y^{(t)}, \theta_{\hat{y}}^{(t)}, \epsilon^{(t)}, \rho^{(t)}, \lambda$)
17:      **return** θ_y ▷ parameter of the clean-label classifier

---

$$R(t) = 1 - \epsilon^{(t)}. \tag{5}$$

In the simplest case, such as Co-teaching [16] or FINE [21], the constraint for θ_y can be written as:

$$L(\theta_y, \epsilon^{(t)}) = \sum_{(x_i, \hat{y}_i) \in \mathcal{S}_{\text{clean}}} \text{KL}\left[\text{Categorical}(Y; \hat{y}) \| \text{Categorical}(Y; f_{\theta_y}(x_i))\right], \tag{6}$$

where:

$$\mathcal{S}_{\text{clean}} = \{(x_i, \hat{y}_i) : (x_i, \hat{y}_i) \in \mathcal{D} \wedge z(x_i, y_i) \leq m\}\},$$
$$m \in \{m : \Pr(z(x_i, y_i) \leq m) \geq R(t) \wedge \Pr(z(x_i, y_i) \geq m) \geq 1 - R(t)\},$$
$$\mathcal{S}_{\text{noisy}} = \mathcal{D} \setminus \mathcal{S}_{\text{clean}}. \tag{7}$$

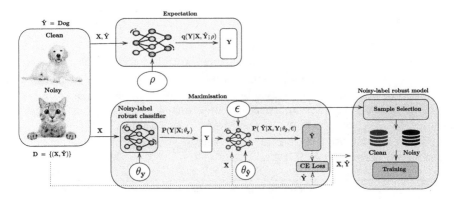

Fig. 3. Our training algorithm uses an existing noisy-label classifier (e.g., DivideMix [26]) parameterised by θ_y as the clean-label model $p(Y|X;\theta_y)$. The generation of noisy-label (given X and Y) is carried out by a model parameterised by noise rate ϵ and $\theta_{\hat{y}}$. The noisy-label classifier is based on a sample-selection mechanism that uses a curriculum $R(t) = 1 - \epsilon^{(t)}$.

with $R(t)$ defined in (5), and $z(x,y)$ representing the score of a criterion (e.g., loss [26,58,59], distance to the largest eigenvectors [21], or KNN scores [11]). Intuitively, the loss in Eq.(6) is simply the cross-entropy loss on the $\lfloor R(t) \times N \rfloor$ samples that have smallest scores (with $\lfloor . \rfloor$ being the floor function). One can also extend to other SOTA models by replacing the loss L accordingly. For example, if DivideMix [26] is used as a base model to constrain θ_y, L will include two additional terms: loss on un-labeled data and regularisation using *mixup* [54].

3.3 Training and Testing

Given the sample selection approach in Sect. 3.2, the M step in Eq. (4) is slightly modified by integrating the constraint in Eq. (6), which can be written as

$$\theta_y^{(t+1)}, \theta_{\hat{y}}^{(t+1)}, \epsilon^{(t+1)} = \arg\max_{\theta_y, \theta_{\hat{y}}, \epsilon} Q\left(\theta_y, \theta_{\hat{y}}, \epsilon, \rho^{(t)}\right) - \lambda L(\theta_y, \epsilon^{(t)}), \qquad (8)$$

where λ is a hyper-parameter and L is defined in Eq. (6). The training procedure is summarised in Algorithm 1 and visualised in Fig. 3. In the implementation, we integrate the proposed method into existing models, such as DivideMix [26] or FINE [21]. Note that the clean-label classifier $f_{\theta_y}(.)$ is also the clean classifier of the base model.

4 Experiments

We show extensive experiments in several noisy-label synthetic benchmarks with CIFAR100 [22], and real-world benchmarks, including CNWL's red mini-ImageNet [17], Clothing1M [45] and mini-WebVision [27]. Section 4.2 describes implementation details. We evaluate our approach by plugging SOTA models into $p(y|x;\theta_y)$, defined in Sect. 3, with results being shown in Sect. 4.3, and ablation studies in Sect. 4.4.

4.1 Dataset Descriptions

We follow the existing literature to generate the IDN labels [44] for **CIFAR100** [22]. The dataset contains 50,000 training images and 10,000 testing images of shape $32 \times 32 \times 3$. The dataset is class-balanced with 100 different categories and the IDN rates considered are 0.2, 0.3, 0.4 and 0.5 [44]. Another important IDN benchmark dataset is the **red mini-ImageNet**, a real-world dataset from CNWL [17]. In this dataset, there are 100 distinct classes, with each class containing 600 images. The images and their corresponding noisy-labels have been crawled from the internet at various controllable label noise rates. For a fair comparison with the existing literature [9,15,46], we have resized the images to 32×32 pixels from the 84×84 original pixel settings. We show results with 40%, 60% and 60% noise rates on this dataset. **Clothing1M** [45] is a real-world clothing classification dataset that contains 1 *million* images with an estimated noise rate of 38.5%. The dataset contains 14 different categories, and the labels are generated from surrounding texts. Images in this dataset are of different sizes, and we follow the resize structure suggested in [21,26]. This dataset also contains $50k$ manually validated clean training images, $14k$ images for the validation set, and $10k$ testing images. We have not used the clean training, validation set, or any extra training images while training. Only the testing set is used for evaluation. **Mini-WebVision** [26] contains 65,944 images with their respective labels from the 50 different initial categories from the WebVision [27], with image size of 256×256 pixels. Following the evaluation process commonly used in this benchmark, 50 categories from the ILSVRC12 [23] dataset are also used for testing.

4.2 Implementation

All methods are implemented in Pytorch [36] and use one NVIDIA RTX 3090 card for training and testing. As mentioned in the original papers, hyperparameter settings are kept the same for the baselines used in the proposed algorithm. All classifier architectures are also kept the same as the baseline models. A random initialisation of noise rate parameter ϵ with the sigmoid as its activation function is employed for all experiments to maintain the fairness of the comparisons with other approaches. The value of λ in Eq. (8) is set to 1 for all the cases. We integrate many SOTA approaches [11,15,21,26,58,59] into our graphical model, as explained in Sect. 3.3. For CIFAR100 [22] with IDN [44], we integrate DivideMix [26], C2D [59], CC [58], and InstanceGM [15] into our model, given their superior performance across various noise rates. Additionally, we also use F-Dividemix from FINE [21], as shown in Fig. 2b. For red mini-ImageNet [17], we test our proposed approach with and without DINO self-supervision [4]. For the implementation without self-supervision,

Table 1. Test accuracy % on *(left)* CIFAR100 [22] under different IDN [45], and *(right)* red mini-ImageNet [17]. Other models' results are from [7,15,46]. Here, we integrate DivideMix [26], C2D [59], CC [58] and InstanceGM [15] into our proposed model for CIFAR100 [22]. For red mini-ImageNet [17], we present the results with and without self-supervision [4]. We integrate DivideMix [26] and InstanceGM [15] into our model, with the latter tested with and without self-supervision.

Method	Noise Rates - IDN			
	0.2	0.3	0.4	0.5
CE [50]	30.42	24.15	21.45	14.42
PartT [44]	65.33	64.56	59.73	56.80
kMEIDTM [7]	69.16	66.76	63.46	59.18
DivideMix [26]	77.03	76.33	70.80	58.61
DivideMix-Ours	77.42	77.21	72.41	64.02
	(+0.39)	(+0.88)	(+1.61)	(+5.41)
C2D [59]	78.61	78.18	72.89	63.19
C2D-Ours	79.07	78.59	73.31	65.28
	(+0.46)	(+0.41)	(+0.42)	(+2.09)
CC [58]	79.61	77.56	76.58	63.19
CC-Ours	79.72	78.71	77.38	67.53
	(+0.11)	(+1.15)	(+0.80)	(+4.34)
InstanceGM [15]	79.69	79.21	78.47	77.19
InstanceGM-Ours	79.61	79.40	79.52	78.21
	(-0.08)	(+0.19)	(+1.05)	(+1.02)

Method	Noise rate		
	0.4	0.6	0.8
CE [46]	42.70	37.30	29.76
MixUp [54]	46.40	40.58	33.58
MentorMix [17]	47.14	43.80	33.46
FaMUS [46]	51.42	45.10	35.50
DivideMix [26]	46.72	43.14	34.50
DivideMix-Ours	50.70	45.11	37.44
	(+3.98)	(+1.97)	(+2.94)
InstanceGM [15]	52.24	47.96	39.62
InstanceGM-Ours	56.61	51.40	43.83
	(+4.37)	(+3.44)	(+4.21)
With self-supervised learning			
PropMix [9]	56.22	52.84	43.42
InstanceGM-SS [15]	56.37	53.21	44.03
InstanceGM-SS-Ours	58.29	53.60	45.47
	(+1.92)	(+0.39)	(+1.44)

Table 2. Final estimated noise rate ϵ for the Table 1, *(left)* CIFAR100 [22] under various IDN [44], and *(right)* red mini-ImageNet [17] with and without self-supervision.

Estimated noise rates				
Method	sActual noise rate			
	0.2	0.3	0.4	0.5
DivideMix-Ours	0.18	0.34	0.47	0.53
C2D-Ours	0.19	0.33	0.38	0.52
CC-Ours	0.21	0.34	0.41	0.54
InstanceGM-Ours	0.23	0.37	0.42	0.47

Estimated noise rates			
Method	Actual noise rate		
	0.4	0.6	0.8
DivideMix-Ours	0.39	0.58	0.73
InstanceGM-Ours	0.38	0.55	0.74
InstanceGM-SS-Ours	0.48	0.53	0.69

we use DivideMix [26] and InstanceGM [15], and for the self-supervised version, we only use InstanceGM [15]. The models for Clothing1M [45] and mini-WebVision [26] (validation on ImageNet [23]) are trained using DivideMix [26], SSR [11], C2D [59] and CC [58]. Additional implementation details, empirical analysis, and computational analysis are presented in the supplementary material in Appendices 1, 2 and 4 respectively.

4.3 Comparison with SOTA

This section compares our approach with SOTA methods on datasets with the IDN settings and noisy real-world settings. The bold text in tables indicate SOTA results, and our results are in the greyed rows.

Synthetic Instance-Dependent Noise. The comparison between various baselines and our proposed method on CIFAR100 [22] with IDN [44] is shown in Table 1 *(left)* with the noise rate ranging from 0.2 to 0.5. It is worth noting that using our proposed model with DivideMix [26], C2D [59], CC [58] and InstanceGM [15] improve their performance in almost all cases, particularly with large noise rates. Table 2 *(left)* also shows the final noise rate ϵ estimated by our model which is reasonably close to the simulated noise rate displayed in the table's header.

Table 3. Test accuracy (%) of competing methods, and final estimated noise rate ϵ on Clothing1M [45]. Also, we have not considered competing models that rely on a clean or validation set whilst training. We integrate DivideMix [26], C2D [59], SSR [11] and CC [58] into our model. DivideMix [26] and CC [58] shows the locally reproduced results. Mentioned noise rate by [45] is 0.385.

Method	Test accuracy (%)		Estimated noise rate
OT-Filter [12]	74.50		
ELR+ with C2D [59]	74.58		
AugDesc [35]	75.11		
NCE [25]	75.30		
DivideMix [26]	74.32		
DivideMix-Ours	74.41	(+0.09)	0.41
C2D [59]	74.58		
C2D-Ours	74.71	(+0.13)	0.41
SSR (class-imbalance) [11]	74.12		
SSR-Ours	74.20	(+0.08)	0.42
CC [58]	75.24		
CC-Ours	**75.31**	(+0.07)	0.41

Real-World Noise. We also evaluate our proposed method on various real-world noisy settings regarding test accuracy and estimated noise rates ϵ in Tables 1, 2, 3 and 4. Similarly to the synthetic IDN in Table 1 *(left)*, the results show that existing noisy-label robust methods can be easily integrated with our model to outperform current SOTA results for real-world noisy-label datasets. Tables 1 and 2 *(right)* shows the results on red mini-ImageNet using two configurations, including cases without self-supervision (top part of the table) and with self-supervision (bottom part of the table). The self-supervision DINO pre-training [4] relies only on images from red mini-ImageNet to enable a fair comparison with existing baselines [9,15]. Results from Table 1 *(right)* demonstrate that our approach improves the performance of SOTA methods by a considerable margin in all cases. In fact, using estimated noise rate ϵ (Table 2 *(right)*) while training InstanceGM [15] without self-supervision shows better performance than existing self-supervised baselines at 0.4 noise rate. Moreover, DivideMix [26], C2D [59], SSR [11], and CC [58] are used as a baselines for Clothing1M [45] as shown in Table 3, with results showing slight improvements with the use of our method. For mini-WebVision [26] (validation on ImageNet [23]), we use DivideMix [26], C2D [59], SSR [11], and CC [58] as baselines (Table 4). It is worth noting that our results are better than the SOTA for all settings in Table 4.

4.4 Ablation

We show an ablation study with testing accuracy *(top)* and training time *(bottom)* of our approach in Table 5 *(left)* on CIFAR100 [22] at 0.3 and 0.5 IDN [44] using DivideMix [26] as baseline. Detailed complexity analysis is shown in the supplementary material, Table 10. Initially, the accuracy result of baseline DivideMix [26] is 76.33%, 58.61% for 0.3 and 0.5 IDN, respectively. In the second row, we fix the noise rate ϵ at 0.3 and 0.5 for DivideMix's sample selection, as explained in Sect. 3.3 (without updating ϵ), then the results improved to 78.14% and 64.44%, which is the ideal case (i.e., a perfect noise rate estimation) that motivated our work. In the third case, we use the proposed graphical model with pre-trained DivideMix [26] that shows accuracy of 75.01% and 52.31%. In the next case, the proposed graphical model is trained together with DivideMix [26] without considering the estimated noise rate ϵ for sample selection, which results in accuracy of 76.20% and 56.30%. In the last row, we show the training of the proposed model with DivideMix [26], together with the estimation of noise rate ϵ, and the selection of samples based on that, with $\approx 1\%$ and $\approx 8\%$ accuracy improvement fo 0.3 and 0.5 IDN, which is quite close to our ideal case (second row). We also study the effect that the initialisation of the noise rate parameter ϵ plays in our method. As described in Sect. 4.2, we randomly initialise ϵ within

Table 4. Test accuracy (%) and final estimated noise rate ϵ on mini-WebVision [26] and validation on ImageNet [23]. We integrate DivideMix [26], Contrast-to-Divide (C2D) [59], SSR [11] and CC [58] into our model, whilst models suffixed with **-Ours** denotes our proposed approach. ImageNet [23] is only considered for validation.

Dataset	mini-WebVision		ImageNet		Estimated noise rate
	Top-1	Top-5	Top-1	Top-5	
BtR [38]	80.88	92.76	75.96	92.20	
NCE [25]	79.50	**93.80**	76.30	94.10	
DivideMix [26]	77.32	91.64	75.20	91.64	
DivideMix-Ours	**78.51**	**92.03**	**76.11**	**93.24**	0.43
	(+1.19)	(+0.39)	(+0.91)	(+1.60)	
C2D [59]	79.42	92.32	78.57	93.04	
C2D-Ours	**80.20**	**92.82**	**79.16**	**93.12**	0.43
	(+0.78)	(+0.50)	(+0.59)	(+0.08)	
SSR [11]	80.92	92.80	75.76	91.76	
SSR-Ours	**81.68**	**93.80**	**76.91**	**93.05**	0.43
	(+0.76)	(+1.00)	(+1.15)	(+1.29)	
CC [58]	79.36	93.64	76.08	93.86	-
CC-Ours	**80.01**	**93.79**	**77.11**	**94.21**	0.44
	(+0.65)	(+0.15)	(+1.03)	(+0.35)	

Table 5. *(left)* Ablation results on CIFAR100 at 0.3 and 0.5 IDN. We display accuracy results under different configurations and compare the training time of our approach against DivideMix [26]. The implementation settings and detailed computational analysis are described in Sect. 4.2 and Appendix 4 respectively. *(right above)* Ablation results for initialisation with different ϵ values on CIFAR100 [22] at a synthetic noise rate of 0.5 IDN [44] with DivideMix-Ours. *(right below)* Ablation graph showing the learning pattern of the noise rate parameter estimation under different initialisation values, i.e., $\epsilon = 0.001$ (blue), $\epsilon = 0.1$ (orange), $\epsilon = 0.9$ (red), when trained on DivideMix-ours case at 0.5 IDN [44] on CIFAR100 [22].

Models	Accuracy (%)	
	0.3	0.5
DivideMix [26]	76.33	58.61
DivideMix with fixed ϵ	78.14	64.44
Our method with pre-trained DivideMix	75.01	52.31
Our method with original DivideMix (no ϵ)	76.20	56.30
DivideMix-Ours	77.21	64.02
Training Time (GPU-hours)		
DivideMix [26]	7.20	
DivideMix-Ours	8.50	
DivideMix-Ours (Mixed Precision)	6.10	

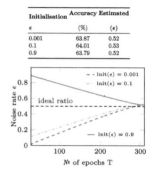

the range $(0, 1)$. To test the robustness of our method to this initialisation, we perform an ablation study by initialising ϵ with three different values, namely: $0.001, 0.1$, and 0.9 on CIFAR100 [24] at 0.5 IDN [44] using the DivideMix-ours approach. Our results, presented in Table 5 *(right above and below)*, demonstrate that the initial value of ϵ has a limited effect on the model's performance, highlighting the stability of our method across different initialisation of ϵ. In the supplementary material, Table 7 *(left)* presents comparative analyses of our model against PartT [44] and SSR [11] on CIFAR100 [22] at 0.5 IDN [44], while Table 7 *(right)* showcases similar comparisons with baselines DivideMix [26] and InstanceGM [15] at a higher noise rate of 0.6 IDN [44]. Additionally, Table 8 provides further insights into the performance of DivideMix [26] on red mini-ImageNet [17] at lower noise rates of 0.1, 0.2 and 0.3. Table 9 illustrates noise rate estimation with standard deviation for our model integrated with DivideMix [26] and InstanceGM [15] on CIFAR100 [22] under IDN [44] settings at 0.2, 0.3, 0.4, and 0.5. Further statistical analysis is detailed in Appendix 5 with confidence interval presented in Table 12. Performance analysis with various architectures (Table 13) and lambda values (Table 14) are mentioned in Appendix 7. Threshold-based comparisons are in Appendix 6 and Table 15.

5 Discussion and Conclusion

In this work, we proposed an innovative graphical model for IDN noisy-label learning, focusing on the estimation of the label noise rate that is leveraged to introduce a robust sample selection curriculum. Given that SOTA IDN noisy-label learning approaches tend to rely on sample selection methods, our method can be seamlessly integrated to them to improve their performance in many synthetic and real-world benchmarks, including CIFAR100 [22], red mini-ImageNet [17], Clothing1M [45], mini-WebVision [26], and ImageNet [23].

From a societal perspective, our methodology offers a positive impact by mitigating biases inherent in noisy data, thereby promoting more fair and accurate machine learning outcomes. It also opens the door to future research avenues, including exploring advanced noise rate estimation methodologies and investigating noise-robust loss functions such as GCE [56] and ELR [28] with it. Our choice of using the same network architecture for clean and noisy labels enables compatibility with co-teaching techniques [16,26]. However, investigating different network structures for clean and noisy labels remains a potential avenue for future research. Moreover, we aim to delve into its dynamics, uncovering substantial improvements for certain models while observing more marginal enhancements for others. For more detailed information, please refer to Sec. 8. Our approach also hints at potential breakthroughs in streamlining data annotation challenges.

References

1. Arpit, D., et al.: A closer look at memorization in deep networks. In: International Conference on Machine Learning, vol. 70, pp. 233–242. PMLR (2017)
2. Bae, H., Shin, S., Jang, J., Na, B., Song, K., Moon, I.C.: From noisy prediction to true label: noisy prediction calibration via generative model. In: International Conference on Machine Learning (2022)
3. Berthelot, D., Carlini, N., Goodfellow, I., Papernot, N., Oliver, A., Raffel, C.A.: Mixmatch: a holistic approach to semi-supervised learning. In: Advances in Neural Information Processing Systems, vol. 32 (2019)
4. Caron, M., et al.: Emerging properties in self-supervised vision transformers. In: International Conference on Computer Vision, pp. 9650–9660 (2021)
5. Chen, P., Ye, J., Chen, G., Zhao, J., Heng, P.A.: Beyond class-conditional assumption: a primary attempt to combat instance-dependent label noise. In: AAAI Conference on Artificial Intelligence, pp. 11442–11450 (2021)
6. Chen, Z., Song, A., Wang, Y., Huang, X., Kong, Y.: A noise rate estimation method for image classification with label noise. J. Phys. Conf. Ser. **2433**(1), 012039 (2023). https://doi.org/10.1088/1742-6596/2433/1/012039
7. Cheng, D., et al.: Instance-dependent label-noise learning with manifold-regularized transition matrix estimation. In: Conference on Computer Vision and Pattern Recognition, pp. 16630–16639 (2022)
8. Cheng, H., Zhu, Z., Li, X., Gong, Y., Sun, X., Liu, Y.: Learning with instance-dependent label noise: A sample sieve approach. In: International Conference on Learning Representations (2020)
9. Cordeiro, F.R., Belagiannis, V., Reid, I., Carneiro, G.: PropMix: hard sample filtering and proportional mixup for learning with noisy labels. In: British Machine Vision Conference (2021)
10. Dempster, A.P., Laird, N.M., Rubin, D.B.: Maximum likelihood from incomplete data via the EM algorithm. J. Roy. Stat. Soc.: Ser. B (Methodol.) **39**(1), 1–22 (1977)

11. Feng, C., Tzimiropoulos, G., Patras, I.: SSR: an efficient and robust framework for learning with unknown label noise. In: British Machine Vision Conference (2022)
12. Feng, C., Ren, Y., Xie, X.: Ot-filter: an optimal transport filter for learning with noisy labels. In: Proceedings of the IEEE/CVF Conference on Computer Vision and Pattern Recognition, pp. 16164–16174 (2023)
13. Frénay, B., Kabán, A.: A comprehensive introduction to label noise. In: European Symposium on Artificial Neural Networks, Computational Intelligence and Machine Learning (2014)
14. Frénay, B., Verleysen, M.: Classification in the presence of label noise: a survey. IEEE Trans. Neural Netw. Learn. Syst. **25**(5), 845–869 (2013)
15. Garg, A., Nguyen, C., Felix, R., Do, T.T., Carneiro, G.: Instance-dependent noisy label learning via graphical modelling. In: Winter Conference on Applications of Computer Vision, pp. 2288–2298 (2023)
16. Han, B., et al.: Co-teaching: robust training of deep neural networks with extremely noisy labels. In: Advances in Neural Information Processing Systems, vol. 31 (2018)
17. Jiang, L., Huang, D., Liu, M., Yang, W.: Beyond synthetic noise: deep learning on controlled noisy labels. In: International Conference on Machine Learning, pp. 4804–4815. PMLR (2020)
18. Jiang, L., Zhou, Z., Leung, T., Li, L.J., Fei-Fei, L.: MentorNet: learning data-driven curriculum for very deep neural networks on corrupted labels. In: International Conference on Machine Learning, pp. 2304–2313. PMLR (2018)
19. Jiang, Z., Zhou, K., Liu, Z., Li, L., Chen, R., Choi, S.H., Hu, X.: An information fusion approach to learning with instance-dependent label noise. In: International Conference on Learning Representations (2022). https://openreview.net/forum?id=ecH2FKaARUp
20. Khosla, P., et al.: Supervised contrastive learning. Adv. Neural. Inf. Process. Syst. **33**, 18661–18673 (2020)
21. Kim, T., Ko, J., Choi, J., Yun, S.Y.: FINE samples for learning with noisy labels. In: Advances in Neural Information Processing Systems, vol. 34 (2021)
22. Krizhevsky, A., Hinton, G.: Learning multiple layers of features from tiny images. University of Toronto, Tech. rep. (2009)
23. Krizhevsky, A., Sutskever, I., Hinton, G.E.: ImageNet classification with deep convolutional neural networks. In: Advances in Neural Information Processing Systems, vol. 25 (2012)
24. Kye, S.M., Choi, K., Yi, J., Chang, B.: Learning with noisy labels by efficient transition matrix estimation to combat label miscorrection. In: European Conference on Computer Vision, pp. 717–738. Springer (2022). https://doi.org/10.1007/978-3-031-19806-9_41
25. Li, J., Li, G., Liu, F., Yu, Y.: Neighborhood collective estimation for noisy label identification and correction. In: European Conference on Computer Vision. pp. 128–145. Springer (2022). https://doi.org/10.1007/978-3-031-20053-3_8
26. Li, J., Socher, R., Hoi, S.C.: DivideMix: learning with noisy labels as semi-supervised learning. In: International Conference on Learning Representations (2020)
27. Li, W., Wang, L., Li, W., Agustsson, E., Gool, L.V.: WebVision Database: Visual learning and understanding from web data. CoRR (2017)
28. Liu, S., Niles-Weed, J., Razavian, N., Fernandez-Granda, C.: Early-learning regularization prevents memorization of noisy labels. In: Advances in Neural Information Processing Systems, vol. 33, pp. 20331–20342 (2020)
29. Liu, Y., Cheng, H., Zhang, K.: Identifiability of label noise transition matrix. In: International Conference on Machine Learning (2023)

30. Ma, X., Huang, H., Wang, Y., Romano, S., Erfani, S., Bailey, J.: Normalized loss functions for deep learning with noisy labels. In: International Conference on Machine Learning, pp. 6543–6553. PMLR (2020)
31. Malach, E., Shalev-Shwartz, S.: Decoupling "when to update" from "how to update". In: Advances in Neural Information Processing Systems, vol. 30 (2017)
32. McLachlan, G.J., Peel, D.: Finite mixture models. Wiley (1996)
33. Neal, R.M., Hinton, G.E.: A view of the EM algorithm that justifies incremental, sparse, and other variants. Learn. Graph. Models, 355–368 (1998)
34. Neyshabur, B., Bhojanapalli, S., McAllester, D., Srebro, N.: Exploring generalization in deep learning. In: Advances in Neural Information Processing Systems, vol. 30 (2017)
35. Nishi, K., Ding, Y., Rich, A., Hollerer, T.: Augmentation strategies for learning with noisy labels. In: Conference on Computer Vision and Pattern Recognition, pp. 8022–8031 (2021)
36. Paszke, A., et al.: PyTorch: an imperative style, high-performance deep learning library. In: Advances in Neural Information Processing Systems, vol. 32 (2019)
37. Shamshad, F., et al.: Transformers in medical imaging: a survey. Med. Image Anal., 102802 (2023)
38. Smart, B., Carneiro, G.: Bootstrapping the relationship between images and their clean and noisy labels. In: Winter Conference on Applications of Computer Vision, pp. 5344–5354 (2023)
39. Song, H., Kim, M., Park, D., Shin, Y., Lee, J.G.: Learning from noisy labels with deep neural networks: a survey. IEEE Trans. Neural Netw. Learn. Syst. (2022)
40. Tanaka, D., Ikami, D., Yamasaki, T., Aizawa, K.: Joint optimization framework for learning with noisy labels. In: Proceedings of the IEEE Conference on Computer Vision and Pattern Recognition (CVPR) (June 2018)
41. Titterington, D.M.: On the likelihood of mixtures of distributions. Biometrika **71**(3), 511–522 (1984). https://doi.org/10.1093/biomet/71.3.511
42. Trummer, I.: From BERT to GPT-3 codex: harnessing the potential of very large language models for data management. VLDB Endowment **15**(12), 3770–3773 (2022)
43. Tu, H., Menzies, T.: Debtfree: minimizing labeling cost in self-admitted technical debt identification using semi-supervised learning. Empir. Softw. Eng. **27**(4), 80 (2022)
44. Xia, X., et al.: Part-dependent label noise: towards instance-dependent label noise. In: Advances in Neural Information Processing Systems, vol. 33, pp. 7597–7610 (2020)
45. Xiao, T., Xia, T., Yang, Y., Huang, C., Wang, X.: Learning from massive noisy labeled data for image classification. In: Conference on Computer Vision and Pattern Recognition, pp. 2691–2699 (2015)
46. Xu, Y., Zhu, L., Jiang, L., Yang, Y.: Faster meta update strategy for noise-robust deep learning. In: Conference on Computer Vision and Pattern Recognition, pp. 144–153 (June 2021)
47. Yang, S., Jiang, L., Liu, Z., Loy, C.C.: Unsupervised image-to-image translation with generative prior. In: Conference on Computer Vision and Pattern Recognition, pp. 18332–18341 (2022)
48. Yang, S., et al.: Estimating instance-dependent bayes-label transition matrix using a deep neural network. In: International Conference on Machine Learning, pp. 25302–25312. PMLR (2022)

49. Yao, Q., Yang, H., Han, B., Niu, G., Kwok, J.T.Y.: Searching to exploit memorization effect in learning with noisy labels. In: International Conference on Machine Learning, pp. 10789–10798. PMLR (2020)
50. Yao, Y., Liu, T., Gong, M., Han, B., Niu, G., Zhang, K.: Instance-dependent label-noise learning under a structural causal model. In: Advances in Neural Information Processing Systems, vol. 34 (2021)
51. Yao, Y., et al.: Dual T: Reducing estimation error for transition matrix in label-noise learning. In: Advances in Neural Information Processing Systems, vol. 33, pp. 7260–7271 (2020)
52. Zhang, C., Bengio, S., Hardt, M., Recht, B., Vinyals, O.: Understanding deep learning requires rethinking generalization. In: International Conference on Learning Representations (2017)
53. Zhang, C., Bengio, S., Hardt, M., Recht, B., Vinyals, O.: Understanding deep learning (still) requires rethinking generalization. Commun. ACM **64**(3), 107–115 (2021)
54. Zhang, H., Cisse, M., Dauphin, Y.N., Lopez-Paz, D.: mixup: beyond empirical risk minimization. In: International Conference on Learning Representations (2017)
55. Zhang, Y., Niu, G., Sugiyama, M.: Learning noise transition matrix from only noisy labels via total variation regularization. In: International Conference on Machine Learning, pp. 12501–12512. PMLR (2021)
56. Zhang, Z., Sabuncu, M.: Generalized cross entropy loss for training deep neural networks with noisy labels. In: Advances in Neural Information Processing Systems, vol. 31 (2018)
57. Zhang, Z., Pfister, T.: Learning fast sample re-weighting without reward data. In: Proceedings of the IEEE/CVF International Conference on Computer Vision, pp. 725–734 (2021)
58. Zhao, G., Li, G., Qin, Y., Liu, F., Yu, Y.: Centrality and consistency: two-stage clean samples identification for learning with instance-dependent noisy labels. In: European Conference on Computer Vision, pp. 21–37. Springer (2022). https://doi.org/10.1007/978-3-031-19806-9_2
59. Zheltonozhskii, E., Baskin, C., Mendelson, A., Bronstein, A.M., Litany, O.: Contrast to divide: Self-supervised pre-training for learning with noisy labels. In: Winter Conference on Applications of Computer Vision, pp. 1657–1667 (2022)
60. Zhu, Z., Liu, T., Liu, Y.: A second-order approach to learning with instance-dependent label noise. In: Conference on Computer Vision and Pattern Recognition, pp. 10113–10123 (June 2021)

Eliminating Warping Shakes for Unsupervised Online Video Stitching

Lang Nie[1,2], Chunyu Lin[1,2](✉), Kang Liao[3], Yun Zhang[4], Shuaicheng Liu[5], Rui Ai[6], and Yao Zhao[1,2]

[1] Institute of Information Science, Beijing Jiaotong University, Beijing, China
[2] Beijing Key Laboratory of Advanced Information Science and Network, Beijing, China
cylin@bjtu.edu.cn
[3] Nanyang Technological University, Singapore, Singapore
[4] Communication University of Zhejiang, Hangzhou, China
[5] University of Electronic Science and Technology of China, Chengdu, China
[6] HAOMO.AI, Beijing, China

Abstract. In this paper, we retarget video stitching to an emerging issue, named ***warping shake***, when extending image stitching to video stitching. It unveils the temporal instability of warped content in non-overlapping regions, despite image stitching having endeavored to preserve the natural structures. Therefore, in most cases, even if the input videos to be stitched are stable, the stitched video will inevitably cause undesired warping shakes and affect the visual experience. To eliminate the shakes, we propose ***StabStitch*** to simultaneously realize video stitching and video stabilization in a unified unsupervised learning framework. Starting from the camera paths in video stabilization, we first derive the expression of stitching trajectories in video stitching by elaborately integrating spatial and temporal warps. Then a warp smoothing model is presented to optimize them with a comprehensive consideration regarding content alignment, trajectory smoothness, spatial consistency, and online collaboration. To establish an evaluation benchmark and train the learning framework, we build a video stitching dataset with a rich diversity in camera motions and scenes. Compared with existing stitching solutions, *StabStitch* exhibits significant superiority in scene robustness and inference speed in addition to stitching and stabilization performance, contributing to a robust and real-time online video stitching system. The codes and dataset are available at https://github.com/nie-lang/StabStitch.

Keywords: image/video stitching · video stabilization · warping shake

Supplementary Information The online version contains supplementary material available at https://doi.org/10.1007/978-3-031-73235-5_22.

1 Introduction

Video stitching techniques are commonly employed to create panoramic or wide field-of-view (FoV) displays from different viewpoints with limited FoV. Due to their practicality, they are widely applied in autonomous driving [23], video

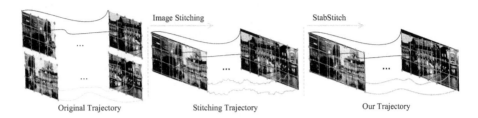

Fig. 1. The occurrence and elimination of warping shakes. Left: stable camera trajectories for input videos. Middle: warping shakes are produced by image stitching, yielding unsmooth stitching trajectories. Right: *StabStitch* eliminates these shakes successfully.

surveillance [37], virtual reality [38], etc. Our work lies in the most common and challenging case of video stitching with hand-held cameras. It does not require camera poses, motion trajectories, or temporal synchronization. It merges multiple videos, whether from multiple cameras or a single camera capturing multiple videos, to create a more immersive representation of the captured scene.

By contrast, image stitching has been studied much more profoundly, which inevitably throws the question of whether existing image stitching solutions can be extended to video stitching. Pursuing this line of thought, we initially leverage existing image stitching algorithms [15,45] to process hand-held camera videos. Although the stitched results for individual frames are remarkably natural, there is obvious content jitter in the non-overlapping regions between temporally consecutive frames, as shown in Fig. 1(mid). It is also important to note that the jitter does not originate from the inherent characteristics of the source video itself. In fact, due to the advancements and widespread adoption of video stabilization in both hardware and software nowadays, the source videos obtained from hand-held cameras are typically stable unless deliberately subjected to shaking. For clarity, we define such content jitter as *warping shake*, which describes the temporal instability of non-overlapping regions induced by temporally non-smooth warps, irrespective of the stability of source videos.

Existing video stitching solutions [10,16,30,47,51] follow a strong assumption that each source video from freely moving hand-held cameras suffers from heavy and independent shakes. Consequently, every source video necessitates stabilization via warping, contradicting the current prevalent reality that video stabilization technology has already been widely integrated into various portable devices (*e.g.*, cellphones, DV cameras, and UAVs). In addition, these approaches, to jointly optimize video stabilization and stitching, often establish a non-linear iterative solving system consisting of various energy terms. Each iteration involves several steps dedicated to optimizing different parameters separately, resulting in a rather slow inference speed and sophisticated optimization.

To solve the above issues, we present the first unsupervised online video stitching framework (termed *StabStitch*) to realize video stitching and video stabilization simultaneously. Building upon the current condition that source videos are typically stable, we simplify this task to stabilize the warped videos by removing warping shakes as illustrated in Fig. 1 (right). To get stable stitching warps, we generate the stitching trajectories drawing on the experience of camera trajectories (*i.e.*, Meshflow [33]) in video stabilization. By ingeniously combining spatial and temporal warps, we derive the formulation of stitching trajectories in the warped video. Next, a warp smoothing model is presented to simultaneously ensure content alignment, smooth stitching trajectories, preserve spatial consistency, and boost online collaboration. Diverging from conventional offline video stitching approaches that require complete videos as input, *StabStitch* stitches and stabilizes videos with backward frames alone. Besides, its efficient designs further contribute to a real-time online video stitching system with only one frame latency.

As there is no proper dataset readily available, we build a holistic video stitching dataset to train the proposed framework. Moreover, it could serve as a comprehensive benchmark with a rich diversity in camera motions and scenes to evaluate image/video stitching methods. Finally, we summarize our principle contributions as follows:

- We retarget video stitching to an emerging issue, termed *warping shake*, and reveal its occurrence when extending image stitching to video stitching.
- We propose *StabStitch*, the first unsupervised online video stitching solution, with a pioneering step to integrating video stitching and stabilization in a unified learning framework.
- We propose a holistic video stitching dataset with diverse scenes and camera motions. The dataset can work as a benchmark dataset and promote other related research work.

2 Related Work

2.1 Image Stitching

Traditional image stitching methods usually detect keypoints [39] or line segments [54] and then minimize the projective errors to estimate a parameterized warp by aligning these geometric features. To eliminate the parallax misalignment [61], the warp model is extended from global homography transformation [2] to other elastic representations, such as mesh [60], TPS [26], superpixel [24], and triangular facet [25]. Meanwhile, to keep the natural structure of non-overlapping regions, a series of shape-preserving constraints is formulated with the alignment objective. For instance, SPHP [3] and ANAP [29] linearized the homography and slowly changed it to the global similarity to reduce projective distortions; DFW [27], SPW [28], and LPC [15] leveraged line-related consistency to preserve geometric structures; GSP [4] and GES-GSP [6] added a global similarity before stitching multiple images together so that the warp of each image resembles a similar transformation as a whole; etc. Besides, Zhang *et al.* [62] re-formulated image stitching with regular boundaries by simultaneously optimizing alignment and rectangling [12,44].

Recently, learning-based image stitching solutions emerged. They feed the entire images into the neural network, encouraging the network to directly predict the corresponding parameterized warp model (*e.g.*, homography [17,42,46], multi-homography [50], TPS [20,45,63], and optical flow [14,22]). Compared with traditional methods based on sparse geometric features, these learning-based solutions train the network parameters to adaptively capture semantic features by establishing dense pixel-wise optimization objectives. They show better robustness in various cases, especially in the challenging cases where traditional geometric features are few to detect.

2.2 Video Stabilization

Traditional video stabilization can be categorized into 3D [31,34], 2.5D [7,32], and 2D [41] [9] [40] methods, according to different motion models. The 3D solutions model the camera motions in 3D space or require extra scene structure for stabilization. The structure is either calculated by structure-from-motion (SfM) [31] or acquired from additional hardware, such as a depth camera [34], a gyroscope sensor [19], or a lightfield camera [49]. Given the intensive computational demands of these 3D solutions, 2.5D approaches relax the full 3D requirement to partial 3D information. To this end, some additional 3D constraints are established, such as subspace projection [32] and epipolar geometry [7]. Compared with them, the 2D methods are more efficient with a series of 2D linear transformations (*e.g.*, affine, homography) as camera motions. To deal with large-parallax scenes, spatially varying motion representations are proposed, such as homography mixture [8], mesh [35], vertex profile [33], optical flow [36], etc. Moreover, some special approaches focus on specific input (*e.g.*, selfie [58,59], 360 [21,53], and hyperlapse [18] videos).

In contrast, learning-based video stabilization methods directly regress unstable-to-stable transformation from data. Most of them are trained with stable and unstable video pairs acquired by special hardware in a supervised manner [55,56,64]. To relieve data dependence, DIFRINT [5] proposed the first unsupervised solution via neighboring frame interpolation. To get a stable interpolated frame, only stable videos are used to train the network. Different from it, DUT [57] established unsupervised constraints for motion estimation and trajectory smoothing, learning video stabilization by watching unstable videos.

2.3 Video Stitching

Video stitching has received much less attention than image stitching. Early works [16,48] stitched multiple videos frame-by-frame, and focused on the temporal consistency of stitched frames. But the input videos were captured by cam-

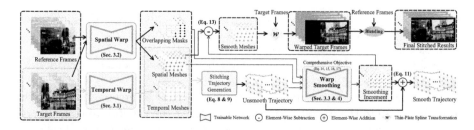

Fig. 2. The overview of *StabStitch*. We first obtain stitching trajectories by integrating spatial and temporal warps. Then the stitching trajectories are optimized by the warp smoothing model to produce unsmooth-to-smooth stitching warps.

eras fixed on rigs. For hand-held cameras with free and independent motions, there is a significant increase in temporal shakes. To deal with it, videos were first stitched and then stabilized in [11], while [30] did it in an opposite way (*e.g.*, videos were firstly stabilized, and then stitched). Both of them accomplished stitching or stabilization in a separate step. Later, a joint optimization strategy was commonly adopted in [10,47,51], where [47] further considered the dynamic foreground by background identification. However, solving such a joint optimization problem regarding stitching and stabilization is fragile and computationally expensive. To this end, we rethink the video stitching problem from the perspective of warping shake and propose the first (to our knowledge) unsupervised online solution for hand-held cameras.

3 StabStitch

We first describe the camera trajectories in video stabilization and then further derive the expression of stitching trajectories in video stitching. Afterward, the unsmooth trajectories are optimized to realize both stitching and stabilization. The pipeline of *StabStitch* is exhibited in Fig. 2.

3.1 Camera Trajectory

Temporal Warp: To obtain camera paths, a temporal warp model is first proposed to represent the temporal motion between consecutive video frames. Different from most video stabilization works [33,35,57] that use point correspondences to estimate the warp, we leverage a convolutional neural network to capture the high-level information in inter-frame motions. This alternative proves to be robust across various scenarios, particularly in low-light and low-texture environments. The network structure is similar to the warp network of UDIS++ [45]. As shown in Fig. 2(left), it takes two consecutive target frames as input and outputs the motions of mesh-like control points $m(t)$ [1]. Due to the temporal continuity of adjacent frames in the video, the estimated motions are often not significant. Consequently, we replace all global correlation layers [43] in UDIS++ with local correlation layers (*i.e.*, cost volume [52]). To improve the efficiency, we substitute the ResNet50 backbone [13] in UDIS++ with ResNet18 and reduce network parameters accordingly. Following UDIS++, our optimization objective also consists of an alignment term and a distortion term, as described in the following equation:

$$\mathcal{L}^{tmp} = \mathcal{L}_{alignment} + \lambda^{tmp}\mathcal{L}_{distortion}. \tag{1}$$

The alignment component leverages photometric errors to implicitly supervise control point motions. The distortion component is composed of an inter-grid constraint and an intra-grid constraint. For brevity, we refer the readers to the supplementary for more details.

Meshflow: The camera paths can be defined as a chain of relative motions, such as Euclidean transformations [34], homography transformations [35], etc. Representing the transformation of the initial frame as an identity matrix $F(1)$, the camera trajectories are written as:

$$C(t) = F(1)F(2)\cdots F(t), \tag{2}$$

where $F(t)$ is the relative transformation from the t-th frame to the $(t-1)$-th frame. Considering that our temporal warp model directly predicts the 2D motions of each control point, we adopt the motion representation of vertex files like MeshFlow [33]. Particularly, we chain the motions of each control point i temporally as the control point trajectory for a more straightforward representation:

$$C_i(t) = m_i(1) + m_i(2) + \cdots + m_i(t), \tag{3}$$

where $m_i(1)$ is set to zero. Note each control point in $m(t)$ is anchored at every vertex in a rigid mesh.

3.2 Stitching Trajectory

Compared with video stabilization, video stitching is more challenging with two or more videos as input and requires the stitched video to possess coherently smooth camera trajectories for the contents from different videos.

Spatial Warp: To obtain the stitching trajectories, in addition to the temporal warp model, we also establish a spatial warp model to represent the spatial motion between different video views, as shown in Fig. 2(left). The spatial warp model has a similar network structure to the temporal warp model except that the first local correlation layer is replaced by a global correlation layer [43] to capture long-range matching (usually longer than half of the image width/height). Considering the significance of spatial warping stability in video stitching, we expect this warp to be as robust as possible, although this network model has been proven to be more robust than traditional methods. To this end, we further introduce a motion consistency term in addition to the basic optimization components of the temporal warp:

$$\mathcal{L}_{consis.} = \frac{1}{(U+1)\times(V+1)} \sum_{i=1}^{(U+1)\times(V+1)} \|m_i(t) - m_i(t-1) - \mu^{spt}\|_2, \quad (4)$$

where μ^{spt} is the maximum tolerant motion difference and $(U+1) \times (V+1)$ denotes the number of control points. We further sum up the total optimization goal as:

$$\mathcal{L}^{spt} = \mathcal{L}_{alignment} + \lambda^{spt}\mathcal{L}_{distortion} + \omega^{spt}\mathcal{L}_{consis.}. \quad (5)$$

Refer to the ablation studies or supplementary for the impact of $\mathcal{L}_{consis.}$.

Stitch-Meshflow: Video stabilization leverages the chain of temporary motions as camera paths, whereas in our video stitching system, how should we represent the stitching paths of a warped video? We dig into this problem by combining the spatial and temporal warp models. With these two models, we first reach the spatial/temporal motions $(m^S/m^T \in \mathbb{R}^{2\times(U+1)\times(V+1)})$ and their corresponding meshes $(M^S/M^T \in \mathbb{R}^{2\times(U+1)\times(V+1)})$ as follows:

$$\begin{aligned} m^T(t) &= TNet(I_{tgt}^{t-1}, I_{tgt}^t) &\Rightarrow M^T(t) &= M^{Rig} + m^T(t), \\ m^S(t-1) &= SNet(I_{ref}^{t-1}, I_{tgt}^{t-1}) &\Rightarrow M^S(t-1) &= M^{Rig} + m^S(t-1), \quad (6) \\ m^S(t) &= SNet(I_{ref}^t, I_{tgt}^t) &\Rightarrow M^S(t) &= M^{Rig} + m^S(t), \end{aligned}$$

where $I_{ref}/I_{tgt} \in \mathbb{R}^{C\times H\times W}$ is the reference/target frame, $SNet/TNet(\cdot, \cdot)$ represents the spatial/temporal warp model, and $M^{Rig} \in \mathbb{R}^{2\times(U+1)\times(V+1)}$ is defined as the 2D positions of control points in a rigid mesh.

Then we need to derive the stitching motion of the warped video from the spatial/temporal meshes. To align the t-th frame with the $(t-1)$-th frame in the warped video, the temporal mesh from the t-th frame to the $(t-1)$-th frame in the source video $(M^T(t))$ should also undergo the same transformation as the spatial warp of the $(t-1)$-th frame $(M^S(t-1))$. Assuming $\mathcal{T}(\cdot)$ is the thin-plate spline (TPS) transformation, the desired stitching motion could be represented as the difference between the desired mesh and the actual spatial mesh $(M^S(t))$:

$$s(t) = \mathcal{T}_{M^{Rig} \to M^S(t-1)}(M^T(t)) - M^S(t). \quad (7)$$

Finally, we attain the stitching paths (we also call it Stitch-Meshflow) by chaining the relative stitching motions between consecutive warped frames as follows:

$$S_i(t) = s_i(1) + s_i(2) + \cdots + s_i(t), \quad (8)$$

where we define $s(1)$ is an all-zero array.

3.3 Warp Smoothing

To get a temporally stable warped video, we need to smooth the stitching trajectories as well as preserve their spatial consistency. Besides, we should also try to prevent the degradation of alignment performance in overlapping areas.

Achitecure: In this stage, a warp smoothing model is designed to achieve the above goals. As depicted in Fig. 2, it takes sequences of (N frames) stitching paths (S), spatial meshes (M^S), and overlapping masks (OP) as input, and outputs a smoothing increment (Δ) as described in the following equation:

$$\Delta = SmoothNet(S, M^S, OP), \tag{9}$$

where $S/M^S/OP \in \mathbb{R}^{2 \times N \times (U+1) \times (V+1)}$. OP are binary mask sequences (1/0 indicates the vertex inside/outside overlapping regions). We calculate it by determining whether each control point in M^S exceeds image boundaries.

The smoothing model first embeds S, M^S, and OP into 32, 24, and 8 channels through separate linear projections, respectively. Then these embeddings are concatenated and fed into three 3D convolutional layers to model the spatiotemporal dependencies. Finally, we reproject the hidden results back into 2 channels to get Δ. The network architecture is designed rather simply to accomplish efficient smoothing inference. In addition, this simple architecture better highlights the effectiveness of the proposed unsupervised learning scheme.

With the smoothing increment Δ, we define the smooth stitching paths as:

$$\hat{S} = S + \Delta. \tag{10}$$

Furthermore, if we expand Eq. 10 based on Eq. 8 and Eq. 7, we obtain:

$$\begin{aligned}\hat{S}(t) &= S(t-1) + s(t) + \Delta(t) \\ &= S(t-1) + \mathcal{T}_{M^{Rig} \to M^S(t-1)}(M^T(t)) - \underbrace{(M^S(t) - \Delta(t))}_{\text{Smooth spatial mesh}}.\end{aligned} \tag{11}$$

In this case, the last term in Eq. 11 can be regarded as the smooth spatial mesh $\hat{M}^S(t)$. Therefore, the sequences of smooth spatial meshes are written as:

$$\hat{M}^S = M^S - \Delta. \tag{12}$$

Objective Function: Given original stitching paths (S) and smooth stitching paths ($\hat{S}$), smooth spatial meshes ($\hat{M}^S$), and overlapping masks (OP), we design the unsupervised learning goal as the balance of different optimization components:

$$\mathcal{L}^{smooth} = \mathcal{L}_{data} + \lambda^{smooth}\mathcal{L}_{smothness} + \omega^{smooth}\mathcal{L}_{space}. \tag{13}$$

Data Term: The data term encourages the smooth paths to be close to the original paths. This constraint alone does not contribute to stabilization. The stabilizing effect of *StabStitch* is realized in conjunction with the data term and the subsequent smoothness term. To maintain the alignment performance of overlapping regions during the smoothing process as much as possible, we further incorporate the awareness of overlapping regions into the data term as follows:

$$\mathcal{L}_{data} = \|(\hat{S} - S)(\alpha OP + 1)\|_2, \tag{14}$$

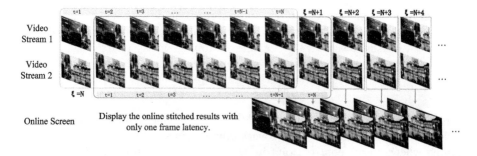

Fig. 3. The online stitching mode. We define a sliding window to process a short sequence and display the last frame on the online screen.

where α is a constant to emphasize the degree of alignment.

Smoothness Term: In a smooth path, each motion should not contain sudden large-angle rotations, and the amplitude of translations should be as consistent as possible. To this end, we constrain the trajectory position at a certain moment to be located at the midpoint between its positions in the preceding and succeeding moments, which implicitly satisfies the above two requirements. Hence, we formulate the smoothness term as:

$$\mathcal{L}_{smoothness} = \sum_{j=1}^{(N-1)/2} \beta_j \|\hat{S}(mid+j) + \hat{S}(mid-j) - 2\hat{S}(mid)\|_2, \quad (15)$$

where mid is the middle index of N (N is required to be an odd number) and β_j is a constant between 0 and 1 to impose varying magnitudes of smoothing constraints on trajectories at different temporal intervals.

Spatial Consistency Term: When there are only data and smoothness constraints, the warping shakes can be already removed. But each trajectory is optimized individually. Actually, our system has $(U+1) \times (V+1)$ control points, which means there are $(U+1) \times (V+1)$ independently optimized trajectories. When these trajectories are changed inconsistently, significant distortions will be produced. To remove the distortions and encourage different paths to share similar changes, we introduce a spatial consistency component as:

$$\mathcal{L}_{space} = \frac{1}{N} \sum_{t=1}^{N} \mathcal{L}_{distortion}(\hat{M}^S(t)), \quad (16)$$

where $\mathcal{L}_{distortion}(\cdot)$ takes a mesh as input and calculates the distortion loss like the spatial/temporal warp model.

4 Online Stitching

Existing video stitching methods [10,16,30,47,51] are offline solutions, which smooth the trajectories after the videos are completely captured. Different from them, *StabStitch* is an online video stitching solution. In our case, the frames after the current frame are no longer available and real-time inference is required.

4.1 Online Smoothing

To achieve this goal, we define a fixed-length sliding window (N frames) to cover previous and current frames, as shown in Fig. 3. Then the local stitching trajectory inside this window is extracted and smoothed according to Sect. 3. Next, the current target frame is re-synthesized using the optimized smooth spatial mesh (Eq. 12). Finally, we blend it with the current reference frame to get a stable stitched frame and display the result when the next frame arrives. With this mode and efficient architectures, *StabStitch* achieves minimal latency with only one frame.

Online Collaboration Term: However, such an online mode could introduce a new issue, wherein the smoothed trajectories in different sliding windows (with partial overlapping sequences) may be inconsistent. This can produce subtle jitter if we chain the sub-trajectories of different windows. Therefore, we design an online collaboration constraint besides the existing optimization goal (Eq. 13):

$$\mathcal{L}_{online} = \frac{1}{N-1} \sum_{t=2}^{N} \|\hat{S}^{(\xi)}(t) - \hat{S}^{(\xi+1)}(t-1)\|_2, \tag{17}$$

where ξ is the absolute time ranging from N to the last frame of the videos. By contrast, t can be regarded as the relative time in a certain sliding window ranging from 1 to N.

4.2 Offline and Online Inference

Offline smoothing takes the whole trajectories as input, outputs the optimized whole trajectories, and then renders all the video frames. It carries on smoothing after receiving whole input videos and can be regarded as a special online case in which the sliding window covers whole videos. By contrast, online smoothing takes local trajectories as input, outputs the optimized local trajectories, and then renders the last frame in the sliding window. The online process smoothes current paths without using any future frames.

5 Dataset Preparation

We establish a dataset, named **StabStitch-D**, for the comprehensive video stitching evaluation considering the lack of dedicated datasets for this task. Our dataset comprises over 100 video pairs, consisting of over 100,000 images, with each video lasting from approximately 5 s to 35 s. To holistically reveal the performance of video stitching methods in various scenarios, we categorize videos into four classes based on their content, including regular (RE), low-texture

Fig. 4. The proposed *StabStitch-D* dataset with a large diversity in camera motions and scenes. We exhibit several video pairs for each category.

(LT), low-light (LL), and fast-moving (FM) scenes. In the testing split, 20 video pairs are divided for testing, with 5 videos in each category. Figure 4 illustrates some examples for each category, where FM is the most challenging case with fast irregular camera movements (rotation or translation). The resolution of each video is resized into 360 × 480 for efficient training, and in the testing phase, arbitrary resolutions are supported.

6 Experiment

6.1 Details and Metrics

Details: We implement the whole framework in PyTorch with one RTX 4090Ti GPU. The spatial warp, temporal warp, and warp smoothing models are trained separately with the epoch number set to 55, 40, and 50, respectively. λ^{tmp}, λ^{spt}, μ^{spt}, and ω^{spt} are defined as 5, 10, 20, and 0.1. The weights for data, smoothness, spatial consistency, and online collaboration terms are set to 1, 50, 10, and 0.1. α, β_1, β_2, and β_3 are set to 10, 0.9, 0.3, and 0.1. The control point resolution and sliding window length are set to $(6 + 1) \times (8 + 1)$ and 7. Moreover, when training the warp smoothing model, we randomly select $N = 7$ frames as the processing window from a larger buffer of 12 frames, which could allow more diverse stitching paths.

Metrics: To quantitatively evaluate the proposed method, we suggest three metrics including **alignment score**, **distortion score**, and **stability score**. Limited by space, we moved the related metric details to the supplementary.

6.2 Compared with State-of-the-Arts

We compare our method with image and video stitching solutions, respectively.

Compared with Image Stitching: Two representative SoTA image stitching methods are selected to compare with our solution: LPC [15] (traditional method) and UDIS++ [45] (learning-based method). The quantitative comparison results are illustrated in Table 1, where '·/·' indicates the PSNR/SSIM values. '-' implies the approach fails in this category (e.g., program crash and extremely severe distortion). The results show our solution achieves comparable alignment performance with SoTA image stitching methods. In fact, our spatial warp model has surpassed UDIS++ as indicated in Table 4. *StabStitch* sacrifices a little alignment performance to reach better temporally stable sequences.

Table 1. Quantitative comparisons with image stitching methods on StabStitch-D dataset. * indicates the model is re-trained on the proposed dataset.

	Method	Regular	Low-Light	Low-Texture	Fast-Moving	Average
1	LCP [15]	24.22/0.812	-	-	23.88/0.813	-
2	UDIS++ [45]	23.19/0.785	31.09/0.936	29.98/0.906	21.56/0.756	27.19/0.859
3	UDIS++ * [45]	24.63/0.829	34.26/0.957	32.81/0.920	**24.78/0.819**	29.78/**0.891**
4	StabStitch	**24.64/0.832**	**34.51/0.958**	**33.63/0.927**	23.36/0.787	**29.89**/0.890

Table 2. User study on the cases that Nie et al. [47] successes, in which the user preference rate is reported. We exclude the failure cases of Nie et al. [47] for fairness.

StabStitch	Nie et al. [47]	No preference
30.47%	6.25%	63.28%

Compared with Video Stitching: We compare our method with Nie et al.'s video stitching solution [47]. To our knowledge, it is the latest and best video stitching method for hand-held cameras. Based on the assumption that the input videos are unstable, it estimates two respective non-linear warps for the reference and target video frame. In contrast, we hold the assumption that currently input videos are typically stable unless deliberately subjected to shaking. Only the target video frame is warped in our system. This difference between Nie et al. [47] and our solution makes the comparison of PSNR/SSIM not completely fair. Therefore, we conduct a user study as an alternative and demonstrate extensive stitched videos in our supplementary video.

User Preference: Nie et al.'s solution [47] is sensitive to different scenes. In our testing set (20 pairs of videos in total), Nie et al. [47] fail in 10 pairs of videos because of program crashes (mainly appearing in the categories of LL and LT).

Fig. 5. Qualitative comparison with Nie *et al.*'s video stitching [47] on a regular case (top) and a fast-moving case (bottom). The numbers below the images indicate the time at which the frame appears in the video. Please zoom in for the best view.

Hence, we exclude these failure cases and conduct the user study only on the successful cases. For a stitched video, different methods may perform differently at different times. So, we segment each complete stitched video into one-second clips (we omit the last clip of a stitched video that is shorter than one second in practice), resulting in 128 clips in total. Then we invite 20 participants, including 10 researchers/students with computer vision backgrounds and 10 volunteers outside this community. In each test session, two clips from different methods are presented in a random order, and every volunteer is required to indicate their overall preference for alignment, distortion, and stability. We average the preference rates and exhibit the results in Table 2. From that, our results are more preferred by users. Besides, we illustrate two qualitative examples in Fig. 5, where our results show much fewer artifacts (refer to our supplementary video for the complete stitched videos).

Table 3. A comprehensive analysis of inference speed (/ms).

SNet	TNet	Trajectory generation	SmoothNet	Warping	Blending	Total
11.5	10	1.1	1	4.4	0.2	28.2

Inference Speed: A comprehensive analysis of our inference speed is shown in Table 3 with one RTX 4090Ti GPU, where 'Blending' represents the average blending. In the example shown in Fig. 5(top), *StabStitch* only takes about 28.2ms to stitch one frame, yielding a real-time online video stitching system. When stitching a video pair with higher resolution, only the time for warping and blending steps will slightly increase. In contrast, Nie *et al.*'s solution [47] takes over 40 min to get such a 26-second stitched video with an Intel i7-9750H 2.60GHz CPU, making it impractical to be applied to online stitching.

6.3 Ablation Study

Table 4. Ablation studies on alignment, distortion, and stability.

	Basic Stitching	$\mathcal{L}_{consis.}$	Warp Smoothing	Alignment ↑	Distortion ↓	Stability ↓
1	✓			30.67/0.902	0.784	81.57
2	✓	✓		**30.75/0.903**	0.804	60.32
3	✓	✓	✓	29.89/0.890	**0.674**	**48.74**

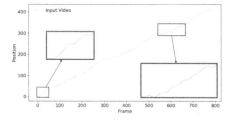 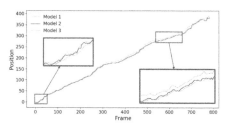

Fig. 6. Left: camera trajectories of the original target video. Right: stitching trajectories of the warped target video from different models (the model index corresponds to the experiment number in Table 4).

Quantitative Analysis: The main ablation study is shown in Table 4, where 'Basic Stitching' (model 1) refers to the spatial warp model without the motion consistency term $\mathcal{L}_{consis.}$. With $\mathcal{L}_{consis.}$ (model 2), the stability is improved. With the warp smoothing model (model 3), both the distortion and stability are significantly optimized at the cost of slight alignment performance, achieving an optimal balance of alignment, distortion, and stability. More experiments can be found in the supplementary materials.

Trajectory Visualization: We visualize the trajectories of the original target video and warped target videos in Fig. 6. Here, the trajectories are extracted from a control point of the example shown in Fig. 5(top) in the vertical direction. It can be observed that even if the input video is stable, image stitching can introduce undesired warping shakes, whereas *StabStitch* (Model 3) minimizes these shakes as much as possible during stitching.

7 Conclusions

Nowadays, the videos captured from hand-held cameras are typically stable due to the advancements and widespread adoption of video stabilization in both hardware and software. Under such circumstances, we retarget video stitching to an emerging issue, *warping shake*, which describes the undesired content instability in non-overlapping regions especially when image stitching technology is directly applied to videos. To solve this problem, we propose the first unsupervised online video stitching framework, *StabStitch*, by generating stitching trajectories and smoothing them. Besides, a video stitching dataset with various camera motions and scenes is built, which we hope can work as a benchmark and promote other related research work. Finally, we conduct extensive experiments to demonstrate our superiority in stitching, stabilization, robustness, and speed.

Acknowledgments. This work was supported by the National Natural Science Foundation of China (No. 62172032), and Zhejiang Province Basic Public Welfare Research Program (No. LGG22F020009).

References

1. Bookstein, F.L.: Principal warps: thin-plate splines and the decomposition of deformations. IEEE TPAMI **11**(6), 567–585 (1989)
2. Brown, M., Lowe, D.G.: Automatic panoramic image stitching using invariant features. IJCV **74**(1), 59–73 (2007)
3. Chang, C.H., Sato, Y., Chuang, Y.Y.: Shape-preserving half-projective warps for image stitching. In: CVPR, pp. 3254–3261 (2014)
4. Chen, Y.-S., Chuang, Y.-Y.: Natural image stitching with the global similarity prior. In: Leibe, B., Matas, J., Sebe, N., Welling, M. (eds.) ECCV 2016. LNCS, vol. 9909, pp. 186–201. Springer, Cham (2016). https://doi.org/10.1007/978-3-319-46454-1_12
5. Choi, J., Kweon, I.S.: Deep iterative frame interpolation for full-frame video stabilization. ACM TOG **39**(1), 1–9 (2020)
6. Du, P., Ning, J., Cui, J., Huang, S., Wang, X., Wang, J.: Geometric structure preserving warp for natural image stitching. In: CVPR, pp. 3688–3696 (2022)
7. Goldstein, A., Fattal, R.: Video stabilization using epipolar geometry. ACM TOG **31**(5), 1–10 (2012)
8. Grundmann, M., Kwatra, V., Castro, D., Essa, I.: Calibration-free rolling shutter removal. In: IEEE International Conference on Computational Photography, pp. 1–8. IEEE (2012)
9. Grundmann, M., Kwatra, V., Essa, I.: Auto-directed video stabilization with robust l1 optimal camera paths. In: CVPR, pp. 225–232. IEEE (2011)
10. Guo, H., Liu, S., He, T., Zhu, S., Zeng, B., Gabbouj, M.: Joint video stitching and stabilization from moving cameras. IEEE TIP **25**(11), 5491–5503 (2016)
11. Hamza, A., Hafiz, R., Khan, M.M., Cho, Y., Cha, J.: Stabilization of panoramic videos from mobile multi-camera platforms. Image Vis. Comput. **37**, 20–30 (2015)
12. He, K., Chang, H., Sun, J.: Rectangling panoramic images via warping. ACM TOG **32**(4), 1–10 (2013)
13. He, K., Zhang, X., Ren, S., Sun, J.: Deep residual learning for image recognition. In: CVPR, pp. 770–778 (2016)
14. Jia, Q., Feng, X., Liu, Y., Fan, X., Latecki, L.J.: Learning pixel-wise alignment for unsupervised image stitching. In: ACM MM, pp. 1392–1400 (2023)
15. Jia, Q., et al.: Leveraging line-point consistence to preserve structures for wide parallax image stitching. In: CVPR, pp. 12186–12195 (2021)

16. Jiang, W., Gu, J.: Video stitching with spatial-temporal content-preserving warping. In: CVPRW, pp. 42–48 (2015)
17. Jiang, Z., Zhang, Z., Fan, X., Liu, R.: Towards all weather and unobstructed multispectral image stitching: algorithm and benchmark. In: ACM MM, pp. 3783–3791 (2022)
18. Joshi, N., Kienzle, W., Toelle, M., Uyttendaele, M., Cohen, M.F.: Real-time hyperlapse creation via optimal frame selection. ACM TOG **34**(4), 1–9 (2015)
19. Karpenko, A., Jacobs, D., Baek, J., Levoy, M.: Digital video stabilization and rolling shutter correction using gyroscopes. CSTR **1**(2), 13 (2011)
20. Kim, M., Lee, Y., Han, W.K., Jin, K.H.: Learning residual elastic warps for image stitching under dirichlet boundary condition. In: WACV, pp. 4016–4024 (2024)
21. Kopf, J.: 360 video stabilization. ACM TOG **35**(6), 1–9 (2016)
22. Kweon, H., Kim, H., Kang, Y., Yoon, Y., Jeong, W., Yoon, K.J.: Pixel-wise warping for deep image stitching. In: AAAI, vol. 37, pp. 1196–1204 (2023)
23. Lai, W.S., Gallo, O., Gu, J., Sun, D., Yang, M.H., Kautz, J.: Video stitching for linear camera arrays. arXiv preprint arXiv:1907.13622 (2019)
24. Lee, K.Y., Sim, J.Y.: Warping residual based image stitching for large parallax. In: CVPR, pp. 8198–8206 (2020)
25. Li, J., Deng, B., Tang, R., Wang, Z., Yan, Y.: Local-adaptive image alignment based on triangular facet approximation. IEEE TIP **29**, 2356–2369 (2019)
26. Li, J., Wang, Z., Lai, S., Zhai, Y., Zhang, M.: Parallax-tolerant image stitching based on robust elastic warping. IEEE TMM **20**(7), 1672–1687 (2017)
27. Li, S., Yuan, L., Sun, J., Quan, L.: Dual-feature warping-based motion model estimation. In: ICCV, pp. 4283–4291 (2015)
28. Liao, T., Li, N.: Single-perspective warps in natural image stitching. IEEE TIP **29**, 724–735 (2019)
29. Lin, C.C., Pankanti, S.U., Natesan Ramamurthy, K., Aravkin, A.Y.: Adaptive as-natural-as-possible image stitching. In: CVPR, pp. 1155–1163 (2015)
30. Lin, K., Liu, S., Cheong, L.F., Zeng, B.: Seamless video stitching from hand-held camera inputs. In: Computer Graphics Forum, vol. 35, pp. 479–487. Wiley Online Library (2016)
31. Liu, F., Gleicher, M., Jin, H., Agarwala, A.: Content-preserving warps for 3d video stabilization. ACM TOG, 1-9 (2009)
32. Liu, F., Gleicher, M., Wang, J., Jin, H., Agarwala, A.: Subspace video stabilization. ACM TOG **30**(1), 1–10 (2011)
33. Liu, S., Tan, P., Yuan, L., Sun, J., Zeng, B.: MeshFlow: minimum latency online video stabilization. In: Leibe, B., Matas, J., Sebe, N., Welling, M. (eds.) ECCV 2016. LNCS, vol. 9910, pp. 800–815. Springer, Cham (2016). https://doi.org/10.1007/978-3-319-46466-4_48
34. Liu, S., Wang, Y., Yuan, L., Bu, J., Tan, P., Sun, J.: Video stabilization with a depth camera. In: CVPR, pp. 89–95. IEEE (2012)
35. Liu, S., Yuan, L., Tan, P., Sun, J.: Bundled camera paths for video stabilization. ACM TOG **32**(4), 1–10 (2013)
36. Liu, S., Yuan, L., Tan, P., Sun, J.: Steadyflow: spatially smooth optical flow for video stabilization. In: CVPR, pp. 4209–4216 (2014)
37. Liu, W., Luo, W., Lian, D., Gao, S.: Future frame prediction for anomaly detection–a new baseline. In: CVPR, pp. 6536–6545 (2018)
38. Lo, I.C., Shih, K.T., Chen, H.H.: Efficient and accurate stitching for 360° dual-fisheye images and videos. IEEE TIP **31**, 251–262 (2021)

39. Lowe, D.G.: Distinctive image features from scale-invariant keypoints. IJCV **60**, 91–110 (2004)
40. Ma, T., Nie, Y., Zhang, Q., Zhang, Z., Sun, H., Li, G.: Effective video stabilization via joint trajectory smoothing and frame warping. IEEE Trans. Visual Comput. Graphics **26**(11), 3163–3176 (2019)
41. Matsushita, Y., Ofek, E., Ge, W., Tang, X., Shum, H.Y.: Full-frame video stabilization with motion inpainting. IEEE TPAMI **28**(7), 1150–1163 (2006)
42. Nie, L., Lin, C., Liao, K., Liu, M., Zhao, Y.: A view-free image stitching network based on global homography. J. Vis. Commun. Image Represent. **73**, 102950 (2020)
43. Nie, L., Lin, C., Liao, K., Liu, S., Zhao, Y.: Depth-aware multi-grid deep homography estimation with contextual correlation. IEEE TCSVT **32**(7), 4460–4472 (2021)
44. Nie, L., Lin, C., Liao, K., Liu, S., Zhao, Y.: Deep rectangling for image stitching: A learning baseline. In: CVPR, pp. 5740–5748 (2022)
45. Nie, L., Lin, C., Liao, K., Liu, S., Zhao, Y.: Parallax-tolerant unsupervised deep image stitching. In: ICCV, pp. 7399–7408 (2023)
46. Nie, L., Lin, C., Liao, K., Zhao, Y.: Learning edge-preserved image stitching from multi-scale deep homography. Neurocomputing **491**, 533–543 (2022)
47. Nie, Y., Su, T., Zhang, Z., Sun, H., Li, G.: Dynamic video stitching via shakiness removing. IEEE TIP **27**(1), 164–178 (2017)
48. Perazzi, F., et al.: Panoramic video from unstructured camera arrays. In: Computer Graphics Forum, vol. 34, pp. 57–68. Wiley Online Library (2015)
49. Smith, B.M., Zhang, L., Jin, H., Agarwala, A.: Light field video stabilization. In: ICCV, pp. 341–348. IEEE (2009)
50. Song, D.Y., Um, G.M., Lee, H.K., Cho, D.: End-to-end image stitching network via multi-homography estimation. SPL **28**, 763–767 (2021)
51. Su, T., Nie, Y., Zhang, Z., Sun, H., Li, G.: Video stitching for handheld inputs via combined video stabilization. In: SIGGRAPH ASIA 2016 Technical Briefs, pp. 1–4 (2016)
52. Sun, D., Yang, X., Liu, M.Y., Kautz, J.: Pwc-net: Cnns for optical flow using pyramid, warping, and cost volume. In: CVPR, pp. 8934–8943 (2018)
53. Tang, C., Wang, O., Liu, F., Tan, P.: Joint stabilization and direction of 360° videos. ACM TOG **38**(2), 1–13 (2019)
54. Von Gioi, R.G., Jakubowicz, J., Morel, J.M., Randall, G.: Lsd: a fast line segment detector with a false detection control. IEEE TPAMI **32**(4), 722–732 (2008)
55. Wang, M., et al.: Deep online video stabilization with multi-grid warping transformation learning. IEEE TIP **28**(5), 2283–2292 (2018)
56. Xu, S.Z., Hu, J., Wang, M., Mu, T.J., Hu, S.M.: Deep video stabilization using adversarial networks. In: Computer Graphics Forum, vol. 37, pp. 267–276. Wiley Online Library (2018)
57. Xu, Y., Zhang, J., Maybank, S.J., Tao, D.: Dut: learning video stabilization by simply watching unstable videos. IEEE TIP **31**, 4306–4320 (2022)
58. Yu, J., Ramamoorthi, R.: Selfie video stabilization. In: ECCV, pp. 551–566 (2018)
59. Yu, J., Ramamoorthi, R., Cheng, K., Sarkis, M., Bi, N.: Real-time selfie video stabilization. In: CVPR, pp. 12036–12044 (2021)
60. Zaragoza, J., Chin, T.J., Brown, M.S., Suter, D.: As-projective-as-possible image stitching with moving dlt. In: CVPR, pp. 2339–2346 (2013)
61. Zhang, F., Liu, F.: Parallax-tolerant image stitching. In: CVPR, pp. 3262–3269 (2014)

62. Zhang, Y., Lai, Y.K., Zhang, F.L.: Content-preserving image stitching with piecewise rectangular boundary constraints. IEEE TVCG **27**(7), 3198–3212 (2020)
63. Zhang, Y., Lai, Y., Lang, N., Zhang, F.L., Xu, L.: Recstitchnet: learning to stitch images with rectangular boundaries. Comput. Visual Media (2024)
64. Zhang, Z., Liu, Z., Tan, P., Zeng, B., Liu, S.: Minimum latency deep online video stabilization. In: ICCV, pp. 23030–23039 (2023)

Vary: Scaling up the Vision Vocabulary for Large Vision-Language Model

Haoran Wei[1](✉), Lingyu Kong[2], Jinyue Chen[2], Liang Zhao[1], Zheng Ge[1], Jinrong Yang[3], Jianjian Sun[1], Chunrui Han[1], and Xiangyu Zhang[1]

[1] MEGVII Technology, Beijing, China
weihaoran18@mails.ucas.ac.cn
[2] University of Chinese Academy of Sciences, Beijing, China
[3] Huazhong University of Science and Technology, Wuhan, China
https://github.com/Ucas-HaoranWei/Vary/

Abstract. Most Large Vision-Language Models (LVLMs) enjoy the same vision vocabulary, *i.e.*, CLIP, for common vision tasks. However, for some special task that needs dense and fine-grained perception, the CLIP-style vocabulary may encounter low efficiency in tokenizing corresponding vision knowledge and even suffer out-of-vocabulary problems. Accordingly, we propose **Vary**, an efficient and productive method to scale up the **V**ision voc**ary**bul**ary** of LVLMs. The procedures of Vary are naturally divided into two folds: the generation and integration of a new vision vocabulary. In the first phase, we devise a vocabulary network along with a tiny decoder-only transformer to compress rich vision signals. Next, we scale up the vanilla vision vocabulary by merging the new with the original one (CLIP), enabling the LVLMs to garner new features effectively. We present frameworks with two sizes: Vary-base (7B) and Vary-toy (1.8B), both of which enjoy excellent fine-grained perception performance while maintaining great general ability.

Keywords: LVLM · Vision vocabulary · Fine-grained perception

1 Introduction

Recently, research into vision dialogue robots [1,23,29,36,63] has been gaining significant traction. These human-like models, mainly relying on two components, *i.e.*, large language models (LLMs) [6,35,38,46,61] and vision vocabulary networks, can not only converse based on user's prompts but also perform well on downstream tasks, such as VQA [26,43], image caption [47], OCR [34], and so on. It is undeniable that large vision-language models (LVLMs) will drive the AI community towards the direction of artificial general intelligence (AGI).

H. Wei—The work was supported by the National Science and Technology Major Project of China (2023ZD0121300). The work was also supported by the Research on AI Terminal Computing Power and End-Cloud Framework (R2411B0R) of China Mobile Group Device Co., Ltd.
L. Kong—Equal contribution.

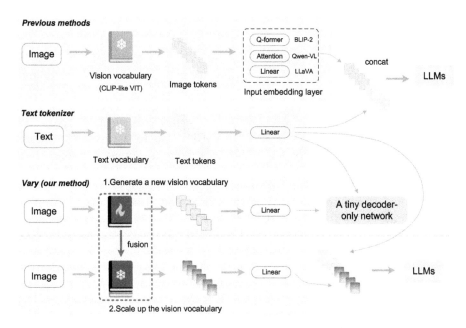

Fig. 1. Previous method *vs.* Vary: Unlike other models that use a ready-made vision vocabulary, the processes of Vary can be divided into two stages: the generation and fusion of vision vocabulary. In the first stage, we use a vocabulary network along with a tiny decoder-only network to produce a powerful new vision vocabulary via autoregression. In the second stage, we fuse the vision vocabulary with the original one to provide new features for the LVLMs efficiently.

Popular GPT-4 [35]-like LVLMs, *e.g.*, BLIP2 [23], MiniGPT4 [63], LLaVA [29], Qwen-VL [4], and *etc.*. [12,57,62] enjoy a stunning performance in multiple aspects with own programming paradigm. Based on an LLM [39,61], BLIP-2 proposes the Q-former, a BERT [11] like network as a vision input embedding layer, aiming to align the image tokens to a text latent. Inherited the structure of BLIP-2, MiniGPT-4 introduces 3500 high-quality image-text pairs as self-supervised fine-tuning (SFT) data, allowing it can "talk" like GPT-4. Unlike BLIP-2, LLaVA utilizes a linear layer as the vision embedding layer, which is similar to the text input embedding layer in text tokenizer [46,61], ensuring consistency in the structure of image and text branches. For Qwen-VL, it utilizes a cross-attention layer to compress and align the image tokens, making the model accept a larger input resolution. Although the above LVLMs' vision input embedding networks are variable (*e.g.*, MLP [29], Qformer, Perceiver [1]), their vision vocabulary is almost identical (a CLIP-based [37] VIT), which we argue maybe a bottle-neck.

It is recognized that CLIP-VIT is a tremendous general vision vocabulary, which is trained via contrastive learning upon more than 400M [40] image-text pairs, covering most natural images and vision scenes. However, for some special scenarios, *e.g.*, high-resolution perception, non-English OCR, document/chart understanding, and so on, the CLIP may regard them as "foreign languages", leading to inefficient tokenizing, i.e., difficulty compress rich vision information into 256×1024 tokens. Although mPlug-Owl [56] and Qwen-VL alleviate the above issues by unfreezing its vision vocabulary network (CLIP) in pretraining, we argue that such a manner may not be reasonable due to four aspects: 1) it may overwrite the knowledge of the original vocabulary; 2) the training efficiency of updating a vision vocabulary upon a relative large LLM (7B) is low; 3) it can not train a dataset with multiple epochs (due to the strong "memory" ability of LLMs) which maybe an important setting in computer vision perception tasks. Therefore, a natural question is: *Is there a strategy that can simplify and effectively intensify the vision vocabulary for an LVLM?*

In this paper, we propose Vary, an efficient and effective approach, to answer the above question. Vary is inspired by the text vocabulary expansion manner in vanilla LLMs [8]. When transferring an English LLM to another foreign language, such as Chinese, it's necessary to expand the text vocabulary to lift the encoding efficiency and model performance under the new language. Intuitively, for the vision branch, if we feed the "foreign language" image to the LVLMs, we also need to scale up the vision vocabulary. In Vary, the process of vocabulary scaling-up can be divided into two steps: 1) generate a new vision vocabulary that can make up the old one (CLIP); 2) integrate the new and old vocabularies to construct a better one. As shown in Fig. 1, we build a small-size pipeline consisting of an encoder network and a tiny decoder-only transformer in the first step to generate the new vocabulary via predicting the next token. It is worth noting that the autoregressive-based process of generating a vocabulary is perhaps more flexible for dense scenes than that based on contrastive learning like CLIP: On the one hand, the next-token way can allow the vision vocabulary to compress longer texts; On the other hand, the data formats that can be used in this manner are more diverse, such as object detection [26,41] data with the prompt. After preparing the new vision vocabulary, we graft it to the vanilla LVLMs to introduce new features. In this process, we freeze both the new and old vocabulary networks to avoid the visual knowledge being overwritten.

Afterward scaling up the vision vocabulary, our LVLM can achieve more fine-grained vision perception, such as document-level Chinese/English OCR, book image to markdown or LaTeX, Chinese/English chart parsing, and so on, while achieving excellent downstream performances. Specifically, Vary-base achieves 79.1% ANLS on DocVQA [33], 66.3% relaxed accuracy on ChartQA [32], 88.6% accuracy on RefCOCO [16]and 36.4% overall accuracy on MMVet [58]. Besides, along with only 1.8B LLM [2], Vary-toy can achieve 65.0% on DocVQA, 59.2% on ChartQA, 88.0% on RefCOCO, and 30.4% on MMVet.

In summary, Vary is a useful strategy to strengthen the vision vocabulary of LVLMs, which can be utilized at arbitrary downstream visual tasks that CLIP is not good at. In addition to the document OCR, chart parsing, and object detection mentioned in this paper, we believe Vary still enjoys more fine-grained tasks. We appeal to researchers to rethink the design ideas of LVLMs from the perspective of vision vocabulary construction.

2 Related Works

2.1 Large Language Models

Over the past year, significant attention has been drawn to large language models (LLMs) in the fields of both natural language processing (NLP) and computer vision (CV). This heightened attention stems from LLMs' outstanding performance in diverse aspects, especially the powerful world knowledge base and universal capabilities. Current LLMs enjoy a unified transformer architecture which is exemplified by BERT [11], GPT-2 [38], T5 [39], *etc.*. Subsequently, researchers have uncovered the concept of an"emergent ability" [50] in LLMs. This implies that as language model sizes reach a certain threshold, there may be a qualitative leap in their capabilities. Motivated by the tremendous success of the decoder-only GPT series, a multitude of other open-source LLMs have emerged, including OPT [61], LLaMA [46], GLM [59], and so on. Building upon these openly available LLMs, numerous tailored fine-tuned models have been introduced to develop LLMs for diverse applications, especially LLaMA-driven models,*e.g.*, Alphaca [44], Vicuna [8], which have become the de-facto component for a Large Vision-Language Model (LVLM).

2.2 LLM-Based Large Vision-Language Models

LLM's robust zero-shot capabilities and logical reasoning make it play the central controller role within an LVLM. There are two primary pipeline styles: plugin-based and end-to-end model. Plugin-based methods [25,42,51,53,55] typically regard LLMs as an agent to invoke various plugins from other foundational or expert models, executing specific functions in response to human instructions. While such methods offer versatility, they have limitations in terms of plugin invocation efficiency and performance. Conversely, end-to-end LVLMs usually rely on a single large multimodal model to facilitate interactions. Following this approach, Flamingo [1] introduces a gated cross-attention mechanism trained on billions of image-text pairs to align vision and language modalities, demonstrating strong performance in few-shot learning. BLIP-2 [23] introduces Q-Former to enhance the alignment of visual features with the language space. More recently, LLaVA [29] proposes using a simple linear layer to replace Q-Former and designed a two-stage instruction-tuning procedure.

Despite the remarkable performance of existing methods, they are confined to the same and limited vision vocabulary – CLIP [37]. For an LVLM, CLIP is a tremendous general vision vocabulary that is trained via contrastive learning upon million-level image-texts pairs, which can cover most nature images and vision tasks, *e.g.*, VQA, caption, scene text OCR. However, some images under special scenarios, *e.g.*, high-resolution image, non-English OCR, document/chart parsing, and so on, will still be regarded as a "foreign language" by CLIP, which may lead to vision "out-of-vocabulary" problem and in turn become a bottleneck.

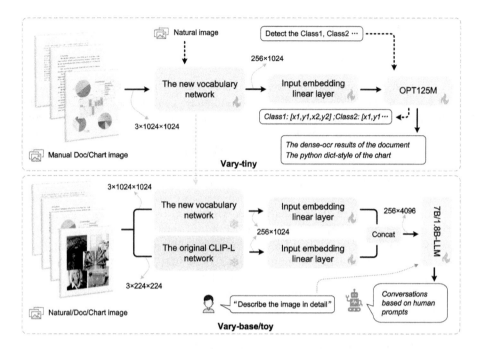

Fig. 2. Overview of the Vary. There are two types of Vary form: Vary-tiny and Vary-base/toy. Vary-tiny is mainly focused on generating a new vision vocabulary while Vary-base/toy is the final Vary body (a new LVLM) aiming to handle various visual tasks based on the new vision vocabulary.

3 Method

3.1 Architecture

Vary enjoys two main conformations: Vary-tiny and Vary-base/toy, as shown in Fig. 2. We devise the Vary-tiny to produce a new vision vocabulary and the Vary-base/toy to make use of the new vocabulary. Specifically, Vary-tiny mainly comprises a vocabulary encoder and a tiny OPT-125M [61] decoder. Between the two modules, we add a linear layer to align the channel dimensions. We hope the new vocabulary network can excel in compressing rich visual signals, *e.g.*, texts within documents/charts, and locations of objects, to compensate for CLIP's shortcomings. Thus we utilize the corresponding three types of data to train the model. Afterward, we extract the new vocabulary network and transplant it to a new language network to build the Vary main body. As shown in the lower half of Fig. 2, the new and original vocabulary (CLIP) networks enjoy independent input embedding layers and are integrated before the LLM. In such a stage, we freeze both weights of two vocabulary networks and unfreeze the weights of others to optimize the whole Vary.

3.2 Towards Generating a New Vision Vocabulary

The New Vocabulary Network We use the SAM [18] initialized ViTDet [24] image encoder (base version with about 80M parameters) as the main body of the new vocabulary network of Vary. Due to the input resolution of the SAM-base encoder being (1024×1024) while the output stride being 16, the feature shape of the last layer is (64×64×256 for H×W×C) that can not be aligned to the output of CLIP-L (256×1024 for N×C). Hence, we introduce two convolution layers, which we found is a good token merging unit, behind the last layer of the SAM encoder, as shown in Fig. 3. The first convolution layer possesses a kernel size of 3 and a stride of 2, aiming to transfer the feature shape to 32×32×512. The setting of the second convolution layer is the same as the first one, which can further compress the output to 16×16×1024. After that, we flattened the output feature to 256×1024 to align the image token shape of CLIP-VIT.

Document Data Engine. We select the high-resolution document image-text pairs as the main dataset used for the new vision vocabulary generation due to the dense OCR task can effectively validate the fine-grained image perception ability of a model. To our knowledge, there is no publicly available dataset of English and Chinese document-text pairs, so we created our own. We first collect pdf-style documents from open-access ArXiv and CC-MAIN-2021-31-PDF-UNTRUNCATED for the English set and assemble from E-books on the Internet for the Chinese set. Then we use *fitz* of PyMuPDF to extract the text information in each pdf page and convert each page into a PNG image via *pdf2image* at the same time. For text ground truth, we concatenate the texts within each block (box) of a page according to the order of top to bottom and left to right. For each image, all pages enjoy the same resolution (96 dpi). During this process, we construct 1M Chinese and 1M English document image-text pairs in total.

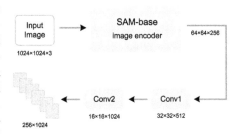

Fig. 3. The structure of new vision vocabulary network of Vary-tiny. We add two convolution layers to convert the output to be similar to CLIP.

Chart Data Engine. We find current LVLMs are not good at chart parsing, so we choose the chart data as another main knowledge that needs to be "written" into the new vocabulary. For chart image-text pair, we all follow a rendering way. We select both the *matplotlib* and *pyecharts* as the rendering tools. For the matplotlib-style chart, we built 250k samples in both Chinese and English. While for pyecharts, we build 500k for both two languages. Besides, we convert the text ground truth of each chart to a python-dict form. The texts used in the chart, *e.g.*, title, x-axis, and y-axis, are randomly selected from the natural language processing (NLP) corpus downloaded from the Internet.

Detection Data Engine. To fully utilize the capacity of the vision vocabulary network and gain the natural image perception ability from SAM initialization, we also introduce object detection data in the vision vocabulary generating process. We gather samples from two large open-source datasets, *i.e.*, Object365 [41] and OpenImage [19]. Due to the low efficiency of coordinate (number texts) tokenizing in OPT's [61] text tokenizer, for images with too many objects, the number of tokens in the ground truth may exceed the maximum token length supported by OPT-125M (although we interpolate it to 4096). Therefore, we re-organize the annotations upon two tasks: 1) **Object Detection**: If there are no more than 30 object-boxes in the image, we will allow the Vary-tiny detect all objects with the prompt: *"Detect all objects in this image"*. 2) **REC**: If the object-box number is over 30, we will regard this image as a REC task using a prompt template: *"Detect class1, class2, ..., in this image"*. The selected classes are random so one image can be used multiple times. Through the above manner, we obtain approximately 3M of detection data.

Input Format. We train all parameters of the Vary-tiny with image-text pairs by autoregression. The input format follows popular LVLMs [14,15], *i.e.*, the image tokens are packed with text tokens in the form of a prefix. We use two special tokens" " and "" to indicate the position of the image tokens. The OPT-125M is interpolated to 4096 for long document texts.

3.3 Towards Scaling up the Vision Vocabulary

The Structure of Vary. After completing the new vocabulary network, we graft it to our LVLM – Vary-base/toy. Specifically, we parallelize the new vision vocabulary with the original CLIP-VIT. Both two vision vocabularies enjoy an individual input embedding layer, *i.e.*, a simple linear layer, as shown in Fig. 2. For Vary-base, the input channel of the linear is 1024 and the output is 2048, ensuring the channel of image tokens after concatenating is 4096, which exactly aligns the input of LLM (Qwen-7B [3] or Vicuna-7B [8]). For Vary-toy, both the input and output channels are 1024 to align the Qwen-1.8B [3].

LaTeX Document Rendering. Except for the collected coarse document data in Sect. 3.2, we also need purified data that enjoys a certain format to support formula and table parsing. To this end, we create new document data through LaTeX rendering. Firstly, we collect abundant *".tex"* source files on Arxiv, and then extracted tables, mathematical formulas, and plain texts using regular expressions. Finally, we re-render these contents with *pdflatex* based on the new templates we prepared. We prepare 10 templates to perform batch rendering. Besides, we transfer the text ground truth of each document page to a *mathpix* markdown style to unify the format. Through the above processes, we acquired about 0.5 million English and Chinese document pages.

Text Associated Chart Rendering. In Sect. 3.2, we batch render chart data to train the new vocabulary network. However, the texts (title, x-axis values, and y-axis values) in those rendered charts suffer low correlation because they are randomly generated. This situation is not a problem in the vocabulary-generating process as we only hope that the new vocabulary can efficiently compress visual information. However, in the training stage of the Vary main body, we hope to use higher quality (strong texts correlated) data due to the LLM being unfrozen. To this end, we use GPT-4 [35] to generate some chart corpus and then we utilize the high-quality chart texts to additionally render 200k chart images for the Vary-base/toy training.

Data Ratio in Pre-training and SFT. For Vary, the pretrain stage is a multi-task training stage, wherein we prepare abundant image-text pairs in various formats. We mainly focus on six types of data in such a stage, containing weakly annotated image caption, PDF-dense OCR, chart parsing, object detection, pure text conversation, and VQA. Specifically, for general natural images, we sample 4M image-text pairs from the Laion-COCO [40] dataset, and we also use the BLIP-558K data proposed in LLaVA [29]. For the PDF image-text pair, we prepare two types of data: one is pure dense text OCR, and the other is a task that converts the PDF image to a markdown format. The previous type of data is randomly sampled from the PDF data used in Vary-tiny and the last one is obtained via LaTeX rendering as aforementioned. For the detection data, images are gathered from the COCO [26] dataset. We sample 50K images with fewer objects for the object detection task and use all train data of RefCOCO for the REC task. We normalize the coordinates of each box and magnify the values to 1000 times. To prevent the language ability of the LLM from deteriorating, we also introduce pure NLP conversation data, including ShareGPT, Baize [52], and Alpaca [44]. For the downstream VQA tasks, we choose two challenge datasets (DocVQA and ChartQA [32]) to monitor the dense text perception and reasoning performance of Vary for artificial data. There are at least 10 prompts made through GPT3.5 [6] for each task.

In the SFT stage, we use the LLaVA-80k or 665k [29] to continue training the model. Both LLaVA-80k and 665k are general SFT datasets with detailed descriptions and prompts produced by GPT4 [29,35].

Conversation Format. When using the Vicuna-7B as the LLM, we prepare the conversation format following Vicuna v1 [8], i.e., USER: "image" "texts input" ASSITANT: "texts output" </s>. When utilizing the Qwen-7B or Qwen1.8B [2], we design the conversations following the LLaVA-MPT [45] format, which can be described as: <—im_start—>user: "imag" "texts input"<—im_end—> <—im_start—>assistant: "texts output" <—im_end—>. The "image" represents 256 image tokens.

4 Experiments

4.1 Datasets and Evaluation Metrics

We evaluate the proposed Vary on multiple datasets, including 1) a document-level OCR test set that we created to explore the performance of dense vision recognition; 2) DocVQA [33], ChartQA [32], and RefCOCO [16] to test the improvement on downstream tasks; 3) MMVet [58] to monitor fluctuations in the general scenes of the model. The document test benchmark contains pure OCR and markdown conversion tasks. In the pure OCR task, the set includes 100 pages with both Chinese and English, which are randomly extracted from Arxiv and E-book. In the markdown conversion task, the test set obtains 200 pages, of which 100 pages contain tables and another 100 pages enjoy formulas.

We report Normalized Edit Distance [5,21] and F1-score along with the page-level precision and recall for the document parsing task. For downstream tasks, e.g., DocVQA, ChartQA, RefCOCO, and MMVet, we use their vanilla metrics for a fair comparison with other LVLMs.

4.2 Implementation Details

During the vision vocabulary generating process, we optimize all parameters of Vary-tiny with a batch size of 512 and train the model for 3 epochs. We utilize the AdamW [31] optimizer and a cosine annealing scheduler [30] along with the learning rate of 5e-5 to train the model.

In the Vary-base/toy training stage, we freeze the weights of both new and vanilla (CLIP-L) vision vocabulary networks and optimize the parameters of input embedding layers and LLM. The initial learning rate is 5e-5 in pretrain while 2e-5 in SFT. Both the pretrain and SFT enjoy a batch size of 256 and an epoch of 1. Other settings are the same as Vary-tiny.

4.3 Fine-Grained Perception Performance

We measure the fine-grained perception performance of Vary through dense text recognition. As shown in Table 1, Vary-tiny gathers both Chinese and English dense OCR ability by the process of vision vocabulary generating. Specifically, it achieves 0.336 and 0.311 edit distance for Chinese and English documents (on plain texts), respectively, proving the new vision vocabulary enjoys good fine-grained text compression capacity. Based on the new vision vocabulary and Qwen-1.8B decoder, Vary-toy further lifts the document OCR accuracy compared to Vary-tiny. For Vary-base, it can achieve an on-par performance with Nougat [5] (a special document parsing model) on English plain-text style documents. Besides, with different prompts (e.g., Convert the image to markdown format), both Vary-toy and Vary-base can realize the document image-markdown format conversion. It is worth noting that in such a task, Vary-base (yields 0.223 edict distance and 80.37% F1 on math and table average) is better than nougat (with 0.245 edict distance and 79.97% F1 on average), which may be due to

Table 1. Fine-grained text perception compared to Nougat. Vary-tiny is the model based on OPT-125M for generating the vision vocabulary, which enjoys pure OCR ability, covering Chinese and English. Vary-base/toy are the models upon Qwen-chat 7B/1.8B upon the new vision vocabulary, enjoying both pure document OCR and markdown format conversation abilities through prompt control.

Method	Forms	Pure OCR		Markdown Conversion		
		Chinese	English	Formula	Table	Average
Nougat [5]	Edit Distance ↓	–	0.126	0.154	0.335	0.245
	F1-score ↑	–	**89.91**	83.97	**75.97**	79.97
	Prediction ↑	–	89.12	82.47	75.21	78.84
	Recall ↑	–	**90.71**	**85.53**	**76.74**	81.14
Vary-tiny	Edit Distance ↓	0.336	0.311	–	–	–
	F1-score ↑	82.89	82.24	–	–	–
	Prediction ↑	83.15	84.37	–	–	–
	Recall ↑	82.63	80.22	–	–	–
Vary-toy	Edit Distance ↓	0.297	0.212	0.182	0.599	0.391
	F1-score ↑	83.84	85.90	79.19	70.67	74.93
	Prediction ↑	84.09	86.74	80.15	72.30	76.23
	Recall ↑	83.60	85.08	78.27	69.12	73.70
Vary-base	Edit Distance ↓	0.201	**0.119**	**0.128**	**0.317**	0.223
	F1-score ↑	86.70	88.10	**85.05**	75.68	80.37
	Prediction ↑	86.46	**89.73**	**85.39**	**77.25**	81.32
	Recall ↑	87.54	86.52	84.72	74.18	79.45

the super strong text correction ability of the 7B LLM decoder. All the above results indicate that by scaling up the vision vocabulary, the new LVLM can gather excellent fine-grained perception performance.

4.4 Downstream Task Performance

We test the performance of Vary on downstream VQA and REC tasks, including DocVQA [33], ChartQA [32], and RefCOCO [16].

For DocVQA and ChartQA, we use the addition prompt: "Answer the following question using a single word or phrase:" [28] to give the model short and precise answers. As shown in Table 2, Vary-base (with Qwen-7B as LLM) can achieve 79.1% (test) and 78.9% (val) ANLS on DocVQA upon LLaVA-80k [29] SFT data. With LLaVA-665k [28] data for SFT, Vary-base can reach 66.3% average accuracy on ChartQA. The performance on both two challenging downstream tasks is much better than Qwen-VL [4], demonstrating the proposed vision vocabulary scaling-up strategy is promising. It is worth noting that along with the only 1.8B language model, Vary-toy can achieve 65.3% ANLS on DocVQA (val) and 59.2% accuracy on ChartQA, further demonstrating the effectiveness of the proposed method.

Table 2. Comparison with popular methods on DocVQA and ChartQA. 80k represents that the SFT data is LLaVA-80k while 665k is the LLaVA-CC665k. The metric of DocVQA is ANLS while the ChartQA is relaxed accuracy following their vanilla papers.

Method	DocVQA		ChartQA		Average
	val	test	human	augmented	
Dessurt [9]	46.5	63.2	–	–	–
Donut [17]	–	67.5	–	–	41.8
Pix2Sturct [20]	–	72.1	30.5	81.6	56.0
mPLUG-DocOwl [56]	–	62.2	–	–	57.4
Matcha [27]	–	–	38.2	90.2	64.2
Qwen-VL [3]	–	65.1	–	–	65.7
Qwen-VL-chat [3]	–	62.6	–	–	66.3
Vary-toy (665k)	65.3	65.0	33.1	85.2	59.2
Vary-base (80k)	78.9	79.1	43.7	87.9	65.8
Vary-base (665k)	78.4	78.2	43.9	88.6	66.3

Table 3. Comparison with popular methods on RefCOCO. Benefiting from the new vision vocabulary, Vary-base/toy can achieve 88.6%/88.0% accuracy on RefCOCO val, which is on par with the Qwen-VL-chat-7B.

Type	Method	Size	RefCOCO		
			val	testA	testB
Traditional	OFA-L [48]	–	80.0	83.7	76.4
	TransVG [10]	–	81.0	82.7	78.4
	VILLA [13]	–	82.4	87.5	74.8
	UniTAB [54]	–	86.3	88.8	80.6
LLM-based	VisionLLM-H [49]	–	–	86.7	–
	Shikra-7B [7]	7B	87.0	90.6	80.2
	Shikra-13B [7]	13B	87.8	91.1	81.7
	Qwen-VL-chat [3]	7B	88.6	92.3	84.5
	Next-chat [60]	7B	85.5	90.0	77.9
	Vary-toy (665k)	1.8B	88.0	90.8	85.1
	Vary-base (665k)	7B	88.6	92.1	84.9

For the REC task, Vary-base/toy can get 88.6%/88.0% accuracy@0.5 on the RefCOCO validation set, which is on par with Qwen-VL-chat (7B) and much better than the Shikra-13B [7]. All the above results show that along with a better vision vocabulary, Vary gathers great natural object perception ability, further proving the effectiveness of using the low-cost Vary-tiny architecture to build a vision vocabulary, allowing us to further reflect on the necessity of CLIP if we add a large amount of weakly labeled image caption data, e.g., Laion-400M [40], during the new vocabulary generating process.

Table 4. Comparison with popular methods on the general vision ability benchmark – MMVet. The abbreviations are as follows: Rec: Recognition; Know: Knowledge; Gen: Language generation; Spat: Spatial awareness.

Method	MM-Vet						
	Rec	OCR	Know	Gen	Spat	Math	Total
BLIP-2 [23]	27.5	11.1	11.8	7.0	16.2	5.8	22.4
LLaVA-7B [29]	28.0	17.1	16.3	18.9	21.2	<u>11.5</u>	23.8
MiniGPT-4 [63]	29.9	16.1	20.4	22.1	22.2	3.8	24.4
Otter [22]	27.3	17.8	14.2	13.8	24.4	3.8	24.7
OpenFlamingo [1]	28.7	16.7	16.4	13.1	21.0	7.7	24.8
LLaVA-13B [29]	**39.2**	22.7	<u>26.5</u>	<u>29.3</u>	29.6	7.7	32.9
LLaVA1.5-7B [28]	–	–	–	–	–	–	30.5
Vary-toy (qwen1.8B)(665k)	37.0	19.6	18.5	20.0	25.1	7.7	30.4
Vary-base (vicuna7B) (665k)	<u>39.5</u>	30.0	24.8	23.0	33.7	**11.5**	36.0
Vary-base (qwen7B) (80k)	36.6	<u>33.4</u>	22.0	22.7	<u>37.2</u>	14.6	<u>36.4</u>

4.5 General Vision Performance

We verify the general vision performance of Vary through MMVet [58] benchmark. As shown in Table 4, with the same LLM (Vicuna-7B) and SFT data (LLaVA-CC665k), Vary lifts 5.5% (36.0% vs. 30.5%) of the total accuracy than LLaVA-1.5. Besides, Vary-toy with only Qwen-1.8B can achieve an on-par 30.4% accuracy with 7B-size LLaVA-1.5. The above results fully demonstrate the proposed Vary can maintain the general visual understanding ability well.

4.6 Complementarity of the New and Old Vision Vocabularies

We apply feature normalization after the input embedding layers (Fig. 2) of both the new and original (CLIP) vision vocabulary networks to monitor whether a branch will collapse before inputting the LLM under some scenarios and analyze the complementarity between them. Specifically, we select four types of common scenarios, including natural, scene text, chart, and document images for experiments. We collect 50 samples for each class to calculate the feature normalization and gain their value ranges. As shown in Fig. 4, for natural images, the feature normalization values of the proposed new vision vocabulary are between 437 and 521 while the CLIP values are from 426 to 552, indicating that the new and old vocabularies are nip and tuck in such class, with the new vocabulary mainly used for object localization and CLIP for image description. For scene text, CLIP values are approximately 100 points higher than the new vision vocabulary due to we do not use such images to train new vision vocabulary, and CLIP plays a major role in this scenario. By contrast, the feature normalization values of new vision vocabulary are much higher than the CLIP for chart/document images, especially for document images which CLIP is not good at, proving the two vocabularies complement each other to a certain extent.

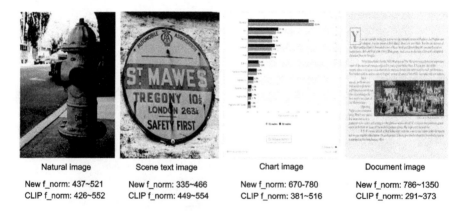

Fig. 4. The values of feature normalization of two vision vocabularies. We choose four commonly used scenarios to test the relationship between two vision branches. For each type of image, we prepare 50 samples, and this figure shows one typical example and the value range of feature normalization.

4.7 Ablation Study

In this section, we conduct ablation analyses to further validate the effectiveness of our designs. We perform ablations from two aspects: the data used to generate the new vocabulary and the selection of the new and original vocabulary.

Datasets in Generating the New Vocabulary. We use document/chart parsing and object detection data in the vision vocabulary generating process. To test the effect of the object detection dataset, we train a new Vary-tiny without detection data. As shown in Table 5, under this setting, the DocVQA only lifts 0.8% (from 65.3% to 66.1%) while the RefCOCO drops 9.7% (from 88.0% to 78.3%). The results demonstrate that: 1) the introduction of detection data has little harm on document text recognition; 2) "writing" general object localization knowledge into the new vision vocabulary is crucial for boosting LVLM's REC performance.

Effectiveness of the Original CLIP. We verify the effect of the old vocabulary (CLIP) of Vary via blocking the new vision vocabulary. As shown in Table 5, when discarding the new vision vocabulary and only keeping the CLIP, the MMVet accuracy is 29.0% (only dropping 0.6%) yet with only 10.9% on DocVQA, proving that the CLIP is crucial for LVLM's general ability, and also indicating that it cannot handle high-resolution document-level VQA task. We also turn off the CLIP branch to further verify its necessity. Under such a setting, we additionally feed 50M image-text pair images sampled from Laion-COCO to make the new vision vocabulary enjoy general vision ability. However, it only yields 16.1% on MMVet. It is difficult to train a general vision vocabulary equivalent to CLIP with limited resources. The above results firmly indicate that our scaling-up vocabulary strategy is very effective at a low cost.

Table 5. Ablation study of Vary. Vary-tiny data means the datasets used in generating the new vocabulary. The "New" represents the new vision vocabulary network. Abbreviations "Doc" and "Det" represent the document and detection data, respectively.

Vary-tiny data			Vary-toy		Performance		
Doc/Chart	Det	Pair	CLIP	New	DocVQA(val)	RefCOCO(val)	MMVet
✓	✓	×	✓	✓	65.3	88.0	30.4
✓	×	×	✓	✓	66.1	78.3	–
–	–	–	✓	×	10.9	78.6	29.0
✓	✓	×	×	✓	65.5	87.2	–
✓	✓	50M	×	✓	65.1	86.9	16.1

5 Limitation and Conclusion

This paper highlights that scaling up the vocabulary in the vision branch for an LVLM is quite significant and we successfully prove such a claim by a simple framework. Profited by the new vision vocabulary we generated, the proposed model – Vary achieves promising performances in multiple tasks. Nevertheless, Vary still suffers limitations such as potential competition between two vision vocabulary networks. We believe that how to effectively scale up the visual vocabulary at a low cost still enjoys much improvement rooms, especially compared to the mature and relatively simple means of text vocabulary expansion. We hope that the useful and efficient design of Vary will attract more research attention to such a direction.

Acknowledgement. The work was supported by the National Science and Technology Major Project of China (2023ZD0121300). The work was also supported by the Research on AI Terminal Computing Power and End-Cloud Framework (R2411B0R) of China Mobile Group Device Co., Ltd.

References

1. Alayrac, J., et al.: Flamingo: a visual language model for few-shot learning. In: NeurIPS (2022)
2. Alibaba: Introducing qwen-7b: Open foundation and human-aligned models (of the state-of-the-arts). https://github.com/QwenLM/Qwen-7B (2023)
3. Bai, J., et al.: Qwen technical report. arXiv preprint arXiv:2309.16609 (2023)
4. Bai, J., et al.: Qwen-vl: a versatile vision-language model for understanding, localization, text reading, and beyond. arXiv preprint arXiv:2308.12966 (2023)
5. Blecher, L., Cucurull, G., Scialom, T., Stojnic, R.: Nougat: Neural optical understanding for academic documents. arXiv preprint arXiv:2308.13418 (2023)
6. Brown, T., et al.: Language models are few-shot learners. Adv. Neural. Inf. Process. Syst. **33**, 1877–1901 (2020)
7. Chen, K., Zhang, Z., Zeng, W., Zhang, R., Zhu, F., Zhao, R.: Shikra: Unleashing multimodal llm's referential dialogue magic. arXiv preprint arXiv:2306.15195 (2023)

8. Chiang, W.L., et al.: Vicuna: an open-source chatbot impressing gpt-4 with 90%* chatgpt quality. https://lmsys.org/blog/2023-03-30-vicuna/ (2023)
9. Davis, B., Morse, B., Price, B., Tensmeyer, C., Wigington, C., Morariu, V.: End-to-end document recognition and understanding with dessurt. In: European Conference on Computer Vision, pp. 280–296. Springer (2022)
10. Deng, J., Yang, Z., Chen, T., Zhou, W., Li, H.: Transvg: end-to-end visual grounding with transformers. In: Proceedings of the IEEE/CVF International Conference on Computer Vision, pp. 1769–1779 (2021)
11. Devlin, J., Chang, M.W., Lee, K., Toutanova, K.: Bert: Pre-training of deep bidirectional transformers for language understanding. arXiv preprint arXiv:1810.04805 (2018)
12. Dong, R., et al.: Dreamllm: Synergistic multimodal comprehension and creation. arXiv preprint arXiv:2309.11499 (2023)
13. Gan, Z., Chen, Y.C., Li, L., Zhu, C., Cheng, Y., Liu, J.: Large-scale adversarial training for vision-and-language representation learning. Adv. Neural. Inf. Process. Syst. **33**, 6616–6628 (2020)
14. Hao, Y., et al.: Language models are general-purpose interfaces. arXiv preprint arXiv:2206.06336 (2022)
15. Huang, S., et al.: Language is not all you need: Aligning perception with language models. arXiv preprint arXiv:2302.14045 (2023)
16. Kazemzadeh, S., Ordonez, V., Matten, M., Berg, T.: Referitgame: referring to objects in photographs of natural scenes. In: Proceedings of the 2014 conference on empirical methods in natural language processing (EMNLP), pp. 787–798 (2014)
17. Kim, G., et al.: Ocr-free document understanding transformer. In: European Conference on Computer Vision, pp. 498–517. Springer (2022)
18. Kirillov, A., et al.: Segment anything. arXiv preprint arXiv:2304.02643 (2023)
19. Kuznetsova, A., et al.: The open images dataset v4: unified image classification, object detection, and visual relationship detection at scale. Int. J. Comput. Vision **128**(7), 1956–1981 (2020)
20. Lee, K., et al.: Pix2struct: screenshot parsing as pretraining for visual language understanding. In: International Conference on Machine Learning, pp. 18893–18912. PMLR (2023)
21. Levenshtein, V.I., et al.: Binary codes capable of correcting deletions, insertions, and reversals. In: Soviet physics doklady, vol. 10, pp. 707–710. Soviet Union (1966)
22. Li, B., Zhang, Y., Chen, L., Wang, J., Yang, J., Liu, Z.: Otter: A multi-modal model with in-context instruction tuning. arXiv preprint arXiv:2305.03726 (2023)
23. Li, J., Li, D., Savarese, S., Hoi, S.: Blip-2: Bootstrapping language-image pre-training with frozen image encoders and large language models. arXiv preprint arXiv:2301.12597 (2023)
24. Li, Y., Mao, H., Girshick, R., He, K.: Exploring plain vision transformer backbones for object detection. In: European Conference on Computer Vision, pp. 280–296. Springer (2022)
25. Liang, Y., et al.: Taskmatrix. ai: Completing tasks by connecting foundation models with millions of apis. arXiv preprint arXiv:2303.16434 (2023)
26. Lin, T.-Y., et al.: Microsoft COCO: common objects in context. In: Fleet, D., Pajdla, T., Schiele, B., Tuytelaars, T. (eds.) ECCV 2014. LNCS, vol. 8693, pp. 740–755. Springer, Cham (2014). https://doi.org/10.1007/978-3-319-10602-1_48
27. Liu, F., et al.: Matcha: Enhancing visual language pretraining with math reasoning and chart derendering. arXiv preprint arXiv:2212.09662 (2022)
28. Liu, H., Li, C., Li, Y., Lee, Y.J.: Improved baselines with visual instruction tuning (2023)

29. Liu, H., Li, C., Wu, Q., Lee, Y.J.: Visual instruction tuning (2023)
30. Loshchilov, I., Hutter, F.: Sgdr: Stochastic gradient descent with warm restarts. arXiv preprint arXiv:1608.03983 (2016)
31. Loshchilov, I., Hutter, F.: Decoupled weight decay regularization. In: ICLR (2019)
32. Masry, A., Long, D.X., Tan, J.Q., Joty, S., Hoque, E.: Chartqa: A benchmark for question answering about charts with visual and logical reasoning. arXiv preprint arXiv:2203.10244 (2022)
33. Mathew, M., Karatzas, D., Jawahar, C.: Docvqa: a dataset for vqa on document images. In: Proceedings of the IEEE/CVF Winter Conference on Applications of Computer Vision, pp. 2200–2209 (2021)
34. Mishra, A., Shekhar, S., Singh, A.K., Chakraborty, A.: Ocr-vqa: visual question answering by reading text in images. In: 2019 International Conference on Document Analysis and Recognition (ICDAR), pp. 947–952. IEEE (2019)
35. OpenAI: Gpt-4 technical report (2023)
36. Ouyang, L., et al.: Training language models to follow instructions with human feedback. In: NeurIPS (2022)
37. Radford, A., et al.: Learning transferable visual models from natural language supervision. In: International Conference on Machine Learning, pp. 8748–8763. PMLR (2021)
38. Radford, A., et al.: Language models are unsupervised multitask learners. OpenAI blog **1**(8), 9 (2019)
39. Raffel, C., et al.: Exploring the limits of transfer learning with a unified text-to-text transformer. J. Mach. Learn. Res. **21**(1), 5485–5551 (2020)
40. Schuhmann, C., et al.: Laion-400m: Open dataset of clip-filtered 400 million image-text pairs. arXiv preprint arXiv:2111.02114 (2021)
41. Shao, S., et al.: Objects365: a large-scale, high-quality dataset for object detection. In: Proceedings of the IEEE/CVF International Conference on Computer Vision, pp. 8430–8439 (2019)
42. Shen, Y., Song, K., Tan, X., Li, D., Lu, W., Zhuang, Y.: Hugginggpt: Solving ai tasks with chatgpt and its friends in huggingface. arXiv preprint arXiv:2303.17580 (2023)
43. Singh, A., et al.: Towards vqa models that can read. In: Proceedings of the IEEE/CVF Conference on Computer Vision and Pattern Recognition, pp. 8317–8326 (2019)
44. Taori, R., et al.: Stanford alpaca: An instruction-following llama model. https://github.com/tatsu-lab/stanford_alpaca (2023)
45. Team, M., et al.: Introducing mpt-7b: A new standard for open-source, commercially usable llms (2023)
46. Touvron, H., et al.: Llama: Open and efficient foundation language models. arXiv preprint arXiv:2302.13971 (2023)
47. Veit, A., Matera, T., Neumann, L., Matas, J., Belongie, S.: Coco-text: Dataset and benchmark for text detection and recognition in natural images. arXiv preprint arXiv:1601.07140 (2016)
48. Wang, P., et al.: Ofa: Unifying architectures, tasks, and modalities through a simple sequence-to-sequence learning framework. In: International Conference on Machine Learning, pp. 23318–23340. PMLR (2022)
49. Wang, W., et al.: Visionllm: Large language model is also an open-ended decoder for vision-centric tasks. arXiv preprint arXiv:2305.11175 (2023)
50. Wei, J., et al.: Emergent abilities of large language models. arXiv preprint arXiv:2206.07682 (2022)

51. Wu, C., Yin, S., Qi, W., Wang, X., Tang, Z., Duan, N.: Visual chatgpt: Talking, drawing and editing with visual foundation models. arXiv preprint arXiv:2303.04671 (2023)
52. Xu, C., Guo, D., Duan, N., McAuley, J.: Baize: An open-source chat model with parameter-efficient tuning on self-chat data. arXiv preprint arXiv:2304.01196 (2023)
53. Yang, R., et al.: Gpt4tools: teaching large language model to use tools via self-instruction. arXiv preprint arXiv:2305.18752 (2023)
54. Yang, Z., et al.: Unitab: Unifying text and box outputs for grounded vision-language modeling. In: European Conference on Computer Vision. pp. 521–539. Springer (2022). https://doi.org/10.1007/978-3-031-20059-5_30
55. Yang, Z., et al.: Mm-react: Prompting chatgpt for multimodal reasoning and action. arXiv preprint arXiv:2303.11381 (2023)
56. Ye, J., et al.: mplug-docowl: Modularized multimodal large language model for document understanding. arXiv preprint arXiv:2307.02499 (2023)
57. Yu, E., et al.: Merlin: Empowering multimodal llms with foresight minds. arXiv preprint arXiv:2312.00589 (2023)
58. Yu, W., et al.: Mm-vet: Evaluating large multimodal models for integrated capabilities. arXiv preprint arXiv:2308.02490 (2023)
59. Zeng, A., et al.: Glm-130b: An open bilingual pre-trained model. arXiv preprint arXiv:2210.02414 (2022)
60. Zhang, A., Zhao, L., Xie, C.W., Zheng, Y., Ji, W., Chua, T.S.: Next-chat: An lmm for chat, detection and segmentation. arXiv preprint arXiv:2311.04498 (2023)
61. Zhang, S., et al.: Opt: Open pre-trained transformer language models. arXiv preprint arXiv:2205.01068 (2022)
62. Zhao, L., et al.: Chatspot: Bootstrapping multimodal llms via precise referring instruction tuning. arXiv preprint arXiv:2307.09474 (2023)
63. Zhu, D., Chen, J., Shen, X., Li, X., Elhoseiny, M.: Minigpt-4: Enhancing vision-language understanding with advanced large language models. arXiv preprint arXiv:2304.10592 (2023)

Merlin: Empowering Multimodal LLMs with Foresight Minds

En Yu[1], Liang Zhao[2], Yana Wei[3], Jinrong Yang[3], Dongming Wu[4], Lingyu Kong[5], Haoran Wei[2], Tiancai Wang[2], Zheng Ge[2], Xiangyu Zhang[2], and Wenbing Tao[1(✉)]

[1] Huazhong University of Science and Technology, Wuhan, China
{yuen,wenbingtao}@hust.edu.cn
[2] MEGVII Technology, Beijing, China
[3] ShanghaiTech University, Shanghai, China
[4] Beijing Institute of Technology, Beijing, China
[5] University of Chinese Academy of Sciences, Beijing, China

Abstract. Humans can foresee the future based on present observations, a skill we term as *foresight minds*. However, this capability remains under-explored within existing MLLMs, hindering their capacity to understand intentions behind subjects. To address this, we integrate the future modeling into MLLMs. By utilizing the **trajectory**, a highly structured representation, as a learning objective, we aim to equip the model to understand spatiotemporal dynamics. Inspired by the learning paradigm of LLMs, we first propose *Foresight Pre-Training (**FPT**)* that jointly learns various tasks centered on trajectories, enabling MLLMs to predict entire trajectories from a given initial observation. Then, we propose *Foresight Instruction-Tuning (**FIT**)* that requires MLLMs to reason about potential future events based on predicted trajectories. Aided by FPT and FIT, we build an unified MLLM named **Merlin** that supports complex future reasoning. Experiments show Merlin's foresight minds with impressive performance on both future reasoning and visual comprehension tasks. Project page: https://ahnsun.github.io/merlin.

Keywords: Multimodal Large Language Model · Future Reasoning

1 Introduction

Human beings can predict future events or outcomes based on current observations, known in neuroscience theory as *predictive processing* [19]. In this paper, we refer to this ability as *foresight minds*, which involves the use of past experiences, knowledge, sensory information, and probabilistic reasoning to generate expectations about future events. In the artificial intelligence (AI) domain, the capability to predict future events is an important topic towards the realization of artificial general intelligence (AGI).

E. Yu and L. Zhao—Equal Contribution.

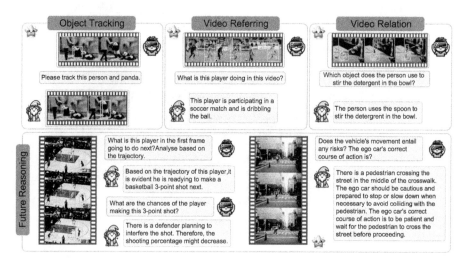

Fig. 1. Demo cases presentation of Merlin. Here we showcase several main capabilities of our built Multimodal Large Language Model (MLLM), **Merlin**. Notably, in the dialogue, the words marked with colors correspond to the trajectory outputs of the targets in the image. To save space, we highlight them using the same colors.

Recent advancements in Multimodal Large Language Models (MLLMs), such as GPT-4V [46] and Bard [2], have shown significant potential in image understanding and logical reasoning. Despite these achievements, these models struggle to foresee future events based on current image observations. Even provided with additional observations, like sequences of multiple frames, the current MLLM models still struggle to adequately analyze and infer specific target behaviors, such as predicting object movements or interactions (shown in Fig. 2). On the contrary, human can reason the future to some extent based on the observed current state [5,52], which shows powerful foresight minds.

To mitigate this existing deficiency in MLLMs, we start from dividing human's process of foreseeing the future into a two-stage system [29,52]: (1) observing the dynamic clues of the subject and then (2) analyzing the behavior pattern and reasoning what might happen according to the observation. For instance, while watching a basketball game, people will first observe the moving players on the court, and then forecast the specific player's forthcoming actions, e.g., shooting, slam-dunking, or passing, by analyzing the current states and movement patterns of the players. Compare this system to current MLLMs, we find that MLLMs can complete the second stage well, thanks to the powerful logical reasoning ability of LLM [48,67]. Therefore the key challenge is the first stage. That is, *how to make MLLM acquire correctly* **spatiotemporal dynamics** *from the multi-image observation?*

Explicitly modeling next frames (*e.g.*, reconstructing next frames [12,71]) can be a straightforward way. However, it can be hard to directly extract dynamic clues from the redundant visual information [24], especially from video sequences.

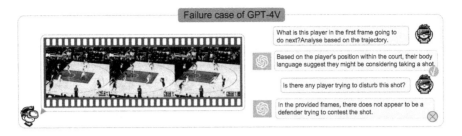

Fig. 2. Failure case of GPT-4V about future reasoning.

It is necessary to construct a suitable learning objective to assist MLLM in obtaining dynamic clues about the specific subjects. To this end, we point out that *trajectory, as a highly structured representation, is a good learning objective which can link the temporal contexts between the past and the future.*

Based on this insight, we propose to model the future to empower existing MLLMs with *"foresight minds"*. Following the modern learning paradigm of LLMs, our future reasoning learning process includes two stages: (1) *Foresight Pre-Training (**FPT**)*, a paradigm that causally models the temporal trajectories, which interleave with multi-frame images. The model starts with the initial observation of one or multiple subjects in the first frame as the query and then is required to predict the whole trajectory. Notably, we introduce various tasks containing richly labeled data [18, 25, 30, 56, 60, 77], including object detection, object tracking, etc., to perform multitask learning. And samples from these tasks are properly formatted to ensure coordinated pre-training. (2) *Future Instruction-Tuning (**FIT**)*, then, considers the trajectory modeling bestowed by FPT as a bridge in the logical chain of future reasoning. Simply put, when querying an MLLM, it must articulate its reasoning in conjunction with the trajectory for each object referenced. This method, as a form of *Trajectory Chain-of-Thought*, effectively narrows the gap between trajectory perception and predictive future reasoning, thereby fully unleashing model's foresight minds.

Aided by the above future modeling technologies, we provide ***Merlin***[1], a novel and unified MLLM capable of handling inputs and outputs of spatial coordinates or tracklets from single image or multiple frames. Moreover, Merlin is adept at performing inductive reasoning about future events based on current observational results. To demonstrate this, we provide several real dialogues between users and Merlin, as displayed in the Fig. 1. Unlike the previous MLLMs [39, 81, 83] which only supported interaction with a single image, Merlin not only provides a richer multi-image interaction, but also on this basis, is capable of executing unique and powerful future reasoning.

We construct a new future reasoning benchmark to evaluate Merlin's logical reasoning and future prediction abilities. The results, which significantly surpass

---

[1] **Merlin** is a legendary character in the tales of King Arthur, renowned as a powerful wizard and a wise counselor in the Arthurian legends. He is depicted as having the power to foresee future events and has a deep understanding of fate and destiny.

previous baselines [10,38,39,72], demonstrate Merlin's stunning performance in future reasoning. We further reveal Merlin's exceptional performance in general visual understanding. Through analysis in scenarios such as VQA (Visual Question Answering) [23,27], comprehensive understanding [40,76], and hallucination [36], we unexpectedly discovered that our proposed novel paradigm of future learning aids MLLMs in gaining a deeper understanding of images. We believe this brings new insights for the training of future MLLMs.

2 Related Work

2.1 Large Language Models

Large Language Models (LLMs) have gained significant attention due to their capabilities in language generation and logical reasoning. Pioneering models like BERT [15], GPT-2 [50], and T5 [51] laid the groundwork, but GPT-3 [8], the first model with a 175 billion parameter size, made notable strides, demonstrating strong zero-shot performance. An emergent ability, wherein model size scaling results in significant language capability improvements, was also observed in LLMs. This was further facilitated by InstructGPT [47] and ChatGPT [45] using Reinforcement Learning with Human Feedback (RLHF) on GPT-3. These advancements led to what's called LLMs' "iPhone moment". Following GPT's success, several open-source LLMs, including OPT [80], LLaMA [62], and GLM [78], have been proposed, showing similar performance to GPT-3. Models like Alpaca [61] and Vicuna [11] illustrate the application of these LLMs, using a self-instruct framework to construct excellent dialogue models.

2.2 Multimodal Large Language Models

The advancements in LLMs [45,62,63] have projected a promising path towards artificial general intelligence (AGI). This has incited interest in developing multimodal versions of these models. Current Multi-modal Large Language Models (MLLMs) harness the potential of cross-modal transfer technologies. These models consolidate multiple modalities into a unified language semantic space, and then employ autoregressive language models as decoders for language outputs. Models like Flamingo [1] have adopted a gated cross-attention mechanism, trained on billions of image-text pairs, to align visual and linguistic modalities, showing impressive performance on few-shot learning tasks. Similarly, BLIP-2 [35] introduced the Q-Former to align visual features more effectively with language space. The LLaVA series [38,39,65] further enhanced this process by using simply a MLP in place of the Q-Former and designing a two-stage instruction-tuning procedure. Apart from creating general MLLMs, techniques have also been developed for visual-interactive multimodal comprehension, involving the precise tuning of referring instructions [10,79,81]. Furthermore, another interesting direction in MLLM research involves integrating MLLMs for cross-modal generation [16,20,31] by using text-to-image models such as Stable Diffusion.

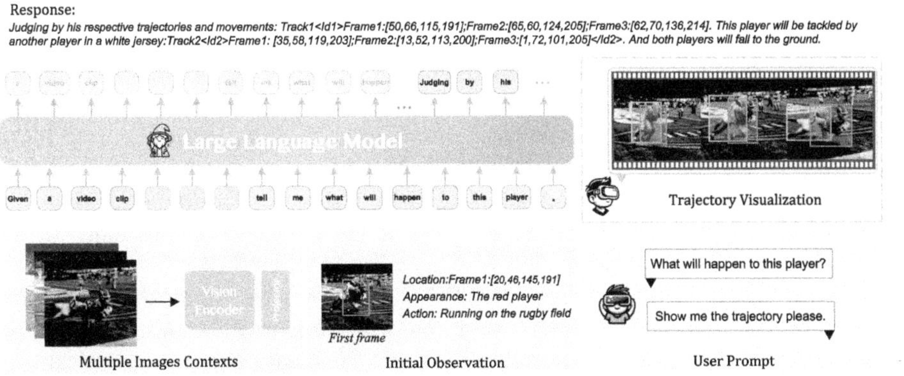

Fig. 3. Overall pipeline of Merlin. The architecture of Merlin consists of three main components: (1) an image encoder, (2) a large language model, and (3) a modality-align projector. **Bottom:** The diverse input format that supports multiple-image contexts, initial observation and the specific user prompt. **Top:** The model response including the predicted trajectory and the future reasoning.

3 Methodology

3.1 Overall Architecture

Merlin is designed to unlock the foresight minds based on observations from single images and multi-frame video clips. In order to accomplish this, images and videos are comprehensively represented through a series of visual tokens, which are then integrated into the language sequence that can be comprehended by Large Language Models (LLMs) in a unified framework. Specifically, Merlin consists of an image encoder, a decoder-only LLM, and a modality alignment block as illustrated in Fig. 3. Following prevalent practice [10,38,39,83], we opt for the pre-trained CLIP [49] ViT-L/14 [17] as the visual encoder and Vicuna-7B v1.5 [11] as the large language decoder. For more details, please refer to our supplementary materials.

To provide enough visual information and details, the input images are resized to a resolution of 448×448. At this juncture, the visual encoder iteratively attends to $(448/14)^2$ uniformly divided image patches, yielding 1024 encoded tokens. Considering the limited context length of LLMs and addressing the substantial computational challenges posed by high resolution and multi-frame context modeling, we simply utilize a 2D convolution to achieve both dimension projection and token aggregation [7,53].

We choose 2D convolution over 1D linear layers [10,38,39] or cross-attention layers [4,35,83] as connector for the following reasons: (1) 2D convolution clusters local visual tokens on a spatial scale [22], effectively achieving a one-step transformation from spatial to channel information; (2) The good convergence properties [28,57] of 2D convolution compared with cross-attention lay a solid foundation for foresight learning in a two-step training approach.

3.2 Foresight Pre-Training

Generative Pre-Training (GPT) [8,45,46] serves as the cornerstone of this generation's Language Models (LLMs). Through learning to predict next token, the model efficiently condenses data, thereby yielding emergent forms of intelligence [66]. In this context, a very natural approach to enhance the model's perception of the dynamic clues across multiple frames is to *explicitly model the next frame* (or image). However, due to the high redundancy in multi-frame visual information, the truly next-frame prediction remains a significant challenge to date. A better approach at this juncture is to *implicitly model high semantic information in the label space* (such as categories, bounding boxes) on a frame-by-frame basis. Temporally, this label information forms a **trajectory**, a highly structured representation. Causally modeling the trajectory in conjunction with each frame of image helps to connect the past and present in time, thus enabling the model to perceive the future.

To this end, we propose the Foresight Pre-Training, a way of *causally modeling the trajectories interleaved with multi-frame images*, to empower the MLLM with the capacity of perceive the dynamic clues, and ultimately achieving future reasoning. Specifically, given a video clip including several frames, we first give the model the observation of the *first frame*, then we require the model to predict the *whole trajectory* of the concerned subject in this video conditioned on the initial observation. Notably, the observation of the first frame can be the description or simple position of the concerned object. Formally,

$$P(Y|X) \sim P(Y|\{X_1, X_2, ...\}, O_{first}), \tag{1}$$

where X_i denotes the ith frame and O_{first} is the first frame observation, Y refers to the trajectory of the subject in O_{first} within the frame sequence. The observation and the raw frames will be regarded as the condition to prompt MLLM to predict the trajectory.

Data Construction. We first aggregate all valuable multimodal information from diverse data resources and then properly organize them for multi-task foresight pre-training. Specifically, for each sample instance **I**, we first collect its multimodal information including consecutive multi-frame images $\{X_1, X_2, ...\}$, subject observations from the *first* frame O_{first}, and subject trajectory Y constructed from *all* frames. Formally,

$$\mathbf{I} = \{\{X_1, X_2, ...\}, O_{first}, Y\}. \tag{2}$$

We categorize observations of one subject of the first frame into three main types: **location** description, **appearance** description and **action** description. Then we *randomly* selected one of these observations of a particular subject in the first frame as the query object. (It is also feasible to select observations of multiple attributes as the query according to the characteristics of the dataset).

To better unleash the powerful generative modeling capacity of LLM, we construct this query process as a type of conversation. Here is an example of the

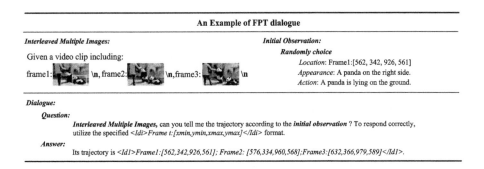

Fig. 4. One example to illustrate the multi-modality pretraining dataset. The top block shows the provided contexts including the multiple images contexts and initial observation (box, appearance and action) about the subject to prompt the LLM. The bottom block shows the dialogue including question and answer.

constructed data shown in Fig. 4. In this case, we want to query the subject—*the panda on the right*—with the randomly select observation, and expect the answer with the movement trajectory of this panda across multiple frames. To model this process, we convert the query to question and trajectory to answer with proper natural language for embellishment.

Overall, the aforementioned process of dialogization roughly follows these three principles: (1) **Precise definition of task prompts and answer formats**. In particular, we use a task prompt to tell MLLM *what specific task to do* (detect or track), and also *specified the answer format* with accurate descriptions in each question. In this way, different types of tasks can be flexibly organized together without compromising the general language ability. (2) **Clear indication of multimodal information.** Concretely, for each group of image tokens, we add a *special frame indicator* in front of then, i.e., frame1:<image> and frame2:<image>, so as to help MLLM better focus on the corresponding image. (3) **Interleaving of frames and observations.** For the same identity, we interleave the frames in which it appears with its positional observations, and enclose them with two ID tokens (i.e. <Idi> and </Idi>) to construct a trajectory. We believe that this interleaved organization helps in *generatively training to model causality within the trajectory*, while the ID tokens ensures that the model can distinguish among different identity objects.

Training Details. The objective in this stage is to initially endow MLLMs with the capacity of modeling the spatiotemporal dynamics across multi-frame images, while ensuring that its general language capabilities do not diminish. Previous practices [4,38,39] typically conducting a separate modality alignment training phase following a multi-task pre-training stage, which however, complicates the training process and data construction. In this paper, we directly incorporate both of them into one stage, and *unfreeze all modules during pre-training*. This is because that we believe the MLLMs are sufficiently powerful to concurrently handle the learning of general multimodal capabilities and multi-task specific abilities *under proper guidance*. Furthermore, we mix a large amount of image-text pairs and rich-annotated conversation data (formatted according to the above method) from diverse data sources [18,25,30,44,56,58,60,77] to conduct multi-task learning. In doing so, not only endows the model with foresight minds but also ensures its multimodal alignment.

3.3 Foresight Instruction Tuning

Although Foresight Pre-Training equips the model with the ability to observe dynamic clues across multiple frames, it still falls short of true foresight minds. This is because models typically struggle to effectively transform such observations into successful future reasoning [64,82].

Recent work [48,82] has highlighted that Chain-of-Thought (CoT) [67] is crucial in bridge the gap between the observations and actions of MLLMs with theory of mind [54,64]. Meanwhile, several prior studies [10,81] have also demonstrated that prompts indicating position (such as bounding boxes or points)—a principle analogous to CoT—can concentrate an MLLM's attention on the relevant area, leading to more accurate dialogues and reducing the likelihood of visual hallucination. Drawing inspiration from these findings, we conduct the Foresight Instruction Training (FIT) building upon the foundation of FPT to further enhance the model's future reasoning capability. In specific, building on the trajectory generating powered by FPT, we further union the trajectories to generatively rationalize the forthcoming events. Mathematically,

$$P(Z|X,Y) \sim P(Z|\{X_1, X_2, ...\}, O_{first}, Y), \qquad (3)$$

where Z refers to the future observation which is deduced from observations in *each* frame. It can be actions, events, trends, or simply likelihoods. In this context, multi-frame images, in conjunction with the first subject observation, and the trajectory of the same subject across all frames, serve as the union condition to prompt MLLM to causally predict the future. This way, akin to ***Trajectory Chain-of-Thought***, effectively bridges the gap between trajectory perception and predictive future reasoning, thereby fully unleashing model's foresight minds.

Data Construction. The specific data construction method is similar to FPT, but on this basis, we also deduce a future observation Z from the information across multiple frames and append it after the trajectory in the answer. Formally,

$$\mathbf{I} = \{\{X_1, X_2, ...\}, O_{first}, Y, Z\}. \qquad (4)$$

Practically, in this paper, we constitute future observations based on multi-frame, multi-target action descriptions combined with human priors, and further process them with GPT-4 [46] to ultimately form reasonable future inferences. More details are provided in the supplementary materials.

Figure 3 provides an illustrative example of FIT, when a user questions Merlin about the future of a player in red attire, Merlin initially presents the observed trajectory of the concerned player, followed by the trajectory of another player in white. Using these trajectories, Merlin deduces that *the player in white is likely to tackle the one in red, resulting in both players falling to the ground.*

Training Details. We freeze the vision encoder and keep the convolutional projector and the LLM unfreezed in this stage. On this basis, we primarily adopt the open-source instruction tuning datasets, *i.e.*, LLaVA-665K [38], for building the basic ability for multi-round visual-language conversation. For further unleashing the foresight minds of model, we first uniformly sample a certain number of multitask dialogues in FPT, in order to maintain the model's capacity of modeling the dynamic clues across multi-frame images. In addiction, we also sample data from three specific scenario datasets [37,43,68] and construct around 30K FIT conversations based on the aforementioned data construction process.

4 Experiment

4.1 Experimental Settings

Datasets. For the foresight pre-training (FPT) stage, we first use 10M image-text pairs sampled from LAION400M [55] to ensure multimodal alignment. On this basis, we gather various open-source datasets with rich annotations to conduct multi-task learning, including (1) **object detection** datasets: Object365 [56] and OpenImage [32]; (2) **tracking** datasets: LaSOT [18], GOT10K [25], MOT17 [44], DanceTrack [60] and SOMPT22 [58]; (3) **grounding** dataset: Ref-COCO [30]; (4) **object relation** dataset: VCR [77]. For these data, as described in Sect. 3.2), we apply strict task definitions and format specifications, and re-organize them in the form of interleaved frames and observations. Ultimately, we obtain approximately 5M question-answer data, which are mixed with 10M paired data for foresight pre-training.

For the foresight instruction-tuning (FIT) stage, we mix approximately 730K conversation data, including (1) open-source instruction-tuning data LLaVA-665K [38], which integrates a series of VQA datasets [59] and multi-round conversation datasets [39]; (2) around 30K FIT multi-frame conversations constructed from three specific scenarios including MultiSports [37], TITAN [43] and STAR [68] based on the data construction method described in Sect. 3.3; (3) nearly 40K randomly sampled FPT multi-task data. For more details of the datasets, please refer to the supplementary materials.

Implementation Details. As outlined in Sect. 3.1, Merlin utilizes the CLIP-ViT-L/14 [49] as its vision encoder for image encoding and the open-source Vicuna-7B v1.5 [11] for foresight decoding. Between them, a 3 × 3 convolution layer with padding set to 1 and a stride of 2 is employed for both dimension projection and token aggregating. During the foresight pre-training, we optimize all parameters of the model, setting the learning rate to $5e-5$ and training for one epoch. In the instruction tuning stage, we freeze the visual encoder and fine-tune the parameters of the projector and LLM. In both stages, we train Merlin using the AdamW [42] optimizer and a cosine annealing scheduler [41] as the learning rate scheduler. The entire training process is conducted on 64 NVIDIA A800 GPUs, with approximately 12 h required for pre-training and 3 h for instruction-tuning. Additional implementation details can be found in the supplementary materials.

Table 1. The Effectiveness of Prediction Reasoning. We mainly select 5 metrics from MMBench develop and test set, respectively, including **OL**: Object localization (Prediction), **PPR**: Physical property reasoning, **FR**: Function reasoning, **IR**: Identity reasoning, and **FP**: Future prediction. **Avg.** denotes the average score. The best and second-best performances are shown in bold font and underlined respectively.

Method	LLM Size	Prediction Reasoning (Dev.)					Prediction Reasoning (Test)				
		Avg.	OL	PPR	FR	IR	FP	Avg.	OL	PPR	FR	IR	FP
InstructBLIP [13]	13B	42.0	14.8	30.7	56.8	88.9	19.0	44.4	5.7	24.0	67.3	<u>92.7</u>	32.4
MiniGPT-4 [83]	13B	43.3	28.4	30.7	49.4	86.7	21.4	48.9	21.0	35.0	67.3	90.2	31.1
OpenFlamingo [3]	7B	5.28	2.5	10.7	8.6	2.2	2.4	11.5	2.9	14.0	9.3	11.0	20.3
MMGPT [40]	7B	19.5	1.2	24.0	9.9	60.0	2.4	16.8	3.8	13.0	12.1	52.4	2.7
MiniGPT-4 [83]	7B	26.8	7.4	14.7	19.8	80.0	11.9	27.9	8.6	13.0	29.9	61.0	27.0
InstructBLIP [13]	7B	34.8	6.2	17.3	51.9	84.4	14.3	39.0	2.9	17.0	52.3	78.0	44.6
LLaVA [39]	7B	38.7	8.6	25.3	53.1	77.8	28.6	39.7	13.3	35.0	48.6	82.9	18.9
mPLUG-Owl [72]	7B	41.0	18.5	18.7	66.7	86.7	14.3	45.9	16.2	23.0	59.8	91.5	39.2
Shikra [10]	7B	51.5	32.1	30.7	63.0	88.9	42.9	<u>60.0</u>	27.6	<u>50.0</u>	<u>70.1</u>	<u>92.7</u>	<u>59.5</u>
Kosmos-2 [26]	1.6B	54.4	38.3	33.3	56.8	91.1	<u>52.4</u>	58.2	<u>40.4</u>	30.0	65.4	89.0	**66.2**
LLaVA-1.5 [38]	7B	<u>59.6</u>	<u>43.2</u>	<u>52.0</u>	71.6	<u>93.3</u>	38.1	-	-	-	-	-	-
Merlin (Ours)	7B	**64.4**	42.0	**54.7**	**72.8**	**97.8**	**54.8**	**66.5**	**41.3**	**51.0**	**83.0**	**97.6**	59.7

4.2 Properties Evaluation of Foresight Minds

In this section, we mainly verify the foresight minds (future reasoning) of Merlin from two aspects, *i.e.*, prediction reasoning and identity association ability, where the former focuses on forecasting and reasoning location, events or behavior based on image observation, and the latter focuses on the model's ability to establish subject identity associations across multiple frames to obtain dynamic clues for future reasoning.

Prediction Reasoning. To evaluate this ability, we probe this ability based on the several sub-tasks of MMBench [40]. MMBench provides a comprehensive evaluation system to assess various capabilities of MLLM, with some metrics focusing on the model's prediction and reasoning capabilities. To this end, we pick out these metrics to establish this new future reasoning benchmark and compare Merlin with the existing SOTA models. As shown in Table 1, Merlin achieves the best overall performance (64.4 average score on the development set and 66.5 average score on the test set). Moreover, it obtains the best in 8/10 indicators and ranks second in all other indicators, which favorably demonstrates Merlin's strong prediction and reasoning ability.

Identity Association. We examine this ability by evaluating the performance of object-tracking tasks [69,73–75], which can comprehensively demonstrate object association and prediction capabilities. To this end, we evaluate Merlin in existing mainstream tracking benchmarks, *i.e.*, LaSOT [18] and GOT10K [25]. It is worth noting that Merlin is the ***first*** **MLLM that can also carry out**

Table 2. Comparison on main tracking benchmarks. Notably, the original LLaVA-1.5 [38] model was incapable of performing tracking tasks. Therefore, we utilized the model configuration of LLaVA-1.5 and trained a version of the model with the same dataset as Merlin for the fair comparison.

Method	LaSOT			GOT10k		
	Success	P_{norm}	P	AO	$SR_{0.5}$	$SR_{0.75}$
Specialist Models						
SiamFC [6]	33.6	42.0	33.9	34.8	35.3	9.8
ATOM [14]	51.5	-	-	55.6	63.4	40.2
SiamRPN++ [34]	49.6	56.9	49.1	51.8	61.8	32.5
SiamFC++ [70]	54.4	62.3	54.7	59.5	69.5	47.9
Generalist Models						
LLaVA-1.5 [38]	19.4	16.5	12.8	23.5	20.2	9.7
Merlin (Ours)	39.8	40.2	38.1	51.4	55.9	42.8

Table 3. Comparison with SOTA methods on main MLLM benchmarks. For VQA tasks, we mainly choose GQA [27] and VisWiz [23] to evaluate the model; For general evaluation, we mainly choose MMBench [40] and MM-Vet [76]. †Includes using in-house data that is not publicly accessible.

Method	VQA Task		Generalist		
	GQA	VisWiz	MMB_d	MMB_t	MM-Vet
BLIP-2 [35]	41.0	19.6	-	-	22.4
InstructBLIP [13]	49.2	34.5	36.0	33.9	26.2
Shikra [10]	-	-	58.8	60.2	-
IDEFICS-9B [33]	38.4	35.5	48.2	45.3	-
IDEFICS-80B [33]	45.2	36.0	54.5	54.6	-
Qwen-VL† [4]	59.3	35.2	38.2	32.2	-
Qwen-VL-Chat† [4]	57.5	38.9	60.6	61.8	-
LLaVA-1.5 [38]	**62.0**	50.0	64.3	59.5	30.5
Merlin (Ours)	60.5	**50.4**	**66.2**	**65.5**	**34.9**

tracking tasks. As shown in Table 2, Merlin achieves comparable performance with expert models and even outperforms on some metrics. Notably, we only *sample a small amount* of tracking data to train Merlin instead of the full amount of data, which means LLM exhibits significant potential in handling temporal tasks, possibly because tracking, as a temporal task, can be viewed as a casually frame-level autoregressive task.

4.3 General Comprehension

In order to showcase the general multi-modal ability, we further benchmark Merlin on various VQA benchmarks and recent benchmarks proposed for evaluating the comprehensive capabilities of MLLMs.

Visual Question Answering (VQA). We first evaluate Merlin on several mainstream VQA benchmarks to reflect the perceptual abilities of MLLMs in understanding image content. As shown in Table 3, Merlin achieves competitive performance compared with existing advanced MLLMs in the selected VQA benchmarks (VQA). The results indicate that Merlin possesses strong image understanding and question-answering capabilities.

Synthetica MLLM Benchmarks. Recently, several benchmarks have been proposed to evaluate the comprehensive performance of MLLMs, encompassing diverse finer-grained scenarios including visual perception, object recognition, optical character recognition (OCR), future reasoning, and so on. In this part, we select several mainstream MLLM benchmarks to evaluate Merlin. As shown in Table 3, We present performance in accuracy on benchmarks including MM-Vet [76] and MMBench [40]. On MMBench, we report results on the both development and test sets. The results show that Merlin significantly outperforms comparative methods, even though many methods utilized a substantial amount of in-house data for pre-training, or employed several times more parameters. This implies that, while introducing foresight minds into MLLMs, we not only preserved their original visual capabilities but even *further enhanced their overall level of visual perception*.

Table 4. Zero-shot object hallucination evaluation on the COCO validation set. "Yes" represents the proportion of positive answers that the model outputs.

Method	LLM Size	Random			Popular			Adversarial		
		Accuracy	F1-Score	Yes	Accuracy	F1-Score	Yes	Accuracy	F1-Score	Yes
LLaVA [39]	13B	64.12	73.38	83.26	63.90	72.63	81.93	58.91	69.95	86.76
MiniGPT-4 [83]	13B	79.67	80.17	52.53	69.73	73.02	62.20	65.17	70.42	67.77
InstructBLIP [13]	13B	88.57	89.27	56.57	82.77	84.66	62.37	72.10	77.32	73.03
Shikra [10]	13B	86.90	86.19	43.26	83.97	83.16	45.23	83.10	82.49	46.50
MultiModal-GPT [21]	7B	50.10	66.71	99.90	50.00	66.67	100.00	50.00	66.67	100.00
mPLUG-Owl [72]	7B	53.97	68.39	95.63	50.90	66.94	98.57	50.67	66.82	98.67
LLaVA [39]	7B	72.16	78.22	76.29	61.37	71.52	85.63	58.67	70.12	88.33
LLaVA-1.5 [38]	7B	83.29	81.33	-	81.88	80.06	-	78.96	77.57	-
Qwen-VL [4]	7B	84.73	82.67	-	84.13	82.06	-	82.26	80.37	-
Merlin (Ours)	7B	**91.58**	**91.66**	49.38	**89.53**	**89.56**	50.27	**84.10**	**84.95**	55.63

Table 5. Ablation study of the proposed strategies in Merlin. (ITP: Image-text pair data, ITD: instruction-tuning data). We mainly report the AO score of GOT10k and the average score of future reasoning.

Pre-Training		Inst.-Tuning		GOT10K	Prediction Rea.
ITP	FPT-Data	ITD	FIT-Data	AO	Average$_{dev}$
✓	✗	✓	✗	-	59.5
✓	✗	✓	✓	-	60.7
✗	✓	✓	✓	15.5	52.8
✓	✓	✓	✗	51.4	61.2
✓	✓	✓	✓	**51.4**	**64.4**

4.4 Object Hallucination

Hallucination presents a significant challenge in existing MLLMs. This term describes the phenomenon where the generated textual content exhibits inconsistencies when compared to its corresponding image content. In this section, we present the experiments from the Polling-Based Object Probing Evaluation (POPE [36]). As demonstrated in Table 4, Merlin surpasses recent SOTA methods with clear margins. More specifically, Merlin achieves optimal performance in all metrics across three scenarios: **Random**, **Popular** and **Adversarial**, with improvements of up to **5** points compared to the highly competitive baseline

Shikra [10]. Surprisingly, in multiple scenarios, the "yes" rate of Merlin is quietly closed to 50%, demonstrating its extraordinary visual perception capabilities.

We analyze this success largely owing to the proposed foresight learning (FPT and FIT). By enabling the model to learn the dynamic correspondence between trajectories across multiple images, the model has gained a *more precise ability to attend to relevant object (trajectories) contexts in the image*, which helps to better avoid misidentification and misalignment of irrelevant targets. We believe that this result will provide new thinking about addressing the issue of hallucinations in MLLM.

Table 6. Ablation studies of the model settings including resolution, vision encoder and projector of Merlin.

Exp	Resolution	Projector	Visual Encoder	Tokens Num	Prediction Rea.	Got-10K
❶	448x	Conv2d	unfrozen	256	**64.4**	**51.4**
❷	336x	Conv2d	unfrozen	256	59.8	47.3
❸	336x	MLP	unfrozen	576	58.1	23.5
❹	448x	Conv2d	frozen	256	60.8	28.4

4.5 Ablative Analysis of FPT and FIT

As introduced in Sects. 3.2 and 3.3, FPT serves as the pre-training strategy to enable MLLM to encapsulate dynamic information across frames by predicting the trajectory of the next frame. FIT is designed to activate the ability of foresight minds in a way of **Trajectory CoT** during instruction fine-tuning. To further explore the effect of FPT and FIT, we conduct an ablation study based on the established future reasoning benchmark and tracking dataset GOT10K [25]. As shown in Table 5, we mainly report the average overlap (AO) of GOT10K and the average score of future reasoning in the development set.

The results show that both FPT and FIT training strategies contribute to the improvement of the metrics. Combining both FPT and FIT, Merlin achieves the best performance which proves the effectiveness of the proposed strategies. Furthermore, we can also observe that the lack of image-text pair data during the pre-training stage considerably hampers the model's general ability. This phenomenon supports our perspective that, during the comprehensive pre-training phase, the integration of image-text pair data is essential for maintaining modality alignment and preventing a decline in combined capabilities.

4.6 Ablative Analysis of Model Configuration

The configuration of model architecture for large-scale models is also a focal point of interest for researchers. In this subsection, we specifically investigate the impact of Merlin's model configuration on performance. As depicted in Table 6, we focus on examining the effects of model input resolution, the visual encoder

This player <Id1> **Frame1**: [461, 278, 651, 976]; **Frame2**: [458, 254, 609, 968]; **Frame3**: [442, 254, 607, 918] </Id1> is dribbling the ball while being pursued by the opponents.

Fig. 5. Attention map visualization. To facilitate the observation, we map the attention between the box responses and the visual tokens of each frame for visualization.

of the model, and the model's projector on the ultimate performance of Merlin. From the experimental outcomes, we can draw the following conclusions:

(i) High-resolution input is more conducive to visual perception and understanding tasks (row ❶ and ❷), particularly for tasks that require precise localization, such as detection and tracking.

(ii) The primary contribution of Conv2d is the ability to compress the number of tokens efficiently and elegantly, which is crucial for supporting high-resolution images. In contrast, MLPs cannot compress tokens. This high token count hinders the training with multiple images. Moreover, more visual tokens does not improve performance in future reasoning tasks (row ❶ and ❸). We speculate that an increased number of visual tokens may lead to the sparsity of supervision.

(iii) During the pre-training phase, the visual encoder should be unfrozen (row ❶ and ❹), which is beneficial for modal alignment and the expansion of the fine-grained spatial information. Similar conclusion is also claimed in [9].

4.7 Visualization Analysis

In this subsection, we visualize the attention map of Merlin to further substantiate the effectiveness of utilizing the proposed strategies. As shown in Fig. 5, we select the output attention map of the middle-level layers of LLM for visualization. We can observe that the word embedding of the output trajectory coordinates can attend to the corresponding object from different frames correctly. This visualization results further prove that the trajectory representation is a good interface to enable MLLM to establish the alignment between the language description and the multi-images dynamic visual contexts. Furthermore, this effectively explains why Merlin possesses a more powerful comprehensive visual capability and a greatly lower level of hallucination compared to previous baselines. Indeed, *the trajectory-driven foresight learning allows the large language model to* **read** *images more profoundly!*

5 Limitation and Conclusion

This study highlighted an obvious deficiency in Multimodal Large-Language Models (MLLMs), specifically their ability to predict future events or outcomes based on current observations, referred as *"foresight minds"*. To address this, we serve as the first to point out that trajectory, as a highly structured representation, is a good learning objective to assist MLLM in obtaining dynamic information from the image observations. Based on this insight, we introduced a unique training method including *Foresight Pre-Training (FPT)* and *Foresight Instruction-Tuning (FIT)*. By synergizing FPT and FIT, we created **Merlin**, a unified MLLM that effectively understands and outputs spatial coordinates or tracklets from single images or multiple frames. Merlin excels at a range of traditional vision-language tasks while demonstrating powerful future reasoning capacities. Despite the substantial advancements made by Merlin, there still are some limitations, particularly in processing long sequential videos and more comprehensive future reasoning evaluation. We aspire for Merlin to guide the enhancement of more advanced MLLMs in the future.

Acknowledgements. This work was supported by the National Natural Science Foundation of China under Grant 62176096 and Grant 61991412. The work was also supported by the Research on AI Terminal Computing Power and End-Cloud Framework (R2411B0R) of China Mobile Group Device Co., Ltd.

References

1. Alayrac, J., et al.: Flamingo: a visual language model for few-shot learning. In: NeurIPS (2022)
2. Anil, R., et al.: PaLM 2 technical report. arXiv preprint arXiv:2305.10403 (2023)
3. Awadalla, A., et al.: OpenFlamingo: an open-source framework for training large autoregressive vision-language models. arXiv preprint arXiv:2308.01390 (2023)
4. Bai, J., et al.: Qwen-vl: a frontier large vision-language model with versatile abilities. arXiv preprint arXiv:2308.12966 (2023)
5. Bates, C., Battaglia, P.W., Yildirim, I., Tenenbaum, J.B.: Humans predict liquid dynamics using probabilistic simulation. In: CogSci (2015)
6. Bertinetto, L., Valmadre, J., Henriques, J.F., Vedaldi, A., Torr, P.H.S.: Fully-convolutional Siamese networks for object tracking. In: Hua, G., Jégou, H. (eds.) ECCV 2016. LNCS, vol. 9914, pp. 850–865. Springer, Cham (2016). https://doi.org/10.1007/978-3-319-48881-3_56
7. Bolya, D., Fu, C., Dai, X., Zhang, P., Feichtenhofer, C., Hoffman, J.: Token merging: your ViT but faster. In: ICLR. OpenReview.net (2023)
8. Brown, T., et al.: Language models are few-shot learners. In: Advances in Neural Information Processing Systems, vol. 33, pp. 1877–1901 (2020)
9. Chen, B., et al.: SpatialVLM: endowing vision-language models with spatial reasoning capabilities. arXiv preprint arXiv:2401.12168 (2024)
10. Chen, K., Zhang, Z., Zeng, W., Zhang, R., Zhu, F., Zhao, R.: Shikra: unleashing multimodal LLM's referential dialogue magic. arXiv preprint arXiv:2306.15195 (2023)
11. Chiang, W.L., et al.: Vicuna: an open-source chatbot impressing GPT-4 with 90%* ChatGPT quality (2023). https://lmsys.org/blog/2023-03-30-vicuna/
12. Cholakov, R., Kolev, T.: Transformers predicting the future. applying attention in next-frame and time series forecasting. arXiv preprint arXiv:2108.08224 (2021)

13. Dai, W., et al.: InstructBLIP: towards general-purpose vision-language models with instruction tuning. arXiv preprint arXiv:2305.06500 (2023)
14. Danelljan, M., Bhat, G., Khan, F.S., Felsberg, M.: ATOM: accurate tracking by overlap maximization. In: CVPR, pp. 4660–4669 (2019)
15. Devlin, J., Chang, M.W., Lee, K., Toutanova, K.: BERT: pre-training of deep bidirectional transformers for language understanding. arXiv preprint arXiv:1810.04805 (2018)
16. Dong, R., et al.: DreamLLM: synergistic multimodal comprehension and creation. arXiv preprint arXiv:2309.11499 (2023)
17. Dosovitskiy, A., et al.: An image is worth 16×16 words: transformers for image recognition at scale. In: ICLR. OpenReview.net (2021)
18. Fan, H., et al.: LaSOT: a high-quality benchmark for large-scale single object tracking. In: Proceedings of the IEEE/CVF Conference on Computer Vision and Pattern Recognition, pp. 5374–5383 (2019)
19. Friston, K.: The free-energy principle: a unified brain theory? Nat. Rev. Neurosci. **11**(2), 127–138 (2010)
20. Ge, Y., Ge, Y., Zeng, Z., Wang, X., Shan, Y.: Planting a seed of vision in large language model. arXiv preprint arXiv:2307.08041 (2023)
21. Gong, T., et al.: Multimodal-GPT: a vision and language model for dialogue with humans. arXiv preprint arXiv:2305.04790 (2023)
22. Goyal, A., Bengio, Y.: Inductive biases for deep learning of higher-level cognition. CoRR abs/2011.15091 (2020)
23. Gurari, D., et al.: VizWiz grand challenge: answering visual questions from blind people. In: Proceedings of the IEEE Conference on Computer Vision and Pattern Recognition, pp. 3608–3617 (2018)
24. He, K., Chen, X., Xie, S., Li, Y., Dollár, P., Girshick, R.: Masked autoencoders are scalable vision learners. In: Proceedings of the IEEE/CVF Conference on Computer Vision and Pattern Recognition, pp. 16000–16009 (2022)
25. Huang, L., Zhao, X., Huang, K.: GOT-10k: a large high-diversity benchmark for generic object tracking in the wild. IEEE Trans. Pattern Anal. Mach. Intell. **43**(5), 1562–1577 (2019)
26. Huang, S., et al.: Language is not all you need: aligning perception with language models. arXiv preprint arXiv:2302.14045 (2023)
27. Hudson, D.A., Manning, C.D.: GQA: a new dataset for real-world visual reasoning and compositional question answering. In: CVPR (2019)
28. Ioffe, S., Szegedy, C.: Batch normalization: accelerating deep network training by reducing internal covariate shift. In: ICML. JMLR Workshop and Conference Proceedings, vol. 37, pp. 448–456. JMLR.org (2015)
29. Kay, K.N., Naselaris, T., Prenger, R.J., Gallant, J.L.: Identifying natural images from human brain activity. Nature **452**(7185), 352–355 (2008)
30. Kazemzadeh, S., Ordonez, V., Matten, M., Berg, T.: ReferItGame: referring to objects in photographs of natural scenes. In: Proceedings of the 2014 Conference on Empirical Methods in Natural Language Processing (EMNLP), pp. 787–798 (2014)
31. Koh, J.Y., Fried, D., Salakhutdinov, R.: Generating images with multimodal language models. arXiv preprint arXiv:2305.17216 (2023)
32. Kuznetsova, A., et al.: The open images dataset v4: Unified image classification, object detection, and visual relationship detection at scale. Int. J. Comput. Vis. **128**(7), 1956–1981 (2020). https://doi.org/10.1007/s11263-020-01316-zD

33. Laurençon, H., et al.: OBELICS: an open web-scale filtered dataset of interleaved image-text documents. In: Thirty-Seventh Conference on Neural Information Processing Systems Datasets and Benchmarks Track (2023)
34. Li, B., Wu, W., Wang, Q., Zhang, F., Xing, J., Yan, J.: SiamRPN++: evolution of siamese visual tracking with very deep networks. In: CVPR, pp. 4282–4291 (2019)
35. Li, J., Li, D., Savarese, S., Hoi, S.: BLIP-2: bootstrapping language-image pre-training with frozen image encoders and large language models. arXiv preprint arXiv:2301.12597 (2023)
36. Li, Y., Du, Y., Zhou, K., Wang, J., Zhao, W.X., Wen, J.R.: Evaluating object hallucination in large vision-language models. arXiv preprint arXiv:2305.10355 (2023)
37. Li, Y., Chen, L., He, R., Wang, Z., Wu, G., Wang, L.: MultiSports: a multi-person video dataset of spatio-temporally localized sports actions. In: Proceedings of the IEEE/CVF International Conference on Computer Vision, pp. 13536–13545 (2021)
38. Liu, H., Li, C., Li, Y., Lee, Y.J.: Improved baselines with visual instruction tuning. arXiv preprint arXiv:2310.03744 (2023)
39. Liu, H., Li, C., Wu, Q., Lee, Y.J.: Visual instruction tuning (2023)
40. Liu, Y., et al.: MMBench: is your multi-modal model an all-around player? arXiv preprint arXiv:2307.06281 (2023)
41. Loshchilov, I., Hutter, F.: SGDR: stochastic gradient descent with warm restarts. arXiv preprint arXiv:1608.03983 (2016)
42. Loshchilov, I., Hutter, F.: Decoupled weight decay regularization. In: ICLR (2019)
43. Malla, S., Dariush, B., Choi, C.: TITAN: future forecast using action priors. In: Proceedings of the IEEE/CVF Conference on Computer Vision and Pattern Recognition, pp. 11186–11196 (2020)
44. Milan, A., Leal-Taixé, L., Reid, I., Roth, S., Schindler, K.: MOT16: a benchmark for multi-object tracking. arXiv preprint arXiv:1603.00831 (2016)
45. OpenAI: ChatGPT (2023). https://openai.com/blog/chatgpt/
46. OpenAI: GPT-4 technical report (2023)
47. Ouyang, L., et al.: Training language models to follow instructions with human feedback. In: NeurIPS (2022)
48. Pi, R., et al.: DetGPT: detect what you need via reasoning. arXiv preprint arXiv:2305.14167 (2023)
49. Radford, A., et al.: Learning transferable visual models from natural language supervision. In: International Conference on Machine Learning, pp. 8748–8763. PMLR (2021)
50. Radford, A., Wu, J., Child, R., Luan, D., Amodei, D., Sutskever, I., et al.: Language models are unsupervised multitask learners. OpenAI Blog **1**(8), 9 (2019)
51. Raffel, C., et al.: Exploring the limits of transfer learning with a unified text-to-text transformer. J. Mach. Learn. Res. **21**(1), 5485–5551 (2020)
52. Ramnani, N., Miall, R.C.: A system in the human brain for predicting the actions of others. Nat. Neurosci. **7**(1), 85–90 (2004)
53. Ryoo, M.S., Piergiovanni, A.J., Arnab, A., Dehghani, M., Angelova, A.: TokenLearner: what can 8 learned tokens do for images and videos? CoRR abs/2106.11297 (2021)
54. Sap, M., Bras, R.L., Fried, D., Choi, Y.: Neural theory-of-mind? On the limits of social intelligence in large LMs. In: EMNLP, pp. 3762–3780. Association for Computational Linguistics (2022)
55. Schuhmann, C., et al.: LAION-400M: open dataset of clip-filtered 400 million image-text pairs. arXiv preprint arXiv:2111.02114 (2021)

56. Shao, S., et al.: Objects365: a large-scale, high-quality dataset for object detection. In: Proceedings of the IEEE/CVF International Conference on Computer Vision, pp. 8430–8439 (2019)
57. Simonyan, K., Zisserman, A.: Very deep convolutional networks for large-scale image recognition. In: ICLR (2015)
58. Simsek, F.E., Cigla, C., Kayabol, K.: SOMPT22: a surveillance oriented multi-pedestrian tracking dataset. In: Karlinsky, L., Michaeli, T., Nishino, K. (eds.) Computer Vision, ECCV 2022 Workshops, ECCV 2022. LNCS, vol. 13805, pp. 659–675. Springer, Cham (2023). https://doi.org/10.1007/978-3-031-25072-9_44
59. Singh, A., et al.: Towards VQA models that can read. In: Proceedings of the IEEE/CVF Conference on Computer Vision and Pattern Recognition, pp. 8317–8326 (2019)
60. Sun, P., et al.: DanceTrack: multi-object tracking in uniform appearance and diverse motion. In: Proceedings of the IEEE/CVF Conference on Computer Vision and Pattern Recognition, pp. 20993–21002 (2022),
61. Taori, R., et al.: Stanford alpaca: an instruction-following LLaMA model (2023). https://github.com/tatsu-lab/stanford_alpaca
62. Touvron, H., et al.: LLaMA : open and efficient foundation language models. arXiv preprint arXiv:2302.13971 (2023)
63. Touvron, H., et al.: LLaMA 2: open foundation and fine-tuned chat models. arXiv preprint arXiv:2307.09288 (2023)
64. Ullman, T.D.: Large language models fail on trivial alterations to theory-of-mind tasks. CoRR abs/2302.08399 (2023)
65. Wei, H., et al.: Small language model meets with reinforced vision vocabulary. arXiv preprint arXiv:2401.12503 (2024)
66. Wei, J., et al.: Emergent abilities of large language models. Trans. Mach. Learn. Res. **2022** (2022)
67. Wei, J., et al.: Chain-of-thought prompting elicits reasoning in large language models. In: Advances in Neural Information Processing Systems, vol. 35, pp. 24824–24837 (2022)
68. Wu, B., Yu, S., Chen, Z., Tenenbaum, J.B., Gan, C.: STAR: a benchmark for situated reasoning in real-world videos. In: Thirty-Fifth Conference on Neural Information Processing Systems Datasets and Benchmarks Track (Round 2) (2021)
69. Wu, D., Han, W., Wang, T., Dong, X., Zhang, X., Shen, J.: Referring multi-object tracking. In: Proceedings of the IEEE/CVF Conference on Computer Vision and Pattern Recognition, pp. 14633–14642 (2023)
70. Xu, Y., et al.: SiamFC++: towards robust and accurate visual tracking with target estimation guidelines. In: AAAI, pp. 140–148 (2020)
71. Yan, W., Zhang, Y., Abbeel, P., Srinivas, A.: VideoGPT: video generation using VQ-VAE and transformers. arXiv preprint arXiv:2104.10157 (2021)
72. Ye, Q., et al.: mplug-owl: modularization empowers large language models with multimodality. arXiv preprint arXiv:2304.14178 (2023)
73. Yu, E., Li, Z., Han, S.: Towards discriminative representation: multi-view trajectory contrastive learning for online multi-object tracking. In: Proceedings of the IEEE/CVF Conference on Computer Vision and Pattern Recognition, pp. 8834–8843 (2022)
74. Yu, E., et al.: Generalizing multiple object tracking to unseen domains by introducing natural language representation. In: Proceedings of the AAAI Conference on Artificial Intelligence, vol. 37, pp. 3304–3312 (2023)

75. Yu, E., Wang, T., Li, Z., Zhang, Y., Zhang, X., Tao, W.: MOTRv3: release-fetch supervision for end-to-end multi-object tracking. arXiv preprint arXiv:2305.14298 (2023)
76. Yu, W., et al.: MM-Vet: evaluating large multimodal models for integrated capabilities. arXiv preprint arXiv:2308.02490 (2023)
77. Zellers, R., Bisk, Y., Farhadi, A., Choi, Y.: From recognition to cognition: visual commonsense reasoning. In: Proceedings of the IEEE/CVF Conference on Computer Vision and Pattern Recognition, pp. 6720–6731 (2019)
78. Zeng, A., et al.: GLM-130B: an open bilingual pre-trained model. arXiv preprint arXiv:2210.02414 (2022)
79. Zhang, S., et al.: GPT4RoI: instruction tuning large language model on region-of-interest. arXiv preprint arXiv:2307.03601 (2023)
80. Zhang, S., et al.: OPT: open pre-trained transformer language models. arXiv preprint arXiv:2205.01068 (2022)
81. Zhao, L., et al.: ChatSpot: bootstrapping multimodal LLMs via precise referring instruction tuning. arXiv preprint arXiv:2307.09474 (2023)
82. Zhou, P., et al.: How far are large language models from agents with theory-of-mind? CoRR abs/2310.03051 (2023)
83. Zhu, D., Chen, J., Shen, X., Li, X., Elhoseiny, M.: MiniGPT-4: enhancing vision-language understanding with advanced large language models. arXiv preprint arXiv:2304.10592 (2023)

ViC-MAE: Self-supervised Representation Learning from Images and Video with Contrastive Masked Autoencoders

Jefferson Hernandez[1]([✉]), Ruben Villegas[2], and Vicente Ordonez[1]

[1] Rice University, Houston, USA
{jefehern,vicenteor}@rice.edu
[2] Google DeepMind, London, UK
https://github.com/jeffhernandez1995/ViC-MAE

Abstract. We propose ViC-MAE, a model that combines both Masked AutoEncoders (MAE) and contrastive learning. ViC-MAE is trained using a global representation obtained by pooling the local features learned under an MAE reconstruction loss and using this representation under a contrastive objective across images and video frames. We show that visual representations learned under ViC-MAE generalize well to video and image classification tasks. Particularly, ViC-MAE obtains state-of-the-art transfer learning performance from video to images on Imagenet-1k compared to the recently proposed OmniMAE by achieving a top-1 accuracy of 86% (+1.3% absolute improvement) when trained on the same data and 87.1% (+2.4% absolute improvement) when training on extra data. At the same time, ViC-MAE outperforms most other methods on video benchmarks by obtaining 75.9% top-1 accuracy on the challenging Something something-v2 video benchmark. When training on videos and images from diverse datasets, our method maintains a balanced transfer-learning performance between video and image classification benchmarks, coming only as a close second to the best-supervised method.

1 Introduction

Recent advances in self-supervised visual representation learning have markedly improved performance on image and video benchmarks [11,15,34,35]. This success has been mainly driven by two approaches: Joint-embedding methods, which encourage invariance to specific transformations-either contrastive [11,15,35] or negative-free [6,18], and masked image modeling which works by randomly masking out parts of the input and forcing a model to predict the masked parts with a reconstruction loss [4,26,34,81]. These ideas have been successfully applied to both images and video (Fig. 1).

Self-supervised techniques for video representation learning have resulted in considerable success, yielding powerful features that perform well across

Supplementary Information The online version contains supplementary material available at https://doi.org/10.1007/978-3-031-73235-5_25.

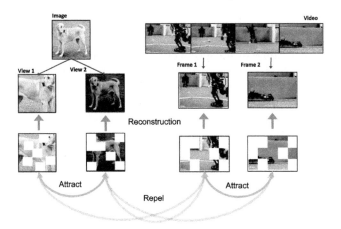

Fig. 1. ViC-MAE operates over video frames and images using masked image modeling at the image and frame level and contrastive learning at the temporal level for videos and under image transformations for images. Our model represents a strong backbone for both image and video tasks.

various downstream tasks [26,27,64,81]. Leveraging image-based models to enhance video feature representations has gained widespread adoption, evidenced by significant advancements in robust video representations [2,47,50]. The reverse—*video-to-image* transfer learning—has not been as successful. This imbalance underscores a nuanced challenge within multimodal learning, and it is not clear how to integrate different modalities. Furthermore, attempts to combine these modalities often result in diminished performance, necessitating tailored adjustments to the underlying architectures or converting one modality (images) into another (repeating images to simulate a video). Learning from video should also yield good image representations since videos naturally contain complex changes in pose, viewpoint, and deformations, among others. These variations can not be simulated through the standard image augmentations used in joint-embedding methods or masked image modeling methods. In this work, we propose a **Vi**sual **C**ontrastive **M**asked **A**uto**E**ncoder (ViC-MAE), a model that learns from both images and video through self-supervision, instead treating short videos as the different views of the same representation, diverging from previous works [29,30]. On transfer experiments, our model also improves *video-to-image* transfer performance while maintaining performance on video representation learning.

Prior work has successfully leveraged self-supervision for video or images separately using either contrastive learning (*i.e.* Gordon *et al.* [31]), or masked image modeling (*i.e.* Feichtenhofer *et al.* [26]). ViC-MAE seeks to leverage the strength of contrastive learning and masked image modeling and seamlessly incorporate images. While trivially this has been done by repeating the image to simulate a still video, ViC-MAE achieves this in the opposite way, treating frames sampled within short intervals (*e.g.* 1 s) as an additional form of temporal data augmentation. Our method uses contrastive learning to align representations across both time-shifted frames and augmented views, and masked image modeling for single video frames or images to encourage learning local features. Diverging from methods that only use a [CLS] token as a global feature, our model aggregates local features using a global pooling layer followed by a contrastive loss to enhance the representation further. This structure is built upon the foundation of the Vision Transformer (ViT) architecture [23], which has become a standard for masked image modeling methods.

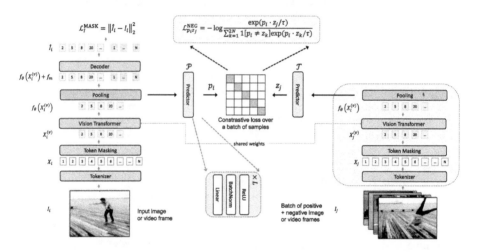

Fig. 2. ViC-MAE inputs two distant frames from a video or two different views of an image within the same batch using a siamese backbone (shared weights), and randomly masks them, before passing them through a ViT model which learns a representation of local features using masked image modeling. A global representation of the video is then constructed by global pooling of the local features learned by the ViT model trained to reconstruct individual patches using an ℓ_2 loss. A standard predictor and a target encoder are used with a contrastive loss. Our use of an aggregation layer before the predictor network aids in avoiding the collapse of the learned global representations.

Closely related to our work is the recently proposed OmniMAE [29] which also aims to be a self-supervised model that can serve as a foundation for image and video downstream tasks. While our experimental evaluations compare ViC-MAE favorably especially when relying on the ViT-L architecture (86% top-1 accuracy on Imagenet vs 84.7%, and 86.8% top-1 accuracy on Kinetics-400 vs 84%), there are also some fundamental differences in the methodology. OmniMAE relies exclusively on masked image modeling and treats images as videos, while ViC-MAE samples frames more sparsely, treating videos within a short time span as the same view. ViC-MAE, leads to reduced training times than video-masked models, while it demands more resources than a basic MAE (which processes 49 visual tokens at a 75% masking rate), it is more efficient (handling 98 tokens at the same rate) than heavier models like OmniMAE or ST-MAE (157 tokens at 90% rate). Surprisingly with these simplifications, ViC-MAE works and achieves high performance on video and image tasks, learning effective temporal representations when finetuning on video. Ultimately, we consider our contributions to be orthogonal and could potentially be integrated to achieve further gains.

Our main empirical findings can be summarized as follows: (i) Treating short videos as augmented views, and then finetuning on regular videos or images yields stronger performance than treating images as videos, while the end models still retain temporal representations, (ii) training with large frame gaps (approx 1.06 s) between sampled frames enhances classification performance, providing the kind of strong augmentation that joint-embedding methods typically require, (iii) including negative pairs in training outperforms negative-free sample training,[1] aligning with other methods that have been successful in *video-to-image* evaluations, and (iv) training with strong image transformations as augmentations is necessary for good performance on images.

Our contributions are as follows: (1) We introduce ViC-MAE, which combines contrastive learning with masked image modeling that works on videos and images by treating short videos as temporal augmentations, unlike previous works; (2) When ViC-MAE is trained only on videos, we achieve state-of-the-art *video-to-image* transfer learning performance on the ImageNet-1K benchmark and state-of-the-art self-supervised performance for video classification on SSv2 [32]; and (3) We demonstrate that ViC-MAE achieves superior transfer learning performance across a wide spectrum of downstream image and video classification tasks, outperforming baselines trained only with masked image modeling. Our source code and model checkpoints are available here.

2 Related Work

Our work is related to various self-supervised learning strategies focusing on video and image data, especially in enhancing image representation through video.

Self-supervised Video Learning. Self-supervised learning exploits temporal information in videos to learn representations aiming to surpass those from static images by designing pretext tasks that use intrinsic video properties such as frame continuity [22,51,54,69,74,75], alongside with object tracking [1,62,79,80]. Contrastive learning approaches on video learn by distinguishing training instances using video temporality [6,18,31,61,82,84]. Recently, Masked Image Modeling (MIM) has used video for pre-training either using the standard design [34] or an asymmetrical siamese design that predicts future masked frames conditioned on present unmasked frames [33]; aiding in transfer learning for various tasks [26,72,81]. Our approach uniquely integrates contrastive learning and masked image modeling into a single pre-training framework suitable for image and video downstream applications.

---

[1] See supplemental material for an evaluation of what we tried and did not work when combining negative-free methods with masked image modeling.

Learning *Video-to-Image* Representations. Several previous models trained only on images have demonstrated remarkable *image-to-video* adaptation [2,47,50]. However, static images lack the dynamism inherent to videos, missing motion cues and camera view changes. In principle, this undermines image-based models for video applications. Recent work has leveraged video data to learn robust image representations to mitigate this. For instance, VINCE [31] shows that natural augmentations found in videos could outperform synthetic augmentations. VFS [84] uses temporal relationships to improve results on static image tasks. CRW [82] employs cycle consistency for inter-video image mapping, allowing for learning frame correspondences. ST-MAE [26] shows that video-oriented masked image modeling can benefit image-centric tasks. VITO [61] develops a technique for video dataset curation to bridge the domain gap between video and images.

Learning General Representations from Video and Images. Research has progressed in learning from video and images, adopting supervised or unsupervised approaches. The recently proposed TubeViT [63] uses sparse video tubes for creating visual tokens across images and video. OMNIVORE [30] employs a universal encoder for multiple modalities with specific heads for each task. PolyViT [48] additionally trains with audio data, using balanced task-training schedules. Expanding on the data modalities, ImageBind [28] incorporates audio, text, and various sensor data, with tailored loss functions and input sequences to leverage available paired data effectively. In self-supervised learning, BEVT [77] adopts a BERT-like approach for video, finding benefits in joint pre-training with images. OmniMAE [29] proposes masked autoencoding for joint training with video and images. OmniVec [70] extends the datasets using in OMNIVORE, creates new task training policies, and adds masked autoencoding as an auxiliary task to learn from multiple modalities. ViC-MAE learns from video and image datasets without supervision by combining masked image modeling and contrastive learning.

Combining Contrastive Methods with Masked Image Modeling. Contrastive learning combined with masked image modeling has been recently investigated. MSN [3] combines masking and augmentations for efficient contrastive learning, using entropy maximization instead of pixel reconstruction to avoid representational collapse, achieving notable few-shot classification performance on ImageNet-1k. CAN [55] uses a framework that combines contrastive and masked modeling, employing a contrastive task on the representations from unmasked patches and a reconstruction plus denoising task on visible patches. C-MAE [39] uses a Siamese network design comprising an online encoder for masked inputs and a momentum encoder for full views, enhancing the discrimination power of masked autoencoders which usually lag in linear or KNN evaluations. C-MAE-V [52] adapts C-MAE to video, showing improvements on Kinetics-400 and Something Something-v2. MAE-CT [43] leverages a two-step approach with an initial masked modeling phase followed by contrastive tuning on the top layers, improving linear classification on masked image modeling-trained models. Our ViC-MAE sets itself apart by effectively learning from images and videos within a unified training approach, avoiding the representational collapse seen in C-MAE through a novel pooling layer and utilizing dual image crops from data augmentations or different video frames to improve modality learning performance.

3 Method

We propose ViC-MAE for feature learning on video and images, which works using contrastive learning at the temporal level (or augmentations on images) and masked image modeling at the image level.

3.1 Background

We provide below some background terminology and review of closely related methods that we build upon.

Masked Image Modeling. This approach provides a way to learn visual representations in a self supervised manner. These methods learn representations by first masking out parts of the input and then training a model to fill in the blanks using a simple reconstruction loss. To do this, these methods rely on an encoder f_θ that takes the non-masked input and learns a representation x, such that a decoder d_ϕ can reconstruct the masked part of the input. More formally, let x be the representation learned by the encoder for masked image I with mask M such that $f_\theta(I \odot M)$. A decoder d is then applied to obtain the first loss over masked and unmasked tokens $d_\phi(x)$. This defines the following reconstruction loss which is only computed over masked tokens:

$$\mathcal{L}_I^{\text{MASK}} = \|d_\phi(f_\theta(I \odot M)) \odot (1 - M) - I \odot (1 - M)\|_2. \quad (1)$$

Contrastive Learning. In common image-level contrastive methods, learning with negatives is achieved by pushing the representation of the positive pairs (different augmented views of the same image) to be close to each other while pulling the representation of negative pairs further apart. More formally, let I and I' be two augmented views of the same image. Contrastive learning uses a siamese network with a prediction encoder $\mathcal{P}$ and a target encoder $\mathcal{T}$ [15,84]. The output of these networks are ℓ_2-normalized: $p = \mathcal{P}(I)/\|\mathcal{P}(I)\|_2$, and $z = \mathcal{T}(I')/\|\mathcal{T}(I')\|_2$. Given a positive pair from a minibatch of size N, the other $2(N-1)$ examples are treated as negative examples. The objective then is to minimize the Info-NCE loss [58]. When learning with negatives, $\mathcal{P}$ and $\mathcal{T}$ typically share the same architecture and model parameters.

3.2 ViC-MAE

We propose a novel approach for learning representations by applying masking image modeling at the individual image level, paired with image-level similarity using either sampled frames or augmented images. Unlike previous methods that inefficiently replicate images to mimic video input, thereby utilizing more computational resources, our methodology treats short video segments as augmented instances of a single view. This perspective not only enhances the efficiency of the learned representations but also significantly broadens the applicability of our model. ViC-MAE offers a versatile "plug and play" solution for image-based tasks. Furthermore, our model can easily be fine-tuned for video tasks and adapted to videos of varying sizes, unlike the traditional 16 frames. Figure 2 shows an overview of our model.

Given a video with T frames $\{I_1, I_2, \cdots, I_T\}$, we sample two frames I_i, I_j as a positive pair input during one training step. We augment single images when they appear in a batch. Notice that our model sees a batch comprising frames and images. After an input image tokenizer layer, we obtain a set of patch-level token representations of X_i and X_j for each frame. Then, we apply token masking by generating a different random mask M_i and M_j and apply them to both of the corresponding input frames to obtain a subset of input visible tokens $X_i^{(v)}$ and $X_j^{(v)}$. These visible token sets are then forwarded to a ViT encoder, which computes a set of representations $f_\theta(X_i^{(v)})$ and $f_\theta(X_j^{(v)})$ respectively. Finally, for the first image, we compute $\hat{I}_i = d_\phi(f_\theta(X_i^{(v)} + f_m))$ where we have added a mask token f_m to let the decoder know which patches were masked and allows to predict patch-shaped outputs through $\hat{I}_i$. These output patches are then trained to minimize the ℓ_2 loss with the true patches in the input image:

$$\mathcal{L}_i^{\text{MASK}} = \|\hat{I}_i - I_i\|_2^2. \tag{2}$$

To apply contrastive pre-training we use a separate prediction branch in the network by applying a global pooling operator Ω over the output representations $f_\theta(X_i^{(v)})$ from the main branch and $f_\theta(X_j^{(v)})$ from the siamese copy of the network. This step simplifies the formulation of our method and avoids using additional losses or the `gradient-stop` operator as in SimSiam [18] to avoid feature representation collapse since the pooled features can not default to the zero vector as they also are being trained to reconstruct patches. We experiment using various aggregation methods, including *mean* pooling, *max* pooling, and *generalized mean* (GeM) pooling [65].

These global representations are then forwarded to a predictor encoder $\mathcal{P}$ and a target encoder $\mathcal{T}$ to obtain frame representations:

$$p_i \triangleq \mathcal{P}(\Omega(f_\theta(X_i^{(v)})))/\|\mathcal{P}(\Omega(f_\theta(X_i^{(v)})))\|_2,$$

and

$$z_j \triangleq \mathcal{T}(\Omega(f_\theta(X_j^{(v)})))/\|\mathcal{T}(\Omega(f_\theta(X_j^{(v)})))\|_2$$

respectively. The predictor network $\mathcal{P}$ and target network $\mathcal{T}$ are symmetrical and we use standard blocks designed for contrastive learning [6,15,18]. These blocks consist of a Linear → BatchNorm1d → ReLU block repeated 2 times.

Table 1. Transfer Learning Results from Video and Image Pre-training to Various Datasets Using the ViT/L-16 Backbone. The pre-training data is a video dataset (MiT, K600, K700, or K400) and/or image dataset (IN1K). All self-supervised methods are evaluated end-to-end with supervised finetuning on IN1K, Kinetics-400, Places365, and SSv2. Best results are in bold. Results of MAE, ST-MAE, and VideoMAE for out-of-domain data were taken from Girdhar et al.[29].

Method	Arch.	Pre-training Data	In-Domain IN1K	In-Domain K400	Out-of-Domain Places-365	Out-of-Domain SSv2
Supervised ViT [23] *ICML'20*	ViT-B	IN1K	82.3	68.5	57.0	61.8
ViT [23] *ICML'20*	ViT-L	IN1K	82.6	78.6	58.9	66.2
COVeR [86] *arXiv'21*	TimeSFormer-SR	JFT-3B + K400 + MiT + IN1K	86.6	87.2	-	70.9
OMNIVORE [30] *CVPR'22*	ViT-B	IN1K + K400 + SUN RGB-D	84.0	83.3	59.2	68.3
OMNIVORE [30] *CVPR'22*	ViT-L	IN1K + K400 + SUN RGB-D	86.0	84.1	-	-
TubeViT [63] *CVPR'23*	ViT-B	K400 + IN1K	81.4	88.6	-	-
TubeViT [63] *CVPR'23*	ViT-L	K400 + IN1K	-	90.2	-	76.1
MAE [34] *CVPR'22*	ViT-B	IN1K	83.4	-	57.9	59.6
MAE [34] *CVPR'22*	ViT-L	IN1K	85.5	82.3	59.4	57.7
ST-MAE [26] *NeurIPS'22*	ViT-B	K400	81.3	81.3	57.4	69.3
ST-MAE [26] *NeurIPS'22*	ViT-L	K400	81.7	84.8	58.1	73.2
VideoMAE [72] *NeurIPS'22*	ViT-B	K400	81.1	80.0	-	69.6
VideoMAE [72] *NeurIPS'22*	ViT-L	K400	-	85.2	-	74.3
OmniMAE [29] *CVPR'23*	ViT-B	K400 + IN1K	82.8	80.8	58.5	69.0
OmniMAE [29] *CVPR'23*	ViT-L	K400 + IN1K	84.7	84.0	59.4	73.4
ViC-MAE	ViT-L	K400	85.0	85.1	59.5	73.7
ViC-MAE	ViT-L	MiT	85.3	84.9	59.7	73.8
ViC-MAE	ViT-B	K400 + IN1K	83.0	80.8	58.6	69.5
ViC-MAE	ViT-L	K400 + IN1K	86.0	86.8	60.0	75.0
ViC-MAE	ViT-B	K710 + MiT + IN1K	83.8	80.9	59.1	69.8
ViC-MAE	ViT-L	K710 + MiT + IN1K	**87.1**	**87.8**	**60.7**	**75.9**

From these representations, we apply the InfoNCE contrastive learning loss as follows:

$$\mathcal{L}^{\text{NEG}}_{p_i, z_j} = -\log \frac{\exp(p_i \cdot z_j / \tau)}{\sum_{k=1}^{2N} \mathbb{1}[p_i \neq z_k] \exp(p_i \cdot z_k / \tau)}, \quad (3)$$

where the denominator includes a set of negative pairs with representations z_k computed for frames from other videos, the same video but at a time longer than the selected time shift and images in the same batch, $\mathbb{1}[p_i \neq z_k] \in \{0, 1\}$ is an indicator function evaluating to 1 when $p_I \neq z_k$ and τ denotes a temperature parameter.

The final loss is $\mathcal{L} = \mathcal{L}^{\text{MASK}} + \lambda \mathcal{L}^{\text{NEG}}$, where λ is a hyperparameter controlling the relative influence of both losses. In practice, we use a schedule to gradually introduce the contrastive loss and let the model learn good local features at the beginning of training.

4 Experiment Settings

We perform experiments to demonstrate the fine-tuning performance of our method on ImageNet-1k and other image recognition datasets. We also evaluate our method on the Kinetics-400 dataset [40] and Something Something-v2 [32] for action recognition to show that our model is able to maintain performance on video benchmarks. Full details are in the supplemental material.

Architecture. We use the standard Vision Transformer (ViT) architecture [23] and conduct experiments fairly across benchmarks and methods using the ViT-B/16 and ViT-L/16 configurations. For masked image modeling, we use a small decoder as proposed by He *et al.* [34]. Finetunig on images requires no changes since this resembles the pre-training configuration. Finetuning on videos is as follows: we initialize the temporal tokenizer by replicating the spatial tokens along the temporal dimension scaled by the length of the video, similarly, we initialize the MHA parameters by replicating them but skip the scaling for them. We use the standard of finetuning on videos of 16 frames, skipping 4.

Pre-training. We adopt Moments in Time [56], Kinetics-400 [40], and ImageNet-1k [21] as our main datasets for self supervised pre-training. They consist of ∼1000K and ∼300K videos of varied length respectively, and ∼1.2M images for Imagenet-1k. We sample frames from these videos using distant sampling, which consists of splitting the video into non-overlapping sections and sampling one frame from each section. Frames are resized to a 224 pixel size, horizontal flipping, and random cropping with a scale range of [0.5, 1], as the only data augmentation transformations on video data. Random cropping (with flip and resize), color distortions, and Gaussian blurring are used for the image modality. For our largest training run, we combine the training sets of Kinetics-400 [40], Kinetics-600 [12], and Kinetics-700 [13], with duplicates removed based on YouTube IDs. We also exclude K400 videos used for evaluation from training to avoid leakage. This process results in a unique, diverse dataset of ∼665K samples, which we label K710, following [76].

Settings. We follow previously used configurations for pre-training [26,34]. We use the AdamW optimizer with a batch size of 512 per device. We evaluate the pre-training quality by end-to-end finetuning. When evaluating on video datasets we follow the common practice of multi-view testing: taking K temporal clips ($K = 7$ on Kinetics) and for each clip taking 3 spatial views to cover the spatial axis (this is denoted as $K \times 3$). The final prediction is the average of all views.

5 Results and Ablations

We first perform experiments to analyze the different elements of the ViC-MAE framework. All the experiments are under the *learning with negative pairs* setting using mean pooling over the ViT features. Linear evaluation and end-to-end finetuning runs are done over 100 epochs for ImageNet-1k, see supplemental material for more details. For our ablations, we restrict ourselves to the ViT-B/16 architecture pre-trained over 400 epochs unless specified otherwise.

5.1 Main Result

Our main result evaluates ViC-MAE on two in-domain datasets that were used during training for most experiments: ImageNet-1K (images) and Kinetics-400 (video), and two out-of-domain datasets that no methods used during training: Places-365 [87] (images) and Something-something-v2 (video). Table 1 shows our complete set of results including comparisons with the state-of-the-art on both supervised representation learning (typically using classification losses), and self-supervised representation learning (mostly using masked image modeling). We consider mostly recent methods building on visual transformers as the most recent TubeViT [63] which relies on this type of architecture.[2]

Our most advanced version of ViC-MAE trained on five datasets (Kinetics-400, Kinetics-600, Kinetics-700, Moments in Time, and Imagenet-1k) using the ViT-Large architecture performs the best across all metrics on all datasets compared to all previous self-supervised representation learning methods and even outperforms the supervised base model OMNIVORE [30] on Imagenet-1k with a top-1 accuracy of 87.1% vs 86%. As well as, COVeR [86] a model trained on a similar data mix, except that it uses more images, COVeR gets 86.6% vs 87.1% on Imagenet-1k. ViC-MAE also comes a close second to other supervised methods and roughly matches the performance of TubeViT [63] which obtains 76.1% top-1 accuracy on Something something-v2 compared to our 75.9% top-1 accuracy. When compared to the current self-supervised state-of-the-art OmniMAE using the same ViT-Large architecture and the same datasets for pre-training (Kinetics-400 and Imagenet-1k), ViC-MAE also outperforms OmniMAE in all benchmarks (Imagenet: 86% $vs.$84.7%, Kinetics-400: 86.8% $vs.$84%, Places-365: 60% $vs.$59.4% and SSv2: 75% vs 73.4%).

Table 2. **Comparison of transfer learning performance of our approach** with supervised baselines across 8 natural image classification datasets. All results correspond to linear evaluation. Best results are shown in bold. ‡MAE trained on MiT and K400 randomly sample a frame from the video to compute a reconstruction loss; these models are trained and evaluated by us. See supplemental material for more evaluation of transfer learning performance.

	Model	Pre-train.	Food	CIFAR10	CIFAR100	Birdsnap	SUN397	VOC2007	DTD	Caltech101
ViT-B/16	MAE [34] ‡	K400	74.54	94.86	79.49	46.51	64.33	83.07	78.01	93.28
	MAE [34] ‡	MiT	76.23	94.47	79.50	47.98	65.32	83.46	78.21	93.08
	ViC-MAE (ours)	K400	76.56	93.64	78.80	47.56	64.75	83.74	78.53	92.27
	ViC-MAE (ours)	MiT	**77.39**	**94.92**	**79.88**	**48.21**	**65.64**	**84.77**	**79.27**	**93.53**
ViT-L/16	MAE [34]	IN1K	77.5	95.0	82.9	49.8	63.2	83.3	74.5	94.8
	OmniMAE [29]	SSv2+IN1K	76.2	94.2	82.2	50.1	62.6	82.7	73.9	94.4
	ViC-MAE (ours)	IN1K+K400	81.9	95.6	85.4	52.8	67.3	84.2	76.8	94.9
	ViC-MAE (ours)	K710+MiT+IN1K	**82.9**	**96.8**	**86.5**	**53.5**	**68.1**	**85.3**	**77.8**	**96.1**

---

[2] Previous methods also use different backbones [31,82,84] $i.e.$ResNet-50. They obtain 54.5%, 33.8%, and 55.6% top-1 accuracies on linear evaluation on ImageNet-1k. Since those works do not use the same setting, we do not include them here.

Another important result is *video-to-image transfer*, where the model is only trained on video but its performance is tested on downstream image tasks. Table 1 shows that when ViC-MAE is trained on the Moments in Time dataset [56], it achieves the best top-1 accuracy of 85.3% for any self-supervised backbone model trained only on video. These results highlight the closing gap in building robust representations that can work seamlessly across image and video tasks.

5.2 Comparison with Other Contrastive Masked Autoencoders

Combining MAE with joint-embedding methods is non-trivial. In our first attempts, we used the [CLS] token as the representation and applied negative free methods such as VicReg [6], and SimSiam [18] with limited success (See supplemental material). When combined with contrastive methods, we found it best to use a pooling operation over the ViT features similar to CAN [55], as we find worse performance when the [CLS] token is used, like in C-MAE [39]. The original MAE [34] is known to have poor linear evaluation performance, obtaining 68% in IN1K linear evaluation when pre-trained on IN1K [34,43]. On the contrary, SimCLR [16] a model trained only using contrastive learning on IN1K achieves 73.5%. Several works have tried to address this by combining contrastive learning with masked image modeling to get the best of both worlds. CAN [55], C-MAE [39] and MAE-CT [43] obtain linear evaluation accuracies of 74.0%, 73.9, 73.4%, respectively when trained on IN1K while ViC-MAE obtains 74.0% trained only on IN1K using ViT/B-16 pre-trained for 800 epochs to make the comparison fair. When using the K400 and IN1K datasets together for pre-training, we get 73.6%, but we highlight that ViC-MAE can now maintain good performance in videos and images using the same pre-trained model.

Table 3. **ViC-MAE ablation experiments** with ViT/B-16. We present linear evaluation results on the ImageNet-1K dataset.

5.3 Transfer Learning Experiments

In this section, we evaluate our pre-trained models from Table 1 for transfer learning on downstream tasks.

Video-to-Image Transfer Learning Performance. We evaluate transfer learning performance of ViC-MAE across a diverse array of 12 downstream image classification tasks [7,9,19,25,41,42,53,57,60,83]. (Due to space constraints, we have shown the six most significant ones. See supplemental material for the full table.) Table 2 shows the results of four models based on a ViT/B backbone. We perform linear evaluation. We train two models using two video datasets. The first model is a baseline MAE model pre-trained on randomly sampled frames from videos on the Moments in Time and Kinetics-400 datasets. The second model is our full ViC-MAE model pre-trained on each of the same two datasets. Our model significantly outperforms the other baselines on 9 out of 12 datasets, whereas the MAE trained on Kinetics is superior on only 3 (i.e. Cars, Aircraft, and Pets). When scaling the size of our models, we see that ViC-MAE surpasses all models, including OmniMAE [29] trained on SSv2+IN1K[3]

Table 4. COCO object detection and segmentation using a ViT-B Mask R-CNN baseline. All entries use data without labels.

Method	pre-train data	AP_{Box}	AP_{Mask}
MAE [34]	IN1K	50.3	44.9
C-MAE [39]	IN1K	52.4	46.5
ViC-MAE	IN1K+K400	52.5	46.5
ViC-MAE	IN1K+K710+MiT	**53.2**	**46.9**

Object Detection and Segmentation. We finetune Mask R-CNN [36] end-to-end on the COCO dataset. We adapted the ViT backbone to be used with the FPN, following the recipe outlined in Li et al. [46]. We apply this approach to ViC-MAE and take the other results from their respective paper. See Table 4. Compared with previous methods, ViC-MAE outperforms other approaches under the same configurations. Specifically, when utilizing the combined IN1K+K400 dataset, ViC-MAE achieves a box AP of 52.5 and a mask AP of 46.5, slightly improving over C-MAE, which stands at 52.4 for box AP and 46.5 for mask AP. More notably, with the expanded dataset of IN1K+K710+MiT, ViC-MAE significantly advances the state-of-the-art, achieving the highest reported scores of 53.2 for box AP and 46.9 for mask AP.

5.4 Ablations

We investigate the effect of scaling the data used to train ViC-MAE, the effect of the ratio of image to videos in pre-training, our choice of frame separation, the choice of pooling operator, and the choice of data augmentations. An extra ablation probing the temporal representation learning of our method can be found in the supplemental material.

Influence of Pre-training Data. We perform an ablation study to the effect of scaling the data points seen by the model. The pre-training data includes

---
[3] These are the only publicly available checkpoints of OmniMAE.

Kinectis-400, ImageNet-1K, Kinectis-600 + Kinectis-700, and the Moments in Time datasets added in that order. We pre-train a ViT/B-16 using ViC-MAE for 400 epochs. As illustrated in Fig. 3a, as we progressively increase the dataset size, our ViC-MAE, shows a steady increase in IN1K top-1 accuracy. This is remarkable when compared to CAN [55], pre-trained on the JFT-300M dataset for 800 epochs that only reaches an accuracy of 84.4%. This shows that our model, when supplied with only about 1.5% of the data that CAN was trained on (4.25M vs. 300M), can achieve comparable accuracy levels.

Contrastive vs Masking-Only Pre-training. We perform an ablation study to the effect of varying the ratio of images to video in the dataset by replicating the entire dataset; notice that the number of training updates changes when doing this. The pre-training data includes Kinectis-400 and ImageNet-1K. We pre-train a ViT/B-16 using ViC-MAE for 400 epochs. As illustrated in Fig. 3b, as we progressively increase the ratio of images to videos, our ViC-MAE, surpasses the OmniMAE model [84], meaning that contrastive plus masking pre-training is better able to use image and video data than masking-only pre-training.

Frame Separation. We aim to explore the effect of frame separation on model performance. We follow the two methods of sampling frames from Xu et al. [84]. Results are shown in Table 3a. The first approach, *Continuous sampling*, involves selecting a start index i and sampling a frame within $(i, i+\delta]$, where δ represents the frame separation, with a separation of 0 meaning identical frames for predictor and target networks. The second, *Distant sampling*, divides the video into n equal intervals, corresponding to the number of frames for contrastive learning, and randomly selects one frame from each interval.

In our experiment, we observe that increasing the frame separation when using *continuous sampling* increases model performance. We observe the best performance using *distant sampling* with $n = 2$ (labeled D in Table 3a). We posit that further increasing frame separation offers potentially stronger augmentations. In the following experiments, we only use strong spatial augmentations combined with distant frame sampling.

Pooling Type. We test which operator Ω used to aggregate local features performs best at producing global features. We report our results in Table 3b. We try common types of pooling (*mean*, *max*) as well as, *generalized mean pooling*. We found *mean* to be more effective in creating a global representation for video, and we use it for all other experiments.

Adding Strong Augmentations to Video Frames. In our ablation study, we investigated the necessity of strong color augmentations for video frames during joint training with the target encoder, as commonly applied to images. The findings, detailed in Table 3c, indicate a performance decrease of over 2% in linear evaluation on the Imagenet dataset when applying solely color augmentations without spatial adjustments. Interestingly, employing color and spatial augmentations does not outperform strong spatial augmentations alone. This diverges from prior approaches that rely heavily on color augmentations for effective contrastive learning, suggesting that the inherent temporal variations in video

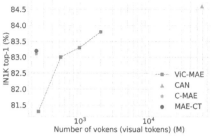 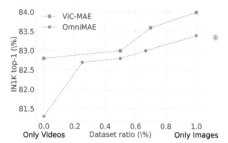

(a) ViC-MAE ViT/B-16 finetuned on IN1K for 100 epochs, compared with CAN pre-trained on JFT-300M, C-MAE, and MAE-CT pre-trained on ImageNet-1K. We increase the amount of data points by adding more video datasets. We can see that our model reaches similar accuracy with ≈ 4.25M data points compared to the 300M of the JFT-300M dataset.

(b) ViC-MAE using the ViT/B-16 architecture finetuned on IN1K for 100 epochs, compared with OmniMAE. We vary the ratio of images vs video in the dataset, from no images to only images. We can see that ViC-MAE can better utilize the videos and images on the dataset compared to masking-only pre-training.

Fig. 3. Additional comparisons with the state-of-the-art and recently proposed methods.

frames may suffice. However, for image datasets, the combination of strong color and spatial augmentations remains necessary.

5.5 Limitations

Our proposed model is able to learn representations from video and image data that transfer to several downstream tasks and surpasses previous models on the same set-up. Given a similar setup, ViC-MAE matches prior results on Kinetics-400, trailing only to supervised models such as TubeViT [63] by 7.1% on ViT/B-16 and 2.4% on ViT/L-16, and MVT [85] slightly on ViT/B-16 but surpasses it by 3.5% on ViT/L-16. It also exceeds MViTv1 [24], TimeSformer [8], and ViViT [2] by margins up to 7.3% on ViT/L-16. Compared to self-supervised models, ViC-MAE falls behind MaskFeat [81] by 0.7% on ViT/B-16 but excels on ViT/L-16 by 3.5%. Our model surpasses V-JEPA [5] by margins up to 1.1% but falls behind VideMAEV2 [76] by 0.8% on ViT/L-16. It is slightly outperformed by DINO [11] and more substantially by models using extra text data or larger image datasets, such as UMT [45], MVD [78], and UniFormerV2 [44], by up to 4.2% on ViT/B-16 and 2.8% on ViT/L-16. Future work could consider leveraging additional weak supervision through other modalities such as text [10,37,66], audio [67,71], 3D geometry [68] or automatically generated data [14,38].

When compared against state-of-the-art ImageNet-pretrained models with comparable computational resources, video-based models, including ours, typically fall short. However, including image and video modalities shows promise in boosting performance. Against models using masked image modeling and contrastive learning, ViC-MAE modestly surpasses MAE [34] by 1.6%, MaskFeat [81] by 1.4% and iBOT [88] by 0.5% with the ViT/L-16 architecture. It also edges out MoCov3 [17] and BeiT [4] by 3% and 1.9% respectively on the same architecture. Yet, it lags behind DINOv2 [59] by 1.2% for ViT/L-16. When compared to supervised models using additional image data, such as DeiT-III [73] and SwinV2 [49] and the distilled ViTs' from [20], our model shows a lag behind of 0.8%, 1.2% and 2.5% respectively on ViT/L-16. These results show that the gap from models pre-trained purely on video still exists, but we believe ViC-MAE pre-trained on image and video data is a step forward in closing that gap.

6 Conclusion

In this work, we introduce ViC-MAE, a method that allows to use unlabeled videos and images to learn useful representation for image recognition tasks. We achieve this by randomly sampling frames from a video or creating two augmented views of an image and using contrastive learning to pull together inputs from the same video and push apart inputs from different videos, likewise, we also use masked image modeling on each input to learn good local features of the scene presented in each input. The main contribution of our work is showing that it is possible to combine masked image modeling and contrastive learning by pooling the local representations of the MAE prediction heads into a global representation used for contrastive learning. The design choices that we have taken when designing ViC-MAE show that our work is easily extensible in various ways. For example, improvements in contrastive learning for images can be directly adapted into our framework. Likewise, pixel reconstruction can be replaced by features important for video representation such as object correspondences or optical flow.

Acknowledgements. The authors would like to thank Google Cloud and the CURe program from Google Research for partially providing funding for this research effort. We are also thankful for support from the Department of Computer Science at Rice University, the National Science Foundation through NSF CAREER Award #2201710, and the Ken Kennedy Institute at Rice University. We also thank anonymous reviewers for their feedback and encouragement.

References

1. Agrawal, P., Carreira, J., Malik, J.: Learning to see by moving. In: Proceedings of the IEEE International Conference on Computer Vision, pp. 37–45 (2015)
2. Arnab, A., Dehghani, M., Heigold, G., Sun, C., Lučić, M., Schmid, C.: ViViT: a video vision transformer. In: Proceedings of the IEEE/CVF International Conference on Computer Vision, pp. 6836–6846 (2021)
3. Assran, M., et al.: Masked siamese networks for label-efficient learning. In: Avidan, S., Brostow, G., Cissé, M., Farinella, G.M., Hassner, T. (eds.) Computer Vision, ECCV 2022. LNCS, vol. 13691, pp. 456–473. Springer, Cham (2022). https://doi.org/10.1007/978-3-031-19821-2_26
4. Bao, H., Dong, L., Piao, S., Wei, F.: BEiT: BERT pre-training of image transformers. In: International Conference on Learning Representations (2021)
5. Bardes, A., et al.: Revisiting feature prediction for learning visual representations from video. arXiv preprint arXiv:2404.08471 (2024)

6. Bardes, A., Ponce, J., Lecun, Y.: VICReg: variance-invariance-covariance regularization for self-supervised learning. In: International Conference on Learning Representations, ICLR 2022 (2022)
7. Berg, T., Liu, J., Woo Lee, S., Alexander, M.L., Jacobs, D.W., Belhumeur, P.N.: Birdsnap: large-scale fine-grained visual categorization of birds. In: Proceedings of the IEEE Conference on Computer Vision and Pattern Recognition, pp. 2011–2018 (2014)
8. Bertasius, G., Wang, H., Torresani, L.: Is space-time attention all you need for video understanding? In: International Conference on Machine Learning, pp. 813–824. PMLR (2021)
9. Bossard, L., Guillaumin, M., Van Gool, L.: Food-101 – mining discriminative components with random forests. In: Fleet, D., Pajdla, T., Schiele, B., Tuytelaars, T. (eds.) ECCV 2014. LNCS, vol. 8694, pp. 446–461. Springer, Cham (2014). https://doi.org/10.1007/978-3-319-10599-4_29
10. Cai, M., et al.: ViP-LLaVA: making large multimodal models understand arbitrary visual prompts. In: Proceedings of the IEEE/CVF Conference on Computer Vision and Pattern Recognition (CVPR), June 2024, pp. 12914–12923 (2024)
11. Caron, M., et al.: Emerging properties in self-supervised vision transformers. In: Proceedings of the IEEE/CVF International Conference on Computer Vision, pp. 9650–9660 (2021)
12. Carreira, J., Noland, E., Banki-Horvath, A., Hillier, C., Zisserman, A.: A short note about kinetics-600. arXiv preprint arXiv:1808.01340 (2018)
13. Carreira, J., Noland, E., Hillier, C., Zisserman, A.: A short note on the kinetics-700 human action dataset. arXiv preprint arXiv:1907.06987 (2019)
14. Cascante-Bonilla, P., et al.: Going beyond nouns with vision & language models using synthetic data. In: Proceedings of the IEEE/CVF International Conference on Computer Vision (ICCV), October 2023, pp. 20155–20165 (2023)
15. Chen, T., Kornblith, S., Norouzi, M., Hinton, G.: A simple framework for contrastive learning of visual representations. In: International Conference on Machine Learning, pp. 1597–1607. PMLR (2020)
16. Chen, T., Kornblith, S., Swersky, K., Norouzi, M., Hinton, G.E.: Big self-supervised models are strong semi-supervised learners. In: Advances in Neural Information Processing Systems, vol. 33, pp. 22243–22255 (2020)
17. Chen, X., Fan, H., Girshick, R., He, K.: Improved baselines with momentum contrastive learning. arXiv preprint arXiv:2003.04297 (2020)
18. Chen, X., He, K.: Exploring simple siamese representation learning. In: Proceedings of the IEEE/CVF Conference on Computer Vision and Pattern Recognition, pp. 15750–15758 (2021)
19. Cimpoi, M., Maji, S., Kokkinos, I., Mohamed, S., Vedaldi, A.: Describing textures in the wild. In: Proceedings of the IEEE Conference on Computer Vision and Pattern Recognition, pp. 3606–3613 (2014)
20. Dehghani, M., et al.: Scaling vision transformers to 22 billion parameters. In: International Conference on Machine Learning, pp. 7480–7512. PMLR (2023)
21. Deng, J., Dong, W., Socher, R., Li, L.J., Li, K., Fei-Fei, L.: ImageNet: a large-scale hierarchical image database. In: 2009 IEEE Conference on Computer Vision and Pattern Recognition, pp. 248–255. IEEE (2009)
22. Diba, A., Sharma, V., Gool, L.V., Stiefelhagen, R.: DynamoNet: dynamic action and motion network. In: Proceedings of the IEEE/CVF International Conference on Computer Vision, pp. 6192–6201 (2019)

23. Dosovitskiy, A., et al.: An image is worth 16×16 words: transformers for image recognition at scale. In: International Conference on Learning Representations (2020)
24. Fan, H., et al.: Multiscale vision transformers. In: Proceedings of the IEEE/CVF International Conference on Computer Vision, pp. 6824–6835 (2021)
25. Fei-Fei, L., Fergus, R., Perona, P.: Learning generative visual models from few training examples: an incremental Bayesian approach tested on 101 object categories. In: 2004 Conference on Computer Vision and Pattern Recognition Workshop, pp. 178–178. IEEE (2004)
26. Feichtenhofer, C., Fan, H., Li, Y., He, K.: Masked autoencoders as spatiotemporal learners. In: Neural Information Processing Systems (NeurIPS) (2022)
27. Feichtenhofer, C., Fan, H., Xiong, B., Girshick, R., He, K.: A large-scale study on unsupervised spatiotemporal representation learning. In: Proceedings of the IEEE/CVF Conference on Computer Vision and Pattern Recognition, pp. 3299–3309 (2021)
28. Girdhar, R., et al.: ImageBind: one embedding space to bind them all. In: Proceedings of the IEEE/CVF Conference on Computer Vision and Pattern Recognition, pp. 15180–15190 (2023)
29. Girdhar, R., El-Nouby, A., Singh, M., Alwala, K.V., Joulin, A., Misra, I.: OmniMAE: single model masked pretraining on images and videos. In: Proceedings of the IEEE/CVF Conference on Computer Vision and Pattern Recognition, pp. 10406–10417 (2023)
30. Girdhar, R., Singh, M., Ravi, N., van der Maaten, L., Joulin, A., Misra, I.: Omnivore: a single model for many visual modalities. In: Proceedings of the IEEE/CVF Conference on Computer Vision and Pattern Recognition, pp. 16102–16112 (2022)
31. Gordon, D., Ehsani, K., Fox, D., Farhadi, A.: Watching the world go by: representation learning from unlabeled videos. arXiv preprint arXiv:2003.07990 (2020)
32. Goyal, R., et al.: The "something something" video database for learning and evaluating visual common sense. In: Proceedings of the IEEE International Conference on Computer Vision, pp. 5842–5850 (2017)
33. Gupta, A., Wu, J., Deng, J., Li, F.F.: Siamese masked autoencoders. In: Oh, A., Naumann, T., Globerson, A., Saenko, K., Hardt, M., Levine, S. (eds.) Advances in Neural Information Processing Systems, vol. 36, pp. 40676–40693. Curran Associates, Inc. (2023). https://proceedings.neurips.cc/paper_files/paper/2023/file/7ffb9f1b57628932518505b532301603-Paper-Conference.pdf
34. He, K., Chen, X., Xie, S., Li, Y., Dollár, P., Girshick, R.: Masked autoencoders are scalable vision learners. In: Proceedings of the IEEE/CVF Conference on Computer Vision and Pattern Recognition, pp. 16000–16009 (2022)
35. He, K., Fan, H., Wu, Y., Xie, S., Girshick, R.: Momentum contrast for unsupervised visual representation learning. In: Proceedings of the IEEE/CVF Conference on Computer Vision and Pattern Recognition, pp. 9729–9738 (2020)
36. He, K., Gkioxari, G., Dollár, P., Girshick, R.: Mask R-CNN. In: Proceedings of the IEEE International Conference on Computer Vision, pp. 2961–2969 (2017)
37. He, R., Cascante-Bonilla, P., Yang, Z., Berg, A.C., Ordonez, V.: Improved visual grounding through self-consistent explanations. In: Proceedings of the IEEE/CVF Conference on Computer Vision and Pattern Recognition, pp. 13095–13105 (2024)
38. He, R., Cascante-Bonilla, P., Yang, Z., Berg, A.C., Ordonez, V.: Learning from models and data for visual grounding (2024). https://arxiv.org/abs/2403.13804
39. Huang, Z., et al.: Contrastive masked autoencoders are stronger vision learners. arXiv preprint arXiv:2207.13532 (2022)

40. Kay, W., et al.: The kinetics human action video dataset (2017). https://arxiv.org/abs/1705.06950
41. Krause, J., Stark, M., Deng, J., Fei-Fei, L.: 3D object representations for fine-grained categorization. In: Proceedings of the IEEE International Conference on Computer Vision Workshops, pp. 554–561 (2013)
42. Krizhevsky, A., Hinton, G.: Learning multiple layers of features from tiny images. Technical report 0, University of Toronto, Toronto, Ontario (2009). https://www.cs.toronto.edu/~kriz/learning-features-2009-TR.pdf
43. Lehner, J., Alkin, B., Fürst, A., Rumetshofer, E., Miklautz, L., Hochreiter, S.: Contrastive tuning: a little help to make masked autoencoders forget. arXiv preprint arXiv:2304.10520 (2023)
44. Li, K., et al.: UniFormerV2: spatiotemporal learning by arming image ViTs with video uniformer. arXiv preprint arXiv:2211.09552 (2022)
45. Li, K., et al.: Unmasked teacher: towards training-efficient video foundation models. arXiv preprint arXiv:2303.16058 (2023)
46. Li, Y., Mao, H., Girshick, R., He, K.: Exploring plain vision transformer backbones for object detection. In: Avidan, S., Brostow, G., Cissé, M., Farinella, G.M., Hassner, T. (eds.) Computer Vision, ECCV 2022. LNCS, vol. 13669, pp. 280–296. Springer, Cham (2022). https://doi.org/10.1007/978-3-031-20077-9_17
47. Li, Y., et al.: MViTv2: improved multiscale vision transformers for classification and detection. In: Proceedings of the IEEE/CVF Conference on Computer Vision and Pattern Recognition, pp. 4804–4814 (2022)
48. Likhosherstov, V., et al.: PolyViT: co-training vision transformers on images, videos and audio. arXiv preprint arXiv:2111.12993 (2021)
49. Liu, Z., et al.: Swin Transformer V2: scaling up capacity and resolution. In: Proceedings of the IEEE/CVF Conference on Computer Vision and Pattern Recognition, pp. 12009–12019 (2022)
50. Liu, Z., et al.: Video Swin Transformer. In: Proceedings of the IEEE/CVF Conference on Computer Vision and Pattern Recognition, pp. 3202–3211 (2022)
51. Lotter, W., Kreiman, G., Cox, D.: Deep predictive coding networks for video prediction and unsupervised learning. In: International Conference on Learning Representations (2016)
52. Lu, C.Z., Jin, X., Huang, Z., Hou, Q., Cheng, M.M., Feng, J.: CMAE-V: contrastive masked autoencoders for video action recognition. arXiv preprint arXiv:2301.06018 (2023)
53. Maji, S., Rahtu, E., Kannala, J., Blaschko, M., Vedaldi, A.: Fine-grained visual classification of aircraft. arXiv preprint arXiv:1306.5151 (2013)
54. Mathieu, M., Couprie, C., LeCun, Y.: Deep multi-scale video prediction beyond mean square error. In: 4th International Conference on Learning Representations, ICLR 2016 (2016)
55. Mishra, S., et al.: A simple, efficient and scalable contrastive masked autoencoder for learning visual representations. arXiv preprint arXiv:2210.16870 (2022)
56. Monfort, M., et al.: Moments in time dataset: one million videos for event understanding. IEEE Trans. Pattern Anal. Mach. Intell. **42**(2), 502–508 (2020)
57. Nilsback, M.E., Zisserman, A.: Automated flower classification over a large number of classes. In: 2008 Sixth Indian Conference on Computer Vision, Graphics & Image Processing, pp. 722–729. IEEE (2008)
58. Oord, A.v.d., Li, Y., Vinyals, O.: Representation learning with contrastive predictive coding. arXiv preprint arXiv:1807.03748 (2018)
59. Oquab, M., et al.: DINOv2: learning robust visual features without supervision. arXiv preprint arXiv:2304.07193 (2023)

60. Parkhi, O.M., Vedaldi, A., Zisserman, A., Jawahar, C.: Cats and dogs. In: 2012 IEEE Conference on Computer Vision and Pattern Recognition, pp. 3498–3505. IEEE (2012)
61. Parthasarathy, N., Eslami, S., Carreira, J., Hénaff, O.J.: Self-supervised video pretraining yields strong image representations. arXiv preprint arXiv:2210.06433 (2022)
62. Pathak, D., Girshick, R., Dollár, P., Darrell, T., Hariharan, B.: Learning features by watching objects move. In: Proceedings of the IEEE Conference on Computer Vision and Pattern Recognition, pp. 2701–2710 (2017)
63. Piergiovanni, A., Kuo, W., Angelova, A.: Rethinking video ViTs: sparse video tubes for joint image and video learning. In: Proceedings of the IEEE/CVF Conference on Computer Vision and Pattern Recognition, pp. 2214–2224 (2023)
64. Qian, R., et al.: Spatiotemporal contrastive video representation learning. In: Proceedings of the IEEE/CVF Conference on Computer Vision and Pattern Recognition, pp. 6964–6974 (2021)
65. Radenović, F., Tolias, G., Chum, O.: Fine-tuning CNN image retrieval with no human annotation. IEEE Trans. Pattern Anal. Mach. Intell. **41**(7), 1655–1668 (2018)
66. Shrivastava, A., Selvaraju, R.R., Naik, N., Ordonez, V.: CLIP-Lite: information efficient visual representation learning with language supervision. In: International Conference on Artificial Intelligence and Statistics, pp. 8433–8447. PMLR (2023)
67. Singh, N., Wu, C.W., Orife, I., Kalayeh, M.: Looking similar sounding different: leveraging counterfactual cross-modal pairs for audiovisual representation learning. In: Proceedings of the IEEE/CVF Conference on Computer Vision and Pattern Recognition, pp. 26907–26918 (2024)
68. Sriram, A., Gaidon, A., Wu, J., Niebles, J.C., Fei-Fei, L., Adeli, E.: HomE: homography-equivariant video representation learning. arXiv preprint arXiv:2306.01623 (2023)
69. Srivastava, N., Mansimov, E., Salakhudinov, R.: Unsupervised learning of video representations using LSTMs. In: International Conference on Machine Learning, pp. 843–852. PMLR (2015)
70. Srivastava, S., Sharma, G.: OmniVec: learning robust representations with cross modal sharing. In: Proceedings of the IEEE/CVF Winter Conference on Applications of Computer Vision, pp. 1236–1248 (2024)
71. Tang, Y., Shimada, D., Bi, J., Xu, C.: AVicuna: audio-visual LLM with interleaver and context-boundary alignment for temporal referential dialogue (2024). https://arxiv.org/abs/2403.16276
72. Tong, Z., Song, Y., Wang, J., Wang, L.: VideoMAE: masked autoencoders are data-efficient learners for self-supervised video pre-training. In: Neural Information Processing Systems (NeurIPS) (2022)
73. Touvron, H., Cord, M., Jégou, H.: DeiT III: revenge of the ViT. In: Avidan, S., Brostow, G., Cissé, M., Farinella, G.M., Hassner, T. (eds.) Computer Vision, ECCV 2022. LNCS, vol. 13684, pp. 516–533. Springer, Cham (2022). https://doi.org/10.1007/978-3-031-20053-3_30
74. Vondrick, C., Pirsiavash, H., Torralba, A.: Anticipating visual representations from unlabeled video. In: Proceedings of the IEEE Conference on Computer Vision and Pattern Recognition, pp. 98–106 (2016)
75. Walker, J., Doersch, C., Gupta, A., Hebert, M.: An uncertain future: forecasting from static images using variational autoencoders. In: Leibe, B., Matas, J., Sebe, N., Welling, M. (eds.) ECCV 2016, Part VII 14. LNCS, vol. 9911, pp. 835–851. Springer, Cham (2016). https://doi.org/10.1007/978-3-319-46478-7_51

76. Wang, L., et al.: VideoMAE V2: scaling video masked autoencoders with dual masking. In: Proceedings of the IEEE/CVF Conference on Computer Vision and Pattern Recognition, pp. 14549–14560 (2023)
77. Wang, R., et al.: BEVT: BERT pretraining of video transformers. In: Proceedings of the IEEE/CVF Conference on Computer Vision and Pattern Recognition, pp. 14733–14743 (2022)
78. Wang, R., et al.: Masked video distillation: rethinking masked feature modeling for self-supervised video representation learning. In: Proceedings of the IEEE/CVF Conference on Computer Vision and Pattern Recognition, pp. 6312–6322 (2023)
79. Wang, X., Gupta, A.: Unsupervised learning of visual representations using videos. In: International Conference on Computer Vision (ICCV) (2015)
80. Wang, X., Jabri, A., Efros, A.A.: Learning correspondence from the cycle-consistency of time. In: Proceedings of the IEEE/CVF Conference on Computer Vision and Pattern Recognition, pp. 2566–2576 (2019)
81. Wei, C., Fan, H., Xie, S., Wu, C.Y., Yuille, A., Feichtenhofer, C.: Masked feature prediction for self-supervised visual pre-training. In: Proceedings of the IEEE/CVF Conference on Computer Vision and Pattern Recognition, pp. 14668–14678 (2022)
82. Wu, H., Wang, X.: Contrastive learning of image representations with cross-video cycle-consistency. In: Proceedings of the IEEE/CVF International Conference on Computer Vision, pp. 10149–10159 (2021)
83. Xiao, J., Hays, J., Ehinger, K.A., Oliva, A., Torralba, A.: Sun database: large-scale scene recognition from abbey to zoo. In: 2010 IEEE Computer Society Conference on Computer Vision and Pattern Recognition, pp. 3485–3492. IEEE (2010)
84. Xu, J., Wang, X.: Rethinking self-supervised correspondence learning: a video frame-level similarity perspective. In: Proceedings of the IEEE/CVF International Conference on Computer Vision, pp. 10075–10085 (2021)
85. Yan, S., et al.: Multiview transformers for video recognition. In: Proceedings of the IEEE/CVF Conference on Computer Vision and Pattern Recognition, pp. 3333–3343 (2022)
86. Zhang, B., et al.: Co-training transformer with videos and images improves action recognition. arXiv preprint arXiv:2112.07175 (2021)
87. Zhou, B., Lapedriza, A., Khosla, A., Oliva, A., Torralba, A.: Places: a 10 million image database for scene recognition. IEEE Trans. Pattern Anal. Mach. Intell. **40**(6), 1452–1464 (2017)
88. Zhou, J., et al.: iBOT: image BERT pre-training with online tokenizer. arXiv preprint arXiv:2111.07832 (2021)

E.T. the Exceptional Trajectories: Text-to-Camera-Trajectory Generation with Character Awareness

Robin Courant[1]([✉])[iD], Nicolas Dufour[1,2][iD], Xi Wang[1][iD], Marc Christie[3][iD], and Vicky Kalogeiton[1][iD]

[1] LIX, Ecole Polytechnique, IP Paris, Paris, France
robin.courant@polytechnique.edu
[2] LIGM, Ecole des Ponts, CNRS, UGE, Paris, France
[3] Inria, IRISA, CNRS, University of Rennes, Rennes, France

Abstract. Stories and emotions in movies emerge through the effect of well-thought-out directing decisions, in particular camera placement and movement over time. Crafting compelling camera trajectories remains a complex iterative process, even for skilful artists. To tackle this, in this paper, we propose a dataset called the Exceptional Trajectories (E.T.) with camera trajectories along with character information and textual captions encompassing descriptions of both camera and character. To our knowledge, this is the first dataset of its kind. To show the potential applications of the E.T. dataset, we propose a diffusion-based approach, named DIRECTOR, which generates complex camera trajectories from textual captions that describe the relation and synchronisation between the camera and characters. To ensure robust and accurate evaluations, we train on the E.T. dataset CLaTr, a Contrastive Language-Trajectory embedding for evaluation metrics. We posit that our proposed dataset and method significantly advance the democratization of cinematography, making it more accessible to common users.

1 Introduction

Cinematography is a collaborative and complex crafting process that mixes technical, artistic and storytelling skills. The ultimate objective is to communicate a distinct message to the audience, at a cognitive (e.g., revealing facts), emotional and aesthetic level, through tasks such as laying out the scene (*mise-en-scène*), setting up the lighting and making decisions to place and move the camera in relation to the characters, their actions or the overall scene content. In this context, the camera is the only window into this staged world and therefore plays a critical role in conveying the director's intention. Through more than a hundred years of practice, cinematography has forged a common language for directors

Supplementary Information The online version contains supplementary material available at https://doi.org/10.1007/978-3-031-73235-5_26.

© The Author(s), under exclusive license to Springer Nature Switzerland AG 2025
A. Leonardis et al. (Eds.): ECCV 2024, LNCS 15082, pp. 464–480, 2025.
https://doi.org/10.1007/978-3-031-73235-5_26

Prompt: The camera trucks right to follow the character.   **Prompt:** The camera performs a push-in to get closer to the character.   **Prompt:** The camera stays static the entire shot.

Fig. 1. Different results generated by our camera trajectory diffusion system. Project page https://www.lix.polytechnique.fr/vista/projects/2024_et_courant.

– the *film grammar* – that prescribes how to place and move the camera to achieve intended effects. Yet mastering camera placements and motions remains challenging, especially for novice users confronted with hundreds of possibilities and little insights into how to generate the best ones.

To lower the barriers in handling camera placement and camera motion, researchers have introduced a variety of methods. These include purely geometric approaches [3,26], optimization- and control-based strategies [8,9], as well as deep learning-grounded methodologies [4,8,16,19] to interactively or automatically compute the parameters of camera trajectories. Typically, these methods address cinematographic tasks as either cinematic-rule-based control [4,9,16] or example-based imitation [18,19,41], conceptually resembling discriminative and regression models or registration and adaptation methods, respectively. Such techniques, however, suffer from the need to either design the underlying geometric model for each type of motion, or to design carefully crafted cost functions for each motion, and are often limited in their capacity to combine mixed motions creatively.

Recent advances in video generation [42,48] enable users to explore more creative possibilities by capturing and reproducing camera motion in their generated videos. Jiang *et al.* [20] followed this path and addressed camera trajectory generation using diffusion models, which incorporate a high degree of controllability. Yet, this work displayed two main drawbacks: first, it relied on a character-centric coordinate system to simplify the problem, thus limiting its generation capabilities, and second its evaluation metrics relied on camera trajectory features with oversimplified assumptions.

In other domains, the generative techniques often rely on the availability of large datasets enriched with textual descriptions, such as language-motion obtained via motion capture (mocap) [11,32] or language-vision [25,36] datasets. Yet in cinematography, there is no movie datasets where crucial cinematic information such as camera and character trajectories are available. Most recent approaches build on synthetic data [18–20], or general videos from streaming platforms (see [16] for drone trajectory generation, or [49] for dedicated real-estate videos) without the cinematic features that conform to the film grammar. Some example-based approaches address cinematic transfer tasks from real film clips [21,41], these approaches only retarget and adapt the camera trajectory with little control or variability in the results and do not encode cinematographic knowledge.

In this work, we propose a new camera trajectory dataset extracted from real movie clips, called *E.T. the Exceptional Trajectories*. It comprises camera trajectories together with textual descriptions of both camera and character trajectory over time (see Fig. 2). E.T. contains more than 11M frames with the corresponding camera and character trajectories, as well as two types of captions: camera-only and camera-character, describing the trajectory of the camera with respect to the trajectory of the character. To our knowledge, E.T. is the first extensive dataset with geometric information on both camera and character trajectories accompanied by textual descriptions.

To exploit this dataset, we also propose DIRECTOR (DiffusIon tRansformEr Camera TrajectORy), a diffusion-based model that generates camera trajectories by leveraging text descriptions and character information, as shown in Fig. 1. This allows us to better encode the correlation between character and camera trajectories. Moreover, unlike previous methods [20] that use a constrained character-relative coordinate system, we propose to use a global coordinate system. DIRECTOR relies on a classical diffusion framework with three distinct architectures for conditioning: in-context, AdaLN and cross-attention settings. Furthermore, we propose a language-trajectory embedding: CLaTr (Contrastive Language-Trajectory), trained at scale using the E.T. dataset. CLaTr serves as a foundation for computing default generative metrics similar to Frechet-Inception-Distance (FID) [13] for generated trajectories. Our experiments show that all three architectures of DIRECTOR successfully leverage the combination of input captions and character trajectories as conditions. Overall, DIRECTOR sets the new state-of-the-art on the camera trajectory generation task.

Our contributions are: (1) We introduce the E.T. camera trajectory dataset extracted from real movie clips. We complement camera trajectories with character trajectories and captions for both camera and character. (2) We present DIRECTOR, a camera trajectory diffusion model that exploits both character trajectories and textual descriptions. It offers higher controllability and granularity for users than existing approaches [20] and achieves state-of-the-art performances. (3) We propose CLaTr, a robust and accurate language-trajectory embedding, which facilitates the evaluation of camera trajectory generation models.

2 Related Work

Camera Control. Over the past twenty years, there have been several paradigm shifts in camera planning and control. Initial studies [3] predominantly focused

[Camera-only]
"The camera **trucks left** throughout the entire shot."

[Camera-character]
"As **the character moves down, the camera trucks left** to follow their motion."

(a)

[Camera-only]
"The camera moves **right along a trucking motion** throughout the entire shot."

[Camera-character]
"While the **character moves right** and then **forward**, the **camera trucks right** to follow their motion."

(b)

Fig. 2. Examples E.T. samples. Each subfigure presents frames from the original movie shot on the left, while the right side depicts the extracted and processed camera and character trajectories. Additionally, the bottom part showcases the generated camera trajectory caption with or without the character trajectory.

on geometric modeling [26] and rule-based trajectory controls [8] to direct and create camera trajectories that comply with either hand-crafted cinematic rules or image-based criteria. With the progress of deep learning, [19] introduced a method to synthesize camera trajectory for 3D animations in two stages: (i) capturing cinematic styles from a reference clip using a Mixture-of-Experts model, and (ii) generating trajectories based on 3D character animations autoregressively. Subsequent research [18], building on this, incorporates keyframing to provide extra control such as positional and velocity constraints. More recently, JAWS [41] pioneered the direction for example-based camera retargeting within a Neural Radiance Field (NeRF) [27] setting, by optimising camera trajectory directly given the 2D reference clip in a 3D NeRF. All these example-based methods share a common limitation: they struggle with generalization because they require carefully selected reference videos to ensure high quality.

Unlike example-based methods, many cinematic-rule-based methods readily integrate with Deep Reinforcement Learning (DRL) and Imitation Learning (IL) techniques, particularly in the drone cinematography domain: [16] exploit optical flow and human poses to guide drone controls via an IL framework. Similarly, [4] use DRL to control drone actions for multiple rewards, including obstacle avoidance, target tracking, shooting style etc. Recently, GAIT [44] employs an aesthetic score-based RL method instead of handcrafted rewards to control the camera in the virtual 3D environment. However, these RL-based camera control approaches also have limitations: (1) they need environment-specific training; (2) they inherently restrict the diversity of results, often leading to collapsed trajectory styles. Instead, we leverage the generalization capabilities of generative models to address the camera control task.

Camera Diffusion. Generative models have recently gained much progress and attention in domains such as textual-conditioned image generation [29,33,35], video synthesis [2,37] and human motion generation [5,38,47]. Among these, diffusion models stand out for their strong ability to produce high-fidelity and diverse generative samples [7,43], making them particularly well-suited for camera trajectory generation tasks.

The first application of diffusion models in camera control is the Cinematographic Camera Diffusion (CCD) [20], which relies on the MDM architecture (human Motion Diffusion Model) [38] and is trained on synthetic data. However, CCD simplifies the task by expressing all the camera trajectories in character-centric relative coordinates. Its small-scale synthetic training dataset also limits the broader application of the method (e.g., only 48-size vocabulary is used during training), thus making it unable to generate camera trajectories from real datasets and, in turn, impractical for common users. In contrast, in our proposed E.T. dataset, we represent camera trajectories in a global coordinate system, distinct from character trajectories. This approach allows for more diverse correlations between character and camera movements. Additionally, E.T. offers a rich vocabulary ($\sim$5.4k) and extensive camera trajectory data.

Recent literature also includes several text-to-video generation techniques that can handle different categories of camera motions [42,48]. These methods, however, assume access to 3D camera trajectories, whereas our approach generates them. Furthermore, they typically overlook the camera's primary targets (i.e., the characters), which are essential for defining camera trajectories. In contrast, our dataset contains character information, and we leverage it to generate camera trajectories that focus on a specific target character.

Camera Trajectory Datasets. Many modern generative methods leverage large multimodal datasets. For instance, in text-to-image generation, the default dataset is LAION [36] with around 400 million image-text pairs. Similarly, in human motion synthesis, the large-scale KIT [32] and HumanML3D [11] datasets offer detailed textual captions that enhance comprehension of human motion. Yet, for camera control, only a few datasets are available [20,49]. This is largely due to the intricacies involved in extracting camera poses from real-world videos, especially in cinematic contexts due to the presence of stylistic elements (e.g. motion blur or depth-of-field). Zhou et al. [49] applied Structure-from-Motion (SfM) methods to YouTube real-estate videos, creating the RealEstate10K dataset. This dataset, designed primarily for 3D reconstruction, comprises solely smooth camera movements and limited scene variation, lacking the nuanced complexity of cinematic camera motion and human presence. More recently Jiang et al. [20] introduced a synthetic cinematic camera trajectory dataset, aiming to circumvent extraction challenges. However, this dataset oversimplifies the intricate cinematic dynamics present in real-world movies.

A recent breakthrough in 3D human pose estimation for videos, termed SLAHMR [10], offers a compelling trade-off between robustness and accuracy by jointly optimizing camera and character trajectory estimations. Motivated by the lack of camera trajectory datasets, the capabilities of SLAHMR and the recent advances in other domains, we propose a new multi-modal camera trajectory dataset E.T. extracted from cinematic content, which we enhance with automatically generated captions for camera and character trajectories.

3 Exceptional Trajectories (E.T.)

We introduce a camera trajectory dataset called *Exceptional Trajectories* (E.T.), extracted from real movies. E.T. is built upon the Condensed Movies Dataset (CMD) [1]. Each *sample* in E.T. represents a camera trajectory at the shot level together with a character trajectory and two types of textual captions: a camera-only caption, which describes the camera motion; and a joint camera-character trajectory caption, which describes the motion of the camera according to the motion of the character (see Fig. 2). Below, we describe the key properties and statistics of E.T. (Sect. 3.1) followed by the creation pipeline (Sect. 3.2).

Table 1. Dataset Comparison. We compare the E.T. dataset to (i) two human motion datasets KIT [32] and HumanML3D [11]; and (ii) camera trajectory datasets RealEstate10K [49] and CCD [20]. Here the notion of sample is common across all datasets and corresponds to data associated with a continuous temporal sequence.

Dataset	#Samples	#Frames	#Hours	Domain	Character		Camera		#Vocabulary
					Traj	#Captions	Traj	#Captions	
KIT Motion-Language [32]	4K	0.8M	11.23	Mocap	✓	6K	-		1,623
HumanML3D [11]	14K	2M	28.59	Mocap	✓	45K	-		**5,371**
RealEstate10k [49]	79K	11M	121	Youtube	-		✓	-	-
CCD [20]	25K	4.5M	50	Synthetic	-		✓	25K	48
E.T. (Ours)	**115K**	**11M**	120	Movie	✓	**115K**	✓	**230K**	1,790

3.1 E.T. Properties and Statistics

The key properties of E.T. are as follows:

Cinematic Content. The camera trajectories in E.T. are both realistic and cinematic, since they are extracted from real-world movies (Table 1). This dual nature allows for effective modelling of various visual styles, in contrast to RealEstate10k's [49] focus on shots characterized by smooth camera trajectories and limited scene variation. Furthermore, by extracting data from real-world movies, E.T. sets itself apart from CCD [20], which only relies on synthetic camera trajectories.

Scale. E.T. is built upon 16,210 different scenes from CMD [1]. It comprises 115K samples spanning 11M frames and totalling 120 h of footage, offering extensive and diverse camera and character (human) trajectories based on real movies. In contrast, existing human motion datasets are much smaller, with only 11.23 h for KIT [32] and 28.59 h for HumanMl3D [11] (see Table 1). When compared against datasets with camera trajectories, it far exceeds CCD [20] in terms of hours, frames and samples. Although its scale is comparable to

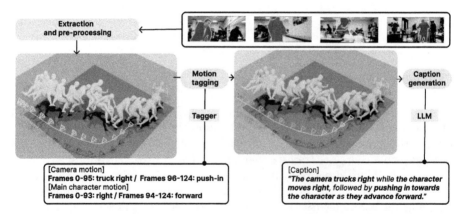

Fig. 3. Dataset creation pipeline. Given RGB frames from a video, we first extract and pre-process camera and character poses, then tag resulting camera and character trajectories (sequence of poses) to obtain rough independent descriptions (middle part). Finally, we translate these descriptions into rich textual captions, aligning the camera trajectory with that of the character (right part).

RealEstate10k [49], it provides additional character trajectories and captions referring to real movies as opposed to RealEstate10k, which focuses only on camera trajectories in another domain.

Controllability. E.T. stands out by comprising not only camera and character trajectories but also camera-only and camera-character captions (see Fig. 2). Incorporating caption information into the model offers multiple advantages: (1) it democratizes the input format for general users; and (2) it adds complementary semantic information to the trajectory data. In comparison, RealEstate lacks captions entirely. CCD's captions are limited by a small vocabulary size and focus only on camera while lacking character information[1]. The richness and complexity of E.T.'s captions are on par in terms of vocabulary size –above a thousand– with human motion datasets such as KIT and HumanML3D, which provide detailed, hand-crafted human motion descriptions[2].

Statistics. Figures 4a and b display the statistics of the E.T. dataset, confirming the diversity and all six degrees of freedom coverage of both camera and character trajectories (see more in Appendix B.1.)

3.2 Dataset Creation Pipeline

E.T. is constructed by a three-step process (see Fig. 3). First, we extract the 3D coordinates of cameras and characters over time, which we further refine to

---

[1] Note that CCD indirectly comprises camera trajectories through the character-relative coordinate system.

[2] Note that E.T. has no overlap with human motion datasets. E.T.'s extracted 3D poses (see Sect. 3.2) are less accurate than the ones in motion capture, while its captions describe camera trajectory relative to character trajectory, as opposed to describing exact human motions targeted by these datasets.

form uniform trajectories. Second, we perform *motion tagging*, i.e. partition each trajectory into segments with each segment comprising a pure camera motion that we label (tag). Third, we generate captions that describe both the camera and the character trajectory over time. We detail each step below

Data Extraction and Pre-processing. To extract camera and character poses, we apply on each shot the joint camera and 3D human poses estimator SLAHMR [46]. Given the complexity of estimating 3D poses from 2D data, the raw outputs tend to be noisy. To address this, we perform various pre-processing steps such as alignment, filtering, smoothing and cropping to a maximum length of 300 frames as in [11]. Refer to the Appendix B.2 for further details.

Motion Tagging. Our objective is to partition camera or character trajectories into segments of pure motion: tags. Besides static, we consider the six fundamental motions across three degrees of freedom. They include lateral movements left, and right; vertical movements up and down; and depth movements forward and backwards. Each trajectory is partitioned into motion tags with one, two, or three pure camera motions, totalling 27 combinations (see Fig. 4a).

We propose a thresholding-based method that uses trajectory velocity for motion tagging: This method consists of two stages: (i) for each dimension (XYZ), we use an initial threshold on velocity to detect whether the camera or character remains static along the dimension; (ii) when multiple dimensions are non-static, we calculate pairwise velocity rates and use a threshold to pinpoint dominant velocities. A dimension is classified as static if its velocity is outmatched. The tag of motion between two points is then determined by the combination of non-static dimensions. Finally, we apply smoothing to avoid noisy and sparse tags and hence enhance the overall trajectory-level tagging.

For *camera trajectory tagging*, we use the rigid body velocity $\in SE(3)$ – derived from rotation and translation– to account for the camera's facing direction. this enables us to differentiate between similar motions, such as 'trucking', where the camera moves along an axis with a perpendicular facing direction, and 'depth', where the facing direction aligns with the movement axis. For *character trajectory tagging*, we assume that characters face the direction of their movement. Hence, we represent character trajectory using only the linear velocity, as derived from the translation of their hip centres.

These result in a coarse description of both camera and character trajectories over time as shown in Fig. 3(left).

Caption Generation. Our objective is to provide rich textual descriptions of the extracted camera trajectories according to the character trajectory. In movie, cameras typically move relative to the subject being filmed, i.e., the main character. Therefore, for each shot, we first identify the main character following [39][3] based on the temporal and spatial coverage of their bounding boxes within the

---

[3] Hitchcock's rule: '*the size of an object in the frame should equal its importance in the story at the moment*' [39].

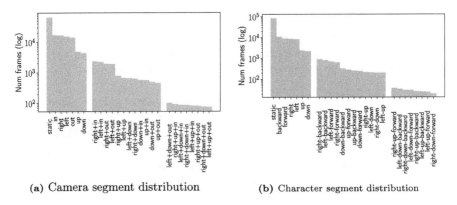

(a) Camera segment distribution (b) Character segment distribution

Fig. 4. E.T. statistics.

shot. Then, for both camera and main character trajectories, we generate captions for each motion tag, as shown in the center of Fig. 3. Then, inspired by [6], our goal is to convert the descriptions obtained via motion tagging for camera and character trajectories into detailed textual annotations. For this, we prompt an LLM –Mistral-7B [17]– to generate camera trajectory captions by referencing the main character's trajectory as anchor points. Our prompt formulation follows a structured approach with context, instruction, constraint, and example. Further details can be found in the Appendix B.3.

This step results in a rich description of both camera and character trajectories over time as shown in Fig. 3(right).

4 Method

Here, we introduce our proposed *DiffusIon tRansformEr Camera TrajectORy* (DIRECTOR) method for camera trajectory generation (Sect. 4.1). DIRECTOR takes as input the character trajectory with the camera-character caption and generates a camera trajectory. Additionally, we present the *Contrastive Language-Trajectory* embedding (CLaTR) that serves as a basis for creating a common space between text and trajectories (Sect. 4.2), enabling the computation of evaluation metrics.

4.1 Camera Trajectory Diffusion

Problem Formulation. We consider a camera trajectory $\mathbf{x}_{1:N}$ as a sequence of N consecutive camera poses. Each camera pose $\mathbf{x} = [\mathbf{R}|\mathbf{t}]$ comprises a rotation $\mathbf{R}$ representing the camera's orientation and a translation $\mathbf{t}$ indicating its position. We aim at generating camera trajectories under two conditions: (i) a target character trajectory $\mathbf{h}_{1:N}$ capturing the 3D positions of the main character; and (ii) a textual description c specifying the desired camera movement relative to the character movement.

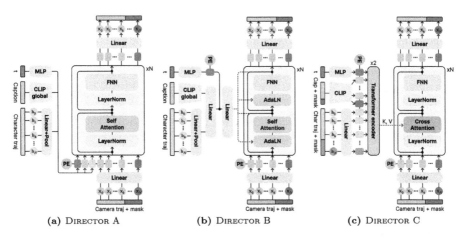

Fig. 5. *DiffusIon tRansformEr Camera TrajectORy* (DIRECTOR). We display 3 variants of our diffusion model DIRECTOR. DIRECTOR A incorporates the conditioning as in-context tokens. DIRECTOR B leverages AdaLN modulation of the transformer block to add the conditioning. DIRECTOR C uses the full text and character trajectory sequences by relying on cross-attention.

Diffusion Framework. We follow the general diffusion paradigm established in EDM [22]. In essence, diffusion models consist of randomly sampling $\mathbf{x}^0 \sim \mathcal{N}(\mathbf{0}, \sigma_{max}^2 \mathbf{I})$, and progressively denoising it to reach the endpoint $\mathbf{x}^K$ of this process, distributed according to the initial data distribution. During the training stage, we perturb an initial data distribution with standard deviation σ_{data}, with i.i.d. Gaussian noise with standard deviation σ. When $\sigma_{\max} \gg \sigma_{\text{data}}$, the noise distribution equivalent to a normal distribution $\mathcal{N}(\mathbf{0}, \sigma_{max}^2 \mathbf{I})$. We use these modified versions of the initial data distribution to train a denoiser module D, which takes as input a sample $\mathbf{x}$ to denoise, the two conditions (character trajectory $\mathbf{h}$ and the caption c), and the corresponding standard deviation σ. Then, D is trained using the denoising score matching loss:

$$\mathcal{L}_{\text{score}} = \big(D(\mathbf{x}, \mathbf{h}, c;\ \sigma) - \mathbf{x}\big)/\sigma^2. \tag{1}$$

During the sampling phase, we apply the 2nd order deterministic sampling introduced in EDM [22] with classifier-free guidance [15].

DIRECTOR *Architecture.* DIRECTOR (*DiffusIon tRansformEr Camera TrajectORy*) takes as input the character trajectory and the caption and generates a camera trajectory. Its architecture is illustrated in Fig. 5. The base of DIRECTOR is a pre-norm Transformer [40,45]. We condition the transformer on the diffusion timestep, the character trajectory, and a textual description that describes the relative movement between the camera and character trajectories (see Fig. 2). The timestep is tokenized using a sinusoidal positional embedding [40] and then mapped with an MLP.

Inspired by the DiT architecture variants [30], we explore three distinct ways to include the conditioning in the denoising process (Fig. 5).

DIRECTOR A (Fig. 5a). The conditioning is added to the ***context*** of the transformer input. We only use the global clip token for the text, and we do a linear embedding of the character trajectories, which in turn gets averaged pooled into a single token.

DIRECTOR B (Fig. 5b). Both conditionings (character trajectory and caption) are concatenated into a single token which gets mapped at each layer into 6 vectors, $\gamma_1, \beta_1, \lambda_1, \gamma_2, \beta_2, \lambda_2$. Then, the layer-norm of the transformer is replaced by the following ***AdaLN*** operation:

$$\text{ADALN}(\gamma, \beta, x) = (1+\gamma)\text{LN}(X) + \beta \quad , \tag{2}$$

where LN refers to the Layer Normalization, γ, β are the scale and bias, respectively. The AdaLN operation is performed before each self-attention and feed-forward layer in the transformer. The output of each self-attention and cross-attention is rescaled by λ. Following [30], we initialize the modulation such that the output is zero.

DIRECTOR C (Fig. 5c). We leverage the full sequence length of the conditioning. We retrieve the CLIP-embedded text sequence and the linearly projected trajectory and concatenate them into a single sequence. We then use 2 layers of transformer encoders to pre-process this sequence, which is then incorporated into the DIRECTOR transformer with a ***cross-attention*** block.

4.2 Contrastive Language-Trajectory Embedding (CLaTr)

Given the scarcity of relevant camera trajectory methods and datasets, the community has not introduced adequate metrics for this task. In the concurrent cinematic camera trajectory diffusion work [20], the authors evaluate their model with metrics from the human motion community. For this, they train a dedicated camera trajectory classifier to extract features. However, their classifier is trained on a simplistic task, comprising only six basic camera motion classes on synthetic data, which fails to capture the true complexity of camera trajectories.

To address this lack of proper evaluation metrics, in this section, we propose to extend existing metrics from text-image-based and text-motion-based generation (which rely on feature embeddings to measure the generation quality) to text-trajectory generation. The main obstacle is that no commonly accepted text-trajectory feature embedding exists. Therefore, we propose to learn a general text-trajectory embedding in a contrastive CLIP-like manner to acquire an accurate and robust feature representation, which can serve as a foundation for computing camera trajectory evaluation metrics.

We introduce *Contrastive Language-Trajectory* embedding (CLaTr) by capitalizing our multi-modal dataset E.T. with a CLIP-like approach [34]. Our language-trajectory embedding follows the methodology outlined in [31], originally designed for human motion. CLaTr consists of a VAE [23] framework with trajectory and text encoders and a shared feature decoder. CLaTr is trained with three losses: (a) a reconstruction loss $\mathcal{L}_R$, quantifying trajectory reconstruction of both trajectory and text features; (b) four KL loss terms $\mathcal{L}_{KL}$, which regularize each modality distribution and also enforce inter-modality similarity; and (c) a cross-modal embedding similarity loss $\mathcal{L}_E$, ensuring alignment between text and trajectory features. See Appendix C for more details.

5 Experiments

Implementation Details. We train DIRECTOR with a batch size of 128 using the AdamW optimizer with a learning rate of 1e−4, $(\beta_1, \beta_2) = (0.9, 0.95)$ and a weight decay of 0.1. We use a cosine decay learning rate scheduler with 5k steps of warmup for a total of 170k steps in bfloat16 mixed precision. The model has 8 layers with a hidden dim of 512 and 16 attention heads. We use dropout and stochastic depth of 0.1. We set the default temporal input size to 300 to match the E.T. sample size (see Sect. 3.2) and use masking to handle inputs with fewer than 300 frames. For the camera trajectory, we use the 6D continuous representation for rotation [50] combined with the 3D translation component. For the character trajectory, we use the 3D position of the character's hip center.

Table 2. Quantitative Results. Comparison of DIRECTOR and concurrent methods on E.T. pure and mixed subsets, evaluating trajectory quality (left) and caption coherence (right). First best and second best .

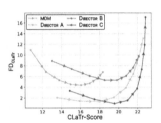

Fig. 6. FD_{CLaTr} vs CLaTr-Score. Guidance range between 0.6 and 2.2 on E.T. mixed subset.

Set	Methods	ω	Camera trajectory quality					Text-camera coherence			
			FD_{CLaTr} ↓	P ↑	R ↑	D ↑	C ↑	CS ↑	C-P ↑	C-R ↑	C-F1 ↑
E.T. pure trajectories	CCD [20]	5.5	31.33	0.79	0.55	0.83	0.72	3.21	0.53	0.28	0.27
	MDM [35]	1.8	6.10	0.77	0.68	0.89	0.80	21.26	0.81	0.75	0.76
	DIRECTOR A	1.6	5.16	0.82	0.67	1.00	0.86	21.88	0.84	0.78	0.80
	DIRECTOR B	1.8	6.61	0.80	0.72	0.92	0.82	23.10	0.85	0.80	0.86
	DIRECTOR C	1.6	4.57	0.83	0.65	1.00	0.87	21.49	0.83	0.78	0.80
E.T. mixed trajectories	CCD [20]	6.0	35.81	0.73	0.55	0.75	0.67	6.26	0.37	0.20	0.17
	MDM [35]	2.0	6.79	0.78	0.65	0.85	0.76	18.32	0.36	0.36	0.34
	DIRECTOR A	1.4	3.88	0.82	0.68	0.98	0.85	20.76	0.43	0.43	0.42
	DIRECTOR B	1.6	6.10	0.78	0.74	0.85	0.78	20.78	0.41	0.40	0.39
	DIRECTOR C	1.4	3.76	0.83	0.67	1.00	0.86	21.05	0.49	0.49	0.48

5.1 Quantitative Results

Metrics. We use two sets of metrics.

First, we assess **camera trajectory quality**, specifically how well the generated camera trajectories match the distribution of the ground truth camera trajectories. For this, we use the CLaTr-based metrics described in Sect. 4.2: the Frechet CLaTr Distance (FD_{CLaTr}) similar to FID [13]), Precision (R), Recall (R), Density (D) and Coverage (C) [28]. As the validation set comprises only a few samples and these metrics need a critical amount of samples (10k+), we compare to the train set as it is common practice in small dataset generative models (e.g. CIFAR image generation [14,24]).

Second, we use **text-camera coherence** metrics, which measure the coherence between the given caption (text) and the generated camera trajectory. For this, we use the CLaTr-Score (CS) (see Sect. 4.2), similar to CLIP-Score [12].

Additionally, we derive Classifier Precision (C-P), Classifier Recall (C-R) and Classifier F1-Score (C-F1) by performing motion tagging (described in Sect. 3.2) on generated camera trajectories and compare them to the ground truth.

Dataset. In our experiments, we train and evaluate our model on two different subsets of the E.T. dataset. First, the *pure camera trajectory subset*, where we only keep the samples having a single camera motion trajectory (e.g. *"the camera trucks right"*). Second, the *mixed camera trajectories subset*, which excludes some static-only camera trajectories to create a balanced subset. In this way, we can both correctly compare against methods suited for simple, pure trajectories and emphasize the difficulty of the mixed compositional camera trajectories. We compare in Table 2 DIRECTOR with concurrent methods on the pure subset (top) and mixed subset (bottom).

Comparison to the State of the Art. We report in Table 2 and Fig. 6 quantitative results of the different DIRECTOR architectures against the previous state-of-the-art CCD [20], and MDM [38], a default modern method in human motion. We observe that overall we outperform both works on all metrics and both subsets. Particularly, in the mixed trajectory subset (bottom of Table 2), we demonstrate superior camera trajectory quality metrics (left section of Table 2) with a margin of -3.0 FD_{CLaTr} against MDM and -32.1 against CCD. Additionally, our method excels in text-camera coherence (right section of Table 2) within the same subset, achieving a substantial improvement of $+3.6$ ClaTR-Score against MDM and $+15.7$ against CCD.

Additionally, we show in Fig. 6 the trade-off between FD_{CLaTr} (trajectory quality) and CLaTr-Score (conditioning coherence) for varying guidance weights. The optimal point is at the bottom right, where FD_{CLaTr} is lowest and CLaTr-Score is highest. We observe that the MDM curve (blue) consistently lies above DIRECTOR's curves, indicating that MDM performs worse.

These results reveal the effectiveness of our method both in generating high-quality camera trajectories and in handling the input caption conditioning.

Ablation of DIRECTOR Architectures. We observe in Table 2 and Fig. 6 that DIRECTOR C outperforms other variants, followed closely by DIRECTOR A. The cross-attention mechanism in DIRECTOR C enables effective incorporation of conditioning into the model, leading to its superior performance. DIRECTOR A offers a compelling balance of efficiency and performance: it exhibits comparable results to DIRECTOR C with a simpler concept and fewer parameters. In contrast, DIRECTOR B excels in text-camera coherence on the pure trajectory subset (top-right of Table 2) but struggles on the mixed trajectory subset (bottom-right of Table 2). We attribute this to the AdaLN's ability to condition the model in simple setups, but its failure to capture sequential complexity in harder scenarios.

5.2 Qualitative Results

Figure 7 shows generated camera trajectories from DIRECTOR (architecture C). Each sub-figure displays the trajectories with pyramid markers for keyframes,

(a) Controllability (b) Diversity (c) Complexity (d) Character-aware

Fig. 7. Qualitative results. Generated camera trajectories with corresponding prompts and character trajectories, highlighting (a) controllability, (b) diversity, (c) complexity, and (d) character awareness. Darker shades indicate later frames.

along with character meshes and corresponding captions. The output trajectories are smooth and consistent with the input conditions. We highlight four key strengths of our method:

Controllability (Fig. 7a). DIRECTOR offers high controllability: by modifying only two words in the caption, the user can generate all kinds of camera trajectories, e.g. "*trucks right*", "*trucks left*", "*booms top*" and "*booms bottom*".

Diversity (Fig. 7b). Given the same input conditions (i.e. character trajectory and caption), DIRECTOR generates diverse camera trajectories, allowing users to explore a wide range of creative and unique outputs.

Complexity (Fig. 7c). DIRECTOR can handle complex input conditions, including character trajectories (e.g., "*moves right*" then "*stops*") and camera trajectories descriptions (e.g., "*stays static and pushes-in*" and "*trucks right and remains static*").

Character-Awareness (Fig. 7d). DIRECTOR effectively considers the character, generating camera trajectories that follow the character's movement when the prompt and character trajectory are mirrored.

6 Conclusion

We designed and implemented E.T., a dataset of camera and character trajectories extracted from movie sequences that we believe will be very beneficial to the community. In addition to their trajectories, E.T. comes with text captions that describe the camera and character trajectories over time. We showed how E.T. can be exploited to train a diffusion-based approach to generate complex camera trajectories from high-level textual descriptions which correlate the trajectory of the camera with the trajectory of the characters. For this, we propose the diffusion-based method DIRECTOR, which sets the new state of the art on camera trajectory generation. In the future, we plan to address the expressiveness of the trajectory captions, by including more information about modifiers and the exact position on the screen where the characters should be located.

Acknowledgements. This work was supported by ANR-22-CE23-0007, ANR-22-CE39-0016, Hi!Paris grant and fellowship, and was granted access to the HPC resources of IDRIS under the allocation 2023-AD011013951 made by GENCI. We would like to thank Hongda Jiang, Mathis Petrovich, Pierre Vassal and the anonymous reviewers for their insightful comments and suggestions.

References

1. Bain, M., Nagrani, A., Brown, A., Zisserman, A.: Condensed movies: story based retrieval with contextual embeddings. In: ACCV (2020)
2. Blattmann, A., et al.: Stable video diffusion: scaling latent video diffusion models to large datasets. arXiv preprint arXiv:2311.15127 (2023)
3. Blinn, J.: Where am I? What am I looking at? (cinematography). IEEE Comput. Graph. Appl. **8**, 76–81 (1988)
4. Bonatti, R., et al.: Autonomous aerial cinematography in unstructured environments with learned artistic decision-making. J. Field Robot. **37**, 606–641 (2020)
5. Chen, X., et al.: Executing your commands via motion diffusion in latent space. In: CVPR (2023)
6. Delmas, G., Weinzaepfel, P., Lucas, T., Moreno-Noguer, F., Rogez, G.: PoseScript: 3D human poses from natural language. In: Avidan, S., Brostow, G., Cissé, M., Farinella, G.M., Hassner, T. (eds.) Computer Vision, ECCV 2022. LNCS, vol. 13666, pp. 346–362. Springer, Cham (2022). https://doi.org/10.1007/978-3-031-20068-7_20
7. Dhariwal, P., Nichol, A.: Diffusion models beat GANs on image synthesis. In: NeurIPS (2021)
8. Drucker, S.M., Galyean, T.A., Zeltzer, D.: CINEMA: a system for procedural camera movements. In: Symposium on Interactive 3D Graphics (1992)
9. Galvane, Q., Christie, M., Lino, C., Ronfard, R.: Camera-on-rails: automated computation of constrained camera paths. In: ACM Motion In Games (2015)
10. Goel, S., Pavlakos, G., Rajasegaran, J., Kanazawa, A., Malik, J.: Humans in 4D: reconstructing and tracking humans with transformers. In: ICCV (2023)
11. Guo, C., et al.: Generating diverse and natural 3D human motions from text. In: CVPR (2022)
12. Hessel, J., Holtzman, A., Forbes, M., Le Bras, R., Choi, Y.: CLIPScore: a reference-free evaluation metric for image captioning. In: EMNLP (2021)
13. Heusel, M., Ramsauer, H., Unterthiner, T., Nessler, B., Hochreiter, S.: GANs trained by a two time-scale update rule converge to a local Nash equilibrium. In: NeurIPS (2017)
14. Ho, J., Jain, A., Abbeel, P.: Denoising diffusion probabilistic models. In: NeurIPS (2020)
15. Ho, J., Salimans, T.: Classifier-free diffusion guidance. In: NeurIPS-W (2021)
16. Huang, C., et al.: Learning to film from professional human motion videos. In: CVPR (2019)
17. Jiang, A.Q., et al.: Mistral 7B. arXiv preprint arXiv:2310.06825 (2023)
18. Jiang, H., Christie, M., Wang, X., Liu, L., Wang, B., Chen, B.: Camera keyframing with style and control. ACM TOG **40**, 1–13 (2021)
19. Jiang, H., Wang, B., Wang, X., Christie, M., Chen, B.: Example-driven virtual cinematography by learning camera behaviors. ACM TOG **39**, 45:1–45:14 (2020)
20. Jiang, H., Wang, X., Christie, M., Liu, L., Chen, B.: Cinematographic camera diffusion model. In: Computer Graphics Forum (2024)

21. Jiang, X., Rao, A., Wang, J., Lin, D., Dai, B.: Cinematic behavior transfer via nerf-based differentiable filming. arXiv preprint arXiv:2311.17754 (2023)
22. Karras, T., Aittala, M., Aila, T., Laine, S.: Elucidating the design space of diffusion-based generative models. In: NeurIPS (2022)
23. Kingma, D.P., Welling, M.: Auto-encoding variational Bayes. Statistics (2014)
24. Krizhevsky, A., et al.: Learning multiple layers of features from tiny images, ON, Canada, Toronto (2009)
25. Lin, T.-Y., et al.: Microsoft COCO: common objects in context. In: Fleet, D., Pajdla, T., Schiele, B., Tuytelaars, T. (eds.) ECCV 2014. LNCS, vol. 8693, pp. 740–755. Springer, Cham (2014). https://doi.org/10.1007/978-3-319-10602-1_48
26. Lino, C., Christie, M.: Intuitive and efficient camera control with the Toric space. ACM TOG **34**, 1–12 (2015)
27. Mildenhall, B., Srinivasan, P.P., Tancik, M., Barron, J.T., Ramamoorthi, R., Ng, R.: NeRF: representing scenes as neural radiance fields for view synthesis. In: Vedaldi, A., Bischof, H., Brox, T., Frahm, J.-M. (eds.) ECCV 2020. LNCS, vol. 12346, pp. 405–421. Springer, Cham (2020). https://doi.org/10.1007/978-3-030-58452-8_24
28. Naeem, M.F., Oh, S.J., Uh, Y., Choi, Y., Yoo, J.: Reliable fidelity and diversity metrics for generative models. In: ICML (2020)
29. Nichol, A.Q., et al.: GLIDE: towards photorealistic image generation and editing with text-guided diffusion models. In: ICML (2022)
30. Peebles, W., Xie, S.: Scalable diffusion models with transformers. In: ICCV (2023)
31. Petrovich, M., Black, M.J., Varol, G.: TMR: text-to-motion retrieval using contrastive 3D human motion synthesis. In: ICCV (2023)
32. Plappert, M., Mandery, C., Asfour, T.: The KIT motion-language dataset. Big Data (2016)
33. Podell, D., et al.: SDXL: improving latent diffusion models for high-resolution image synthesis. arXiv preprint arXiv:2307.01952 (2023)
34. Radford, A., et al.: Learning transferable visual models from natural language supervision. In: ICML (2021)
35. Rombach, R., Blattmann, A., Lorenz, D., Esser, P., Ommer, B.: High-resolution image synthesis with latent diffusion models. In: CVPR (2022)
36. Schuhmann, C., et al.: LAION-400M: open dataset of clip-filtered 400 million image-text pairs. arXiv preprint arXiv:2111.02114 (2021)
37. Singer, U., et al.: Make-a-video: text-to-video generation without text-video data. arXiv preprint arXiv:2209.14792 (2022)
38. Tevet, G., Raab, S., Gordon, B., Shafir, Y., Cohen-Or, D., Bermano, A.H.: Human motion diffusion model. In: ICLR (2023)
39. Truffaut, F., Scott, H.: Hitchcock/Truffaut. Revised Edition. Simon and Schuster (1985)
40. Vaswani, A., et al.: Attention is all you need. In: NeurIPS (2017)
41. Wang, X., Courant, R., Shi, J., Marchand, E., Christie, M.: JAWS: just a wild shot for cinematic transfer in neural radiance fields. In: CVPR (2023)
42. Wang, Z., et al.: MotionCtrl: a unified and flexible motion controller for video generation. arXiv preprint arXiv:2312.03641 (2023)
43. Xiao, Z., Kreis, K., Vahdat, A.: Tackling the generative learning trilemma with denoising diffusion GANs. In: ICLR (2021)
44. Xie, D., et al.: GAIT: generating aesthetic indoor tours with deep reinforcement learning. In: ICCV (2023)
45. Xiong, R., et al.: On layer normalization in the transformer architecture. In: ICML (2020)

46. Ye, V., Pavlakos, G., Malik, J., Kanazawa, A.: Decoupling human and camera motion from videos in the wild. In: CVPR (2023)
47. Zhang, M., et al.: MotionDiffuse: text-driven human motion generation with diffusion model. IEEE TPAMI **46**, 4115–4128 (2024)
48. Zhao, R., et al.: MotionDirector: motion customization of text-to-video diffusion models. arXiv preprint arXiv:2310.08465 (2023)
49. Zhou, T., Tucker, R., Flynn, J., Fyffe, G., Snavely, N.: Stereo magnification: learning view synthesis using multiplane images. ACM TOG **37**, 1–12 (2018)
50. Zhou, Y., Barnes, C., Lu, J., Yang, J., Li, H.: On the continuity of rotation representations in neural networks. In: CVPR (2019)

OphNet: A Large-Scale Video Benchmark for Ophthalmic Surgical Workflow Understanding

Ming Hu[1,2,3], Peng Xia[1,3], Lin Wang[5], Siyuan Yan[1,2], Feilong Tang[1,3], Zhongxing Xu[7], Yimin Luo[6], Kaimin Song[3], Jurgen Leitner[2], Xuelian Cheng[1], Jun Cheng[8], Chi Liu[9], Kaijing Zhou[4(✉)], and Zongyuan Ge[1,2,3(✉)]

[1] Faculty of IT, AIM Lab, Melbourne, Australia
[2] Faculty of Engineering, Monash University, Melbourne, Australia
[3] Airdoc-Monash Research, Airdoc, Shanghai, China
zongyuan.ge@monash.edu
[4] Eye Hospital, Wenzhou Medical University, Wenzhou, China
zhoukaijing@wmu.edu.cn
[5] Bosch Corporate Research, Shanghai, China
[6] King's College London, London, UK
[7] Cornell University, Ithaca, USA
[8] Institute for Infocomm Research, A*STAR, Singapore, Singapore
[9] Faculty of Data Science, City University of Macau, Macao, Macao, Special Administrative Region of China

Abstract. Surgical scene perception via videos is critical for advancing robotic surgery, telesurgery, and AI-assisted surgery, particularly in ophthalmology. However, the scarcity of diverse and richly annotated video datasets has hindered the development of intelligent systems for surgical workflow analysis. Existing datasets face challenges such as small scale, lack of diversity in surgery and phase categories, and absence of time-localized annotations. These limitations impede action understanding and model generalization validation in complex and diverse real-world surgical scenarios. To address this gap, we introduce OphNet, a large-scale, expert-annotated video benchmark for ophthalmic surgical workflow understanding. OphNet features: 1) A diverse collection of 2,278 surgical videos spanning 66 types of cataract, glaucoma, and corneal surgeries, with detailed annotations for 102 unique surgical phases and 150 fine-grained operations. 2) Sequential and hierarchical annotations for each surgery, phase, and operation, enabling comprehensive understanding and improved interpretability. 3) Time-localized annotations, facilitating temporal localization and prediction tasks within surgical workflows. With approximately 285 h of surgical videos, OphNet is about

M. Hu, P. Xia and L. Wang—Equal contribution.

Supplementary Information The online version contains supplementary material available at https://doi.org/10.1007/978-3-031-73235-5_27.

20 times larger than the largest existing surgical workflow analysis benchmark. Code and dataset are available at: https://minghu0830.github.io/OphNet-benchmark/.

Keywords: Surgical Workflow Understanding · Ophthalmic Surgery · Medical Image Analysis · Video Benchmark

1 Introduction

As surgical robot platforms such as the da Vinci® surgical system become increasingly sophisticated, there is growing interest in integrating enhanced intelligence into scenarios like minimally invasive surgery [20,32,68]. The advancements in machine vision perception empower these robotic systems to autonomously recognize and adapt to the intricacies of surgical environments, without relying on binary instrument usage signals, RFID tags, sensor data from tracking devices, or other signal information that necessitates laborious manual annotations or additional equipment installations [28]. This autonomy includes the capacity to identify anatomical structures, detect anomalies, and adjust surgical plans in real-time, which is crucial in dynamic and unpredictable surgical settings. In recent years, especially in endoscopy and ophthalmic surgery, the application of deep learning has demonstrated considerable promise in bolstering these autonomous capabilities. This encompasses the analysis of surgical workflows [6,28,74], segmentation of instruments and anatomy [4,24,43], and depth estimation [78], among others.

Automatic video surgical workflow understanding is a fundamental yet challenging problem for developing computer-assisted and robotic-assisted surgery, which can be divided into internal (e.g., laparoscopic and endoscopic [28,47,57]) and external (e.g., operating room and nursing procedure [41]) analysis. In addition to promoting the development of intelligent surgery, it also greatly benefits surgical documentation, education, and training [11,12,65]. Baret et al. [6] showed networks, like Inflated 3D ConvNet (I3D) [10] that utilize spatiotemporal convolutions, require a relatively extensive dataset for effective training. In their study, the model achieves an accuracy exceeding 80% when trained on 100 videos, with a progressive improvement as the sample size surpasses 700. However, the highly efficient and rapidly evolving deep learning technologies for surgical workflow analysis are currently limited by the following shortcomings in current video benchmarks: **1) Small-scale:** the majority of video datasets contain no more than 100 videos. For example, the CATARACTS [23] and CatRelDet [23] datasets contain only 50 and 21 surgical videos, respectively. These datasets are relatively small, insufficient for large-scale validation. **2) Limited categories of surgeries and phases:** almost all ophthalmic surgical video datasets only include cataract surgery and do not further classify specific types of surgeries. Additionally, the number of phase categories is also limited, like CatRelDet [23] only contains 4 different phase labels, which is insufficient to meet the requirements for evaluation in real clinical environments. **3)**

Table 1. The statistics comparison among existing workflow analysis datasets and our OphNet. Compared to other datasets, OphNet focuses on more comprehensive coverage of various surgery, phase and operation categories, collects a large number of videos, totaling 284.8 h, and also enables a variety of recognition, localization and prediction tasks. OphNet demonstrates considerable competitiveness in both its scale and the richness of its labels. For instance, Cholec120 [45], Cholec80 [65], m2cai-workflow and Lap-Chole [62] form one series, whereas CholecT50 [46], CholecT45 [46], and CholecT40 [44] comprise another series. We have excluded the following scenarios from our comparison: (1) non-open-source datasets such as Bypass170 [66], ESD [30], Yu's [74], etc.; (2) a superset of multiple open-source or non-open-source datasets, like Cholec207 [6], etc.; (3) datasets employed for lesion, anatomy, and instrument classification and segmentation, such as SUN-SEG [27], CVC-ClinicDB [7], ROBUST-MIS [52], Mesejo's [40], Cata7 [43], etc., anomaly detection such as PolypDiag [63] (from Hyper-Kvasir [9] and LDPolypVideo [38]), Kvasir-Capsule [59], etc., and other datasets not dedicated to workflow analysis. It's worth mentioning that even in comparison with the above datasets, OphNet demonstrates considerable competitiveness in both its scale and the richness of its labels. *Endo&Lap* denotes the endoscopic and laparoscopic protocol, *OphScope* denotes the ophthalmic microscope protocol. We choose the latest version for comparison in cases where datasets have multiple supplementary updates.

Protocol	Datasets	Dataset Properties					Tasks			
		No. of Videos	No. of Action Segments	No. of Surgery Categories	No. of Action Categories	Total Duration	Multi-Surgery Presence Recognition	Phase Recognition	Phase Localization	Phase Prediction
Endo&Lap	Cholec120 [45]	120	-	1	7	76.2h	✗	✓	✗	✗
	SurgicalActions160 [54]	160	160	1	16	0.2h	✗	✓	✗	✗
	HeiCo [39,50]	30	-	3	14	2.8h	✗	✓	✓	✓
	EndoVis 2021 [67]	33	250	1	7	22.0h	✗	✓	✗	✗
	PitVis [3]	25	287	1	17	33.3h	✗	✓	✗	✗
	CholecT50 [46]	50	-	1	10	44.7h	✗	✓	✓	✓
	AutoLaparo [70]	21	300	1	7	23.1h	✗	✓	✓	✓
OphScope	LensID [22]	100	2,440	1	2	11.7h	✗	✓	✗	✗
	Cataract-101 [55]	101	1,266	1	10	14.0h	✗	✓	✓	✓
	CatRelDet [23]	21	2,400	1	4	2.0h	✗	✓	✗	✗
	CATARACTS [4]	50	1,536	1	19	20.0h	✗	✓	✓	✓
	Cataract-1K [21]	1,000	931	1	12	118.7h	✗	✓	✓	✓
	OphNet(Ours)	**2,278**	**9,795**	**66**	**150**	**284.8h**	✓	✓	✓	✓

Coarse-grained annotation: due to annotation costs, existing benchmarks often have coarse-grained action definitions. For example, adhesive injection may occur in two different phases: main incision and capsulorhexis, so it may be classified into different phase categories. Coarse-grained action definitions may lead to annotation bias. **4) Single time-boundary annotation:** they only annotate designated phases in the videos, ignoring the continuity across different stages of ophthalmic surgery, as well as the hierarchical relationship between surgery, phase, and operation. Simpler datasets, such as LensID [22], are limited to binary classification tasks distinguishing lens implantation from other irrelevant phases. **5) Uniform domain:** the videos are meticulously collected, and while this ensures video quality, the uniform style is not conducive to testing the model's domain generalization ability.

While some works have explored semi-supervised and self-supervised learning strategies [8,51,73,75] to alleviate the cost of annotations or use only a small fraction of available labels, these approaches still lack competitiveness in performance compared to fully supervised learning. This deficiency in performance is impeding the widespread clinical application of these strategies. To address the shortage of sufficient labeled datasets, we construct OphNet, a large-scale and expert-level video benchmark with high diversity, for ophthalmic surgical workflow understanding. The main advantages of OphNet are as follows:

- **Largest scale and diversity:** to the best of our knowledge, OphNet is currently the largest and most richly labeled dataset for surgical workflow analysis. It contains a number of videos 20 times greater than the current largest benchmark in ophthalmic surgery and far exceeds datasets in more established fields such as endoscopy. Additionally, OphNet includes the greatest variety of different types of surgeries, encompassing 66 different surgeries such as cataract, glaucoma, and corneal surgeries, along with 102 unique surgical phases and 150 distinct operations. This diversity significantly surpasses that of previous research.
- **Fine-grained, sequential and hierarchical annotation:** we have meticulously selected a subset of videos for annotation localization, with each video being annotated for an average of 22 operations. Additionally, we provide exquisite annotations at the levels of surgery, phase, and operation, catering to the requirements for training specific challenge models. This annotation design aims to offer a multifaceted understanding of surgical protocols, accommodate the nuances of each distinct surgery, and enhance the usability and interpretability of our dataset.
- **Expert-level manual annotation:** the annotation work for OphNet was completed by ten experienced ophthalmologists and five individuals with ophthalmic experience, encompassing, but not limited to, standardization of definitions for surgery, phase, and operation labels, video filtering, classification and localization annotations, and secondary verification, among others. Expert-level annotators ensure the quality and professionalism of OphNet.

2 Related Work

Surgical Workflow Understanding. Beyond its therapeutic advantages, minimally invasive surgery also provides the capability for operative video recording. These videos can be stored and later utilized for various purposes such as cognitive training, skill assessment, and surgical workflow analysis [76,77]. Techniques derived from the broader field of video content analysis and representation are increasingly being incorporated into the surgical realm [6,36,74]. A typical surgical workflow can be defined by a sequence of tasks or events, including patient positioning, incision, dissection, and suturing. These events are influenced not only by the specific type of surgery but also by the individual surgeon's proficiency and technique. Consequently, a comprehensive understanding of the surgical workflow necessitates a thorough examination of the temporal, spatial, and contextual facets of these tasks.

textbfWeakly-Supervised Video Learning. Substantial pioneering work has been undertaken in the realms of video understanding [5,16,33,53,69,71,72,79]. Since even a small amount of videos easily comprises several million frames, methods that do not rely on a frame-level annotation are of special importance. Weakly-supervised video learning makes use of loosely labeled data to train models, thereby obviating the necessity for exhaustively annotated training data. These weak labels might manifest in a variety of forms, including video-level labels or partial labels, which are less specific than frame-level or pixel-level annotations. The objective of weakly-supervised learning is to utilize these coarse labels to produce models capable of delivering fine-grained predictions, such as temporally precise action localization or detailed semantic segmentation [15,35]. The large-scale annotation of surgical videos demands a significant investment of invaluable medical time resources. In this context, weakly supervised learning emerges as a viable solution to this bottleneck.

3 Dataset Construction

In this section, we detail the construction of our dataset, which involved meticulous data collection and preprocessing. YouTube was leveraged as a primary source to circumvent privacy issues while ensuring a broad representation of ophthalmic surgeries. Our selection criteria aimed to capture a wide range of video qualities and styles, specifically targeting cataract, glaucoma, and corneal surgeries due to their prevalence in clinical settings. We refined the dataset by excluding videos of inadequate quality or those depicting non-human subjects. The annotation process was designed to reflect the complex nature of eye surgeries, incorporating hierarchical classification to account for multiple conditions often treated within a single procedure. This was complemented by detailed localization annotations, delineating the distinct phases and techniques characteristic of ophthalmic operations, all undertaken by a dedicated team of ophthalmologists to ensure accuracy and relevance.

3.1 Data Collection and Preprocessing

Collection. Medical data exhibit unique privacy considerations and tend to be of smaller data volumes, characteristics that are particularly salient in the case of video data. By capitalizing on the wealth of surgical videos available on YouTube, we are able to obviate potential ethical and privacy concerns while concurrently enabling the rapid procurement of a substantial corpus of videos

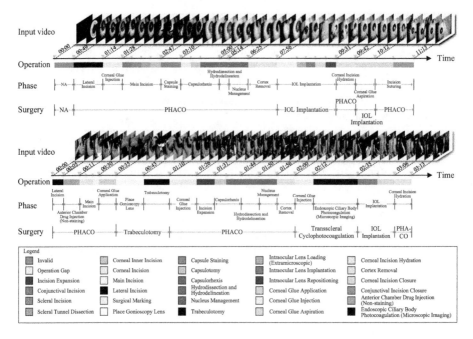

Fig. 1. The figure shows two combined surgical videos, *PHACO + IOL implantation* and *PHACO + Trabeculotomy + Transscleral Cyclophotocoagulation + IOL implantation*. For each frame marked in color, we provide time-boundary annotations at surgical, phase and operation levels.

for further screening [1,2,17,29]. To fulfill this objective, we deploy text-based search algorithms to probe each surgery on YouTube, obtaining videos with titles that incorporate the requisite surgical keywords. We select cataract, glaucoma, and corneal surgery-three of the most commonly performed ophthalmic surgeries in actual clinical environments-as the central subjects of our research. To expand our video collection, we bolster our search queries by integrating synonyms and abbreviations associated with each type of ophthalmic surgery. For instance, *cataract surgery* encompasses various types: *Phacoemulsification (abbr. PHACO), Intraocular Lens implantation (abbr. IOL)*, and *Extracapsular Cataract Extraction (abbr. ECCE)*, etc.

Preprocessing. Given the nature of text-based retrieval and the varying quality, style, and filming methods of surgical videos influenced by different YouTube sources, some videos may exhibit poor quality or deviations in surgical representation. We filter out low-resolution videos, black-and-white color schemes, and animated demonstrations. Furthermore, surgical videos featuring non-human eyes, such as pig eyes, rabbit eyes, or pseudo-human eyes, are also outside the scope of our annotation. The data preprocessing was jointly completed by five professionals with experience in ophthalmology for initial screening, followed by a review conducted by six attending ophthalmologists.

3.2 Data Annotation

Classification Annotation. Different from the single-label classification in most natural video datasets [17,25,58,60,80], in ophthalmic surgery, physicians typically consider and implement multiple types of surgeries and phases based on the specific conditions and needs of the patient. This multifaceted approach stems from the complexity of the eye as an organ, where numerous diseases often coexist. Consequently, multiple ocular problems may need to be addressed in a single surgery. As shown in Fig. 1, a patient with cataracts may also have glaucoma, necessitating the simultaneous treatment of both conditions within one surgery. Hence, we initially classified all videos based on the primary surgical categories into three principal groups: cataract, glaucoma, and corneal surgeries. Subsequently, these were further allocated for the detailed annotation of primary and secondary surgery. It is important to note that there exists only a single type of primary surgery, whereas multiple secondary surgery may be present. The categorization and annotation were meticulously completed by a team comprised of 8 experienced ophthalmologists.

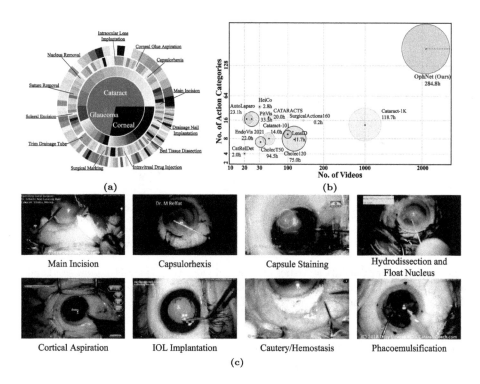

Fig. 2. OphNet's composition, comparison with other datasets for the same task, and some phase examples: (a) an overview of the composition ratios at the levels of surgery, phase, and operation; (b) comparison among existing open-source ⬤ laparoscopic & endoscopic, and ◯ ophthalmic microscope workflow analysis video datasets and ⬤ our OphNet. OphNet stands as the largest real-world video dataset for ophthalmic surgical workflow understanding, featuring the highest number of videos, longest duration, and diverse categories of surgeries and phases; (c) eight phase examples in OphNet.

Hierarchical Localization Annotation. A single surgery is often multi-faceted, involving a series of intricate phases, each requiring distinct techniques and instruments, and the transition between surgeries entails high precision and coordination. Routine cataract surgery involves several steps. It begins with the administration of anesthesia, followed by a small incision in the cornea. The surgeon then creates an opening in the lens capsule and uses ultrasonic vibrations to break up and remove the cataract. Afterward, an artificial intraocular lens (IOL) is inserted into the lens capsule, and the incision is sealed without stitches. We define the phases and operation for various surgeries based on the textbook *Ophthalmic Surgery: Principles and Practice* [61]. To ensure the quality of localization annotations, we assign each annotator videos of 2 different types of primary surgeries. For each action time-boundary annotation, we annotate at three different levels of granularity: surgery, phase, and operation. We also employed the complete linkage algorithm [13] to cluster and merge various temporal boundaries into stable boundaries that received multiple agreements. It's worth noting that an individual video may have multiple separate instances of the same or different phases, thereby leading to multiple boundary definitions. A team of 15 ophthalmologists completes the localization annotation.

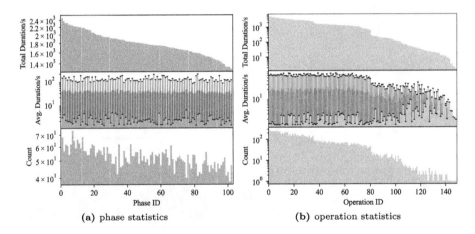

Fig. 3. We present the data statistics of trimmed videos at the levels of phase and operation, including the number of trimmed videos, average duration, and total duration. The IDs and corresponding names can be found in the appendix.

3.3 Dataset Statistics and Analysis

OphNet includes 2,278 surgical videos (284.8 h), demonstrating 66 different types of ophthalmic surgeries: 13 types of cataract surgery, 14 types of glaucoma surgery, and 39 types of corneal surgery. There are 102 phases and 150 operations for recognition, detection and prediction tasks, summarized in Table 1. Over 77% of videos have high-definition resolutions of 1280 × 720 pixels or higher. To facilitate algorithm development and evaluation, we selected 523 videos for localization annotations. Additionally, we trim videos according to annotated action boundaries, resulting in 7,320 phase instances and 9,795 operation instances, totaling 51.2 h. The average duration of trimmed videos is 32 s, while untrimmed videos average 337 s (Fig. 3).

4 Experiments

In our study, we explore four potential tasks using the OphNet dataset: 1) primary surgery presence recognition, 2) phase and operation recognition, 3) phase localization, and 4) phase anticipation. To establish robust baselines, we employ state-of-the-art models known for their effectiveness in human action recognition, detection and anticipation. For each task, we provide a detailed problem formulation and evaluate the baseline models' performance. Our findings offer valuable insights into video understanding within sequences and fine granularity, contributing to the field's knowledge of video tasks in medical contexts.

4.1 Surgery Presence, Phase and Operation Recognition

Task Description. Surgery presence recognition focuses on identifying various surgical types in untrimmed videos through weakly supervised methods. This requires the model to discern and capture distinct surgical action features within extensive video footage. In OphNet, all types of surgeries occurring in each video are annotated, but only a subset of the videos have time-boundary annotations, as detailed in Sect. 3.2. We identify the primary surgery based on the key objectives and durations of the surgeries. To streamline the process, our experiments are limited to recognizing the presence of these primary surgeries. Additionally, phase recognition segments the surgery into distinct phases using visual cues and movements, such as incision, lens removal, and implantation. Operation recognition involves identifying finer-grained surgical actions.

Baselines. We compare the performance of I3D [10], SlowFast [19], X3D [18], and MViT V2 [31] models on this task. These models are evaluated in two versions: 1) random initialization training and 2) pre-training with weights from Kinetics 400 [29], which is a human action recognition dataset. In addition, we also explored the classification performance of X-CLIP [42] and ViFi-CLIP [49], two CLIP [48]-based models. We also compared the effects of different numbers of input frames on the models' perfor-

Table 2. Per-class Top-1 and Top-5 accuracy (%) for the primary surgery presence recognition on untrimmed videos. The best performance for each split has been highlighted in **bold**.

Baselines	Primary Surgery Classification							
	Cataract		Glaucoma		Cornea		All	
	Top-1	Top-5	Top-1	Top-5	Top-1	Top-5	Top-1	Top-5
I3D [10]	38.9	86.3	43.6	71.7	42.7	78.2	29.8	53.2
SlowFast [19]	45.3	82.1	44.6	72.3	45.5	77.3	27.2	54.4
X3D [31]	42.1	87.4	44.6	74.2	43.8	76.6	28.5	62.7
MViT V2 [18]	43.2	84.2	45.5	81.8	45.5	75.9	29.1	60.1
I3D [10]*	36.8	84.7	48.2	81.8	48.6	76.5	27.2	50.6
SlowFast*	49.0	83.7	47.3	80.9	49.2	75.6	27.2	50.6
X3D*	47.4	86.3	46.4	81.5	48.3	78.2	35.4	61.4
MViT V2*	44.2	85.3	49.1	81.8	47.7	77.3	28.5	63.3
X-CLIP$_{16}$ [42]	58.5	**94.7**	51.8	**92.8**	61.4	**88.6**	40.5	79.0
X-CLIP$_{32}$	**60.6**	92.6	**53.5**	83.7	56.8	84.1	**58.9**	**81.0**
ViFi-CLIP$_{16}$ [49]	59.6	88.3	52.4	80.2	61.4	81.8	58.9	79.8
ViFi-CLIP$_{32}$	59.6	88.3	51.5	84.8	50.0	75.0	56.3	77.2

pared the effects of different numbers of input frames on the models' performance, where the subscript $_{16}$ represents an input of 16 frames, and $_{32}$ represents an input of 32 frames.

Setup. The dataset was randomly partitioned to ensure a balanced representation of examples for each surgery category. Specifically, we allocated 70% of the data for training (1,449 surgical videos), 10% for validation (205 surgical videos), and 20% for testing (424 surgical videos). For the phase and operation recognition experiments, we followed the same settings, with 70% of the data used for training (5,024 phase segments, 6,856 operations segments), 10% for validation (730 phase segments, 975 operations segments), and 20% for testing (1,566 phase segments, 1,964 operations segments). The input for the surgery presence recognition experiment is untrimmed videos, while for the phase and operation recognition experiments, the input consists of trimmed segments. In all classification experiments, we set up the analysis and comparison from four perspectives: cataract surgery, glaucoma surgery, corneal surgery, and all surgical videos.

Results. We summarize the results in Tables 2 and 3. In primary surgery classification, X-CLIP [42] achieved the highest overall Top-1 accuracy at 58.9%, leading in cataract and glaucoma surgeries with accuracies of 60.6% and 53.5% respectively. For corneal surgeries, X-CLIP also recorded the highest Top-1 and Top-5 accuracies of 61.4% and 88.6%. In phase classification, ViFi-CLIP [49] led with the highest Top-1 accuracy, particularly in cataract surgeries at 75.9%, and exhibited the best performance in both glaucoma and corneal surgeries. Furthermore, in operation classification, ViFi-CLIP outperformed all other models in all categories, especially noted in cataract and corneal surgeries with Top-1 accuracies of 75.1% and 83.7% and Top-5 accuracies of 93.8% and 85.2% respectively. Overall, ViFi-CLIP showed superior performance in both phase and operation classifications across various surgery types. Besides, in the phase and operation classification experiments, a higher number of input frames generally had a positive effect on the model. In Fig. 4, we present the heatmap visualization of four examples from the test set of phase recognition experiments using the ViFi-CLIP model. It can be observed that in a series of frame images, the model focuses on surgical instruments and the operated eye area, which is consistent with human experience. The instruments used in the same phase of ophthalmic surgery are often similar.

4.2 Phase Localization

Task Description. Phase localization in ophthalmic surgical workflow analysis refers to the task of pinpointing the exact moments or time intervals within a surgical video where specific phases of the surgery begin and end. This involves the detailed temporal segmentation of the entire surgical procedure into its constituent phases based on visual cues, surgeon's actions, and the progression of the surgery. The objective of phase localization is to accurately identify the start and end times of different surgical stages, such as pre-operative preparation, incision, and removal of the lens facilitating a granular and precise understanding of the surgery timeline. This task is crucial for detailed surgical documentation, efficient surgical training, and the development of targeted interventions during specific stages of the surgery, enhancing overall surgical management and post-operative analysis.

Table 3. Per-class Top-1 and Top-5 accuracy (%) for the primary surgery presence recognition on untrimmed videos and phase recognition on trimmed videos. * denotes the initialization from the model pre-trained on Kinetics 400 [29]. For the two CLIP models, we chose ViT-B/16 as the backbone and compared the performance of two different input frame numbers, 16 and 32. The best performance for each split has been highlighted in **bold**.

Baselines	Phase Classification								Operation Classification							
	Cataract		Glaucoma		Cornea		All		Cataract		Glaucoma		Cornea		All	
	Top-1	Top-5	Top-1	Top-5	Top-1	Top-5	Top-1	Top-5	Top-1	Top-5	Top-1	Top-5	Top-1	Top-5	Top-1	Top-5
I3D [10]	27.2	55.7	24.1	57.5	18.9	52.1	25.7	58.2	26.8	54.9	23.5	56.0	18.0	51.2	25.0	57.1
SlowFast [19]	26.5	56.5	23.1	56.5	24.2	49.1	26.7	60.1	25.8	55.2	22.9	45.9	23.5	48.5	26.0	59.0
X3D [31]	27.0	58.3	21.0	55.5	21.8	28.5	26.6	62.3	26.4	47.2	20.5	44.6	21.3	27.8	26.1	61.5
MViT V2 [18]	26.2	54.9	21.0	53.4	26.0	46.8	27.0	59.8	25.9	43.8	20.5	42.7	25.5	45.9	26.5	58.9
I3D*	29.5	68.9	22.1	58.6	26.0	50.9	30.2	71.2	28.8	67.5	21.8	47.9	25.7	50.3	29.5	60.0
SlowFast*	30.6	72.3	25.2	54.7	30.7	59.8	31.7	61.8	29.9	71.1	24.8	43.9	29.5	58.7	30.5	60.9
X3D*	27.2	72.9	22.1	59.6	30.7	61.5	33.5	63.2	26.5	71.8	21.7	48.8	29.9	60.2	32.8	62.1
MViT V2*	34.2	76.5	23.3	52.0	38.4	65.1	28.3	60.2	33.5	75.2	22.8	41.5	37.9	64.0	27.8	59.5
X-CLIP$_{16}$ [42]	68.3	92.2	47.3	89.8	53.0	77.4	63.4	85.3	67.5	91.0	46.5	78.9	82.2	76.1	62.5	84.0
X-CLIP$_{32}$	69.1	**94.0**	48.7	81.7	54.8	80.4	62.7	85.8	68.0	93.0	**47.9**	**80.5**	84.0	79.5	62.0	84.7
ViFi-CLIP$_{16}$ [49]	**75.9**	93.7	40.4	85.4	66.6	81.6	66.1	**88.4**	74.5	92.5	42.8	74.5	**85.0**	80.5	**65.0**	**87.5**
ViFi-CLIP$_{32}$	73.0	92.9	**49.6**	**82.7**	**57.7**	**81.6**	**68.4**	87.2	**75.1**	**93.8**	43.2	80.2	83.7	**85.2**	64.8	86.5

Table 4. The results for phase detection. ActionFormer and TriDet are state-of-the-art models for human action detection tasks, and we use three different backbones for feature extraction and report mAP at the IoU thresholds of [0.1:0.2:0.9]. Average mAP is computed by averaging different IoU thresholds. The best performance for each split has been highlighted in **bold**.

Baselines	Backbones	mAP (%)				
		0.1	0.3	0.5	0.7	Avg.
ActionFormer [79]	CSN [64]	53.7	50.1	40.6	24.5	42.5
	SwinViviT [34]	59.3	54.7	43.3	26.3	46.4
	SlowFast [19]	60.0	55.9	45.1	26.0	47.5
TriDet [56]	CSN	56.1	53.0	43.1	29.4	46.2
	SwinViviT	61.0	**57.1**	**47.1**	**33.1**	**50.4**
	SlowFast	**61.3**	56.0	45.6	30.4	48.6

Table 5. The results for phase anticipation. We report top-1 accuracy at the observation ratios [0.1:0.2:0.9]. Average top-1 accuracy is computed by averaging different observation ratios. The best performance for each split has been highlighted in **bold**.

Baselines	Top-1 Acc. (%)				
	0.1	0.3	0.5	0.7	Avg.
I3D [10]	26.5	42.2	49.8	51.3	47.3
SlowFast [19]	25.4	42.6	48.9	52.2	47.2
MViT V2 [18]	25.6	43.7	49.3	52.3	47.5
I3D*	27.3	43.5	50.1	51.4	47.6
SlowFast*	27.5	43.2	49.9	**52.3**	47.8
MViT V2*	**27.8**	**43.8**	**50.5**	51.7	**48.2**

Baselines. We conducted the experiments using phase-level labels and used two baseline models, ActionFormer [79] and TriDet [56], with backbone networks configured as CSN [64], SwinViviT [34], and SlowFast [19]. Data split follows the setup of the primary surgery classification experiment.

Setup. We excluded *Operation Gap* and *Invalid*, and filtered out tags with fewer than 20 segments. We used two baseline models, ActionFormer [79] and TriDet [56], with backbone networks configured as CSN [64], SwinViviT [34], and SlowFast [19]. To extract features from the videos, we first extracted RGB frames from each video at a rate of 25 frames per second. We also extracted optical flow using the TV-L1 [26,37] algorithm. We then fine-tuned an I3D [10] model that had been pre-trained on the ImageNet [14] dataset, and used it to generate features for each RGB and optical flow frame. Because each video has a variable duration, we performed uniform interpolation to generate 100 fixed-length features for each video. Finally, we concatenated the RGB and optical flow features into a 2048-dimensional embedding, which served as the input for our model.

Results. The experimental results for phase localization are presented in the Table 4, showcasing the performance of different baseline models with various backbones in terms of mean Average Precision (mAP) at different Intersection over Union (IoU) thresholds [0.1:0.2:0.9]. The models evaluated include Action-Former with SwinViviT and SlowFast backbones, and TriDet with CSN, SwinViviT, and SlowFast backbones. The results indicate that the TriDet model with a SwinViviT backbone outperforms other combinations, achieving the highest mAP scores across most IoU thresholds, with notable scores of 61.0% (IoU = 0.1), 57.1% (IoU = 0.3), 47.1% (IoU = 0.5), and 33.1% (IoU = 0.7), resulting in an average mAP of 50.4%. This indicates that the TriDet model, especially when combined with the SwinViviT backbone, is particularly effective for phase localization in surgical videos (Fig. 5). On the other hand, the TriDet model with a SlowFast backbone shows competitive performance, particularly achieving the highest mAP of 61.3% at the lowest IoU threshold (0.1). However, it falls slightly behind in performance at higher IoU thresholds compared to the SwinViviT backbone.

4.3 Phase Anticipation

Task Description. This task requires the analysis of real-time or recorded video data to foresee the sequence of events based on current and past surgical activities. By understanding the typical progression of ophthalmic surgeries and recognizing patterns in the surgeon's actions and the use of instruments, the system aims to forecast the next phase of the surgery, allowing for proactive preparation and response. The objective of phase anticipation is to enhance the efficiency and safety of surgical procedures by providing the surgical team with advanced notice of upcoming steps, enabling better resource allocation, timing for critical tasks, and overall coordination within the operating room.

Setup. Following previous approaches of the primary surgery classification experiment in Sect. 4.1, We randomly mask phase sequences in the test video with different observation ratios.

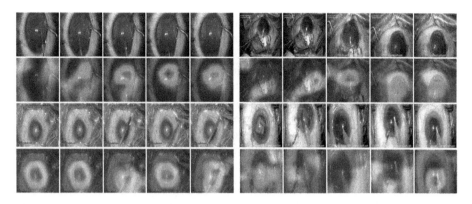

Fig. 4. Attention map visualizations of ViFi-CLIP [49] on four examples from OphNet's test set in the phase recognition task.

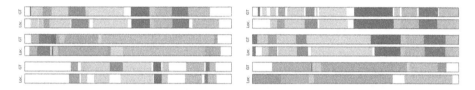

Fig. 5. Phase localization visualization of TriDet [56]. *GT* represents the ground truth visualization for phases, while *Loc.* visualizes the model's highest confidence phase category and the time-boundary results. Blank segments denote invalid segments or operation gaps.

Baselines. We evaluate our datasets with the baseline models such as I3D [10], SlowFast [19], and MViT V2 [31]. For each model, we also adopted two training approaches: random initialization training and using pre-trained weights from Kinetics 400 [29].

Results. The phase detection results are illustrated in Table 5. The results demonstrate that the baseline models pretrained on Kinetics 400 [29] generally outperform their original counterparts in terms of Top-1 accuracy across different observation ratios for phase anticipation. Specifically, the modified MViT V2* model exhibits the highest improvement, achieving the best average Top-1 accuracy of 48.2%. Moreover, while all models show increased accuracy with higher observation ratios, indicating that more observed data contributes to better performance, the consistent improvement across all ratios for the enhanced models suggests effective modifications.

5 Limitations

Dataset Bias. OphNet's videos are sourced from YouTube and exhibit diverse styles, clarity, and screen elements. This diversity can aid detection models in generalization but may affect their effectiveness and performance. Some videos in the dataset include subtitles or additional video windows, such as a little subtitle or watermark shown in Fig. 2. Similarly, additional video windows offer another perspective but can make the scene chaotic, making it harder to recognize primary surgical actions. The presence of these factors in OphNet reflects the complexity of real-world surgical environments, because an ophthalmic microscope may inherently display different windows or show parameters during recording. While they pose challenges, they also present opportunities for developing models that can better handle variability and unpredictability, which are crucial aspects of real-world surgical scenarios.

Annotation Bias. OphNet is entirely annotated by ophthalmologists, there is a distinct possibility of annotation bias reflecting specific regional practices, terminologies, and interpretations. Despite the universal nature of many ophthalmic procedures, subtle differences in surgical techniques, procedural preferences, and clinical terminologies could lead to inconsistencies in how surgeries are categorized and described across different regions. For instance, the terminology used to describe certain procedures might differ, with one region referring to a procedure as *anterior vitrectomy* while another uses *pars plana vitrectomy*. To reduce the possibility of biases in precise annotations, we have taken great care to establish a unified definition prior to describing the surgery, phase and operation. However, potential biases arising from regional variations and individual surgical practices are inevitable.

6 Conclusion

In response to the current challenges in ophthalmology, a surgical field apt for automation and remote control, we introduce OphNet, a large-scale, diverse, and expert-level video benchmark for understanding ophthalmic surgical workflows. OphNet is the most extensive dataset of its kind, containing a broad range of cataract, glaucoma, and corneal surgeries and detailed annotations for distinct surgical phases. OphNet comprises 2,278 surgical videos (284.8 h), 7,320 phase segments and 9,795 operation segments (51.2 h), showcasing 66 different types of ophthalmic surgeries: 13 cataract, 14 glaucoma, and 39 corneal. It is annotated with 102 phases and 150 operations. With OphNet, we explored primary surgery presence recognition, phase localization and phase anticipation on untrimmed videos, phase and operation recognition on trimmed videos. We employed state-of-the-art models to establish robust baselines and provided valuable insights into video understanding within sequences and fine granularity. Our work contributes to the broader understanding of surgical video tasks in medical contexts and promotes the integration of deep learning technologies into ophthalmic surgical procedures.

References

1. Fair use on Youtube. https://support.google.com/youtube/answer/9783148?hl=en#:~:text=If%20the%20use%20of%20copyright,copyright%20removal%20request%20to%20YouTube
2. Youtube's copyright exception policy. https://www.youtube.com/howyoutubeworks/policies/copyright/#copyright-exceptions
3. Adrito, D., et al.: PitVis: workflow recognition in endoscopic pituitary surgery
4. Al Hajj, H., et al.: CATARACTS: challenge on automatic tool annotation for cataract surgery. Med. Image Anal. **52**, 24–41 (2019). https://doi.org/10.1016/j.media.2018.11.008. https://www.sciencedirect.com/science/article/pii/S136184151830865X
5. Alwassel, H., Giancola, S., Ghanem, B.: TSP: temporally-sensitive pretraining of video encoders for localization tasks. In: Proceedings of the IEEE/CVF International Conference on Computer Vision Workshops, pp. 3173–3183 (2021)
6. Bar, O., et al.: Impact of data on generalization of AI for surgical intelligence applications. Sci. Rep. **10**(1), 22208 (2020). https://doi.org/10.1038/s41598-020-79173-6
7. Bernal, J., Sánchez, F.J., Fernández-Esparrach, G., Gil, D., Rodríguez, C., Vilariño, F.: WM-DOVA maps for accurate polyp highlighting in colonoscopy: validation vs. saliency maps from physicians. Comput. Med. Imag. Graph. **43**, 99–111 (2015). https://doi.org/10.1016/j.compmedimag.2015.02.007. https://www.sciencedirect.com/science/article/pii/S0895611115000567
8. Bodenstedt, S., et al.: Unsupervised temporal context learning using convolutional neural networks for laparoscopic workflow analysis. arXiv preprint arXiv:1702.03684 (2017)
9. Borgli, H., et al.: HyperKvasir, a comprehensive multi-class image and video dataset for gastrointestinal endoscopy. Sci. Data **7**(1), 283 (2020). https://doi.org/10.1038/s41597-020-00622-y
10. Carreira, J., Zisserman, A.: Quo Vadis, action recognition? A new model and the kinetics dataset. In: Proceedings of the IEEE Conference on Computer Vision and Pattern Recognition, pp. 6299–6308 (2017)
11. Czempiel, T., et al.: TeCNO: surgical phase recognition with multi-stage temporal convolutional networks. In: Martel, A.L., et al. (eds.) MICCAI 2020. LNCS, vol. 12263, pp. 343–352. Springer, Cham (2020). https://doi.org/10.1007/978-3-030-59716-0_33
12. Czempiel, T., Paschali, M., Ostler, D., Kim, S.T., Busam, B., Navab, N.: OperA: attention-regularized transformers for surgical phase recognition. In: de Bruijne, M., et al. (eds.) MICCAI 2021, Part IV 24. LNCS, vol. 12904, pp. 604–614. Springer, Cham (2021). https://doi.org/10.1007/978-3-030-87202-1_58
13. Defays, D.: An efficient algorithm for a complete link method. Comput. J. **20**(4), 364–366 (1977). https://doi.org/10.1093/comjnl/20.4.364
14. Deng, J., Dong, W., Socher, R., Li, L.J., Li, K., Fei-Fei, L.: ImageNet: a large-scale hierarchical image database. In: 2009 IEEE Conference on Computer Vision and Pattern Recognition, pp. 248–255. IEEE (2009)
15. Dong, S., Hu, H., Lian, D., Luo, W., Qian, Y., Gao, S.: Weakly supervised video representation learning with unaligned text for sequential videos. In: Proceedings of the IEEE/CVF Conference on Computer Vision and Pattern Recognition, pp. 2437–2447 (2023)

16. Duong, H.T., Le, V.T., Hoang, V.T.: Deep learning-based anomaly detection in video surveillance: a survey. Sensors **23**(11) (2023). https://doi.org/10.3390/s23115024. https://www.mdpi.com/1424-8220/23/11/5024
17. Heilbron, F.C., Escorcia, V., Ghanem, B., Niebles, J.C.: ActivityNet: a large-scale video benchmark for human activity understanding. In: Proceedings of the IEEE Conference on Computer Vision and Pattern Recognition, pp. 961–970 (2015)
18. Feichtenhofer, C.: X3D: expanding architectures for efficient video recognition. In: Proceedings of the IEEE/CVF Conference on Computer Vision and Pattern Recognition, pp. 203–213 (2020)
19. Feichtenhofer, C., Fan, H., Malik, J., He, K.: SlowFast networks for video recognition. In: Proceedings of the IEEE/CVF International Conference on Computer Vision, pp. 6202–6211 (2019)
20. Forslund Jacobsen, M., Konge, L., Alberti, M., la Cour, M., Park, Y.S., Thomsen, A.S.S.: Robot-assisted vitreoretinal surgery improves surgical accuracy compared with manual surgery: a randomized trial in a simulated setting. Retina **40**(11), 2091–2098 (2020)
21. Ghamsarian, N., et al.: Cataract-1K: cataract surgery dataset for scene segmentation, phase recognition, and irregularity detection. arXiv preprint arXiv:2312.06295 (2023)
22. Ghamsarian, N., Taschwer, M., Putzgruber-Adamitsch, D., Sarny, S., El-Shabrawi, Y., Schoeffmann, K.: LensID: a CNN-RNN-based framework towards lens irregularity detection in cataract surgery videos. In: de Bruijne, M., et al. (eds.) MICCAI 2021. LNCS, vol. 12908, pp. 76–86. Springer, Cham (2021). https://doi.org/10.1007/978-3-030-87237-3_8
23. Ghamsarian, N., Taschwer, M., Putzgruber-Adamitsch, D., Sarny, S., Schoeffmann, K.: Relevance detection in cataract surgery videos by spatio-temporal action localization. In: 25th International Conference on Pattern Recognition, ICPR 2020, Virtual Event, Milan, Italy, 10–15 January 2021, pp. 10720–10727. IEEE (2020). https://doi.org/10.1109/ICPR48806.2021.9412525
24. Grammatikopoulou, M., et al.: CaDIS: cataract dataset for image segmentation. arXiv preprint arXiv:1906.11586 (2019)
25. Gu, C., et al.: AVA: a video dataset of spatio-temporally localized atomic visual actions (2018)
26. Horn, B.K., Schunck, B.G.: Determining optical flow. Artif. Intell. **17**(1–3), 185–203 (1981)
27. Ji, G.P., et al.: Video polyp segmentation: a deep learning perspective. Mach. Intell. Res. **19**(6), 531–549 (2022). https://doi.org/10.1007/s11633-022-1371-y
28. Jin, Y., et al.: SV-RCNet: workflow recognition from surgical videos using recurrent convolutional network. IEEE Trans. Med. Imaging **37**(5), 1114–1126 (2018)
29. Kay, W., et al.: The kinetics human action video dataset. arXiv preprint arXiv:1705.06950 (2017)
30. Li, J., et al.: Imitation learning from expert video data for dissection trajectory prediction in endoscopic surgical procedure. In: Greenspan, H., et al. (eds.) Medical Image Computing and Computer Assisted Intervention, MICCAI 2023. LNCS, vol. 14228, pp. 494–504. Springer, Cham (2023). https://doi.org/10.1007/978-3-031-43996-4_47
31. Li, Y., et al.: MViTv2: improved multiscale vision transformers for classification and detection. In: Proceedings of the IEEE/CVF Conference on Computer Vision and Pattern Recognition, pp. 4804–4814 (2022)

32. Lin, S., et al.: Semantic-super: a semantic-aware surgical perception framework for endoscopic tissue identification, reconstruction, and tracking. In: 2023 IEEE International Conference on Robotics and Automation (ICRA), pp. 4739–4746 (2023). https://doi.org/10.1109/ICRA48891.2023.10160746
33. Lin, T., Liu, X., Li, X., Ding, E., Wen, S.: BMN: boundary-matching network for temporal action proposal generation. In: Proceedings of the IEEE/CVF International Conference on Computer Vision, pp. 3889–3898 (2019)
34. Liu, Z., et al.: Video swin transformer. arXiv preprint arXiv:2106.13230 (2021)
35. Long, F., Yao, T., Qiu, Z., Tian, X., Luo, J., Mei, T.: Bi-calibration networks for weakly-supervised video representation learning. Int. J. Comput. Vis. **131**(7), 1704–1721 (2023). https://doi.org/10.1007/s11263-023-01779-w
36. Loukas, C.: Video content analysis of surgical procedures. Surg. Endosc. **32**(2), 553–568 (2018). https://doi.org/10.1007/s00464-017-5878-1
37. Lucas, B.D., Kanade, T., et al.: An iterative image registration technique with an application to stereo vision, vol. 81, Vancouver (1981)
38. Ma, Y., Chen, X., Cheng, K., Li, Y., Sun, B.: LDPolypVideo benchmark: a large-scale colonoscopy video dataset of diverse polyps. In: de Bruijne, M., et al. (eds.) MICCAI 2021. LNCS, vol. 12905, pp. 387–396. Springer, Cham (2021). https://doi.org/10.1007/978-3-030-87240-3_37
39. Maier-Hein, L., et al.: Heidelberg colorectal data set for surgical data science in the sensor operating room. Sci. Data **8**(1), 101 (2021)
40. Mesejo, P., et al.: Computer-aided classification of gastrointestinal lesions in regular colonoscopy. IEEE Trans. Med. Imaging **35**(9), 2051–2063 (2016). https://doi.org/10.1109/TMI.2016.2547947
41. Ming, H., et al.: NurViD: a large expert-level video database for nursing procedure activity understanding. In: Thirty-Seventh Conference on Neural Information Processing Systems Datasets and Benchmarks Track (2023)
42. Ni, B., et al.: Expanding language-image pretrained models for general video recognition (2022)
43. Ni, Z.-L., et al.: RAUNet: residual attention U-Net for semantic segmentation of cataract surgical instruments. In: Gedeon, T., Wong, K.W., Lee, M. (eds.) ICONIP 2019. LNCS, vol. 11954, pp. 139–149. Springer, Cham (2019). https://doi.org/10.1007/978-3-030-36711-4_13
44. Nwoye, C.I., et al.: Recognition of instrument-tissue interactions in endoscopic videos via action triplets. In: Martel, A.L., et al. (eds.) MICCAI 2020, Part III 23. LNCS, vol. 12263, pp. 364–374. Springer, Cham (2020). https://doi.org/10.1007/978-3-030-59716-0_35
45. Nwoye, C.I., Padoy, N.: Data splits and metrics for benchmarking methods on surgical action triplet datasets. arXiv preprint arXiv:2204.05235 (2022)
46. Nwoye, C.I., et al.: Rendezvous: attention mechanisms for the recognition of surgical action triplets in endoscopic videos. Med. Image Anal. **78**, 102433 (2022)
47. Pan, X., Gao, X., Wang, H., Zhang, W., Mu, Y., He, X.: Temporal-based Swin Transformer network for workflow recognition of surgical video. Int. J. Comput. Assist. Radiol. Surg. **18**(1), 139–147 (2023). https://doi.org/10.1007/s11548-022-02785-y
48. Radford, A., et al.: Learning transferable visual models from natural language supervision. In: International Conference on Machine Learning, pp. 8748–8763. PMLR (2021)
49. Rasheed, H., Khattak, M.U., Maaz, M., Khan, S., Khan, F.S.: Fine-tuned CLIP models are efficient video learners. In: The IEEE/CVF Conference on Computer Vision and Pattern Recognition (2023)

50. Ross, T., et al.: Comparative validation of multi-instance instrument segmentation in endoscopy: results of the ROBUST-MIS 2019 challenge. Med. Image Anal. **70**, 101920 (2021)
51. Ross, T., et al.: Exploiting the potential of unlabeled endoscopic video data with self-supervised learning. Int. J. Comput. Assist. Radiol. Surg. **13**, 925–933 (2018)
52. Roß, T., et al.: Comparative validation of multi-instance instrument segmentation in endoscopy: results of the ROBUST-MIS 2019 challenge. Med. Image Anal. **70**, 101920 (2021). https://doi.org/10.1016/j.media.2020.101920. https://www.sciencedirect.com/science/article/pii/S136184152030284X
53. Sato, F., Hachiuma, R., Sekii, T.: Prompt-guided zero-shot anomaly action recognition using pretrained deep skeleton features. In: Proceedings of the IEEE/CVF Conference on Computer Vision and Pattern Recognition, pp. 6471–6480 (2023)
54. Schoeffmann, K., Husslein, H., Kletz, S., Petscharnig, S., Münzer, B., Beecks, C.: Video retrieval in laparoscopic video recordings with dynamic content descriptors. Multim. Tools Appl. **77**(13), 16813–16832 (2018). https://doi.org/10.1007/s11042-017-5252-2
55. Schoeffmann, K., Taschwer, M., Sarny, S., Münzer, B., Primus, M.J., Putzgruber, D.: Cataract-101: video dataset of 101 cataract surgeries. In: César, P., Zink, M., Murray, N. (eds.) Proceedings of the 9th ACM Multimedia Systems Conference, MMSys 2018, Amsterdam, The Netherlands, 12–15 June 2018, pp. 421–425. ACM (2018). https://doi.org/10.1145/3204949.3208137
56. Shi, D., Zhong, Y., Cao, Q., Ma, L., Li, J., Tao, D.: TriDet: temporal action detection with relative boundary modeling. In: Proceedings of the IEEE/CVF Conference on Computer Vision and Pattern Recognition, pp. 18857–18866 (2023)
57. Shi, X., Jin, Y., Dou, Q., Heng, P.A.: LRTD: long-range temporal dependency based active learning for surgical workflow recognition. Int. J. Comput. Assist. Radiol. Surg. **15**(9), 1573–1584 (2020)
58. Sigurdsson, G.A., Varol, G., Wang, X., Farhadi, A., Laptev, I., Gupta, A.: Hollywood in homes: crowdsourcing data collection for activity understanding. In: Leibe, B., Matas, J., Sebe, N., Welling, M. (eds.) ECCV 2016. LNCS, vol. 9905, pp. 510–526. Springer, Cham (2016). https://doi.org/10.1007/978-3-319-46448-0_31
59. Smedsrud, P.H., et al.: Kvasir-Capsule, a video capsule endoscopy dataset. Sci. Data **8**(1), 142 (2021). https://doi.org/10.1038/s41597-021-00920-z
60. Soomro, K., Zamir, A.R., Shah, M.: UCF101: a dataset of 101 human actions classes from videos in the wild (2012)
61. Spaeth, G., Danesh-Meyer, H., Goldberg, I., Kampik, A.: Ophthalmic Surgery: Principles and Practice. Elsevier Health Sciences (2011). E-Book. https://books.google.com.hk/books?id=wHWMUGH-5csC
62. Stauder, R., Ostler, D., Kranzfelder, M., Koller, S., Feußner, H., Navab, N.: The TUM LapChole dataset for the M2CAI 2016 workflow challenge. arXiv preprint arXiv:1610.09278 (2016)
63. Tian, Y., et al.: Contrastive transformer-based multiple instance learning for weakly supervised polyp frame detection. In: Wang, L., Dou, Q., Fletcher, P.T., Speidel, S., Li, S. (eds.) Medical Image Computing and Computer Assisted Intervention, MICCAI 2022. LNCS, vol. 13433, pp. 88–98. Springer, Cham (2022). https://doi.org/10.1007/978-3-031-16437-8_9
64. Tran, D., Wang, H., Torresani, L., Feiszli, M.: Video classification with channel-separated convolutional networks (2019)
65. Twinanda, A.P., Shehata, S., Mutter, D., Marescaux, J., de Mathelin, M., Padoy, N.: EndoNet: a deep architecture for recognition tasks on laparoscopic videos.

IEEE Trans. Med. Imaging (TMI) **36**, 86–97 (2016). https://api.semanticscholar.org/CorpusID:5633749
66. Twinanda, A.P., Yengera, G., Mutter, D., Marescaux, J., Padoy, N.: RSDNet: learning to predict remaining surgery duration from laparoscopic videos without manual annotations. IEEE Trans. Med. Imaging **38**(4), 1069–1078 (2019). https://doi.org/10.1109/TMI.2018.2878055
67. Wagner, M., et al.: Comparative validation of machine learning algorithms for surgical workflow and skill analysis with the HeiChole benchmark. Med. Image Anal. **86**, 102770 (2023)
68. Wang, T., Li, H., Pu, T., Yang, L.: Microsurgery robots: applications, design, and development. Sensors **23**(20) (2023). https://doi.org/10.3390/s23208503. https://www.mdpi.com/1424-8220/23/20/8503
69. Wang, X., Zhang, S., Qing, Z., Shao, Y., Gao, C., Sang, N.: Self-supervised learning for semi-supervised temporal action proposal. In: CVPR (2021)
70. Wang, Z., et al.: AutoLaparo: a new dataset of integrated multi-tasks for image-guided surgical automation in laparoscopic hysterectomy. In: Wang, L., Dou, Q., Fletcher, P.T., Speidel, S., Li, S. (eds.) Medical Image Computing and Computer Assisted Intervention, MICCAI 2022. LNCS, vol. 13437, pp. 486–496. Springer, Cham (2022). https://doi.org/10.1007/978-3-031-16449-1_46
71. Wu, W., Sun, Z., Ouyang, W.: Revisiting classifier: transferring vision-language models for video recognition (2023)
72. Wu, W., Wang, X., Luo, H., Wang, J., Yang, Y., Ouyang, W.: Bidirectional cross-modal knowledge exploration for video recognition with pre-trained vision-language models. In: Proceedings of the IEEE/CVF Conference on Computer Vision and Pattern Recognition (2023)
73. Yengera, G., Mutter, D., Marescaux, J., Padoy, N.: Less is more: surgical phase recognition with less annotations through self-supervised pre-training of CNN-LSTM networks. arXiv preprint arXiv:1805.08569 (2018)
74. Yu, F., et al.: Assessment of automated identification of phases in videos of cataract surgery using machine learning and deep learning techniques. JAMA Netw. Open **2**(4), e191860–e191860 (2019). https://doi.org/10.1001/jamanetworkopen.2019.1860
75. Yu, T., Mutter, D., Marescaux, J., Padoy, N.: Learning from a tiny dataset of manual annotations: a teacher/student approach for surgical phase recognition. arXiv preprint arXiv:1812.00033 (2018)
76. Yuan, K., Srivastav, V., Navab, N., Padoy, N.: HecVL: hierarchical video-language pretraining for zero-shot surgical phase recognition. arXiv preprint arXiv:2405.10075 (2024)
77. Yuan, K., et al.: Learning multi-modal representations by watching hundreds of surgical video lectures (2023)
78. Zha, R., Cheng, X., Li, H., Harandi, M., Ge, Z.: EndoSurf: neural surface reconstruction of deformable tissues with stereo endoscope videos. In: Greenspan, H., et al. (eds.) Medical Image Computing and Computer Assisted Intervention, MICCAI 2023. LNCS, vol. 14228, pp. 13–23. Springer, Cham (2023). https://doi.org/10.1007/978-3-031-43996-4_2

79. Zhang, CL., Wu, J., Li, Y.: ActionFormer: localizing moments of actions with transformers. In: Avidan, S., Brostow, G., Cissé, M., Farinella, G.M., Hassner, T. (eds.) Computer Vision, ECCV 2022. LNCS, vol. 13664, pp. 492–510. Springer, Cham (2022). https://doi.org/10.1007/978-3-031-19772-7_29
80. Zhou, L., Xu, C., Corso, J.J.: Towards automatic learning of procedures from web instructional videos. In: AAAI Conference on Artificial Intelligence, pp. 7590–7598 (2018). https://www.aaai.org/ocs/index.php/AAAI/AAAI18/paper/view/17344

Author Index

Ai, Rui 390
Anderson, Stuart 206

Bagautdinov, Timur 206
Bi, Mengxiao 300
Black, Michael J. 246

Carneiro, Gustavo 372
Cendra, Fernando Julio 188
Chae, Yeongnam 229
Chan, Antoni B. 19
Chen, Jinyue 408
Chen, Junxi 92
Chen, Xiaokang 1
Chen, Zhaoxi 1
Cheng, Jun 481
Cheng, Xuelian 481
Chitta, Kashyap 57
Christie, Marc 464
Courant, Robin 464
Cui, Ziteng 37

D'Alessandro, Adriano 75
Dai, Bo 112
Dauner, Daniel 57
Deng, Jiajun 131
Do, Thanh-Toan 372
Dong, Jiangxin 336
Dong, Junhao 92
Du, Dalong 131
Dufour, Nicolas 464

Felix, Rafael 372

Garg, Arpit 372
Ge, Zheng 408, 425
Ge, Zongyuan 481
Geiger, Andreas 57
Guo, Qing 319

Hamarneh, Ghassan 75
Han, Chunrui 408
Han, Kai 188
Harada, Tatsuya 37
Hernandez, Jefferson 444
Hong, Fangzhou 112
Hu, Ming 481
Huang, Hui 19

James, Austin 206
Jesslen, Artur 264
Jin, Xin 131, 300

Kalantidis, Yannis 353
Kalogeiton, Vicky 464
Kang, Xueyang 149
Khirodkar, Rawal 206
Khoshelham, Kourosh 149
Kim, Mijung 229
Kong, Lingyu 408, 425
Koniusz, Piotr 92
Kortylewski, Adam 264

Lan, Yushi 112
Larlus, Diane 353
Lee, Sehyung 229
Leitner, Jurgen 481
Li, Bohan 131
Li, Chenxin 168
Li, Ke 282
Li, Lincheng 300
Li, Xiaoguang 319
Liang, Zhujin 131
Liao, Kang 390
Lin, Chunyu 390
Liu, Chi 481
Liu, Shuaicheng 390
Liu, Xinyu 168
Liu, Yifan 168, 300

Author Index

Liu, Ziwei 1
Loy, Chen Change 112
Luan, Zhaoliang 149
Lucas, Thomas 353
Luo, Yimin 481
Lv, Xintao 300

Ma, Wufei 264
Mahdavi-Amiri, Ali 75
Martinez, Julieta 206
Meng, Xuyi 112

Nguyen, Cuong 372
Nie, Lang 390

Ong, Yew-Soon 92
Ordonez, Vicente 444

Pan, Jinshan 336
Pan, Xingang 112
Peng, Xue Bin 246

Rempe, Davis 246

Saito, Shunsuke 206
Sarıyıldız, Mert Bülent 353
Selednik, Peter 206
Shao, Jing 168
Shao, Zedian 282
Song, Kaimin 481
Song, Liang 282
Stenger, Björn 229
Sun, Jianjian 408
Sun, Peng 282

Tang, Feilong 481
Tang, Jiaxiang 1
Tang, Jinhui 336
Tao, Wenbing 425
Thies, Justus 246

Villegas, Ruben 444

Wang, Angtian 264
Wang, Bing 149
Wang, Cheng 168

Wang, Lin 481
Wang, Song 319
Wang, Tengfei 1
Wang, Tiancai 425
Wang, Xi 464
Wei, Haoran 408, 425
Wei, Yana 425
Weinzaepfel, Philippe 353
Wu, Dongming 425
Wu, Shuwen 300

Xia, Peng 481
Xiao, Dongling 282
Xu, Congsheng 300
Xu, Liang 300
Xu, Zhongxing 481

Yan, Siyuan 481
Yan, Yichao 300
Yang, Dingkang 282
Yang, Jinrong 408, 425
Yang, Kun 282
Yang, Shuai 112
Yang, Xiaokang 300
Yang, Yixin 336
Yi, Hongwei 246
Yu, En 425
Yu, Weihao 168
Yuan, Yixuan 168
Yuille, Alan 264

Zeng, Gang 1
Zeng, Wenjun 131, 300
Zhang, Canyu 319
Zhang, Guofeng 264
Zhang, Kaiyi 19
Zhang, Qi 19
Zhang, Wenyao 131
Zhang, Xiangyu 408, 425
Zhang, Yun 390
Zhao, Bingchen 188
Zhao, Liang 408, 425
Zhao, Yao 390
Zhaoen, Su 206
Zhou, Kaijing 481
Zhou, Shangchen 112

Printed in the USA
CPSIA information can be obtained
at www.ICGtesting.com
CBHW060915101024
15650CB00005B/40